REVI

VOLUME 17

THERMODYNAMIC MODELING OF GEOLOGICAL MATERIALS: MINERALS, FLUIDS AND MELTS

EDITORS: I.S.E. Carmichael & H.P. Eugster

AUTHORS:

Robert C. Newton
Dept of the Geophysical Sciences
University of Chicago
Chicago, Illinois 60637

David A. Crerar
Alexandra Navrotsky
Dept of Geological & Geophysical Sciences
Princeton University
Princeton, New Jersey 08544

Bernard J. Wood
Dept of Geological Sciences
Northwestern University
Evanston, Illinois 60201

Kenneth S. Pitzer
Department of Chemistry
University of California, Berkeley
Berkeley, California 94720

John H. Weare
Department of Chemistry
University of California, San Diego
La Jolla, California 92093

Lukas Baumgartner
Hans P. Eugster
John M. Ferry
Dimitri A Sverjensky
Dept of Earth & Planetary Sciences
The Johns Hopkins University
Baltimore, Maryland 21218

John R. Holloway
Depts of Chemistry and Geology
Arizona State University
Tempe, Arizona 85287

George H. Brimhall
Ian S.E. Carmichael
Dept of Geology & Geophysics
University of California, Berkeley
Berkeley, California 94720

Robert G. Berman
Thomas H. Brown
Dept of Geological Sciences
University of British Columbia
Vancouver, BC, Canada V6T 2B4

Mark S. Ghiorso
Dept of Geological Sciences
University of Washington
Seattle, Washington 98195

SERIES EDITOR: Paul H. Ribbe
Department of Geological Sciences
Virginia Polytechnic Institute & State University
Blacksburg, Virginia 24061

MINERALOGICAL SOCIETY of AMERICA

Printed by BookCrafters, Inc., Chelsea, Michigan

REVIEWS in MINERALOGY

(Formerly: SHORT COURSE NOTES)

ISSN 0275-0279

Volume 17: *Thermodynamic Modeling of Geological Materials: Minerals, Fluids and Melts*

ISBN 0-939950-21-9

FOREWORD

The editors and authors of this volume presented a short course, entitled "Thermodynamic Modeling of Geological Materials: Minerals, Fluids amd Melts," October 22-25, 1987, at the Wickenburg Inn near Phoenix, Arizona. This was the fourteenth in a series of such courses sponsored by the Mineralogical Society of America since 1974, and this is the eighteenth book published under the banner, *Reviews in Mineralogy* [Volume 9 was issued in two parts -- see list of available titles on the opposite page].

The text of this volume was assembled from author-prepared, camera-ready copy -- thus the wide variety in style and font types represented. Mrs. Marianne Stern patiently and skillfully did most of the paste-up of Volume 17.

Paul H. Ribbe
Series Editor
Blacksburg, VA

PREFACE

When Van't Hoff calculated the effect of solution composition on the gypsum-anhydrite transition a century ago, he solved a significant geochemical problem (Hardie, 1967). Other well known examples of the early use of chemical thermodynamics in geology are Bowen's calculations of the plagioclase melting loop and the diopside-anorthite eutectic (Bowen, 1913, 1928). Except for a few specialists, however, these techniques were largely ignored by earth scientists during the first half of the 20th century. The situation changed dramatically by the 1950's when more and better thermodynamic data on geologic materials became available, and when thermodynamic arguments of increasing sophistication began to permeate the petrologic and geochemical literature. This rejuvenation was spearheaded by D.S. Korzhinskii, H. Ramberg, J.B. Thompson, J. Verhoogen and others. Today a graduating petrologist or geochemist can be expected to have a thorough grounding in geological thermodynamics.

Rapid intellectual growth in a field brings with it the difficulty of keeping abreast of parallel and diverging specialties. In order to alleviate this problem, we asked a group of active researchers to contribute up-to-date summaries relating to their specialties in the thermodynamic modeling of geological materials, in particular minerals, fluids and melts. Whereas each of these topics could fill a book, by covering the whole range we hope to emphasize similarities as much as differences in the treatment of various materials. For instance, there are useful parallels to be noted between Margules parameters and Pitzer coefficients. The emphasis here is on modeling, after the required data have been collected, and the approach ranges form theoretical to empirical.

We deliberately imposed few restrictions on the authors. Some chose to interpret modeling in the rigorous thermodynamic sense, while others approached their topics from more general geochemical viewpoints. We hope that any lack of unity and balance is compensated for by a collection of lively and idiosyncratic essays in which students and professionals will find new ideas and helpful hints. If the selection appears tilted towards fluids, it is because other recent summaries have emphasized minerals and melts.

This volume could not have been assembled without the dedication, cooperation and understanding of every author and his or her typist(s), and without the unselfish efforts of the Series Editor and current President of MSA, Paul Ribbe. As a tribute to the foremost geochemist of the century, we wish to dedicate this volume to the memory of V.M. Goldschmidt, on the eve of his 100th birthday.

References:

Bowen, N.L. (1913) Melting phenomena of the plagioclase feldspars. Amer. J. Sci. 35, 577-590. Bowen, N.L. (1928) The Evolution of the Igneous Rocks. Princeton Univ. Press, Princeton, NJ, 322 p. Hardie, L.A. (1967) The gypsum-anhydrite equilibrium at one atmosphere pressure. Amer. Mineral. 52, 171-200.

I.S.E. Carmichael
Berkeley, California

H.P. Eugster
Baltimore, Maryland

August 1987

Thermodynamic Modeling of Geological Materials: Minerals, Fluids and Melts

TABLE OF CONTENTS

Page

Chapter 1 **Robert C. Newton**

THERMODYNAMIC ANALYSIS OF PHASE EQUILIBRIA IN SIMPLE MINERAL SYSTEMS

Chapter 2 **Alexandra Navrotsky**

MODELS OF CRYSTALLINE SOLUTIONS

Chapter 3 — Bernard J. Wood

THERMODYNAMICS OF MULTICOMPONENT SYSTEMS CONTAINING SEVERAL SOLID SOLUTIONS

Chapter 4 — Kenneth S. Pitzer

A THERMODYNAMIC MODEL FOR AQUEOUS SOLUTIONS OF LIQUID-LIKE DENSITY

Chapter 5 — John H. Weare

MODELS OF MINERAL SOLUBILITY IN CONCENTRATED BRINES WITH APPLICATION TO FIELD OBSERVATIONS

Chapter 6 — Dimitri A. Sverjensky

CALCULATION OF THE THERMODYNAMIC PROPERTIES OF AQUEOUS SPECIES AND THE SOLUBILITIES OF MINERALS IN SUPERCRITICAL ELECTROLYTE SOLUTIONS

Chapter 7 **John R. Holloway**

IGNEOUS FLUIDS

Chapter 8 **George H. Brimhall** and **David A. Crerar**

ORE FLUIDS: MAGMATIC TO SUPERGENE

Part III: Formation of Primary and Secondary Ore Deposits

Chapter 9 **John M. Ferry** and **Lukas Baumgartner**

THERMODYNAMIC MODELS OF MOLECULAR FLUIDS AT THE ELEVATED PRESSURES AND TEMPERATURES OF CRUSTAL METAMORPHISM

Chapter 10 — Hans P. Eugster and Lukas Baumgartner

Mineral Solubilities and Speciation in Supercritical Metamorphic Fluids

Chapter 11 — R. G. Berman and T. H. Brown

Development of Models for Multicomponent Melts: Analysis of Synthetic Systems

Chapter 12 **Mark S. Ghiorso**

MODELING MAGMATIC SYSTEMS: THERMODYNAMIC RELATIONS

Chapter 13 **Mark S. Ghiorso** and **Ian S.E. Carmichael**

MODELING MAGMATIC SYSTEMS: PETROLOGIC APPLICATIONS

Chapter 1

Robert C. Newton

THERMODYNAMIC ANALYSIS OF PHASE EQUILIBRIA IN SIMPLE MINERAL SYSTEMS

INTRODUCTION

Knowledge of the thermodynamic properties of minerals has become of great importance in prediction of the physical and chemical conditions of the formation of mineral assemblages and as a guide to the synthesis of minerals. The three major sources of information are thermochemical and thermophysical measurements, derivation from experimental phase equilibrium data at elevated temperatures and pressures, and attempted reproduction of the expected temperatures and pressures of natural assemblages, under the assumption of "frozen-in" equilibrium (method of paragenetic analysis). All of these methods have made important contributions to the body of thermodynamic data, and comparison of results from the three independent sources is valuable in cross-checking, systemization and extending of the data sets.

The most important properties of minerals for stability calculations are heat capacity, entropy, enthalpy of formation at one bar and molar volume as a function of temperature and pressure. This information is needed for end-member substances and for solid solutions. The volume properties are relatively easily and accurately measured by X-ray diffraction, but the thermal properties are much harder to evaluate accurately for complex substances. The most direct information is supplied by calorimetry, both thermophysical (heat capacity measurements) and thermochemical (enthalpy of solution measurements). Less direct but increasingly important sources of information are spectroscopy, such as infrared absorption, especially at low temperatures (Kieffer, 1981), electrochemical measurements, as in the solid electrolyte cell (Sato, 1971), and vapor pressure measurement, as with the high-temperature Knudsen cell (Rammansee and Fraser, 1982).

The most extensively-used source of thermodynamic data of minerals remains derivation from experimental phase equilibrium diagrams, although the thermochemical and thermophysical methods approach the phase equilibrium method in comprehensiveness and potential accuracy. The precision of enthalpy of solution work is not quite at a level which allows generally accurate calculation of phase equilibrium. Nevertheless, constraints provided by thermochemical measurements are very important in initializing phase equilibrium calculations, and, in turn, phase equilibrium data are of utmost importance in optimizing thermodynamic parameters within the uncertainty limits of calorimetry, and in discriminating between conflicting sets of thermochemical measurements.

The simplest kind of experimental equilibrium for deriving thermodynamic data is the univariant P-T curve of a reaction among pure substances. An example is the decarbonation of magnesite ($MgCO_3$) to periclase (MgO):

$$MgCO_3 = MgO + CO_2. \qquad (A)$$

The general equation of equilibrium for simple devolatilization is

$$\Delta G(P,T) = 0 = \Delta G°(T) + \int_1^P \Delta V_S dP + nRT \ln f_V \qquad (1)$$

where ΔG is the free energy change, ΔV_S the difference in volume between the solid products and reactants, n is the coefficient of the volatile species in the balanced equation, f_V is the fugacity of the volatile species, in this case CO_2, and R is the gas constant (8.3144 J/K). Experiments locating the P-T curve of Reaction (A) define, with the aid of equation (1), the most important property of magnesite, its free energy of formation from the oxides (ΔG_f°) at one bar pressure by:

$$\Delta G_f^\circ \text{ (magnesite)} = -\Delta G_A^\circ. \quad (2)$$

The free energy of reaction may be expressed further as:

$$\Delta G^\circ = \Delta H^\circ - T\Delta S^\circ. \quad (3)$$

ΔH° and ΔS° are, respectively, the changes in standard enthalpy and entropy of the reaction, likewise expressible as differences in standard enthalpies and entropies of formation of the products and reactants. The standard enthalpy of a substance at any temperature is related to that at any reference temperature, usually taken as 298.15 K, by integration of the heat capacity, C_P:

$$H(T) - H(298.15) = \int^T_{298.15} C_P dT. \quad (4)$$

Heat capacities above 298 K can usually be expressed as polynomial functions of temperature. A form convenient for representation of measured data is (Robie et al., 1978):

$$C_P = a + bT + c/T^2 + d/\sqrt{T} + eT^2. \quad (5)$$

Truncated forms lacking some of the terms are commonly used. The standard entropy of a substance may be expressed as a Third Law entropy plus a configurational term for those substances with possibility of significant residual atomic disorder at zero Kelvin, such as Al,Si disorder of feldspar:

$$S(T) = \int_0^T (C_P/T)dT + S \text{ (conf.)}. \quad (6)$$

The entropy of disorder of a mineral can sometimes be characterized by some physical measurement, such as X-ray diffraction, as

$$S \text{ (conf.)} = -R\Sigma X_i \ln X_i \quad (7)$$

for each atomic site, where X_i are the molar fractions of the atoms mixing on a site.

The free energies of compression, $-RT \ln f_V$, of H_2O and CO_2 are well measured in the ranges 0-1300 K and 1-10000 bars (Burnham et al., 1969; Shmulovich and Shmonov, 1978). The fugacities of other volatiles, and of volatile mixtures, are much less well known, but can be approximated from theoretical considerations, such as the hard-sphere Modified Redlich-Kwong (MRK) equation of state (Holloway, 1977; Kerrick and Jacobs, 1981). Extrapolation beyond measured pressure ranges by the MRK procedure may not be trustworthy (Haselton et al., 1978).

Very accurate work at elevated temperatures and pressures above a few kbar requires use of expansivities, a, and compressibilities, β, of the solids:

$$V(T,P) = V(298.15)[1 + a(T-298) - \beta P]. \quad (8)$$

Since a is, in general, a function of pressure, and β of temperature, and since few high temperature and high pressure measurements of these quantities exist, some authors ignore a and β in derivation of thermodynamic data from phase equilibria (Connolly and Kerrick, 1984; Berman et al., 1985).

In a few propitious reactions, such as magnesite decarbonation, thermodynamic parameters of minerals may be determined accurately from one or a few experimental brackets of univariant equilibria. Where several or numerous experimental brackets of a single equilibrium are available, various statistical treatments may be applied, and where a number of experimental brackets of several equilibria involving minerals in common are available, multiple regression (Gasparik and Newton, 1984) or Monte Carlo (Perkins et al., 1981) methods may be used. In the latter method, large numbers of trial parameters are inserted into the thermodynamic equations, and the combination of parameters yielding minimal total deviation from experimental points or bracket midpoints is sought. In the method of linear parametric analysis ("linear programming"), as developed by Gordon (1973), allowable ranges of some parameters, such as standard entropy and enthalpy changes of a reaction, are defined by the composite restriction of experimental points.

The most important thermodynamic parameter to be derived by analysis of phase equilibria is the enthalpy of formation, $\Delta H^\circ_{f,298}$, from the oxides or elements, because data obtainable from physicochemical measurements, principally solution calorimetry, are usually not precise enough for accurate calculations. Heat capacities are often fixed by thermophysical data, though they may sometimes be optimized within small ranges (Halbach and Chatterjee, 1984). Standard entropies are sometimes fixed by precise low-temperature heat capacity measurements, but are often uncertain to the extent of a configurational entropy. A practical problem is that samples too large to be synthesized by ordinary methods are often required for adiabatic calorimetry. Measurements on near-end-member natural samples require uncertain corrections for impurities. For these reasons standard entropies may be derived from phase equilibria even where heat capacity measurements exist. Molar volumes are almost always fixed by X-ray diffraction measurements.

Thermodynamic analysis of univariant equilibria in oxide-silicate systems is conveniently referred to a generalization of expressions (1), (3), (4) and (6) (cf. Day and Kumin, 1980):

$$\Delta H^\circ_{f,S,298} - T\Delta S^\circ_{f,S,298} - \int^T_{298} dT \int^T_{298} \frac{\Delta Cp_{f,s}}{T} dT + \int_1^P \Delta V_S dP + nRT \ln f_V \qquad (9)$$

$$= 0 \quad (a)$$

$$> 0 \quad (b)$$

$$< 0 \quad (c)$$

$\Delta H^\circ_{f,S,298}$ and $S^\circ_{f,S,298}$ are, respectively, the differences in enthalpy and entropy of formation from the oxides of the solid products and reactants:

$$\Delta H^\circ_{f,s,298} = \Sigma n_i \Delta H^\circ_{f,298,i} \text{ (solid products) } - \Sigma m_i \Delta H^\circ_{f,298,i} \text{ (solid reactants)}, \qquad (10)$$

with a similar expression for $\Delta S^\circ_{f,S,298}$, where n and m are the stoichiometric coefficients.

At (T,P) points within the field of stability of the condensed (left-hand) assemblage, (9b) applies, and for the (T,P) region where devolatilization occurs, (9c) applies. Regression methods to obtain $\Delta H^\circ_{f,S,298}$ and $\Delta S^\circ_{f,S,298}$ with (9a) commonly use midpoints of isothermal or isobaric experimental brackets or points on empirical curves drawn through the brackets. In this method, if $\Delta H^\circ_{f,S,298}$ and $\Delta S^\circ_{f,S,298}$ are best-fit quantities to all of the brackets, then, at a bracket mid-point (T_i,P_i), Equation (9a) becomes

$$\Delta H^\circ_{f,S,298} - T_i \Delta S^\circ_{f,S,298} + k_i(T_i,P_i) = \delta_i, \qquad (11)$$

where δ_i is the free energy deviation of the bracket midpoint from the best-fit curve. Optimal $\Delta H^\circ_{f,S,298}$ and $\Delta S^\circ_{f,S,298}$ are found by minimizing the sum of the squared deviations:

$$\frac{\partial(\Sigma\delta_i^2)}{\partial(\Delta H^\circ_{f,s,298})} = 0; \quad \frac{\partial(\Sigma\delta_i^2)}{\partial(\Delta S^\circ_{f,s,298})} = 0. \tag{12}$$

These expressions lead to two equations in the two unknowns $\Delta H^\circ_{f,s,298}$ and $\Delta S^\circ_{f,s,298}$:

$$p\Delta H^\circ_{f,s,298} - (\Sigma T_i)\Delta S^\circ_{f,s,298} + (\Sigma k_i) = 0,$$
$$-(\Sigma T_i)\Delta H^\circ_{f,s,298} + (\Sigma T_i^2)\Delta S^\circ_{f,s,298} - (\Sigma T_i k_i) = 0, \tag{13}$$

where p is the number of brackets. Methods using the inequalities (9b) and (9c), principally linear parametric analysis, make use of individual experimental data points. Examples of the use of these methods are given below.

Since, in general, there can be at most j - 1 independent reactions among j phases, the number of conditions (10) will not be sufficient to determine all of the $\Delta H^\circ_{f,298}$ quantities, one or more of which must be evaluated by other means, such as solution calorimetry, if superior measurements exist. These "anchor phases" (Holland and Powell, 1985) are used to derive the less well known or unknown properties of the other phases. Alternatively, a set of parameters from thermochemical measurements is used as initial input in multiple regression (Halbach and Chatterjee, 1984). The parameters are allowed to vary within the stated limits of the thermochemical measurements, and the set yielding the smallest composite residuals from experimental data is found.

Free energy, enthalpy and entropy of mixing in solid solutions are important in many calculations of mineral equilibria. These quantities are especially hard to define accurately by thermochemical and thermophysical methods, and are often derived from experimental phase equilibrium data. Discussion of solid solutions is deferred, for the most part, to Chapters 2 and 3 by A. Navrotsky and B.J. Wood.

Serious problems with retrieval of thermodynamic parameters from phase equilibria have become apparent, especially where solid solutions are involved and where ordinary regression techniques are applied without regard to constraints from thermochemical measurements (see discussion by Cohen, 1986). Three major sources of error are: (a) too many parameters regressed from too few data points, (b) high correlation of derived parameters, as in the simultaneous retrieval of Fe,Mg mixing parameters of coexisting garnet and olivine (O'Neill and Wood, 1979), and (c) insufficient temperature, pressure, or composition baselines of experimental data. Temperature dependences of standard and solid solution free energies have characteristically been overestimated from the last source of error.

Rigorous definition of uncertainties in derived thermodynamic properties is very difficult, or sometimes virtually impossible; the cumulative uncertainties in experimental observations, input thermochemical data, unknown validity of assumptions and approximations such as ideal solution behavior, uncertainties in volatile fugacities, especially where extended by theoretical equations of state, and neglect of solubility of the solids in the vapor, as in "pure water" calculations, tend to discourage rigorous treatments. Somewhat arbitrary criteria may be adopted, such as the extreme slopes of regression lines which can be drawn through plotted data points or the mean deviations of derived parameters from thermochemically measured ones (Day and Kumin, 1980; Halbach and Chatterjee, 1984).

Discussions of error analysis in derivation of thermodynamic properties are given in Bird and Anderson (1973), Anderson (1977), Zen (1977) and Demarest and Haselton (1981). The last paper demonstrates that, for isothermal or isobaric pairs of runs which bracket an equilibrium curve, if the half-width of the bracket is greater than the experimental temperature or pressure uncertainties of the runs, which is usually the case, the brackets are not subject to Gaussian error distribution and, hence, cannot be used in rigorous error propagation.

Another problem is that error analysis based on individual brackets undoubtedly overestimates uncertainties in reaction parameters. The net constraint of many experimental brackets of a single univariant curve is often very stringent: a practically unique line may sometimes be demanded by a set of experimental brackets covering large P-T ranges, subject to the additional constraint of constant sign of curvature (no inflections). Thus, the composite constraint exacted by a number of brackets must considerably reduce the uncertainty in the free energy of the reaction. To the author's knowledge, no rigorous treatment of the multiple-bracket effect is available in the literature. For these reasons, the present paper, following the majority of authors (cf. Berman et al., 1986), does not attempt definition of standard errors of derived parameters, but concentrates instead on internal consistency and optimization.

The purpose of this chapter is to illustrate some of the methods of deriving thermodynamic properties of some common rock-forming minerals, principally silicates, to show the correspondence between measured and derived quantities, and to indicate some of the strengths and limitations of the thermochemical, experimental and paragenetic methods.

MgO-Al_2O_3-SiO_2 PERIDOTITE MINERALS

General approach

A comprehensive thermodynamic dataset is most conveniently built up by considering simple subsystems, such as MgO-Al_2O_3-SiO_2, for which there are many univariant equilibria with good experimental coverage. In this way complicating factors of solid solution are kept to a minimum. Standard properties of minerals may be derived by a step-wise process, in which the least ambiguous and most reliable data are selected to define the properties of a few minerals, and these properties in turn define those of other minerals through more complex reactions (Powell, 1978; Helgeson et al., 1978). Alternatively the thermodynamic properties of a mineral subset are derived simultaneously by regression of experimental data available for all reactions in the subset (Holland and Powell, 1985). The former method has the advantage that the best-determined mineral properties serve as a foundation, whereas the latter method tends to distribute the uncertainties more broadly. The latter method has the advantage that it does not run the risk of biasing the entire dataset through misjudgment of the most reliable experimental data. The step-wise approach is illustrated in the SiO_2-undersaturated portion of the system MgO-Al_2O_3-SiO_2. This model peridotite system includes the phases forsterite, enstatite, pyrope, spinel and Mg-cordierite. Some of the most rigorously constraining data consist of single brackets, which favors the step-wise approach.

Enstatite and forsterite

A convenient starting point is the reactions of enstatite and forsterite with magnesite. Several workers have calculated the free energy of formation of magnesite from the reversed experimental data of Harker and Tuttle (1955) on Reaction (A), using Equation (1). The most comprehensive analysis is that of Robinson et al. (1982), which is accepted here. The input parameters ΔV_S and f_V are very accurately known at the temperatures (600°-900°C) and pressures (100-2800 bars) of the equilibrium and contribute almost negligible uncertainties. The uncertainty from all sources in $\Delta G°_f$ (magnesite) is of the order of 350 J. From $\Delta G°_f$ (magnesite), the corresponding quantities for enstatite and forsterite are obtained from the reactions:

Table 1

Enthalpy of formation (oxides), entropy, heat capacity coefficients, volume, thermal expansion and compressibility of $MgO-Al_2O_3-SiO_2$ phases

Mineral	Formula	$\Delta H^\circ_{f,298}$ kJ	S°_{298} J/K	a J/K	b $x10^3$	c $x10^{-5}$	d $x10^{-2}$	V°_{298} cm^3	$\alpha x10^5$ K^{-1}	$\beta x10^7$ bar^{-1}
Enstatite	$MgSiO_3$	-32.69[a]	66.27[c]	188.76[c,e]	-5.332	.00865	-18.129	31.32[b]	2.9	8.7
Forsterite	Mg_2SiO_4	-58.42[a]	94.11[d]	27.98	3.414	-8.990	-17.45	43.79[b]	3.5	7.5
Pyrope	$Mg_3Al_2Si_3O_{12}$	-74.19[a]	266.27[e]	544.95	20.680	-83.312	-22.8	113.13	2.7	6.0
Spinel	$MgAl_2O_4$	-19.59[a]	87.25[a,f]	222.91	6.127	-16.857	-15.512	39.71[b]	2.7	4.1
Cordierite	$Mg_2Al_4Si_5O_{18}$	-48.67	416.53[a,d]	821.34[f]	43.339	-82.112	-50.003	233.55	0.7	5.1
MgTs	$MgAl_2SiO_6$	+1.61[a]	132.63	346.12[c,f]	-4.613	-18.960	-28.009	58.36[g]	2.9	8.7
Corundum	Al_2O_3	0	50.92[f]	157.36[f]	0.719	-18.969	-9.880	25.58[f]	2.6	3.6
Periclase	MgO	0	26.94[f]	65.211	-1.270	-4.619	-3.872			
Quartz	SiO_2	0	41.46[i]	104.40	6.070	0.340	-10.700			

Sources: a Present study; b Charlu et al., 1975; c Krupka et al., 1979a; d Robie et al., 1982; e Haselton, 1979; f Robie et al., 1978; g Gasparik and Newton, 1984; h Clark, 1966; i Holland and Powell, 1985 (composite of α- and β-quartz). All compressibilities and thermal expansions from g, except for cordierite (h, with β modelled by beryl).

Table 2

Enthalpy of formation determinations of forsterite and enstatite, kJ/mol (oxides)

FORSTERITE			ENSTATITE		
Author	Type of Determination	$\Delta H^\circ_{f,298}$ [b]	Author	Type of Determination	$\Delta H^\circ_{f,298}$ [b,c]
Torgesen and Sahama (1948)	HF sol'n calorimetry[a] 346 K	-63.26±1.05	Torgesen and Sahama (1948)	HF sol'n calorimetry[c] 346 K	-35.73±0.63
King et al. (1967)	HF sol'n calorimetry 346 K	-57.99±1.26	Shearer and Kleppa (1973)	$Pb_2B_2O_5$ sol'n calorimetry[c] 970 K	-35.73±0.88
Navrotsky (1971)	$Pb_2B_2O_5$ sol'n calorimetry 965 K	-61.63±4.18	Charlu et al. (1975)	$Pb_2B_2O_5$ sol'n calorimetry 970 K	-36.90±0.71
Charlu et al. (1975)	$Pb_2B_2O_5$ sol'n calorimetry 970 K	-61.50±1.13	Kiseleva et al. (1979)	$Pb_2B_2O_5$ sol'n calorimetry 1170 K	-34.18±1.92
Kiseleva et al. (1979)	$Pb_2B_2O_5$ sol'n calorimetry 1170 K	-58.20±2.59	Chatillon-Colinet et al. (1983a)	$(Na,Li)BO_2$ sol'n calorimetry 1170 K	-35.98±0.88
Brousse et al. (1984)	$(Na,Li)BO_2$ sol'n calorimetry 1073 K	-57.82±1.84	Brousse et al. (1984)	$(Na,Li)BO_2$ sol'n calorimetry 1073 K	-33.93±1.76
Present	Phase equilibrium analysis	-58.42	Present	Phase equilibrium analysis	-32.69

a Extrapolation based on measurements on natural olivines.

b Correction of ΔH°_f to 298 K based on heat capacities of $MgSiO_3$ (Haselton, 1979), Mg_2SiO_4 (Robie et al., 1982), MgO and SiO_2 (Robie et al., 1978).

c Clinoenstatite used in study; ΔH°_f of orthoenstatite derived by adding 0.63 kJ to ΔH°_f of clinoenstatite, based on enstatite-clinoenstatite equilibrium data (Boyd and England, 1965).

$$\underset{\text{magnesite}}{MgCO_3} + \underset{\text{quartz}}{SiO_2} = \underset{\text{enstatite}}{MgSiO_3} + CO_2 \qquad (B)$$

$$\underset{\text{magnesite}}{MgCO_3} + \underset{\text{enstatite}}{MgSiO_3} = \underset{\text{forsterite}}{Mg_2SiO_4} + CO_2 \qquad (C)$$

Equilibrium temperatures of 791 K for Reaction (B) and 831 K for Reaction (C) at 2000 bars were determined by Johannes (1969). The equilibrium temperatures are short projections from data with H_2O-CO_2 fluids but seem well determined. Uncertainties are estimated to be ± 10 K and ± 50 bars. The 15-parameter polynomial function constructed by Holland (1981) from the P-V-T data of CO_2 of Shmulovich and Shmonov (1978) gives $f(CO_2)$ (791, 2000) = 3448 bars, and $f(CO_2)$ (831, 2000) = 3468 bars. The standard free energy of formation of magnesite is 18.58 kJ at 791 K and 26.77 kJ at 831 K, interpolating from Table 115 of Robinson et al. (1982). The 791 K values may be used with the other data of Table 1 to get the free energy of formation of enstatite from the relation:

$$\Delta G_B = 0 = \Delta G°_{f,791}(en) - \Delta G°_{f,791}(mag) + \int_1^{2000}(V_{en}-V_{mag}-V_{qtz})dP$$
$$+ 791\ R\ \ln f(CO_2)\ (791,\ 2000) \qquad (14)$$

This gives $\Delta G°_{f,791}$ (enst) = -31.00 kJ and, from 3), $\Delta H°_{f,791}$ (enst) = -32.93 kJ. A similar procedure yields the free energy and enthalpy of formation of forsterite. Maximum uncertainties in these quantities are of the order of one kJ. The enthalpies of formation corrected to 298 K are given in Table 1.

The standard enthalpies of formation of enstatite and forsterite from phase equlibrium analysis are compared with the available determinations by enthalpy of solution measurements in Table 2. The later determinations with larger uncertainties (Kiseleva et al., 1979; Brousse et al., 1984) on synthetic phases are in agreement with the present derived quantities. However, the uncertainties from calorimetry are considerably larger than those obtainable from phase equilibrium analysis, a general result noted previously by Anderson (1977) and Helgeson et al. (1978). Earlier enthalpy of solution determinations have underestimated the uncertainties. The oxide melt calorimetry of Charlu et al. (1975) yielded $\Delta H°_f$ of enstatite too negative to be compatible with high temperature, high pressure phase equilibrium relations of pyrope and enstatite, as noted by several workers (Perkins et al., 1981; Gasparik and Newton, 1984).

Pyrope

The reaction:

$$\underset{\text{pyrope}}{2Mg_3Al_2Si_3O_{12}} = \underset{\text{enstatite}}{3Mg_2Si_2O_6} + \underset{\text{corundum}}{2Al_2O_3} \qquad (D)$$

defines the standard properties of pyrope. The enstatite formula is doubled for convenience. This univariant equilibrium is stable over a narrow interval of temperature and pressure near 850°C and 16.3 kbar (Fig. 1), as first noted by Schreyer and Seifert (1970). It was reversed hydrothermally in the low-friction 1"-diameter NaCl-medium piston-cylinder apparatus by Gasparik and Newton (1984). At these conditions, enstatite in equilibrium with pyrope has 5.75 wt. percent of Al_2O_3 in solid solution, or 11.33 molar percent of $MgAl_2SiO_6$ (Mg-Tschermak pyroxene). Assuming an ideal solution of $MgAl_2SiO_6$ (MgTs) and $Mg_2Si_2O_6$ (En) (Wood and Banno, 1973) yields a small correction to the free energy expression:

$$G_D\ (T,P) = 0 = 3\Delta G^\circ_f\ (\text{enst}) - 2\Delta G^\circ_f\ (\text{py})$$
$$+ \int_1^{16250} (2V_{cor} + 3V_{enst} - 2V_{py})\ dP + 3RT \ln X_{En}. \quad (15)$$

Expression (15) uses expressions (3) and (7), implies that ΔH of mixing is zero (ideal solution), and assumes that there is cation mixing on only one site of the pyroxene (the M1 site), which is approximately valid for small amounts of Al_2O_3, according to IR absorption studies (Smirnova, 1986). Deviation from ideal mixing would not affect the derived standard properties of pyrope by more than a few hundred joules because of the high concentration of $Mg_2Si_2O_6$. The C_P of pyrope from low-temperature adiabatic calorimetry and high-temperature drop calorimetry of Haselton (1979) is accepted. The 298.15 K formation properties of pyrope are given in Table 1. Uncertainty in ΔH°_f is of the order of 2.5 kJ, mainly from the uncertainty in ΔH°_f (enst).

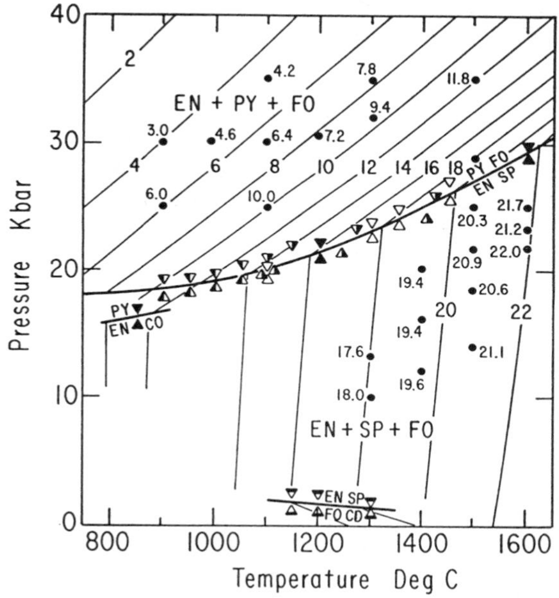

Figure 1. Peridotite equilibria in the system MgO-Al_2O_3-SiO_2. Letter symbols: EN = enstatite, PY = pyrope, FO = fosterite, SP = spinel, CO = corundum, CD = cordierite. Numbers: mol % $MgAl_2SiO_6$ (MgTs) in enstatite, according to Perkins et al. (1981) [EN + PY + FO field] and Gasparik and Newton (1984) [EN + SP + FO field]. Triangles are reversal runs of univariant equilibria (bold curves): solid: Gasparik and Newton, 1984; left-shaded: Danckwerth and Newton, 1978; right-shaded: Perkins et al., 1981; open: Haselton, 1979; base-shaded: Herzberg, 1983. Light lines are least-squares fits to enstatite isopleth data (see text).

The derived standard enthalpy of formation of pyrope is less negative by 4.8 kJ than that found by the oxide melt solution calorimetry of Charlu et al. (1975). Among the available derived datasets, those of Holland and Powell (1985) and Berman et al. (1985) give values close to the present value.

Spinel - a disordered phase

Spinel remains essentially stoichiometric in systems with forsterite, though it can acquire excess Al_2O_3 in solid solution in more aluminous systems. Hafner and Laves (1961) discovered through IR absorption studies of heat-treated natural spinel that some Mg,Al disorder occurs, according to the schematic reaction $Mg^{IV} + Al^{VI} = Mg^{VI} + Al^{IV}$, where the superscripts denote Al coordination numbers. Thus, although good low- and high-temperature heat capacity data exist, the entropy of spinel is uncertain to the extent of a temperature-dependent configurational entropy.

Thermodynamic properties of spinel can be calculated from the univariant "haploperidotite" reaction:

$$\underset{\text{pyrope}}{Mg_3Al_2Si_3O_{12}} + \underset{\text{forsterite}}{Mg_2SiO_4} = \underset{\text{spinel}}{MgAl_2O_4} + \underset{\text{enstatite}}{2Mg_2Si_2O_6}. \qquad \text{(E)}$$

This reaction has been determined experimentally several times. The collection of reversals determined under well-controlled conditions with aluminous enstatite starting material is shown in Figure 1. The Al_2O_3 content of enstatite along the reaction curve was determined by Perkins et al. (1981). A quadratic curve fitted to the midpoints of reversal brackets by least squares is $P(\text{kbar}) = 30.459 - .02641\ T(K) + 1.3772 \times 10^{-5}T^2$ from which the entropy of the reaction and ultimately, of spinel, is given by the Clausius-Clapeyron equation:

$$dP/dT = \Delta S/\Delta V = (2S_{En} - 2R \ln X_{En}+S_{Sp}-S_{Fo}-S_{Py})/(2V_{En}+V_{Sp}-V_{Fo}-V_{Py}). \qquad (16)$$

Ideal solution of $Mg_2Si_2O_6$ is again assumed. The standard entropy of spinel is found to exceed the Third Law entropy by a nearly constant 6.6 J/K in the range 800°-1300°C. Entropy functions of enstatite and pyrope, which have been measured only to 1000°C, cannot be safely extrapolated above 1300°C. The free energy of formation of spinel is determined in a manner analogous to that of pyrope in expression (15). The standard enthalpy of formation at 298.15 K, derived from expression (3), is in agreement with the oxide melt solution calorimetry of Charlu et al. (1975), within the broad uncertainties.

The nearly constant non-Third Law entropy increment of spinel corresponds to only about 20% of the configurational entropy of complete Mg,Al disorder according to Equation (7). However, in a NMR spectroscopy study of fast-quenched synthetic spinel, Wood et al. (1986) found considerably greater amounts of disorder in spinels heated at temperatures greater than 900°C, and concluded that disordering goes rapidly to completion above that temperature, but is largely non-quenchable. A possible explanation of the nearly constant entropy of disorder inferred by the present analysis is that spinel along the univariant curve (E) is completely disordered above about 1000°C, and that the entropy of disorder is largely registered in the measured high-temperature heat capacities based on drop calorimetry, except for about 6.6 J/K, which represents quenchable configurational entropy. Chapters 2 and 3 by Navrotsky and Wood give further discussion of the problem of spinel disorder. Several workers have found that spinel heat capacities and entropies should be increased by variable amounts above the Third Law values to give results consistent with phase equilibrium calculations in simple systems (cf. Saxena and Chatterjee, 1986).

<u>MgTs - a fictive substance</u>

P-T contours of constant Al_2O_3 content (isopleths) of enstatite coexisting with pyrope can be constructed from analysis of the divariant reaction, first written by Wood and Banno (1973):

$$\underset{\text{pyrope}}{Mg_3Al_2Si_3O_{12}} = \underset{\text{enstatite}}{Mg_2Si_2O_6} + \underset{\text{MgTs}}{MgAl_2SiO_6} \qquad \text{(F)}$$

In the treatment of Gasparik and Newton (1984), $Mg_2Si_2O_6$ and $MgAl_2SiO_6$ end members are considered to form ideal solid solutions at all pressures and temperatures. Thus, lines of constant enstatite composition should obey the equation:

$$\Delta G_F = 0 = \Delta G°_f\,(\text{enst}) + \Delta G°_f\,(\text{MgTs}) - \Delta G°_f\,(\text{py})$$

$$+ \int_1^P (V_{En} + V_{MgTs} - V_{Py})dP + RT \ln (X_{En} \cdot X_{MgTs}) \quad (17)$$

The volume of MgTs is found by extrapolation of the initial trend of molar volume of enstatite with Al_2O_3 content to the $MgAl_2SiO_6$ composition. Isopleths of Al_2O_3 in enstatite coexisting with spinel and forsterite may be similarly constructed from the divariant reaction:

$$\underset{\text{enstatite}}{Mg_2Si_2O_6} + \underset{\text{spinel}}{MgAl_2O_4} = \underset{\text{MgTs}}{MgAl_2SiO_6} + \underset{\text{forsterite}}{Mg_2SiO_4} \quad (G)$$

Fitting the enstatite compositions determined by reversed experiments (Perkins et al., 1981; Gasparik and Newton, 1984) (Fig. 1) simultaneously to Equation 17) and its spinel field analog by Equations (10) to (13) defines the standard free energy of formation of a fictive MgTs orthopyroxene. Properties at 298 K (Table 1) are derived by modelling the MgTs heat capacity as the sum of those of $MgSiO_3$ and Al_2O_3. A very small pressure effect on the entropies can be ignored, because the thermal expansions of pyrope and enstatite are nearly identical. In this analysis $\Delta H°_f$ and S° of MgTs are adjustable parameters which allow a self-consistent dataset. However, some physical reality to the ideal MgTs model is suggested by the plausible standard entropy, which exceeds that of $MgSiO_3 + Al_2O_3$ by 15.4 J/K. This is in the range of the ΔS of Al coordination change from six-fold to four-fold shown by silicate reactions and polymorphism, as in the kyanite-sillimanite inversion. This is consonant with the Al^{IV} of MgTs.

Other fictive Al_2O_3-bearing pyroxene end members can be used with apparent success. Ganguly and Ghose (1979) suggested a pyrope formula pyroxene, $Mg_3Al^{VI}Al^{IV}Si_3O_{12}$, and Perkins et al. (1981) found an equally good fit to their experimental data with this component. Saxena (1981) used a fictive pyroxene-structure Al_2O_3 component with consistent results, though with a smaller dataset of pyroxene compositions.

Orthopyroxene compositions in the system $MgO\text{-}Al_2O_3\text{-}SiO_2$ are useful for geothermometry and geobarometry of garnet and spinel peridotites. Information on the pressures of the upper mantle were first derived from enstatite compositions in garnet peridotite xenoliths by Chatterjee (1969) and Boyd (1973), based on experimental isopleth data in the simple system. Spinel-field enstatite compositions are nearly pressure-independent, and useful for geothermometry of spinel lherzolites (Gasparik and Newton, 1984).

Cordierite in peridotites

Magnesian cordierite, $Mg_2Al_4Si_5O_{18}$, is stable with forsterite at pressures below those of the reaction

$$\underset{\text{enstatite}}{5Mg_2Si_2O_6} + \underset{\text{spinel}}{2MgAl_2O_4} = \underset{\text{cordierite}}{Mg_2Al_4Si_5O_{18}} + \underset{\text{forsterite}}{5Mg_2SiO_4}. \quad (H)$$

Figure 1 shows the reversed determinations of Herzberg (1983) using aluminous enstatite starting material. The dP/dT slope is negative, but not closely constrained. Thermodynamic properties of cordierite derived from H) and the properties of the other substances involve large uncertainties. The derived $\Delta H°_f$ and S° of cordierite given in Table 1 are based on an analysis of hydrous cordierite presented below.

Cordierite has the possibility of Al,Si tetrahedral disorder, and, although a few natural samples have some disorder (Gibbs, 1966), most well-crystallized natural samples are highly ordered. The derived entropy of high-temperature synthetic cordierite exceeds the Third-Law value (Weller and Kelley, 1963) by 9.20 J/K. The inferred moderate amount of configurational entropy is consistent with the fact that nearly pure natural cordierite has nearly the same enthalpy of solution as high-temperature synthetic cordierite (Charlu et al., 1975). The dP/dT slope of reaction H) with this S^{conf} is -0.71 bar/K.

Structural H_2O has a substantial effect in stabilization of cordierite to higher pressures, as discussed below. The standard properties derived here apply to calculations in very dry systems. The high temperature magnesian cordierite peridotite from Zabargad in the Red Sea rift (Kurat et al., 1982) is such a hot and dry paragenesis. The experimental results suggest that the cordierite formed at pressures less than 3 kbar by subsolidus recrystallization.

ALUMINUM SILICATES

The Al_2SiO_5 system has wide application to petrology and has been very intensively studied. Of the three stable polymorphs, kyanite (density 3.67 gm/cm^3) and andalusite (density 3.15) are crystallographically well-defined and highly ordered phases, whereas sillimanite (density 3.25), with one tetrahedral Al per formula, has the possibility of Al,Si order disorder (Greenwood, 1972; Navrotsky et al., 1973). There is evidence also that substantial amounts of defect or stacking fault energy in sillimanite, produced by mechanical deformation or, possibly, non-equilibrium growth conditions, could affect the stability relations with the other polymorphs (Salje and Wernecke, 1982). Cation disorder or defects may bear on the petrologic role of the textural variant, fibrolite.

Three decades of discussion of Al_2SiO_5 stability have greatly focussed the problem. In this discussion, experimental measurements, thermochemical measurements, and paragenetic observations (field data) have all played important roles. Two somewhat different experimental stability studies remain in contention, those of Richardson et al. (1969) and Holdaway (1971). The P-T location of the kyanite-andalusite (K-A) equilibrium is well-defined by all data. The kyanite-sillimanite (K-S) P-T curve is less well constrained, and the andalusite-sillimanite (A-S) curve is considerably different in the two studies. This lack of agreement is probably the result of the extreme sluggishness of the A-S reaction, the different experimental methods, and, quite possibly, differences in the starting materials or treatment of them. Holdaway's hydrothermal experiments on A-S used the sensitive method of weight gain or loss of a polished crystal of natural andalusite immersed in a fine powder of natural sillimanite. He found an andalusite field considerably contracted over that of Richardson et al. (1969), who used comparison of X-ray peak heights before and after hydrothermal runs on powder mixes, a method used in earlier studies (Newton, 1966a). Holdaway's method was repeated with similar results by Kerrick and Heninger (1984).

Holdaway (1971) ascribed the disparate results on A-S to the presence of metastable fibrolite in the natural sample of sillimanite used by Richardson et al. (1969). Another factor could have been the vigorous grinding of their starting materials. The latter possibility is supported by theoretical analysis of heat capacity measurements on severely ground sillimanite by Salje (1986).

Figure 2 shows the reported phase equilibrium data derived under well-defined conditions. Gordon (1973) and Day and Kumin (1980) analyzed these data by the method of linear parametric analysis to extract self-consistent thermodynamic parameters. In Day and Kumin's analysis an individual data point, for instance a run in the andalusite field adjacent to the K-A boundary (point 2, Fig. 2), corresponds to an expression of the form of (9c) without the volatile fugacity term.At the T and P of data point 2, plotted at the extreme limit of stated uncertainties in experimental conditions, following Day and Kumin (1980), expression (9c) becomes $\Delta H°_{298} - T\Delta S°_{298} < C$, where C is a constant.

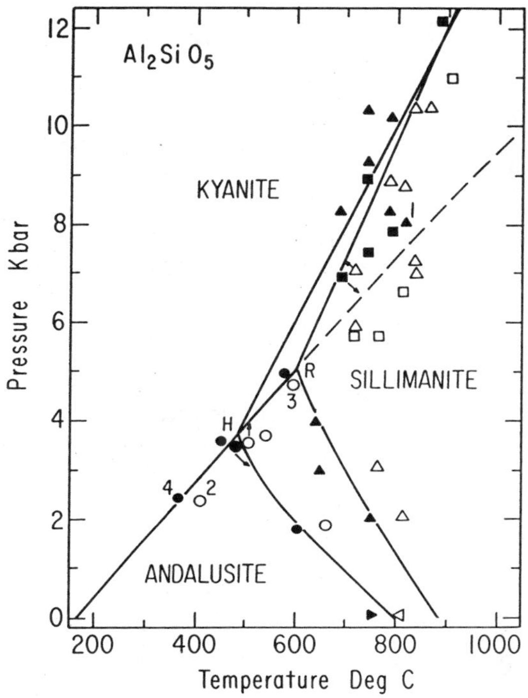

Figure 2. Al_2SiO_5 equilibria satisfying measured thermophysical and volumetric data and andalusite-sillimanite reversal data of either Richardson et al. (1969) leading to R triple point, or Holdaway (1971), leading to H triple point. Also satisfied are reversal data of kyanite-andalusite and kyanite-sillimanite equilibria. Shaded symbols: relatively low-T assemblages stable. Triangles: Richardson et al. (1968; 1969); circles: Holdaway (1971); squares: Newton (1966a, 1966b, and unpublished data); tilted triangles: Weill (1966, unreversed). Runs are plotted at extremes of P and T uncertainties.

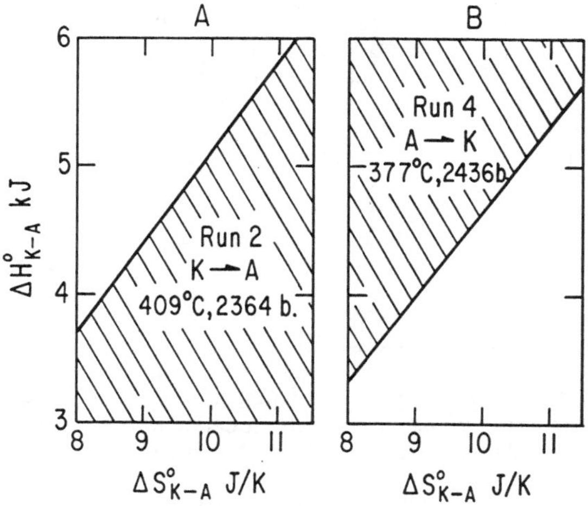

Figure 3. Half-spaces (shaded) of possible standard enthalpy-standard entropy changes of the kyanite-andalusite reaction in accord with two experimental runs of Figure 2, according to the inequalities (9b) and (9c) of the text.

This inequality defines a half-space of allowable ΔH°_{298} and ΔS°_{298}, as shown in Figure 3a. For data point 4, a similar inequality exists, but of the opposite sign, since the run reacted to kyanite ($\Delta G > 0$). Figure 3b shows the allowable half-space of point 4. In principle, any three or more data points on opposite sides of the reaction may constrain ΔH°_{298} and ΔS°_{298} to a triangular, quadrilateral, etc. element if a mutually accessible space exists. The minimum and maximum values of ΔH°_{298} and ΔS°_{298} consistent with all of the data points form apices of the smallest feasible ΔH°_{298}-ΔS°_{298} multilateral. The parametric method has the useful property of identifying inconsistent runs which contradict the accessible space defined by the others. Experimental data for all three Al_2SiO_5 reactions may be analyzed simultaneously by this method to extract three sets of self-consistent ΔH° and ΔS° parameters, subject to the conditions that $\Sigma\Delta H^\circ = 0$ and $\Sigma\Delta S^\circ = 0$.

The smallest accessible ΔH°_{298}-ΔS°_{298} area determined by any four kyanite-andalusite reversals is shown in Figure 4. The extreme corners of the feasible ΔH°_{298}-ΔS°_{298} space might be considered to define the uncertainties of the standard parameters. This is not a valid assessment, however, because the uncertainties so determined are highly correlated, rather than independent. To estimate uncertainties in the standard reaction parameters, Day and Kumin (1980) introduced the quantity

$$Z = \left|\frac{\Delta H - \Delta H'}{U(\Delta H')}\right| + \left|\frac{\Delta S - \Delta S'}{U(\Delta S')}\right| \qquad (18)$$

where ΔH and ΔS are a pair of accesible solutions of the inequalities 9b) and 9c), $\Delta H'$ and $\Delta S'$ are corresponding calorimetrically-determined quantities, and $U(\Delta H')$ and $U(\Delta S')$ are the calorimetric uncertainties in these quantities. The pair of parameters, $\Delta H^{\circ *}$ and $\Delta S^{\circ *}$, in best agreement with calorimetry corresponds to the minimum Z for any pair within the accessible space, and the quantities ΔH^*-$\Delta H'$ and ΔS^*-$\Delta S'$ are measures of the uncertainties.

The standard deviation of the entropy measurements alone may be added as a constraint on feasible values of ΔH°_{298} because heat capacity measurements are generally more precise than heat of solution measurements. Using the precise low temperature C_P data of Robie and Hemingway (1984), ΔS°_{298} (K-A) is 9.09 ± 0.21 J/K, and Figure 4 gives ΔH°_{298} (KA) = 4.15 kJ with an uncertainty of no more than 0.25 kJ. This agrees, within the uncertainties, with the value of 3.77 ± 1.13 kJ which follows from the solution calorimetry of Anderson and Kleppa (1969) and Anderson et al. (1977), when corrected from the calorimeter temperature of 700°C. The two solution calorimetry studies, though published eight years apart, are internally consistent in that they were made at the same time in the same batch of $Pb_2B_2O_5$ solvent. The close agreement between Day and Kumin's (1980) rigorous analysis of the K-A equilibrium and the results of solution calorimetry is an encouraging but rare instance of the convergence that should be expected in the application of different thermodynamic methods to simple mineral systems.

Day and Kumin's (1980) analysis shows that the Richardson et al. (1969) and the Holdaway (1971) A-S data are equally compatible with all other experimental Al_2SiO_5 data, except for the unreversed cryolite solubility measurement of Weill (1966) at one bar, which favors the Holdaway diagram. The Robie and Hemingway (1984) C_P data of the Al_2SiO_5 phases show that the Holdaway diagram can be calculated without recourse to disordering entropy of sillimanite. Salje (1986) showed, however, that heat capacities in the range 320-400 K of fibrolitic sillimanite are significantly higher than those of coarse-grained sillimanite, and that the heat capacities can be increased significantly by ultra-grinding. By theoretical modelling of the heat capacity data, he was able to calculate a range of triple point positions encompassing both the Holdaway (1971) and Richardson et al. (1969) versions, depending generally on the grain size and crystallinity of the sillimanite. This observation is generally consistent with the suggestion that fibrolite in Richardson et al's experimental sillimanite and the grinding they gave it might have influenced their triple point location.

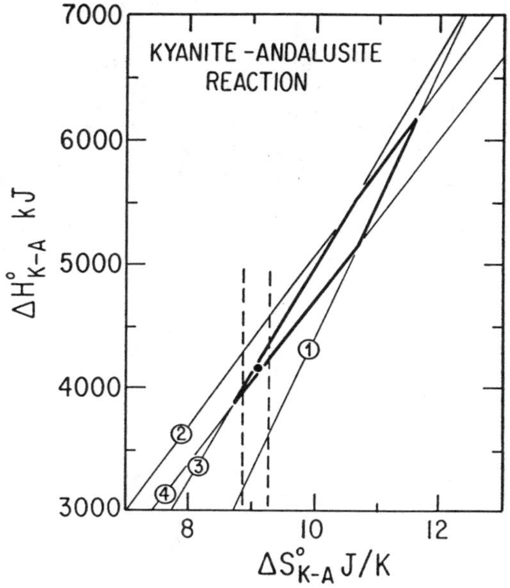

Figure 4. Linear parametric analysis of the kyanite-andalusite reaction based on runs 1-4 of Figure 2 defining a feasible space of ΔH°_{K-A}-ΔS°_{K-A} values (enclosed by bold lines). Dashed lines show 2ß error limits of ΔS°_{K-A} at 298.15 K based on adiabatic calorimetry (Robie and Hemingway, 1984). Their intersection with the experimental feasible space defines ΔH°_{K-A} of 4.15 kJ at 298.15 K.

Table 3

Enthalpy of formation (oxides), entropy, and molar volume at 298 K, heat capacity coefficients, thermal expansion and compressibility of the Al_2SiO_5 polymorphs

Mineral	ΔH°_f (kJ)	σ_H	S° (J/K)	σ_S	a (J/K)	b x10^2	c x10^{-5}	d x10^{-3}	V (cm^3)	αx10^5 (k^{-1})	βx10^7 (bar^{-1})	Refs.
Kyanite												
Measured	-5.90	0.59	82.30	0.13	303.9	-1.339	-8.952	-2.9043	44.15	2.54	4.30	1-5
Derived	-7.91											6
Andalusite												
Measured	-1.13	0.75	91.39	0.14	290.4	-1.052	-11.09	-2.6278	51.52	2.47	5.43	1-5
Derived	-3.76											6
Sillimanite (Lo S)												
Measured	+1.34	1.26	95.79	0.14	226.1	1.407	-24.40	-1.3760	49.86	1.44	6.22	1-5
Derived	+0.32											6
Sillimanite (Hi S)												
Derived	+2.13		97.25									6

References: 1: Kiseleva et al., 1983; 2: Robie and Hemingway, 1984; 3: Pankratz and Kelley, 1964 (refitted by Holland and Powell, 1985); 4: Skinner et al., 1961; 5: Brace et al., 1969; 6: Present study
Heat capacity coefficients valid 298-1600 K.

Figure 2 shows two Al_2SiO_5 diagrams which are compatible with either the Richardson et al. (1969) or the Holdaway (1971) A-S data points and with other reversed data and the heat capacity measurements. The diagram with the lower triple point coinciding with Holdaway's diagram assumes zero disorder entropy of sillimanite. The diagram of the higher triple point, resembling the diagram of Richardson et al., assumes a 1.46 J/K entropy increment of sillimanite above the Robie and Hemingway (1984) value, resulting from Al,Si disorder, defects, or hyperfine crystals. The heat capacity and standard entropy data are collected in Table 3, along with enthalpy of formation data compatible with the two diagrams of Figure 2. Calorimetric enthalpy of formation measurements of the Al_2SiO_5 polymorphs are fragmentary and somewhat inconsistent (Charlu et al., 1975; Kiseleva et al., 1983). More consistent values are gotten using a paragenetic approach. Although corundum and quartz are not uncommonly found together in high temperature metamorphic rocks (Steefel and Atkinson, 1984; Powers and Bohlen, 1986), textural criteria suggest that they are unstable relative to sillimanite. The criterion that the least stable form of sillimanite ("Hi S" form, Table 3), corresponding to the "R" triple point of Figure 2 is marginally stable relative to corundum and quartz at 1000 K and 5 kbar ($\Delta G°_f = 0$) leads to the $\Delta H°_{f,298}$ values in Table 3.

The most important contribution of the modern Al_2SiO_5 diagrams to petrology is to show emphatically that many assemblages in metamorphic rocks represent "frozen-in" equilibrium from conditions at elevated temperature and pressure, and that lithostatic burial pressure is the principal source of metamorphic pressure. Various field observers (e.g. Johnson, 1963) had criticized earlier experimental Al_2SiO_5 diagrams arguing that the maximum possible burial pressures estimated from known stratigraphic and structural controls were considerably less than the pressures of over 8 kbar required to stabilize kyanite according to some of the diagrams. Indeed, some early estimates of the triple point pressure and temperature based on purely geologic criteria are remarkably close to Holdaway's triple point (Schuiling, 1957; Hietanen, 1967). This coincidence virtually rules out the possibility that kyanite or other high pressure minerals commonly formed under sporadic "overpressure", either local or regional (Clark, 1961; Coleman and Lee, 1963).

Geologic criteria are invoked to discriminate between the Holdaway (1971) and Richardson et al. (1969) diagrams. The existence of the apparently stable natural assemblages chloritoid-sillimanite (Holdaway, 1978; Milton, 1986) and paragonite-quartz-sillimanite (Grambling, 1984) are compatible with the Holdaway diagram, but not with the Richardson et al. diagram, as shown in Figure 5. On the other hand, the presence of apparently primary magmatic andalusite in pegmatites (Fyfe, 1969) and granites (Clark et al., 1976) are allowed for by the Richardson et al. diagram, but not the Holdaway diagram. Thus, discrimination by paragenetic analysis is at a stalemate. Despite the remaining controversy over 1.5 kbar and 130°C differences in the triple points of the competing diagrams, the Al_2SiO_5 diagram has been of great value in lending credence to the paragenetic approach to mineral physical chemistry.

CALCIUM-ALUMINUM SILICATES

The system CaO-Al_2O_3-SiO_2-H_2O is the most-discussed several-component system in experimental and theoretical mineralogy because of the many natural phases with compositions close to the simple system and because of experimental accessibility of many reactions among these phases. Reliable reversed experimental data are available on reactions among the phases anorthite ($CaAl_2Si_2O_8$), zoisite ($Ca_2Al_3Si_3O_{12}[OH]$), grossular ($Ca_3Al_2Si_3O_{12}$), margarite ($CaAl_4Si_2O_{10}[OH]_2$), lawsonite ($CaAl_2Si_2O_7[OH]_2 \cdot H_2O$), CaTs pyroxene ($CaAl_2SiO_6$), gehlenite ($Ca_2Al_2SiO_7$), wollastonite ($CaSiO_3$) and the hydrous and anhydrous aluminum silicates. These reactions have been analyzed by regression analysis (Helgeson et al., 1978; Haas et al., 1981) and

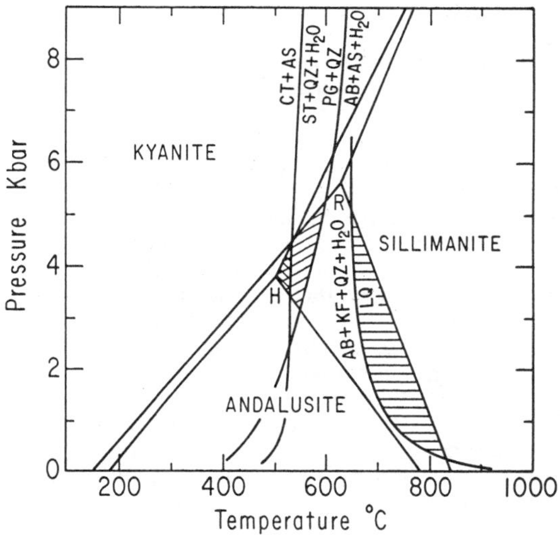

Figure 5. Experimental univariant equilibria used with natural parageneses to test the experimental Al_2SiO_5 diagrams of Richardson et al. (1969) (triple point at "R") and of Holdaway (1971) (triple point at "H"). Reaction of chloritoid (CT) and Al_2SiO_5 (AS) to staurolite (ST), quartz (QZ) and H_2O from Ganguly (1969); reaction of paragonite (PG) and quartz (QZ) to albite (AB), (AS) and H_2O from Chatterjee (1973); reaction of AB, K-feldspar (KF), QZ and H_2O to granite liquid (LQ) from Luth et al. (1964). Right-slanted ruling shows P-T area compatible with the observed occurrence of chloritoid-sillimanite and the Holdaway (1971) diagram. Left-slanted ruling shows P-T area compatible with the observed assemblage paragonite-quartz-sillimanite and the Holdaway (1971) diagram. Horizontal ruling shows P-T area compatible with the occurrence of andalusite-bearing pegmatites and granites and the Richardson et al. (1969) diagram.

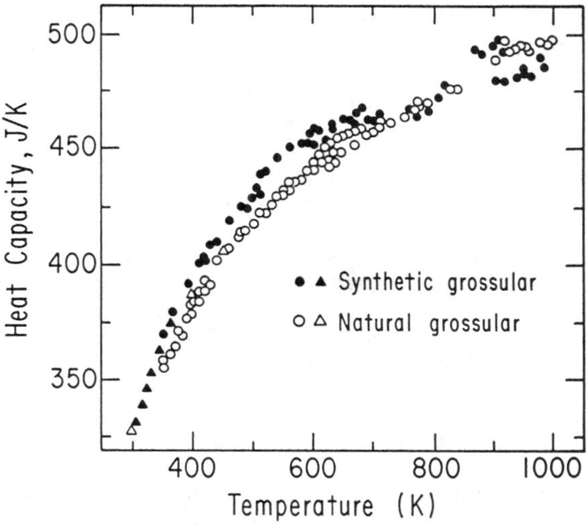

Figure 6. Heat capacity data of synthetic and natural grossular by adiabatic calorimetry (triangles) and differential scanning calorimetry (circles). Data for synthetic grossular are from Haselton and Westrum (1980) and Krupka et al. (1979b); data for natural grossular are from Westrum et al. (1979) and Krupka et al. (1979b).

linear parametric analysis (Gordon, 1973; Halbach and Chatterjee, 1984; Berman et al., 1985) to extract thermodynamic parameters. Several adiabatic calorimetry and solution calorimetry studies have been carried out on synthetic and nearly end-member natural samples. The derived and measured thermodynamic properties are in encouraging agreement.

Many of the phases have the possibility of variable cation disorder, including anorthite, margarite, CaTs and gehlenite, and the configurational entropies and appropriate standard states of these phases are, to some extent, ambiguous. Grossular is a well-defined substance which can serve as a base for analysis. Four studies of heat capacity (Kolesnik et al., 1979; Westrum et al., 1979; Krupka et al., 1979b; Haselton and Westrum, 1980), and three of enthalpy of solution (Charlu et al., 1978; Kiseleva, 1984; Stolyarova, 1986) exist. There are significant discrepancies in heat capacities between measurements on synthetic and natural samples, as shown in Figure 6. Synthetic grossular differs from natural grossular in having slightly higher C_P and $S°$. The discrepancy in $S°_{298}$ may result from impurities in the natural samples or inadequate impurity correction procedures (Westrum et al., 1979) or from some subtle differences in crystal structure. Differences in thermodynamic parameters of Ca-Al silicates derived from experimental phase equilibria by various authors may often be traced to the C_P dataset adopted for grossular.

A convenient starting point for analysis is the breakdown reaction of anorthite at high pressures:

$$\underset{\text{grossular}}{Ca_3Al_2Si_3O_{12}} + \underset{\text{kyanite}}{2Al_2SiO_5} + \underset{\text{quartz}}{SiO_2} = \underset{\text{anorthite}}{3CaAl_2Si_2O_8}. \qquad (I)$$

Numerous experimental studies of this reaction exist. The most precise data available are those of Koziol and Newton (1986) on synthetic material using a molten Li_2MoO_4 intergranular flux (Ito,1975). Their reversals are shown in Figure 7. The dP/dT slope of 22.47 ± 0.30 bars/K defined by the experimental data can be rationalized in two contrasting ways by the heat capacity measurements. If the data on synthetic grossular are accepted, a significant amount of configurational entropy in anorthite, presumably from about 5% of Al,Si tetrahedral disorder, is necessary (Haselton, 1979; Wood and Holloway, 1984; Gasparik, 1984), in order to reproduce the experimental dP/dT slope. If the heat capacity data of natural grossular are accepted, the slope is reproduced using the Third Law entropy of anorthite without a configurational increment. Further insight into the problem of anorthite disorder requires analysis of other equilibria.

Experimental data on three reactions involving anorthite, grossular and zoisite are shown in Figure 8. The three univariant reactions are:

$$\underset{\text{zoisite}}{2Ca_2Al_3Si_3O_{12}(OH)} + \underset{\text{kyanite}}{Al_2SiO_5} + \underset{\text{quartz}}{SiO_2} = \underset{\text{anorthite}}{4CaAl_2Si_2O_8} + H_2O, \qquad (J)$$

$$\underset{\text{zoisite}}{4Ca_2Al_3Si_3O_{12}(OH)} + \underset{\text{quartz}}{SiO_2} = \underset{\text{anorthite}}{5CaAl_2Si_2O_8} + \underset{\text{grossular}}{Ca_3Al_2Si_3O_{12}} + 2H_2O, \qquad (K)$$

$$\underset{\text{zoisite}}{6Ca_2Al_3Si_3O_{12}(OH)} = \underset{\text{anorthite}}{6CaAl_2Si_2O_8} + \underset{\text{grossular}}{2Ca_3Al_2Si_3O_{12}} + \underset{\text{corundum}}{Al_2O_3} + 3H_2O, \qquad (L)$$

The P-T curves of these equilibria are tightly constrained by numerous reversed experimental data determined under well-defined conditions.

The method of linear parametric analysis based on expression (9) may be used to determine the standard reaction parameters of the dehydration equilibria. The procedure is similar to that used by Chatterjee (1977). In Reaction (K), for example, $\Delta H°_{f,s}$

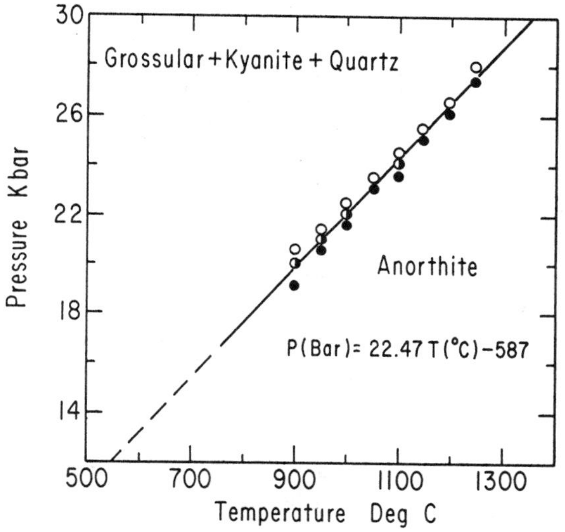

Figure 7. Reversed experimental data of the reaction 3 anorthite = grossular + 2 kyanite + quartz of Koziol and Newton (1986) and A.M. Koziol (personal communication, 1987). Line through data is based on regression analysis of the data of this and three other CASH reactions (see text).

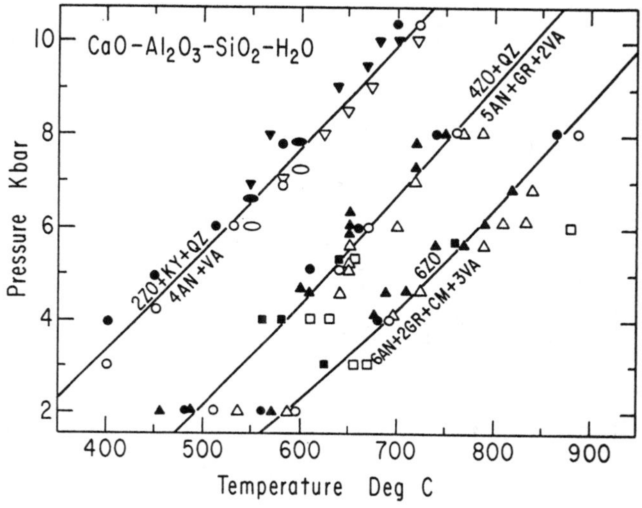

Figure 8. Reversed experimental data of three CASH reactions involving zoisite (ZO), anorthite (AN), grossular (GR), kyanite (KY), quartz (QZ), corundum (CM) and an aqueous vapor (VA). Shaded symbols indicate zoisite-forming reactions. Open symbols indicate anorthite-forming reactions. Triangles: Newton (1965; 1966c). Inverted triangles: Goldsmith (1981). Ovals: Jenkins et al. (1984). Circles: Chatterjee et al. (1984). Squares: Boettcher (1970). Reaction lines are calculated from data in Table 4.

Table 4

Enthalpies of formation (oxides), entropies, heat capacity coefficients, volumes, expansivities and compressibilities of $CaO-Al_2O_3-SiO_2-H_2O$ phases and component oxides

Mineral Source	Ideal formula Impurities, wt. %	$\Delta H^\circ_{f,298}$ kJ/Mol	σ_H	S°_{298} J/K-Mol	σ_S	$a \cdot 10^{-2}$ J/K	$b \cdot 10^{2}$	$c \cdot 10^{-6}$	$d \cdot 10^{-3}$	$e \cdot 10^{5}$	V°_{298} cm^3	$\alpha \cdot 10^{5}$ K^{-1}	$\beta \cdot 10^{7}$ bar^{-1}	Refs.
GROSSULAR	$Ca_3Al_2Si_3O_{12}$													
Natural*	FeO 1.81, MgO .24, MnO .65, TiO_2 .25			255.50	0.51	16.333	-75.99	9.113	-20.873	26.690				1,2
Natural**	Fe_2O_3 .30, MgO .01, MnO .10	-333.46	1.51	258.77	0.20									3,4
Synthetic		-320.73	2.80	260.12	0.26	5.1633	3.2184	-9.9419	-1.4031		125.30	2.66	6.4	5,6,7
Derived		-316.91												8
ANORTHITE	$CaAl_2Si_2O_8$													
Synthetic		-96.48	1.30	199.	0.30	5.1683	-9.2492	-1.4085	-4.5885	4.1883	100.79	1.42	12.9	9,10,11
Derived		-97.91	1.21	202.52										8
ZOISITE	$Ca_2Al_3Si_3O_{12}(OH)$													
Natural	MgO .10, V_2O_3 .23, rest .17			295.98	0.30	8.3462	-3.9689		-8.1488		136.52	2.80	5.9	12,13,14
Derived		-259.58												8
CLINOZOISITE	$Ca_2Al_3Si_3O_{12}(OH)$	-265.12												8,15
Derived				305.31										
KYANITE	Al_2SiO_5	-7.96		82.30		4.3612	-13.576		-4.8027	4.7236				8,10,11,16,17
LIME	CaO	0		38.21	0.13	0.5244	0.3673	-0.7507	-0.05099					10
CORUNDUM	Al_2O_3	0		50.92	0.10	1.5736	0.07190	-1.8969	-0.9880		25.58	2.50	3.5	10,11
α-QUARTZ†	SiO_2	0		41.46	0.20	0.4460	3.7754	-1.0018			22.69	5.91	26.7	10,11,18
β-QUARTZ††	SiO_2	0				0.5893	1.0031				23.31	-0.77	17.7	10,11,18
STEAM	H_2O	0		188.83	0.04	0.07368	2.7468	-0.22316	0.36174	-0.48117				10

* 0.19 wt. % H_2O; ** 0.10 wt. % H_2O; † C_p valid 298-844 K; †† C_p valid 844-1800 K; $\Delta H_{\alpha-\beta} = 0.48$ kJ; $S_{\alpha-\beta} = 0.55$ J/K; 1: Westrum et al., 1979; 2: Krupka et al., 1979b; 3: Kolesnik et al., 1979; 4: Kiseleva, 1984; 5: Charlu et al., 1978; 6: Haselton, 1979; 7: Hazen and Finger, 1978; 8: Present work; 9: Newton et al., 1980; 10: Robie et al., 1978; 11: Clark, 1966; 12: Perkins et al., 1980; 13: Haas et al., 1981; 14: Holland and Powell, 1985; 15: Jenkins et al., 1984; 16: Robie and Hemingway, 1984; 17: Brace et al., 1969; 18: Simmons and Wang, 1971. $C_p = a + bT + c/T^2 + d/\sqrt{T} + eT^2$. $V(T,P) = V^\circ[1 + \alpha(T-298) - \beta P]$.

denotes $5\Delta H°_f$ (anorthite) + $\Delta H°_f$ (grossular) - $4\Delta H°_f$ (zoisite), with formation from the oxides, including steam. Similar expressions exist for $\Delta S°_{f,s}$ and $\Delta C_{P f,s}$. The "greater than" sign applies to a data point which indicates reaction to zoisite, and the "less than" sign applies to an anorthite-forming reaction. A plot of the points given by the last three terms of the inequality versus T yields a narrow region confining a straight line of slope $\Delta S°_{f,s,298}$ and intercept at zero Kelvin of $-\Delta H°_{f,s,298}$.

The reaction most tightly constrained by the reversal data is (K). The plot of expression (9) in Figure 9 has almost no uncertainty in defining a $\Delta S°_{f,s,298}$ of 427.05 J/K, using the heat capacities in Table 4, including the C_P equation of Haselton (1979) for synthetic grossular. The C_P of zoisite must be extrapolated beyond the 730 K limit of the measurements by Perkins et al. (1980). The extrapolation of Haas et al. (1981), used here, and those of Holland and Powell (1985) and Perkins et al. (1980) are closely similar. The C_P data for natural grossular of Westrum et al. (1979) yield $\Delta S°_{f,s,298}$ (K) of 430.11 J/K. The maximum uncertainties in these estimates are ± 5 J/K. In contrast, the entropy of formation changes computed from the 298 K Third Law entropies are 406.8 J/K using the C_P data for natural grossular and 410.8 J/K using the C_P data for synthetic grossular. These values are much smaller than $\Delta S°_{f,s,298}$ calculated from expression (9), which apparently calls for a disordering entropy increment of anorthite of about 3.2 J/K, using the Haselton (1979) C_P data of grossular. This anorthite disorder entropy is similar to that derived from phase equilibrium analysis by Wood and Holloway (1984) and Gasparik.

A similar analysis applied to the experimental reactions (I)-(L) yields, with relations of the type of (10), values of $\Delta H°_{f,298}$ (anorthite), $\Delta H°_{f,298}$ (zoisite), $\Delta H°_{f,298}$ (grossular) and S^{conf} (anorthite). The measured and derived thermodynamic quantities are given in Table 4. The derived enthalpy of formation of anorthite and grossular agree fairly well with measurements by solution calorimetry.

The present formation properties of the CASH minerals are compared in Table 5 with those of several other internally consistent datasets. Additional experimental reactions do not improve the reliability of the derived properties, since additional minerals, many having order-disorder complications, are necessarily introduced also. Differences among the derived properties of Table 5 arise largely from use of different numbers of experimental reactions involving additional CASH minerals.

The method of paragenetic analysis is illustrated by the polymorphic reaction

$$\underset{\text{clinozoisite}}{Ca_2Al_3Si_3O_{12}(OH)} = \underset{\text{zoisite}}{Ca_2Al_3Si_3O_{12}(OH)}. \qquad (M)$$

The clinozoisite polymorph has been stated to be the high-temperature form by Ackermand and Raase (1973) and the low-temperature form by Holdaway (1972). The natural phases always contain significant amounts of ferric iron, which undoubtedly has a stabilizing effect, and the two polymorphs commonly coexist in low- and middle-grade calcareous and basic metamorphic rocks. Clinozoisite invariably contains more iron than does zoisite in these occurrences.

Jenkins et al. (1984) reversed Reaction (J) in the range 500°C-600°C using both synthetic Fe-free zoisite and clinozoisite. They found that the equilibrium with clinozoisite lies at pressures 1.0 ± 0.4 kbar higher than that with zoisite at 550°C and therefore that zoisite is more stable than clinozoisite by $\Delta G°_M$ = -4.35 ± 1.80 kJ at this temperature.

The effect of ferric iron on the equilibrium may be evaluated from analyses of natural coexisting pairs and a simple ideal solution model. Using extrapolated end member components of $Ca_2Al_3Si_3O_{12}(OH)$ = ZS and $Ca_2Fe^{3+}Al_2Si_3O_{12}(OH)$ = PS, the equilibrium

Table 5

Enthalpy of formation (oxides) and entropy of anorthite, grossular and zoisite according to several authors

AUTHOR	SUBSTANCE AND PARAMETERS, 298 K						METHODS
	Anorthite		Grossular		Zoisite		LPA = Linear parametric analysis
	ΔH°_f (kJ)	S° (J/K)	ΔH°_f (kJ)	S° (J/K)	ΔH°_f (kJ)	S° (J/K)	
Helgeson et al., 1978	-98.48	205.43	-326.07	254.68	-263.53	295.98	Regression, 5 reactions
Haas et al., 1981	-95.63 ±1.12	199.29 ±0.15	-323.25 ±3.22	255.97 ±2.95	-254.36 ±0.88	295.89 ±0.66	Regression, 7 reactions
Halbach & Chatterjee, 1984	-103.41	199.60	-330.02	259.60	-266.78	296.45	LPA, 13 reactions
Holland & Powell, 1985	-101.65 ±1.49	201.0	-323.86 ±2.80	260.1	-263.58 ±2.46	295.9	Regression, 17 reactions
Berman et al., 1985	-97.21	200.44	-320.63	255.00	-255.64	295.44	LPA, 32 reactions
Present	-97.91	202.52	-316.91	260.12	-259.58	295.98	Regression, 4 reactions

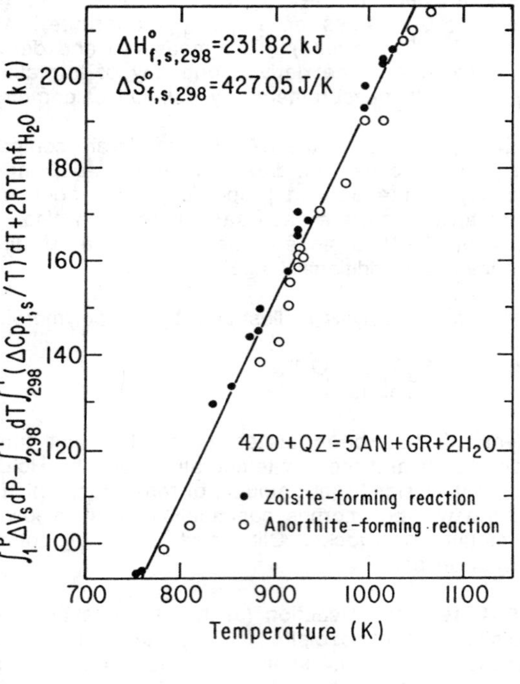

Figure 9. Linear parametric analysis of the reaction 4 zoisite + quartz = 5 anorthite + grossular + $2H_2O$ based on the data points of Figure 8 and inequalities (9b) and (9c) of text. Volume and heat capacity data are from Table 4 and H_2O fugacity (f_{H2O}) data are from Burnham et al. (1969).

relation becomes (Jenkins et al., 1984):

$$-\Delta H^\circ_M/RT + \Delta S^\circ_M/R \approx \ln X_{ZS}^{ZO}/X_{ZS}^{CZO}. \quad (19)$$

The approximation sign is used because small differences in volumes and heat capacities of the polymorphs are neglected and because of the ideal activity-composition. Figure 10 shows a plot of the right-hand side of the above equation versus reciprocal temperature, based on some analyses of natural coexisting zoisite-clinozoisite pairs and broad assumed temperature ranges of the parageneses (Jenkins et al., 1984). In addition, the Jenkins et al. (1984) 550°C point and an unreversed experimental composition determination of Prunier and Hewitt (1985) are plotted. The plot suggests that Fe-free clinozoisite is stable relative to Fe-free zoisite below about 200°C. The intercept at 1/T = 0 (infinite T) gives an estimate of the entropy of inversion of 9.33 J/K. The inferred properties of clinozoisite are given in Table 4. As with the crystallochemically similar polymorphic relationship of enstatite and clinoenstatite, the more symmetrical orthorhombic form has the higher entropy. The deductions from natural parageneses may be tested by calorimetric measurements on synthetic Fe-free clinozoisite, if suitable material in sufficiently large quantities can be synthesized.

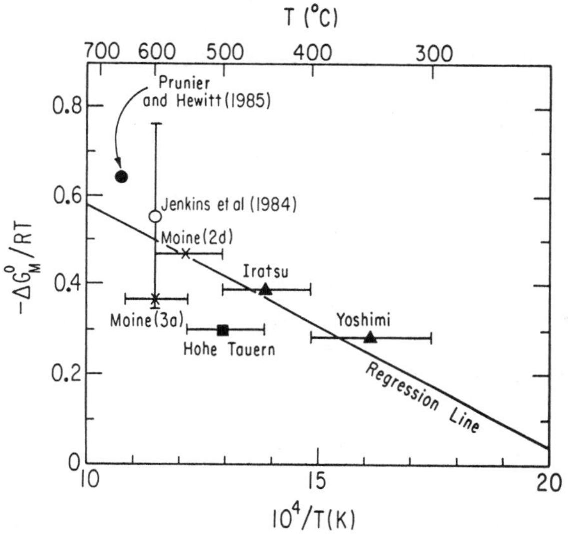

Figure 10. Temperature-normalized standard free energy of the polymorphic reaction zoisite = clinozoisite versus reciprocal temperature, based on two experimental data (Prunier and Hewitt, 1985; Jenkins et al., 1984) and analyses of Fe^{3+} in coexisting natural zoisite-clinozoisite pairs from Japan, Scotland and Austria with assumed temperature ranges (sources in Jenkins et al., 1984). A linear regression of the data gives an extrapolated equilibrium temperature near 200°C (see text).

The near agreement of the thermochemical quantities derived from phase equilibrium diagrams and those measured by calorimetry in the CASH system is very encouraging, considering the completely different lines of approach. The laboratory results, guided and extended by deductions from natural parageneses, can be applied effectively to the study of high-temperature and high-pressure calc-silicate assemblages from a variety of natural environments.

CONTINUOUS DEHYDRATION REACTIONS - HYDROUS CORDIERITE

Certain minerals, including cordierite and some zeolites, have the property of continuously variable hydration states. Water molecules occupy channels and cages in aluminosilicate framework structures and are more loosely bound than hydroxyl groups in other silicates, accounting for ease of dehydration and variable hydration state dependent on $P(H_2O)$ and T. The partial molal enthalpy of H_2O is therefore small and the partial molal entropy is high (Zen, 1972), giving rise to peculiar phase equilibria in reactions with other minerals, such as the retrograde (decreasing temperature) dehydration (Fig. 11) of hydrous cordierite to talc, kyanite, quartz and vapor (Newton, 1972). Analogous retrograde dehydrations are those of laumontite to lawsonite, quartz and vapor (Liou, 1971) and of analcime to jadeite and vapor (Manghnani, 1970).

The thermodynamics of cordierite dehydration can be analyzed by the divariant reaction,

$$Mg_2Al_4Si_5O_{12} + nH_2O \quad = \quad Mg_2Al_4Si_5O_{18}{\cdot}nH_2O. \qquad \text{(N)}$$

This expression supposes two endmembers, anhydrous Mg-cordierite and a somewhat fictive fully hydrated endmember. The geometry of the structural cages enclosing H_2O suggests only one H_2O molecule per cage (Smith and Schreyer, 1960), so that n should be unity in the hydrous end member, corresponding to 2.99 wt % H_2O. However, experimental measurements of the hydration state of Mg-cordierite at various $P(H_2O)$ and T (Mirwald and Schreyer, 1977) indicates that $n > 1$ is possible. Helgeson et al. (1978) and Kurepin (1979) fitted an ideal solution of the end-members to the isohydron data, with n = 1 for the hydrous component, whereas Newton and Wood (1979) used n = 1.2. Ideal solution assumes random mixing of H_2O and voids over the cage sites, which must be an oversimplification, since Aines and Rossman (1984) found two discrete crystallographic orientations of the H_2O molecules in cordierite by IR absorption studies.

According to the ideal solution analysis, the equation of equilibrium N) is

$$\Delta H^\circ_N - T\Delta S^\circ_N + RT \ln (X_{HCd}/X_{Cd}{\cdot}f^n_{H2O}) = 0, \qquad (20)$$

where the X's are the mole fractions of the assumed end members and $f(H_2O)$ is the fugacity of H_2O. A solid volume term is neglected because Cd and HCd have nearly identical molar volumes. A least-squares fit of the hydration data for n = 1 yields $\Delta H^\circ = -38.55$ kJ and $\Delta S^\circ = -96.44$ J/K. The Mirwald and Schreyer (1977) points of wt % $H_2O > 3.0$ were not used in the regression. The quality of the fit is shown in Figure 11. It will be noted that the calculated isohydrons and some of the experimental data points extend metastably beyond the cordierite field boundaries. The assumption of n = 1.2 (Fig. 12) gave $\Delta H^\circ = -51.46$ kJ and $\Delta S^\circ = -137.53$ J/K, or -42.88 kJ and -114.61 J/K per mol of H_2O absorbed, which values are not greatly different from those given by n = 1. Newton and Wood (1979) did not use the hydration data below 3 kbar in their regression.

The enthalpy and entropy of sorption of H_2O into cordierite are close to the values for condensation of steam to liquid water, which suggests that sorption is dominantly by hydrogen bonding (Langer and Schreyer, 1976). As a result of weak bonding, the partial molal entropy of H_2O in the cordierite structure is 85% to 95% of the molar entropy of free H_2O at the same conditions. This high H_2O entropy makes possible retrograde dehydration curves. The unusual dehydration behavior of cordierite creates complications in the P-T reaction-curve topologies of middle-grade metamorphism, with the possibility of retrograde dehydration in some common cordierite-forming reactions in metapelites (Newton and Wood, 1979). Reduction of $P(H_2O)$ in metamorphism would have a substantial effect of destabilizing cordierite, as illustrated by the difference in breakdown pressure of cordierite to enstatite, sillimanite and quartz in the wet (w) and dry (d) systems, shown in Figure 12, based on the expression (Newton and Wood, 1979):

$$P_w - P_d = 1.2RT \ln (1-X_{HCd})/\Delta V_s, \qquad (21)$$

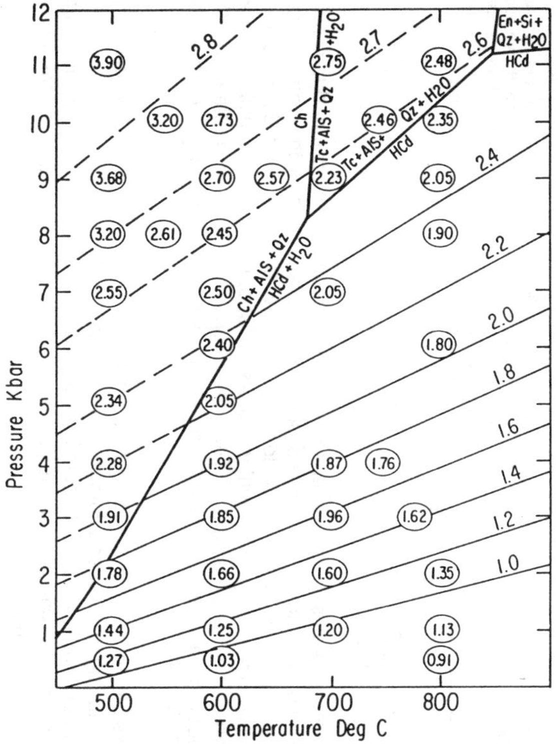

Figure 11. Isohydrons in wt % H_2O of hydrous Mg-cordierite (HCd) from $P(H_2O) = P_{total}$ (light lines, dashed where metastable) calculated from ideal Cd-HCd mixing theory with $n(H_2O) = 1$ by regression of experimental points of Mirwald and Scheyer (1977) (ovals). Symbols: Ch = clinochlore Tc = talc; AlS = Al_2SiO_5; En = enstatite; Si = sillimanite; Qz = quartz. Univariant equilibria (bold lines) discussed in Newton and Wood (1979). Note retrograde (decreasing temperature) dehydration of cordierite to talc, Al_2SiO_5, and quartz.

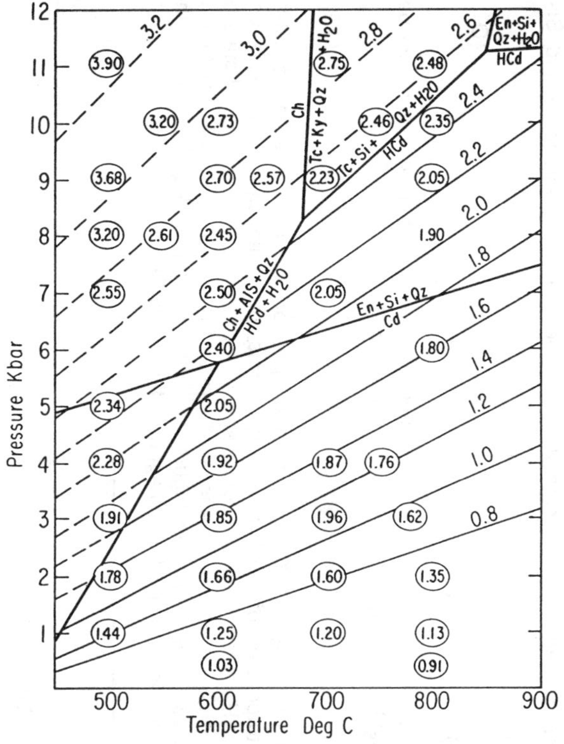

Figure 12. Isohydrons of H_2O in cordierite for $P(H_2O) = P_{total}$ calculated with ideal solution model and $n(H_2O) = 1.2$ (Newton and Wood, 1979). Position of anhydrous cordierite breakdown reaction to En + Si + Qz and standard properties of anhydrous cordierite calculated from the analogous HCd breakdown (upper right hand side) with the hydration theory. Note that H_2O in Mg-cordierite stabilizes it by more than three kbar.

where X_{HCd} is the mole fraction of hydrous cordierite (n = 1.2) at the breakdown pressure in a system of $P(H_2O) = P_{total}$.

A different sort of hydration theory starts with the reaction (Martignole and Sisi, 1981)

$$\underset{\text{(in cordierite)}}{H_2O} = \underset{\text{(in vapor).}}{H_2O} \tag{O}$$

Defining the hydration number, $n(H_2O)$, as the number of moles of H_2O in cordierite per mole of $Mg_2Al_4Si_5O_{18}$, the activity, $a(H_2O)$ in cordierite is

$$a(H_2O) = f(H_2O)/k = \gamma(H_2O) \cdot n(H_2O), \tag{22}$$

where $f(H_2O)$ is the fugacity of H_2O in the coexisting vapor, k is a Henry's Law constant determined by extrapolation of the hydration isotherms from infinitely low pressure and H_2O content, and $\gamma(H_2O)$ is the activity coefficient. Selection of a fictive standard state of $n(H_2O) = 1$ and $a(H_2O) = 1$ gives k for each isotherm as $f°(H_2O)$, the fugacity of H_2O at T and the extrapolated P, as shown in Figure 13. The activity of $Mg_2Al_4Si_5O_{18}$, the anhydrous component, is obtained from integrating the Gibbs-Duhem equation.

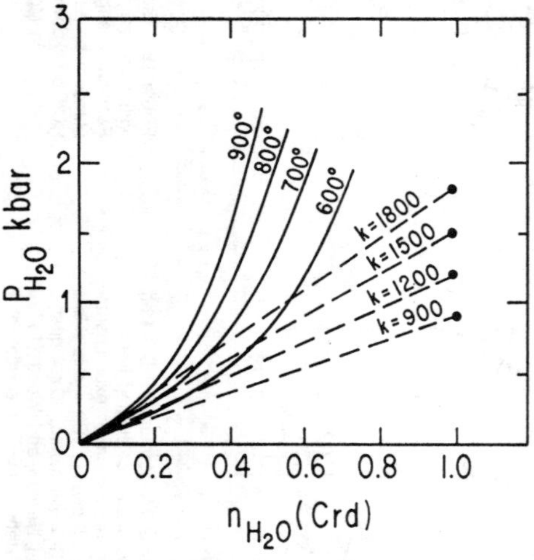

Figure 13. Hydration isotherms constructed by Martignole and Sisi (1981) from experimental hydration data of Schreyer and Yoder (1964). Henry's Law constants obtained by extrapolation to $n(H_2O) = 1$ of low pressure tangents are shown (see text).

The great advantage of this method is that it is not dependent on an assumed molecular model. The measured isohydrons themselves determine the activities of the components. Martignole and Sisi (1981) used the older, low-pressure hydration data of Schreyer and Yoder (1964), determined in slow-quenching cold-seal apparatus, with possible failure to retrieve equilibrium hydration states. The isohydrons are not well defined by the Schreyer and Yoder data, and long extrapolation of strongly curving isotherms was necessary to get H_2O activities.

It is quite possible that the method of Martignole and Sisi (1981) will be the most useful method of making phase equilibrium calculations of hydrous cordierite, when a large body of reversed hydration data becomes available. It may be necessary to define separately two extrapolated standard states for two P-T regions separated by the phase transition of Mirwald (1982).

FERROUS IRON MINERALS

Calorimetry of ferrous iron minerals has lagged behind work on other substances because of difficulties of controlling oxidation states in solution calorimetry and because of difficulty of synthesizing large quantities of suitable material. However, available phase equilibrium data make possible reliable calculations of the standard properties of some important ferrous iron minerals.

The "anchor" substances are ilmenite, $FeTiO_3$, and fayalite, Fe_2SiO_4. The free energy of formation of ilmenite was derived by Anovitz et al. (1985) from the experimental data on the oxidation reaction

$$\underset{\text{metal}}{Fe} + \underset{\text{rutile}}{TiO_2} + {}^1/_2O_2 = \underset{\text{ilmenite}}{FeTiO_3}. \qquad (P)$$

The free energy of formation from the elements is given by

$$\Delta G^\circ_{f,el}\ (\text{ilmenite}) = \Delta G^\circ_{f,el}\ (\text{rutile}) + {}^1/_2\ RT\ \ln f(O_2). \qquad (23)$$

A similar method gives the free energy of formation of fayalite from iron, quartz and oxygen, based on experimental determination of the oxygen pressure of fayalite decomposition (Myers and Eugster, 1983).

Precise experimental studies at high pressures and temperatures are available to obtain the formation properties of almandine from the reaction

$$\underset{\text{ilmenite}}{3FeTiO_3} + \underset{\text{sillimanite}}{Al_2SiO_5} + \underset{\text{quartz}}{2SiO_2} = \underset{\text{almandine}}{Fe_3Al_2Si_3O_{12}} + \underset{\text{rutile}}{3TiO_2} \qquad (Q)$$

(Bohlen et al., 1983). The standard entropy of almandine at 298 K has been measured by low-temperature adiabatic calorimetry of a synthetic sample (Metz et al., 1983). $\Delta H^\circ_{f,298}$ of almandine from the oxides calculated from phase equilibrium is -37.91 ± 1.67 kJ and the measured S°_{298} is 340.16 J/K with a probable uncertainty of ± 0.1%, based on precision of similar measurements given by this group. Heat capacity coefficients derived from the experimental data for Reaction (Q) are:

$$a = 487.44\ \text{J/K};\quad b = 5.6649 \times 10^{-3};\quad c = 1.5202 \times 10^{7};\quad d = -2.4137 \times 10^{2}.$$

The large entropy and numerically small enthalpy of formation of almandine are noteworthy. Early attempts to derive these properties from experimental stability work in hydrous systems at low pressures (Hsu, 1968) yielded low entropies in the range 272-314 J/K (Zen, 1973) and much more negative enthalpies of formation, of the order of -100 kJ/mol (Zen, 1973; Ahrens, 1972). Overestimation of the temperature dependence of the reaction free energies is a common problem when experimental baselines are short. The only available calorimetric detemination of the enthalpy of formation, that of Chatillon-Colinet et al. (1983b) giving $\Delta H^\circ_{f,298,ox}$ = -53.89 ± 6.02 kJ, is not in good agreement with the calculation from phase equilibria, which is preferred over the calorimetry.

SUMMARY

Some aspects of the derivation of standard thermodynamic properties of minerals from phase equilibria have been presented, including the combined use of measured thermodynamic data and phase equilibrium data, the use of regression methods, treatment of disordered phases, treatment of devolatilization reactions, including continuous dehydration of zeolites, the use of fictive components and their reference states, and the use of deductions from natural parageneses. It is concluded that the most important properties which can be evaluated are non-Third-Law configurational entropies and enthalpies of formation. The latter quantity can, under favorable circumstances of supporting measurements, be evaluated with considerably greater accuracy than obtainable by solution calorimetry.

The present discussion has merely touched upon a very important aspect of the thermodynamical analysis of mineral systems, namely that of solid solutions. Extendec development of this subject is presented in Chapter 2. Chapter 3 illustrates practical calculations of mineralogic phase equilibrium pertinent to problems of the deep crust and upper mantle. These calculations show the interplay of optimized standard formation properties, concepts of order-disorder of minerals, and mixing properties of solid solutions.

ACKNOWLEDGMENTS

The author's research is supported by a grant from the National Science Foundation, #EAR-8411192. I thank H.W. Day, J.M. Ferry, H.T. Haselton and B.J. Wood for critica readings of the manuscript, A.M. Koziol for use of her unpublished experimental data, and Benjamin Newton for programming some of the calculations.

REFERENCES

Ackermand, D. and Rasse, P. (1973) Coexisting zoisite and clinozoisite in biotite schists from the Hohe Tauern, Austria. Contrib. Mineral. Petrol. 42, 333-341.

Ahrens, T.J. (1972) The mineralogic distribution of iron in the upper mantle. Phys. Earth Plan. Int. 5, 267-281.

Aines, R.D. and Rossman, G.R. (1984) The high temperature behavior of water and carbon dioxide in cordierite and beryl. Amer. Mineral. 69, 319-327.

Anderson, G.M. (1977) The accuracy and precision of calculated mineral dehydration equilibria. In: Thermodynamics in Geology, D.G. Fraser, ed., Dordrecht, Holland: Reidel, p. 115-136.

Anderson, P.A.M. and Kleppa, O.J. (1969) The thermochemistry of the kyanite-sillimanite equilibrium. Amer. J. Sci. 267, 285-290.

Anderson, P.A.M., Newton, R.C. and Kleppa, O.J. (1977) The enthalpy change of the andalusite-sillimanite reaction and the Al_2O_3 diagram. Amer. J. Sci. 277, 585-593.

Anovitz, L.M., Treiman, A.H., Essene, E.J., Hemingway, B.S., Westrum, E.F. Jr., Wall, V.J., Burriel, R. and Bohlen, S.R. (1985) The heat-capacity of ilmenite and phase equilibria in the system Fe-Ti-O. Geochim. Cosmochim. Acta 49, 2027-2040.

Berman, R.B., Brown, T.H. and Greenwood, H.J. (1985) An internally consistent thermodynamic data base for minerals in the system Na_2O-K_2O-CaO-MgO-FeO-Fe_2O_3-Al_2O_3- SiO_2-TiO_2-H_2O-CO_2. Atomic Energy of Canada, Ltd., Technical Rep't. TR-377, 1-62.

Berman, R.G., Engi, M., Greenwood, H.J. and Brown, T.H. (1986) Derivation of internally-consistent thermodynamic data by the technique of mathematical programming: a review with application to the system MgO-SiO_2-H_2O. J. Petrol. 27, 1331-1364.

Bird, G.W. and Anderson, G.M. (1973) The free energy of formation of magnesian cordierite and phlogopite. Amer. J. Sci. 273, 84-91.

Boettcher, A.L. (1970) The system CaO-Al_2O_3-SiO_2-H_2O at high pressures and temperatures. J. Petrol. 11, 337-379.

Bohlen, S.R., Wall, V.J. and Boettcher, A.L. (1983) Experimental investigations and geological applications of equilibria in the system FeO-TiO-Al_2O_3-SiO_2-H_2O.

Boyd, F.R. (1973) A pyroxene geotherm. Geochim. Cosmochim. Acta 37, 2533-2546.

Boyd, F.R. and England, J.L. (1965) The rhombic enstatite-clinoenstatite inversion. Carnegie Inst. Wash. Yrbk. 64, 117-120.

Brace, W.F., Scholz, C.H. and LaMori, P.N. (1969) Isothermal compressibility of kyanite, andalusite and sillimanite from synthetic aggregates. J. Geophys. Res. 74, 2089-2098.

Brousse, C., Newton, R.C. and Kleppa, O.J. (1984) Enthalpy of formation of forsterite, enstatite, akermanite, monticellite and merwinite at 1073 K determined by alkali borate solution calorimetry. Geochim. Cosmochim. Acta 48, 1081-1088.

Burnham, C.W., Holloway, J.R. and Davis, N.F. (1969) Thermodynamic properties of water to 1,000°C and 10,000 bars. Geol. Soc. Amer. Spec. Paper 132, 96 pp.

Charlu, T.V., Newton, R.C. and Kleppa, O.J. (1975) Enthalpies of formation at 970 K of compounds in the system $MgO\text{-}Al_2O_3\text{-}SiO_2$ by high temperature solution calorimetry. Geochim. Cosmochim. Acta 39, 1487-1497.

Charlu, T.V., Newton, R.C. and Kleppa, O.J. (1978) Enthalpy of formation of some lime silicates by high-temperature solution calorimetry, with discussion of high pressure phase equilibria. Geochim. Cosmochim. Acta 42, 367-375.

Chatillon-Colinet, C., Newton, R.C., Perkins, D. and Kleppa, O.J. (1983) Thermochemistry of $(Fe^{2+},Mg)SiO_3$ orthopyroxene. Geochim. Cosmochim. Acta 47, 1597-1603.

Chatillon-Colinet, C., Kleppa, O.J., Newton, R.C. and Perkins, D. (1983) Enthalpy of formation of $Fe_3Al_2Si_3O_{12}$ (almandine) by high temperature alkali borate solution calorimetry. Geochim. Cosmochim. Acta 47, 439-444.

Chatterjee, N.D. (1969) Aus welchen Erdtiefen stammen die diamantführenden Kimberlite? N. Jahrb. Mineral., Mh., 289-305.

Chatterjee, N.D. (1973) Low-temperature compatibility relations of the assemblage quartz-paragonite and the thermodynamic status of the phase rectorite. Contrib. Mineral. Petrol. 42, 259-271.

Chatterjee, N.D. (1977) Thermodynamics of dehydration equilibria. In Thermodynamics in Geology, D.G. Fraser, ed., Dordrecht-Holland, Reidel, p. 137-160.

Chatterjee, N.D., Johannes, W. and Leistner, H. (1984) The system $CaO\text{-}Al_2O_3\text{-}SiO_2\text{-}H_2O$: new phase equilibria data, some calculated phase relations, and their petrological applications. Contrib. Mineral. Petrol. 88, 1-13.

Clark, S.P. (1961) A redetermination of equilibrium relations between kyanite and sillimanite. Amer. J. Sci. 259, 641-650.

Clark, S.P. (1966), ed., Handbook of Physical Constants. Geol. Soc. Amer. Memoir 97, 587 pp.

Clarke, D.B., McKenzie, C.B., Muecke, G.K. and Richardson, S.W. (1976) Magmatic andalusite from the South Mountain Batholith, Nova Scotia. Contrib. Mineral. Petrol. 56, 279-287.

Cohen, R.E. (1986) Thermodynamic solution properties of aluminous clinopyroxenes: Non-linear least squares refinements. Geochim. Cosmochim. Acta 50, 563-576.

Coleman, R.G. and Lee, D.E. (1962) Metamorphic aragonite in the glaucophane schists of Cazadero, California. Amer. J. Sci. 260, 577-595.

Connolly, J.A.D. and Kerrick, D.M. (1984) Analylsis of thermodynamic data and phase equilibria for the system lime-alumina-silica-water-carbon dioxide. Geol. Soc. Amer. Abstr. w/ Prog. 16, 476.

Danckwerth, P. A. and Newton, R.C. (1978) Experimental determination of the spinel peridotite to garnet peridotite reaction in the system $MgO\text{-}Al_2O_3\text{-}SiO_2$ in the range 900°-1100°C and Al_2O_3 isopleths of enstatite in the spinel field. Contrib. Mineral. Petrol. 66, 189-201.

Day, H.W. and Kumin, H.J. (1980) Thermodynamic analysis of the aluminum silicate triple point. Amer. J. Sci. 280, 265-287.

Demarest, H.H. Jr. and Haselton, H.T. Jr. (1981) Error analysis for bracketed phase equilibrium data. Geochim. Cosmochim. Acta 45, 217-224.

Fyfe, W.S. (1969) Some second thoughts on $Al_2O_3\text{-}SiO_2$. Amer. J. Sci. 267, 291-296.

Ganguly, J. (1969) Chloritoid stability and related parageneses: theory, experiments, and applications. Amer. J. Sci. 267, 910-944.

Ganguly, J. and Ghose, S. (1979) Aluminous orthopyroxene: order-disorder, thermodynamic properties, and petrologic implications. Contrib. Mineral. Petrol. 69, 375-385.

Gasparik, T. (1984) Experimental study of subsolidus phase relations and mixing properties of pyroxene in the system $CaO\text{-}Al_2O_3\text{-}SiO_2$. Geochim. Cosmochim. Acta 48, 2537-2545.

Gasparik, T. and Newton, R.C. (1984) The reversed alumina contents of orthopyroxene in equilibrium with spinel and forsterite in the system $MgO\text{-}Al_2O_3\text{-}SiO_2$. Contrib. Mineral. Petrol. 85, 186-196.

Gibbs, G.V. (1966) The polymorphism of cordierite I: the crystal structure of low cordierite. Amer. Mineral. 51, 1068-1087.

Goldsmith, J.R. (1981) The join $CaAl_2Si_2O_8$-H_2O (anorthite-water) at elevated pressures and temperatures. Amer. Mineral. 66, 1183-1188.

Gordon, T.M. (1973) Determination of internally consistent thermodynamic data from phase equilibrium experiments. J. Geol. 81, 199-208.

Grambling, J.A. (1984) Coexisting paragonite and quartz in sillimanitic rocks from New Mexico. Amer. Mineral. 69, 79-87.

Greenwood, H.J. (1972) Al^{IV}-Si^{IV} disorder in sillimanite and its effect on phase relations of the aluminum silicate minerals. Geol. Soc. Amer. Mem. 132, 553-571.

Haas, J.L. Jr., Robinson, G.R. Jr. and Hemingway, B.S. (1981) Thermodynamic tabulations for selected phases in the system CaO-Al_2O_3-SiO_2-H_2O. U.S. Dept. Int. Geol. Surv., Open-file Rep't. 80-908, 1-135.

Hafner, S. and Laves, F. (1961) Ordnung/Unordnung und Ultrarotabsorbtion III Die Systeme $MgAl_2O_4$-Al_2O_3 und $MgAl_2O_4$-$LiAl_5O_8$. Kristallogr. 115, 321-330.

Halbach, H. and Chatterjee, N.D. (1984) An internally consistent set of thermodynamic data for twentyone CaO-Al_2O_3-SiO_2-H_2O phases by linear parametric programming. Contrib. Mineral. Petrol. 88, 14-23.

Harker, R.I. and Tuttle, O.F. (1955) Studies in the system CaO-MgO-CO_2. Pt. 1, The thermal dissociation of calcite, dolomite, and magnesite. Amer. J. Sci. 253, 209-224.

Haselton, H.T. (1979) Calorimetry of synthetic pyrope-grossular garnets and calculated stability relations. Ph.D. thesis, Univ. of Chicago, 98 pp.

Haselton, H.T., Sharp, W.E. and Newton, R.C. (1978) CO_2 fugacity at high temperatures and pressures from experimental decarbonation reactions. Geophys. Res. Lett. 5, 753-756.

Haselton, H.T. and Westrum, E.F. (1980) Low-temperature heat capacities of synthetic pyrope, grossular, and $pyrope_{60}grossular_{40}$. Geochim. Cosmochim. Acta 44, 701-709.

Hazen, R.M. and Finger, L.W. (1978) Crystal structures and compressibilities of pyrope and grossular to 60 kbar. Amer. Mineral. 63, 297-303.

Helgeson, H.C., Delany, J.M., Nesbitt, H.W. and Bird, D.K. (1978) Summary and critique of the thermodynamic properties of rock-forming minerals. Amer. J. Sci. 278-A, 1-299.

Herzberg, C.T. (1983) The reaction forsterite + cordierite = aluminous orthopyroxene + spinel in the system MgO-Al_2O_3-SiO_2. Contrib. Mineral. Petrol. 84, 84-90.

Hietanen, A. (1967) On the facies series in various types of metamorphism. J. Geol. 75, 187-214.

Holdaway, M.J. (1971) Stability of andalusite and the aluminum silicate phase diagram. Amer. J. Sci. 97-131.

Holdaway, M.J. (1972) Thermal stability of Al-Fe epidote as a function of $f(O_2)$ and Fe content. Contrib. Mineral. Petrol. 37, 307-340.

Holdaway, M.J. (1978) Significance of chloritoid-bearing rocks in the Picuris Range, New Mexico. Geol. Soc. Amer. Bull. 89, 1404-1414.

Holland, T.J.B. (1981) Thermodynamic analysis of simple mineral systems. In Thermodynamics of Minerals and Melts, R.C. Newton, A. Navrotsky and B.J. Wood, eds., New York: Springer, p. 19-34.

Holland, T.J.B. and Powell, R. (1985) An internally consistent thermodynamic dataset with uncertainties and correlations: 2. Data and results. J. Meta. Geol. 3, 343-370.

Holloway, J.R. (1977) Fugacity and activity of molecular species in supercritical fluids. In Thermodynamics in Geology, D.G. Fraser, ed., Dordrecht, Holland: Reidel, p. 161-182.

Hsü, L.C. (1968) Selected phase relationships in the system Al-Mn-Fe-Si-O-H; a model for garnet equilibria. J. Petrol. 9, 40-83.

Ito, J. (1975) High temperature solvent growth of orthoenstatite, $MgSiO_3$, in air. Geophys. Res. Lett. 2, 533-536.

Jenkins, D.M., Newton, R.C. and Goldsmith, J.R. (1984) Relative stability of Fe-free zoisite and clinozoisite. J. Geol. 93, 663-672.

Johannes, W. (1969) An experimental investigation of the system MgO-SiO_2-H_2O-CO_2. Amer. J. Sci. 267, 1083-1104.

Johnson, M.R.W. (1963) Some time relations of movement and metamorphism in the Scottish Highlands. Geologie en Mijnbouw 42, 121-142.

Kerrick, D.M. and Heninger, S.G. (1984) The andalusite-sillimanite equilibrium revisited. Geol. Soc. Amer. Abstr. w/ Prog. 16, 558.

Kerrick, D.M. and Jacobs, G.K. (1981) A modified Redlich-Kwong equation for H_2O-CO_2, and H_2O-CO_2 mixtures at elevated pressures and temperatures. Amer. J. Sci. 281, 735-767.
Kieffer, S. (1981) Thermodynamics and lattice vibrations of minerals, 4, Application to chain and sheet silicates and orthosilicates. Rev. Geophys., Space Phys. 18, 862-886.
Kiseleva, I.A. (1984) The enthalpy of formation of grossular. Geochem. Int'l. 21, 138-141.
Kiseleva, I.A., Ogodorova, L.P., Topor, L.P. and Chigareva, O.G. (1979) Thermochemical studies of CaO-MgO-SiO_2 system. Geokhimiya 1979, 1821-1825.
Kiseleva, I.A., Ostapenko, G.T., Ogorodova, L.P., Topor, N.D. and Timoshkova, L.P. (1983) High-temperature calorimetric data on the andalusite-kyanite-sillimanite-mullite phase equilibria. Geokhimiya 1983, 1247-1256.
Kolesnik, Yu.N., Nogteva, V.V., Arkhipenko, D.K., Orekhov, B.A. and Pankov, I.E. (1979) Grossular heat capacity under the 13-300 K temperature interval and thermodynamics of solid solution pyrope-grossular. Geokhimiya 1979, 1811-1820.
Koziol, A.M. and Newton, R.C. (1986) Definition of anorthite = grossular + kyanite + quartz in the range 650°-1250°C. Geol. Soc. Amer. Abstr. w/ Prog. 18, 661.
Krupka, K.M., Kerrick, D.M. and Robie, R.A. (1979) Heat capacities of synthetic orthoenstatite and natural anthophyllite from 5 to 1000 K. Trans. Amer. Geophys. Union, EOS 60, 405.
Krupka, K.M., Robie, R.A. and Hemingway, B.S. (1979) High temperature heat capacities of corundum, periclase, anorthite, $CaAl_2Si_2O_8$ glass, muscovite, pyrophyllite, $KAlSi_3O_8$ glass, grossular, and $NaAlSi_3O_8$ glass. Amer. Mineral. 64, 86-101.
Kurat, G., Niedermayr, G., and Prinz, M. (1982) Peridot von Zabargad, Rotes Meer. Aufschluss 33, 169-182.
Kurepin, V.A. (1979) Thermodynamics of hydrous cordierite, and mineral equlibria involving it. Geochem. Int'l. 16, 34-44.
Langer, K. and Schreyer, W. (1976) Apparent effects of molecular water on the lattice geometry of cordierite: a discussion. Amer. Mineral. 61, 1036-1040.
Liou, J.G. (1971) P-T stabilities of laumontite, wairakite, lawsonite, and related minerals in the system $CaAl_2Si_2O_8$-SiO_2-H_2O. J. Petrol. 12, 379-411.
Luth, W.C., Jahns, R.H. and Tuttle, O.F. (1964) The granite system at pressures of 4 to 10 kilobars. J. Geophys. Res. 69, 759-774.
Manghnani, M.H. (1970) Analcite-jadeite phase boundary. Phys. Earth, Plan. Int. 3, 456-461.
Martignole, J. and Sisi, J-C. (1981) Cordierite-garnet-H_2O equilibrium: a geological thermometer, barometer and water fugacity indicator. Contrib. Mineral. Petrol. 77, 38-46.
Metz, G.W., Anovitz, L.M., Essene, E.J., Bohlen, S.R., Westrum, E.F. and Wall, V.J. (1983) The heat capacity and phase equilibria of almandine. Trans. Amer. Geophys. Union, EOS 64, 346-347.
Milton, D.J. (1986) Chloritoid-sillimanite assemblage from North Carolina. Amer. Mineral. 71, 891-894.
Mirwald, P.W. (1982) A high-pressure phase transition in cordierite. Amer. Mineral. 67, 277-283.
Mirwald, P.W. and Schreyer, W. (1977) Die stabile und metastabile Abbaureaktion von Mg-cordierit in Talk, Disthen und Quarz und ihre Abhängigkeit vom Gleichgewichts - wassergehalt des Cordierits. Fortsch. Mineral. 55, 95-97.
Myers, J. and Eugster, H.P. (1983) The system Fe-Si-O: oxygen buffer calibrations to 1,500K. Contrib. Mineral. Petrol. 82, 75-90.
Navrotsky, A., Newton, R.C. and Kleppa, O.J. (1973) Sillimanite-disordering enthalpy by calorimetry. Geochim. Cosmochim. Acta 37, 2497-2508.
Navrotsky, A. and Kleppa, O.J. (1967) The thermodynamics of cation distributions in simple spinels. J. Inorg. Nucl. Chem. 29, 2201-2214.
Newton, R.C. (1965) The thermal stability of zoisite. J. Geol. 73, 431-441.
Newton, R.C. (1966a) Kyanite-sillimanite equilibrium at 750°C. Science 151, 1222-1225.
Newton, R.C. (1966b) Kyanite-andalusite equilibrium from 700°-800°C. Science 153, 170-172.
Newton, R.C. (1966c) Some calc-silicate equilibrium relations. Amer. J. Sci. 264, 204-222.
Newton, R.C. (1972) An experimental determination of the high pressure stability limits of magnesian cordierite under wet and dry conditions. J. Geol. 80, 398-420.
Newton, R.C. and Wood, B.J. (1979) Thermodynamics of water in cordierite and some petrologic consequences of cordierite as a hydrous phase. Contrib. Mineral. Petrol. 68, 391-405.

Newton, R.C., Charlu, T.V. and Kleppa, O.J. (1980) Thermochemistry of the high structural state plagioclases. Geochim. Cosmochim. Acta 44, 933-941.

O'Neill, H.St.C. and Wood, B.J. (1979) An experimental study of Fe-Mg partitioning between garnet and olivine and its calibration as a geothermometer. Contrib. Mineral. Petrol. 70, 59-70.

Pankratz, L.B. and Kelley, K.K. (1964) High-temperature heat contents and entropies of andalusite, kyanite and sillimanite. U.S. Bur. Mines Rep't. #6370, 1-7.

Perkins, D., Holland, T.J.B. and Newton, R.C. (1981) The Al_2O_3 contents of enstatite in equilibrium with garnet in the system MgO-Al_2O_3-SiO_2 at 15-40 kbar and 900°-1600°C. Contrib. Mineral. Petrol. 78, 99-109.

Perkins, D., Westrum, E.F. and Essene, E.J. (1980) The thermodynamic properties and phase relations of some minerals in the system CaO-Al_2O_3-SiO_2-H_2O. Geochim. Cosmochim. Acta 44, 61-84.

Powell, R.(1978) Equilibrium Thermodynamics in Geology. London: Harper and Row, 284 pp.

Powers, R.E. and Bohlen, S.R. (1985) The role of synmetamorphic igneous rocks in the metamorphism and partial melting of metasediments, Northwest Adirondacks. Contrib. Mineral. Petrol. 90, 401-409.

Prunier, A.R. and Hewitt, D.A. (1985) Experimental observations on coexisting zoisite-clinozoisite. Amer. Mineral. 70, 375-378.

Rammansee, W. and Fraser, D.G. (1982) Determination of activities in silicate melts by Knudsen cell mass spectrometry - I. The system $NaAlSi_3O_8$-$KAlSi_3O_8$. Geochim. Cosmochim. Acta 46, 2269-2278.

Richardson, S.W., Bell, P.M. and Gilbert, M.C. (1968) Kyanite-sillimanite equilibrium between 700°C and 1500°C. Amer. J. Sci. 266, 513-541.

Richardson, S.W., Gilbert, M.C. and Bell, P.M. (1969) Experimental determination of kyanite-andalusite and andalusite-sillimanite equilibria; the aluminum silicate triple point. Amer. J. Sci. 267, 259-272.

Robie, R.A. and Hemingway, B.S. (1984) Entropies of kyanite, andalusite, and sillimanite: additional constraints on the pressure and temperature of the Al_2SiO_5 triple point. Amer. Mineral. 69, 298-306.

Robie, R.A., Hemingway, B.S. and Fisher, J.R. (1978) Thermodynamic properties of minerals and related substances at 298.15°K and 1 bar (10^5 pascals) pressure and at higher temperatures. U.S. Geol. Surv. Bull. 1452, 456 pp.

Robie, R.A., Hemingway, B.S. and Takei, H. (1982) Heat capacities and entropies of Mg_2SiO_4, Mn_2SiO_4, and Co_2SiO_4 between 5 and 380 K. Amer. Mineral. 67, 470-482.

Robinson, G.P., Haas, J.L., Schafer, C.M. and Haselton, H.T. (1982) Thermodynamic and thermophysical properties of selected phases in the MgO-SiO_2-H_2O-CO_2, CaO-Al_2O_3-SiO_2-H_2O-CO_2, and Fe-FeO-Fe_2O_3-SiO_2 chemical systems, with special emphasis on the properties of basalts and their mineral components. U.S. Geol. Survey Open File Rep't. 83-79, 1-429.

Salje, E. (1986) Heat capacities and entropies of andalusite and sillimanite: The influence of fibrolitization on the phase diagram of the Al_2SiO_5 polymorphs. Amer. Mineral. 71, 1366-1371.

Salje, E. and Werneke, Chr. (1982) The phase equilibrium between sillimanite and andalusite as determined from lattice vibrations. Contrib. Mineral. Petrol. 79, 56-67.

Sato, M. (1971) Electrochemical measurements and control of oxygen fugacity and other gaseous fugacities with solid electrolyte systems. In Research Techniques for High Pressure and High Temperature, G. Ulmer, ed., New York: Springer-Verlag, p. 43-100.

Saxena, S.K. (1981) The MgO-Al_2O_3-SiO_2 system: free energy of pyrope and Al_2O_3-enstatite. Geochim. Cosmochim. Acta 45, 821-825.

Saxena, S.K. and Chatterjee, N. (1986) Thermochemical data on mineral phases: the system CaO-MgO-Al_2O_3-SiO_2. J. Petrol. 27, 827-842.

Schreyer, W. and Seifert, F. (1970) Pressure dependence of crystal structures in the system MgO-Al_2O_3-SiO_2-H2O at pressures up to 30 kilobars. Phys. Earth, Plan. Int. 3, 422-430.

Schreyer, W. and Yoder, H.S. (1964) The system Mg-cordierite-H_2O and related rocks. N. Jb. Mineral. Abh. 101, 271-342.

Schuiling, R.D. (1957) A geo-experimental phase diagram of Al_2SiO_5. Proc. Koninkl. Nederl. Akad. van Wetenschappen, B series, 60, 220-226.

Shmulovich, K.I. and Shmonov, V.M. (1978) Tables of thermodynamic properties of gases and liquids (carbon dioxide). Gosdarst. Sluzhba Standart. Dannykh, 1-165.

Simmons, G. and Wang, H. (1971) Single Crystal Elastic Constants and Calculated Aggregate Properties: A Handbook, 2nd Ed. Cambridge, Mass.: M.I.T. Press.

Skinner, B.J., Clark, S.P. Jr. and Appleman, D.E. (1961) Molar volumes and thermal expansions of andalusite, kyanite and sillimanite. Amer. J. Sci. 259, 651-668.

Smirnova, N.S. (1986) IR data on isomorphous substitution in natural orthopyroxenes. Geochem. Int'l. 23, 102-106.

Smith, J.V. and Schreyer, W. (1960) Location of argon and water in cordierite. Mineral. Mag. 33, 226-236.

Steefel, C.I. and Atkinson, W.W. (1984) Hydrothermal andalusite and corundum in the Elkhorn District, Montana. Econ. Geol. 79, 573-579.

Stolyarova, T.A. (1986) The enthalpy of dissolution of grossular $Ca_3Al_2(SiO_4)_3$ in hydrofluoric acid. Geochem. Int'l. 23, 67-70.

Weill, D.F. (1966) Stability relations in the Al_2O_3-SiO_2 system calculated from solubilities in the Al_2O_3-SiO_2-Na_3AlF_6 system. Geochim. Cosmochim. Acta 30, 223-237.

Weller, W.W. and Kelley, K.K. (1963) Low-temperature heat capacities and entropies at 298.15°K of akermanite, cordierite, gehlenite, and merwinite. U.S. Bur. Mines Rep't. 6343.

Westrum, E., Essene, E.J. and Perkins, D. (1979) Thermophysical properties of the garnet, grossular: $Ca_3Al_2Si_3O_{12}$. J. Chem. Thermodyn. 11, 57-66.

Wood, B.J. and Banno, S. (1973) Garnet-orthopyroxene and orthopyroxene- clinopyroxene relationships in simple and complex systems. Contrib. Mineral. Petrol. 42, 109-124.

Wood, B.J. and Holloway, J.R. (1984) A thermodynamic model for subsolidus equilibria in the system CaO-MgO-Al_2O_3-SiO_2. Geochim. Cosmochim. Acta 48, 159-176.

Wood, B.J., Kirkpatrick, R.J. and Montez, B. (1986) Order-disorder phenomena in $MgAl_2O_4$ spinel. Amer. Mineral. 71, 999-1006.

Zen, E-An (1972) Gibbs free energy, enthalpy, and enthalpy of ten rock-forming minerals: calculations, discrepancies, implications. Amer. Mineral. 57, 524-553.

Zen, E-An (1973) Thermochemical parameters of minerals from oxygen-buffered hydrothermal equilibrium data: method, application to annite and almandine. Contrib. Mineral. Petrol. 39, 65-80.

Zen, E-An (1977) The phase-equilibrium calorimeter, the petrogenetic grid, and a tyranny of numbers. Amer. Mineral. 62, 189-204.

 Alexandra Navrotsky

MODELS OF CRYSTALLINE SOLUTIONS

INTRODUCTION

The complexity and richness of natural minerals arises not only from the large number of phases and crystal structures, but also from the variety of solid solutions, complex elemental substitutions, order-disorder relations, and exsolution phenomena one encounters. Indeed, a pure mineral of end-member composition is the exception rather than the rule. Elemental substitutions in minerals may be classified in terms of the crystallographic sites on which they occur and the formal charges of the chemical elements (ions) which participate. Most common rock-forming silicates can be said to have a general formula $A_aM_mT_tO_oY_y$, where the capital letters refer to types of sites and the small letters to their multiplicity per formula unit. "A" represents large 8-12 coordinated cations (alkalis, alkaline earths, sometimes rare earths or other "incompatible" elements). "M" represents smaller divalent or trivalent cations in approximately octahedral coordination. These include Mg^{2+}, Fe^{2+}, Fe^{3+}, Cr^{3+}, Al^{3+}, and other transition metals. "T" represents smaller ions in tetrahedral coordination, chiefly Si^{4+} and Al^{3+}, sometimes B^{3+}, Be^{2+}, and P^{5+}, as well as smaller to trace amounts of Fe^{3+}, Ga^{3+}, Ge^{4+}. "O" represents oxygen, and "Y" various substituting anions, chiefly F^- and OH^-. Carbonates, sulfates, halides, and sulfides follow their own crystal chemical patterns and ionic substitutions.

Within each classification above, there may be several crystallographically distinct sites, e.g. the M1 and M2 sites of olivine or the several T sites of the feldspars. The substitution of atoms (or ions - I shall not argue about formal versus actual charges in this context) in a solid solution may proceed in a number of different ways. These are summarized in Table 1. All substitutions must maintain electroneutrality (balance of formal charges) in the crystal. The substitutions may involve one or more sets of crystallographic sites. They may involve ions of like charge, charge-coupled substitutions on one sublattice or on several, or the creation or filling of vacant sites.

This structural complexity must be reflected in the thermodynamics of solid solution formation. The basic theoretical question is to relate the microscopic details of atomic interactions to site occupancies and macroscopic thermodynamic parameters. The basic practical problem is to obtain useful expressions for the thermodynamic mixing properties of multicomponent solid solutions which can be applied to petrologic systems. These practical models should be simple enough to incorporate into computer codes describing phenomena such as the thermal evolution of a magma. They should be robust enough to withstand interpolation and some extrapolation in (P,T,X) space, even in the hands, heads, and computers of those who are neither crystallographers nor thermodynamicists.

At first glance, the theoretical and practical goals appear almost contradictory. The main part of this review will deal with how one tries to serve both taskmasters. The formalism and examples presented are meant to emphasize three points. (1) Any expression one writes for a free energy of mixing, an activity, or an activity coefficient has buried in it some assumptions about the mixing process on an atomistic scale. Therefore one must make sure those assumptions are reasonable. (2) Equations, whose form is constrained by theoretical considerations and whose parameters are physically reasonable, are more reliably

Table 1. Substitutions in Solid Solutions

Substitutions	Examples
$Mg^{2+}_M = Fe^{2+}_M$ = transition metal$_M$	olivines, pyroxenes, spinels, amphiboles, micas, carbonates (M = octahedral or distorted octahedral site)
$Mg^{2+}_M = Ca^{2+}_M$	pyroxenes, carbonates
$Mg^{2+}_{M'} = Ca^{2+}_{M'} = Fe^{2+}_{M'} = Mn^{2+}_{M'}$	garnets (M' = 8-fold site)
$Al^{3+}_M = Ca^{3+}_M = Fe^{3+}_M$	garnets, spinels
$OH^-_X = F^-_X$	amphiboles, micas, clays
$Na^+_A = K^+_A$	amphiboles, micas, clays
$Si^{4+}_T = Al^{3+}_T + Na^+_A$	stuffed silica derivatives, amphiboles, micas, clays
$Na^+_A + Si^{4+}_T = Ca^{2+}_A + Al^{3+}_T$	feldspars
$Mg^{2+}_{M1} + Si^{4+}_{M2} = Al^{3+}_{M1} + Al^{3+}_T$	pyroxenes
$Mg^{2+}_{M1} + Ca^{2+}_{M2} = Al^{3+}_{M1} + Na^+_{M2}$	pyroxenes, amphiboles

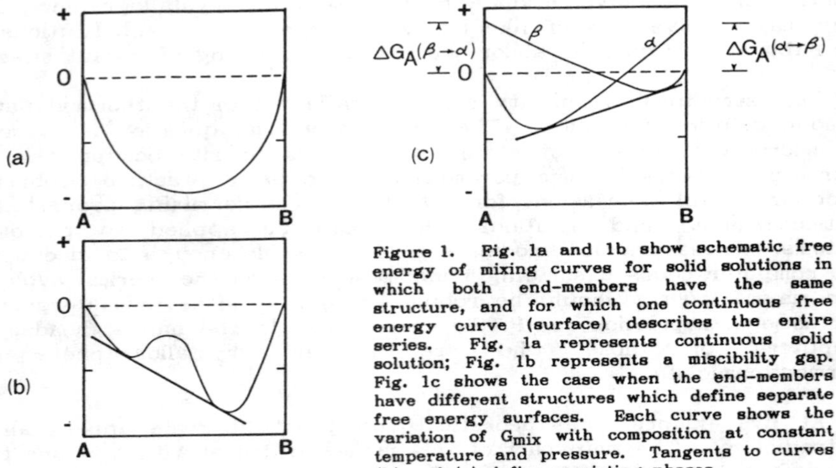

Figure 1. Fig. 1a and 1b show schematic free energy of mixing curves for solid solutions in which both end-members have the same structure, and for which one continuous free energy curve (surface) describes the entire series. Fig. 1a represents continuous solid solution; Fig. 1b represents a miscibility gap. Fig. 1c shows the case when the end-members have different structures which define separate free energy surfaces. Each curve shows the variation of G_{mix} with composition at constant temperature and pressure. Tangents to curves (b) and (c) define coexisting phases.

extrapolated than arbitrary polynomials fit to data in a small (P,T,X) range. (3) Models which simultaneously include constraints imposed by phase equilibria, calorimetry, and crystal chemistry are more reliable than those based on any one source alone.

This chapter stresses the development of thermodynamic models consistent with crystal chemical complexity. Chapter 3 shows how such models can be simplified, while keeping at least some essence of the microscopic detail, for petrologic applications. Both chapters stress the perils of arbitrary models using large numbers of purely empirical parameters.

SOME THERMODYNAMIC FORMALISM

Consider the equation:

$$x\ A + (1-x)\ B = A_xB_{1-x}\ , \tag{1}$$

where A and B are end-members of a solid solution series which mix to form an intermediate phase of mole fraction, x. If both end members have the same crystal structure, a continuous free energy curve can be drawn to represent the substitution, see Fig. 1a and 1b. If the space group of the structure of one end-member is a subgroup of that of the other, complex free energy relations may lead to higher order phase transitions. If the two structures are not so related, then each represents a different free energy surface, see Fig. 1c. That being the case, complete miscibility cannot occur, but common tangents to the two free energy curves (or appropriate generalizations to many dimensions) define the mutual solubilities.

At atmospheric pressure and temperature T, the free energy per mole of an isostructural solid solution can be written as

$$G^\circ_T(A_xB_{1-x}) = x\ G^\circ_T(A) + (1-x)_T G^\circ(B) + \Delta G^\circ_{mix}\ , \tag{2}$$

where $\Delta G^\circ(A_xB_{1-x})$ is the standard free energy of the solid solution, $G^\circ_T(A)$ and $G^\circ_T(B)$ are the standard free energies of the end members A and B, and ΔG°_{mix} is a free energy of mixing term. Note that throughout this chapter, the superscipt zero refers to an atmospheric pressure standard state.

The ΔG°_{mix} term consists of an enthalpy and entropy term

$$\Delta G^\circ_{mix} = \Delta H^\circ_{mix} - T\Delta S^\circ_{mix}\ . \tag{3}$$

The enthalpy of mixing term contains the energetics of the interactions; a positive ΔH°_{mix} generally indicates clustering of like species and a tendency toward exsolution, while a negative ΔH°_{mix} generally suggests ordering of unlike species and a tendency toward formation of an intermediate compound. Both tendencies may occur simultaneously in the same system; see the examples of carbonates and feldspars below.

The free energy in Eq. (2) can be written in terms of chemical potentials of components A and B

$$G^{\circ}_{T}(A_xB_{1-x}) = x\ \mu^{\circ}_{T}(A) + (1-x)\ \mu^{\circ}_{T}(B) + x\ \Delta\mu^{\circ}(A) + (1-x)\ \Delta\mu^{\circ}(B)\ . \tag{4}$$

The $\mu^{\circ}_{T}(A)$ and $\mu^{\circ}_{T}(B)$ terms are the free energies per mole of pure A and B and identical to the $G^{\circ}_{T}(A)$ and $G^{\circ}_{T}(B)$ terms in Eq. (2). The $\Delta\mu^{\circ}$ terms represent partial molar free energies of mixing. Therefore

$$\Delta G^{\circ}_{mix} = x\ \Delta\mu^{\circ}_{T}(A) + (1-x)\ \Delta\mu^{\circ}_{T}(B)\ . \tag{5}$$

The thermodynamic activity is defined as

$$\Delta\mu^{\circ}_{T}(A) = RT\ \ln a(A)\ , \quad \Delta\mu^{\circ}(B) = RT\ \ln a(B)\ . \tag{6}$$

The changes in chemical potential on mixing can be related to partial molar enthalpies and entropies of mixing:

$$\Delta\mu^{\circ}_{T}(A) = \Delta\bar{h}^{\circ}_{T}(A) - T\Delta\bar{s}^{\circ}_{T}(A)\ , \quad \Delta\mu^{\circ}_{T}(B) = \Delta\bar{h}^{\circ}_{T}(B) - T\Delta\bar{s}^{\circ}_{T}(B)\ . \tag{7}$$

The two chemical potentials (and other partial molar quantities) are not independent of each other but are related by the Gibbs-Duhem equation, a form of which is

$$x\ \frac{\partial\Delta\mu^{\circ}_{T}(A)}{\partial x} + (1-x)\ \frac{\partial\Delta\mu^{\circ}_{T}(B)}{\partial x} = 0\ . \tag{8}$$

Pressure affects the above equations through the volume term, since

$$(\partial G/\partial P)_T = V\ , \quad (\partial\Delta G_{mix}/\partial P)_T = \Delta V_{mix}\ , \quad (\partial\Delta\mu_A/\partial P)_T = \Delta\bar{v}(A)\ . \tag{9}$$

The enthalpy of mixing at P and T is related to the enthalpy of mixing at 1 atm and T by a pressure-volume term, with $\Delta\bar{V}_{mix}$, $\Delta V(A)$ dependent, in general, on P and T

$$\begin{aligned} \Delta H_{mix} &= \Delta H^{\circ}_{mix} + \int_{1\ atm}^{P} \Delta V_{mix} dP\ , \\ \Delta\bar{h}(A) &= \Delta\bar{h}^{\circ}(A) + \int_{1\ atm}^{P} \Delta\bar{v}(A) dP\ . \end{aligned} \tag{10}$$

The above equations can all be generalized to multicomponent systems.

THE IDEAL SOLUTION - THE ENTROPY OF MIXING TERM

There are three possible contributions to the entropy of mixing: (a) changes in the vibrational entropy term, (b) terms from changes in magnetic and electronic entropy, and (c) the configurational or statistical term arising from the occupancy of equivalent sites by different chemical species. The last is usually the most important. It arises through statistical mechanics and is fundamentally related to the greater randomness (or loss of information about which ion is on a particular site) in the solid solution compared to the pure end member.

If Eq. (1) represents the random mixing of a total of one mole of species over a total of one mole of sites, then (and only then) the entropy of mixing is given by

$$\Delta S_{mix} = -R[x \ln x + (1-x) \ln (1-x)] \tag{11}$$

and

$$\Delta\bar{s}(A) = -R \ln x \ , \quad \Delta\bar{s}(B) = -R \ln (1-x) \ . \tag{12}$$

If, furthermore, the enthalpy of mixing is zero, then the simple ideal solution results, with

$$\Delta G^{\circ}_{mix} = RT[x \ln x + (1-x) \ln (1-x)] \ ; \tag{13}$$

$$\Delta\mu^{\circ}(A) = RT \ln x \ , \quad \Delta\mu^{\circ}(B) = RT \ln (1-x) \ ; \tag{14}$$

$$a(A) = x \ , \quad a(B) = 1-x \ . \tag{15}$$

Equation (15) is familiar as Raoult's Law, often used as a first approximation when little information is available. Note that this formulation predicates the mixing of one mole of species per formula unit of solid solution. Thus it is directly applicable to the following cases as examples: NiO-MgO rocksalt solid solutions, $MgAl_2O_4$-$ZnAl_2O_4$ spinel solid solutions if one assumes Ni and Zn mix on tetrahedral sites only, and $MgSiO_3$-$FeSiO_3$ pyroxene solid solutions if one assumes that M1 and M2 sites are energetically indistinguishable, with Fe and Mg distributed at random. In each of the cases above, ideality would mean activity equal to mole fraction, i.e. a(NiO) = x, a(MgO) = 1-x or $a(MgAl_2O_4)$ = x, $a(ZnAl_2O_4)$ = 1-x or $a(MgSiO_3)$ = x, $a(FeSiO_3)$ = 1-x. More complex expressions, even for "ideal mixing" (in the sense of a random distribution of species with no preferred energetics) arise when stoichiometry dictates several species per formula unit. Thus for olivines, Mg_2SiO_4-Fe_2SiO_4, there are two moles of (Fe+Mg) per mole of $(Mg,Fe)_2SiO_4$. One can proceed in either of two ways: One can redefine the formula unit as $(Mg,Fe)Si_{0.5}O_2$. Then, for ideal mixing, $a(MgSi_{0.5}O_2)$ = x, $a(FeSi_{0.5}O_2)$ = 1-x. This is a convenient formulation for application to element distribution equilibria. For example, the distribution of Fe and Mg between olivine and orthopyroxene can be written as

$$MgSi_{0.5}O_2(ol) + FeSiO_3(opx) = FeSi_{0.5}O_2(ol) + MgSiO_3(opx) \ ; \tag{16}$$

$$K_D = \frac{[a(FeSi_{0.5}O_2,\ ol)]\ [a(MgSiO_3,\ opx)]}{[a(MgSi_{0.5}O_2,ol)]\ [a(FeSiO_2,\ opx)]} \ . \tag{17}$$

Then $-RT \ln K_D$ is the standard free energy change of Reaction (16), given as

$$\Delta G^\circ = -RT \ln K_D = \Delta G_f^\circ(FeSi_{0.5}O_2,ol) + \Delta G_f^\circ(MgSiO_3,opx)$$
$$-\Delta G_f^\circ(MgSi_{0.5}O_2,ol) - \Delta G_f^\circ(FeSiO_3,opx) \ , \tag{18}$$

or

$$\Delta G^\circ = 1/2\Delta G_f^\circ(Fe_2SiO_4,ol) + \Delta G_f^\circ(MgSiO_3,opx)$$
$$-1/2\Delta G_f^\circ(Mg_2SiO_4,ol) - \Delta G_f^\circ(FeSiO_3,opx) \ . \tag{19}$$

where the ΔG_f° terms are appropriate free energy of formation terms, all taken with respect to elements or to oxides. Note the factors of "1/2" which appear because Eq. (16) is written involving one mole of exchangeable species per formula unit.

If one assumes ideal mixing in olivine and orthopyroxene, then $a(MgSiO_{0.5}O_2,ol) = x$, $a(FeSi_{0.5}O_2,ol) = 1-x$, $a(MgSiO_3,opx) = x'$, $a(FeSiO_3,opx) = 1-x'$, where $x = (Mg/(Mg+Fe))_{ol}$, $x' = (Mg/(Mg+Fe))_{opx}$ and

$$K_D = \frac{x'(1-x)}{x(1-x')} \ . \tag{20}$$

The second procedure is to leave the formula unit of olivine as $(Fe,Mg)_2SiO_4$. Then, for the random mixing of xMg_2SiO_4 and $(1-x)Fe_2SiO_4$,

$$\Delta S_{mix} = -2R[x \ln x + (1-x) \ln (1-x)] \tag{21}$$

with the factor of two arising because entropy is an extensive parameter and two moles of ions are being mixed. If $\Delta H^{mix} = 0$, then

$$\Delta\mu(Mg_2SiO_4,ol) = -2RT \ln x \ , \quad \Delta\mu(Fe_2SiO_4,ol) = -2RT \ln (1-x) \ , \tag{22}$$

and

$$a(Mg_2SiO_4,ol) = x^2 \ , \ a(Fe_2SiO_4,ol) = (1-x)^2 \ , \tag{23}$$

while, from the discussion above

$$a(MgSi_{0.5}O_2,\ ol) = x \ , \quad a(FeSi_{0.5}O_2,\ ol) = (1-x) \ . \tag{24}$$

Similarly, if one considers $Mg_3Al_2Si_3O_{12}$-$Fe_3Al_2Si_3O_{12}$ garnet solid solutions, with Mg and Fe mixing ideally in 8-fold sites

$$a(Mg_3Al_2Si_3O_{12}) = x^3 \ , \quad a(Fe_3Al_2Si_3O_{12} = (1-x)^3 \ , \tag{25}$$

or

$$a(MgAl_{2/3}SiO_4) = x \ , \quad a(FeAl_{2/3}SiO_4) = (1-x) \ . \tag{26}$$

In the garnet series $Mg_3Al_2Si_3O_{12}$-$Mg_3Cr_2Si_3O_{12}$, ideal mixing of Al and Cr on octahedral sites means that

$$a(Mg_3Al_2Si_3O_{12}) = x^2 \ , \quad a(Mg_3Cr_2Si_3O_{12}) = (1-x)^2 \ , \tag{27}$$

or

$$a(Mg_{1.5}AlSi_{1.5}O_6) = x \ , \quad a(Mg_{1.5}CrSi_{1.5}O_6) = (1-x) \ . \tag{28}$$

Finally for garnets $Mg_3Al_2Si_3O_{12}$-$Mg_3Al_2Ge_3O_{12}$, with ideal mixing of Si and Ge on tetrahedral sites,

$$a(Mg_3Al_2Si_3O_{12}) = x^3 \ , \quad a(Mg_3Al_2Ge_3O_{12}) = (1-x)^3 \ , \tag{27}$$

or

$$a(MgAl_{2/3}SiO_4) = x \ , \quad a(MgAl_{2/3}SiO_4) = (1-x) \ . \tag{28}$$

These concepts of ideal solution can be extended to involve simultaneous mixing of ions on several sets of sites. If no altervalent substitutions occur, and each species can be unambiguously assigned to a given sublattice, this extension is quite straightforward. Consider, for example, a natural garnet containing Mg^{2+}, Fe^{2+}, Ca^{2+}, Fe^{3+}, Mn^{2+}, Fe^{3+}, Cr^{3+}, Al^{3+}, and Si^{4+} as major constituent. Its formula may be written as $(Mg,Fe^{2+},Ca,Mn^{2+})_3(Al,Cr,Fe^{3+})_2Si_3O_{12}$. The activity of pyrope $Mg_3Al_2Si_3O_{12}$ in such a complex garnet may then be estimated as

$$a(Mg_3Al_2Si_3O_{12}) = x^3(x')^2 \ , \tag{29}$$

where x is the fraction of 8-fold sites occupied by Mg, $x = Mg/(Mg+Fe^{2+}+Ca^{3+}+Mn^{2+})$, and x' is the mole fraction of octahedral sites occupied by Al, $x' = Al/(Al+Cr+Fe^{3+})$.

Similarly, the activity of diopside, $CaMgSi_2O_6$, in a clinopyroxene can be calculated from the site occupancy. If one knows a composition from electron microprobe analysis and can estimate the Fe^{2+}/Fe^{3+} ratio, crystal chemical reasoning suggests the following. All Na^+, Mn^{2+} and Ca^{2+} enter the large M2 sites. Enough Cr^{3+}, Fe^{3+} and, if needed, Al^{3+} enter the small M1 sites to balance the Na^+ charge (jadeite-like substitution). The remaining Al is distributed equally between M1 and T sites (Tschermak substitution). The Mg^{2+} and Fe^{2+} are then distributed approximately randomly between M1 and M2. The structural formula is then $(Mg,Fe^{2+},Al,Cr,Fe^{3+})(Ca,Mg,Fe^{2+},Na)(SiAl)_2O_6$, and the activity of $CaMgSi_2O_6$ is approximated as

$$a(CaMgSi_2O_6) = [Ca/(Ca+Mg+Fe^{2+}+Na]_{M2} \cdot$$

$$[Mg/(Mg+Fe^{2+}+Al+Cr+Fe^{3+})]_{M1} \cdot [Si/(Si+Al)]_T^2 \tag{30}$$

Other trace elements may also be included. Nonstoichiometry (deviation from totals of two M cations, and two T cations per six oxygens) may be a complication.

This discussion has given specific examples rather than general formulas involving "the j-th species on the i-th sublattice". This loss of generality is purposeful; one is better off tailoring a thermodynamic model to a specific crystal chemical situation. The entropy model then need not be needlessly cumbersome, yet it must count all the important contributions to the entropy of mixing. In applications to silicate melts, similar statistical counting of site occupancies for framework and nonframework cations has been termed the "two-lattice" model (Weill et al., 1980; Hon et al., 1981; Henry et al., 1982). When end-member compositions also contain configurational disorder, additional care must be taken to correctly compute the entropy of mixing (see below).

REGULAR, SUBREGULAR AND GENERALIZED MIXING MODELS

A starting point for thermodynamic models is the useful but inherently contradictory assumption (see Chapter 3) that though the heats of mixing are not zero, the configurational entropies of mixing are those of a random solid solution. The excess free energy of mixing is then

$$\Delta G^{\circ}_{mix,ex} = \Delta H^{\circ}_{mix} - T\Delta S^{\circ}_{mix,ex} \ . \tag{31}$$

$\Delta S^{\circ}_{mix,ex}$ arises from two sources. First, it is the nonconfigurational (vibrational, magnetic, electronic) contribution to the entropy of mixing. This may be positive or negative. Second, it is the correction to the configurational entropy arising from deviation from a random distribution of ions. Whether this deviation arises from ordering or from clustering, it diminishes the configurational entropy from its maximum (random) value. For a two-component system, the simplest formulation for ΔG_{excess} is

$$\Delta G_{excess} = \Delta H_{mix} = \lambda \ x(1-x) = W_H \ x(1-x) \ , \tag{32}$$

where λ (in the chemistry literature) or W_H (in the geology literature) is the enthalpy interaction parameter. If this parameter is truly independent of temperature, then Eq. (32) represents the "strictly regular" solution.
For this case:

$$\begin{aligned} \Delta\mu^{\circ}(A) &= \lambda \ (1-x)^2 + RT \ln x \ , \\ \Delta\mu^{\circ}(B) &= \lambda \ x^2 + RT \ln (1-x) \ , \end{aligned} \tag{33}$$

where, as before, A is the component with mole fraction x, B with mole fraction (1-x). Then

$$a(A) = x \exp(\frac{\lambda(1-x)^2}{RT}) \ , \quad a(B) = (1-x) \exp(\frac{\lambda(x)^2}{RT}) \ , \tag{34}$$

or, if one defines an activity coefficient, γ, such that $a = \gamma x$,

$$\gamma(A) = \exp\ (\frac{\lambda(1-x)^2}{RT}) \ , \quad \gamma(B) = \exp\ (\frac{\lambda x^2}{RT}) \ . \tag{35}$$

For λ independent of temperature, λ/RT decreases in magnitude as T increases and the solution approaches ideal behavior. At low temperature, if λ is positive, a symmetric miscibility gap develops with a critical temperature $T_c = \lambda/2R$.

The strictly regular model can be generalized in two ways. First, λ can depend on temperature:

$$\lambda = a - bT$$

or

$$W_G = W_H - TW_S \ . \tag{36}$$

In effect, this introduces a nonzero excess entropy of mixing to the model. Secondly, the form of Eq. (32) can be generalized such that λ depends on composition.

$$\Delta G_{excess} = x(1-x)\ (\sum_{i=0}^{n} a_i x^n) \ . \tag{37}$$

If one takes only the first two terms in Eq. (37) then

$$\Delta G_{mix,ex} = x\ (1-x)\ (a_o + a_1 x) \ . \tag{38}$$

This can be written as

$$\Delta G_{excess} = x\ (1-x)\ [W_{G2}\ x + W_{G1}\ (1-x)] \ . \tag{39}$$

The parameters W_{G1} and W_{G2} can depend on temperature:

$$W_{G1} = W_{H1} - TW_{S1} \ , \quad W_{G2} = W_{H2} - TW_{S2} \ . \tag{40}$$

Equation (37) is known as the generalized Margules formulation. Equations (39, 40) are of forms popularized by J.B. Thompson and coworkers for mineralogical applications (see Thompson, 1967). The four parameters (W_{H1}, W_{H2}, W_{S1}, W_{S2}) are generally sufficient for petrologic applications. Indeed, more problems arise from using too many parameters than from using too few. The reader is also referred to the discussion in Chapter 3 of the limitations of the ideal configurational entropy expression for regular and subregular solutions.

Immiscibility occurs when positive W_G terms outweigh the configurational entropy. The miscibility gaps are asymmetric. The conditions for equilibrium between two phases (α and β) are the simultaneous equalities of chemical potentials or activities:

$$\mu(\text{A, phase } \alpha) = \mu(\text{A, phase } \beta)$$

and (41)

$$\mu(\text{B, phase } \alpha) = \mu(\text{B, phase } \beta)$$

or

$$a(\text{A, phase } \alpha) = a(\text{A, phase } \beta)$$

and (42)

$$a(\text{B, phase } \alpha) = a(\text{B, phase } \beta) \ .$$

In general, such coexisting compositions need to be found by a numerical process rather than by a closed form algebraic solution.

SYSTEMATICS IN MIXING PROPERTIES

In solid solutions without order-disorder phenomena, enthalpies of mixing are generally zero or positive. At low temperature, a positive enthalpy of mixing will outweigh the $T\Delta S$ term and result in a miscibility gap. Three main factors affect enthalpy of mixing and therefore the range and stability of solid solution. The first and usually most important is the size difference of the ions or atoms being mixed. This results in a strain energy which is larger the greater the difference in size. For a given size difference, it is often easier to put a smaller atom into a larger site than vice versa. The second factor is differences in bonding character. In cases where the ionicity (however defined) of the species being mixed differs significantly, solubility can be limited, even when the ions are similar in size. Examples are NaCl-AgCl and NaBr-AgBr, which show large positive heats of mixing. Similar effects are expected in sulfides, though differences in crystal structure are complications. Specific electronic effects resulting from incompletely filled "d" shells may influence the thermodynamics of solid solutions containing transition metal ions. The third factor is charge. The higher the charge when species of the same charge are mixed, the more positive the heat of mixing for comparable size difference. When species of different charges mix in one sublattice (e.g. Fe^{2+}, Fe^{3+}, and Ti^{4+} in the iron-titanium oxides), short range order may dominate the thermodynamics.

Of these factors, the role of size mismatch has been most thoroughly explored. Fancher and Barsch (1969a,b) modeled eight alkali halide systems using the classical Born theory of ionic crystals. They found that the heats of mixing could be explained well by a calculated strain energy term. In addition, vibrational excess entropies were found to play an important role.

Urusov (1968) also used a "Born-type" approach, but ionicity differences were taken into account. He proposed that the mixing enthalpy of a given system is a result of differences in size and effective charge of the substituents and a consequence of deformation of the crystal structure and of changing chemical interactions. For cases in which the electronegativity difference is small, Urusov related ΔH_{mix} to a bond-length mismatch parameter. He obtained

$$\Delta H_{mix} = cx\ (1-x)\ nz^2\ \delta^2 \quad , \tag{43}$$

with c an empirical parameter, n the coordination number, z the charge, and

$$\delta = \frac{(V_A)^{1/3}-(V_B)^{1/3}}{x(V_A)^{1/3}+(1-x)(V_B)^{1/3}} \quad , \tag{44}$$

A and B being the two components, V the molar volume.

Semiempirical treatments of solid solution formation in oxide systems with rocksalt structure have been given by Driessens (1968) and Navrotsky (1974). Driessens attempted to calculate the contribution to the interaction parameter from cation pair interactions by subtracting, from the experimental parameter, calculated contributions from lattice energy and crystal field stabilization. Navrotsky also used a regular solution approach in the modeling of deviations from ideality in some oxide solid solutions with the rocksalt structure. An approximately linear dependence was found for λ, the interaction parameter, on δ^2, the square of a normalized size difference parameter, with, for component oxides A and B, with metal-oxygen bond lengths r_{AO} and r_{BO}

$$\delta = \frac{r_{AO} - r_{BO}}{r_{AO} + r_{BO}} \quad . \tag{45}$$

Differences in bond lengths were also used by O'Neill and Navrotsky (1984) in models of spinel solid solution energetics.

However, for many minerals, the ions being mixed occupy only a small fraction of the total volume of the structure. If one compares solid solubility in CaO-MgO (very limited), $CaCO_3$-$MgCO_3$ (extensive) and $Ca_3Al_2Si_3O_{12}$- $Mg_3Al_2Si_3O_{12}$ (complete), one is drawn to the conclusion that greater size mismatch can be tolerated by a structure with larger molar volume in which the ions being mixed are embedded in a matrix which can itself change geometry slightly to absorb the strain. In a thermodynamic sense, the volume of a phase is a more unique macroscopic parameter than any individual bond length. With these considerations in mind, Davies and Navrotsky (1983) sought a correlation between the thermodynamics of mixing (excess free energy in a regular or subregular solution model) and a term describing the volume mismatch for a variety of oxide, chalcogenide, halide, and silicate systems. Values of the excess free energy (W_G or W_{G1} and W_{G2}) were obtained from (a) measured heats of mixing, (b) measured activity-composition relations, or (c) solid solubilities in systems showing a miscibility gap. A survey of the literature yielded about sixty halide, oxide, chalcogenide, and silicate systems for which these parameters can be determined. A volume mismatch term was defined as follows. For a system showing regular solution behavior, W_G be correlated with $\Delta V = (V_A-V_B)/(0.5(V_A+V_B))$, which is an average volume mismatch with $V_A > V_B$, where V_A and V_B are molar volumes. For subregular behavior W_{GA} is correlated with $(V_B-V_A)/V_B$ and W_{GB} with $(V_B-V_A)/V_A$. Thus the limiting interaction parameter for the dissolution of a component in the opposite end-member is correlated with the volume mismatch for substitution into that end-member. The resulting correlations are shown in Figs. 2, 3 and 4. Several conclusions may be drawn. The use of a volume mismatch term rather than a bond mismatch enables one to group together many diverse systems in which divalent ions are being mixed (oxides, chalcogenides, spinels, garnets, olivines, other silicates) into one correlation, which gives, for $\lambda = W_G$, W_{GA}, or W_{GB}, as defined above:

$$\lambda = 100.8\Delta V - 0.4 \text{ kJ mol}^{-1} \quad . \tag{46}$$

Within this correlation are points for cation mixing on 4-, 6-, and 8-coordinated sites and for anion mixing. Thus, once the effect of coordination number is included in the molar volume of the phases involved, it does not appear explicitly in the operations. These correlations also suggest that the data for cation mixing and for anion mixing follow the same trend although data on anion mixing are generally rather limited for systems other than alkali halides. Figure 4 shows all the correlations for alkali halides, rock-salt oxides and chalcogenides, and tungstates and molybdates. The correlation segregates these systems into three distinct groups. The alkali halide systems as a group show much smaller positive deviations from ideality than the oxide and chalcogenide systems. The different slopes of these two correlations confirm the expectation that more highly charged ions mix less easily than ions of lower charge. A few points (e.g. Cr_2O_3-Al_2O_3, TiO_2-SnO_2) for trivalent and tetravalent ions also support this trend. The molybdate and tungstate systems appear anomalous for unknown reasons.

When the volume difference is large, there is asymmetry in the interaction parameters and in resulting miscibility gaps. The W_G parameters imply that it is easier to dissolve the component of smaller volume into a host with larger volume than vice versa, generally in accord with experimental observation.

The main value of such empirical correlations is that they can be used to estimate deviations from ideality in phases where such measurements are difficult. For example, Table 2 lists predicted W_H parameters (based on the regular solution approximation) for a number of (Mg,Fe) mineral solid solutions including some high pressure phases for which mixing data are not readily available. The calculations suggest the following. "FeO"-MgO shows the largest positive deviation from ideal mixing. In reality this is compounded by nonstoichiometry, which is much larger in this solid solution than in any iron-bearing silicate solid solution. Olivine, pyroxene, β-phase, and silicate spinels are all predicted to have small but significant positive heats of mixing (W near 5 kJ). Garnet (both aluminous and Al-free) and ilmenite are predicted to have smaller heats of mixing (W near 2 kJ) mainly because the molar volume difference between Mg- and Fe-endmembers is smaller. As Table 2 shows, these predictions agree reasonably well with available experimental data.

PHASES WITH DIFFERENT STRUCTURES

Partial solid solution can exist among endmembers of different structure. Consider two substances, A with structure "α" and B with

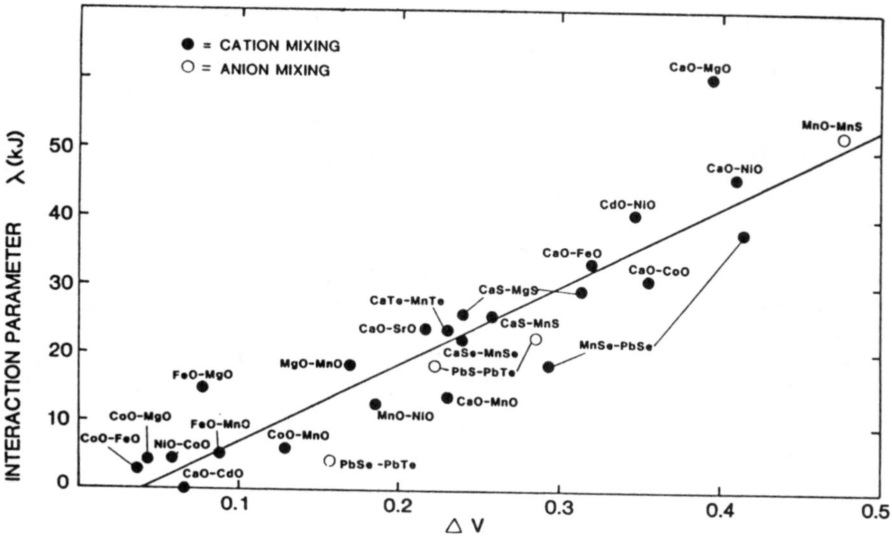

Figure 2. Interaction parameter, λ, versus volume mismatch, ΔV, for oxide and chalcogenide solid solutions with rocksalt structure. From Davies and Navrotsky (1983).

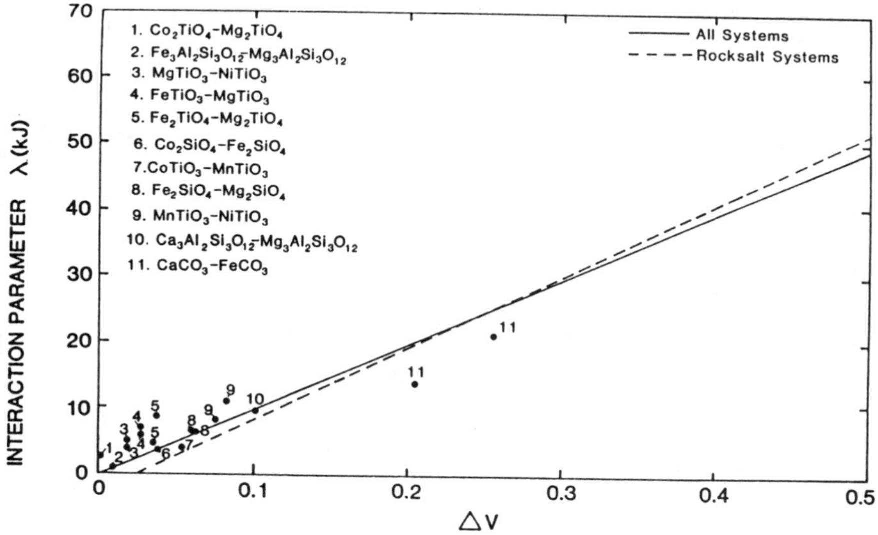

Figure 3. Interaction parameter, λ, for other solid solutions in which divalent ions are being mixed. The solid line indicates the best correlation using both rocksalt and other systems. From Davies and Navrotsky (1983).

Table 2. Predicted and Experimental Values of Energetics of Fe,Mg Mixing

System	Structure	Volume (cm^3/mol)[a] Mg	Fe	W (kJ/mol) calc[b]	expt.
MgO-FeO	rocksalt	11.25	12.25	8.2	15.9[c]
1/2(Mg_2SiO_4-Fe_2SiO_4)	olivine	21.84	23.14	5.4	3 to 7[d], 6.3[e]
1/2(Mg_2SiO_4-Fe_2SiO_4)	β-phase	20.26	21.61	6.1	6.3[e]
1/2(Mg_2SiO_4-Fe_2SiO_4)	spinel	19.83	21.02	5.5	6.3[e]
$MgSiO_3$-$FeSiO_3$	orthopyroxene	31.33	32.96	4.7	4.0[f]
$MgSiO_3$-$FeSiO_3$	garnet	28.53	29.38	2.6	
$MgSiO_3$-$FeSiO_3$	ilmenite	26.35	26.85	1.5	
$MgSiO_3$-$FeSiO_3$	perovskite	24.46	25.49	3.8	
1/3($Mg_3Al_2Si_3O_{12}$-$Fe_3Al_2Si_3O_{12}$	garnet	37.73	38.41	1.4	

a. Volume per one mol Mg or Fe, volume data from Jeanloz and Thompson (1983)

b. Calculated from Eq. (46), $W = \lambda = 100.8\ \Delta V - 0.4$, $\Delta V = (V_{Fe}-V_{Mg})/(0.5(V_{Fe}+V_{Mg}))$

c. Hahn and Muan (1962). Complicated by nonstoichiometry.

d. Consistent with several sets of calorimetry, phase equilibria, ignoring asymmetry; Newton et al. (1981).

e. Bina and Wood (1987), averaging asymmetry in γ,α phases, from phase diagram fitting.

f. Chatillon-Colinet et al. (1983), alkali borate solution calorimetry, order-disorder a complication.

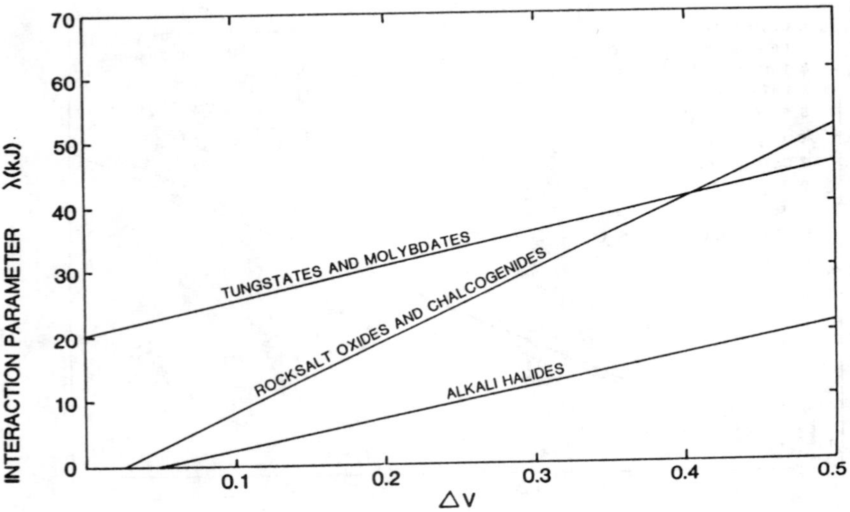

Figure 4. Linear correlations of interaction parameters and volume mismatch for alkali halide, oxide, chalcogenide, tungstate, and molybdate systems. From Davies and Navrotsky (1983).

structure "β". Writing formula units corresponding to one mole of ions being mixed, one has

$$
\begin{aligned}
\mu(A,\alpha) &= \mu^{\circ}(A,\alpha) + RT \ln a(A,\alpha) \\
\mu(A,\beta) &= \mu^{\circ}(A,\alpha) + \Delta\mu(A,\alpha\rightarrow\beta) + RT \ln a(A,\beta) \\
\mu(B,\alpha) &= \mu^{\circ}(B,\beta) + \Delta\mu(B,\beta\rightarrow\alpha) + RT \ln a(B,\alpha) \\
\mu(B,\beta) &= \mu^{\circ}(B,\beta) + RT \ln a(B,\beta) \ ,
\end{aligned}
\tag{47}
$$

where $\mu^{\circ}(A,\alpha)$ is the free energy per mole of pure A in structure "α", $\Delta\mu^{\circ}(A(\alpha\rightarrow\beta))$ is the free energy change for the transformation of A from structure "α" to structure "β", $a(A,\alpha)$ refers to the activity of A in structure "α" (unity for pure A) and $a(A,\beta)$ refers to the activity of A in structure "β" (also unity for pure A, with structure "β" taken as standard state). Analogous definitions hold for the terms in component B. Pure A in structure "β" can be a real phase at some P and T (e.g. $MgSi_{0.5}O_2$ in the spinel structure) or a physically unrealizable "cryptomodification" or "fictive phase" (see Chapter 1) unstable at all P and T (e.g. $FeSi_{0.5}O_2$ in the modified spinel structure). The limiting solubilities are given by equating chemical potentials:

$$
\begin{aligned}
\mu(A,\alpha) &= \mu(A,\beta) \ , \\
\mu(B,\alpha) &= \mu(B,\beta) \ .
\end{aligned}
\tag{48}
$$

The calculated phase relations as a function of pressure among the α, β, and γ polymorphs in the system Mg_2SiO_4-Fe_2SiO_4 are shown in Fig. 5. The parameters used in the calculation are shown in Table 3. The first set of parameters correspond to those used by Navrotsky and Akaogi (1984). They assumed ideal mixing of $FeSi_{0.5}O_2$ and $MgSi_{0.5}O_2$ in all three phases (olivine (α), modified spinel (β) and spinel (γ)) and neglected the effect of compressibility and thermal expansion. The second set of parameters correspond to those used by Bina and Wood (1987), who included both energetic nonideality and the dependence of volume of transition on pressure. In either case, the equations to be solved (e.g. for the α-β transition) are

$$
\begin{aligned}
0 = {} & \Delta H^{\circ}(FeSi_{0.5}O_2,\alpha\rightarrow\beta) - T\Delta S^{\circ}(FeSi_{0.5}O_2,\alpha\rightarrow\beta) \\
& + \int_{1\ atm}^{P} \Delta V(FeSi_{0.5}O_2,\alpha\rightarrow\beta)dP + RT \ln a(FeSi_{0.5}O_2,\alpha) \\
& - RT \ln a(FeSi_{0.5}O_2,\beta) \ ,
\end{aligned}
\tag{49}
$$

and

$$
\begin{aligned}
0 = {} & \Delta H^{\circ}(MgSi_{0.5}O_2,\alpha\rightarrow\beta) - T\Delta S^{\circ}(MgSi_{0.5}O_2,\alpha\rightarrow\beta) \\
& + \int_{1\ atm}^{P} \Delta V(MgSi_{0.5}O_2,\alpha\rightarrow\beta)dP + RT \ln a(MgSi_{0.5}O_2,\alpha) \\
& - RT \ln a(MgSi_{0.5}O_2,\beta) \ .
\end{aligned}
\tag{50}
$$

In each equation, the standard state for activity is the pure endmember (A or B) of the given structure (α or β).

Table 3. Parameters Used for Mg_2SiO_4–Fe_2SiO_4 Phase Diagram Calculations.

Parameters	Bina and Wood Mg_2SiO_4	Bina and Wood Fe_2SiO_4	Navrotsky and Akaogi Mg_2SiO_4	Navrotsky and Akaogi Fe_2SiO_4
Olivine				
V°_{298} (cm^3/mol)	43.67	46.27	43.67	46.27
$\alpha \times 10^6$ (K^{-1})	26.08	26.18	—	—
$(d\alpha/dT) \times 10^6$ (K^{-2})	0.015	0.012	—	—
K_O (GPa)	129	138	—	—
K_O'	5.2	5.2	—	—
W_G (J)	8368	4184	—	—
W_V (J/GPa)	125	125	—	—
Modified Spinel (β)				
V°_{298} (cm^3/mol)	40.52	43.22	40.52	43.22
$\alpha \times 10^6$ (K^{-1})	20.63	22.1	—	—
$(d\alpha/dT) \times 10^6$ (K^{-2})	0.017	0.011	—	—
K_O (GPa)	174	189	—	—
K_O'	4.8	4.8	—	—
W_G (J)	6276	6276	—	—
W_V (J/GPa)	0	0	—	—
$\Delta H^\circ_{1000}(\alpha\to\beta)$ J/mol	23200	14221	29957	16351
$\Delta S^\circ_{1000}(\alpha\to\beta)$ J/mol·K	-8.45	-8.53	-10.46	-11.59
$\Delta C_p(\alpha\to\beta)$ J/mol·K	-3.34	-0.50	—	—
Spinel (γ)				
V°_{298} (cm^3/mol)	39.65	42.02	39.54	41.92
$\alpha \times 10^6$ (K^{-1})	18.63	21.16	—	—
$(d\alpha/dT) \times 10^6$ (K^{-2})	0.017	0.011	—	—
K_O (GPa)	184	199	—	—
K_O'	4.8	4.8	—	—
W_G (J)	4180	4180	—	—
W_V (J/GPa)	0	0	—	—
ΔH°_{1000} ($\alpha\to\gamma$) J/mol	33589	11142	36777	2941
ΔS°_{1000} ($\alpha\to\gamma$) J/mol·K	-10.67	-10.88	-16.74	-17.57
ΔC_p ($\alpha\to\gamma$) J/mol·K	-7.78	-1.17	—	—

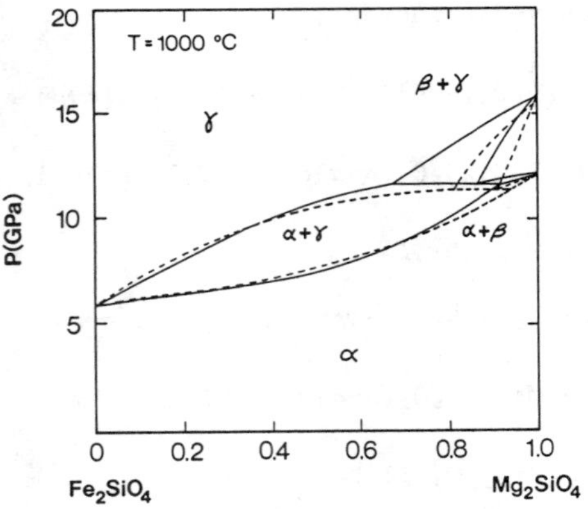

Figure 5. Mg_2SiO_4–Fe_2SiO_4 phase diagram at 1000°C as calculated by Bina and Wood (1987) (solid lines) and Navrotsky and Akaogi (1984) (dotted lines).

Knowing the ΔH°, ΔS°, ΔV, and activity terms, one can solve for coexisting compositions at a given P and T. Both calculated diagrams show similar topologies. The α-β loop is intersected at high Mg_2SiO_4 content by the α-β and β-γ loops. The resulting "rabbit ears" phase diagram combines the stable portions of all three binary loops. The width of the two-phase regions, however, is very sensitive to the input data. A very narrow α-β loop, consistent with a sharp seismic discontinuity near 400 km, is obtained when the parameters of Bina and Wood (1987) are used.

ORDER-DISORDER IN SOLID SOLUTIONS

General Comments.

Ordering phenomena complicate the thermodynamic behavior of solid solutions in several ways.
(1) If the order-disorder process is slow, several series of solid solutions, each having a differing degree of order can be prepared or observed in nature. Then the thermodynamics of mixing of other species (e.g. Na,K in feldspars) will depend on the degree of order in the aluminosilicate framework.
(2) The extent of ordering (be it Al,Si in feldspar, Mg,Fe in pyroxene, or Fe,Ti in spinels) depends on temperature in a complex fashion. Thus the equilibrium properties of a solid solution have a complex dependence on temperature which is not modelled easily using conventional formulation Margules parameters. Extrapolation of measured properties to higher or lower temperatures is far more risky than in "simple" systems. Thermodynamic models which specifically include order-disorder must be developed.
(3) The observed degree of ordering is often kinetically controlled. At low temperatures, changes in the degree of order are hindered, thus both ordering and disordering occur slowly, and many metastable states are preserved. At intermediate temperatures, equilibrium is attained and that state can be quenched to ambient conditions. At high temperatures, equilibrium is attained, but it may be impossible to quench the full extent of disorder back to ambient conditions. The effect of pressure on the rates of order-disorder reactions has only begun to be explored. Thus the interpretation of experimental observations must be made with both thermodynamics and kinetics in mind. Specifically, the extent to which order-disorder contributes to experimentally measured heat capacities poses a complex problem.
(4) A strong tendency to ordering is manifested in significant negative heats of mixing and sometimes leads to compound formation, (e.g. dolomite, omphacitic pyroxenes, etc.). Both the heat and entropy of mixing in such systems may depend strongly on the degree of order, and therefore on temperature. In addition, destabilizing energetic effects from mixing of ions in the disordered state may be present at the same time as the stabilizing effect of ordering, leading to complex interplay between ordering and exsolution.
(5) Order-disorder reactions often involve a change in the symmetry of the structure. Such changes may be ordinary first-order phase transitions, or they may be higher order phase transitions leading to some unconventional phase diagram topologies. The following sections will describe some thermodynamic models for order-disorder transitions and present thermodynamic analyses for several systems.

Cation interchange equilibria, especially in spinels.

Many minerals (e.g. spinel, olivine, pyroxene, feldspar, melilite, pseudobrookite) have two or more crystallographically nonequivalent sites. If these sites represent positions of very different energy for the ions in the crystal, a completely ordered structure results. If the energetics of cation interchange are not prohibitive, a partially or completely randomized structure will result, and the degree of disorder can vary with thermal history of the sample. Since the sites remain crystallographically distinct, even in the fully disordered state, the disordering process is nonconvergent (Thompson, 1969). Octahedral-tetrahedral cation disorder in spinels is an example of this situation.

The spinel unit cell is face-centered cubic, space group Fd3m. Since the anions occupy general positions, both the lattice parameter, "a", and a positional parameter, generally designated "u" and known in oxide spinels as the oxygen parameter, are required for a complete description. The cation to anion distances, R, are given by

$$R_{tet} = (1.732(u - 1/8)) \,, \tag{51}$$

$$R_{oct} = a(3u^2 - 2u + 3/8)^{1/2} \,. \tag{52}$$

Thus the octahedral and tetrahedral bond lengths can be used to determine the two parameters, a and u, and vice versa. A perfectly normal spinel is one in which the single A cation of the formula unit occupies the tetrahedral site, and the two B cations the two equivalent octahedral sites. If the symbols ( ) and [ ] denote tetrahedral and octahedral sites, respectively, then this distribution may be written $(A)[B_2]O_4$. The alternative distribution $(B)[AB]O_4$, called inverse, is also possible. In fact all distributions falling between these two extremes may be realized, and one can define an additional parameter, x, the degree of inversion, which is the fraction of B ions occupying tetrahedral sites. Thus x may vary between 0 for the perfectly normal case and 1 for the perfectly inverse case. Of special note is the completely random arrangement, $(A_{1/3}B_{2/3})[A_{2/3}B_{4/3}]O_4$, with $x = 0.667$. Note that in this section I use "x" to indicate degree of inversion, N to indicate mole fraction in a solid solution.

A thermodynamic treatment of cation distributions in simple spinels was presented by Callen et al. (1956) and by Navrotsky and Kleppa (1967). The configurational entropy (per mole AB_2O_4) is given, assuming completely random mixing of ions on each site by:

$$S_{conf} = -R[x \ln x + (1-x) \ln (1-x) + x \ln (x/2) + (2-x) \ln (1-x/2)] \,. \tag{53}$$

If the change in enthalpy on disordering is $\Delta H_D = x\Delta H_{int}$ (the extent of reaction times interchange energy) and the change in the nonconfigurational entropy as ΔS_{non}, then the change in free energy on disordering, ΔG_D, is

$$\Delta G_D = x\Delta H_{int} - T(\Delta S_{conf} + \Delta S_{non}) \,, \tag{54}$$

where ΔS_{conf} is the change in configurational entropy on disordering, equal to S_{conf}, if the ordered state is taken as $x = 0$.

At equilibrium

$$\partial \Delta G_D/\partial x = 0 \tag{55}$$

and

$$-RT \ln \left[\frac{x^2}{(1-x)(2-x)}\right] = \Delta H_{int} - T\Delta S_{non} \ . \tag{56}$$

Navrotsky and Kleppa (1967) have shown that the term inside the logarithm on the left hand side of Eq. (56) is equivalent to an equilibrium constant, K, for the interchange reaction

$$(A) + [B] = (B) + [A] \ . \tag{57}$$

If, furthermore, the change in nonconfigurational entropy on disordering can be neglected, then the simple model proposed by Navrotsky and Kleppa (1967) results. The interchange enthalpy for Reaction (57) can be regarded as a difference of site preference energies ΔE_A and ΔE_B for ions A and B, and

$$-RT \ln \left[\frac{x^2}{(1-x)(2-x)}\right] = (\Delta E_A - \Delta E_B) = \Delta H_{int} \ . \tag{58}$$

Thus the interchange enthalpy uniquely defines the equilibrium cation distribution as a function of temperature and, conversely, knowledge of the equilibrium cation distribution at one temperature is sufficient to estimate ΔH_{int}.

Though the simple treatment above is useful, greater complexity is seen in the cation distributions of some spinels and especially in solid solutions between largely normal and largely inverse spinels. In particular, the Madelung constant and lattice energy of a spinel depends on its lattice constant, a, oxygen parameter, u, and cation distribution, x. Both a and u are uniquely determined by a combination of the octahedral and tetrahedral bond lengths (Eqs. (51) and (52)). Hill et al. (1979) showed that, one could account for 96.9% of the variation of bond lengths in 149 oxide spinels by using the ionic radii of Shannon and Prewitt (1969).

O'Neill and Navrotsky (1983) used this observation of constant bond lengths, R_{oct} and R_{tet}, to compute the change of a and u, and hence of the electrostatic energy with x. They argued that the change in electrostatic energy may be the dominant term in the disordering enthalpy and that the most appropriate form for the interchange enthalpy is quadratic in x, giving:

$$\Delta G_D = \alpha x + \beta x^2 - T\Delta S_{non} + RT\,[x \ln x + (1-x) \ln (1-x) + x \ln (x/2) + (2-x) \ln (1-x/2)] \ . \tag{59}$$

At equilibrium, $(dG_D/dx) = 0$, and (neglecting the nonconfigurational entropy)

$$-RT \ln \left[\frac{x^2}{(1-x)(2-x)}\right] = \alpha + 2\beta x \ . \tag{60}$$

The coefficients α and β are generally of approximately equal magnitude and opposite sign. The effects of crystal field stabilization energies are implicitly included in this model, since they can be accommodated in the term linear in x. The coefficients α and β will depend mainly on the differences in radii and charge of the two ions, but these coefficients must be determined empirically from observed cation distributions.

O'Neill and Navrotsky (1984) applied this model to spinel solid solutions as follows. After examining data for a number of spinels, they concluded that the value of the coefficient β depended mainly on the charge type, with β = ~-20 kJ/mol for 2-3 spinels and β = ~-60 kJ/mol for 2-4 spinels. Assuming these constant values of β, a series of α's, which maintain a meaning of cation site preference energies, could be calculated and are shown in Fig. 6. For a spinel solid solution at any given temperature, the cation distribution may be calculated by solving a set of cation distribution equilibria involving several pairs of cations. The lattice parameters in the solid solution can be calculated from the site occupancies using Eq. (51) and (52). Activity-composition relations can then be calculated at different temperatures from the cation distributions at each temperature, the change in configurational entropy for solid solution formation at that temperature, the negative change in enthalpy resulting because the site occupancy is not a simple average of the site occupancies of the end-members, and finally an empirical regular solution enthalpy of mixing term reflecting destabilization from size mismatch.

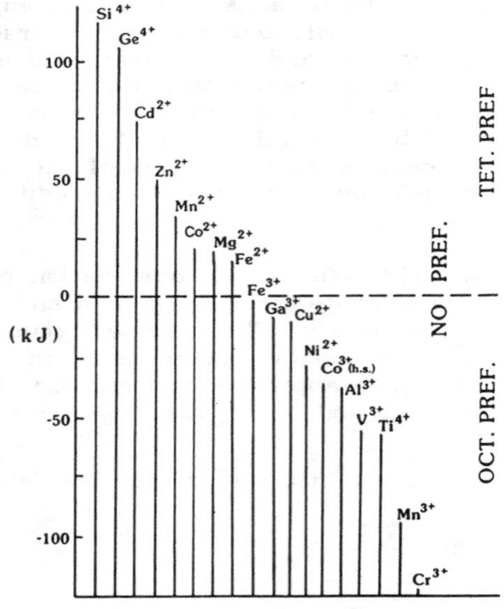

Figure 6. Site preference energies in spinels, from O'Neill and Navrotsky (1984).

Several general conclusions were drawn.

(1) The enthalpy and entropy of mixing depend strongly on temperature because the degree of order is temperature dependent.

(2) The lattice parameter can vary rather nonlinearly with composition, especially at low temperature and when the end members have different cation distributions. A sigmoid variation of the lattice parameter with composition is commonly seen in solid solutions between a normal and an inverse spinel (e.g. $FeCr_2O_4$-Fe_3O_4).

(3) The calculated entropies of mixing can be quite asymmetric, especially for the mixing of a normal and an inverse spinel. This results in significantly asymmetric free energies of mixing, not because one invokes Margules parameters and asymmetric enthalpies and nonconfigurational entropies of mixing, but because the configurational entropy term itself departs from symmetrical behavior and does not resemble $-nR\ [x \ln x+(1-x) \ln (1-x)]$ in form. At low temperature this asymmetry is pronounced and a solvus between a largely normal and a largely inverse spinel develops. Again I stress that such asymmetry can arise from the form of the entropy of mixing and not from strongly asymmetric energies.

O'Neill and Navrotsky (1983, 1984) cautioned that the general trends in their model were not totally applicable to two very common spinels, namely Fe_3O_4 and $MgAl_2O_4$. In the former, Fe^{2+}-Fe^{3+} electron exchange on the octahedral sublattice is rapid, which complicates the thermodynamics, perhaps introducing an excess entropy term (Wu and Mason, 1981; O'Neill and Navrotsky, 1984). In addition, the high temperature disorder in magnetite is almost certainly not quenchable, and the extent to which a cation distribution can be quenched in an Fe_3O_4-containing solid solution probably depends strongly on composition. The behavior of $MgAl_2O_4$ disordering also appears anomalous, and it may be impossible to completely quench disorder from temperatures above ~1000°C (Wood et al., 1986; Navrotsky, 1986). Navrotsky (1986) studied three spinel solid solutions ($MgFe_2O_4$-$MgAl_2O_4$, $ZnFe_2O_4$-$ZnAl_2O_4$, and $NiAl_2O_4$-$ZnAl_2O_4$) by high temperature solution calorimetry. The enthalpies of mixing do not follow the O'Neill-Navrotsky systematics based on bond-length differences. In addition, in the range 25-700°C, the system $MgAl_2O_4$-$MgFe_2O_4$ shows a small negative excess heat capacity whereas the other two systems show no excess. Navrotsky (1986) also reports enthalpies of annealing associated with re-equilibration of cation distribution in $NiAl_2O_4$, $MgAl_2O_4$, and $ZnAl_2O_4$. These enthalpies and independently measured cation distribution data taken from the literature support the general form of the O'Neill-Navrotsky model, but suggest somewhat different values for the parameters proposed. Thus one must be cautious about using the model for quantitative prediction of mixing properties for individual systems. Rather, for maximum accuracy, data for each individual system must be fit to the model equations. The equations and these fitted parameters then provide an accurate formalism for both interpolation and extrapolation.

The system Fe_3O_4-Fe_2TiO_4 (magnetite-ulvospinel) is one of the most widespread and interesting natural spinel series. Iron-titanium oxide equilibria involving this series and the Fe_2O_3-$FeTiO_3$ (hematite-ilmenite) series are used as petrologic temperature and oxygen fugacity indicators and figure strongly in paleomagnetism. Yet the cation distribution equilibria and solid solution thermodynamics of these spinels have been poorly understood. A major complication in this system arises because the Fe^{2+}-Fe^{3+} equilibrium involves electron transfer rather than the movement of cations, and samples quenched from high temperatures probably do not preserve their cation distributions. Nonstoichiometry (oxidation toward a γ-Fe_2O_3 or $Fe^{3+}_{8/3}Fe^{2+}_{1/3}O_4$ spinel) presents additional complexity.

Recently Mason and coworkers (Wu and Mason, 1981; Trestman-Matts et al., 1983; Mason, 1987) have shown that Fe^{2+}-Fe^{3+} distributions in iron-bearing minerals can be obtained by electrical measurements (Seebeck coefficient or Seebeck coefficient combined with electrical conductivity). These methods measure the site occupancies at high temperature directly and avoid problems of redistribution during the quench. Trestman-Matts et al. made measurements on titanomagnetites. For the series (N) Fe_2TiO_4-(1-N) Fe_3O_4) the site occupancies may be written as follows, assuming all Ti remains octahedral:

	Tet	Oct	Sum
Fe^{2+}	1-x	N+x	1+N
Fe^{3+}	x	2-2N-x	2-2N
Ti^{4+}	0	N	N
	1	2	3

(61)

Then the equilibrium constant K is

$$K = \frac{x(N+x)}{(1-x)(2-2N-x)} \ . \tag{62}$$

Using Seebeck coefficient data for Fe_3O_4 and for solid solutions with N = 0.19, 0.37, 0.58 and 0.69, Trestman-Matts et al. found that, in the range 600-1300 °C, with T in degrees Kelvin, -RT ln K in J/mol.

$$-RT \ln K = -9610 + 30.8\ T\ x + 25.6\ T \ . \tag{63}$$

This equation is similar to the O'Neill-Navrotsky model but with α = -9610 + 25.6T, β = 15.4T. The temperature dependence of α suggests a small negative excess entropy of disordering, that of β is a novel finding. The cation distribution data do appear to need three rather than two parameters to describe their temperature- and composition-dependence. The parameters chosen probably are not a unique interpretation but appear reasonable. The strong temperature dependence of α and β may reflect electronic entropy effects related to Fe^{2+}-Fe^{3+} electron hopping or to some degree of short range order on the octahedral sublattice which contains Fe^{2+}, Fe^{3+}, and Ti^{4+}. Nevertheless, the parameters chosen do maintain an essential feature of the O'Neill-Navrotsky model, namely that the Fe^{2+}-Fe^{3+} distribution energetics depend strongly on the average charge on the octahedral sublattice, which in this case is influenced strongly by the titanium content.

The model above can be used to calculate the distribution at different temperatures; the results are shown in Fig. 7. Fe_3O_4, an inverse spinel at low temperature, tends to a largely random distributions at high temperature. Fe^{3+}-Fe^{3+} distribution at high

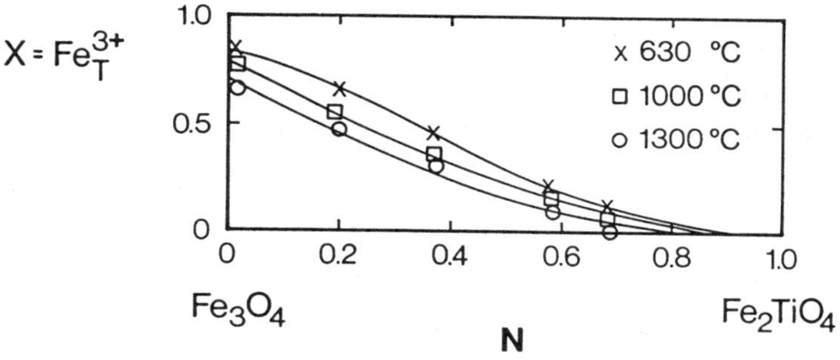

Figure 7. Experimental cation distributions and those fit by Eqs. (62) and (63) (solid lines) for (N)Fe_2TiO_4-(1-N)Fe_3O_4 from Trestman-Matts et al. (1983).

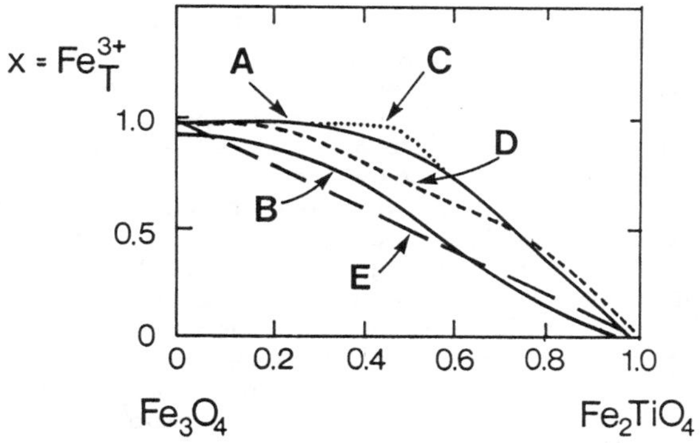

Figure 8. Cation distributions at 25°C (A), 300°C (B) calculated from Eq. (62) and (63) compared to those suggested by Akimoto (1954) (C), Neel (1955) and Chevallier et al. (1955) (D) and O'Reilly and Banerjee (1965) (E).

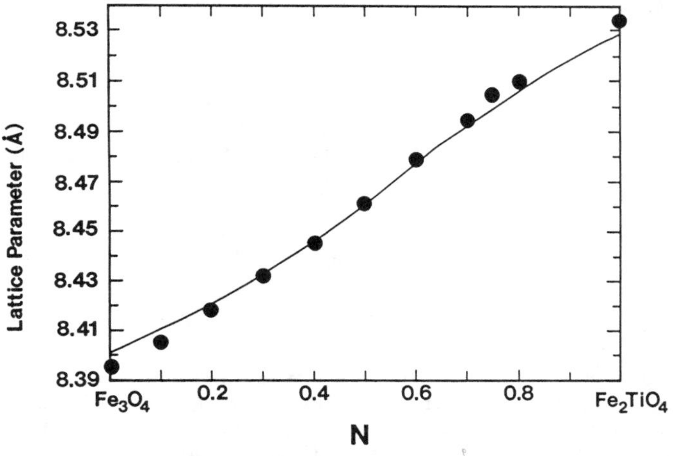

Figure 9. Lattice parameters of NFe_2TiO_4-(1-N)Fe_3O_4 solid solutions. Points are from Wechsler et al. (1984), curve is calculated for 300°C (assumed, somewhat arbitrary, temperature of frozen cation equilibrium) using cation distribution model.

temperature vary smoothly with composition, with Fe^{3+} showing less tetrahedral preference with increasing Fe_2TiO_4 content (N). The cation distribution predicted from the high temperature data and Eqs. (62) and (63) for $(N)Fe_2TiO_4$-$(1-N)Fe_3O_4$ at several temperatures is shown in Fig. 8 and compared with the prediction of several previously proposed models based on saturation magnetization data on quenched samples. Although the data show a more gradual variation than the models, the low temperature limit is similar to that proposed by Neel (1955).

The lattice parameters may be calculated using Eq. (51) and (52) and the radii given by O'Neill and Navrotsky (1983). The results are compared to the lattice parameters obtained by Wechsler et al. (1983), for an assumed equilibration temperature of 300°C (see Fig. 9). The agreement is quite satisfactory. The calculations reproduce the noticeably sigmoid variation of lattice parameter with composition, a consequence of the changing Fe^{2+}-Fe^{3+} site preference.

Activity-composition relations in Fe_3O_4-Fe_2TiO_4 were determined by Taylor (1964) at 1300°C and Webster and Bright (1961) at 1200°C. They show (see Fig. 10a) almost Raoultian behavior, with $a_{Fe_3O_4}$ slighly larger than (1-N) at Fe_3O_4-rich compositions and $a_{Fe_3O_4}$ slightly less than (1-N) at Fe_2TiO_4-rich compositions. This seemingly "ideal" one-site-mixing behavior is surprising because in $Fe_2^{2+}TiO_4$-$Fe^{2+}Fe_2^{3+}O_4$, the substitution involves a total of two moles of ions ($Fe^{2+}+Ti^{4+} = 2Fe^{3+}$) being mixed. Thus for totally random mixing of these species one might expect $a_{Fe_3O_4} = (1-N)^2$ as a limiting case. In fact, the complex cation distribution equilibrium discussed above will determine the configurational entropy, and short range order may further complicate the situation. Using the cation distribution model and the values of x calculated from it (Fig. 8), one can calculate ΔG_{mix}. At a given mole fraction, N, and temperature, T, with x = the cation distribution parameter for the solid solution with mole fraction Fe_2TiO_4 = N, and x_o = the cation distribution parameter for Fe_3O_4 (N = 0)

$$
\begin{aligned}
\Delta G_{mix} = {} & (-9610 + 25.6\ T)(x_o - x) \\
& + 15.4\ (x_o^2 - x^2) \\
& + 8.314\ T\ [x \ln x + (1-x) \ln (1-x) \\
& + (N+x) \ln (0.5\ (N+x)) + (2-2N-x) \ln (1-N-0.5x)] \\
& - 16.628\ NT \ln 2 \\
& - 8.314\ (1-N)\ T[x_o \ln x_o + (1-x_o) \ln (1-x_o) \\
& + x_o \ln (x_o/2) + (2-x_o) \ln (1-x_o/2)] \ . \qquad (64)
\end{aligned}
$$

This equation takes into account the difference in free energy, arising from both nonconfigurational and configurational terms between the solid solution and N moles of Fe_2TiO_4 (with Fe and Ti distributed at random on octahedral sites) and (1-N) moles of Fe_3O_4 with its equilibrium degree of disorder x_o. The result of this calculation (curve B in Fig. 10) is compared with the experimental free energy of mixing

Table 4. Order Parameters and Enthalpy of Disordering and of Mixing in Dolomites.

Quench Temp. (°C)	Order Parameter s^a	Enthalpy (kJ per mole $MM'(CO_3)_2$) Disordering[b]	Mixing[c]
$CdMg(CO_3)_2$ [d]			
600	1	0	−5.6
750	0.656±0.036	6.5	+0.9
775	0.620±0.028	6.7	+1.1
800	0.052±0.078	13.8	+8.1
$CaMg(CO_3)_2$ [e]			
natural sample	1	0	−11.5
1250 (15 kbar)	0 (?)	14.0	+2.5
$CaMn(CO_3)_2$ [f]			
600	0	(≥5.3)	−5.7

a. s = long range order parameter; s = 1 for ordered, s = 0 for disordered structure

b. For reaction $MM'(CO_3)_2$ (ordered) = $MM'(CO_3)_2$ (partially or totally disordered)

c. For reaction $MCO_3 + M'CO_3 = MM'(CO_3)_2$

d. From Navrotsky and Capobianco (1987)

e. From Capobianco et al. (1987)

f. From Capobianco and Navrotsky (1987)

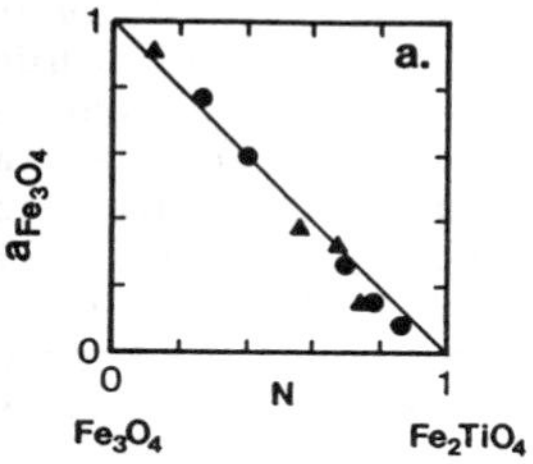

Figure 10. (a) Activity of Fe_3O_4 in $(1-N)Fe_3O_4-(N)Fe_2TiO_4$ solid solutions. Triangles are data of Webster and Bright (1961), at 1200°C, circles are data of Taylor (1964) at 1300°C, line is Raoult's Law, $a_{Fe_3O_4} = N_{Fe_3O_4}$. (b) Free energy of mixing at 1300°C. (A) Ideal mixing of one mole of ions, (B) Calculated from cation distribution model (no positive contribution from size mismatch), (C) Experimental data from a fit to activities in (A), (D). Ideal mixing of one mole of ions.

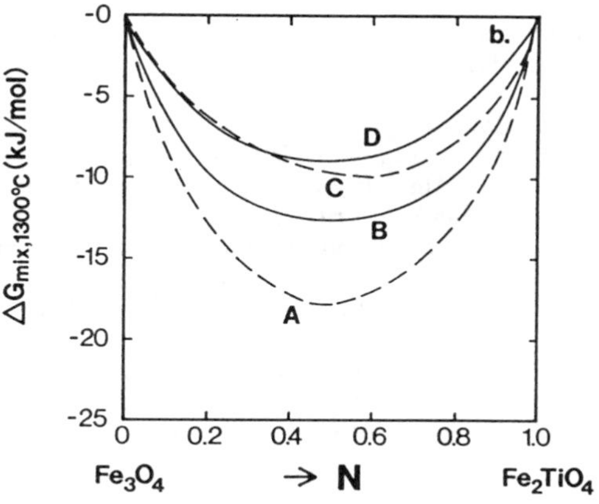

(curve C), the curve for ideal mixing of one mole of ions (curve D) and that for ideal mixing of two moles of ions (curve A). The experimental curve is less negative than that calculated from the cation distribution model. This may suggest a positive strain energy related to size mismatch, as suggested for other spinel solid solutions by O'Neill and Navrotsky (1983). The magnitude of this positive interaction parameter suggested by Trestman-Matts et al. (1983) follows the systematic trends suggested by Davies and Navrotsky (1982) and O'Neill and Navrotsky (1983). However, because the free energy of mixing calculated from the cation distribution is almost symmetric about N = 0.5, the strain energy must be asymmetric to produce an asymmetric solvus. In this respect Fe_3O_4-Fe_2TiO_4 is somewhat different from the solid solution between a normal 2-3 spinel (e.g. $FeCr_2O_4$) and an inverse 2-3 spinel (e.g. Fe_3O_4) discussed by O'Neill and Navrotsky (1983) for which the change in cation distribution itself produces both a sigmoid variation in lattice parameters and marked asymmetry in the thermodynamics of mixing. However, at present the delineation of the Fe_2TiO_4-Fe_3O_4 solvus is not very accurate, and experimental enthalpies of mixing in the Fe_3O_4-Fe_2TiO_4 solid solution series have not been measured. Thus the exact calculation of mixing thermodynamics at low temperature must await further measurements. I stress, however, that the apparently almost Raoultian behavior of the system at 1200-1300 °C is probably fortuitous, cannot be extrapolated to lower or higher temperatures, and does not represent ideal one-site mixing.

Carbonates - calcite and dolomite structures.

The calcite and dolomite crystal structures commonly occur in nature for divalent metal carbonate minerals (usually $CaCO_3$ and $CaMg(CO_3)_2$). The major crystal-chemical difference between the calcite and dolomite structures is that calcite contains only one distinct metal layer while dolomite has two. The chemical difference between these layers results in two geometrically inequivalent cation sites. Furthermore, a continuous range of structures between the ordered dolomite and disordered calcite structures is possible because the space group of dolomite ($R\bar{3}$) is a subgroup of that of calcite ($R\bar{3}c$). Indeed, some natural dolomites (i.e., $CaMg(CO_3)_2$) are reported to have disordered structural states (Reeder, 1984) while synthetically disordered samples may be obtained by annealing ordered material at high temperature. However, the high temperatures involved and possible problems in quenching the disordered phase make detailed study of the order-disorder transition in $CaMg(CO_3)_2$ difficult.

The system $CdCO_3$-$MgCO_3$ is analogous to the $CaCO_3$-$MgCO_3$ system in that similar ordered and disordered phases exist. Because chemical homogenization of distinct Cd and Mg layers proceeds within an experimentally accessible range of pressure and temperature (1-10 kbar $P(CO_2)$ and 600 to 850°C), samples can be prepared which, on quenching, possess greatly different states of order. Phase diagrams for both the $CaCO_3$-$MgCO_3$ and the $CdCO_3$-$MgCO_3$ systems are shown in Fig. 11.

Recent studies (Navrotsky and Capobianco, 1987; Capobianco et al., 1987) provide new data on the dependence of long range order parameter and of enthalpy of mixing on quench temperature (see Table 4).

The Cd,Mg disordering reaction is convergent (Thompson, 1969), i.e.

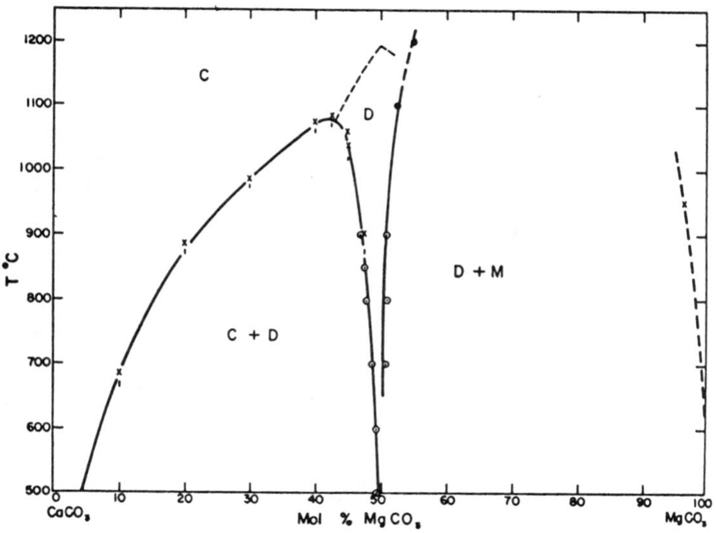

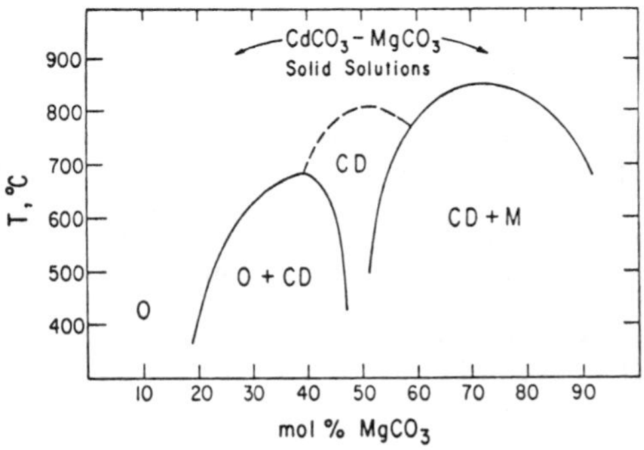

Figure 11. Phase diagrams for $CaCO_3$-$MgCO_3$ and $CdCO_3$-$MgCO_3$ (from Goldsmith and Heard, 1961; Goldsmith, 1972; Goldsmith, 1983). C = calcite, M = magnesite, D = dolomite, O = oldhamite, CD = cadmium dolomite.

alternate cation layers are crystallographically distinct only when the cations are ordered. The simplest model (see Capobianco et al., 1987) which can be applied to the order-disorder reaction is of the Bragg-Williams type in which the long range order parameter, s, is related to one energy parameter, W, by the equation (at the stoichiometric dolomite composition)

$$s = \tanh\,(-Ws/2R)\ . \qquad (65)$$

W is negative and its numerical value reflects the energy of ordering per mole $CdMg(CO_3)_2$.

The enthalpy of mixing (formation of $CdMg(CO_3)_2$ from $CdCO_3$ and $MgCO_3$) is then

$$\Delta H_{mix} = W\,[0.5 - 0.25(1-s^2)]\ , \qquad 0.5\,W \le \Delta H_{mix} \le 0.25W\ . \qquad (66)$$

The critical temperature at which s becomes zero is given by T_c = $-W/2R$. It is about 830°C for $CdMg(CO_3)_2$ and about 1150°C for

$CaMg(CO_3)_2$ (Goldsmith and Heard, 1961; Goldsmith, 1972).

A striking feature of both systems is that the ordered dolomite phase has a negative enthalpy of mixing from the end-member carbonates, while the disordered phase at dolomite composition has a positive enthalpy of mixing. This cannot be accounted for by a simple Bragg-Williams model because its one parameter, W, must be negative for ordering to take place at low temperatures. Thus even the disordered phase would show a negative heat of mixing.

To include positive heats of mixing in the disordered phase, one can apply a generalized Bragg-Williams, or generalized point approximation (PA), see discussion by Capobianco et al. (1987), and Burton (1987). In such a model there are two distinct interactions: (1) an attractive (interlayer) interaction promoting unlike pair formation (parameterized as W_{ier}) and (2) a repulsive (intralayer) interaction promoting segregation (parameterized as W_{ira}).

The Gibbs free energy of mixing (formation of $(Cd_{1-x}Mg_x)_2(CO_3)_2$ from $CdCO_3$ and $MgCO_3$) is given by, per formula unit containing two carbonates:

$$\Delta G_{mix} = W_{ier}[x - x^2(1 - s^2)] + W_{ira}[x - x^2(1 + s^2)] + RT \sum_i \sum_j x_{ij} \ln x_{ij} \ , \qquad (67)$$

where x_{ij} is the cation site occupancy of the "j-th" atom on the "i-th" sublattice. Thus for $i = \alpha$ (layer preferred by Mg), β (layer preferred by Cd) and j = Mg, Cd, $x_{\alpha,Mg} = x + xs$, $x_{\alpha,Cd} = 1 - x - xs$, $x_{\beta,Mg} = x - xs$, and $x_{\beta,Cd} = 1 - x + sx$. Here x is the mole fraction of $MgCO_3$ and s, as before, is the long range order parameter. As in the original Bragg-Williams model, the equilibrium value of s is found by minimizing ΔG_{mix} with respect to variations in s, and the critical temperature for disordering is obtained by setting the second derivative of ΔG_{mix} to zero for $s = 0$. For $W_{ier} < 0$ and $W_{ira} > 0$, intersite ordering decreases ΔH_{mix} while intrasite mixing increases it, and at low temperature the ordered phase is stable. Above T_C, $s = 0$, so that ΔH_{mix}, given by $(W_{ier} + W_{ira})(x - x^2)$ is positive for $W_{ira} > |W_{ier}|$. W_{ier} and/or W_{ira} may depend on composition, leading to asymmetry in solution properties.

A model based on the tetrahedral approximation (TA) in the cluster variation method (CVM) also can be applied (see Burton, 1987). The CVM approach includes the effects of short range order (which the point or Bragg-Williams approximations neglect) by considering the possible arrangements of clusters of atoms (tetrahedra of cations in this case). The calculated equilibrium proportions of the chemically distinct tetrahedra are constrained by the total solution composition and the energetic interactions between the atoms of the tetrahedron. These calculated proportions then provide information on both long-range and short-range order. Because of short-range order, the configurational entropy in the CVM calculations is generally smaller than in the previous models.

Calculations using both the PA and TA models correctly predict the topology and qualitatively fit the observed phase boundaries in both the

$CaCO_3$-$MgCO_3$ and $CdCO_3$-$MgCO_3$ systems (Burton, 1987; Capobianco et al., 1987). Such models account qualitatively for positive heats of mixing in the disordered phases. However, quantitative fit to the structural and thermochemical data in $CdCO_3$-$MgCO_3$ is poor (Capobianco et al., 1987). This failure may be related to several factors. First, the carbonate group may undergo shifts of position and small deviations from a planar configuration as a result of Cd,Mg ordering. Such involvement of CO_3 groups might give the transition more cooperative character and compress it, as observed, into a smaller temperature interval. Second, the compositional asymmetry of inter- and intra-layer interactions may be more complex than considered in the above models. Third, differences in lattice vibrations between ordered and disordered phases may result in excess heat capacities, excess entropies, and temperature dependent enthalpies not considered in simple models.

Nevertheless, it is important to note the essential features of a model which can qualitatively explain the variation of long-range-order with temperature, the enthalpy of mixing, and the topology of the phase diagram. Such a model must incorporate three main features: a negative (stabilizing) interaction for ordering of cations between layers, a positive (destabilizing) interaction for the mixing of cations within a layer, and compositional asymmetry for at least one parameter. These features lead to a strong temperature dependence of the order parameters and of the enthalpy and entropy of mixing. Empirical models based on a Margules-formulation simply cannot build in the appropriate temperature dependence and cannot account for the phase diagram topology.

The system $CaCO_3$-$MnCO_3$ is interesting because an ordered dolomite-like phase ($Ca,Mn(CO_3)_2$, kutnahorite) exists only at low temperature and a continuous solid solution with no long range order exists above 500°C. Calorimetric and phase equilibrium study of this system (Capobianco and Navrotsky, 1987) shows that the system at 600–700°C has positive excess free energies of mixing (relative to ideal mixing) in $(Mn_XCa_{1-X})CO_3$ but that the enthalpies of mixing are complex, being slightly positive at $x < 0.35$ and negative at $x > 0.35$, with a minimum near $x = 0.6$. Thus even the long-range-disordered phase of kutnahorite composition appears energetically stabilized relative to $CaCO_3$ and $MnCO_3$. The smaller difference in size between Ca and Mn than between Ca and Mg or Cd and Mg may diminish the magnitudes of the positive intralayer interactions.

Feldspar solid solutions.

The ability of the feldspar structure to undergo fairly rapid alkali exchange at low temperature while preserving a state of Al-Si order enables one to study the influence of the latter on the energetics of Na,K mixing. The results (Hovis and Waldbaum, 1977) are shown in Table 5. For a series with constant Al,Si order, the enthalpy of Na,K mixing becomes less positive with increasing framework disorder, i.e. in the series maximum microcline-orthoclase-high sanidine. The slightly larger volumes of the disordered feldspars, coupled with a larger variety of local environments within the framework, may make it energetically easier for Na and K to find local geometries which minimize the strain. It is interesting that when the constraints of long range order are removed entirely, namely in $NaAlSi_3O_8$-$KAlSi_3O_8$ glasses, the enthalpies of mixing are exothermic (Hervig and Navrotsky, 1984).

Several approaches to the thermodynamics of $NaAlSi_3O_8$-$CaAl_2Si_2O_8$ solid solutions have been discussed in "Microscopic to Macroscopic" (Volume 14 of this series - Navrotsky, 1985; Carpenter, 1985). Two sorts of experimental data are available: activity-composition in plagioclase of "high" structural state derived from hydrothermal and ion-exchange equilibria (Orville, 1972; Blencoe et al., 1982) and solution calorimetric determinations of enthalpies of mixing in solid solutions of several states of Al,Si order (Newton et al., 1980; Carpenter et al., 1985). The interpretation of these data depends on the thermodynamic model one assumes. Saxena and Ribbe (1972) and Blencoe et al. (1982) assumed that high plagioclases represent a continuous solid solution represented by one free energy surface and derived empirical polynomials to describe the excess free energy, relative to "ideal" (Raoultian) behavior. These polynomials fit the observed positive deviations from Raoult's Law at 700-1000°C and extrapolate to ideal one site mixing ($a = x$) at the high temperature limit.

However, the stoichiometry and crystallography preclude simple one-site mixing in plagioclase, since Na and Ca mix on A-sites and Si and Al mix on T-sites. The nature of the configurational entropy depends on the degree of Al,Si order and the extent of local charge balance observed in the Na + Si = Ca + Al substitution (Kerrick and Darken, 1975). Newton et al. (1980) and Henry et al. (1982) pointed out that the activity data could be described reasonably well by combining the observed positive heats of mixing with entropies of mixing calculated assuming aluminum avoidance in the high plagioclases. This leads to positive deviations from Raoult's law at 600-1000°C , virtually Raoultian behavior at 1000-1200°C, and negative deviations from Raoult's Law at higher temperature. This crossover occurs because positive heats of mixing are balanced by entropies of mixing larger than $-R(x \ln x + (1-x) \ln (1-x))$.

In his original study of ion exchange equilibria, Orville chose to separate his activity plots into three regions. (a) $0 \leq x_{An} < {\sim}0.5$, $\gamma_{An} = 1.27$, $\gamma_{Ab} = 1$ (Henry's Law for An, Raoult's Law for Ab), (b) $0.5 < x_{An} < {\sim}0.9$, both γ_{Ab} and γ_{An} change with composition, and (c) ${\sim}0.9 < x_{An} < 1$, $\gamma_{An} = 1$, $\gamma_{Ab} = 1.89$. Though Orville did not explicitly discuss the reason for this choice, the separation into two Henry's Law regions separated by an intermediate region implies that at albite-rich compositions, the anorthite component is substituting into a rather constant albite-like environment, at anorthite-rich compositions an anorthite-like environment persists, and a fairly rapid change in structure and thermodynamic properties occurs at intermediate compositions which are, in fact, close to those for the $C\bar{1}$ - $I\bar{1}$ transition (Carpenter, 1985). This transition, which is not first order, separates albite-rich feldspars with a largely disordered Al,Si distribution from anorthite-rich feldspars with a largely ordered Al,Si distribution. The effect of this transition on thermodynamics of the solid solutions is shown qualitatively in Fig. 12 (Carpenter and Ferry, 1984).

A quantitative description of the thermodynamics of mixing requires knowledge of the extent of both long and short range order and the energetics of both mixing and ordering. A starting point for such a model may be the order parameter theory developed for sodium feldspar by Salje (Salje, 1985; Salje et al., 1985). Two simultaneously occurring processes that reduce the symmetry of Na-feldspar are known: very sluggish Al,Si ordering and rapid distortive transformation. As a result,

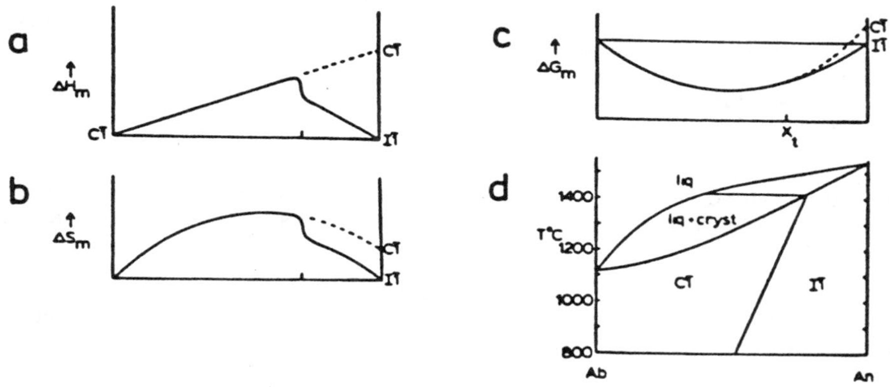

Figure 12. Schematic form of mixing parameters for the plagioclase solid solution at an arbitrary high temperature, using $C\bar{1}$ (disordered) and $I\bar{1}$ (ordered) anorthite end members and ignoring melt relations. The solid solution is shown as consisting of two approximately ideal segments, with $C\bar{1}$ and $I\bar{1}$ symmetry, related by a non-first order order/disorder transformation at X_t. A fictive $C\bar{1}$ pure anorthite end member is shown as the extrapolation of the $C\bar{1}$ solid solution to An_{100}. The equilibrium $C\bar{1}/I\bar{1}$ transformation line shown as a function of temperature and composition and its effect on melting relations. From Carpenter and Ferry (1984).

Table 5. Regular Solution Parameters for Enthalpies of Mixing in $(Na,K)AlSi_3O_8$

	W_H (kJ)
low albite - microcline	29.6 ± 0.4[a]
analbite - sanidine	23.8 ± 0.1[a]
glass	10.6 ± 3.4[b]

a. Hovis and Waldbaum (1977)

b. Hervig and Navrotsky (1984)

a two order parameter theory using Q_{od} and Q (representing the above processes, respectively) was derived from fundamental Landau theory. Based on symmetry considerations, Q and Q_{od} are not independent of each other due to their common strain components. An investigation of this coupling mechanism demonstrated that if one order parameter was known, the other could be calculated if the crystal was in thermal equilibrium. If, however, the crystal was in a metastable state, Q_{od} would be independent of temperature and the observed thermodynamic behavior would be controlled entirely by the temperature dependence of Q.

It was further shown that the symmetry breaking transition ($C2/m \rightarrow C\bar{1}$) under thermal equilibrium was a function of both Q and Q_{od}, and was accompanied by a smooth crossover between high and low albite. Finally, Salje derived simple formulae as functions of the order parameters, Q and Q_{od}, and temperature T, from which any thermodynamic functions, lattice parameters, etc. could be derived. One can then evaluate the thermodynamic behavior of sodium feldspar through investigations of the heat capacities of albite, analbite, ordered and disordered Or_{31}. By deriving numerical parameters for their equations from experimental data, Salje et al. (1985) could calculate expressions for the free energy, entropy, and enthalpy of sodium feldspar in thermal equilibirum and in metastable states as a function of the two order parameters, Q and Q_{od}, and the temperature T. Then the lattice strains of all the stable and metastable states were calculated and a phase diagram could be constructed.

An attractive feature of this order-theory is that its parameters can be determined from independent thermodynamic and physical measurements, rather than simply varied in computer simulations and obtained by a "best fit" to a phase diagram.

Recently Salje (Salje, 1987; Redfern and Salje, 1987) has formulated a Landau theory approach to the $I\bar{1} \rightarrow P\bar{1}$ phase transition in anorthite and in Ca-rich plagioclases. The theory uses two coupled order parameters, one related to rapid lattice distortion at the phase transition itself (which occurs at 510 K in anorthite) the other related to the degree of "frozen in" Al,Si disorder arising from the high temperature $C\bar{1}$-$P\bar{1}$ equilibrium near the melting point. Thus the low temperature transition is sensitive to plagioclase composition and thermal history and behavior at low temperature offers a tool for studying the fairly inaccessible high temperature disordering. Such an approach can be expected to lead toward a full description of plagioclase thermodynamics in the future. In the meantime, the most petrologically useful simplified description of plagioclase solid solution behavior is still in doubt (see Chapter 3 for one interpretation).

An interesting sidelight to plagioclase thermodynamics is that at temperatures of interest to igneous petrology (900-1300 °C), the numerical deviations of activity-composition relations from Raoultian behavior are in fact rather small. This leads to the use of "ideal solution" models ($a_{NaAlSi_3O_8} = x_{NaAlSi_3O_8}$) in petrologic applications (e.g. Loomis, 1979). The discussion above suggests that such behavior is fortuitous and results from the approximate cancellation of complex and temperature dependent heats and entropies within a restricted range. However their extrapolation to lower or higher temperature or to high pressure may lead to significant errors.

CONCLUSIONS

Thermodynamic analyses of complex solid solutions should be based on appropriate crystal chemical features of the substitutions which often occur over several crystallograhpic sites. By making the macroscopic punishment fit the microscopic crime, one avoids the pitfalls of arbitrary parameters and uncertain extrapolations.

ACKNOWLEDGMENTS

A. Navrotsky's work has been supported by the National Science Foundation and the Department of Energy. R.C. Newton, B.J. Wood and H.P. Eugster provided thoughtful reviews of this chapter. J. Bialkowski typed the manuscript and drafted most of the figures. P. H. Ribbe made the final editing easy.

REFERENCES

Akimoto, S. (1954) Thermomagnetic study of ferromagnetic minerals contained in igneous rocks. J. Geomagnetism Geoelectricity 6, 1-14.

Bina, C.R. and Wood, B.J. (1987) The olivine-spinel transitions: experimental and thermodynamic constraints and implications for the nature of the 400 km seismic discontinuity. J. Geophys. Res. (in press).

Blencoe, J.G., Merkel, G.A., and Seil, M.K. (1982) Thermodynamics of crystal-fluid equilibria, with applications to the system $NaAlSi_3O_8$-$CaAl_2Si_2O_8$-SiO_2-NaCl-$CaCl_2$-H_2O. In: Advances in Physical Geochemistry, Vol. 2, S.K. Saxena, Ed. Springer Verlag, N.Y., p. 191-222.

Burton, B.P. (1987) Theoretical analysis of cation ordering in binary rhombohedral carbonate systems. Am. Mineral. 72, 329-336.

Callen, H.B., Harrison, S.E. and Kriessman, C.J. (1956) Cation distributions in ferrospinels - theoretical. Phys. Rev. 103, 851-856.

Capobianco, C. and Navrotsky, A. (1987) Solid solution thermodynamics in $CaCO_3$- $MnCO_3$. Am. Mineral. 72, 312-318.

Capobianco, C., Burton, B.P., Davidson, P.M., and Navrotsky, A. (1987) Structural and calorimetric studies of order-disorder in $CdMg(CO_3)_2$. J. Solid State Chem. (in press).

Carpenter, M.A. (1985) Order-disorder transformations in mineral solid solutions. In "Microscopic to Macroscopic", S.W. Kieffer and A. Navrotsky, Eds., Reviews in Mineralogy 13, 187-224.

Carpenter, M.A. and Ferry, J.M. (1984) Constraints on the thermodynamic mixing properties of plagioclase feldspars. Contrib. Mineral. Petrol. 87, 138-148.

Carpenter, M.A., McConnell, J.D.C. and Navrotsky, A. (1985) Enthalpies of Al-Si ordering in the plagioclase feldspar solid solution. Geochim. Cosmochim. Acta 49, 947-966.

Chatillon-Colinet, C., Newton, R.C., Perkins, D. and Kleppa, O.J. (1983) Thermochemistry of $(Fe^{2+},Mg)SiO_3$ orthopyroxenes. Geochim. Cosmochim. Acta 47, 1597-1603.

Chevallier, R., Bolfa, J. and Mathieu, S. (1955) Titanomagnetites et ilmenites ferromagnetiques. Bull. Soc. france Mineral. Crist. 78, 307-346.

Davies, P.K. and Navrotsky, A. (1983) Quantitative correlations of deviations from ideality in binary and pseudo-binary solid solutions. J. Solid State Chem. 46, 1-22.

Driessens, F.C.M. (1968) Thermodynamics and defect chemistry of some oxide solid solutions. Ber. Bunsengesellschaft Phys. Chem. 72, 754-764.

Fancher, D.L. and Barsch, G.L. (1969a) Lattice theory of alkali halide solid solutions - I. Heat of formation. J. Phys. Chem. Solids. 30, 2501-2516.

Fancher, D.L. and Barsch, G.L. (1969b) Lattice theory of alkali halide solid solutions - II. Entropy of mixing and solid solubility. J. Phys. Chem. Solids 30, 3517-3525.

Goldsmith, J.R. (1972) Cadmium dolomite and the system $CdCO_3$-$MgCO_3$. J. Geol. 80, 617-626.

Goldsmith, J.R. (1983) Phase relations in rhombohedral carbonates. In "Carbonates: Mineralogy and Chemistry", R.J. Reeder, Ed., Reviews in Mineralogy 11, 49-76.

Goldsmith, J.R. and Heard, H.C. (1961) Subsolidus phase relations in the system $CaCO_3$-$MgCO_3$. J. Geol. 69, 45-74.

Hahn, W.C. and Muan, A. (1962) Activity measurements in oxide solid solutions: the system "FeO"-MgO in the temperature range 1100 to 1300°C. Trans. Met. Soc. Amer. Inst. Met. Eng. 224, 416-420.

Henry, D.J., Navrotsky, A., and Zimmermann, H.D. (1982) Thermodynamics of plagioclase-melt equilibria in the system albite-anorthite-diopside. Geochim. Cosmochim. Acta 46, 381-391.

Hervig, R.L. and Navrotsky, A. (1984) Thermochemical study of glasses in the system $NaAlSi_3O_8$-$KAlSi_3O_8$-Si_4O_8, and the join $Na_{1.6}Si_{2.4}O_8$-$K_{1.6}Al_{1.6}Si_{2.4}O_8$. Geochim. Cosmochim. Acta 48, 513-522.

Hill, R.J., Craig, J.R., and Gibbs, G.V. (1979) Systemaics of the spinel structure type. Phys. Chem. Minerals 4, 317-340.

Hon, R., Henry, D.J., Navrotsky, A., and Weill, D.F. (1981) A thermochemical calculation of the pyroxene saturation surface in the system diopside-albite-anorthite. Geochim. Cosmochim. Acta 45, 157-161.

Hovis, G.L. and Waldbaum, D.R. (1977) A solution calorimetric investigation of K-Na mixing in a sanidine-analbite ion exchange series. Am. Mineral. 62, 680-686.

Kerrick, D.H. and Darken, L.S. (1975) Statistical thermodynamic models for ideal oxide and silicate solid solutions, with applications to high temperature plagioclase. Geochim. Cosmochim. Acta 39, 1431-1442.

Loomis, T.F. (1979) An empirical model for plagioclase equilibria in hydrous melts. Geochim. Cosmochim. Acta 43, 1753-1759.

Mason, T.O. (1987) Cation intersite distributions in iron-bearing minerals via electrical conductivity / Seebeck effect. Phys. Chem. Minerals 14, 156-162.

Navrotsky, A. and Kleppa, O.J. (1967) The thermodynamics of cation distributions in simple spinels. J. Inorg. Nucl. Chem. 29, 2701-2714.

Navrotsky, A. (1974) Thermodynamics of binary and ternary transition metal oxides in the solid state. MTP International Reviews of Science, Inorganic Chemistry, Series 2, Vol. 5, D.W.A. Sharp, Ed. Butterworths- University Park Press, Baltimore, MD, 29-70.

Navrotsky, A. (1985) Crystal chemical constraints on thermochemistry of minerals. In "Microscopic to Macroscopic", S.W. Kieffer and A. Navrotsky, Eds., Reviews in Mineralogy 13, 225-276.

Navrotsky, A. (1986) Cation distribution energetics and heats of mixing in $MgFe_2O_4$-$MgAl_2O_4$, $ZnFe_2O_4$-$ZnAl_2O_4$, and $NiAl_2O_4$-$ZnAl_2O_4$ spinels: study by high temperature calorimetry. Am. Mineral. 71, 1160-1169.

Navrotsky, A. and Akaogi, M. (1984) α-β-γ phase relations in Fe_2SiO_4-Mg_2SiO_4 and Co_2SiO_4-Mg_2SiO_4: calculation from thermochemical data and geophysical applications. J. Geophys. Res. 89, 10135-10140.

Navrotsky, A. and Capobianco, C. (1987) Enthalpies of formation of dolomite and of magnesian calcites. Am. Mineral (in press).

Neel, L. (1955) Some theoretical aspects of rock magnetism. Advances in Physics 4, 191-243.

Newton, R.C., Charlu, T.V. and Kleppa, O.J. (1980) Thermochemistry of the high structural state plagioclases. Geochim. Cosmochim. Acta 44, 933-941.

Newton, R.C., Wood, B.J., and Kleppa, O.J. (1981) Thermochemistry of silicate solid solutions. Bull. Mineral. 104, 162-171.

O'Neill, H.St.C., and Navrotsky, A. (1983) Simple spinels: crystallographic parameters, cation radii, lattice energies, and cation distributions. Am. Mineral. 68, 181-194.

O'Neill, H.St.C. and Navrotsky, A. (1984) Cation distributions and thermodynamic properties of binary spinel solid solutions. Am. Mineral. 69, 733-755.

O'Reilly, W. and Banerjee, S.K. (1965) Cation distributions in titanomagnetites $(1-x)Fe_3O_4$- x Fe_2TiO_4. Phys. Lett. 17, 237-238.

Orville, P.M. (1972) Plagioclase cation exchange equilibria with aqueous chloride solution: results at 700°C and 2000 bars in the presence of quartz. Am. J. Sci. 272, 236-272.

Redfern, S.A.T. and Salje, E. (1987) Thermodynamics of plagioclase II: temperature evolution of the spontaneous strain at the $I\bar{1}$-$P\bar{1}$ phase transition in anorthite. Phys. Chem. Minerals. 14, 189-195.

Reeder, R.J. (1984) Crystal chemistry of the rhombohedral carbonates. In "Carbonates: Mineralogy and Chemistry", R.J. Reeder, Ed., Reviews in Mineralogy 11, 1-48.

Salje, E. (1985) Thermodynamics of sodium feldspar, I: order parameter treatment and strain induced coupling effects. Phys. Chem. Minerals. 12, 93-98.

Salje, E. (1987) Thermodynamics of plagioclases I: theory of $I\bar{1}$-$P\bar{1}$ phase transition in anorthite and Ca-rich plagioclases. Phys. Chem. Minerals. 14, 181-188.

Salje, E., Kuscholke, B., Wruck, B. and Kroll, H. (1985) Thermodynamics of sodium feldspar, II: experimental results and numerical calculations. Phys. Chem. Minerals. 12, 132-140.

Saxena, S.K. and Ribbe, P.H. (1972) Activity-composition relations in feldspars. Contrib. Mineral. Petrol. 37, 131-138.

Shannon, R.D. and Prewitt, C.T. (1969) Effective ionic radii in oxides and fluorides. Acta Cryst. B25, 925-946.

Taylor, R.W. (1964) Phase equilibria in the system FeO-Fe_2O_3-TiO_2 at 1300°C. Am. Mineral. 49, 1016-1030.

Thompson, J.B. (1967) Thermodynamic properties of simple solutions. In: Researches in Geochemistry, P.H. Abelson, Ed., John Wiley and Sons, NY, 340-361.

Thompson, J.B. (1969) Chemical reactions in crystals. Am. Mineral. 54, 341-375.

Trestman-Matts, A., Dorris, S.E., Kumarakrishnan, S., and Mason, T.O. (1983) Thermoelectric determination of cation distribution in Fe_3O_4-Fe_2TiO_4. J. Amer. Ceram. Soc. 66, 829-834.

Urusov, V.S. (1968) Energies of solid solutions. Geokhimiya 9, 1033-1043.

Webster, A.H. and Bright, F.H. (1961) The system iron-titanium-oxygen at 1200°C and oxygen partial pressures between 1 atm and 2 x 10^{-14} atm. J. Amer. Ceram. Soc. 44, 110-116.

Wechsler, B.A., Lindsley, D.H., and Prewitt, C.T. (1984) Crystal structure and cation distribution in titanomagnetites ($Fe_{3-X}Ti_XO_4$). Am. Mineral. 69, 754-770.

Weill, D. F., Hon, R. and Navrotsky, A. (1980) The igneous system $CaMgSi_2O_6$-$CaAl_2Si_2O_8$-$NaAlSi_3O_8$: Variations on a classic theme by Bowen. In: Physics of Magmatic Processes, R.B. Hargraves, Ed., Princeton Univ. Press, 49-92.

Wood, B.J., Kirkpatrick, R.J. and Montez, B. (1986) Order-disorder in $MgAl_2O_4$ spinel. Am. Mineral. 71, 999-1008.

Wu, C.C. and Mason, T.O. (1981) Thermopower measurement of the cation distribution in magnetite. J. Amer. Ceram. Soc. 64, 237-242.

Chapter 3

Bernard J. Wood

THERMODYNAMICS OF MULTICOMPONENT SYSTEMS CONTAINING SEVERAL SOLID SOLUTIONS

In the chapter dealing with pure solids (Newton, this volume) it was seen how thermodynamic data may be extracted from experimental determinations of univariant equilibria. Navrotsky (this volume) continued with a discussion of some of the theories which have been used to predict and explain solid solution behavior and finished with several applications of these theories to binary solutions. One of the main goals of experimental and theoretical petrology is, of course, to predict partitioning under any desired pressure and temperature conditions. The object of this chapter is to show that currently available thermodynamic data are sufficiently precise to enable some complex systems to be modelled with considerable confidence. As an example, subsolidus relations in the system Na_2O-FeO-CaO-MgO-Al_2O_3–SiO_2 (NFCMAS) have been calculated for a number of compositions which approximate natural basalts. Before these results are discussed, however, it is necessary to describe briefly the methods used to calculate equilibrium in multiphase, multicomponent systems.

COMPUTATION OF MULTICOMPONENT, MULTIPHASE EQUILIBRIA

In a six-component system involving multicomponent solid solutions there is no way to solve explicitly for the equilibrium phase assemblage and the equilibrium mineral compositions. These can, however, be obtained numerically if the bulk composition of the system and pressure and temperature are fixed. From the geologic standpoint the first comprehensive use of numerical methods to solve this type of problem was described by Brown and Skinner (1974). Their method can be described as the "rocking tangent-plane" approach and is, in principle, general to any number of components and phases. It is illustrated for a two-component, two-phase case (BC) in Figure 1 and may be summarized as follows:

1) Express the total free energy of each phase in terms of μ, ΔS_{mix} and ΔG_{XS} (Fig. 1A).
2) Place an n-dimensional hyperplane in free energy space well below all of the free energy surfaces of all the phases (Fig. 1B).
3) Bring the plane up in free energy space until it touches the free energy surface of one of the phases. This then prohibits upward movement of the plane at that point in composition space (Fig. 1C).
4) Continue to increase the free energy of the fixed bulk composition by rocking the plane on the intersected free energy surface (Fig. 1C,D).
5) As further intersections are made, the increase in free energy of the bulk composition becomes more constrained until the plane becomes completely fixed (Fig. 1D).
6) Compute amounts of each phase when plane is fixed. If any phase is negative in amount, remove it from consideration and return to step 5). When all amounts are positive the equilibrium assemblage and phase compositions have been found.

Although it is not practical to attempt multicomponent calculations other than by computer, the numerical methods can be illustrated for the binary, two-phase case by assuming that both phases α and β are ideal solutions of A and B. The free energy of any bulk composition is given by

$$G_{BC} = \sum_i n_i \mu_i \ , \tag{1}$$

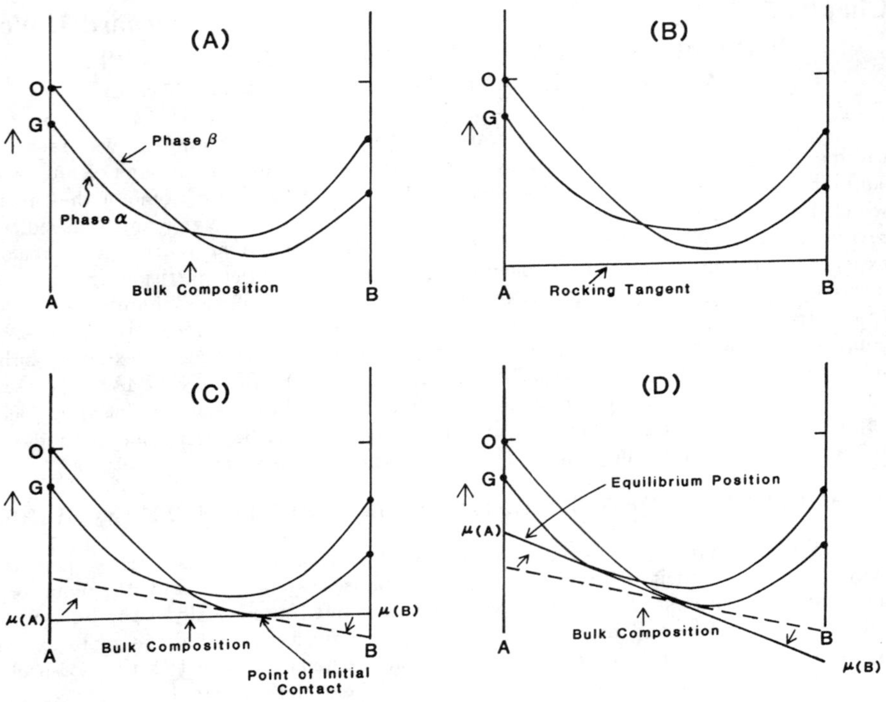

Figure 1. Equilibrium between the phases α and β is attained when their free energy surfaces have a common tangent as in 1D. Methods of calculating the composition of equilibrium assemblages are discussed in the text.

where n_i is the number of moles of i and μ_i is the chemical potential of i. On a one mole basis this reduces, in the binary case, to

$$G_{BC} = X_A^{BC}\ \mu(A) + X_B^{BC}\ \mu(B)\ , \tag{2}$$

where X_A^{BC} and X_B^{BC} are the mole fractions of A and B in the bulk composition. The molar free energies of phases α and β (ideal one site solution) are given, with standard states at the P and T of interest by:

$$G(\alpha) = X_A^\alpha\ \mu^\circ(A,\alpha) + X_B^\alpha\ \mu^\circ(B,\alpha) + RT(X_A^\alpha \ln X_A^\alpha + X_B^\alpha \ln X_B^\alpha) \tag{3}$$

$$G(\beta) = X_A^\beta\ \mu^\circ(A,\beta) + X_B^\beta\ \mu^\circ(B,\beta) + RT(X_A^\beta \ln X_A^\beta + X_B^\beta \ln X_B^\beta)\ ,$$

where X_A^α, X_B^β, etc. refer to mole fractions in phases α and β.

When the tangent touches the free energy curve it is found (Navrotsky, this volume) that $\mu(A)$ is related to $\mu^\circ(A,\alpha)$ and X_A^α as follows:

$$\mu(A) = \mu^\circ(A,\alpha) + RT \ln X_A^\alpha\ . \tag{4}$$

Therefore, for each of the phases we have a relationship

$$\mu(B) - \mu(A) = \mu^\circ(B,\alpha) - \mu^\circ(A,\alpha) + RT \ln X_B^\alpha - RT \ln X_A^\alpha\ . \tag{5}$$

Let us consider a specific case in which the values of μ° are as follows, at 1000 K:

$$\mu^\circ(A,\alpha) = -2000 \text{ J} \qquad \mu^\circ(A,\beta) = 0 \text{ J}$$

$$\mu^\circ(B,\alpha) = -3000\ J \qquad \mu^\circ(B,\beta) = -6000\ J$$

Starting with the free energy "tangent" horizontal (Fig. 1B) and below all free energy surfaces (say at –10,000 J) Equations (5) are solved for the compositions which would be first touched by the horizontal line ($\mu(B) - \mu(A) = 0$) as it is moved upwards:

$$0 = -6000 + RT \ln X_B^\beta - RT \ln X_A^\beta = f(X_B^\beta)\ , \tag{6}$$

$$0 = -1000 + RT \ln X_B^\alpha - RT \ln X_A^\alpha = f(X_B^\alpha)\ .$$

These nonlinear equations can be solved iteratively by Newton's method using initial estimates of those compositions with horizontal tangents to the free energy surface. Newton's method gives the (n+1)th estimate of the root of a nonlinear equation from the nth estimate as follows (e.g., Gerald and Wheatley, 1984, p. 15):

$$(X_B^\beta)_{n+1} = (X_B^\beta)_n - \frac{f(X_B^\beta)}{f'\ (X_B^\beta)}\ . \tag{7}$$

The derivatives $f'\ (X_B^\beta)$ are given by (since $dX_A^\beta/dX_B^\beta = -1$)

$$f'\ (X_B^\beta) = \frac{RT}{X_B^\beta} + \frac{RT}{X_A^\beta}\ . \tag{8}$$

Starting with an initial estimate of X_B^β of 0.70 one obtains

$$X_B^\beta = 0.7 - \frac{1044}{39590} = 0.674\ .$$

The next iteration for the root yields X_B^β of 0.673, indicating rapid convergence. Similarly, for phase α it can be shown that the horizontal tangent is at X_B^α of 0.53. Substituting the X's and μ°'s in Equation (3) it is found that the horizontal tangent will touch phase β first and that this occurs at $G_{BC} = \mu(A) = \mu(B) = -9293$ J. The tangent is now raised to this position.

In order to rock the tangent to its equilibrium position the free energy of the bulk composition is raised (Fig. 1C,D) and the composition of phase β has to shift towards A (Fig. 1C,D). The extent of the compositional shift for a given change in G_{BC} (ΔG_{BC}) is obtained by differentiating Equation (2) with respect to G_{BC} and writing down an appropriate Gibbs-Duhem equation for phase β:

$$1 = X_A^{BC} \frac{d\mu(A)}{dG_{BC}} + X_B^{BC} \frac{d\mu(B)}{dG_{BC}}\ , \tag{9}$$

$$0 = X_A^\beta \frac{d\mu(A)}{dG_{BC}} + X_B^\beta \frac{d\mu(B)}{dG_{BC}}\ . \tag{10}$$

These simultaneous equations enable solution for $d\mu(A)/dG_{BC}$ and $d\mu(B)/dG_{BC}$. For example, if X_B^{BC} is 0.4 and X_A^{BC} is 0.6 and a horizontal tangent in contact with β at X_B^β of 0.673 and X_A^β of 0.327, one obtains:

$$\frac{d\mu(A)}{dG_{BC}} = +2.465\ , \quad \frac{d\mu(B)}{dG_{BC}} = -1.198\ .$$

The free energy of the bulk composition is now incremented by a small amount ΔG_{BC} (on the order of 5-10 J) and new values of $\mu(A)$ and $\mu(B)$ obtained by expanding μ as a Taylor Series and truncating after the first derivative:

$$\mu(A,\ new) = \mu(A,\ old) + \frac{d\mu(A)}{dG_{BC}} \cdot \Delta G_{BC}\ . \tag{11}$$

For an increment ΔG_{BC} of +10 J, one obtains

$$\mu(A) = -9293 + 24.7 = -9268.3\ J\ , \quad \mu(B) = -9293 - 12.0 = -9305\ J\ .$$

Substituting back into Equation (5) yields:

$$-36.7 = -6000 + RT \ln X_B^\beta - RT \ln X_A^\beta ,$$

which can be solved by Newton's method to give an X_B^β of 0.672. This procedure is repeated by solving Equations (9), (10), (11) and (5) until X_B^β reaches 0.47, at which point the free energy curve for phase α becomes tangent to the same line as phase β and equilibrium is reached. Note that progression towards equilibrium is slow but that larger step sizes can introduce numerical errors because Equation (11) has been truncated after the first derivative.

A completely different method was used by Crerar (1975) and Nicholls (1977) (see also Van Zeggeren and Storey, 1970) to calculate stable mineral and solution compositions in multicomponent systems. This solves the problem using equilibrium constant data rather than free energies, so that for the binary case discussed above, one has to consider the two relevant equilibria:

$$\underset{\text{phase } \alpha}{B} \rightleftarrows \underset{\text{phase } \beta}{B} \tag{12}$$

$$\underset{\text{phase } \alpha}{A} \rightleftarrows \underset{\text{phase } \beta}{A} \tag{13}$$

with the conditions of equilibrium:

$$\exp \frac{-\Delta G_{12}^o}{RT} - \frac{a(B,\beta)}{a(B,\alpha)} = 0 \ , \ \exp \frac{-\Delta G_{13}^o}{RT} - \frac{a(A,\beta)}{a(A,\alpha)} = 0 \ , \tag{14}$$

and the mass balance requirements:

$$n_A^{BC} = N_\beta X_A^\beta + N_\alpha X_A^\alpha \ , \quad n_B^{BC} = N_\beta X_B^\beta + N_\alpha X_B^\alpha \ , \tag{15}$$

where n_A^{BC} and n_B^{BC} are the number of moles of A and B in the bulk composition and N_α and N_β are the number of moles of the α and β phases. In practice, activity coefficient expressions must be introduced into (14) to correct for activity to mole fractions, but ideality will be assumed for the purpose of illustration. In the binary case there are two linear Equations (15) and the two nonlinear equations:

$$\exp \frac{-\Delta G_{12}^o}{RT} - \frac{X_B^\beta}{X_B^\alpha} = 0 = f \ , \tag{16}$$

$$\exp \frac{-\Delta G_{13}^o}{RT} - \frac{(1-X_B^\beta)}{(1-X_B^\alpha)} = 0 = g \ .$$

The nonlinear Equations (16) are solved iteratively by the Newton (otherwise known as the Newton-Raphson) method. This is done by expanding the functions, as before, in a Taylor series about the point of an initial estimate for the roots, $X_B^\beta(1)$ and $X_B^\alpha(1)$, in terms of the differences between the actual roots and the estimated values (ΔX_B^β, ΔX_B^α). The series is truncated after the first derivative term:

$$f = 0 = f(X(1)) + f'_{X_B^\beta} \Delta X_B^\beta + f'_{X_B^\alpha} \Delta X_B^\alpha \tag{17}$$

$$g = 0 = g(X(1)) + g'_{X_B^\beta} \Delta X_B^\beta + g'_{X_B^\alpha} \Delta X_B^\alpha \ .$$

Initial estimates of X are made and Equations (17) are evaluated to obtain ΔX_B^β and ΔX_B^α. For example, if one uses initial estimates of X_B^β of 0.5 and X_B^α of 0.4 then the following values of f and g are obtained by substitution into (16) ($\Delta G_{12}^o = -3000$ J, $\Delta G_{13}^o = +2000$ J, T = 1000 K):

$$f(X(1)) = 0.1845 \ , \qquad g(X(1)) = -0.047 \ .$$

The derivatives ($f'_{X_B^\beta}$, etc.) of f and g with respect to X_B^β and X_B^α can be obtained explicitly in this case. Generally, however, they are obtained numerically by finite difference which involves incrementing the variable by a small amount δ and re-evaluating the function f and g, i.e.:

$$f'_X \approx \frac{f(x+\delta) - f(x)}{\delta} .$$

If X_B^α and X_B^β are incremented by 0.001 to recalculate f and g, one obtains:

$$f'_{X_B^\beta} = -2.5 \; , \; f'_{X_B^\alpha} = 3.117 \; , \; g'_{X_B^\beta} = 1.67 \; , \; g'_{X_B^\alpha} = -1.387 .$$

Equations (17) then reduce to two simultaneous equations with the unknowns ΔX_B^β, ΔX_B^α which can readily be solved. In larger problems, with more unknowns, this is done using Gaussian elimination, but in this case it is a trivial matter to obtain $\Delta X_B^\beta = -0.06$ and $\Delta X_B^\alpha = -0.11$. For the next iteration, therefore, values of X_B^β of 0.44 and X_B^α of 0.29 are used. In a matter of three iterations the problem converges to equilibrium compositions of X_B^β of 0.47 and X_B^α of 0.33. For a bulk composition containing 0.4 moles of B and 0.6 moles of A, Equation (15) may then be solved to show that there are 0.5 moles of α and 0.5 moles of β in the equilibrium assemblage.

The final method which is commonly used (Saxena and Ericksson, 1983, 1985; Wood and Holloway, 1984, for example) is that of free energy minimisation by the method of steepest descent. The direction of steepest descent is obtained using Lagrangian multipliers, the standard method of solving problems when a maximum or minimum of a function is required subject to a constraining relationship between the variables. In this case one starts with an initial estimate of the numbers of moles of A and B in the two phases, n_A^α, n_B^α, etc. and introduces a search parameter l to find the direction of steepest descent. Starting with Equation (1):

$$G = \sum_i n_i \, \mu_i ,$$

the most negative (minimum) value of dG/dl is required:

$$\frac{dG}{dl} = \sum_i \mu_i \frac{dn_i}{dl} = \text{minimum} \tag{18}$$

subject to the conditions of mass balance which, in the binary case, are

$$\frac{dn_A^\alpha}{dl} + \frac{dn_A^\beta}{dl} = 0 \, , \; \frac{dn_B^\alpha}{dl} + \frac{dn_B^\beta}{dl} = 0 . \tag{19a}$$

In order to prevent the n_i from becoming negative during iteration it is usual to transform the variable as follows:

$$n_i \equiv \exp(\eta_i) \tag{19b}$$

so that (18) and (19) become

$$\frac{dG}{dl} = \sum_i \mu_i \, n_i \, \frac{d\eta_i}{dl} = \text{minimum} \tag{20}$$

$$n_A^\alpha \frac{d\eta_A^\alpha}{dl} + n_A^\beta \frac{d\eta_A^\beta}{dl} = 0 \tag{21a}$$

$$n_B^\alpha \frac{d\eta_B^\alpha}{dl} + n_B^\beta \frac{d\eta_B^\beta}{dl} = 0 , \tag{21b}$$

with the normalization condition

$$\sum_i \left(\frac{d\eta_i}{dl}\right)^2 = 1 . \tag{21c}$$

In order to minimise dG/dl with respect to the search parameter it is necessary to differentiate (20) with respect to l and set the result equal to zero. The constraints (21a,b,c) may, however, be added directly into the problem by multiplying each of them by some factor (Lagrangian multiplier), subtracting from (or adding to) (20) and minimising the new function, i.e.:

minimise

$$\sum_i \mu_i n_i \frac{d\eta_i}{dl} - \nu\left[\sum_i(\frac{d\eta_i}{dl})^2 - 1\right] - \lambda_A\left[n_A^\alpha \frac{d\eta_A^\alpha}{dl} + n_A^\beta \frac{d\eta_A^\beta}{dl}\right] - \lambda_B\left[n_B^\alpha \frac{d\eta_B^\alpha}{dl} + n_B^\beta \frac{d\eta_B^\beta}{dl}\right],$$

which gives:

$$\sum_i \mu_i n_i \frac{d^2\eta_i}{dl^2} - 2\nu\sum_i(\frac{d\eta_i}{dl})(\frac{d^2\eta_i}{dl^2}) -$$

$$\lambda_A\left[n_A^\alpha \frac{d^2\eta_A^\alpha}{dl^2} + n_A^\beta \frac{d^2\eta_A^\beta}{dl^2}\right] - \lambda_B\left[n_B^\alpha \frac{d^2\eta_B^\alpha}{dl^2} + n_B^\beta \frac{d^2\eta_B^\beta}{dl^2}\right] = 0 . \quad (22)$$

Equation (22) is now separated into four equations, one in each of the components A in phase α, A in phase β, etc.:

$$\mu_A^\alpha n_A^\alpha \frac{d^2\eta_A^\alpha}{dl^2} - 2\,\nu \frac{d\eta_A^\alpha}{dl}\frac{d^2\eta_A^\alpha}{dl^2} - \lambda_A n_A^\alpha \frac{d^2\eta_A^\alpha}{dl^2} = a'$$

$$\mu_B^\alpha n_B^\alpha \frac{d^2\eta_B^\alpha}{dl^2} - 2\,\nu \frac{d\eta_B^\alpha}{dl}\frac{d^2\eta_B^\alpha}{dl^2} - \lambda_B n_B^\alpha \frac{d^2\eta_B^\alpha}{dl^2} = b' \quad (23)$$

$$\mu_A^\beta n_A^\beta \frac{d^2\eta_A^\beta}{dl^2} - 2\,\nu \frac{d\eta_A^\beta}{dl}\frac{d^2\eta_A^\beta}{dl^2} - \lambda_A n_A^\beta \frac{d^2\eta_A^\beta}{dl^2} = c' .$$

(The fourth of Eqns. (23) can be derived by symmetry and the second derivative terms cancel out.) The three Lagrangian multipliers ν, λ_A, λ_B are permitted to take any value appropriate to the solution of the problem. They are arbitrarily given values (as yet undetermined) so that when each Equation (23) is multiplied by n_A^α, n_A^β, etc. as appropriate, and terms A and B are collected separately, the following two conditions apply:

$$\mu_A^\alpha n_A^{\alpha 2} + \mu_A^\beta n_A^{\beta 2} - \lambda_A(n_A^{\alpha 2} + n_A^{\beta 2}) = 0 , \quad (24a)$$

$$\mu_B^\alpha n_B^{\alpha 2} + \mu_B^\beta n_B^{\beta 2} - \lambda_B(n_B^{\alpha 2} + n_B^{\beta 2}) = 0 . \quad (24b)$$

Note that the terms in 2 ν cancel out because one obtains:

$$-2\,\nu\left[n_A^\alpha \frac{d\eta_A^\alpha}{dl} + n_A^\beta \frac{d\eta_A^\beta}{dl}\right],$$

and from (21) the term in brackets is zero.

Equations (24a,b) are now solved for λ_A and λ_B using the initial values of n_A^α, n_A^β etc. The results are substituted into (23) to obtain the four values of $d\eta_i/dl$, e.g., for component A in phase α,

$$\frac{d\eta_A^\alpha}{dl} = \frac{n_A^\alpha}{2\nu}(\mu_A^\alpha - \lambda_A) . \quad (25)$$

The Lagrangian multiplier ν is obtained simply by setting it equal to 1/2, determining all of the values of $d\eta_i/dl$, and then forcing them to obey the normalisation conditions (21c). New values of η_i for the next iteration are found, as before, with a truncated Taylor series,

In order to minimise dG/dl with respect to the search parameter it is necessary to differentiate (20) with respect to l and set the result equal to zero. The constraints (21a,b,c) may, however, be added directly into the problem by multiplying each of them by some factor (Lagrangian multiplier), subtracting from (or adding to) (20) and minimising the new function, i.e.:

minimise

$$\sum_i \mu_i n_i \frac{d\eta_i}{dl} - \nu \left(\sum_i (\frac{d\eta_i}{dl})^2 - 1 \right) - \lambda_A \left(n_A^\alpha \frac{d\eta_A^\alpha}{dl} + n_A^\beta \frac{d\eta_A^\beta}{dl} \right) - \lambda_B \left(n_B^\alpha \frac{d\eta_B^\alpha}{dl} + n_B^\beta \frac{d\eta_B^\beta}{dl} \right),$$

which gives:

$$\sum_i \mu_i n_i \frac{d^2\eta_i}{dl^2} - 2\nu \sum_i (\frac{d\eta_i}{dl})(\frac{d^2\eta_i}{dl^2}) -$$
$$\lambda_A \left(n_A^\alpha \frac{d^2\eta_A^\alpha}{dl^2} + n_A^\beta \frac{d^2\eta_A^\beta}{dl^2} \right) - \lambda_B \left(n_B^\alpha \frac{d^2\eta_B^\alpha}{dl^2} + n_B^\beta \frac{d^2\eta_B^\beta}{dl^2} \right) = \text{other terms} . \quad (22)$$

Equation (22) is now separated into four equations, one in each of the components A in phase α, A in phase β, and so on. The three Lagrangian multipliers ν, λ_A, λ_B are permitted to take any value appropriate to the solution of the problem. They are arbitrarily given values (as yet undertermined) so that the L.H.S. of each equation (23) is equal to zero:

$$\mu_A^\alpha n_A^\alpha \frac{d^2\eta_A^\alpha}{dl^2} - 2\,\nu \frac{d\eta_A^\alpha}{dl} \frac{d^2\eta_A^\alpha}{dl^2} - \lambda_A n_A^\alpha \frac{d^2\eta_A^\alpha}{dl^2} = 0$$
$$\mu_B^\alpha n_B^\alpha \frac{d^2\eta_B^\alpha}{dl^2} - 2\,\nu \frac{d\eta_B^\alpha}{dl} \frac{d^2\eta_B^\alpha}{dl^2} - \lambda_B n_B^\alpha \frac{d^2\eta_B^\alpha}{dl^2} = 0 \quad (23)$$
$$\mu_A^\beta n_A^\beta \frac{d^2\eta_A^\beta}{dl^2} - 2\,\nu \frac{d\eta_A^\beta}{dl} \frac{d^2\eta_A^\beta}{dl^2} - \lambda_A n_A^\beta \frac{d^2\eta_A^\beta}{dl^2} = 0 \; .$$

(The fourth of Eqns. (23) can be derived by symmetry and the second derivatives can be cancelled out.) Multiplying each equation (23) by n_B^α, n_A^β, etc. and collecting terms in A and B, we have:

$$\mu_A^\alpha n_A^{\alpha 2} + \mu_A^\beta n_A^{\beta 2} - \lambda_A (n_A^{\alpha 2} + n_A^{\beta 2}) = 0 \, , \quad (24a)$$
$$\mu_B^\alpha n_B^{\alpha 2} + \mu_B^\beta n_B^{\beta 2} - \lambda_B (n_B^{\alpha 2} + n_B^{\beta 2}) = 0 \, . \quad (24b)$$

Note that the terms in 2 ν cancel out because one obtains:

$$-2\,\nu \left(n_A^\alpha \frac{d\eta_A^\alpha}{dl} + n_A^\beta \frac{d\eta_A^\beta}{dl} \right),$$

and from (21) the term in brackets is zero.

Equations (24a,b) are now solved for λ_A and λ_B using the initial values of n_A^α, n_A^β etc. The results are substituted into (23) to obtain the four values of $d\eta_i/dl$, e.g., for component A in phase α,

$$\frac{d\eta_A^\alpha}{dl} = \frac{n_A^\alpha}{2\nu} (\mu_A^\alpha - \lambda_A) \; . \quad (25)$$

The Lagrangian multiplier ν is obtained simply by setting it equal to 1/2, determining all of the values of $d\eta_i/dl$, and then forcing them to obey the normalisation conditions (21c). New values of η_i for the next iteration are found, as before, with a truncated Taylor series,

$$\eta_i(j+1) = \eta_i(j) + (\frac{d\eta_i}{dl})\ \Delta l\ , \quad (26)$$

and new values of n_i obtained from (19b). Returning to the binary of Figure 1, let us begin by making initial estimates of equilibrium compositions for a bulk composition $n_A^{BC} = 1.2$, $n_B^{BC} = 0.8$, i.e., $X_A^{BC} = 0.6$ as before:

$$n_A^\alpha = 0.75\ ,\ n_B^\alpha = 0.35\ ,\ n_A^\beta = 0.45,\ n_B^\beta = 0.45\ .$$

Using ideal solution relationships (4) one obtains at 1000 K the following μ values:

$$\mu_A^\alpha = -5184\ J\ ,\ \mu_B^\alpha = -12520\ J\ ,\ \mu_A^\beta = -5763\ J,\ \mu_B^\beta = -11763\ J\ .$$

The Lagrangian multipliers λ are obtained from (24):

$$\lambda_A = -5337\ ,\ \lambda_B = -12048\ ,$$

and from (25) with the normalisation condition applied:

$$\frac{d\eta_A^\alpha}{dl} = +0.376\ ,\ \frac{d\eta_B^\beta}{dl} = -0.627\ ,\ \frac{d\eta_B^\alpha}{dl} = -0.539\ ,\ \frac{d\eta_B^\beta}{dl} = +0.418\ .$$

Since there is no requirement that the step parameter Δl be positive, it is necessary to determine which sign gives the direction of descent as opposed to ascent of the free energy surface. In this case a negative sign gives descent as will be shown from calculating the total initial free energy: $\sum \mu_i n_i = -16157$ J. Setting dl at –0.1, the new compositions are computed from (26) with substitution into (19b):

$$n_A^\alpha = 0.7223\ ,\ n_B^\alpha = 0.3694\ ,\ n_A^\beta = 0.4791,\ n_B^\beta = 0.4316\ .$$

Note that perfect mass balance is not maintained because of truncation errors, and that small changes in the total amounts of A and B are obtained. It is necessary, therefore, to force mass balance before calculating free energy:

$$n_A^\alpha = 0.7215\ ,\ n_B^\alpha = 0.3689\ ,\ n_A^\beta = 0.4785,\ n_B^\beta = 0.4311\ .$$

The new chemical potential values are:

$$n_A^\alpha = -5433\ ,\ n_B^\alpha = -12010\ ,\ n_A^\beta = -5340,\ n_B^\beta = -12208\ ,$$

and the total free energy $G_{BC} = -16168$ J. Continued iteration will proceed towards the equilibrium values of X_B^β of 0.47 and X_B^α of 0.33 discussed earlier.

It is commonly stated that there are no really good ways of solving systems of nonlinear equations so one has to trade-off between computational efficiency and time spent for programming. Of the three methods outlined above, that involving Newton-Raphson iteration to solve the equilibrium constant equations is generally the most computationally efficient. For problems involving large numbers of components and multicomponent phases, however, free energy minimisation is the easiest method to program (see Storey and Van Zeggeren, 1964; Van Zeggeren and Storey, 1970).

In a multicomponent system containing solid phases the bulk composition is generally expressed in terms of M component oxides. Each phase is made up of one or more oxides so that there are M mass balance equations of type

$$N_j = \sum_{i=1}^{S} a_{ji}\ n_i\ , \quad (27)$$

where N_j is the total number of moles of the jth oxide in the sample, S is the total number of complex components ($Mg_2Si_2O_6$, $CaMgSi_2O_6$, etc.), and a_{ji} is the number of moles of j in the ith complex component, e.g., for oxide SiO_2, a_{ji} is 2 for $Mg_2Si_2O_6$ and 0 for $MgAl_2O_4$.

Modification of Equation (24) to take account of the new mass balance Equations (27) gives an equation for each of the M oxides:

$$\sum_{j=1}^{M} \left(\sum_{i=1}^{S} a_{ki}\, a_{ji} n_i^2\right) \lambda_j = \sum_{i=1}^{S} \mu_i a_{ki} n_i^2 , \quad k = 1,2,3 \cdots M . \tag{28}$$

The M unknowns, the λ_j are obtained from the M simultaneous linear equations (e.g., by Gaussian elimination) and the change in amount of the ith component obtained from:

$$\frac{d\eta_i}{dl} = \frac{n_i}{2\,\nu}\left(\mu_i - \sum_{j=1}^{M} \lambda_j a_{ji}\right) . \tag{29}$$

In all other respects the equations are unchanged from those used in the 2-component case.

The discussion will proceed with a brief description of the data base used for calculation in the Na_2O-FeO-CaO-MgO-Al_2O_3-SiO_2 system and the assumptions associated with its use.

METHOD OF APPROACH

As far as possible, the data base was built up with the methods described by Newton (this volume) using univariant equilibria to refine the properties of single component phases in the systems Al_2O_3-SiO_2, Na_2O-Al_2O_3-SiO_2, MgO-Al_2O_3-SiO_2, CaO-Al_2O_3-SiO_2 and FeO-SiO_2. The only points of note are the following assumptions.

(a) The value of ΔCp for all solid-solid equilibria was assumed to be constant, independent of temperature. Measured Cp's at 1000 K were used wherever possible. With the constant ΔCp assumption, the standard state (pure phase at P and T of interest) free energy change of a reaction at 1 bar pressure and T°K becomes

$$\Delta G^{\circ}_{1,T} = \Delta H^{\circ}_{1,1000} + \Delta Cp_{1,1000}(T - 1000) - T(\Delta S^{\circ}_{1,1000} + \Delta Cp_{1,1000} \ln T/1000) . \tag{30}$$

(b) The dependencies of volume on pressure and temperature have been assumed to be linear:

$$V(P,T) = V(1,298)[1 + \alpha(T - 298)](1 - \beta P) , \tag{31}$$

where α and β are coefficients of thermal expansion and compressibility, respectively. This limits the approach to pressures below about 50 kbar.

(c) No systematic attempt has been made to attach uncertainties to the derived thermodynamic parameters. This is because an internally consistent data base can, by definition, be used as an ensemble to reproduce all of the available calorimetric and experimental data. It is, however, a "house of cards" in that alteration of one of the data requires changes in all of the others. With many different types of calorimetric and experimental data being used, it is not obvious that it is realistic or even useful to assign rigid uncertainties to H and S (e.g., Berman et al., 1986).

TREATMENT OF SOLID SOLUTIONS

The important link between thermodynamic data derived for pure phases and the computation of multicomponent phase relationships is the treatment of complex solid solutions. Navrotsky's chapter dealt with solid solution theories of progressively greater complexity which therefore have increasing ability to describe atomic interactions in solid solutions. In this chapter the approach has to be rather different, since there are insufficient data to describe, for example, a four- or five-component pyroxene solution with multivalent substitution (e.g., AlAl $\longleftrightarrow$ MgSi) by anything other than a very simple model. The use of simple models has several advantages: they are computationally simple; they minimise the use of arbitrary adjustable parameters; and extrapolation outside the experimental temperature range gives physically reasonable results. The disadvantage is that they often reproduce the experimental data poorly.

Partial molar entropy of mixing

For all solid solutions except plagioclase, the partial molar entropy of mixing was calculated from the multisite ideal equations (see Navrotsky, this volume) assuming complete disorder on each individual sublattice, e.g., for $(Ca,Mg,Fe)_3Al_2Si_3O_{12}$ garnet,

$$\overline{S}^{\,mix}_{Mg_3Al_2Si_3O_{12}} = -\,R\ln {X^{c}_{Mg}}^{3} ,$$

where X^{c}_{Mg} is the atomic fraction of Mg on the 8-coordinate cubic site in the garnet. For $(Ca,Mg,Fe)(Mg,Fe)Si_2O_6$ clinopyroxene,

$$\overline{S}^{\,mix}_{CaFeSi_2O_6} = -R\ln X^{M2}_{Ca}\cdot X^{M1}_{Fe} \;,\; \overline{S}^{\,mix}_{Mg_2Si_2O_6} = -R\ln X^{M2}_{Mg}\cdot X^{M1}_{Mg} \,. \qquad (32)$$

Excess free energies of mixing

The symmetric model (Navrotsky, this volume) was used wherever possible, with extension to asymmetric being made for plagioclase, garnet and olivine solutions. Note that for a multicomponent case the symmetric model yields (Wohl, 1946):

$$RT\ln\gamma_A = \sum_{\substack{J=1\\J\neq A}}^{N} X_J^2 W_{AJ} + \sum_{J=1}^{N-1}\sum_{\substack{Q=2\\(J<Q)\\J,Q\neq A}}^{N} X_J X_Q (W_{AJ} + W_{AQ} - W_{JQ}) \,, \qquad (33)$$

where W_{AJ}, W_{AQ}, etc. are the binary interaction parameters and γ_A is the activity coefficient of component A. For a ternary (A,B,C) asymmetric solution the activity coefficient expression is (Andersen and Lindsley, 1981)

$$\begin{aligned} RT\ln\gamma_A = {} & W_{AB}[X_B^2(1-2X_A)] + W_{BA}[2X_AX_B(1-X_A)] \qquad (34)\\ & + W_{AC}[X_C^2(1-2X_A)] + W_{CA}[2X_AX_B(1-X_A)] \\ & - 2X_C[X_CX_BW_{BC} + X_B^2W_{CB}] + WSUM(\tfrac{1}{2}X_BX_C - X_AX_BX_C) - X_BX_C(1-2X_A)W_{ABC} \,, \end{aligned}$$

where WSUM is the sum of all of the binary interaction parameters and W_{AB} refers to the molar excess free energy for component A when it is dissolved in an infinite matrix of B. W_{BA} refers to the reverse case, B being dissolved in an infinite matrix of A.

Additional free energy terms arise in some multisite solid solutions due to "reciprocal" or "cross-site" interactions (Wood and Nicholls, 1978). If, for example, one considers an $(Mg,Fe)(Al,Cr)_2O_4$ spinel solid solution, then it is necessary to take account of the internal equilibrium:

$$MgAl_2O_4 + FeCr_2O_4 = MgCr_2O_4 + FeAl_2O_4 \,. \qquad (35)$$

For Equilibrium (35) the standard free energy change is given by

$$\Delta G^{o}_{R} = G^{o}_{MgCr_2O_4} + G^{o}_{FeAl_2O_4} - G^{o}_{MgAl_2O_4} - G^{o}_{FeCr_2O_4} \,,$$

where the standard state is taken to be the pure phase at the pressure and temperature of interest. Wood and Nicholls (1978) have shown that the free energies of reciprocal equilibria such as (35) add the following terms to the activity coefficients of, for example, $MgAl_2O_4$ and $FeAl_2O_4$:

$$RT \ln \gamma_{MgAl_2O_4} = X_{Fe}^{tet} \cdot X_{Cr}^{oct} \Delta G_R^o + \text{site terms} , \quad (36)$$

$$RT \ln \gamma_{FeAl_2O_4} = - X_{Mg}^{tet} \cdot X_{Cr}^{oct} \Delta G_R^o + \text{site terms} ,$$

where X_{Mg}^{tet}, X_{Cr}^{oct}, etc. refer to mole fractions on tetrahedral and octahedral sites, respectively. As indicated in Equation (36), these reciprocal terms derive from the properties of the end members and are added to the symmetric or asymmetric terms which describe the interactions between atoms on each of the individual sublattices.

The discussion will continue with a consideration of the experimental and calorimetric data for some of the simple systems which bound NFCMAS.

THE SYSTEMS AS, MAS, CAS and NAS

Most of the data relevant to these three systems have been discussed at length by Newton (this volume). The important parameters which need to be considered when attempting to fit calorimetric and phase equilibrium data together are those dealing with disorder in $MgAl_2O_4$ spinel, $CaAl_2SiO_6$ pyroxene, anorthite, albite, and sillimanite. In most cases the contributions of disorder to enthalpies and entropies of these phases have not been measured calorimetrically, but they do affect phase equilibrium relationships. Hence comparison of phase equilibrium and calorimetric data enable some estimate of the effects of disorder to be made.

Albite

Holland (1980) obtained reversed phase equilibrium data on the univariant reaction:

$$\underset{\text{albite}}{NaAlSi_3O_8} = \underset{\text{jadeite}}{NaAlSi_2O_6} + \underset{\text{quartz}}{SiO_2} \quad (37)$$

Enthalpies of solution of high albite (Newton et al., 1980) and of jadeite (Wood et al., 1980) in lead borate melt at 970 K require the enthalpy change of reaction at this temperature (with high albite) to be –8.5 J/K, approximately. A linear least-squares fit to Holland's (1980) data (with high albite) is consistent with a 1 bar enthalpy change for Reaction (24) of –10040 J, in reasonable agreement with calorimetry. The derived entropy change, –43.39 J/K, indicates, however, that Al-Si disorder in high albite is not complete. Use of the measured heat capacity of albite (Robie et al., 1978) without taking account of disorder would give a standard entropy change for (24) of –29.83 J/K at 1000 K. The value determined from phase equilibrium measurements indicates that the entropy of Al-Si disorder in high albite is about 13.56 J/K (to give ΔS_{1000}^o of –43.39 J/K) rather than the 18.75 J/K required for complete Al-Si disorder. A certain amount of short range Al-Si order in high albite (and hence reduced entropy) is consistent with the observations of Mazo (1977) and Kroll and Ribbe (1980).

$CaAl_2SiO_6$ pyroxene and anorthite

The position and slope of the pyroxene breakdown reaction

$$\underset{\text{pyroxene}}{3CaAl_2SiO_6} = \underset{\text{grossular}}{Ca_3Al_2Si_3O_{12}} + \underset{\text{corundum}}{2Al_2O_3}$$

indicates, as shown by Wood and Holloway (1984) that $CaAl_2SiO_6$ pyroxene has about 10 J/K of entropy due to Al-Si disorder. This is, once again, slightly less than the 11.55 J/K that would be expected for random Al-Si mixing. Similarly, the positions and slopes of subsolidus reactions involving anorthite are consistent with, for this phase, about 2.5 J/K (Newton, this volume, obtained 3.22 J/K) of additional, non-calorimetric entropy due to the observed small amounts of Al-Si disorder (Bruno et al., 1976).

$MgAl_2O_4$ spinel

Wood et al. (1986) have shown that the $MgAl_2O_4$ can be readily disordered at low temperatures, attaining about 39% inversion at 900°C. Analysis of phase equilibrium data on reactions such as

$$\text{enstatite} + \text{spinel} = \text{pyrope} + \text{forsterite} \tag{38}$$

have shown, however, that the residual entropy of disorder is much smaller than that predicted by the data of Wood et al. (1986), being close to 5 J/K (see Newton, this volume.) The small residual entropy and enthalpy were, for simplicity, added to spinel in an ad hoc fashion by fitting Reaction (38) (see Wood and Holloway, 1984) to a non-convergent disordering model (Navrotsky and Kleppa, 1967). The internal equilibrium, Mg(tet) + Al(oct) = Mg(oct) + Al(tet), was treated as if it had a standard state enthalpy change ΔH° of 83.7 kJ/mol and the hypothetical degree of inversion x calculated from

$$\frac{-10065}{T} = \ln\left(\frac{x^2}{(1-x)(2-x)}\right) .$$

This results in a contribution to the free energy of disorder G_{dis} calculated from

$$G_{dis} = x\,\Delta H^\circ + RT[x \ln x + (1-x)\ln(1-x) + x\ln(\tfrac{x}{2}) + (2-x)\ln(1-\tfrac{x}{2})] \;. \tag{39}$$

Although an unsatisfactory representation of the microscopic data, this model yields appropriate values of the free energy of $MgAl_2O_4$ disorder. Although the extent of Al-Si disorder in sillimanite is also uncertain (Robie and Hemingway, 1984), the Al_2SiO_5 phase diagram was fitted by assuming a similar model to that described above (Wood and Holloway, 1984). The fraction x of Al on tetrahedral-Si sites is estimated from

$$\frac{-9060}{T} = \ln\frac{x^2}{(1-x)^2} .$$

The free energy of sillimanite disorder may then be obtained from

$$G_{dis} = 75320x + 16.627T[x \ln x + (1-x)\ln(1-x)] \text{ J/mol} .$$

For both $MgAl_2O_4$ spinel and sillimanite, the "best-fit" enthalpy and entropy values have been recalculated to zero disorder in Table 1. At any temperature of concern, G_{dis} is simply added to the calorimetric free energy derived from the data in Table 1.

CMAS SYSTEM

Pyroxenes

For simplicity, and because of its limited stability field, the pigeonite phase was ignored. This enables $CaMgSi_2O_6$-$Mg_2Si_2O_6$ orthopyroxene and clinopyroxene solutions to be treated as symmetric (Holland et al., 1979, 1980; Lindsley et al., 1981) and

Table 1. Data Base for the System CaO-MgO-Al_2O_3-SiO_2 (Wood and Holloway, 1984)

Phase	Component	V (c.c)	$\alpha \cdot 10^5$	$\beta \cdot 10^6$	Cp (J/K)	ΔH_f^o (KJ/mol)	S (J/K)
clinopyroxene	$CaMgSi_2O_6$	66.1	2.92	0.82	249.99	-147.78	401.66
clinopyroxene	$Mg_2Si_2O_6$	62.64	2.92	1.01	252.21	-63.66	394.00
clinopyroxene	$CaAl_2SiO_6$	63.62	2.7	0.8	250.04	-76.99	406.27
orthopyroxene	$Mg_2Si_2O_6$	62.64	2.92	1.01	252.21	-70.46	391.25
orthopyroxene	$MgAl_2SiO_6$	58.93	2.92	1.01	251.04	-3.85	387.94
orthopyroxene	$CaMgSi_2O_6$	66.1	2.92	0.82	249.99	-147.78	401.66
olivine	Mg_2SiO_4	43.79	4.14	0.79	175.35	-63.35	277.11
garnet	$Mg_3Al_2Si_3O_{12}$	113.20	2.57	0.47	485.39	-84.94	777.81
garnet	$Ca_3Al_2Si_3O_{12}$	125.24	2.34	0.54	494.21	-324.30	783.24
spinel	$MgAl_2O_4$	39.71	2.75	0.49	178.28	-28.03	264.47[a]
kyanite	Al_2SiO_5	44.09	2.74	0.5	195.73	-7.28	285.22
sillimanite	Al_2SiO_5	49.90	2.02	0.8	193.43	-1.88	295.06[a]
plagioclase	$CaAl_2Si_2O_8$	100.79	1.48	1.1	319.20	-101.42	532.96
corundum	Al_2O_3	25.575	2.61	0.37	124.93	0	180.16
quartz	SiO_2	22.688	6.0	2.5	68.75	0	115.56

All enthalpies and entropies are 1000 K, 1 atmosphere values. Enthalpies are relative to constituent oxides.
[a] Values for $MgAl_2O_4$ and sillimanite are recalculated to zero disorder (see text).

Table 2. Thermodynamic Properties of Fe- and Na-bearing phases

Phase	Component	V (c.c.)	$\alpha \cdot 10^5$	$\beta \cdot 10^6$	C_P J/K	ΔH_f^o KJ/mol	S J/K
Clinopyroxene	$CaFeSi_2O_6$	67.88	3.89	0.82	255.35	-124.06	435.97
Clinopyroxene	$Fe_2Si_2O_6$	65.98	3.89	1.01	264.93	-28.87	459.99
Clinopyroxene	$NaAlSi_2O_6$	60.4	2.66	0.95	244.05	-75.73	385.43
Orthopyroxene	$NaAlSi_2O_6$	60.4	2.66	0.75	244.05	-25.10	385.43
Orthopyroxene	$CaFeSi_2O_6$	67.88	3.89	0.82	255.35	-124.06	435.97
Orthopyroxene	$Fe_2Si_2O_6$	65.98	3.89	1.01	264.93	-29.00	459.99
Garnet	$Fe_3Al_2Si_3O_{12}$	115.30	2.44	0.53	504.46	-42.47	879.35
Olivine	Fe_2SiO_4	46.39	3.19	0.91	188.07	-27.45	349.24
Plagioclase	$NaAlSi_3O_8$	100.43	3.44	1.6	312.25	-65.69	544.38
Spinel	$FeAl_2O_4$	40.75	2.54	0.49	180.12	-16.42	296.10[a]

Enthalpy, entropy and heat capacity are 1000K values.
[a] Enthalpy and entropy recalculated to complete order (see text).

Table 3. Mixing Properties of Solid Solutions (kJ/mol) (Olivine and garnet values are pressure-dependent; see text.)

Atom Pair	Garnet	Olivine	Spinel	Plagioclase
CaMg	46.90	-		-
MgCa	13.40	-		-
CaFe	4.20	-		-
FeCa	12.55	-		-
MgFe	12.55	16.74	1.25	-
FeMg	2.50	8.37	1.25	-
CaNa	-	-		8.47
NaCa	-	-		28.23

nonideal parameters were adopted from Holland et al. (1979, 1980; Table 4).

Table 4. Mixing Properties of Pyroxenes (kJ/mol)

Regular Solution Terms

	Clinopyroxene		Orthopyroxene	
Pair	M2 Site	M1 Site	M2 Site	M1 Site
Ca-Mg	24.74 + 0.105P	-	34.0	-
Ca-Fe	20.04	-	22.80	
Mg-Fe	8.37	0.42	6.28	0
Mg-Al	-	7.53	-	0
Fe-Al	-	0	-	0
Na-Ca	0	-	0	-
Na-Fe	0	-	0	-
Na-Mg	0	-	0	-

Reciprocal Terms

	Clinopyroxene	Orthopyroxene
NaAl-MFe	30.33	30.33
NaAl-MMg	22.80	22.80
MAl_2–MFe	0	-5.9

Mixing of the aluminous pyroxene components $MgAl_2SiO_6$ and $CaAl_2SiO_6$ in clinopyroxene and orthopyroxene were assumed to be as molecular units (Wood and Banno, 1973; Newton, this volume) with end-member (Table 1) and mixing (Table 4) properties taken from Wood and Holloway (1984).

The free energy of the internal (reciprocal) equilibrium,

$$CaMgSi_2O_6 + MgAl_2SiO_6 = CaAl_2SiO_6 + Mg_2Si_2O_6 \ ,$$

was set at zero for both ortho- and clinopyroxene (Wood and Holloway, 1984).

Garnets

Newton et al. (1977) found considerable asymmetry in the excess enthalpies of mixing of synthetic Ca-Mg garnets. These results were used by Wood and Holloway (1984) with excess volume data to show that Ca-Mg garnets with low mole fractions of grossular ($\simeq$0.15) have near-zero excess entropies. Recent activity measurements (Wood, unpublished) tend to confirm this observation. The Ca-Mg mixing properties shown in Table 3 are therefore the calorimetric values adopted by Wood and Holloway (1984).

At high pressure the activity coefficients were corrected as follows using the partial molar volume relationships ($\overline{V}_i$) given by Haselton and Newton (1980),

$$[\ln \gamma_i]_{P,T,X} = [\ln \gamma_i]_{1bar,T,X} + \frac{P(\overline{V}_i - V_i^{o})}{RT} \ .$$

$FeO–Al_2O_3–SiO_2$ SYSTEM

Fayalite

The heat capacity of fayalite at low temperatures has recently been measured by Robie et al. (1982). These data also yield fayalite entropy and, when combined with high temperature heat of formation and heat content data (Robie et al., 1978) result in fayalite being the best characterized of the ferrous silicates with a heat of formation from oxides of –27450 J/mol at 1000 K and an entropy of 349.24 J/K at 1000 K (Table 2).

Ferrosilite

Wood and Kleppa (1981) determined the standard state enthalpy change of the reaction

$$\begin{array}{cccc} Fe_2SiO_4 + & SiO_2 & = & Fe_2Si_2O_6 \\ \text{fayalite} & \text{quartz} & & \text{ferrosilite} \end{array}$$

to be –2340±880 J at 1000 K and 1 bar. Experimental reversals of the equilibrium boundary by Bohlen et al. (1980) lead to ΔH° of –1550 J and ΔS° of –4.85 J/K at 1000 K. Taken in conjunction with the fayalite and quartz data (Tables 1, 2) these constrain 1000 K enthalpy and entropy of $Fe_2Si_2O_6$ to values close to –29000 J and 460.0 J/K, respectively.

Almandine

Recent adiabatic and DSC calorimetric measurements (G. Metz and E. J. Essene, pers. commun.) yield an entropy of almandine of 879 J/K at 1000 K. This result may be used in conjunction with several phase equilibrium studies to estimate the heat of formation of almandine. Newton et al. (1987) used the results of Bohlen et al. (1983a) on the reaction ilmenite + sillimanite + quartz = almandine + rutile to obtain ΔH_f° (almandine) of –44.35 $kJ \cdot mol^{-1}$ at 1000 K. In this study a combination of the activity measurements of Cressey et al. (1978) and the data of Bohlen et al. (1983b) on the reaction

$$\text{fayalite} + \text{anorthite} = \text{garnet}(Ca_1Fe_2) \qquad (30)$$

were used to estimate an enthalpy of formation of almandine of –43.5 $kJ \cdot mol^{-1}$ at 1000 K with entropy fixed at 879.35 J/K at the same temperature. These agree, within experimental uncertainty, with the values derived by Newton et al. (1987) and require small positive deviations from ideality of CaFe garnets Table 3 and Geiger et al., 1987).

Although the phase equilibria data are in good agreement as far as ΔH_f° (almandine) is concerned, the results do not agree with the calorimetric measurements of Chatillon-Colinet et al. (1983c) who obtained –59.0±5.1 kJ/mol at 1000 K. At present the reason for the discrepancy is unclear, but the Chatillon-Colinet et al. data cannot be brought into agreement with the phase equilibrium results unless the entropy of almandine is lowered by several J/K.

Hercynite

The entropy of $FeAl_2O_4$ hercynite was obtained from electrochemical measurements of the equilibrium

$$\begin{array}{ccccc} 2Fe & + O_2 + & 2Al_2O_3 & = & 2FeAl_2O_4 \\ \text{metal} & \text{gas} & \text{corundum} & & \text{hercynite} \end{array}$$

by Chan et al. (1973). This yields an entropy of about 300 J/K at 1000 K and includes the effects of Fe-Al disorder. Iron-aluminum disorder was treated in the same manner and, for simplicity, with the same disordering enthalpy as $MgAl_2O_4$ (83.7 kJ/mol). Taking account of this effect, a hypothetical completely ordered $FeAl_2O_4$ spinel should have an entropy of about 296.2 J/K at 1000 K (Table 2). This value was adopted and supplemented with the entropy of Fe-Al disorder in the same way as described earlier for $MgAl_2O_4$ spinel. The 1000 K enthalpy of $FeAl_2O_4$ was obtained by fitting the Fe-Mg partitioning data of Engi (1983) for coexisting olivine and spinel. $MgAl_2O_4$-$FeAl_2O_4$ spinels were assumed to be analogous to other Fe-Mg solid solutions and to exhibit small positive deviations from ideality. Thus, a reasonable fit to the data can be obtained with W^{Sp}_{FeMg} of +1250 J/mol and $H_{FeAl_2O_4}$ (recalculated to complete order) of -16420 J/mol at 1000 K. Although not constrained to fit the 1300°C olivine-spinel partitioning data of Jamieson and Roedder (1984) these thermodynamic properties also reproduce their results within experimental uncertainty.

COMPLEX SOLID SOLUTIONS

Olivine

Olivine was treated as a binary asymmetric solution using the interaction parameters of Wood and Kleppa (1981). This gives, for example, for the activity of Fe_2SiO_4 at temperature T (K) and pressure (bars):

$$RT \ln a_{Fe_2SiO_4} = RT \ln X_{Fe}^2 + X_{Mg}^2[W_{FeMg} + 2X_{Fe}(W_{MgFe} - W_{FeMg})] ,$$

where a is activity, X_{Fe} is the mole fraction of fayalite in the solid solution and both W's increase by 0.025 J/bar because of the excess volumes of Fe-Mg olivine solutions.

Garnet

Garnet was treated as an asymmetric ternary Ca-Mg-Fe solid solution with six temperature-independent mixing parameters (Eqn. 34, Table 3). The Ca-Mg and Ca-Fe nonidealities were obtained from phase equilibrium and calorimetric data in the manner described above. Iron-magnesium mixing in garnet is more ideal than in olivine at the Mg-end of the pyrope-almandine series (O'Neill and Wood, 1979) but less ideal at the Fe end (Ganguly and Saxena, 1984; Bohlen et al., 1983a; Geiger et al., 1987). Values of W_{FeMg} of +2500 J (three-site basis) and W_{MgFe} of 12550 J fit the Fe-Mg partitioning experiments of O'Neill and Wood (1979) and are in reasonable agreement with the excess enthalpies of Geiger et al. (1987).

Plagioclase

The free energies of mixing of plagioclases are difficult to model because of the structural transformation from $C\bar{1}$ to $I\bar{1}$ in intermediate compositions (Navrotsky, this volume). The aluminum avoidance model of Newton et al. (1980) has been used for high temperature plagioclases (Navrotsky, this volume). This is a simple model which generates activity coefficients in reasonable agreement (within 20% in γ) with those derived from the reversed experimental data of Goldsmith (1982).

Pyroxenes

Interaction parameters involving Ca-, Mg- and Al- components were derived in the manner discussed earlier. The data of Lindsley (1981) on the $CaFeSi_2O_6$-$Fe_2Si_2O_6$ join can be fitted reasonably well to the symmetric model (ignoring pigeonite as

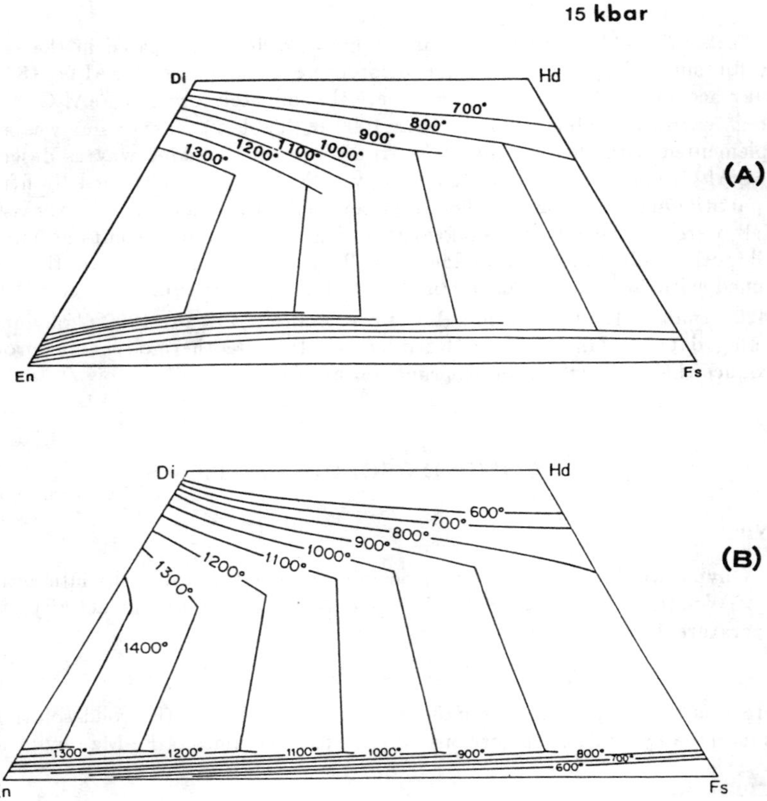

Figure 2. Comparison of calculated orthopyroxene-clinopyroxene relations at 15 kb pressure (2a) with Lindsley's (1983) experimentally-based phase diagram (2b). Note good agreement between calculated and observed Fe-Mg partitioning (tie lines) but that calculated Ca contents of orthopyroxene are slightly greater than observed. Pigeonite field has been neglected in the calculations.

Table 5. Quartz Tholeiite (A) at 11.5 kbar / 800 ° C

	Before			After	
Phase	Component	# Moles	μ (kJ/mol)	# Moles	μ (kJ/mol)
Cpx	$NaAlSi_2O_6$	0.427	-446.91	0.804	-441.81
Cpx	$CaMgSi_2O_6$	5.190	-506.81	7.349	-506.47
Cpx	$CaFeSi_2O_6$	0.051	-527.05	0.065	-528.92
Cpx	$Mg_2Si_2O_6$	0.050	-424.53	0.038	-424.42
Cpx	$Fe_2Si_2O_6$	1.212	-464.80	1.280	-468.95
Cpx	$CaAl_2SiO_6$	0.286	-466.14	0.445	-465.42
Opx	$NaAlSi_2O_6$	0.001	-511.70	0.0016	-475.10
Opx	$CaFeSi_2O_6$	0.762	-526.16	0.157	-527.95
Opx	$MgAl_2SiO_6$	0.495	-385.74	0.110	-382.82
Opx	$Mg_2Si_2O_6$	3.956	-426.30	0.873	-424.60
Opx	$Fe_2Si_2O_6$	2.512	-464.81	0.392	-468.78
Opx	$CaMgSi_2O_6$	0.094	-503.90	0.026	-502.43
Gt	$Mg_3Al_2Si_2O_{12}$	0.113	-812.58	1.607	-808.30
Gt	$Fe_3Al_2Si_2O_{12}$	0.173	-871.22	1.738	-874.71
Gt	$Ca_3Al_2Si_2O_{12}$	0.060	-1054.10	0.647	-1053.69
Plag	$NaAlSi_3O_8$	3.425	-541.11	3.047	-539.06
Plag	$CaAl_2Si_2O_8$	5.823	-560.82	2.404	-562.68
Quartz	SiO_2	2.113	-97.27	5.91	-97.27

before). These yield the Ca-Fe interaction parameters shown in Table 4 and a free energy of transformation of ferrosilite (orthopyroxene structure) to ferrosilite (C2/c clinopyroxene) of 125 J/mol.

In the $CaMgSi_2O_6R$–$RCaFeSi_2O_6$-$Fe_2Si_2O_6$-$Mg_2Si_2O_6$ quadrilateral the M2 sites in both ortho- and clinopyroxenes become ternary Ca-Fe-Mg solutions. In addition, the Fe and Mg also partition unequally between M1 and M2 sites (Saxena and Ghose, 1971; Besancon, 1981).

The orthopyroxene equilibrium

$$\mathrm{Fe(M2) + Mg(M1) = Fe(M1) + Mg(M2)} \tag{40}$$

has an enthalpy change of about 29 kJ/gm·atom (Besancon, 1981), a value which explains almost all of the calorimetric enthalpy of disordering measured by Chatillon-Colinet et al. (1983b). The orthopyroxene site occupancy data of Saxena and Ghose (1971), when taken together with this enthalpy change, yield the following approximate temperature dependence:

$$\ln \frac{X_{Fe}^{M1} \cdot X_{Mg}^{M2}}{X_{Fe}^{M2} \cdot X_{Mg}^{M1}} = \frac{-3520}{T} + 1.51 \; . \tag{41}$$

The pyroxene quadrilateral presented by Lindsley (1983) was used to estimate Fe-Mg mixing parameters for ortho- and clinopyroxene by a process of trial and error. Calcium was assigned to the M2 sites in ortho- and clinopyroxene and the partitioning of Fe and Mg between M1 and M2 sites assumed, in both cases, to obey (41) This latter constraint takes account of the similar Fe-Mg site preferences in both ortho- and clinopyroxene (McCallister et al., 1976; Saxena and Ghose, 1971). The resultant calculated phase diagram for coexisting ortho- and clinopyroxene is shown in Figure 2 where it may be compared with Lindsley's (1983) figure. As can be seen, the clinopyroxene limb of the miscibility gap and the orthopyroxene-clinopyroxene tie line orientation can be reproduced very well with the simple model described here. Apart from the absence of the pigeonite field, the calculated orthopyroxene composition is consistently 2-3 mol % richer in $CaSiO_3$ component than is observed. The entropic and enthalpic effects of this additional solubility are extremely small (50-100 J only), however, and may be neglected safely for most phase equilibrium or geothermobarometric calculations.

With the addition of sodium as $NaAlSi_2O_6$ component the pyroxene M2 sites become quaternary Ca-Mg-Na-Fe solutions and the M1 sites Mg-Fe-Al solutions. The dependence of aluminum solubility in orthopyroxene on Fe/Mg ratio (Wood, 1974; Harley, 1984) requires nonideal Fe-Al mixing. This can either be treated as arising from an on-site W_{FeAl}^{M1} mixing parameter of from a nonzero standard free energy change for the internal reaction

$$Fe_2Si_2O_6 + MgAl_2SiO_6 = FeAl_2SiO_6 + MgFeSi_2O_6 \; . \tag{42}$$

Assuming, for convenience, that it is all due to the latter effect, a ΔG_{42}^{o} of –5.9 kJ is obtained. Nonideal mixing on the clinopyroxene join $CaMgSi_2O_6$-$NaAlSi_2O_6$ has been observed in both calorimetric (Wood et al., 1980) and phase equilibrium experiments (Holland, 1983) and may be ascribed in part to non-ideal Al-Mg interactions (Table 4). These are insufficient to explain all of the observed nonideality, however. Since the data of Gasparik (1985) on $CaAl_2SiO_6$-$NaAlSi_2O_6$ pyroxenes preclude large Ca-Na nonidealities, a large contribution was assumed to arise from a reciprocal term due to the internal equilibrium

$$CaMgSi_2O_6 + NaAlSi_2O_6 = CaAlSi_2O_6 + NaMgSi_2O_6 \; . \tag{43}$$

Since the sum of ΔG_{43}^{o} and W_{MgAl} has to be about 30.3 kJ in order to fit the observed nonidealities (Wood et al., 1980), a ΔG_{43}^{o} of 22.8 kJ is obtained (Table 4). In the absence of any data identical values were adopted for analogous reactions involving Fe^{2+} components (Table 4).

CALCULATION of COMPLEX PHASE DIAGRAMS

As an example of the use of the data base discussed above the transformation of gabbro to eclogite has been calculated for the quartz tholeiite (A) composition studied experimentally by Green and Ringwood (1967). After excluding the minor components Fe_2O_3, TiO_2 and MnO the bulk composition was recalculated into initial estimates of mole fractions of each of the components (Table 5) and then scaled to a large total number of moles (about 25). (The total mass of the system is purely a matter of convenience and this total number of moles gives reasonable convergence.) The free energy of the system was then minimised using the steepest descent method at a pressure of 11.5 kbar and temperature of 800°C (Table 5). As can be seen from the concentrations of components at the beginning and end of free energy minimisation, the major change which occurs is the production of garnet through the complex reactions

(A) orthopyroxene + clinopyroxene + plagioclase → garnet + quartz ,

(B) orthopyroxene + plagioclase → garnet + quartz .

At the same time the clinopyroxene becomes richer in jadeite through the breakdown of the albite component in plagioclase

(C) plagioclase → jadeitic clinopyroxene + quartz .

The net result is that orthopyroxene virtually disappears, garnet increases from 0.35 to 4.0 moles and plagioclase decreases from 9.25 to 5.45 moles while its composition changes from An_{63} to An_{44}.

The test of convergence is that two successive iterations give a negligible change in free energy (–0.005 J). With this constraint, an approach to equilibrium can be independently verified by calculating the free energy changes of a number of equilibria. Some results for this particular minimisation are shown in Table 6. In general, equilibrium for major components is approached within a few tens of joules, but minor components are largely ignored by the minimisation since they have little effect on the total free energy and the increments are exponential (19b). As an example, one may see in Table 5 that $NaAlSi_2O_6$ (orthopyroxene) for which an essentially arbitrary free energy was employed (Table 2) did not approach equilibrium with $NaAlSi_2O_6$ (clinopyroxene). Because of the use of exponential increments and decrements (19b), phases which are unstable do not completely disappear but simply reach a low level at which the exponential changes have negligible effects on the total free energy. The instability of a phase (e.g., orthopyroxene) then needs to be verified independently by calculating the free energies of equilibria such as those shown in Table 6. As can be seen from Table 6, the equilibria controlling orthopyroxene ($Mg_2Si_2O_6$ and $Fe_2Si_2O_6$) disappearance both have negative values of ΔG implying that at this point the high pressure stability limit of orthopyroxene has been reached. Computer runs at higher pressures do not completely remove orthopyroxene for the reasons discussed above, but the calculated free energies of the two last equilibria in Table 6 decrease by 0.24 kJ for every 100 bars increase in pressure. Thus it is possible to define the disappearance of orthopyroxene within 100 to 200 bars.

The calculated phase diagram for quartz tholeiite (A) is shown in Figure 3 where it may be compared with Green and Ringwood's (1967) experimental data. Agreement between calculated and observed positions for the upper (plagioclase-out) and lower

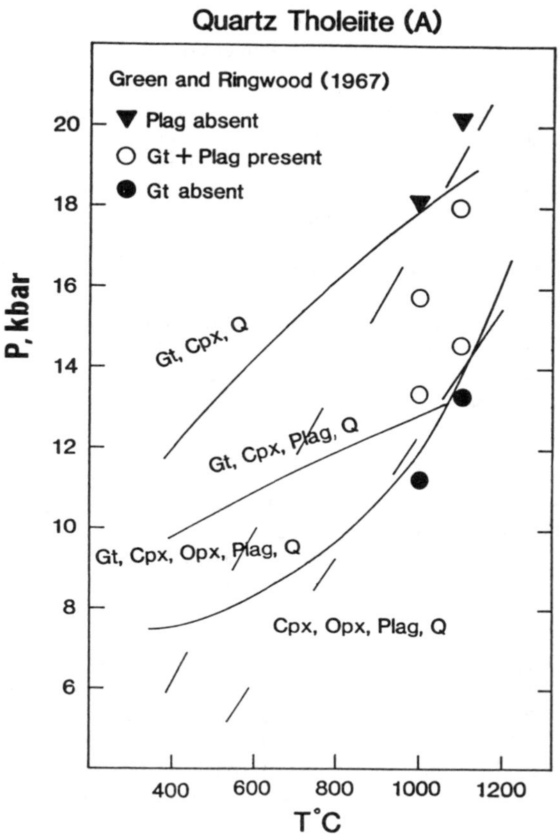

Figure 3. Calculated transition from gabbro (cpx, opx, plag, qtz) to eclogite (gt, cpx, qtz) for the quartz tholeiite (A) composition of Green and Ringwood (1967). Symbols are experimental data. Solid lines are (with increasing pressure) calculated garnet-in, orthopyroxene-out and plagioclase-out curves. Dashed lines are extrapolations of Green and Ringwood (1967).

Table 6. Tests of Equilibrium at 11.5 kbar / 800 °C

Equilibrium	ΔG (start) (kJ)	ΔG (end) (kJ)
$NaAlSi_2O_6$(cpx)+Q=Albite	+3.07	+0.02
$CaAl_2SiO_6$(cpx)+Q=An	+2.59	+0.01
$Mg_2Si_2O_6$(opx)+1/3 Alm(gt) =1/3Pyr(gt)+1/2$Fe_2Si_2O_6$(opx)	+0.29	+0.05
$Mg_2Si_2O_6$(opx)+An =1/3gro(gt)+2/3Pyr(gt)+qtz	-3.24	-0.09
$Fe_2Si_2O_6$(opx)+An =1/3Gro(gt)+2/3Pyr(gt)+Q	-3.82	-0.18

(garnet-in) curves is excellent despite the fact that the thermodynamic data were not, in any way, forced to fit these particular experimental results.

Of particular interest in Figure 3 is the fact that extrapolation of Green and Ringwood's data to lower temperatures is a more complex problem than they anticipated. In their discussion of eclogite stability at low temperatures and pressures Green and Ringwood made a simple and reasonable assumption. They assumed that the slopes of garnet-in and plagioclase-out reactions are subparallel to analogous reactions in simple systems. The calculations demonstrate, however, that the boundaries must actually be flatter and more curved than Green and Ringwood believed. The curvature and flattening are due to entropy of mixing effects in the complex phases involved in the reactions. The garnet-in curve flattens because the pyroxenes on the low pressure sides of Equations (A) and (B) decrease in mutual solubility and hence in entropy of mixing as temperature is lowered (Fig. 3). This raises the overall entropy of the transition and, since ΔV is negative, lowers its slope. The plagioclase-out curve is also flatter than Green and Ringwood's extrapolated transition boundary, which is subparallel to the end-member reaction

$$\text{albite} = \text{jadeite} + \text{quartz} \; . \qquad (46)$$

The reason that the actual plagioclase-out boundary is flatter than the end-member reaction is that the natural clinopyroxene contains only about 25 mol % jadeite and hence the partial molar entropy of jadeite is substantially higher than that of pure jadeite under the same P,T conditions. In contrast, plagioclases at the plagioclase-out boundary are much closer to pure albite ($\sim Ab_{70}$) than is the pyroxene to pure jadeite. The entropic effect on the plagioclases is hence smaller than on the pyroxenes and the net result is to flatten the boundary. The implication of these results is that, in contrast to Green and Ringwood's (1967) suggestion, eclogite is not stable in continental crust of normal thickness and geothermal gradient and that anhydrous mafic compositions should produce intermediate pressure granulite facies assemblages at the base of the continental crust. The thermodynamic approach provides a rational and consistent basis for calculation of the phase diagram at temperatures where reactions rates are too low to enable equilibrium to be obtained on a laboratory time scale.

Although the gabbro-eclogite transition provides a good test of the data base and method, the experimental data do not include equilibrium compositions of minerals. A more rigorous test would therefore involve reproducing equilibrium mineral compositions in addition to the phase boundaries. Although there are no data available for such a complete test, well analyzed suites of natural rocks provide a means of investigating how accurately mineral compositions may be calculated under independently estimated pressures and temperatures of equilibration. Granulite facies metamorphic rocks from the Adirondacks are suitable for this purpose.

Adirondack granulites

Bohlen and Essene (1979, 1980) and Johnson and Essene (1982) have obtained mineral compositions for quartz-bearing and olivine-bearing mafic granulites of the Adirondacks. Temperatures and pressures of metamorphism are well constrained by feldspar and oxide geothermometry (Bohlen and Essene, 1979; Bohlen et al., 1980) and pyroxene-garnet-plagioclase geobarometry (Newton, 1983; Bohlen et al., 1983c). As a test of the data base the compositions reported for the coexisting minerals in a number of these granulites were taken and the phases "re-equilibrated" in the computer at a range of pressures and temperatures. The results were used to construct lines in P-T space along which calculated plagioclase, clinopyroxene, garnet and olivine compositions correspond to those observed in the rocks.

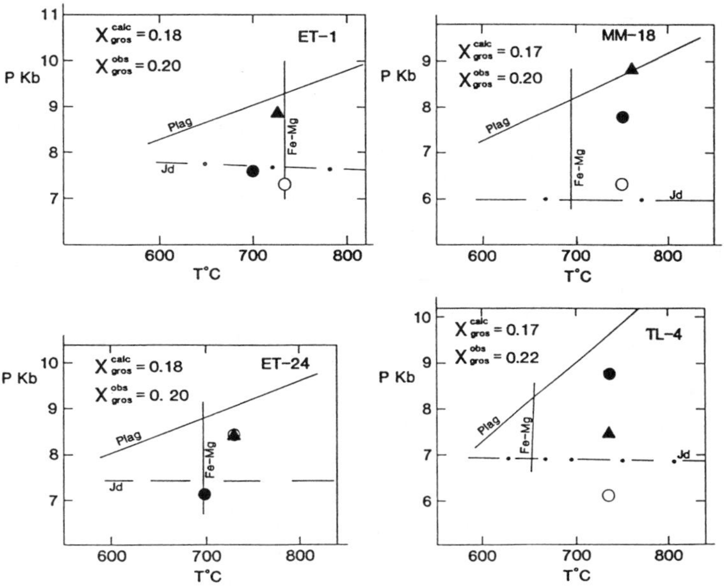

Figure 4. Comparison of calculated and observed mineral compositions for 4 quartz-bearing mafic granulites from the Adirondacks. Lines labelled "plag" and "Jd" refer to those P-T lines along which calculated An content of plagioclase and $NaAlSi_2O_6$ content of clinopyroxene correspond to those observed. Lines labelled "Fe-Mg" correspond to P-T lines of calculated garnet-clinopyroxene Fe-Mg partitioning which correspond to observed partitioning. "X_{gros}" refers to $Ca_3Al_2Si_3O_{12}$ content of garnet. Independent geobarometry: △ Newton (1983) based on An content of plagioclase, • Bohlen et al. (1983c) based on An content of plagioclase, ○ Newton (1983) based on jadeite content of clinopyroxene.

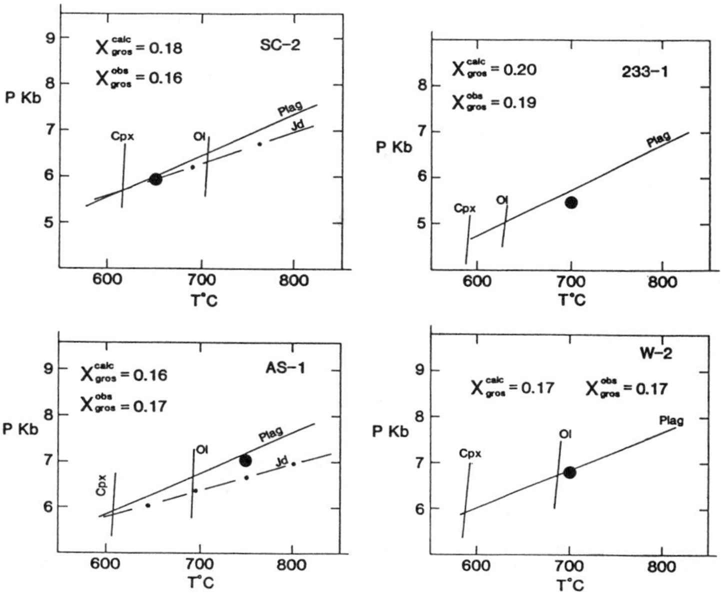

Figure 5. Comparison of calculated and observed mineral compositions for 4 olivine-bearing metagabbros from the Adirondacks. Lines labelled "plag" and "Jd" have same significance as in Figure 7. Lines labelled "cpx" and "ol" refer to lines along which calculated Fe-Mg partitioning between clinopyroxene and garnet and olivine and garnet, respectively, correspond to the observed partitioning. • calculated P and T from Bohlen et al. (1983c). Dashed box indicates approximate uncertainty in the geothermobarometry.

Figure 4 shows the calculated P-T lines for which plagioclase, clinopyroxene and garnet compositions correspond to those present in four quartz-clinopyroxene-orthopyroxene-garnet-plagioclase assemblages. As can be seen, the anorthite contents of the plagioclases (0.2-0.35 mole fraction) produce P-T lines which are in very good agreement with pressure estimates from pyroxene-plagioclase-garnet geobarometry (Newton, 1983; Bohlen et al., 1983c). The jadeite contents of the clinopyroxenes are also close to independently determined P-T ranges, but produce slightly lower pressures, possibly due to analytical difficulties at low Na (0.036-0.047 mole fraction $NaAlSi_2O_6$) contents of the clinopyroxenes (Newton, 1983). The calculated grossular contents of the garnets stable along the plagioclase isopleths are very close to those observed, the largest discrepancy, 0.05 mole fraction, coming from the most Fe- and Mn-rich composition, TL-4, in which the observed grossular, almandine and spessartine contents are 0.22, 0.65 and 0.07, respectively (Bohlen and Essene, 1980). The discrepancy may, in part, be due to uncertainties in the mixing properties at Fe-rich compositions (Ganguly and Saxena, 1984; Geiger et al., 1987).

The near-vertical lines labelled Fe-Mg refer to correspondence between calculated and observed Fe-Mg partitioning between clinopyroxene and garnet. As can be seen, there is good agreement between garnet-clinopyroxene relationships and independent feldspar and oxide thermometry. In sum, one must conclude that the data obtained from simple systems enable subsolidus equilibria involving complex pyroxene, garnet, plagioclase and quartz to be calculated with considerable confidence in the P-T regime of the granulite facies. The results provide additional support for the validity of the gabbro-eclogite calculations presented earlier.

Calculated mineral compositions in olivine-plagioclase-orthopyroxene-clinopyroxene-garnet assemblages are shown in Figure 5. As with the quartz-bearing assemblages the calculated P-T line for the equilibrium plagioclase is in very good agreement with independent geobarometry and the calculated grossular contents of garnets are also in good agreement with those observed. Iron-magnesium partitioning between garnet and olivine appears to be in better agreement with oxide and feldspar thermometry than is that between garnet and clinopyroxene. A possible explanation of the low apparent garnet-clinopyroxene temperatures (apart from accumulated errors in the thermodynamic data) is that of partial re-equilibration. Recent re-calibration of the garnet-clinopyroxene Fe-Mg thermometer (Patterson and Newton, 1987) indicates that granulites record low garnet-clinopyroxene temperatures (Newton, pers. comm.) in agreement with the calculations of Figures 4 and 5. Despite this uncertainty, the general agreement between calculated and observed equilibria (Figs. 3-5) is extremely encouraging.

SUMMARY

A thermodynamic data base has been developed for calculation of subsolidus equilibria in the system Na_2O-FeO-CaO-MgO-Al_2O_3-SiO_2 This is particularly useful for the calculation of phase relations outside the P-T range where experiments can be performed on a laboratory time scale (Fig. 3). Comparison with Green and Ringwood's (1967) experimental data on the gabbro to eclogite transformation demonstrates that the appearance and disappearance of phases in mafic compositions can be calculated with considerable confidence. Eclogite appears not to be stable in continental crust of normal thickness. Comparison of calculated phase compositions with those observed in mafic granulites at about 7.5 kbar/725°C confirms the general validity of the data base and solution models which have been derived.

A FORTRAN 77 program which performs the calculations illustrated in Table 5 is available on request from the author.

ACKNOWLEDGEMENTS

Drs. A. Navrotsky and R.C. Newton are thanked for their corrections on the first draft of this paper. The data base and program was originally developed with the support of NASA grant NAGW-182 and later extended with support from NSF EAR-8212502 and 8416793.

REFERENCES

Andersen, D.J. and Lindsley, D.H. (1981) A valid Margules formulation for an asymmetric ternary solution: revision of the olivine-ilmenite thermometer, with applications. Geochim. Cosmochim. Acta 45, 847-853.

Besancon, J.R. (1981) Rate of cation ordering in orthopyroxenes,. Am. Mineral. 66, 965-973.

Bohlen, S.R. and Essene, E.J. (1979) A critical evaluation of two-pyroxene thermometry in Adirondack granulites. Lithos 12, 335-345.

—— and —— (1980) Evaluation of coexisting garnet-biotite, garnet-clinopyroxene and other Mg-Fe exchange thermometers in Adirondack granulites. Geol. Soc. Amer. Bull. 91, Pt. II, 685-719.

——, —— and Hoffman, K.S. (1980) Update on feldspar and oxide thermometry in the Adirondack Mountains, New York. Geol. Soc. Amer. Bull. 91, Pt. I, 110-113.

——, Wall, V.J. and Boettcher, A.L. (1983a) Experimental investigations and geological applications of equilibria in the system $FeO-TiO_2-Al_2O_3-SiO_2-H_2O$. Amer. Mineral. 68, 1049-1058.

——, —— and —— (1983b) Experimental investigation and application of garnet granulite equilibria. Contrib. Mineral. Petrol. 83, 52-61.

——, —— and —— (1983c) Geobarometry in granulites. In: Advances in Physical Geochemistry 3, S.K. Saxena, ed. Springer-Verlag, New York.

Brown, T.H. and Skinner, B.J. (1974) Theoretical prediction of equilibrium phase assemblages in multicomponent systems. Amer. J. Sci. 274, 961-986.

Bruno, E., Chiari, G. and Facchinelli, A. (1976) Anorthite quenched from 1530 ° C. I. Structure refinements. Acta Cryst. B32, 3270-3280.

Chatillon-Colinet, C., Kleppa, O.J., Newton, R.C. and Perkins, D., III (1983a) Enthalpy of formation of $Fe_3Al_2Si_3O_{12}$ (almandine) by high temperature alkali borate solution calorimetry. Geochim. Cosmochim. Acta 47, 439-444.

——, Newton, R.C., Perkins, D., III and Kleppa, O.J. (1983b) Thermochemistry of $(Fe^{2+},Mg)SiO_3$ orthopyroxene. Geochim. Cosmochim. Acta 47, 1597-1603.

Crerar, D.A. (1975) A method for computing multicomponent chemical equilibria based on equilibrium constants. Geochim. Cosmochim. Acta 39, 1375-1384.

Cressey, G., Schmid, R. and Wood, B.J. (1978) Thermodynamic properties of almandine-grossular garnet solid solutions. Contrib. Mineral. Petrol. 67, 397-404.

Engi, M. (1983) Equilibria involving Al-Cr spinel: Mg-Fe exchange with olivine: experiments, thermodynamic analysis and consequences for geothermometry,. Amer. J. Sci. 283-A, 29-71.

Ganguly, J. and Saxena, S.K. (1984) Mixing properties of aluminosilicate garnets: Constraints from natural and experimental data and applications to geothermometry. Amer. Mineral. 69, 88-97.

Gasparik, T. (1985) Experimental study of subsolidus phase relations and mixing properties of pyroxene and plagioclase in the system $Na_2O–CaO–Al_2O_3–SiO_2$. Contrib. Mineral. Petrol. 89, 346-357.

Geiger, G.A., Newton, R.C. and Kleppa, O.J. (1987) Enthalpy of mixing of synthetic almandine-grossular and almandine-pyrope garnets from high temperature solution calorimetry. Geochim. Cosmochim. Acta, in press.

Gerald, C.F. and Wheatley, P.O., (1984) "Applied Numerical Analysis," Addison-Wesley Co., Reading, Mass., 579 p.

Goldsmith, J.R. (1982) Plagioclase stability at elevated temperatures and water pressures,. Amer. Mineral. 67, 653-675.

Green, D.H. and Ringwood, A.E. (1967) An experimental investigation of the gabbro to eclogite transformation and its petrological applications. Geochim. Cosmochim. Acta 31, 767-833.

Harley, S.L. (1984) The solubility of aluimina in orthopyroxene coexisting with garnet in the system FeO-MgO-Al_2O_3-SiO_2 and CaO–FeO–MgO–Al_2O_3-SiO_2. J. Petrol. 25, 665-696.

Haselton, H.T. and Newton, R.C. (1980) Thermodynamics of pyrope-grossular garnets and their stabilities at high temperatures and high pressures. J. Geophys. Res. 85, 6973-6982.

Holland, T.J.B. (1980) The reaction albite=jadeite+quartz determined experimentally in the range 600-1200 °C. Amer. Mineral. 65, 129-134.

—— (1983) The experimental determination of activities in disordered and short-range ordered jadeitic pyroxenes. Contrib. Mineral. Petrol. 82, 214-220.

—— Navrotsky, A. and Newton, R.C. (1979) Thermodynamic parameters of $CaMgSi_2O_6$-$Mg_2Si_2O_6$ pyroxenes based on regular and cooperative disorder models. Contrib. Mineral. Petrol. 69, 337-334.

——, ——, and —— (1980) Thermodynamic parameters of $CaMgSi_2O_6$–$Mg_2Si_2O_6$ pyroxenes based on regular solution and cooperative disordering models. Reply. Contrib. Mineral. Petrol. 75, 305-306.

Jamieson, H.E. and Roedder, P.L. (1984) The distribution of Mg and Fe^{2+} between olivine and spinel at 1300 ° C. Amer. Mineral. 69, 283-291.

Johnson, C.A. and Essene, E.J. (1982) The formation of garnet in olivine-bearing metagabbros from the Adirondacks. Contrib. Mineral. Petrol. 81, 240-251.

Lindsley, D.H. (1981) The formation of pigeonite on the join hedenbergite-ferrosilite at 11.5 and 15 kbar: experiments and a solution model. Amer. Mineral. 66, 1175-1182.

—— (1983) Pyroxene thermometry. Amer. Mineral. 68, 477-493.

——, Grover, J.E. and Davidson, P.M. (1981) The thermodynamics of the $Mg_2Si_2O_6$-$CaMgSi_2O_6$ join: a review and an improved model. In: Advances in Physical Geochemistry 1, R.C. Newton, A. Navrotsky and B.J. Wood, eds. Springer-Verlag, New York, p. 149-175.

Mazo, R.M. (1977) Statistical mechanical calculation of aluminum-silicon disorder in albite. Amer. Mineral. 62, 1232-1237.

McCallister, R.H., Finger, I.W. and Ohashi, Y. (1976) Intracrystalline Fe^{2+}-Mg equilibria in three natural Ca-rich clinopyroxenes. Amer. Mineral. 61, 671-676.

Navrotsky, A. and Kleppa, O.J. (1967) The thermodynamics of cation distributions in simple spinels. J. Inorg. Nucl. Chem. 29, 2701-2714.

Newton, R.C. (1983) Geobarometry of high-grade metamorphic rocks. Amer. J. Sci. 283-A, 1-28.

—— Charlu, T.V. and Kleppa, O.J. (1977) Thermochemistry of high pressure garnets and clinopyroxenes in the system CaO-MgO-Al_2O_3-SiO_2. Geochim. Cosmochim. Acta 41, 369-377.

——, ——, and —— (1980) Thermochemistry of the high structural state plagioclases. Geochim. Cosmochim. Acta 44, 933-941.

——, Geiger, C.A. and Kleppa, O.J. (1987), Thermodynamics of $(Fe,Mg,Ca)_3Al_2Si_3O_{12}$ garnet: A review. In: Advances in Physical Geochemistry (submitted).

Nicholls, J. (1977) The calculation of mineral compositions and modes of olivine-two pyroxene-spinel assemblages: Problems and possibilities. Contrib. Mineral. Petrol. 60, 119-142.

O'Neill, H.St.C. and Wood, B.J. (1979) An experimental study of Fe-Mg partitioning between garnet and olivine and its calibration as a geothermometer. Contrib. Mineral. Petrol. 70, 59-70.

Pattison, D.R.M. and Newton, R.C., (1987) Garnet-clinopyroxene K_D (Fe-Mg) variations due to X_{Ca}^{Grt} and pressure. EOS, Trans. Amer. Geophys. Union 68, 461.

Robie, R.A. and Hemingway, B.S. (1984) Entropies of kyanite, andalusite and sillimanite: Additional constraints on the pressure and temperature of the Al_2SiO_5 triple point. Amer.

Mineral. 69, 298-306.
——, Finch, C.B. and Hemingway, B.S. (1982) Heat capacity and entropy between 5.1 and 383 Kelvin of fayalite (Fe_2SiO_4): comparison of calorimetric and equilibrium values for the QFM buffer reaction. Amer. Mineral. 67, 463-469.
——, Hemingway, B.S. and Fisher, J.R. (1978) Thermodynamic properties of minerals and related substances at 298.15 K and 1 bar (10^5 Pa) pressure and at higher temperature. U.S. Geol. Surv. Bull. 1452, 456 pp.
Saxena, S.K. and Ericksson, G. (1983) Theoretical computation of mineral assemblages in pyrolite and lherzolite. J. Petrol. 24, 538-555.
—— and —— (1985) Anhydrous phase equilibria in earth's upper mantle,. J. Petrol. 26, 378-390.
—— and Ghose, S. (1971) Mg^{2+}-Fe^{2+} order-disorder and the thermodynamics of the orthopyroxene solution. Amer. Mineral. 56, 532-559.
Storey, S.H. and Van Zeggeren, F. (1964) Computation of chemical equilibrium compositions. Can. J. Chem. Eng. 42, 54-55.
Van Zeggeren, F. and Storey, S.H. (1970) "The Computation of Chemical Equilibria," Cambridge Univ. Press, Cambridge, England. 176 p.
Wohl, K. (1946) Thermodynamic evaluation of binary and ternary liquid systems. Trans. Amer. Inst. Chem. Eng. 42, 215-249.
Wood, B.J. (1974) The solubility of alumina in orthopyroxene coexisting with garnet. Contrib. Mineral. Petrol. 46, 1-15.
—— and Banno, S. (1973) Garnet-orthopyroxene and orthopyroxene-clinopyroxene relationships in simple and complex systems. Contrib. Mineral. Petrol. 42, 109-124.
—— and Holloway, J.R. (1984) A thermodynamic model for subsolidus equilibria in the system CaO-MgO-Al_2O_3-SiO_2. Geochim. Cosmochim. Acta 48, 159-176.
——, Holland, T.J.B., Newton, R.C. and Kleppa, O.J. (1980) Thermochemistry of jadeite-diopside pyroxenes. Geochim. Cosmochim. Acta 44, 933-941.
—— and Kleppa, O.J. (1981) Thermochemistry of forsterite-fayalite olivine solutions. Geochim. Cosmochim. Acta 45, 569-581.
—— and Nicholls, J. (1978) The thermodynamic properties of reciprocal solid solutions. Contrib. Mineral. Petrol. 66, 389-400.
——, Holland, T.J.B., Newton, R.C. and Kleppa, O.J. (1980) Thermochemistry of jadeite-diopside pyroxenes. Geochim. Cosmochim. Acta 44, 933-941.
——, Kirkpatrick, R.J. and Montez, B. (1986) Order-disorder phenomena in $MgAl_2O_4$ spinel. Amer. Mineral. 71, 999-1006.

Chapter 4 **Kenneth S. Pitzer**

A Thermodynamic Model for Aqueous Solutions of Liquid-Like Density

INTRODUCTION

There are many geologically important natural systems involving multicomponent aqueous solutions; other similar systems are important in steam power generation, chemical processing, and other industrial operations. Thus it is important to have an accurate and convenient model for the prediction of the thermodynamic properties of such aqueous solutions since it would be very burdensome to make experimental measurements for each composition at all of the temperatures and pressures of interest. Aqueous solutions are also of theoretical interest, and existing theory provides a general structure for the desired model as well as the precise form of certain terms, but other terms are best evaluated empirically from appropriate experiments. The model and its applications are described and discussed in the present chapter and the following one by Weare.

This model was initially developed (Pitzer, 1973, 1975; Pitzer and Kim, 1974) for solutions near room temperature, but it has been found to be applicable to aqueous systems up to 300°C or a little higher temperature. A liquid-like density and relatively small compressibility are assumed. Thus the present model is not applicable close to the critical point of water at 374°C. The requirements for a theory applicable to ionic solutions in the critical region were discussed by Levelt Sengers et al. (1986). Progress has been made toward models valid for aqueous systems at above critical temperatures and pressures (Sverjensky, a chapter in this volume; Pitzer and Li, 1983, 1984; Bischoff and Pitzer, 1985), but much more research is needed for these more extreme conditions.

A typical application is the prediction of the equilibrium between an aqueous phase (brine) and one or more solid phases (minerals). There are many published examples of solubility calculations; especially pertinent are those of Harvie and Weare (1980), Krumgalz and Millero (1983), and Harvie et al. (1984) for 25°C, and for high temperatures that of Pabalan and Pitzer (1987b). The following chapter by Weare emphasizes solubility calculations. The vapor phase may also be present and important. A complete calculation of either type involves several steps which can be considered separately as follows:

1. <u>The standard state chemical potential or molar Gibbs energy of formation of each substance involved</u>. For aqueous species the standard state is a hypothetical state with ideal properties at one molal which is actually evaluated in a limiting process at infinite dilution. General tables of standard state chemical potentials for solids, gases, and solution species at the reference temperature, 25°C = 298.15 K, and the reference pressure of 1 bar are available from several sources (Robie et al., 1978; Wagman et al., 1982). Our interest is in accurate values for particular differences, and the tables of Harvie and Weare (1980), Harvie et al. (1984), and in the following chapter of this volume have been prepared to best meet this need.

2. <u>The excess Gibbs energy for the solution, which yields the activity coefficients of all solute species and the osmotic coefficient</u>. The model for these quantities is described below.

NOTATION

a	activity; also a negative ion; also parameter in eq. (46)
A	Helmholtz energy
A_ϕ	Debye-Hückel parameter
B_{ij}	second virial coefficient or binary ion interaction parameter for interaction i-j (also B^ϕ_{ij})
b	parameter in extended Debye-Hückel expression; also parameter in eq. (46)
c	positive ion; also parameter in eq. (46)
C_{ij}	(also C^ϕ_{ij}) third virial coefficient
d_w	density of water
e	electronic charge
f(I)	extended Debye-Hückel function
F	collection of terms for the activity coefficient (eq. 32,33)
g(x)	defined function (eq. 19)
g'(x)	defined function (eq. 20)
G^{ex}	excess Gibbs energy
H	enthalpy
I	ionic strength
J_{ij}	defined function (eq. A11)
k	Boltzmann constant
K	equilibrium constant
ℓ	electrostatic length (eq. A9)
m_i	molality of i
M	positive ion
n_w	number of kg of water
N_o	Avogadro's number
P	pressure
q_{ij}	defined function (eq. A12)

NOTATION

r	interparticle distance
R	gas constant
S	entropy
t	temperature in °C
T	absolute temperature
X	negative ion
z_i	charge on ion i
Z	total charge molality (eq. 35)
u_{ij}	short range potential
v_{ij}	interionic potential of mean force

Greek letters

α	parameter in second virial coefficient expression (also α_1, α_2)
$\beta_{ij}^{(0)}$	(also $\beta_{ij}^{(1)}$, $\beta_{ij}^{(2)}$) second virial coefficient for interaction i-j
γ_i	activity coefficient of ion i
γ_{MX}	mean activity coefficient for electrolyte MX
θ_{ij}	second virial coefficient contribution from short-range forces
${}^E\theta_{ij}$	same but for long-range force
κ	Debye length (eq. A3)
λ_{ij}	second virial coefficient for interaction i-j
μ_i	standard chemical potential of species i
μ_{ijk}	third virial coefficient for interaction i-j-k
ν_i	number of ions of type i in the complete salt
ν	total number of ions in salt MX (also ν_{MX})
ϕ	osmotic coefficient
Φ_{ij}	total second virial coefficient for ions i-j of the same sign
ψ_{ijk}	third virial coefficient for mixing effects

Additional information concerning the theoretical basis for the model is given in Appendix A. This model was first tested at 25°C and the most extensive array of parameters pertain to that temperature. Also there have been a very extensive array of successful applications at 25°C, some of which are described below and others by Weare in the next chapter. Recently, this model has been extended to 200°C or above for a number of the geologically important solutes and these results are also described below. Appendix B gives numerical values for several solutes at elevated temperatures.

3. For high temperature applications, the changes of standard-state chemical potential values from 25° to the temperature of interest. This calculation is normally based on values for the standard-state entropy (or enthalpy) at 25°C and for the standard-state heat capacity over a range of temperature from 25°C upward. Sample calculations of this type are discussed and a summary of entropy values and heat capacity equations are given below.

Since the model is based on a general equation for the Gibbs energy of the aqueous fluid, any thermodynamic property can be obtained from the appropriate derivative. Phase equilibria with several solid minerals are especially interesting applications which are discussed by Weare in the following chapter. Thermal properties of the brine are sometimes required, and these are discussed below, as are equilibria of the aqueous fluid with a vapor phase.

There are, of course, other models which with similar accuracy treat some of the properties and systems here considered. But none of the other models has shown a comparably wide range of applicability and accuracy. Hence, we shall not burden this paper with extensive discussions of other models and comparisons of results. Such comparisons are available elsewhere; many publications here cited contain such comparisons.

EXCESS GIBBS ENERGY; ACTIVITY AND OSMOTIC COEFFICIENTS

Basic equation

In constructing a model for aqueous ionic solutions all valid and useful theory should be included. The statistics of ideal solution behavior are first included in the definitions of the excess Gibbs energy and the activity and osmotic coefficients. Next in importance and now fully confirmed is the limiting law of Debye and Hückel; this is the leading term for departure from ideality for any ionic solution. It depends only on the concentrations of the electrically charged species, the temperature, and the macroscopic properties of the solvent, the density and dielectric constant. This type of term arises purely from long-range electrostatic forces. There are higher order terms of the same type, depending only on electrical forces, but in many cases these are indistinguishable from terms arising from short-range forces between solute species. These short-range forces are complex and not practically predictable at present; hence, their effects are treated empirically.

There is one higher-order electrostatic term that yields effects quite different from those from short-range forces. It arises only when ions of different charge of the same sign are mixed, i.e., Ca^{2+} and Na^{+} or SO_4^{2-} and Cl^{-}. It can be calculated from theory and the properties of the solvent.

Neutral solute species interact only through short-range forces. These effects are easily included in any general equation for the excess Gibbs energy. The remaining questions relate to the choice of variables for the basic equation. The basic theory is developed in terms of the total volume and the concentrations of solute species as well as the temperature. But for most practical systems pressure is a more appropriate variable than total volume. Also composition variables such as molality or mole fraction, which are independent of pressure and temperature, are much more convenient than concentration. For electrolytes, molality is so widely used that it seems desirable to adopt it. It is possible to make rigorous transformations from volume to pressure and from concentration to molality, but these are very cumbersome in general. Thus, it is convenient and satisfactory for most purposes to set up the basic equation with the variables T, P, m_i and with guidance from theory for the details of certain terms and the general structure of the equation. The empirical terms can be accepted as such or can be related to the corresponding theoretical terms in the T, V, c_i equation, if desired. But, it must be remembered that the volume to pressure transformation breaks down with the infinite compressibility at the solvent critical point. Hence, the present treatment has this limitation and should not be used above 350°C, except at pressures far above critical. Most of the applications of the present model have been at 300°C or lower temperatures, and in that range the pressure basis is fully satisfactory.

There are alternate presentations of the basic statistical mechanics of multicomponent ionic fluids which suggest the basic equation adopted for the present model. Appendix A describes the most rigorous analysis which, however, is abstract. In the original presentation and an early summary of this model (Pitzer, 1973, 1979), a basic but approximate treatment was presented which suggests the same form of equation and gives an easier understanding of the general physical picture. The approximations of the latter treatment do not carry over to the basic equation of the model, however. The model is postulated for empirical use and is to be justified by its empirical success. This equation (A7 in Appendix A) expresses the excess Gibbs energy as follows:

$$G^{ex}/n_w RT = f(I) + \sum_i \sum_j m_i m_j \lambda_{ij}(I) + \sum_i \sum_j \sum_k m_i m_j m_k \mu_{ijk} + \cdots \tag{1}$$

Here n_w is the number of kg of water and m_i, m_j,... are the molalities of all solute species. The ionic strength is given by

$$I = \tfrac{1}{2} \sum_i m_i z_i^2 \quad ,$$

where z_i is the number of charges on the ith solute. The first term on the right in Equation (1) includes the Debye-Hückel limiting law, but is an extended form chosen for empirical effectiveness. Note that f(I) depends only on the ionic strength and not on individual ionic molalities or other solute properties.

The quantity $\lambda_{ij}(I)$ represents the short range interaction in the presence of the solvent between solute particles i and j. This binary interaction parameter or second virial coefficient does not itself have any composition dependence for neutral species, but for ions it is dependent on the ionic strength; it does depend, of course, on the particular solute species i and j and the temperature and pressure. The similar quantity for triple interaction is μ_{ijk}; in principle it might be ionic strength dependent, but with a single possible exception (Phutela and Pitzer, 1986a), there is no indication of such dependence. Hence, we shall write our equations without considering any I-dependence for μ. Fourth or higher order interactions could be included, but we will not do so in this presentation. They are needed only for extremely concentrated solutions (Ananthaswany and Atkinson, 1985), and then alternate methods may be preferable (Pitzer and Simonson, 1986).

The standard definitions and thermodynamic transformations yield the equations for the activity and osmotic coefficients:

$$\begin{aligned} ln\gamma_i &= [\partial(G^{ex}/n_w RT)/\partial m_i]_{n_w} \\ &= (z_i^2/2)f' + 2\sum_j \lambda_{ij} m_j + (z_i^2/2)\sum_j\sum_k \lambda'_{jk} m_j m_k \\ &\quad + 3\sum_j\sum_k \mu_{ijk} m_j m_k + \cdot\cdot\cdot \end{aligned} \tag{2}$$

$$\begin{aligned} \phi - 1 &= -(\partial G^{ex}/\partial n_w)_{n_i}/kT \sum_i m_i \\ &= (\sum_i m_i)^{-1}[If' - f) + \sum_i\sum_j (\lambda_{ij} + I\lambda'_{ij}) m_i m_j \\ &\quad + 2\sum_i\sum_j\sum_k \mu_{ijk} m_i m_j m_k + \cdot\cdot\cdot] \quad . \end{aligned} \tag{3}$$

Here f' and λ' are the ionic strength derivatives of f and λ. Also $m_i = n_i/n_w$ with n_i the number of moles of species i. The multiple sums in equations (1, 2, 3) are unrestricted, i.e., each sum covers all solute species. Also, we note the definition of the excess Gibbs energy in the molality system and another useful relationship:

$$G^{ex} = G - n_1\bar{G}_1^0 - \sum n_i[\bar{G}_i^0 - RT\,(1 - ln m_i)] \quad , \tag{4}$$

$$G^{ex}/n_w RT = \sum m_i(ln\gamma_i + 1 - \phi) \quad . \tag{5}$$

Here $\bar{G}_1^0$ is the molar Gibbs energy of water and $\bar{G}_i^0$ the partial molar Gibbs energy for species i, each in its standard state.

For solutions containing ions, the requirement of electrical neutrality makes it impossible to evaluate certain individual ionic quantities. This becomes more explicit as one derives the working equation for an electrolyte with a single solute.

Pure electrolytes

If the electrolyte MX has ν_M positive ions of charge z_M in its formula and ν_X negative ions of charge z_X; neutrality requires $z_M\nu_M = |z_X|\nu_X$; also we take $\nu = \nu_M + \nu_X$. For a salt molality m, the ion molalities are $m_M = \nu_M m$ and $m_X = \nu_X m$. The osmotic coefficient becomes

$$\phi-1 = (\nu m)^{-1}\{(If'-f) + m^2[2\nu_M\nu_X(\lambda_{MX} + I\lambda'_{MX}) + \nu_M^2(\lambda_{MM} + I\lambda'_{MM}) \tag{6}$$

$$+ \nu_X^2(\lambda_{XX} + I\lambda'_{XX})] + m^2(6\nu_M^2\nu_X\mu_{MMX} + 6\nu_M\nu_X^2\mu_{MXX} + 2\nu_M^3\mu_{MMM} + 2\nu_X^3\mu_{XXX}) + \ldots\}.$$

From the experimentally measured properties of the pure electrolyte, together with the Debye-Hückel term (If'-f), one can evaluate only the bracketed term in λ's, a function of ionic strength, and the final term in parentheses involving μ's. Thus we define f^ϕ, $B^\phi(I)$ and C^ϕ as follows:

$$f^\phi = (f' - f/I)/2 \tag{7}$$

$$B_{MX}^\phi = \lambda_{MX} + I\lambda'_{MX} + (\nu_M/2\nu_X)(\lambda_{MM} + I\lambda'_{MM}) + (\nu_X/2\nu_M)(\lambda_{XX} + I\lambda'_{XX}) \tag{8}$$

$$C_{MX}^\phi = [3/(\nu_M\nu_X)^{\frac{1}{2}}](\nu_M\mu_{MMX} + \nu_X\mu_{MXX}) \quad . \tag{9}$$

At this point we could have included the terms in μ_{MMM} and μ_{XXX} in the definition of C^ϕ, but we shall neglect them later so omit them now. These terms relate to short-range interactions of three ions all of the same sign. Since electrical repulsions make it unlikely that three ions of the same sign are often close together, these terms are expected to be very small, and no indication has arisen that they need to be included. Equation (6) now reduces to the simple form

$$\phi-1 = |z_M z_X| f^\phi + m(2\nu_M\nu_X/\nu)B_{MX}^\phi + m^2[2(\nu_M\nu_X)^{3/2}/\nu]C_{MX}^\phi \quad . \tag{10}$$

For a 1-1 electrolyte all of the coefficients became unity.

In originally developing this model (Pitzer, 1973), two choices were made at this point: the extended form of the Debye-Hückel term f^ϕ and the form for the ionic strength dependence of B_{MX}^ϕ. All combinations of the most likely forms were tested with an array of accurately measured experimental osmotic coefficients for several 1-1, 2-1, and 1-2 type salts at 25°C, and the best results were obtained for the forms:

$$f^\phi = -A_\phi I^{\frac{1}{2}}/(1 + bI^{\frac{1}{2}}) \quad , \tag{11}$$

$$B_{MX}^\phi = \beta_{MX}^{(0)} + \beta_{MX}^{(1)} \exp(-\alpha I^{\frac{1}{2}}) \quad . \tag{12}$$

The general pattern of ionic strength dependence was indicated theoretically for each function, but alternate forms were equally plausible, and the choice was made for empirical effectiveness. Here b is a universal parameter with the value 1.2 $kg^{\frac{1}{2}}\cdot mol^{-\frac{1}{2}}$ and α has the value 2.0 $kg^{\frac{1}{2}}\cdot mol^{-\frac{1}{2}}$ for all of the salts in the test set. It will be possible to use a different value of α for other salts or salts of other charge types; this will be discussed below. The parameters $\beta_{MX}^{(0)}$ and $\beta_{MX}^{(1)}$ are specific to the salt MX. It was expected that these parameters representing short-range interactions would be specific to the interacting ions.

As salts of other valence types were studied, it was found that Equation (12) with α = 2.0 served well for 3-1 and even 4-1 salts but not for 2-2 salts such as $MgSO_4$ (Pitzer and Mayorga, 1973, 1974). The 2-2 type salts show an electrostatic ion pairing effect which has usually been represented by considering the ion pair as a separate solute species. Introduction of such a species in equilibrium with other species complicates the calculations considerably, however. We found that good agreement with observed properties was obtained for the 2-2 salts if one simply added another term to B_{MX}^{ϕ} as follows:

$$B_{MX} = \beta_{MX}^{(0)} + \beta_{MX}^{(1)} \exp(-\alpha_1 I^{\frac{1}{2}}) + \beta_{MX}^{(2)} \exp(-\alpha_2 I^{\frac{1}{2}}) \quad . \qquad (13)$$

The values α_1 = 1.4 $kg^{\frac{1}{2}}\cdot mol^{-\frac{1}{2}}$ and α_2 = 12 $kg^{\frac{1}{2}}\cdot mol^{-\frac{1}{2}}$ were satisfactory for all 2-2 electrolytes at 25°C. The parameter $\beta_{MX}^{(2)}$ is negative and is related to the association equilibrium constant.

In a very extensive investigation of aqueous HCl Holmes et al. (1987) found that the $\beta^{(2)}$ term was unnecessary below 250°C. From 250 to 375°C they obtained a good fit with a $\beta^{(2)}$ term and with α_1 = 1.4 and $\alpha_2 = 6.7\ A_\phi$. There is theoretical support for the proportionality of α_2 to the Debye-Hückel parameter A_ϕ.

Finally, we note the theoretical expression for the Debye-Hückel parameter,

$$A_\phi = (1/3)\ (2\pi N_o d_w/1000)^{1/2} (e^2/\varepsilon kT)^{3/2} \quad , \qquad (14)$$

with N_o Avogadro's number, d_w the density of water, e the electronic charge, k Boltzmann's constant, and ε the dielectric constant or the relative permittivity of water. For SI units ε is multiplied by $4\pi\varepsilon_o$ with ε_o the permittivity of free space. In many papers the symbol D is used instead of ε for the dielectric constant.

Next we transform B^{ϕ}_{MX} and C^{ϕ}_{MX} as defined by Equations (8) and (9) to the corresponding forms for the excess Gibbs energy as follows:

$$B_{MX} = \lambda_{MX} + (\nu_M/2\nu_X)\lambda_{MM} + (\nu_X/2\nu_M)\lambda_{XX} \quad ; \qquad (15)$$

$$C_{MX} = (3/2)(\mu_{MMX}/z_M + \mu_{CXX}/|z_X|) \quad . \qquad (16)$$

These quantities are used for the equations for mixed electrolytes. Introduction of the selected forms from Equations (11) and (13) yields

$$f = -(4IA_\phi/b)\,ln(1+bI^{\frac{1}{2}}) \quad ; \qquad (17)$$

$$B_{MX} = \beta_{MX}^{(0)} + \beta_{MX}^{(1)}\ g(\alpha_1 I^{\frac{1}{2}}) + \beta_{MX}^{(2)}\ g(\alpha_2 I^{\frac{1}{2}}) \quad ; \qquad (18)$$

$$g(x) = 2[1-(1+x)\exp(-x)]/x^2 \quad . \tag{19}$$

Further manipulations yield the useful relationships:

$$B'_{MX} = [\beta^{(1)}_{MX} g'(\alpha_1 I^{\frac{1}{2}}) + \beta^{(2)}_{MX} g'(\alpha_2 I^{\frac{1}{2}})]/I \quad ; \tag{20}$$

$$g'(x) = -2[1-(1+x+x^2/2)\exp(-x)]/x^2 \quad ; \tag{21}$$

$$C_{MX} = C^{\phi}/2|z_M z_X|^{\frac{1}{2}} \quad . \tag{22}$$

For the activity coefficient equation, it is useful to define the following:

$$f^{\gamma} = -A_{\phi}[I^{\frac{1}{2}}/(1+bI^{\frac{1}{2}}) + (2/b)ln(1+bI^{\frac{1}{2}})] \quad ; \tag{23}$$

$$B^{\gamma}_{MX} = B_{MX} + B^{\phi}_{MX} \quad ; \tag{24}$$

$$C^{\gamma}_{MX} = 3C^{\phi}_{MX}/2 \quad . \tag{25}$$

The mean activity coefficient for a salt is defined as

$$ln\gamma_{\pm} = (\nu_M ln\gamma_M + \nu_X ln\gamma_X)/\nu \quad , \tag{26}$$

and for the present model with a single salt this becomes

$$ln\gamma_{\pm} = |z_M z_X| f^{\gamma} + m(2\nu_M \nu_X/\nu)B^{\gamma}_{MX} + m^2[2(\nu_M \nu_X)^{3/2}/\nu]C^{\gamma}_{MX} \quad . \tag{27}$$

Equations (10) and (27) were applied to the very extensive array of data for 25°C with excellent agreement to about 6 mol·kg^{-1} for various types of electrolytes (Pitzer and Mayorga, 1973, 1974). The resulting parameters are discussed in a subsequent section.

Mixed electrolytes

In order to treat mixed electrolytes it is desirable to rewrite Equation (1) in terms of the experimentally determinable quantities B and C instead of the individual ion quantities λ and μ. Appropriate transformations yield for the excess Gibbs energy

$$\begin{aligned} G^{ex}/(n_w RT) = {} & f(I) + 2\sum_c \sum_a m_c m_a [B_{ca} + (\sum_c m_c z_c)C_{ca}] \\ & + \sum_{c<c'}\sum m_c m_{c'}[2\Phi_{cc'} + \sum_a m_a \psi_{cc'a}] + \sum_{a<a'}\sum m_a m_{a'}[2\Phi_{aa'} + \sum_c m_c \psi_{caa'}] \\ & + 2\sum_n \sum_c m_n m_c \lambda_{nc} + 2\sum_n \sum_a m_n m_a \lambda_{na} + 2\sum_{n<n'}\sum m_n m_{n'} \lambda_{nn'} + \cdot\cdot\cdot \; , \end{aligned} \tag{28}$$

where the sums are over the various cations c,c' and over the anions a,a'. If neutral solute species n, n' are present, the terms from Equation (1) in λ_{ij} and μ_{ijk} are retained; only those in λ_{ij} are shown in the last three sums. Difference combinations of λ's and μ's arise which are defined as follows:

$$\Phi_{cc'} = \lambda_{cc'} - (z_{c'}/2z_c)\lambda_{cc} - (z_c/2z_{c'})\lambda_{c'c'} \quad ; \tag{29}$$

$$\psi_{cc'a} = 6\mu_{cc'a} - (3z_{c'}/z_c)\mu_{cca} - (3z_c/z_{c'})\mu_{c'c'a} \tag{30}$$

Analogous expressions for $\Phi_{aa'}$ and $\psi_{caa'}$ arise from permutation of the indices. These quantities account for interactions between ions of like sign, which arise only for mixed solutions, and can best be determined from simple common-ion mixtures.

In terms of various quantities defined above, the following equations give the osmotic coefficient of the mixed electrolyte and the activity coefficients of cation M and anion X, respectively.

$$(\phi-1) = (2/\sum_i m_i)[-A_\phi I^{3/2}/(1 + bI^{1/2}) + \sum_c \sum_a m_c m_a (B^\phi_{ca} + Z\, C_{ca})$$
$$+ \sum_{c<c'} \sum m_c m_{c'} (\Phi^\phi_{cc'} + \sum_a m_a \psi_{cc'a}) + \sum_{a<a'} \sum m_a m_{a'} (\Phi^\phi_{aa'} + \sum_c m_c \psi_{caa'})$$
$$+ \sum_n \sum_c m_n m_c \lambda_{nc} + \sum_n \sum_a m_n m_a \lambda_{na} + \sum_{n<n'} \sum m_n m_{n'} \lambda_{nn'} + \cdots] \tag{31}$$

$$ln\gamma_M = z_M^2 F + \sum_a m_a (2B_{Ma} + Z\, C_{Ma}) + \sum_c m_c (2\Phi_{Mc} + \sum_a m_a \psi_{Mca})$$
$$+ \sum_{a<a'} \sum m_a m_{a'} \psi_{Maa'} + |z_M| \sum_c \sum_a m_c m_a C_{ca} + 2\sum_n m_n \lambda_{nM} + \cdots \tag{32}$$

$$ln\gamma_X = z_X^2 F + \sum_c m_c (2B_{cX} + Z\, C_{cX}) + \sum_a m_a (2\Phi_{Xa} + \sum_c m_c \psi_{cXa})$$
$$+ \sum_{c<c'} \sum m_c m_{c'} \psi_{cc'X} + |z_X| \sum_c \sum_a m_c m_a C_{ca} + 2 \sum_n m_n \lambda_{nX} + \cdots \tag{33}$$

The third virial terms for neutrals are omitted in Equations (31)-(33).

The quantity F includes the Debye-Hückel term and other terms as follows:

$$F = -A_\phi [I^{\frac{1}{2}}/(1 + bI^{\frac{1}{2}}) + (2/b)\, ln(1 + bI^{\frac{1}{2}})] + \sum_c \sum_a m_c m_a B'_{ca}$$
$$+ \sum_{c<c'} \sum m_c m_{c'} \Phi'_{cc'} + \sum_{a<a'} \sum m_a m_{a'} \Phi'_{aa'} \quad . \tag{34}$$

Also, Φ' is the ionic strength derivative of Φ, and

$$Z = \sum_i m_i |z_i| \quad ; \tag{35}$$

$$\Phi^\phi_{cc'} = \Phi_{cc'} + I\Phi'_{cc'} \quad . \tag{36}$$

It should be remembered that single ion activity coefficients are not measurable by ordinary thermodynamic methods because of space charge limitations. Also in the transformation from Equation (2) to (32) and (33) certain terms in λ's and μ's remain which cancel for any neutral combination of ions; see Pitzer (1979) for more details. Thus there is no pretense that Equations (32) and (33) yield absolute values of single-ion activity coefficients; rather these are practical values for use in all practical thermodynamic calculations. For complex mixed electrolytes, the use of the single ion activity coefficients is much more convenient than the use of mean activity coefficients and electrically neutral differences of activity coefficients, although the final results are identical.

Since like charged ions repel one another, we expect their short-range interactions to be small and that $\lambda_{cc'}$ and $\lambda_{cc'}$, etc. are all small. We further note that $\Phi_{cc'}$ and $\psi_{cc'a}$ are differences between these small quantitites; hence, they should certainly be small. Indeed both Bronsted (1922) and Guggenheim (1935) neglected these terms completely. We do not neglect these quantities, but we do find them to be small in most cases. There is an exception, noted above, where the long-range electrical forces yield a term that appears in $\Phi_{cc'}$ in this formulation. It appears only for unsymmetrical mixing, i.e., where the charges on c and c' (or a and a') differ. This term is given by theory (Pitzer, 1975, 1983). The complete expressions for Φ_{ij} are:

$$\Phi_{ij} = \theta_{ij} + {}^{E}\theta_{ij}(I) \quad , \tag{37}$$

$$\Phi'_{ij} = {}^{E}\theta'_{ij}(I) \quad , \tag{38}$$

$$\Phi^{\phi}_{ij} = \theta_{ij} + {}^{E}\theta_{ij}(I) + I\,{}^{E}\theta'_{ij}(I) \quad , \tag{39}$$

where ${}^{E}\theta(I)$ and ${}^{E}\theta'(I)$ account for these electrostatic unsymmetrical mixing effects and depend only on the charges of the ions i and j, the total ionic strength, and on the solvent properties ε and d_w (hence, on the temperature and pressure). The theory and equations for calculating these terms are given in Appendix A, which also describes a method of numerical calculation by Chebyshev approximations devised by Harvie (1981). The remaining term θ_{ij} arising from short-range forces is taken as a constant for any particular c,c' or a,a' at a given T and P. Its ionic strength dependence is very small and is usually neglected as we have done here, but this effect may be significant in some cases as shown by recent calculations of heats of mixing by Phutela and Pitzer (1986a).

Thus starting from theoretical considerations, the ion-interaction model gives an expression for the activity and osmotic coefficients of electrolyte mixtures in terms of the empirical parameters, $\beta^{(0)}_{MX}$, $\beta^{(1)}_{MX}$, $\beta^{(2)}_{MX}$, C^{ϕ}_{MX}, θ_{ij}, and ψ_{ijk}. Provided that their temperature and pressure dependencies are known, these coefficients permit the calculation of solubilities in binary, ternary and more complex mixtures at different temperatures and pressures.

It is important to recognize that the same equation for total excess Gibbs energy yields other thermodynamic functions such as excess volumes, enthalpies, entropies, and heat capacities by appropriate differentiation. These other functions are directly obtainable from experiment and their evaluation yields accurate data on the pressure or temperature-dependencies of the ion-interaction coefficients. Also the entropy or enthalpy of the fluid is required for analysis of some important processes, and it can be calculated from the temperature derivative of the Gibbs energy. Since these relationships are standard thermodynamics, we will not burden this paper with the detailed equations for mixtures. The equations for pure electrolytes were given by Pitzer et al. (1984).

There is, therefore, a wide array of experimental data from which the ion-interaction parameters and their temperature functions can be determined. This includes measurements of

1) freezing point depression	*6) enthalpy of dilution
*2) boiling point elevation	7) enthalpy of mixing
*3) vapor pressure	*8) enthalpy of solution
*4) isopiestic concentrations	*9) heat capacity
5) E.M.F.	*10) solubility

Those marked with an asterisk are those that so far have been most useful at temperatures greater than 100°C. Enthalpies of dilution measured at 25°C are also important for their relation to the temperature dependency of the Gibbs energy.

<u>Neutral solutes</u>

The situation is much simpler for uncharged solute species, and there is no need to rearrange the terms in λ_{ij} and μ_{ijk} in the basic Equations (1), (2), and (3). The terms for neutral species were included in Equations (32) and (33) for the activity coefficients of ions. The corresponding equation for the activity coefficient of a neutral species is

$$ln\gamma_N = 2(\sum_c m_c\lambda_{Nc} + \sum_a m_a\lambda_{Na} + \sum_n m_n\lambda_{N,n}) \qquad (40)$$

Third virial terms from Equation (2) can be added to Equation (40), if needed. Even for neutral molecules interacting with ions, the forces are short ranged, and there is no need to modify the λ's and μ's. Electrical neutrality limits the evaluation of λ's and μ's to electrically neutral sums and differences, but it does not seem worthwhile to define new quantities. Since Setchenow in 1892, the departure of neutral-species activity coefficients from unity has traditionally been described in terms equivalent to the λ's of these equations. An extensive review of neutral solutes in aqueous salt solutions was presented by Long and McDevit (1952).

Empirically there is no evidence for an ionic-strength dependence for the λ's for neutral species. Thus all λ'_{ij} terms in Equations (2) and (3) can be omitted for both neutral-neutral and neutral-ion interactions. There is some theoretical basis for a small ionic strength effect for interactions of ions with neutrals having large dipole moments (or higher-order electric moments); this is described in Appendix A. But this effect is so small that there is no reason to complicate the equations at present.

Also the question may be asked whether the dielectric constant should be that of a mixed solvent, including neutral solute species, instead of the value for water. It is possible to set up equations for mixed solvents, and this is necessary if the solvent composition varies over a wide range, e.g., from pure water to pure methanol. But the present equations assume a pure solvent, and its dielectric constant must be used. The effect of neutral-molecule solutes on interionic effects via changes in dielectric constant are included along with other effects in the second and third virial coefficients including a possible ionic-strength dependence of the ion-neutral second virial coefficient. Such an ionic-strength dependence has not been detected up to the present, but the possibility should be kept in mind.

Silica dissolves in water to form a neutral species, presumably $Si(OH)_4$; this is an interesting example to discuss briefly. Silica solubility, both in pure water and in salt solutions, has been measured by various investigators including Chen and Marshall (1981).

For the three-component system SiO_2-salt-H_2O with salt molality m, they summarized their results with the equation

$$\log \gamma = Dm + Em^2 \quad , \qquad (41a)$$

where γ is the activity coefficient of the silica. If S represents silica and the salt MX has ν_M and ν_X ions, respectively, our Equation (2) reduces in this case to

$$ln\gamma_S = 2m(\nu_M\lambda_{SM} + \nu_X\nu_{SX}) + 3m^2(\nu_M^2\mu_{SMM} + 2\nu_M\nu_X\mu_{SMX} + \nu_X^2\mu_{SXX}) \quad . \qquad (41b)$$

Evidently

$$2.303D = 2(\nu_M\lambda_{SM} + \nu_X\lambda_{SX}) \text{ and } 2.303E = 3(\nu_M^2\mu_{SMM} + 2\nu_M\nu_X\mu_{SMX} + \nu_X^2\mu_{SXX}) \quad .$$

Chen and Marshall fit their equation to their data both with and without the term in E and then give equations for the parameters over the range of temperature 25-300°C. Even for solutions extending to as high molality as 7, the effect of the E term was rather small for the salts NaCl, Na_2SO_4, $MgCl_2$, and $MgSO_4$.

In an accompanying paper Marshall and Chen (1981) restated their results (with E = 0) in terms of ion molalities. Since all of the individual ion parameters cannot be evaluated, they set the parameter for Na^+ to zero. Their parameter D_i now becomes just (2/2.303) λ_{Si} in our Equation (2). Marshall and Chen also made solubility measurements for silica in mixed salt solutions and verified that the ionic effects were additive as implied by Equation (2).

Association equilibria

Up to this point we have assumed that the selection of solute species was unambiguous and that electrical neutrality was the only supplementary relationship between solute molalities. But there may be association equilibria such as $H^+ + HCO_3^- = CO_2(aq) + H_2O$ which relate one solute molality to other molalities. The chemical thermodynamics of each such equilibrium is straightforward with an equilibrium constant relationship involving molalities and activity coefficients. For the carbonic acid case this is

$$K_{assoc.} = a_{H_2O} m_{CO_2} \gamma_{CO_2} \; / \; m_{H^+} m_{HCO_3^-} \gamma_{H^+} \gamma_{HCO_3^-} \quad . \tag{42}$$

Each such equilibrium adds one or more relationships between the molalities and an equilibrium-constant equation all of which must be satisfied simultaneously in the complete solution of the problem.

For carbonates and many other cases the association constants are large (i.e., the dissociation constants are very small) and there is no question about the need to recognize the associated species. Examples treated using the present model include Na^+ - HCO_3^- - CO_3^{2-} - Cl^- - CO_2 - H_2O (Peiper and Pitzer, 1982), H^+ - HSO_4^- - SO_4^{2-} - H_2O (Pitzer et al., 1977), and H^+ - K^+ - $H_2PO_4^-$ - H_3PO_4 - H_2O (Pitzer and Silvester, 1976). The last example is interesting in that it models the phosphoric acid to such high molality that a third virial coefficient for triple interaction of the neutral species is required.

In some cases the association, although significant, is not strong, and the fraction in the associated form is never large. In the dilute range the degree of association increases with concentration. But at higher concentration, the activity coefficients of the ions decrease, and the degree of association levels off and may even decrease. This last effect is particularly strong for multiple charged ions such as the divalent metal sulfates. In the case of the M^{2+} - SO_4^{2-} solutions it was found (Pitzer and Mayorga, 1974) that the associated species could be omitted provided an additional ionic-strength dependent term with a large exponent α_2 was added to the second virial coefficient, see Equation (13). The coefficient in this term $\beta_{MX}^{(2)}$ is negative and, in the limit of low molality, is related to the association constant K by $\beta^{(2)} = -K/2$. Also the exponent α_2 is related to the Debye-Hückel parameter A_ϕ. Indeed, the equation without the MSO_4 neutral species and with the $\beta^{(2)}$ term represented the properties to high molality without difficulty. In contrast, association treatments with the simple inclusion of the ion-pair association equilibrium constant are not successful at high molality; further terms are required - either virial coefficients involving the MSO_4 species or association equilibria to triple or quadruple ions. Thus the treatment with the $\beta^{(2)}$ term and without the association equilibrium has many advantages.

With increase in temperature the dielectric constant of water decreases rather rapidly and one expects ion pairing to become stronger. Archer and Wood (1985) found this to be the case in a general treatment of $MgSO_4$(aq). They included not only the neutral ion pair but also the triplets M_2X^+, MX_2^- and a sextuplet M_3X_3 with association constants at 25°C of 126.4, 557.3, and 3.813×10^6, respectively, and enthalpies of association of approximately 6, 6, and 27 $kJ \cdot mol^{-1}$. They also adjusted heat capacities of association for each reaction as well as a simple pattern of temperature-dependent second and third virial coefficients to account for repulsive interactions and obtained good agreement with the available data up to 3 $mol \cdot kg^{-1}$ and 150°C. Soon thereafter Phutela and Pitzer (1986b) presented high-temperature heat capacity measurements for $MgSO_4$(aq) and a comprehensive treatment of that system without association equilibria but with temperature-dependent $\beta^{(0)}$, $\beta^{(1)}$, $\beta^{(2)}$ and C^{ϕ}. They found that $-\beta^{(2)}$ increased with temperature, as expected. They also investigated the effect of change of the exponent α_2 with temperature proportionally to A_{ϕ}. The temperature-dependent α_2 gave better agreement below 0.1 $mol \cdot kg^{-1}$, but there was no difference above 0.1 $mol \cdot kg^{-1}$. Thus for mixed electrolytes at ionic strength above 0.4 $mol \cdot kg^{-1}$ the simple treatment with constant α_2 is fully satisfactory. The general agreement for various properties for the two treatments was comparable; the treatment without association equilibria gave agreement for the osmotic coefficient at 110°C to higher molality (5 $mol \cdot kg^{-1}$) and better agreement at 140°C. The association treatment gives a better fit to the heats of dilution at high temperatures in the dilute range below 0.03 $mol \cdot kg^{-1}$. The more recent heat capacity measurements were not available at the time of Archer and Wood's treatment; as expected, the Phutela and Pitzer treatment fits these data much more accurately.

In estimating the need to introduce associated species, these results for $MgSO_4$ give the best guide for 2-2 electrolytes. For less highly charged ions, the ion pairs must be recognized for somewhat smaller values of the association constant because the activity coefficient for the ions decreases less rapidly with increase of molality. The case of aqueous HCl was investigated very thoroughly by Holmes et al. (1987). They found no need to include a $\beta^{(2)}$ term below 250°C. For higher temperatures up to 375°C, they obtained good agreement with an equation including a $\beta^{(2)}$ term. In the range of their data $-\beta^{(2)}$ is as large as 32 corresponding to an association constant of 64.

At room temperature most 1-1 electrolytes are either unambiguously associated, such as acetic acid or ammonia, or are clearly strong electrolytes where the present equations are adequate without the $\beta^{(2)}$ term. The situation for 2-1 or 1-2 electrolytes at 25°C was studied carefully by Harvie et al. (1984); they concluded that association should be recognized for cases with association constants greater than 20 (or dissociation constants less than 0.05). This general topic is also discussed by Weare in the following chapter.

TEMPERATURE AND PRESSURE EFFECTS ON STANDARD-STATE PROPERTIES

For the calculation of equilibria one requires the chemical potentials of all substances at the temperature and pressure of interest. We first consider the change of the chemical potential from the reference temperature, normally 25°C, to another temperature. The entropy and heat capacity determine this difference as follows:

$$(\partial \mu_i^{\circ}/\partial T)_P = -S_i^{\circ} \quad , \tag{43}$$

or

$$\mu_{i,T}^{\circ} - \mu_{i,T_r}^{\circ} = -(T-T_r)S_{i,T_r}^{\circ} + \int_{T_r}^{T} C_{p,i}^{\circ} dT' - T \int_{T_r}^{T} (C_{p,i}^{\circ}/T')dT' \quad . \tag{44}$$

Here S_i° represents the standard absolute entropy of component _i_, and $C_{p,i}^\circ$ is the standard state heat capacity (at constant pressure) which itself is expressed as a function of temperature. T_r and T represent the reference temperature and temperature of interest, respectively. An analogous but somewhat more complex expression for the temperature dependence of the chemical potential can be written in terms of the enthalpy and heat capacity.

For aqueous species the heat capacities become large as the critical temperature of water is approached; for ions the heat capacity becomes negative but for neutral species it can be positive. At the critical pressure this trend becomes a divergence to plus or minus infinity at the critical point.

For solids the heat capacities remain moderate in magnitude except at certain solid phase transitions. In the case of transitions, the integration is divided into ranges and the entropy change for the transition is included explicitly.

The effect of pressure on the chemical potential is given by the molar or partial molar volume and can be determined from density data. At saturation pressure these effects are usually small, particularly for the solids, but can be significant at higher pressures (Rogers and Pitzer, 1982; Pitzer, 1986).

The basic equation for pressure dependency is

$$\mu_{i,P}^\circ - \mu_{i,P_r}^\circ = \int_{P_r}^{P} \bar{V}_i^\circ dP \quad , \tag{45}$$

where $\bar{V}_i^\circ$ is the partial molar volume of solute species i in its standard state at the temperature of interest. The data base for volumes is discussed below.

DATA BASE

Standard state values for 25°C

There are many sources of standard-state chemical potentials (molar Gibbs energies) of formation, entropies, heat capacities, and volumes at 25°C (298.15 K) and 1 bar for substances of geological interest. Wagman et al. (1982) give very extensive tables while Robie et al. (1978) consider geological interest in their selection. For solubility calculations, the accuracy of particular differences is especially important and the tables of Harvie and Weare (1980), of Harvie et al. (1984), and of Weare in the following chapter have been prepared to meet this need. Table 1 contains, for convenience, an abbreviated list of chemical potentials and enthalpies of formation, and of entropies and parameters for heat capacity equations. The sources of the entropies and heat capacities will be discussed below; the chemical potentials are from Harvie et al. (1984).

Standard-state enthalpies, entropies, heat capacities, and volumes

The conversion of standard-state chemical potentials to other temperatures requires entropy values for the reference temperature and heat capacities as a function of temperature. The entropy values are

Table 1: The Chemical Potentials, Enthalpies of Formation, and Entropies at 298.15 K of the Species and Minerals of the Na-K-Mg-Cl-SO_4-OH-H_2O System and the Temperature Functions of the Heat Capacity of the Solids

Substance	Formula	$-\mu_f^\circ/RT$	$-\Delta_f H^\circ/RT$	S°/R	$C_p/R = a + bT + cT^{-2}$			
					a	$10^3 b$	$10^{-5} c$	T(K)range
Water	$H_2O(\ell)$	95.6635	115.304	8.409				
Hydroxide ion	$OH^-(aq)$	63.452	92.780	-1.293				
Chloride ion	$Cl^-(aq)$	52.955	67.432	6.778				
Sulfate ion	$SO_4^{2-}(aq)$	300.386	366.800	2.42				
Magnesium ion	$Mg^{2+}(aq)$	183.468	188.329	-16.64				
Calcium ion	$Ca^{2+}(aq)$	233.30	218.98	-6.39				
Sodium ion	$Na^{2+}(aq)$	105.651	96.865	7.096				
Potassium ion	$K^+(aq)$	113.957	101.81	12.33				
Arcanite	$K_2SO_4(c)$	532.39	580.01	21.12	14.48	11.98	-2.14_4	298 - 856
Bischofite	$MgCl_2 \cdot 6H_2O(c)$	853.1	1008.11	44.03	29.08	29.56	-	298 - 385
Epsomite	$MgSO_4 \cdot 7H_2O(c)$	1157.74	(1366.27)	44.79	11.8	118	-	~273 - ~473
Halite	NaCl(c)	154.99	165.88	8.676	5.525	1.96_3	-	298 - 1073
Hexahydrite	$MgSO_4 \cdot 6H_2O(c)$	1061.37	(1244.79)	41.87	10.9	104	-	~273 - ~473
Kieserite	$MgSO_4 \cdot H_2O(c)$	579.18	649.34	(14.99)	6.89	31.05	-	~273 - ~473
Leonhardtite	$MgSO_4 \cdot 4H_2O(c)$	868.55^4	1007.13	(30.64)	9.39	74.8	-	~273 - ~473
Magnesium chloride	$MgCl_2(c)$	238.74	258.71	10.78	9.511	0.714_6	-1.03_7	298 - 987
Magnesium chloride hydrate	$MgCl_2 \cdot H_2O(c)$	347.66	389.94	16.505	10.95	9.788	-	298 - 650
Magenesium chloride dihydrate	$MgCl_2 \cdot 2H_2O(c)$	451.06	516.24	21.64	15.05	13.74	-	298 - 500
Magnesium chloride tetrahydrate	$MgCl_2 \cdot 4H_2O(c)$	654.93	766.06	31.75	22.56	21.65	-	298 - 450
Magnesium sulfate	$MgSO_4(c)$	472.26	518.33	11.02	6.71	16.30	-	298 - 700
Mirabilite	$Na_2SO_4 \cdot 10H_2O(c)$	1471.15	1475.75	71.21	10.65	196.0	-	~200 - ~450
Pentahydrite	$MgSO_4 \cdot 5H_2O(c)$	965.13	-	(35.7)	10.2	89.3	-	~273 - ~473
Sylvite	KCl(c)	164.84	176.034	9.934	5.575	2.011	-	~298 - ~700
Thenardite	$Na_2SO_4(c)$	512.39	559.55	17.99	13.16	13.70	-1.666	270 - 700

generally available from the tables of Wagman et al. (1982) or Robie et al. (1978). In some cases the entropy was obtained indirectly from the enthalpy and the Gibbs energy (chemical potential), and in such cases one should revise the entropy to be consistent with any change in the Gibbs energy. The properties of solids do not ordinarily depend appreciably on crystal size, but there are cases where extremely small crystals are obtained with significantly larger molar entropies and chemical potentials. Kieserite ($MgSO_4 \cdot H_2O$) is an example where the data of Ko and Daut (1979) for the enthalpy and Frost et al. (1957) for the heat capacity and entropy show significant differences with crystal size.

In Table 1 parameters are given for the following equation for the heat capacity of a solid:

$$C_p/R = a + bT + cT^{-2} \quad . \tag{46}$$

The range of validity is also indicated. These heat capacity parameters were taken mostly from Kelley (1960). The coefficients given by Kelley (1960) for KCl(s) are incorrect, and the values listed are from Holmes and Mesmer (1983). The values for $Na_2SO_4 \cdot 10\ H_2O(s)$ and $Na_2SO_4(s)$ were fitted by Pabalan and Pitzer (1987b) to data from Brodale and Giauque (1958, 1972), while those for $MgSO_4(s)$ were derived from JANAF data (Stull and Prophet, 1971). The C_p coefficients for $MgSO_4 \cdot H_2O(s)$ and $MgSO_4 \cdot 6H_2O(s)$ were derived from linear fits to the low temperature data of Frost et al. (1957) and Cox et al. (1955), respectively.

Heat capacity data for $MgSO_4 \cdot 4H_2O(s)$, $MgSO_4 \cdot 5H_2O(s)$, and $MgSO_4 \cdot 7H_2O(s)$ are not available. However, the contribution of each water molecule to the entropy or heat capacity of a hydrated solid is expected to be about the same. Pabalan and Pitzer (1987b) present isothermal plots of the measured entropies and heat capacities of $MgCl_2 \cdot nH_2O(s)$ and $MgSO_4 \cdot nH_2O(s)$ versus the number of hydration waters which show a linear trend. Thus the unknown C_p temperature functions for $MgSO_4 \cdot 4H_2O(s)$, $MgSO_4 \cdot 5H_2O(s)$, and $MgSO_4 \cdot 7H_2O(s)$ were estimated on this basis.

The S°/R values in Table 1 are mostly from Wagman et al. (1982). Values in parentheses were calculated from $\mu_f°/RT$ and $\Delta H_f°/RT$. For pentahydrite, whose entropy is not known independently, its value was estimated from the entropy values of the other $MgSO_4$ hydrous salts. With the exception of $MgSO_4 \cdot H_2O(s)$, values of $\Delta H_f°/RT$ values given in the table were calculated from $\mu_f°/RT$ and S°/R or were taken from Wagman et al. (1982) when the value of S°/R is in parenthesis. For $MgSO_4 \cdot H_2O(s)$, the enthalpy of formation at 25°C was taken from Ko and Daut (1979).

Numerical equations for the aqueous standard state heat capacities as a function of temperature are given along with the solution properties for several solutes in Appendix B. Table 2 includes several other solutes for which this information is available (sometimes over a more limited range) together with appropriate references.

Table 2: Binary Electrolyte Solutions with Available High Temperature Data

	Range of T and P		Ref.
	T/°C	P/kbar	
	1:1 Electrolytes		
HCl	0 - 375	0 - 0.4	a,b
LiCl	0 - 250	-	c
NaCl	0 - 300	0 - 1.0	d
NaI	25 - 100	0 - 0.1	a
NaOH	0 - 350	0 - 0.4	e
KCl	0 - 250	-	c
CsF	25 - 100	0 - 0.1	a
CsCl	0 - 250	0 - 0.1	a,c
CsI	25 - 100	0 - 0.1	a
	1:2 or 2:1 Electrolytes		
Li_2SO_4	0 - 225	-	f
Na_2SO_4	0 - 225	0 - 0.1	f,j
K_2SO_4	0 - 225	-	f
Cs_2SO_4	0 - 225	-	f
$MgCl_2$	25 - 200	0 - 0.1	g,h
$CaCl_2$	25 - 250	0 - 0.1	k,ℓ
$SrCl_2$	25 - 200	0 - 0.1	g
	2:2 Electrolytes		
$MgSO_4$	25 - 200	0 - 0.1	i,j

[a] Saluja et al. (1986).
[b] Holmes et al. (1987).
[c] Holmes and Mesmer (1983).
[d] Pitzer et al. (1984).
[e] Pabalan and Pitzer (1987a).
[f] Holmes and Mesmer (1986).
[g] Phutela et al. (1987).
[h] Holmes et al. (1978)
[i] Phutela and Pitzer (1986b).
[j] Phutela and Pitzer (1986c).
[k] Ananthaswamy and Atkinson (1985).
[ℓ] Møller (submitted).

Table 3: Single Electrolyte Solution Parameter Values for 25°C[a]

Cation	Anion	$\beta^{(0)}_{ca}$	$\beta^{(1)}_{ca}$	$\beta^{(2)}_{ca}$	C^{ϕ}_{ca}
H	Cl	0.1775	0.2945	-	0.0008
H	SO_4	0.0298	-	-	0.0438
H	HSO_4	0.2065	0.5556	-	-
Li	Cl	0.1494	0.3074	-	0.00359
Na	Cl	0.0765	0.2644	-	0.00127
Na	Br	0.0973	0.2791	-	0.00116
Na	SO_4	0.01958	1.113	-	0.00497
Na	HSO_4	0.0454	0.398	-	-
Na	OH	0.0864	0.253	-	0.0044
Na	HCO_3	0.0277	0.0411	-	-
Na	CO_3	0.0399	1.389	-	0.0044
Na	SiO_4H_3[a]	0.043	0.024	-	-
K	Cl	0.04835	0.2122	-	-0.0084
K	Br	0.0569	0.2212	-	-0.00180
K	SO_4	0.04995	0.7793	-	-
K	HSO_4	-0.0003	0.1735	-	-
K	OH	0.1298	0.320	-	0.0041
K	HCO_3	0.296	-0.013	-	-0.008
K	CO_3	0.1488	1.043	-	-0.0015
Cs	Cl	0.0300	0.0558	-	0.00038
NH_4	Cl	0.0522	0.1918	-	-0.00301
Mg	Cl	0.35235	1.6815	-	0.00519
Mg	Br	0.43268	1.75275	-	0.00312
Mg	SO_4	0.2210	3.343[b]	-37.23[c]	0.0025
Mg	HSO_4	0.4746	1.729	-	-
Mg	OH	-	-	-	-
Mg	HCO_3	0.329	0.6072	-	-
MgOH	Cl	-0.010	1.658	-	-
Ca	Cl	0.3159	1.614	-	-0.00034
Ca	Br	0.3816	1.61325	-	-0.00257
Ca	SO_4	0.20	3.1973[b]	-54.24[c]	-
Ca	HSO_4	0.2145	2.53	-	-
Ca	OH	-0.1747	-0.2303	-5.72[c]	-
Ca	HCO_3	0.004	2.977	-	-
UO_2	Cl[b]	0.4274	1.644	-	-0.03686
UO_2	SO_4	0.322	1.827[b]	(-40)[c]	-0.0176
Al	Cl	0.6993	5.845	-	0.00273
La	Cl	0.5889	5.600	-	-0.0238
Th	Cl	1.0138	13.331	-	-0.1034

[a] From Pitzer (1979) or Harvie et al. (1984) except $Na(SiO_4H_3)$ from Hershey and Millero (1986); note that $C_{ca} = C^{\phi}_{ca}/2|z_c z_a|$.

[b] $\alpha_1 = 1.4\ kg\cdot mol^{-1}$ (otherwise $\alpha_1 = 2.0\ kg\cdot mol^{-1}$).

[c] $\alpha_2 = 12\ kg\cdot mol^{-1}$.

The volumetric properties of many aqueous electrolytes have been measured to high pressure at 25°C or over the 0-50°C range. This information is of oceanographic interest. Above 50°C the volumetric information is much more limited. NaCl(aq) has been investigated over a wide range of temperature and pressure and general correlations are available (Rogers and Pitzer, 1982). Gates (1985) has made further measurements by an additional method. Information on aqueous HCl is now extensive (Holmes et al., 1987). The densities of a number of other aqueous systems have been measured at moderate pressure to 200°C and in additional cases to 100°C. Equations have been fitted to these data yielding standard state volumes; the particular solutes are listed in Table 2 which includes the pertinent references.

Since compressibilities of aqueous solutions decrease with increase in pressure, the assumption of a constant molar volume in Equation (45) usually overestimates the effect of pressure on the chemical potential. In any case, there is no divergence of the partial molar volume at high pressure as there is with temperature for the heat capacity as the critical point is approached. Thus one can make reasonable estimates of pressure effects to higher pressures than those listed in Table 2.

There have been several general correlations of high-temperature volumetric data with use of approximate theory. A very recent study is by Tanger and Helgeson (1987), while an earlier paper was presented by Zarembo and L'vov (1982); these papers give references to additional experimental measurements. Thus it is possible to estimate the change of chemical potential with pressure even in cases where there are no experimental measurements for a particular solute.

Pure electrolyte parameters for 25°C

The activity and osmotic coefficients at 25°C of many electrolytes as well as neutral solutes in water were measured quite accurately many years ago. Robinson and Stokes with their collaborators made many of these measurements, and their book (Robinson and Stokes, 1965) includes excellent tables of selected values over wide ranges of molality. The parameters for the present equations were derived by least squares regression for a very wide range of electrolytes by Pitzer and Mayorga (1973, 1974) using more recent measurements as well as the selected values of Robinson and Stokes. Subsequently, Pitzer et al. (1978) obtained values for the parameters for the rare earth chlorides, nitrates, and perchlorates from the extensive measurements of Spedding and associates. Other research has contributed new or improved values for particular solutes. Table 3 contains values selected for possible geological interest.

Very recently, Kodytek and Dolejs (1986) refitted the rare-earth-salt data including a $\beta^{(2)}$ term. Their values of $\beta^{(2)}$ are positive; thus there is no indication of ion-pairing. But the extra flexibility does give a better fit for the low molalities. Their treatment emphasized the dilute range, and the older values may be preferable at high molalities.

Pure electrolyte parameters for high temperature

In an earlier section the various measurements were listed from which the pure electrolyte parameters can be derived. The isopiestic method, which is so useful at 25°C, has provided valuable data up to about 250°C. The reference solution is usually NaCl which has been thoroughly studied by other methods including the vapor pressure relative to that of pure water. Provided the heat of dilution as well as the excess Gibbs energy are known at 25°C, measurements of the heat capacity at higher temperatures provide the required information for both the solution parameters and the standard state entropy and chemical potential. In practice two or more methods are usually used, and the parameters determined from a least-squares optimization of the parameters to fit all of the accurate measurements.

There is now an extensive array of data for the most important aqueous solutes extending upward in temperature as indicated in Table 2. But in many cases further work would be welcome to extend the temperature range further, to better account for the effect of pressure, and to increase accuracy. Appendix B gives a brief listing of the numerical parameters in the temperature dependency expressions for both standard-state heat capacities and nonideality properties for several pure electrolytes and includes remarks concerning the Debye-Hückel parameter.

For NaCl(aq), there is an extensive array of thermodynamic data as reported by Pitzer, Peiper, and Busey (1984). Their evaluation of these data yield a complete set of parameters valid in the region 0-300°C and saturation pressure to 1 kbar. Solubility data were not used in the general regression; hence a comparison of calculated solubilities with experimentally measured values is a check on a prediction. As shown in Figure 1, there is excellent agreement between calculated and experimentally determined values with a maximum deviation of 1.5% at 275°C. This general equation is also consistent with the very recent heat capacity measurements of Gates et al. (1987); most values agree within the uncertainty stated for the equation. For KCl(aq), Holmes and Mesmer (1983a) combined their isopiestic vapor pressure measurements to 250°C with other literature data to yield a thermodynamically consistent set of parameters for KCI solutions.

In the case of $MgCl_2$ there is no fully satisfactory general treatment. The ion-interaction parameters of de Lima and Pitzer (1983) were based on isopiestic measurements of Holmes et al. (1978) at high temperature and Rard and Miller (1981) at 25°C. Pabalan and Pitzer (1987b) adjusted the trend of C^{ϕ} at the higher temperatures to better fit the solubility of the various hydrates of $MgCl_2$ which rises to 14 $mol \cdot kg^{-1}$ at 200°C. This adjustment affects the agreement with the osmotic coefficients of Holmes et al., which extend only to 3.5 $mol \cdot kg^{-1}$, but the agreement is still good with standard deviations less than 0.003. The standard state heat capacity is taken from Phutela et al. (1987). This investigation also reported ion-interaction parameters, but their validity is limited to about 2 $mol \cdot kg^{-1}$; hence, they are not useful for solubility calculations. The final comparison of calculated and observed solubilities is shown on Figure 2.

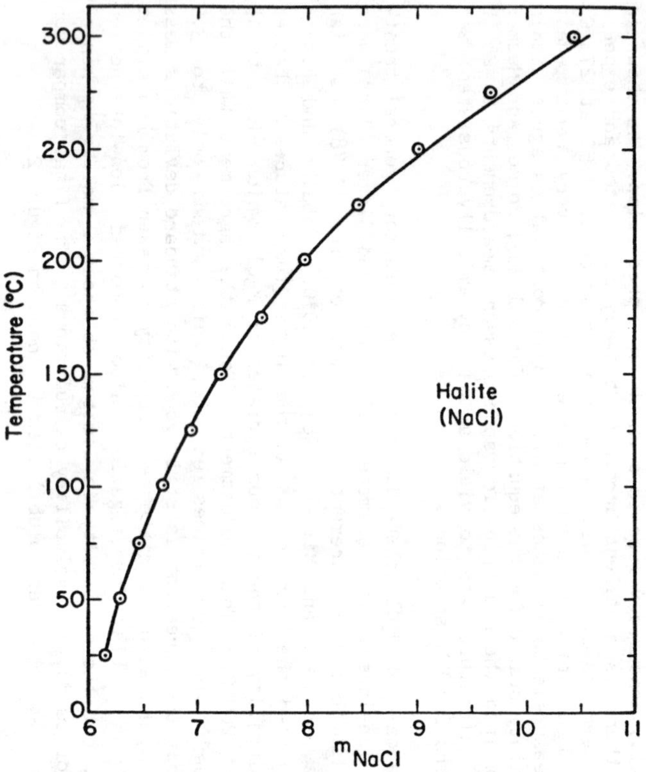

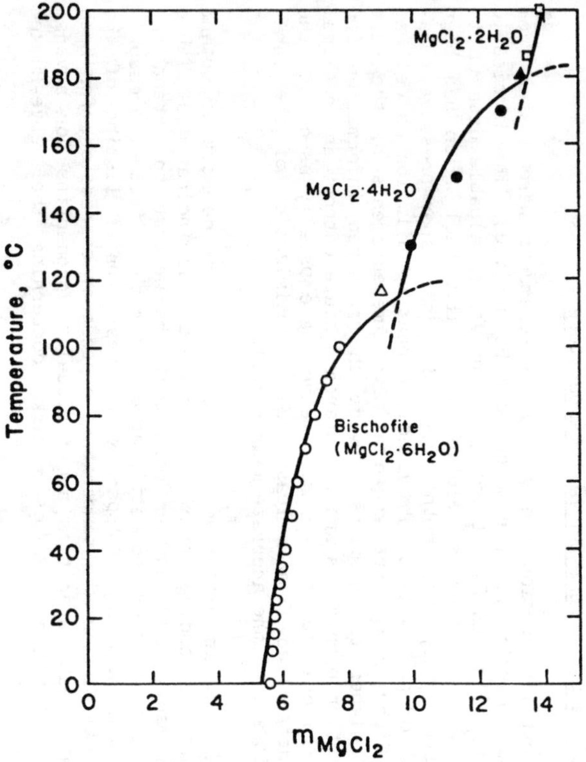

Figure 1 (left). Calculated halite (NaCl) solubilities in the binary $NaCl\text{-}H_2O$ system compared to experiment. The solubility data from 75-300°C are from Liu and Lindsay (1972), and those below 75°C are from Linke and Seidell (1965).

Figure 2 (right). Calculated solubilities in the $MgCl_2\text{-}H_2O$ binary system compared to experimental data from Linke and Seidell (1965). The triangles represent experimental data at the triple points. The C_{MX} function above 100°C was adjusted to fit the solubilities.

For $CaCl_2$ Phutela and Pitzer (1983) gave equations valid to 200°C and 4.3 $mol \cdot kg^{-1}$ based on osmotic-coefficient and heat-of-dilution data. Above 5 $mol \cdot kg^{-1}$ the usual form of equation terminating with the third virial coefficient is inadequate, and Ananthaswamy and Atkinson (1985) used a form extended through the sixth virial coefficient. Their treatment is valid to 100°C. Very recently Møller (submitted) has considered vapor pressure data for $CaCl_2(aq)$ above 200°C and derived equations for the ion-interaction parameters valid to 250°C. Møller (submitted) also considered $CaSO_4$ and reported appropriate parameters. Although it is not necessary to recognize the neutral $CaSO_4$ species at 25°C, Møller found that the equilibrium forming this species should be included at higher temperature.

For $Na_2SO_4(aq)$ and $K_2SO_4(aq)$, one has the equations of Holmes and Mesmer (1986) which were fit to data up to 225°C including heat capacities of Na_2SO_4 to 200°C from Rogers and Pitzer (1981). Pabalan and Pitzer (1987b) made solubility comparisons based on these equations and found excellent agreement to 175°C for Na_2SO_4 but some deviation at higher temperature. For K_2SO_4 the comparison showed some differences in the 100-215°C interval, but the agreement was reasonably good over the entire range.

In the case of $MgSO_4(aq)$ solutions, a comprehensive regression of heat capacity, enthalpy, and osmotic coefficient data by Phutela and Pitzer (1986b) yielded parameters that are valid from 25-200°C. Figure 3 shows that calculated solubilities are in very good agreement with experimental data to 200°C.

Holmes et al. (1987) presented a comprehensive treatment for HCl(aq). They gave three sets of parameters for the present model. The first, valid to 523 K and 7 $mol \cdot kg^{-1}$, involves only the usual terms for a 1-1 electrolyte and is summarized in Appendix B. A second treatment adds a fourth virial coefficient and is valid to 16 $mol \cdot kg^{-1}$. For temperatures from 523 to 648 K and from 0 to 7 $mol \cdot kg^{-1}$, they include a $\beta^{(2)}$ term.

For NaOH(aq) the available volumetric and vapor pressure data to 10 $mol \cdot kg^{-1}$ were evaluated by Pabalan and Pitzer (1987a) who give equations for the virial coefficients extending to 350°C. Standard state heat capacities could not be obtained from these data, but this is not a deficiency for solubility calculations since the solubility of NaOH exceeds the range of validity of the solution model. These parameters are useful, however, for the calculation of the effect of NaOH on mixed solution properties including the solubility of other salts.

Various forms have been used to describe the temperature dependency of the ion-interaction coefficients. No attempt has been made to find a singular form to describe this dependency; the parameters obtained in the cited studies are listed in Appendix B for the solutes discussed above. There are equations covering less extensive temperature ranges for a few other solutes as noted in Table 2 where references are listed. The initial change with temperature above 25°C for an even wider range of solutes is given by Silvester and Pitzer (1978).

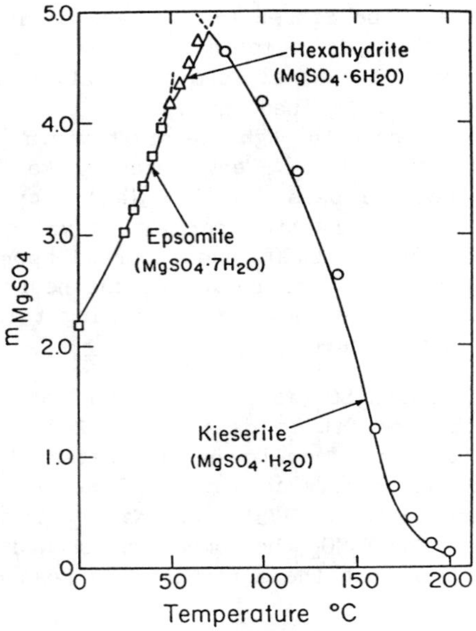

Figure 3. Calculated solubilities in the system $MgSO_4$-H_2O compared to experimental data from Linke and Seidell (1965).

Table 4: Mixing Parameters for 25°C

c°	c'	$\theta_{cc'}$	$\psi_{cc'Cl}$	$\psi_{cc'SO_4}$	$\psi_{cc'HSO_4}$	$\psi_{cc'OH}$	$\psi_{cc'HCO_3}$	$\psi_{cc'CO_3}$
Na	K	-0.012	-0.0018	-0.010	-	-	-0.003	0.003
Na	Ca	0.07	-0.007	-0.055	-	-	-	-
Na	Mg	0.07	-0.012	-0.015	-	-	-	-
Na	MgOH	-	-	-	-	-	-	-
Na	H	0.036	-0.004	-	-0.0129	-	-	-
K	Ca	0.032	-0.025	-	-	-	-	-
K	Mg	0.	-0.022	-0.048	-	-	-	-
K	MgOH	-	-	-	-	-	-	-
K	H	0.005	-0.011	0.0197	-0.0265	-	-	-
Ca	Mg	0.007	-0.012	0.024	-		-	-
Ca	MgOH	-	-	-	-	-	-	-
Ca	H	0.092	-0.015	-	-	-	-	-
Mg	MgOH	-	0.028	-	-	-	-	-
Mg	H	0.010	-0.011	-	-0.0178	-	-	-

a	a'	$\theta_{aa'}$	$\psi_{aa'Na}$	$\psi_{aa'K}$	$\psi_{aa'Ca}$	$\psi_{aa'Mg}$	$\psi_{aa'MgOH}$	$\psi_{aa'H}$
Cl	SO_4	0.030	0.000	-0.005	-0.002	-0.008	-	-
Cl	HSO_4	-0.006	-0.006	-	-	-	-	0.013
Cl	OH	-0.050	-0.006	-0.006	-0.025	-	-	-
Cl	HCO_3	0.003	-0.015	-	-	-0.096	-	-
Cl	CO_3	-0.002	0.0085	0.004	-	-	-	-
SO_4	HSO_4	-	-0.0094	-0.0677	-	-0.425	-	-
SO_4	OH	-0.013	-0.009	-0.050	-	-	-	-
SO_4	HCO_3	0.001	-0.005	-	-	-0.161	-	-
SO_4	CO_3	0.002	-0.005	-0.009	-	-	-	-
OH	CO_3	0.010	-0.017	-0.001	-	-	-	-
HCO_3	CO_3	-0.004	0.002	0.012	-	-	-	-

Mixing parameters

The mixing parameters θ_{ij} and ψ_{ijk} for many simple ions at 25°C were derived from the available measurements of activity and osmotic coefficients by Pitzer and Kim (1974). These parameters are best evaluated from data for common-ion mixtures. For example, the values of $\theta_{Na,K}$ and $\psi_{Na,K,Cl}$ may be determined from osmotic coefficient data for NaCl-KCl solutions. The θ_{ij} parameters are independent of the oppositely charged ions. Thus, all the data for Na^+ - K^+ mixing are considered simultaneously (NaCl-KCl, Na_2SO_4-K_2SO_4, $NaNO_3$-KNO_3, etc.) to determine a single value of $\theta_{Na,K}$ and values of $\psi_{Na,K,Cl}$, ψ_{Na,K,SO_4}, etc. Solubility measurements also contribute valuable information concerning these parameters, and they were not considered by Pitzer andKim. Harvie and Weare (1980) and Harvie et al. (1984) obtained additional values and recommended some changes, primarily of the third virial ψ_{ijk}, on the basis of solubility data. Since the effect of the term in ψ_{ijk} increases with the square of the molality, this quantity is best determined from measurements at the highest molalities, i.e., in saturated solutions where the solubility of a solid is measured.

Table 4 lists values of θ_{ij} and ψ_{ijk} for 25°C for various interionic interactions while Table 5 lists values of λ_{ij} for ion-neutral interactions on the basis of zero for the interaction with H^+. The values in Tables 4 and 5 are the same as those of Harvie et al. (1984), except for the parameters involving Cl-SO_4 interactions, where slightly different values from Downes and Pitzer (1976) and Pabalan and Pitzer (1987b) are listed.

In their 1974 work Pitzer and Kim obtained reasonable results without the higher-order electrostatic terms for cases of unsymmetrical 2-1 mixing of ions of one sign and a common ion of the opposite sign. But they concluded that the more extreme 3-1 mixing showed clear deviation, and this was explained by the higher-order terms (Pitzer, 1975). Subsequently, the case of $CaSO_4$ solubility in NaCl, with 2-1 mixing of both cations and anions, clearly required the higher-order terms (Harvie and Weare, 1980). The result is that two sets of values are often given for θ_{ij}: one to be used with the higher-order electrostatic functions ${}^E\theta_{ij}$ and ${}^E\theta'_{ij}$, and the other to be used without these terms. It now seems best always to include the ${}^E\theta$ and ${}^E\theta'$ terms for all cases of unsymmetrical mixing, and all values in Table 4 are for that basis. Once a modern computer has a subroutine for the calculation of these terms, the further complications are minimal. Thus, it is easier and less likely to lead to error if these terms are always included than if choices are made in each case to include or exclude them.

Both θ_{ij} and ψ_{ijk} undoubtedly vary with temperature, and there are heat of mixing data which give their temperature derivatives at 25°C (Phutela and Pitzer, 1986a). Until heat of mixing measurements become generally available at higher temperature, however, we must depend primarily on solubility data for the values of θ_{ij} and ψ_{ijk} at high temperatures. Pabalan and Pitzer (1987b) found that the mineral solubilities in many systems could be fitted with constant θ's at their 25°C values together with ψ's either constant or varying with temperature in

Table 5: Neutral-ion Parameter Values for 25°C

i	$\lambda_{CO_2,i}$	$\lambda_{CaCO_3,i}$	$\lambda_{MgCO_3,i}$
H	0.0	-	-
Na	.100	-	-
K	.051	-	-
Ca	.183	-	-
Mg	.183	-	-
MgOH	-	-	-
Cl	-.005	-	-
SO_4	.097	-	-
HSO_4	-.003	-	-
HCO_3	-	-	-
CO_3	-	-	-

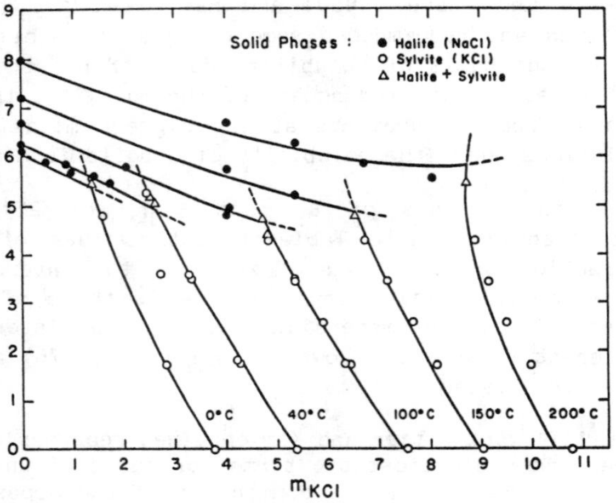

Figure 4 (above). Calculated solubilities in the ternary mixture $NaCl$-KCl-H_2O compared with experimental data taken from Linke and Seidell (1965). The intersections of isothermal curves represent calculated triple points. One parameter in the $\psi_{Na,K,Cl}$ function was adjusted to fit the solubilities above 25°C.

Figure 5 (below). Calculated solubilities in the $NaCl$-$MgCl_2$-H_2O system. Experimental data below 100°C are from Linke and Seidell (1965), while those above 100°C are from Akhumov and Vasil'ev (1932). The dashed curves are extrapolations of the solubilities of either $MgCl_2 \cdot nH_2O$ or NaCl into the supersaturated solution concentration of the other solid. The intersection of the curves denotes the calculated triple point. One parameter in the $\psi_{Na,Mg,Cl}$ function was adjusted to fit the solubilities above 25°C.

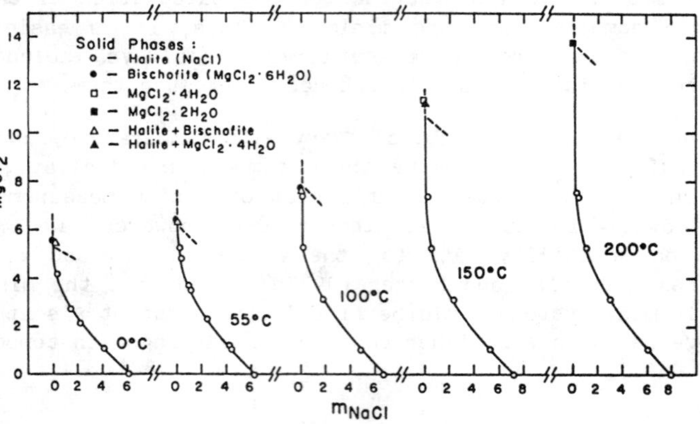

a simple manner. Figures 4 and 5 show the results for the systems NaCl-KCl-H_2O and NaCl-$MgCl_2$-H_2O, respectively. In each case the parameter giving the temperature dependence of $\psi_{Na,K,Cl}$ or $\psi_{Na,Mg,Cl}$ was determined to best fit the array of solubility measurements above 25°C. Møller (submitted) obtained excellent agreement for the solubility of NaCl in the NaCl-$CaCl_2$-H_2O system with a temperature-independent $\psi_{Na,Ca,Cl}$. Table 6 includes the θ's and ψ's valid at elevated temperatures for several systems of interest.

APPLICATIONS

Solubilities of solids

Since solubility calculations are discussed in detail by Weare in his chapter, only brief comments are included here. A multicomponent example at high temperature is shown in Figure 6. Here the solubilities of both NaCl and of KCl at fixed $MgCl_2$ molality are indicated for the four-component system NaCl-KCl-$MgCl_2$-H_2O. Since all parameters were evaluated in simpler systems, the curves on Figure 6 are independently determined predictions of the solubilities shown. While the agreement is not perfect, it approaches the accuracy of the solubility measurements. This calculation was first presented by Pabalan and Pitzer (1987b).

There are many calculations for 25°C of mineral solubilities in complex, concentrated brines. In addition to those of Harvie and Weare (1980) and Harvie et al. (1982, 1984), other notable examples include Krumgalz and Millero (1982, 1983), Gueddari et al. (1983), Monnin and Schott (1984), and Langmuir and Melchior (1985).

Complex-ion equilibria

Many heavy metals form a series of complex ions with chloride and other anions. Multiple charged cations also form various complex ions with hydroxide. Various other ions, if present in the solution, have specific effects on these complex-ion equilibria. A good example of the use of the present model is the treatment of lead chloride complexes by Millero and Byrne (1984). They consider the effects of H^+, Na^+, Mg^{2+}, and Ca^{2+} on the effective complexation constants.

Vapor-phase equilibria

Within the criterion of the present model, water of liquid-like density as solvent, one can calculate the fugacities of vapor species in equilibrium. The activity of water is obtained from the osmotic coefficient.

$$\ln a_1 = -\phi M_1 \sum m_i/1000 \quad , \tag{47}$$

where M_1 is the molecular mass of water and the sum is over all solute species. The activity of a volatile solute species is given by

$$a_i = m_i \gamma_i \quad , \tag{48}$$

provided that species is recognized as a solute in the liquid phase. In some cases a solute is so fully dissociated in the liquid that one

Table 6:

Mixed Electrolyte Parameters for High Temperatures

i	j	k	θ_{ij}	ψ_{ijk}
Na	K	Cl	-0.012	-0.0068 + 1.68E-5T
Na	Mg	Cl	0.07	0.0199 - 9.51/T
Na	Ca	Cl	0.05	-0.003
K	Mg	Cl	0	0.0259 - 14.27/T
Cl	SO_4	Na	0.030	0.00
		K		-0.005
		Mg		-0.1175 + 32.63/T
Cl	OH	Na	-0.050	0.0273 - 9.93/T
SO_4	OH	Na	-0.013	0.0302 - 11.69/T

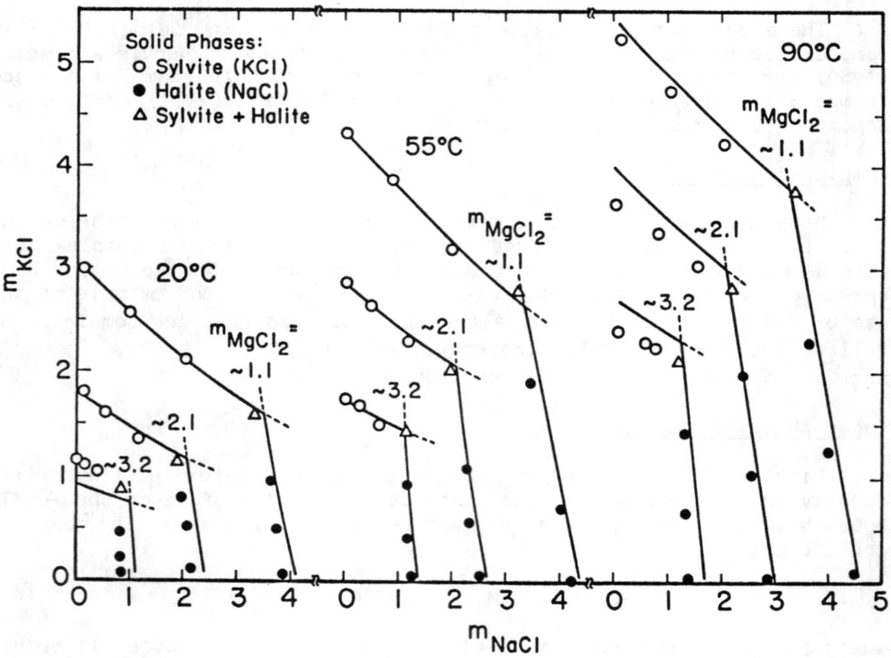

Figure 6. Predicted and experimental solubilities of halite and/or sylvite in the quaternary system $NaCl-KCl-MgCl_2-H_2O$ at 20, 55, and 90°C and at $MgCl_2$ molalities of approximately 1.1, 2.1, and 3.2. Experimental data are from Kayser (1923).

considers a direct relationship between the ν_M cations and ν_X anions of the ionized solute and the associated vapor, whereupon

$$ln\ a_i = \nu_M\ ln(m_M\gamma_M) + \nu_X\ ln(m_X\gamma_X) \quad . \qquad (49)$$

Given the activities of the volatile species, one can calculate the vapor fugacities from the Henry's constants and the vapor fugacity of water. Then the pressure and composition of the vapor phase is easily calculated on an ideal-gas basis. For moderately imperfect gases, the fugacity coefficients are readily obtained if the pertinent gas-phasesecond virial coefficients are known or can be estimated by a method such as that of Tsonopoulos (1974). More substantial departure from the ideal gas requires the use of a more comprehensive equation of state to determine the fugacities (see Holloway, 1977). These methods are steadily improving and are often reported in the chemical engineering journals.

An interesting application of the present solution model concerns the vapor in equilibrium with atmospheric aerosols where droplets contain mixed electrolytes. When the relative humidity decreases below 75%, these solutions become very concentrated with ionic strength above 10 $mol{\cdot}kg^{-1}$. Clegg and Brimblecomb (submitted) show that strong acids such as HNO_3 and HCl develop appreciable vapor pressure under these conditions. They have extended these studies to include the vapor pressures of various acids with marine seasalt aerosols (Brimblecomb and Clegg, submitted).

Another set of interesting applications involving neutral as well as ionized solutes and vapor equilibria were presented by Chen et al. (1979). One example involves vapor-liquid equilibrium for the system $K^+ - CO_3^{2-} - HCO_3^- - CO_2 - H_2O$ at temperatures to 140°C and concentrations to 40 wt. % carbonate.

Thermal properties

The enthalpy, entropy, and heat capacity of the liquid are all given by appropriate temperature derivatives of the Gibbs energy. These are standard thermodynamic relationships. Since temperature dependent expressions are given in Appendix B for various parameters of the present model, these derivatives are readily taken. Indeed, it is the standard-state heat capacity that is given, and it must be integrated to obtain the enthalpy or entropy. The total entropy of the liquid as a function of temperature and pressure is required for reversible flow processes where entropy is conserved. Similarly the enthalpy is required for irreversible, throttled processes. In other cases where heat transfer is recognized, the enthalpy of the fluid is also needed.

The total fluid entropy and enthalpy for $NaCl-H_2O$ are tabulated by Pitzer et al. (1984) for the range to 6 $mol{\cdot}kg^{-1}$, 300 C, and 1 kbar. For calculations concerning seawater and related fluids, it is often sufficient to take the thermal properties of an appropriate molality of NaCl in H_2O. The parameters of the present model, however, allow the explicit inclusion of other substantial components if desired.

SUPPLEMENTARY COMMENTS

With increasing accuracy and precision of experimental measurements, empirical functions that first seemed adequate will need to be improved or replaced. Some comments concerning the desirable procedure for improvement may be appropriate. Since there is now a very extensive base of parameters for various interactions, their continued validity should be retained as far as possible. Since the extended Debye-Hückel term f(I) or its derivative enters the equations for all properties, it should remain unchanged. It is shown in Appendix A that modifications of the ionic-strength dependence of this term will only make corresponding changes in the ionic-strength dependence of the second virial coefficients. Thus one can accomplish an improvement of this type by a change in the second virial coefficients instead of f(I).

Implicit in f(I) is the Debye-Hückel parameter A_ϕ and in turn the dielectric constant or relative permittivity. At 25°C the uncertainty is so small, that no problem seems likely, but the temperature dependence of the dielectric constant is not as accurately known as would be desired. Most of the high-temperature correlations within the present model have used the equation of Bradley and Pitzer (1979) for the dielectric constant. This equation is valid to 350°C and 1 kbar. Until an alternate equation is clearly established as superior, it seems desirable to retain the Bradley and Pitzer equation.

At present, the other equation in significant use is that of Uematsu and Franck (1980). It is somewhat more complex but has the advantage of validity to higher temperature, 550°C. In the range below 350°C it is not clear which equation better represents the true property of water, since measurements disagree by more than the difference between the equations. The values from 0 to 100°C recommended by the International Union of Pure and Applied Chemistry (Kienitz and Marsh, 1981), fall closer to those given by the Bradley and Pitzer equation. Thus, for work below 350°C there is no need to change at this time. When a change to a definitely superior equation is made, it will be necessary to present alternate sets of values of those other parameters that are very accurately known for use with each dielectric constant equation until the entire array is available on the new basis.

Fortunately, activity and osmotic coefficients and as a result, solubilities and vapor pressures, depend only on the dielectric constant itself, and the uncertainties are quite small. The problem is more serious where the temperature derivative is required for enthalpies and entropies, and it is much more serious for heat capacities which involve the second temperature derivative.

In contrast to changes in f(I) which have a general effect, changes can be made in the ionic-strength dependence of the second virial coefficient for a particular ion interaction or even for a subcategory of ions without any effect on the rest of the model. Thus Holmes and Mesmer (1986) recommend the use of $\alpha_1 = 1.4$ instead of 2.0 for the alkali metal sulfates. Kodytek and Dolejs (1986) found that inclusion of a $\beta^{(2)}$ term (with $\alpha_2 = 6$) for 3-1 electrolytes gave an appreciably better fit. Since their $\beta^{(2)}$ values are positive, this is not an indication of ion pairing in the lanthanide salts; rather it represents just a more flexible empirical expression for the ionic-strength dependence in this case. Improvements of this type can be made without disturbing the previously determined expressions for other parameters.

ACKNOWLEDGMENTS

I thank all of my collaborators in research on this electrolyte model and especially Dr. Roberto Pabalan whose advice and assistance was particularly related to parts of this chapter. It has been a pleasure to cooperate with Dr. John Weare in this area of research. This work was supported by the Director, Office of Energy Research, Office of Basic Energy Sciences, Division of Engineering, and Geosciences of the U.S. Department of Energy under Contract No. DE-AC03-76SF00098.

APPENDIX A: THEORETICAL BACKGROUND

There are alternate formulations of rigorous statistical mechanics for multicomponent fluid systems. The McMillan-Mayer (1945) system is appropriate where a solvent, in our case water, is always the most abundant component. In this system the interactions between solute species are given by potentials of mean force in the solvent and the detailed interaction of individual solvent molecules can largely be ignored. The excess Helmholtz energy can be expressed in a power series in concentrations c_i, c_j, .. of solute species

$$A^{ex}/VkT = \sum_i \sum_j c_i c_j B^\circ_{ij} + \sum_i \sum_j \sum_k c_i c_j c_k C^\circ_{ijk} + \ldots \tag{A1}$$

The quantities B°_{ij}, C°_{ijk}, etc. arise from the binary, tertiary, etc. solute-solute interactions in the presence of the solvent and in the limit of low solute concentration; they depend on the solvent and the temperature but not on the solute concentrations. They can be calculated from the potentials of mean force and can be called the second, third, etc. virial coefficients.

When ions are present, with long-range (R^{-1}) interparticle potentials, the integrals for B°_{ij}, C°_{ijk}, etc. for interionic interactions diverge. Mayer (1950) showed how the calculation could be rearranged to avoid this divergence and Friedman (1962) developed further this method. Friedman's Equations (6.10) and (13.44), with minor changes in symbols, give for the excess Helmholtz energy

$$A^{ex}/VkT = -\kappa^3/12\pi + \sum_i \sum_j c_i c_j B_{ij}(\kappa) + \sum_i \sum_j \sum_k c_i c_j c_k C_{ij}(\kappa) + \ldots \tag{A2}$$

The first term on the right is just the Debye-Hückel limiting law with the reciprocal length κ defined by

$$\kappa^2 = (4\pi e^2/\varepsilon kT) \sum_i c_i z_i^2 \quad . \tag{A3}$$

Here ε is the dielectric constant or relative permittivity of the solvent, e the electronic charge, and z_i the number of charges on particle i. The sum in Equation (A3) is clearly related to the ionic strength.

In Equation (A2) the virial coefficients B_{ij}, C_{ijk}, etc. differ from the corresponding B°_{ij}, C°_{ijk}, etc. in equation (A1) by the omission of the terms which, when rearranged, became the Debye-Hückel term. This transformation gives B_{ij}, C_{ijk}, etc. a dependence on κ in addition to their dependence on solvent properties, etc.

In proceeding to the general equation of the model, one shifts to the Gibbs energy and to molalities instead of concentrations. The Debye-Hückel term can be transformed exactly. It is assumed that other effects of this transformation can be absorbed in the virial coefficients which will be determined empirically. One then has

$$G^{ex}/n_w RT = -\alpha I^{3/2} + \sum_i \sum_j m_i m_j \beta_{ij}(I) + \ldots \qquad \text{(A4)}$$

where n_w is number of kg of water and α is the Debye-Hückel parameter. If one sums only over interactions of ions of opposite sign and ignores the ionic strength dependency of β_{ij}, this is equivalent to the equation proposed by Bronsted (1922). One now finds that the ionic strength dependency of β_{ij} is very great, but that it can be reduced by replacing the Debye-Hückel limiting law by an extended term. This was first suggested on an empirical basis by Guggenheim (1935). A test of several extended forms for the Debye-Hückel term (Pitzer, 1973) led to the choice $(\alpha/b)I\ ln\ (1 + bI^{\frac{1}{2}})$ with b an empirical constant. This can be rearranged

$$\begin{aligned} \alpha I\ ln\ (1 + bI^{\frac{1}{2}})/b &= \alpha(I^{3/2} - bI^2/2 + b^2 I^{5/2}/3 \ldots.) \\ &= \alpha I^{3/2} - I^2 \alpha q(I) \\ &= \alpha I^{3/2} - (\sum_{ij} \sum m_i m_j z_i^2 z_j^2) q(I)(\alpha/4) \quad , \qquad \text{(A5)} \end{aligned}$$

where q(I) is a function of ionic strength but not of individual molalities. Thus, in addition to the limiting law term $\alpha I^{3/2}$, the remaining contribution of the extended D-H term has exactly the same molality dependence as the second virial coefficient term of equation (A4) and these can be combined. The particular form for the extended D-H term was chosen to minimize the ionic strength dependence of the resulting second virial coefficient term,

$$\lambda_{ij}(I) = \beta_{ij}(I) + z_i^2 z_j^2\ q(I)(\alpha/4) \quad , \qquad \text{(A6)}$$

at high ionic strength.

The higher-order electrostatic term for unsymmetrical mixing of ions of the same sign, which was identified by Friedman (1962), appears as a special term within the ionic strength dependent second virial coefficient. Hence it does not need to be recognized separately at this point. Its evaluation on the molality basis was given by Pitzer (1975, 1983) and is discussed below. Our final form of equation can then be written as

$$G^{ex}/n_w RT = f(I) + \sum_i \sum_j m_i m_j \lambda_{ij}(I)$$

$$+ \sum_i \sum_j \sum_k m_i m_j m_k \mu_{ijk}(I) + \ldots \quad \text{(A7)}$$

The Debye-Hückel term f(I) includes the limiting law and depends only on the ionic strength. The second virial coefficients $\lambda_{ij}(I)$ are functions of ionic strength and include terms for mixing of ions of the same sign but different charge as appropriate. With increasing molalities of solutes, additional terms will be required in this power series to attain a given accuracy. In the semi-empirical application, the required number of terms will become apparent.

An alternate presentation of basic theory (Pitzer, 1973, 1979) is less abstract and gives more of a physical picture and an estimate of the pattern of ionic strength dependency of the second virial coefficient. These aspects may be useful to readers. But this alternate presentation includes approximations which the Mayer-Friedman treatment avoids and which are not intrinsic to the form of Equation (A7).

We return now to the higher-order electrostatic term for unsymmetrical mixing (Pitzer, 1975, 1983). The second virial coefficient $B_{ij}(\kappa)$ of equation (A2) is shown by Friedman (1962) to be given by

$$B_{ij}(\kappa) = (2\pi z_i z_j \ell/\kappa^2) J_{ij}(\kappa, z_i, z_j \ldots) \quad \text{(A8)}$$

with the electrostatic length

$$\ell = e^2/\varepsilon kT \quad . \quad \text{(A9)}$$

We note that the interionic potential of mean force can be written as

$$v_{ij} = u_{ij} + kTz_i z_j \ell/r \quad , \quad \text{(A10)}$$

where the second term is the electrostatic interaction and u_{ij} a function of the interionic distance r, is the short-range potential. Then the function J_{ij} of (A8) is

$$J_{ij} = -(\kappa^2/z_i z_j \ell) \int_0^\infty [\exp(q_{ij} - u_{ij}/kT) - 1 - q_{ij} - q_{ij}^2/2] r^2 dr \quad \text{(A11)}$$

with

$$q_{ij} = -(z_i z_j \ell/r) \exp(-\kappa r) \quad . \quad \text{(A12)}$$

The integral in Equation (A11) cannot be evaluated, in general, without knowledge of the short-range potential u_{ij}. Since that quantity is not known accurately, the entire second virial coefficient is treated as an empirical quantity. But for the particular case of ions of the same sign, an approximation yields useful results.

Ions of the same sign repel one another strongly enough that they seldom approach one another closely; hence the short-range potential should have little or no effect. This can be seen mathematically in Equation (A11). If q_{ij} is large and negative for the range of r for which u_{ij} differs from zero, then the value of $\exp(q_{ij})$ is extremely small throughout this range. Thus, provided u_{ij} is positive (or if negative, is small), the effect of u_{ij} will be negligible.

In view of this situation, one can evaluate the effect of electrostatic forces on the difference terms Φ_{ij} without making any detailed assumption about short-range forces. We write

$$\Phi_{ij} = \theta_{ij} + {}^{E}\theta_{ij}(I) \quad , \tag{A13}$$

where the first term on the right arises from the combined effects of short-range forces acting directly or through the solvent, of the use of molalities instead of concentration, and of the difference in the Debye-Hückel term in Equation (A2) from that in (A7) or (28). The second term ${}^{E}\theta_{MN}$ will be calculated from the corresponding terms of the cluster-integral theory with the omission of short-range forces. From the definition of Φ_{MN} we have

$${}^{E}\theta_{MN} = {}^{E}\lambda_{MN} - (z_N/2z_M)\,{}^{E}\lambda_{MM} - (z_M/2z_N)\,{}^{E}\lambda_{NN} \quad , \tag{A14}$$

$${}^{E}\lambda_{ij} = (z_i z_j/4I)J_{ij} \text{ with } u_{ij} = 0 \quad , \tag{A15}$$

$$J_{ij} = \frac{\kappa^2}{z_i z_j \ell} \int_0^\infty (1 + q_{ij} + \tfrac{1}{2}q_{ij}^2 - e^{q_{ij}})r^2\, dr \quad . \tag{A16}$$

With the substitutions

$$y = \kappa r \quad , \tag{A17}$$

$$x = z_i z_j \ell\kappa \quad , \tag{A18}$$

$$q = -(x/y)e^{-y} \quad , \tag{A19}$$

$$J(x) = x^{-1} \int_0^\infty (1 + q + \tfrac{1}{2}\, q^2 - e^{q})y^2\, dy \quad . \tag{A20}$$

In our working units:

$$x_{ij} = 6z_i z_j A_\phi I^{\frac{1}{2}} \quad , \tag{A21}$$

where for ions of the same sign x_{ij} is always positive. Also

$${}^{E}\theta_{MN} = (z_M z_N/4I)[J(x_{MN}) - \tfrac{1}{2}J(x_{MM}) - \tfrac{1}{2}J(x_{NN})] \quad . \tag{A22}$$

We also need the temperature derivative of ${}^{E}\theta$ and therefore of J. If $J' = \partial J/\partial x$, we find for ${}^{E}\theta'$ the expression

$${}^{E}\theta'_{MN} = -\,{}^{E}\theta_{MN}/I + (z_M z_N/8I^2)[x_{MN}J'(x_{MN}) - \tfrac{1}{2}x_{MM}J'(x_{MM}) - \tfrac{1}{2}x_{NN}J'(x_{NN})] \quad . \tag{A23}$$

For J the integrals of the second and third terms in the parentheses in Equation (A20) are straightforward with the results:

$$J = \frac{1}{4}\, x - 1 + J_2 \quad , \tag{A24}$$

$$J' = \frac{1}{4} - (J_2/x) + J_3 \quad , \tag{A25}$$

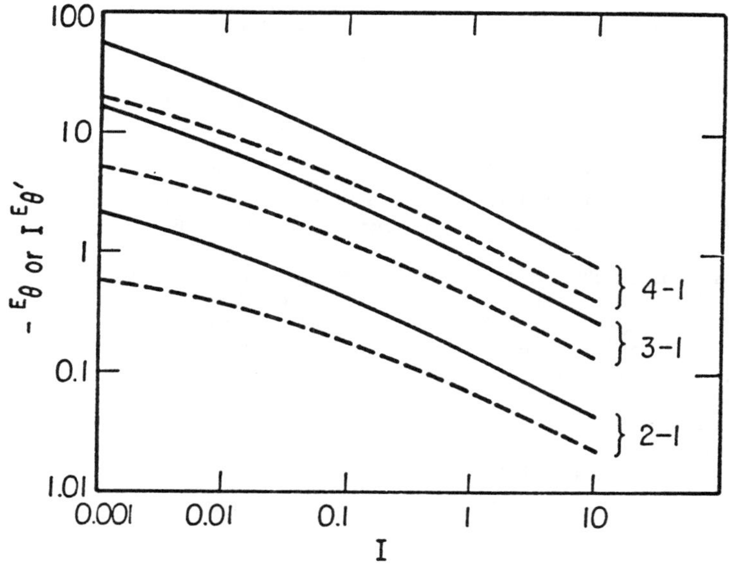

Figure A1. The functions $-^{E}\theta$ (solid curves) and $I^{E}\theta'$ (dashed curves) for mixing ions of charge types 2-1, 3-1, and 4-1.

Table A1.
Numerical arrays for calculating J(x) and J'(x)

k	a_k^{I}	a_k^{II}
0	1.925154014814667	.628023320520852
1	-.060076477753119	.462762985338493
2	-.029779077456514	.150044637187895
3	-.007299499690937	-.028796057604906
4	.000388260636404	-.036552745910311
5	.000636874599598	-.001668087945272
6	.000036583601823	.006519840398744
7	-.000045036975204	.001130378079086
8	-.000004537895710	-.000887171310131
9	.000002937706971	-.000242107641309
10	.000000396566462	.000087294451594
11	-.000000202099617	.000034682122751
12	-.000000025267769	-.000004583768938
13	.000000013522610	-.000003548684306
14	.000000001229405	-.000000250453880
15	-.000000000821969	.000000216991779
16	-.000000000050847	.000000080779570
17	.000000000046333	.000000004558555
18	.000000000001943	-.000000006944757
19	-.000000000002563	-.000000002849257
20	-.000000000010991	.000000000237816

$$J_2 = x^{-1} \int_0^{\infty} (1 - e^q)y^2 dy \quad , \qquad (A26)$$

$$J_3 = x^{-1} \int_0^{\infty} \exp(q - y)y dy \quad . \qquad (A27)$$

There are no simple integrals for J_2 and J_3 but they are readily evaluated numerically with modern computers. The resulting functions $^E\theta$ and $^E\theta'$ for 2-1, 3-1, and 4-1 mixing are shown in Figure A1.

For more efficient compution of these functions than numerical integration, several methods have been proposed (Pitzer, 1975), (Roy et al., 1983), (Harvie, 1981). Harvie's method uses two Chebyshev polynomial approximations, one for $x \le 1$ and the other for $x \ge 1$. The appropriate equations for these regions follow:

Region I. $x \le 1$

$$z = 4\, x^{1/5} - 2 \qquad (A28)$$

$$\frac{dz}{dx} = \frac{4}{5}\, x^{-4/5} \qquad (A29)$$

$$b_k = z\, b_{k+1} - b_{k+2} + a_k^{I} \qquad (A30)$$

$$d_k = b_{k+1} + z\, d_{k+1} - d_{k+2} \qquad (A31)$$

$k = 0, 20$ (for A30 and A31)

Region II. $x \ge 1$

$$z = \frac{40}{9}\, x^{-1/10} - \frac{22}{9} \qquad (A32)$$

$$\frac{dz}{dx} = -\frac{40}{90}\, x^{-11/10} \qquad (A33)$$

$$b_k = z\, b_{k+1} - b_{k+2} + a_k^{II} \qquad (A34)$$

$$d_k = b_{k+1} + z\, d_{k+1} - d_{k+2} \qquad (A35)$$

$k = 0, 20$ (for A34 and A35)

Using the calculated values for the b_k and the d_k, $J(x)$ and $J'(x)$ can be calculated from the following formulas:

$$J(x) = \frac{1}{4} x - 1 + \frac{1}{2} (b_0 - b_2) \quad , \qquad (A36)$$

$$J'(x) = \frac{1}{4} + \frac{1}{2} \frac{dz}{dx} (d_0 - d_2) \quad . \qquad (A37)$$

Some discussion with regard to the calculation of the arrays b_k and d_k is appropriate. The coefficients a_k^{I} and a_k^{II} are given in Table A1. By definition $b_{21} = b_{22} = d_{21} = d_{22} = 0$. Therefore, by using Equation (A30) or (A34) the numbers b can be generated in decreasing sequence. Similar arguments apply to the array d. The values $J(1) = 0.116437$ and $J'(1) = 0.160527$ can be used to check a program for this calculation.

Another theoretical topic concerns the possible ionic-strength dependence of second virial coefficients for interactions of ions with neutral molecules containing dipole or higher electrical moments. The work of Kirkwood (1934) pertains to this question, but it considers only electrical effects subject to a distance of closest approach and ignores all other effects of short range forces which are normally the dominant terms. For charge-dipole effects on the activity coefficient of the dipolar molecule i, Kirkwood's Equation (21) yields

$$\delta\, ln\, \gamma_i = - \frac{3\pi e^2 \mu_i^2 \, \Sigma\, c_j z_j^2}{2a\varepsilon^2 k^2 T^2 (1 + \kappa a + \kappa^2 a^2/3)} , \qquad (A38)$$

with μ_i the dipole moment of i and c_j and z_j the concentration and charge on ion j. Also a is the distance of closest approach of an ion to the molecule and κ is the Debye reciprocal length which is related to the ionic strength. For each term in the sum, equation (A38) yields the electrical contribution to a second virial coefficient λ_{ij}. The appearance of κ in the denominator indicates an ionic strength dependence. But for typical values of the dipole moment and other quantities, this term is very small compared to that for short-range forces. For quadrupole or higher moments the corresponding term is even smaller. Hence there is no present indication that an ionic-strength dependence need be considered for the second virial coefficients for neutral-ion interactions.

APPENDIX B: NUMERICAL PARAMETERS FOR TEMPERATURE DEPENDENCY EXPRESSIONS

The following are the temperature functions for the parameters of the present solution model and for the standard state heat capacities of aqueous electrolytes reported by various investigators. The pressure dependencies of the ion-interaction coefficients are currently neglected except for NaCl(aq), HCl(aq), and NaOH(aq), which have the requisite PVT data available in sufficient quantity to allow their evaluation. In the case of standard state heat capacities, only NaCl(aq) and HCl(aq) have both P- and T-dependent functions. The standard state heat capacities of NaOH(aq) are not included since present information is inadequate. This lack is not important for the solubility of NaOH, however, since it is beyond the valid concentration range of the present model. In all the equations, T_R and T refer to the reference temperature of 298.15 K and the temperature of interest, respectively. Pressures are designated by P and are in bars; standard state heat capacities are in units of J K^{-1} mol^{-1}.

<u>NaCl(aq)</u>

Ion-interaction parameters - Pitzer et al. (1984), 273-573 K and saturation pressure to 1 kbar. P refers to pressure in bars.

$$f(T) = Q1/T + Q2 + Q3\,P + Q4\,P^2 + Q5\,P^3 + Q6\,ln(T) + (Q7 + Q8\,P + Q9\,P^2 + Q10\,P^3)T + (Q11 + Q12\,P + Q13\,P^2)T^2 + (Q14 + Q15\,P + Q16\,P^2 + Q17\,P^3)/(T-227) + (Q18 + Q19\,P + Q20\,P^2 + Q21\,P^3)/(680-T)$$

	$\beta^{(0)}$	$\beta^{(1)}$	$C^{\phi} = 2C$
Q1	-656.81518	119.31966	-6.1084589
Q2	24.86912950	-0.48309327	4.0217793E-1
Q3	5.381275267E-5	0	2.2902837E-5
Q4	-5.588746990E-8	0	0
Q5	6.589326333E-12	0	0
Q6	-4.4640952	0	-0.075354649
Q7	0.01110991383	1.4068095E-3	1.531767295E-4
Q8	-2.657339906E-7	0	-9.0550901E-8
Q9	1.746006963E-10	0	0
Q10	1.046261900E-14	0	0
Q11	-5.307012889E-6	0	-1.538600820E-8
Q12	8.634023325E-10	0	8.6926600E-11
Q13	-4.178596200E-13	0	0
Q14	-1.579365943	-4.2345814	0.3531041360
Q15	2.202282079E-3	0	-4.3314252E-4
Q16	-1.310550324E-7	0	0
Q17	-6.381368333E-11	0	0
Q18	9.706578079	0	-0.09187145529
Q19	-0.02686039622	0	5.1904777E-4
Q20	1.534474401E-5	0	0
Q21	-3.215398267E-9	0	0

Standard state heat capacity - the following equation was fit to values from 273-573 K and at 1 bar or saturation pressure tabulated by Pitzer et al. (1984), Table A-4. For values at other pressures, the reader is referred to the tables and equations given by Pitzer et al. (1984) which are valid in the range 273-573 K and to 1 kbar pressure.

$$C_p^\circ = -1.848175E6 + 4.411878E7/T + 3.390654E5 \ln(T) - 8.893249E2T + 4.005770E\text{-}1T^2 - 7.244279E4/(T-227) - 4.098218E5/(647-T).$$

<u>KCl(aq)</u>

Ion-interaction parameters - Holmes and Mesmer (1983a), 273-523 K.

$$f(T) = Q1 + Q2(1/T-1/T_R) + Q3 \ln(T/T_R) + Q4(T-T_R) + Q5(T^2-T_R^2) + Q6 \ln(T-260).$$

	$\beta^{(0)}$	$\beta^{(1)}$	$C^{\phi} = 2C$
Q1	0.04808	0.0476	-7.88E-4
Q2	-758.48	303.9	91.270
Q3	-4.7062	1.066	0.58643
Q4	0.010072	0	-0.0012980
Q5	-3.7599E-6	0	4.9567E-7
Q6	0	0.0470	0

Standard state heat capacity - Holmes and Mesmer (1983a), 273-523 K.

$$C_p^\circ = -991.51 + 5.56452T - 0.00852996T^2 - 686/(T-270).$$

$MgCl_2(aq)$

Ion-interaction parameters - de Lima and Pitzer (1983b), with equation for C^{ϕ}_{MX} modified to fit the solubility data, 298-473 K.

$$f(T) = Q1\ T^2 + Q2\ T + Q3$$

	$\beta^{(0)}$	$\beta^{(1)}$	$C^{\phi} = 2^{3/2}\ C$
Q1	5.93915E-7	2.60169E-5	2.41831E-7
Q2	-9.31654E-4	-1.09438E-2	-2.49949E-4
Q3	0.576066	2.60135	5.95320E-2

Standard state heat capacity - Phutela et al. (1987), 298-453 K

$$C^{\circ}_p = -7.39872E6/T + 7.96487E4 - 3.25868E2T$$
$$+ 5.98722E\text{-}1T^2 - 4.21187E\text{-}4T^3.$$

$CaCl_2(aq)$

Ion-interaction parameters - Møller (submitted), 298-523 K, 0-4 mol·kg^{-1}.

$$f(T) = Q1 + Q2\ T + Q3/T + Q4\ \ln T + Q5/(T-263) + Q6\ T^2 + Q7/(680-T)$$

	$\beta^{(0)}$	$\beta^{(1)}$	$C^{\phi} = 2^{3/2}\ C$
Q1	-9.41895832E1	3.4787	-3.03578731E1
Q2	-4.0475002E-2	-1.5417E-2	-1.36264728E-2
Q3	2.34550368E3	0	7.64582238E2
Q4	1.70912300E1	0	5.50458061
Q5	-9.22885841E-1	0	-3.27377782E-1
Q6	1.51488122E-5	3.1791E-5	5.69405869E-6
Q7	-1.39082000	0	-5.36231106E-1

A simpler equation valid to 473 K and 4.3 mol·kg^{-1} is given by Phutela and Pitzer (1983).

Standard state heat capacity - Phutela et al. (1987), 298-373 K.

$$C^{\circ}_p = -1.26721E6/T + 7.41013E3 - 11.5222\ T.$$

$Na_2SO_4(aq)$

Ion-interaction parameters - Holmes and Mesmer (1986), 273-498 K. Note: these parameters are consistent with using a value of $\alpha_1 = 1.4$, instead of the usual value of 2.0.

$$F(T) = Q1 + Q2(T_R - T_R^2/T) + Q3(T^2 + 2T_R^3/T - 3T_R^2) + Q4(T + T_R^2/T - 2T_R)$$
$$+ Q5[\ln(T/T_R) + T_R/T - 1] + Q6\{1/(T-263) + (263-T_R^2)/[T(T_R-263)^2]\}$$
$$+ Q7\{1/(680-T) + (T_R^2-680T)/[T(680-T_R)^2]\}.$$

	$\beta^{(0)}$	$\beta^{(1)}$	$C^{\phi} = 2^{3/2}\ C$
Q1	-1.727E-2	0.7534	1.1745E-2
Q2	1.7828E-3	5.61E-3	-3.3038E-4
	9.133E-6	-5.7513E-4	1.85794E-5
Q4	0	1.11068	-3.9200E-2
Q5	-6.552	-378.82	14.2130
Q6	0	0	0
Q7	-96.90	1861.3	-24.950

Standard state heat capacity - Holmes and Mesmer (1986), 273-498 K.

$$C_p^\circ = -1206.2 + 7.6405T - 1.23672E\text{-}2T^2 - 6045/(T-263).$$

$K_2SO_4(aq)$

Ion-interaction parameters - Holmes and Mesmer (1986), 273-498 K. Note: these parameters are consistent with using a value of $\alpha_1 = 1.4$, instead of the usual value of 2.0.

$$f(T) = Q1 + Q2(T_R - T_R^2/T) + Q3(T^2 + 2T_R^3/T - 3T_R^2) + Q4(T + T_R^2/T - 2T_R)$$
$$+ Q5[\ln(T/T_R) + T_R/T - 1] + Q6\{1/(T-263) + (263T - T_R^2)/[T(T_R-263)^2]\}$$
$$+ Q7\{1/(680-T) + (T_R^2 - 680T)/[T(680-T_R)^2]\}.$$

	$\beta^{(0)}$	$\beta^{(1)}$	$C^{\phi} = 2^{3/2}\ C$
Q1	0	0.6179	9.1547E-3
Q2	7.476E-4	6.85E-3	0
Q3	0	5.576E-5	0
Q4	4.265E-3	-5.841E-2	-1.81E-4
Q5	-3.088	0	0
Q6	0	-0.90	0
Q7	0	0	0

Standard state heat capacity:

$$C_p^\circ(K_2SO_4) = 2C_p^\circ(KCl) + C_p^\circ(Na_2SO_4) - 2C_p^\circ(NaCl).$$

$MgSO_4(aq)$

Ion-interaction parameters - Phutela and Pitzer (1986), 298-473 K. Note that the final column gives the temperature coefficients for C. For $MgSO_4$, this is related to C^ϕ by: $C^\phi = 4C$ (see Eqn. 22).

$$f(T) = Q1(T/2 + 298^2/2T - 298) + Q2(T^2/6 + 298^3/3T - 298^2/2)$$
$$+ Q3(T^3/12 + 298^4/4T - 298^3/3) + Q4(T^4/20 + 298^5/5T - 298^4/4)$$
$$+ (298 - 298^2/T)Q5 + Q6.$$

	$\beta^{(0)}$	$\beta^{(1)}$	$\beta^{(2)}$	$C = C^\phi/4$
Q1	-1.0282	-2.9596E-1	-1.3764E-1	1.0541E-1
Q2	8.4790E-3	9.4564E-4	1.2121E-1	-8.9316E-4

Q3	-2.33667E-5	0	-2.7642E-4	2.51E-6
Q4	2.1575E-8	0	0	-2.3436E-9
Q5	6.8402E-4	1.1028E-2	-2.1515E-1	-8.7899E-5
Q6	0.21499	3.3646	-32.743	0.006993

Standard state heat capacity - Phutela and Pitzer (1986), 298-473 K.

$$C_p^\circ = -6.2543E6/T + 6.5277E4 - 2.6044E2T + 4.6930E\text{-}1T^2 - 3.2656E\text{-}4T^3.$$

<u>HCl(aq)</u>

Ion-interaction parameters - Holmes et al. (1987), 273-523 K. The equation and parameters listed are valid to 7 $mol\cdot kg^{-1}$, but this paper also includes more complex equations valid to 648 K and to 16 $mol\cdot kg^{-1}$; also note that the equations use ρ, the density in $kg\cdot m^{-3}$ of pure water at the particular P and T and that they include pressure dependence to 400 bars.

$$f(T) = Q1 + Q2\ ln(\rho/997) + Q3(\rho-997) + Q4(T-T_R) + Q5(P-1).$$

	$\beta^{(0)}$	$\beta^{(1)}$	$C = C^\phi/2$
Q1	0.17690	0.2973	0.362E-3
Q2	-9.140E-2	16.147	0
Q3	0	-1.7631E-2	0
Q4	-4.034E-4	0	-3.036E-5
Q5	6.20E-6	7.20E-5	0

Standard State heat capacity - Holmes et al. (1987), 273-648 K.

$$C_p^\circ = 17.93 - 16.79T/(T-240) + 6.4579E5\ TX_P,$$

where

$$X_p = [(\partial^2 ln\varepsilon/\partial T^2)_P - (\partial ln\varepsilon/\partial T)_P^2]/\varepsilon,$$

with ε the dielectric constant (relative permittivity)

<u>NaOH(aq)</u>

Ion-interaction parameters - Pabalan and Pitzer (1987a), 0-350°C and saturation pressure to 400 bars. P refers to pressure in bars.

$$f(T) = Q1 + Q2\ P + (Q3 + Q4\ P)/T + Q5\ ln(T) + (Q6 + Q7\ P)T$$
$$+ (Q8 + Q9\ P)T^2 + Q10/(T-227) + (Q11 + Q12\ P)/(647-T).$$

	$\beta^{(0)}$	$\beta^{(1)}$	$C^\phi = 2\ C$
Q1	2.7682478E+2	4.6286977E+2	-1.6686897E+1
Q2	-2.8131778E-3	0	4.0534778E-4
Q3	-7.3755443E+3	-1.0294181E+4	4.5364961E+2
Q4	3.7012540E-1	0	-5.1714017E-2
Q5	-4.9359970E+1	-8.5960581E+1	2.9680772
Q6	1.0945106E-1	2.3905969E-1	-6.5161667E-3
Q7	7.1788733E-6	0	-1.05530373E-6

Q8	-4.0218506E-5	-1.0795894E-4	2.3765786E-6
Q9	-5.88474E-9 0	8.9893405E-10	
Q10	1.1931144E-1	0	-6.8923899E-1
Q11	2.4824963 0	-8.1156286E-2	
Q12	-4.8217410E-3	0	0

Debye-Hückel parameter

Most calculations for the present model have used the equation of Bradley and Pitzer (1979) for the dielectric constant and the density of water from the equation of Haar et al. (1984). Tables of values of A_ϕ and of the corresponding parameters for enthalpy, heat capacity, and volume were given by Bradley and Pitzer (1979) and are also available in other papers including Pitzer et al. (1984) and Ananthaswamy and Atkinson (1984). Møller (submitted) gives a seven-constant equation for A_ϕ at saturation pressure to 300°C.

REFERENCES

Akhumov, E.I. and Vasil'ev, B.B. (1932) A study of aqueous solutions at elevated temperatures. II. Zh. Obs. Khimii, 2, 282-289 (in Russian).

Ananthaswamy, J. and Atkinson, G. (1984) Thermodynamics of concentrated electrolyte mixtures. 4. Pitzer-Debye-Hückel limiting slopes for water from 0 to 100°C and from 1 atm to 1 kbar. J. Chem. Engr. Data, 29, 81-87.

Ananthaswamy, J. and Atkinson, G. (1985) Thermodynamics of concentrated electrolyte mixtures. 5. A review of the thermodynamic properties of aqueous calcium chloride in the temperature range 273.15-373.15 K. J. Chem. Engr. Data, 30, 120-128.

Archer, D.G. and Wood, R.H. (1985) Chemical equilibrium model applied to aqueous magnesium sulfate solutions. J. Solution Chem., 14, 757-780.

Bradley, D.J. and Pitzer, K.S. (1979) Thermodynamics of electrolytes. 12. Dielectric properties of water and Debye-Hückel parameters to 350°C and 1 kbar. J. Phys. Chem. 83, 1599-1603.

Brimblecombe, P. and Clegg, S.L. (submitted) The solubility and behavior of acid gases in the marine aerosol. J. Atmos. Chem.

Brodale, G.E. and Giauque, W.F. (1958) The heat of hydration of sodium sulfate. Low temperature heat capacity and entropy of sodium sulfate decahydrate. J. Amer. Chem. Soc. 80, 2042-2044.

Brodale, G.E. and Giauque, W.F. (1972) The relationship of crystalline forms I, III, IV, and V of anhydrous sodium sulfate as determined by the third law of thermodynamics. J. Phys. Chem. 76, 737-743.

Bronsted, J.N. (1922) Studies on solubility. IV. The principle of specific interaction of ions. J. Amer. Chem. Soc. 44, 877-898.

Chen, C.-C., Britt, H.I., Boston, J.F., and Evans, L.B. (1979) Extension and application of the Pitzer equation for vapor-liquid equilibrium of aqueous electrolyte systems with molecular solutes. Amer. Inst. Chem. Engin. J. 25, 820-831.

Chen, C.-T.A. and Marshall, W.L. (1981) Amorphous silica solubilities IV. Behavior in pure water and aqueous sodium chloride, sodium sulfate, magnesium chloride, and magnesium sulfate solutions up to 350°C. Geochim. Cosmochim. Acta 46, 279-287.

Clegg, S.L. and Brimblecombe, P. (submitted) Equilibrium partial pressures of strong acids over concentrated saline solutions. Part

I. HNO_3; Part II. HCl. Atmos. Environment.
Cox, W.P., Hornung, E.W., and Giauque, W.F. (1955) The spontaneous transformation from macrocrystalline to microcrystalline phases at low temperatures. The heat capacity of $MgSO_4 \cdot 6H_2O$. J. Amer. Chem. Soc. 77, 3935-3938.
de Lima, M.C.P. and Pitzer, K.S. (1983) Thermodynamics of saturated electrolyte mixtures of NaCl with Na_2SO_4 and $MgCl_2$. J. Solution Chem. 12, 187-199.
Downes, C.J. and Pitzer, K.S. (1976) Thermodynamics of electrolytes. Binary mixtures formed from aqueous NaCl, Na_2SO_4, $CuCl_2$ and $CuSO_4$ at 25°C. J. Solution Chem. 5, 389-398.
Friedman, H.L. (1962) Ionic Solution Theory. Wiley-Interscience, New York, 265 pp.
Frost, G.B., Breck, W.G., Clayton, R.N., Reddoch, A.H., and Miller C.G. (1957) The heat capacities of the crystalline and vacuum dehydrated form of magnesium sulphate monohydrate. Can. J. Chem. 35, 1446-1453.
Gates, J.A. (1985) Thermodynamics of Aqueous Electrolyte Solutions at High Temperatures and Pressures, Ph.D. Dissertation, University of Delaware, Newark, Delaware, 340 pp.
Gates, J.A., Tillett, D.M., White, D.E., and Wood, R.H. (1987) Apparent molar heat capacities of aqueous NaCl solutions from 0.05 to 3.0 $mol \cdot kg^{-1}$, 350 to 600 K, and 2-18 MPa. J. Chem. Thermodyn. 19, 131-146.
Gueddari, M., Monnin C., Perret, D., Fritz, B., and Tardy, Y. (1983) Geochemistry of the brines of Chott El Jerid in Southern Tunisia: application of Pitzer's equations. Chem. Geol. 39, 165-178.
Guggenheim, E.A. (1935) The specific thermodynamic properties of aqueous solutions of strong electrolytes. Phil. Mag. 19, 588-643.
Haar, L., Gallagher, J.S., and Kell, G.S. (1984) NBS/NRC Steam Tables: Thermodynamic and Transport Properties and Computer Programs for Vapor and Liquid States of Water in SI Units. Hemisphere Publishing, Washington, D.C., 318 pp.
Harvie, C.E. (1981) Theoretical investigations in geochemistry and atom surface scattering. Ph.D. dissertation, University of California at San Diego. Univ. Microfilm Int., Ann Arbor, MI # AAD82-03026.
Harvie, C.E. and Weare, J.H. (1980) The prediction of mineral solubilities in natural waters: the Na-K-Mg-Ca-Cl-SO_4-H_2O system from zero to high concentration at 25°C. Geochim. Cosmochim. Acta 44, 981-997.
Harvie, C.E., Eugster, H.P., and Weare, J.H. (1982) Mineral equilibria in the six-component seawater system, Na-K-Mg-Ca-SO_4-Cl-H_2O at 25°C. II. Compositions of the saturated solutions. Geochim. Cosmochim. Acta 46, 1603-1618.
Harvie, C.E., Møller, N., and Weare, J.H. (1984) The prediction of mineral solubilities in natural waters: The Na-K-Mg-Ca-H-Cl-SO_4-OH-HCO_3-CO_3-CO_2-H_2O system to high ionic strengths at 25°C. Geochim. Cosmochim. Acta 48, 723-751.
Hershey, J.P. and Millero, F.J. (1986) The dependence of the acidity constants of silicic acid on NaCl concentration using Pitzer's equations. Marine Chem. 18, 101-105.
Holloway, J.R. (1977) Fugacity and activity of molecular species in supercritical fluids. In: Thermodynamics in Geology, D. G. Fraser, ed., D. Reidel, Dordrecht, Holland, 161-181 pp.
Holmes, H.F., Baes, C.F., and Mesmer, R.E. (1978) Isopiestic studies of aqueous solutions at elevated temperatures: 1. KCl, $CaCl_2$ and $MgCl_2$. J. Chem. Thermodyn. 10, 983-996.

Holmes, H.F., Busey, R.H., Simonson, J.M., Mesmer, R.E., Archer, D.G., and Wood, R.H. (1987) Thermodynamic properties of HCl(aq) to 648 K. J. Chem. Thermodyn. 19, xxxx.

Holmes, H.F. and Mesmer, R.E. (1983) Thermodynamic properties of aqueous solutions of the alkali metal chlorides to 250°C. J. Phys. Chem. 87, 1242-1255.

Holmes, H.F. and Mesmer, R.E. (1986) Thermodynamics of aqueous solutions of the alkali metal sulfates. J. Solution Chem. 15, 495-518.

Kayser, E. (1923) Die ersatzzahlen inkonstanter Lösungen über Kaliumchlorid und Natriumchlorid. Kali 17, 1-9 (in German).

Kelley, K.K. (1960) Contributions to the data on theoretical metallurgy. XIII. High-temperature heat-content, heat-capacity, and entropy data for the elements and inorganic compounds. U.S. Bur. Mines Bull. 584, 232 pp.

Kienitz, H. and Marsh, K.N. (1981) Recommended reference materials for realization of physiochemical properties: Permittivity. Pure and Appl. Chem. 53, 1847-1862.

Kirkwood, J.G. (1934) Theory of solutions of molecules containing widely separated charges with special application to zwitterions. J. Chem. Phys. 2, 351-361.

Ko, H.C. and Daut G.E. (1979) Enthalpies of Formation of α- and β-Magnesium Sulfate and Magnesium Sulfate Monohydrate. U.S. Bureau of Mines, Rep. Investig. 8409, 8 pp.

Kodytek, V. and Dolejs, V. (1986) On the Pitzer equation for the osmotic coefficient of aqueous solutions of 3:1 rare earth electrolytes. Z. phys. Chem., Leipzig, 267, 743-746.

Krumgalz, B.S. and Millero, F.J. (1982) Physico-chemical study of Dead Sea Waters. I. Activity coefficients of major ions in Dead Sea water. Marine Chem. 11, 209-222.

Krumgalz, B.S. and Millero, F.J. (1983) Physico-chemical study of Dead Sea waters. III. On gypsum saturation in Dead Sea waters and their mixtures with Mediterranean Sea water. Marine Chem. 13, 127-139.

Langmuir, D. and Melchior, D. (1985) The geochemistry of Ca, Sr, Ba, and Ra sulfates in some deep brines from the Palo Duro Basin, Texas. Geochim. Cosmochim. Acta 49, 2423-2432.

Levelt Sengers, J.M.H., Everhart, C.M., Morrison, G., and Pitzer, K.S. (1986) Thermodynamic anomalies in near-critical aqueous NaCl solutions. Chem. Engr. Commun. 47, 315-328.

Linke, W.F. and Seidell, A. (1965) Solubilities of Inorganic and Metal Organic Compounds, 4th ed. Amer. Chem. Soc.

Liu, C. and Lindsay, W.T., Jr. (1972) Thermodynamics of sodium chloride solutions at high temperatures. J. Solution Chem. 1, 45-69.

Long, F.A. and McDevit, W.F. (1952) Activity coefficients of nonelectrolyte solutes in aqueous salt solutions. Chem. Rev. 51, 119-169.

Marshall, W.L. and Chen, C.-T.A. (1981) Amorphous silica solubilities. V. Predictions of solubility behavior in aqueous mixed electrolyte solutions to 300°C. Geochim. Cosmochim. Acta 46, 289-291.

Mayer, J.E. (1950) The theory of ionic solutions. J. Chem. Phys. 18, 1426-1436.

McMillan, W.G. and Mayer, J.E. (1945) The statistical thermodynamics of multicomponent systems. J. Chem. Phys. 13, 276-303.

Millero, F.J. and Byrne, R.H. (1984) Use of Pitzer's equations to determine the media effect in the formation of lead chloro complexes. Geochim. Cosmochim. Acta 48, 1145-1150.

Møller, N. (submitted) The prediction of mineral solubilities in natural waters: a chemical equilibrium model for the $CaSO_4$-NaCl-$CaCl_2$-H_2O system to high temperature and concentration. Geochim. Cosmochim. Acta.

Monnin, C. and Schott, J. (1984) Determination of the solubility products of sodium carbonate minerals and an application to trona deposition in Lake Magadi (Kenya). Geochim. Cosmochim. Acta 48, 571-581.

Pabalan, R.T. and Pitzer, K.S. (1987a) Thermodynamics of NaOH(aq) in hydrothermal solutions. Geochim. Cosmochim. Acta, 51, 829-837.

Pabalan, R.T. and Pitzer, K.S. (1987b) Thermodynamics of concentrated electrolyte mixtures and the prediction of mineral solubilities to high temperatures for mixtures in the system Na-K-Mg-Cl-SO_4-OH-H_2O. Geochim. Cosmochim. Acta 51, xxxx.

Peiper, J.C. and Pitzer, K.S. (1982) Thermodynamics of aqueous carbonate solutions including mixtures of sodium carbonate, bicarbonate, and chloride. J. Chem. Thermodynamics, 14, 613-638.

Phutela, R.C. and Pitzer, K.S. (1983) Thermodynamics of aqueous calcium chloride. J. Solution Chem. 12, 201-207.

Phutela, R.C. and Pitzer, K.S. (1986a) Thermodynamics of electrolyte mixtures. Enthalpy and the effect of temperature on the activity coefficient. J. Solution Chem. 15, 649-662.

Phutela, R.C. and Pitzer, K.S. (1986b) Heat capacity and other thermodynamic properties of aqueous magnesium sulfate to 473 K. J. Phys. Chem. 90, 895-901.

Phutela, R.C. and Pitzer, K.S. (1986c) Densities and apparent molar volumes of aqueous magnesium sulfate and sodium sulfate to 473 K and 100 bar. J. Chem. Engr. Data, 31, 320-327.

Phutela, R.C., Pitzer, K.S., and Saluja, P.P.S. (1987) Thermodynamics of aqueous magnesium chloride, calcium chloride and strontium chloride at elevated temperatures. J. Chem. Eng. Data 32, 76-80.

Pitzer, K.S. (1973) Thermodynamics of electrolytes. I. Theoretical basis and general equations. J. Phys. Chem. 77, 268-277.

Pitzer, K.S. (1975) Thermodynamics of electrolytes. V. Effects of higher-order electrostatic terms. J. Solution Chem. 4, 249-265.

Pitzer, K.S. (1979) Theory: Ion interaction approach. In Activity Coefficients in Electrolyte Solutions, R. Pytkowicz, ed., CRC Press, Boca Raton, Florida, 157-208 pp.

Pitzer, K.S. (1983) Thermodynamics of unsymmetrical electrolyte mixtures: Enthalpy and heat capacity. J. Phys. Chem. 87, 2360-2364.

Pitzer, K.S. (1986) Theoretical considerations of solubility with emphasis on mixed electrolytes. Pure Appl. Chem. 58, 1599-1610.

Pitzer, K.S. and Kim, J.J. (1974) Thermodynamics of electrolytes. IV. Activity and osmotic coefficients for mixed electrolytes. J. Amer. Chem. Soc. 96, 5701-5707.

Pitzer, K.S. and Li, Y.-g. (1983) Thermodynamics of aqueous sodium chloride to 823 K and 1 kbar. Proc. Nat'l Acad. Sci. U.S.A. 80, 7689-7693.

Pitzer, K.S. and Li, Y.-g. (1984) Critical phenomena and thermodynamics of dilute aqueous sodium chloride. Proc. Nat'l Acad. Sci. U.S.A. 81, 1268-1271.

Pitzer, K.S. and Mayorga, G. (1973) Thermodynamics of electrolytes. II. Activity and osmotic coefficients for strong electrolytes with one or both ions univalent. J. Phys. Chem. 77, 2300-2308.

Pitzer, K.S. and Mayorga, G. (1974) Thermodynamics of electrolytes. III. Activity and osmotic coefficients for 2-2 electrolytes. J. Solution Chem. 3, 539-546.

Pitzer, K.S., Peiper, J.C., and Busey, R.H. (1984) Thermodynamic properties of aqueous sodium chloride solutions. J. Phys. Chem. Ref. Data 13, 1-102.
Pitzer, K.S., Peterson, J.R., and Silvester, L.F. (1978) Thermodynamics of electrolytes. IX. Rare earth chlorides, nitrates, and perchlorates, J. Solution Chem. 7, 45-56.
Pitzer, K.S., Roy, R.N., and Silvester, L.F. (1977) Thermodynamics of electrolytes. 7. Sulfuric Acid. J. Amer. Chem. Soc. 99, 4930-4936.
Pitzer, K.S. and Silvester, L.F. (1976) Thermodynamics of electrolytes. VI. Weak electrolytes including H_3PO_4. J. Solution Chem. 5, 269-278.
Pitzer, K.S. and Simonson, J.M. (1986) Thermodynamics of Multicomponent, miscible, ionic systems: theory and equations. J. Phys. Chem. 90, 3005-3009.
Rard, J.A. and Miller D.G. (1981) Isopiestic determination of the osmotic and activity coefficients of aqueous $MgCl_2$ solutions at 25°C. J. Chem. Eng. Data 26, 38-43.
Robie, R.A., Hemingway, B.S., and Fischer, J.R. (1978) Thermodynamic properties of minerals and related substances at 298.15 K and 1 bar pressure and at higher temperatures. U.S. Geol. Survey Bull. 1452, 456 pp.
Robinson, R.A. and Stokes, R.H. (1965) Electrolyte Solutions, 2nd ed, revised, Butterworths, London, 571 pp.
Rogers, P.S.Z. and Pitzer, K.S. (1981) High-temperature thermodynamic properties of aqueous sodium sulfate solutions. J. Phys. Chem. 85, 2886-2895.
Rogers, P.S.Z. and Pitzer, K.S. (1982) Volumetric properties of aqueous sodium chloride solutions. J. Phys. Chem. Ref. Data, 11, 15-81.
Roy, R.N., Gibbons, J.J., Peiper, J.C., and Pitzer, K.S. (1983) Thermodynamics of the unsymmetrical mixed electrolyte HCl-$LaCl_3$. J. Phys. Chem. 87, 2365-2369.
Saluja, P.P.S., Pitzer, K.S., and Phutela, R.C. (1986) High-temperature thermodynamic properties of several 1:1 electrolytes. Can. J. Chem. 64, 1328-1335.
Silvester, L.F. and Pitzer, K.S. (1978) Thermodynamics of electrolytes, X. Enthalpy and the effects of temperature on the activity coefficients. J. Solution Chem. 7, 327-337.
Stull, D.R. and Prophet, H. (1971) JANAF Thermochemical Tables, 2nd ed. Nat'l Stand. Ref. Data Ser., U.S. Nat'l Bur. Stand. 37, 1141 pp.
Tanger, J.C., IV, and Helgeson, H.C. (1987) Calculation of the thermodynamic and transport properties of aqueous species at high pressures and temperatures: Revised equations of state for the standard partial molal properties of ions and electrolytes. Amer. J. Sci., in press.
Tsonopoulos, C. (1974) An empirical correlation of second virial coefficients. Am. Inst. Chem. Engin. J. 20, 263-272.
Uematsu, M. and Franck, E.U. (1980) Static dielectric constant of water and steam. J. Phys. Chem. Ref. Data 9, 1291-1305.
Wagman, D.D., Evans, W.H., Parker, V.B., Schumm, R.H., Halow, I., Bailey, S.M., Churney, K.L., and Nuttall, R.L. (1982) The NBS tables of chemical thermodynamic properties. Selected values for inorganic and C_1 and C_2 organic substances in SI units. J. Phys. Chem. Ref. Data 11, Suppl. 2, 392 pp.
Zarembo, V.I. and L'vov, S.N. (1982) Limiting partial volumes of individual ions in aqueous solution. Geochim. Int'l 19, 57-69.

Chapter 5

John H. Weare

MODELS OF MINERAL SOLUBILITY IN CONCENTRATED BRINES WITH APPLICATION TO FIELD OBSERVATIONS

INTRODUCTION

Experimentally determined phase equilibria have played an important role in the analysis of field observations in evaporite deposits (see for example Braitsch, 1971). These studies, while providing insights into the formation and transformation of evaporite minerals, have been limited by the availability of phase equilibria data appropriate for the systems under study. Equilibria data for systems of more than two components are difficult to obtain in the laboratory. With the large numbers of components typical of natural systems enough data points to adequately describe the phase equilibria even over the limited region of composition that may represent a particular natural processes are seldom available.

The availability of the solubility data to define a phase diagram varies from system to system depending somewhat on the difficulties associated with measurements. A complete data base rarely exists for temperatures other than 25°C. Solubility data as a function of brine composition is scarce for even such important systems as $CaCO_3-H_2O$ and $CaSO_4-H_2O$. On the other hand, for studies of important processes such as formation of minerals via the evaporation of a dilute ground water or seawater the region of composition variation required to describe the process may be very large.

Fortunately, for some systems which are particularly important and common in nature, for example brine compositions related to evaporated seawater, enough data exist to adequately describe the portion of the required phase diagrams describing the formation and transformation of common minerals (Braitsch, 1971). However, even for these relatively well-determined systems another problem is encountered. Analysis using phase diagrams is limited by the number of degrees of freedom (compositions, temperature, and pressure) which may be represented simultaneously on a diagram. While this approach is essentially exact for simple systems it cannot be generalized to include all the compositional degrees of freedom typical of natural environments. Temperature and pressure may influence the processes of interest and also need to be included in the analysis. For quantitative studies phase diagrams typically are limited to two or three degrees of freedom. For example, reporting solubility of $CaSO_4$ in $NaCl-H_2O$ requires two variables. To represent the solubility of $CaSO_4$ in KCl, NaCl, H_2O mixtures requires more variables than can be easily represented on a two dimensional diagram. But a typical seawater evaporite formation process problem may require six or more components. Various projections can be introduced and assumptions can be made as to the important species, but these approaches introduce uncertainties in the analysis and may even lead to incorrect results (Harvie et al., 1980; Eugster et al., 1980).

These problems would be eased by the development of a theoretical model of the chemical behavior of the system. A theoretical model can provide a means of interpolating relatively simple binary and ternary data into more complicated quaternary and higher systems, thereby reducing the dependence on direct measurements in complex systems and removing one of the most significant limitations of the phase diagram approach. Computer models are also able to deal with many variables, thereby removing the second difficulty with the application of phase diagrams.

Theoretical models of natural waters at low concentrations (total dissolved solids less than 0.2m) were introduced by Garrels and Thompson (1962). Models of this kind represent phase equilibria by introducing mineral solubility products appropriate for the solid phase and an expression for activity coefficients of the aqueous species based on Debye-Hückel behavior. Mineral equilibria or supersaturation is calculated by solving the mass action equations constrained by mass balance on a computer. For systems at low concentration, these models have been generalized to treat a very wide variety of species for a very large range of pressure and temperature by Helgeson and coworkers (1970).

These models are theoretically appropriate for systems at very low concentration. However, many natural brines have fairly high concentration of dissolved solids. Extensions of the models to systems of high concentration requires important modifications in the treatment of the activity coefficients as a function of concentration.

Recently, we have shown, (Harvie and Weare, 1980; Harvie et al., 1984[1]; Felmy and Weare, 1986) that, in principle, models of sufficient accuracy and complexity to describe most natural near equilibrium water processes can be developed provided the necessary experimental data is available. The limitation of models to low concentration is created by the choice of equations used to describe the concentration behavior of the activity coefficients of the species in the aqueous phase (Nordstrom et al., 1979). These equations are not well-suited to treat concentrated systems. However, recent developments in the phenomenology of electrolyte solutions have led to highly reliable models of electrolyte behavior for even very concentrated solutions. In the work of Weare and coworkers, one of the most successful of these approaches has been adapted to treat geochemical systems. The details of this model have been discussed by Pitzer in chapter 4 of this volume. In this chapter, we will describe some aspects of the parameterization of this model for natural water solubility calculations. The most significant limitation to the wider application of this approach is the lack of experimental data.

The models of Weare and coworkers have been successfully applied to interpret geochemical processes in a variety of well characterized natural systems; studies of ancient (Zechstein II: Harvie et al., 1980; Eugster et al., 1980) and modern (Bocana de Virrila: Brantley, 1984) marine evaporite systems, studies of the solar evaporation of seawater (Møller in press), studies of mineral precipitation in lakes (Great Salt Lake: Spencer et al., 1985; Searles Lake with borate species added: Felmy and Weare, 1987), and in fluid inclusions (collaboration with B. Lazar and H.D. Holland). The agreement between the calculations of the model and field measurements has been remarkably detailed and has led to the identification of various geochemical controls which may be operating on the systems. Some of these applications will be reviewed here.

---

[1] In the following, the model in Harvie et al. (1984) will be referred to as the HMW model. The parameters for this model have been collected in the Appendix. The equations containing these parameters are compiled in Chapter 4.

OVERVIEW OF THE MODEL

Because of the great importance of highly concentrated brines to natural and industrial processes, there has been a continuing effort to develop models of their behavior as a function of the intensive variables pressure, temperature, and composition. Unfortunately, present first principle models of brine thermodynamics are very far from providing a description of complex brines that would be convenient enough for application to geochemical problems. On the other hand, recently developed phenomenological equations, which have some basis in the current theory of electrolyte solutions, appear to have the required flexibility. These models contain parameters which must be evaluated from experimental data. Because of this, it is important to choose a phenomenology which will have the required generality and accuracy while introducing a minimum number of parameters. A variety of approaches have been suggested in the literature (Bromley, 1973; Meissner, 1972; Pitzer, Chapter 4). Most of these approaches, while providing the necessary accuracy in binary systems, have not been generalized to accurately treat complex mixtures which are typical of natural water problems. The approach introduced by Pitzer and coworkers (discussed in Chapter 4) appears to satisfy both the mixing requirement and to introduce a minimum of parameters. In a number of articles (Harvie and Weare, 1980; Harvie et al., 1984; Felmy and Weare, 1986; Møller, in press), we have demonstrated that these equations can be used to describe complex aqueous systems to high concentration and for a range of temperatures up to 250°C (see also Pabalan and Pitzer, in press).

Our model development begins with the well known free energy minimum criterion giving the equilibrium concentration for a system at constant temperature and pressure:

$$\text{minimize } G = \sum_j n_j \mu_j \tag{1}$$

subject to mass balance constraints

$$\sum_j a_{ij} n_j = b_i \quad i = 1,M \tag{2}$$

$$n_j \geq 0. \tag{3}$$

In Equation 1, G is the Gibbs free energy. M is the number of components. n_j is number of moles of species j, and μ_j is the chemical potential of species j.

The method of solving the minimization problem, Equations 1-3, is an important aspect of a model. The algorithm chosen must be efficient and reliable. Problems with reliability become much more severe for highly concentrated very nonideal systems. In our model, the free energy is minimized by an algorithm which has been developed to treat the special problems of phase equilibrium calculations. This algorithm:

1. Incorporates stable numerical methods, providing not only rapid convergence but also guarding against failure because of a poorly conditioned problem or inadequate selection of initial concentrations.

2. Is capable of adding and removing mineral, gas, liquid, and solid solution phases. It selects the correct phase assemblage for an arbitrary set of composition, temperature, and pressure conditions.

3. Generates its own stable reaction paths which are automatically chosen to emphasize major species
4. Can precipitate and dissolve many phases simultaneously in systems constrained by mass balance. Again, it independently chooses the most stable phase assemblage.
5. Can handle nonconvex solution models with some instructions given by the user.

An article with an extensive description of this algorithm has recently been published (Harvie et al., 1987).

In addition to Equation 1 for aqueous phases it is generally convenient to further define measurable properties in terms of an ideal solution reference. In order to do this the activity, a_i, is defined by the equation

$$\mu_i = \mu_i^o + RT \ln a_i \,. \tag{4}$$

The relation to ideal solutions is further emphasized by defining the activity coefficient, γ_i, for solute species by the equation

$$a_i = \gamma_i m_i \,. \tag{5}$$

For the solvent, water, the activity is written in terms of the osmotic coefficient, ϕ, defined by

$$\ln a_{H_2O} = \frac{-W}{1000}(\sum_i m_i)\phi \,. \tag{6}$$

In Equations 5 and 6, m_i is the molality of species i. W is molecular weight of water. The sum in Equation 6 is over all solute species.

Given Equations 4-6 the free energy of the solution phase is defined provided the dependence of the activity coefficient and osmotic coefficients are given as a function of concentration. The equations of Pitzer and coworkers (see Chapter 4), when parameterized using experimental data, provide the necessary concentration dependence to fully define the chemical potentials for all the species in an aqueous phase.

For systems in which there is more than one phase, free energies equations must be specified for each phase. For gases, this is usually done by introducing the fugacity via the equation

$$\mu_i = \mu_i^o + RT \ln f_i \,. \tag{7}$$

In this case the fugacity must be defined as a function of gas phase composition, pressure, and temperature. For low total pressure applications in geochemistry it is often sufficient to assume that the fugacity of species i is equal to the partial pressure of gas phase species i. However, for complex gas mixtures at high pressure, such as that found in crustal environments, this assumption clearly is not adequate (see Ferry and Baumgartner, Chapter 9).

At sufficiently low concentrations, the activity of ions in an aqueous solution obey universal behavior which is not specific to the identity of the ion. This is the well known Debye-Hückel limiting region. At higher concentrations, however, the chemical identity of the species begin to play an important role in the thermodynamics of the system. While some of this behavior can be described by a phenomenological ionic strength dependent extension of the Debye-Hückel equation (e.g. the Davies equation), important contributions which are specific to the chemistry of the species are evident in measurements even at fairly low ionic strength. For example for $I = 0.01$, the mean activity coefficient of $CaCl_2$ is 0.727 and for $Ca(OH)_2$ is 0.671. At higher concentrations, these differences become much larger. To correctly predict solubilities in complex systems, phenomenologies must be able to treat these specific interactions. For complex concentrated aqueous systems there may be a large number of minerals or combinations of minerals which are candidates for the equilibrium assemblage coexisting with the brine. In a typical situation various mineral associations have similar stabilities requiring highly accurate models of the solution phase to correctly select among alternatives (see Table 8 and discussion). Even small errors in the solution free energy may yield qualitatively incorrect prediction of mineral paragenesis.

In the Pitzer approach (Chapter 4), this specific interaction behavior of the ions is described by expressing the activity coefficients of the aqueous species in virial expansion form

$$\ln \gamma_i = \ln \gamma_i^{DH} + \sum_j D_{ij}(I)\, m_j + \sum_{jk} E_{ijk}\, m_j\, m_k \,..... \quad . \tag{8}$$

γ_i^{DH} is a modified Debye-Hückel coefficient which is equal to the "z_i^2F" term in $\ln\gamma$ in Chapter 4. $D_{ij}(I)$ and E_{ijk} are specific for each ion interaction. As indicated, D is a function of ionic strength. The form given by Pitzer for $D_{ij}(I)$ is different for like-like (e.g. $i = K^+$, $j = Na^+$), like-unlike (e.g. $i = K^+, j = Cl^-$), and ion-neutral (e.g. $i = Na^+, j = CO_2(aq)$) interactions reflecting the expected difference in these interactions in solution. The functional form for D given in Chapter 4 (the coefficient of the term linear in m in Pitzer's expression for $\ln\gamma$) for like-unlike interactions contains various parameters (e.g. $\beta^{(0)}$, $\beta^{(1)}$ and $\beta^{(2)}$, and C). Each D_{ij} for like-like interactions contains one parameter for θ_{ij} while all E_{ijk} (related to Pitzer's C and ψ) are constants. All these parameters are evaluated by comparison to experimental data. (A compilation of the parameters for the HMW model is included in the Appendix.)

A significant simplification in the parameterization process is shown by Equation 8. The most important contributions to the concentration behavior of the model comes from the binary terms (e.g. D_{ij} where i and j refer to oppositely charged species). These terms may be determined from binary thermodynamic measurements. For example, in the system $CaSO_4 - NaCl - H_2O$, the Ca-Cl parameters contained in $D(I)_{Ca,Cl}$ can be evaluated from binary osmotic data in the $CaCl_2 - H_2O$ system, as can certain triple interaction terms such as $E_{Ca,Ca,Cl}$ which is related to Pitzer's $C_{Ca,Cl}$ (Mϕller, in press; Holmes et al., 1978; Pitzer, Chapter 4). At this point, these parameters are fixed and additional parameters which must be evaluated in ternary systems are evaluated from ternary measurements (e.g. $D_{Ca,Na}$ from solubility measurements of NaCl in $CaCl_2$ brines). Once these parameters have been established, they are used in Equation 8 to describe systems of arbitrary composition.

If the intention is to use the model at high concentration, it is important to parameterize both binary and ternary data at high concentration. This is because binary third virial terms (e.g. C_{CaCl}) and ternary second and third virial terms (θ^E's and ψ's) are insensitive to low concentration data. Evaluating them from low concentration measurements, no matter how accurate, may provide a poor extrapolation to high concentration.

For moderately concentrated systems ($I < 1m$) a model of a complex brine using only binary information gives quite reliable solubility predictions. For these systems, accurate estimates of the D_{ij} for unlike charged species (e.g. D_{NaCl}), which can be obtained from binary measurements, will often be sufficient to describe the concentration dependence. For higher reliability and for systems which are more concentrated, it is necessary to evaluate the additional parameters in Equation 8. The solubility calculations that we have performed and which are described in the following sections have shown that for most systems of geologic interest up to the solubility limit the series represented by Equation 8 may be terminated at the triplet interaction (m^2) terms which may be evaluated from ternary data. Once all the binary and ternary interaction parameters have been evaluated, a system of arbitrary complexity may be treated. This results in a significant simplification when considering highly complex systems.

The parameters in the Pitzer equation may be degenerate. That is, for a limited and not very high quality data set, several sets of parameters may fit the data with roughly the same accuracy. For example, in evaluating Pitzer parameters from binary cell measurements with a limited number of data points a range of β^0 and β^1 may give agreement within experimental error. Usually these binary parameters will need to be refined when ternary data is included. These degeneracy problems tend to be more severe when evaluating mixing behavior. This is partly because ternary data is less available. In such situations, it may be an advantage to set some parameters equal to zero or to some value that is taken from another similar system. It is always important to control the size of the parameters. Models which have large ternary parameters are apt to be poorly behaved in regions of high concentration. A good guide to the acceptable size of parameters is found from the parameters collected in HMW and given here in the Appendix. Degeneracy is not limited to the Pitzer approach, but is common to any complicated model. It may be more severe in ion association models (see next section and Nordstrom et al., 1979) because of the strong nonlinear dependencies introduced by speciation.

The above equations have been written in terms of single ion activities. As discussed in HMW, because of the condition of electric neutrality, single ion activities cannot be measured. From a computational point of view, however, they provide a significant simplification. A measured property corresponds to some neutral combination of ions, for example, the product of the activity coefficient of Na^+ and Cl^-. Such products are called mean ion activities. Separation of activities into single ion values implies some convention. Any way of separating the mean activity product into single ion contributions is correct as long as the value of the mean ion activity coefficient is not changed. The separation of the Pitzer equations into single ion contributions as in HMW amounts to a convention. This, of course, does not cause a problem provided single ion activities in different conventions are not compared.

MODELS FOR SYSTEMS SHOWING STRONG ASSOCIATION: IONS PAIRS VS. SPECIFIC INTERACTION

One point about the application of the above equations deserves special emphasis. The Pitzer equations provide a means to interpolate behavior from regions of low concentration to regions of high concentration. However, they cannot replace the relations describing the complex low concentration chemistry that many important systems display (e.g. Al, B). This has led to some confusion. The first step in any modeling program must be to identify the important chemical behavior of the system at low concentration. For example, the pH behavior in the Al system is influenced by the formation of various hydration species $AlOH^{2+}$, $Al(OH)_2^+$ etc. (Baes and Mesmer, 1976). The Pitzer equations, as discussed in the previous section (Eqn. 8), cannot be used to describe these strong association interactions (see Chapter 4). For aqueous solutions, strong association effects are usually adequately described in terms of association equilibria (e.g. $Al^{3+} + OH^- \leftrightarrows AlOH^{2+}$). Once this speciation chemistry has been identified, the specific interaction equations may be used to interpolate the thermodynamics into the concentrated region.

The situation is complicated by changes in the chemical behavior with composition, temperature, and pressure. Usually ion association becomes stronger at high temperature. In favorable cases, there may be spectroscopic evidence for associated species in the aqueous phase. In Figure 1, taken from Irish et al. (1984), we show experimental evidence of this kind. The appearance of the second peak in the raman spectra of dissolved $MgSO_4$ ($\lambda \approx 990\ cm^{-1}$) is interpreted as the formation of an associated species (Irish et al., 1984; Archer and Wood, 1985). Unfortunately, while spectroscopic evidence is extremely valuable to identify the species present, it does not necessarily provide an accurate means to evaluate an association constant (Dawson et al., 1986).

Once the strong speciation chemistry at low concentration has been identified, a phenomenological approach must be chosen to extrapolate the behavior into the concentrated region. For aqueous solutions, two approaches have been utilized: approaches based on virial expansion (e.g. the Pitzer equations) and approaches based on introducing further speciation to describe weaker association effects (e. g. the introduction of ion pairs such as $NaSO_4^-$(aq)).

Unfortunately, the situation may not be well defined. For species such as aqueous Ca^{++} ions and $SO_4^=$ ions, which definitely show some association, it is not always clear whether the specific interaction approach or the ion pair species approach will provide the most efficient parameterization. As discussed in considerable detail in HMW and in Chapter 4, Pitzer and coworkers have suggested the inclusion of an additional specific interaction parameter, β_2, to describe such situations. The alternative approach, also suggested by Pitzer, but used more often in other models introduces ion associated species (e.g. $CaSO_4$(aq) ion pair) (for a review of ion pairing models see Nordstrom, 1979). In situations where the activity of a particular species is less than the activity predicted by an extended Debye-Hückel expression (a negative deviation), ion association at least offers the possibility of providing the correct behavior. However, for situations where the deviations from Debye-Hückel activity are positive, this approach cannot be used. Many systems in going from high to low concentration display both positive and negative deviations.

In the models we have developed, the choice between these two alternative phenomenological descriptions is made on the basis of which is the most efficient representation of thermodynamic data. This is well illustrated by contrasting the parameterizations of $CaSO_4$ solubility by HMW with the more recent work in the same system at high temperature by Møller (in press). In developing the HMW model, all the data in the $Na-Ca-Cl-SO_4-H_2O$ system were included in a data base. As usual in our modeling approach, as many binary parameters (e.g. parameters in $D_{Ca,Cl}$,D_{Na,SO_4} etc, see Chapter 4) as possible were evaluated from binary data. In order to describe the $CaSO_4$ system, both the specific interaction approach (i.e. the addition of β_2) and an ion pairing approach (inclusion of an ion pair, $CaSO_4$ (aq)) were tried. In the interest of efficiency, no additional parameters describing the activity of the neutral $CaSO_4$ (aq) species were included in the ion pairing approach. This means that only one additional parameter was included in each approach (both approaches included the usual $\beta^{(0)}_{Ca,SO_4}$ and $\beta^{(1)}_{Ca,SO_4}$ parameters). At this point, both models were optimized to give the best agreement with experimental data. Both models were in good agreement with the data, but it was found that the specific interaction approach (addition of β_2) was somewhat better over the entire data set (approximately 5% improvement in σ (see HMW)). It is important to recognize, of course, that the inclusion of more parameters in the ion association model would improve the accuracy of that approach, but would be a less efficient method.

On the other hand, Mϕller, using the same modeling techniques, found that at higher temperatures in the $CaSO_4$ system ion association did provide a more efficient representation of the data. Typically ion association effects become strong at higher temperatures (see Fig. 1). When these effects become sufficiently strong, an approach based solely on specific interaction (e.g. the inclusion of a β_2) will not provide a good representation of the data. At this point ion associated species must be introduced. Typically when these species are introduced, $\beta^{(2)}$ parameters are no longer necessary. For some situations it may be desirable to include ion association even though it is not necessary to describe the data. However, in such situations more parameters may have to be introduced to describe positive deviations resulting in a less efficient model. As an example, Mϕller found that in order to continue her speciation model to lower temperatures (e.g. 25°C), an additional parameter was necessary.

At 25°C, $Ca–SO_4$ interactions represent an intermediate level of association. The association behavior at 25°C is almost as well described by introducing a neutral associated species as by specific interaction. For other ions, such as $Na-Cl$ and $Na-SO_4$ interactions, ion association is simply not evident in the equilibrium thermodynamic data at 25°C. Inclusion of ion association in such systems neglecting specific interaction will not provide an accurate representation of the data.

On the other hand, for some important systems (e.g. the carbonate system) ion association effects are very strong. This situation is illustrated by the formation of the bicarbonate species (HCO_3^-) which may be considered as an ion association of H^+ with $CO_3^=$. If a specific interaction approach is taken for such systems, very large and negative $\beta^{(2)}$ parameters will be obtained and the resulting description ot the thermodynamic data will be poor. The large size of the $\beta^{(2)}$ parameters is an indication that the virial expansion is not quickly convergent. If such parameters are retained, incorrect behavior (i.e. non convex free energy) may result. In our approach, we introduce ion speciation equilibria when virial parameters become too negative (a good indication of ion association). A guide to the size limits we tolerate for the β_2 parameters may be found in the Appendix and in HMW. There are very few negative parameters in Tables A1-A3. For the seawater system at 25°C, ion association does not play an important role.

The choice of which parameterization to use for a given system is of central importance to the development of a model . Some guidelines may be found from tabulations of ion association equilibria. For example, HMW were able to give a criteria for the inclusion of ion associated species which appears to be reliable. For ion associations of the form $A^{2+} + C^- \rightarrow AC^+$ they suggest that when $K_a > 100.$ a speciation model will be more efficient than a specific interaction model. This criterion appears to work for the H/SO_4 interactions. HMW chose K_a for this system to be 100. and the ion association model of the thermodynamics appears to give a better representation of the thermodynamics of the system than a specific interaction model. For K_a much less then 100., replacement of ion pairs by a specific interaction parameter ($\beta^{(2)}$) is likely to provide a more accurate description. For systems in which association is weak, the data often does not uniquely define a value for the association constant.

The borate system recently studied by Felmy and Weare (1986) illustrates model parameterization in chemically very complex systems. In aqueous solutions boron is known to form highly associated polymeric species. The species present in the solution are a strong function of the composition. The concentration and complexity of the polymeric species increases with increasing boron concentration. This situation is not unique to boron, but is true of many other important elements such as Al and Pb. Since borate minerals are quite soluble a model of high concentration behavior is necessary to describe their mineralogy. This implies that a useful model of the borate equilibria will have to describe both polymerization and specific interaction. To illustrate this, Table 1 shows the borate speciation in a solution saturated with respect to borax ($Na_2B_4O_7 \bullet 10H_2O$). In this example there are four different borate species with roughly the same concentration. In order to provide an accurate model, all of these species and the interactions between them must be described.

At low concentrations, boron exists in aqueous solutions as the monomeric boric acid $B(OH)_3$ species and as the hydrolization product of this species $B(OH)_4^-$ via the reaction

$$B(OH)_3 + H_2O \leftrightarrows B(OH)_4^- + H^+ \quad . \tag{9}$$

Such a simple speciation equilibrium is typical of many important species found in natural waters such as the $HCO_3^-/CO_3^=$ pair and the $HSO_4^-/SO_4^=$ pair. As such , there are no special problems with parameterization (see for example HMW). However, in the borate system, as the concentration of total boron is increased (beyond approximately 0.03 m), the acid base equilibria no longer shows the simple behavior expected from Equation 9. This effect has been accounted for by the introduction of several sets of polymeric species. While each of these sets of species works for an individual data set, it is much more demanding to find a speciation scheme which describes the entire data set. Felmy and Weare (1986) found that it was necessary to include only the two polymeric species $B_3O_3(OH)_4^-$ and $B_4O_5(OH)_4^{-2}$ to describe the entire range of data in the literature to the desired accuracy. Once a speciation scheme has been decided upon, the many other parameters which describe the interaction between species in solution must be evaluated.

The detailed process for the borate system has been described in Felmy and Weare (1986). The simplest systems with the fewest parameters are parameterized first. For example, boric acid solubility in various salt solutions were used to evaluate neutral boric acid ion interaction parameters. After finding the neutral ion interactions, the interaction parameters for the orthoborate ion were obtained from osmotic measurements in salt systems. For example, the $Na^+ - B(OH)_4^-$ interaction was obtained from sodium metaborate osmotic data (Platford, 1968). The accuracy of these parameters were checked by comparison of prediction with the emf data of Owen and King (1943).

In order to describe strong ion association with the divalent cations in the seawater system, $CaB(OH)_4^+$ and $MgB(OH)_4^+$ species were included in the model and were obtained from emf data (Hershey et al., 1985). Specific ion interactions of the $MgB(OH)_4^+$ ion with the Cl^- ion were necessary.

When all these parameters had been established at low boron concentration, emf and solubility data for higher boron concentration were used to evaluate polymerization reactions. An example of the accuracy of the phase description possible in such a complicated system is given in Figure 2. The accuracy represented by this figure is

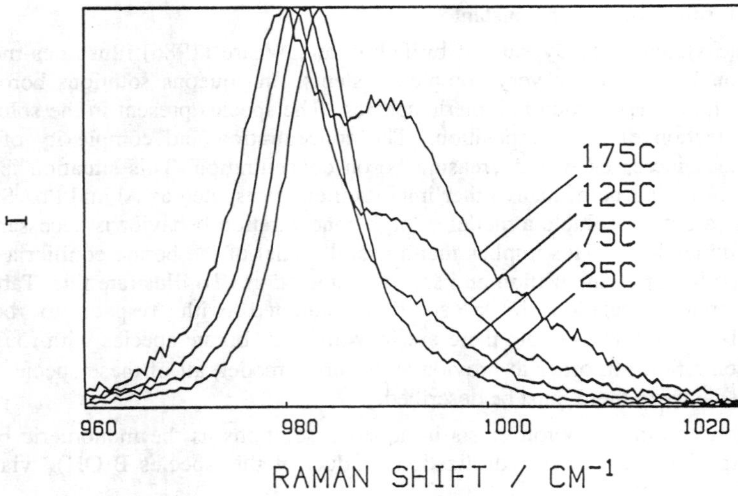

Figure 1. Raman spectrum (Irish et al., 1984) of an aqueous solution of $MgSO_4$ (2.18 m) illustrating ion association at various temperatures between 25 and 175°C (intensity vs. wavenumber).

Table 1. Calculated speciation in pure water saturated with respect to borax ($Na_2B_4O_7{\bullet}10H_2O$) at 25°C.

Species	Concentration (m)
Na^+	.323
$B(OH)_3$	.081
$B(OH)_4^-$	.131
$B_3O_3(OH)_4^-$	.050
$B_4O_5(OH)_4^{2-}$	.071

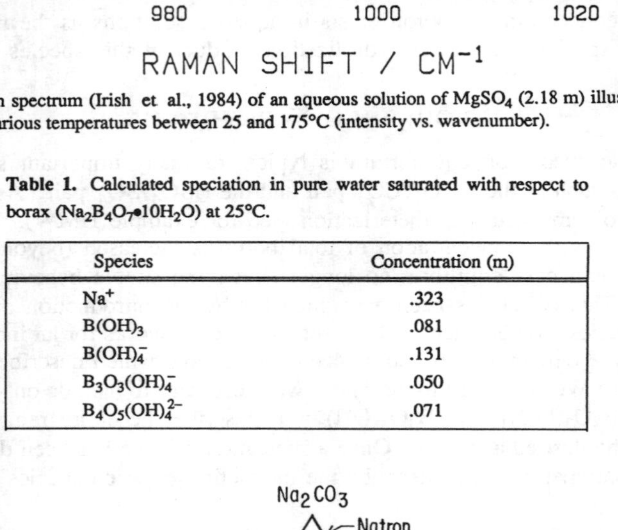

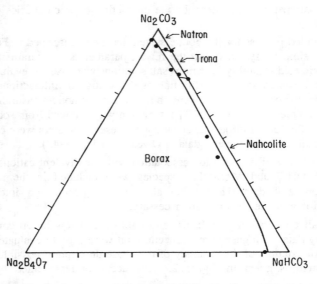

Figure 2. Janëcke projection for the system $Na_2B_4O_7-NaHCO_3-Na_2CO_3$ at 25°C. Experimental data taken from Linke (1965).

encouraging in view of the complications involved with this system.

It is sometimes desirable to transfer parameters form one model to another. For example, to use ion association constants from measurements of one researcher in a model developed by another worker. Such transfers should be viewed with caution. The models generated this way must be extensively tested before they can be applied with confidence. In a particular model, even the standard properties are likely to be a function of the model used to extrapolate to the standard state. For example, solubility products for soluble minerals (e.g. NaCl) must be measured in concentrated solutions. In order to obtain a thermodynamic equilibrium constant, calculated ion activities must be used to extrapolate to zero concentration. Therefore, the K_{sp} obtained is a function of the model used to calculate activities. In order to retain consistency, this same model for activities should be used when predicting solubility. Similar care must be taken with other standard properties (e.g. ion association constants, etc.) that are extracted from solution behavior. Many measurements are not taken to low enough concentration to provide a unique extrapolation to zero concentration. The situation is similar for the various parameters in the Pitzer equations. On the other hand, utilizing parameters from different sets of measurements can be very valuable. An excellent example is the work of Monnin and Schott (1983) who synthesized a temperature dependent model for the sodium carbonate system by using combinations of parameters from Peiper and Pitzer (1982) and HMW. However, these authors were particularly careful to validate their synthesized model against experimental data. Also, the models of Pitzer and of HMW have the same limiting behavior.

INCLUSION OF TEMPERATURE AND PRESSURE AS VARIABLES

The modeling approach I have discussed considers the evaluation of parameters which describe the compositional variation of the free energy at a given temperature and pressure. In order to describe the behavior of the system for a range of temperature and pressure, the pressure and temperature dependence of the parameters must be given. This can pose a difficult problem for model building because the speciation in a system can change dramatically as a function of temperature (see Fig. 1). In the seawater system, electrolytes such as NaCl and KCl do not seem to show a change in speciation as a function of temperature. However, the divalent electrolytes may show strong variation.

To address the development of models with temperature and pressure variation, it is useful to distinguish between parameters describing the limiting behavior of a system, for example, standard chemical potentials and their related thermodynamic equilibrium constants and those describing the extension of the model to high ionic strengths. For the standard properties Helgeson and coworkers (see Chapter 6) have developed an elaborate estimation scheme based on limited data and analogy with other systems. The reliability of such extrapolation is typically about a 0.1 pK unit. While this kind of accuracy may be sufficient to describe problems where there is a large difference between the free energies of the possible choices of coexisting phases, there are many situations in highly concentrated brines where better accuracy is necessary (see Table 8 and discussion). Because of this, the recent extension of the model of HMW (Møller, in press) and the work of Pabalan and Pitzer (in press). have relied on a elaborate fit of both the standard properties and the binary interaction parameters to the thermodynamic data as a function of temperature (see also the work of Holmes and Mesmer, 1986). From this work it

Table 2. Invariant Points[a] for the Quarternary System, NaCl-Na_2SO_4–$CaSO_4$–H_2O at 110°C. (Experimental values[b] are given in parentheses)

Point	Solution Composition m_{Na}	m_{Ca}	m_{Cl}	m_{SO_4}	Solid Phases	a_{H_2O}
1[a]	6.734	.0259	6.734	.0259	H,A	.7413
2	6.930 (6.934)	.00637 (.00590)	6.664 (6.716)	.1391 (.1150)	H,A,Gl	.7401
3	7.381	.00121	6.509	.4374	H,Th,Gl	.7364
4	7.382 (7.614)	.0 (.0)	6.590 (6.586)	.4366 (.5143)	H,Th	.7365
5	5.976 (5.838)	.00371 (.00312)	.0 (.0)	2.992 (2.922)	Th,Gl	.9040
6	3.292 (2.586)	.00794 (.00835)	.0 (.0)	1.654 (1.301)	A,Gl	.9445

a. Point no. 1 is a maximum in a_{H_2O} along the univariant line of halite-anhydrite coexistence.

b. Gromova (1960). See Stephens and Stephens (1963), p. 914.

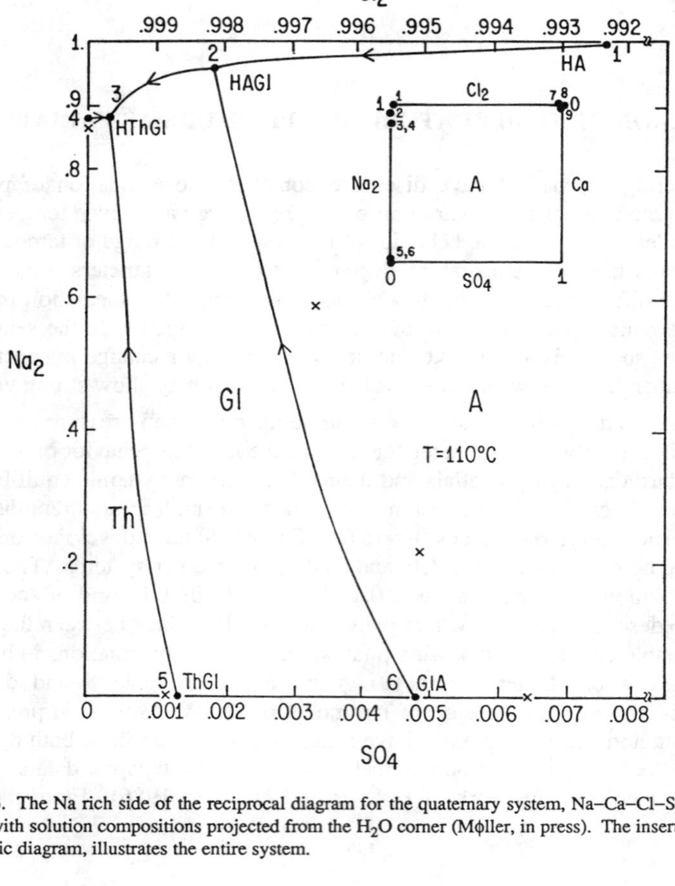

Figure 3. The Na rich side of the reciprocal diagram for the quaternary system, Na–Ca–Cl–SO_4–H_2O, at 110°C with solution compositions projected from the H_2O corner (Mϕller, in press). The insert, which is a schematic diagram, illustrates the entire system.

appears that ternary parameters are adequately described by a simple temperature dependence. This amounts to an important simplification since the temperature dependence of the parameters (Holmes and Mesmer, 1983, 1986; Chapter 4) is already available for a number of important binaries. An example of the solubility calculations of Mϕller (in press) in a quaternary system at 110°C is given in Figure 3 and Table 2. The accuracy and flexibility of the Pitzer phenomenology appears to be sufficient to describe systems at high temperature and pressure. The real limitation for modeling such systems is the lack of sufficient data. References to high temperature and pressure data and available models may be found in Chapter 4 (also Mϕller, in press, Greenberg and Mϕller, in preparation, and Millero, 1979, for data in the seawater system). Recent pressure data and models have been reported by Millero et al. (1985) and Monnin (1987). Pressure and temperature data for the important carbonate system have been measured by Patterson et al. (1982, 1984) and by Plummer and Busenberg (1982).

MODELS FOR POORLY DETERMINED SYSTEMS

Often a model of a system for which there is very little data is necessary. Recently, Mϕller (in press) investigated the use of the Pitzer phenomenology to describe the solubility of $BaSO_4$ in NaCl rich brine system. Rather than reparameterize all the binaries in this system (e.g. obtain Ba/Cl interactions form osmotic data) Mϕller took most parameters from the $CaSO_4$–NaCl–H_2O system. Examples of her results is given in Figure 4. These results indicate that an adequate model may be obtained even for very high concentration by the adjustment of only a single parameter (in this case $\theta_{Ba,Na}$). The reliability of this approach is still somewhat in question, however, since this model has not been tested against mixing data in common ion systems. As discussed in the following in connection with Figure 5, data in the common ion system is extremely sensitive to the detailed description of mixing since the solubility is directly related to the strength of the association constant. This data does not exist for the $BaSO_4$ system.

COMPARISON OF THE HMW MODEL TO OTHER MODELS

The approach that we have taken to the evaluation of parameters is similar to that of other groups which have emphasized the Pitzer phenomenology (Chapter 4; Holmes et al., 1978; Holmes and Mesmer, 1983, 1986). It does differ in one important aspect. We place a larger emphasis on solubility data than other groups. Our model is intended for prediction of the solubility of minerals in natural waters. Since the Pitzer phenomenology is not exact, it is sometimes necessary to relax some of the accuracy (give less weight) in fitting non-solubility measurements (e.g. emf and osmotic measurements) in order to obtain the required accuracy in predicting solubility. These losses in accuracy are always very small. When using data other than solubility data to estimate parameters (e.g. heat data, osmotic data, emf data) it is important to recognize that parameters established at very low concentration may not give good extrapolation for higher concentration solubility predictions. For example, using low concentration heat capacity measurements ($I < 1m$) to estimate β^0, β^1, and C^ϕ in Pitzer's equation can yield a model that is not convex at the high concentrations where solid Na_2SO_4 precipitates (Mϕller, in press).

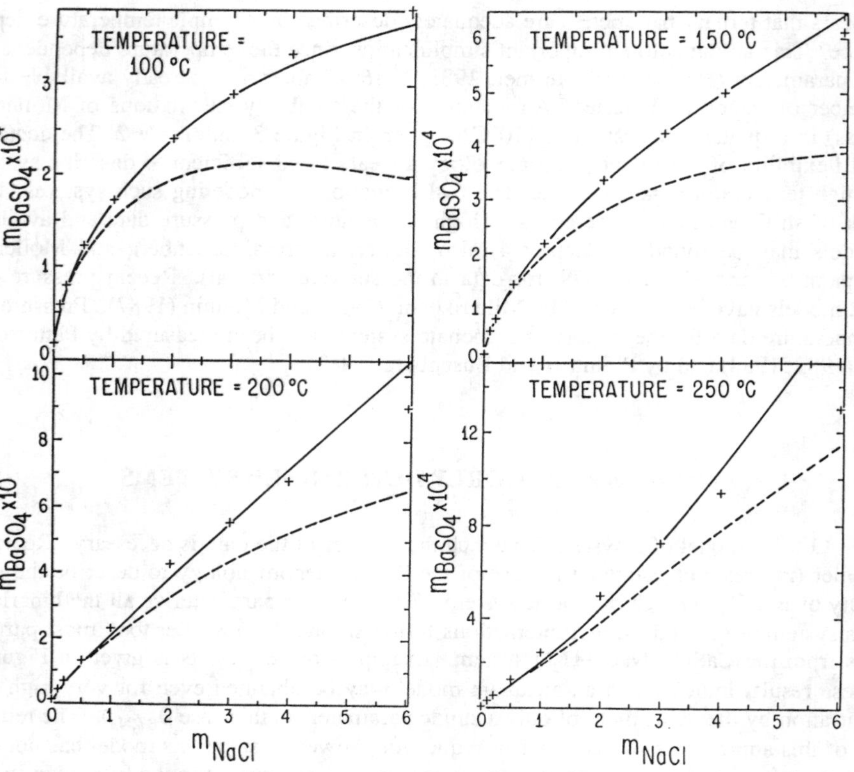

Figure 4. The prediction of barite solubility in aqueous NaCl solutions at 100°C, 150°C, 200°C and 250°C using parameters for the $BaSO_4$ system taken from the $CaSO_4$(T) model and: (1) the chemical potential (μ^o/RT) for barite determined from a single $BaSO_4$–H_2O point (dashed line) or (2) the chemical potential for barite determined in step 1 and $\theta_{Na,Ba}$ fit from the 3.m NaCl point (solid line) (Møller, in press). The data (+) are those of Blount (1977).

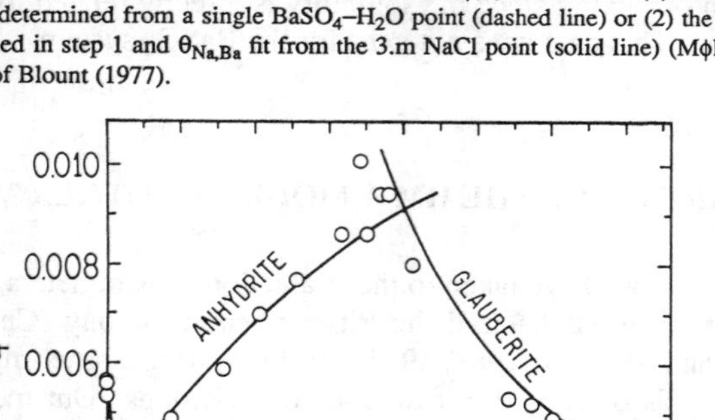

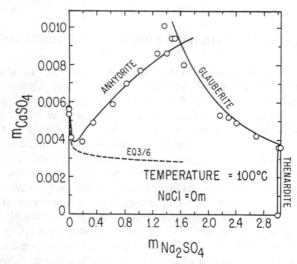

Figure 5. A comparison of model calculations (solid lines) of anhydrite, glauberite and thenardite solubilities in $CaSO_4$–$NaCl$–Na_2SO_4–H_2O solutions at 100°C (Møller, in press) with experimental data (NaCl = 0. m). The dashed line is predicted by EQ3/6.

Table 3 illustrates the complexity of the data set necessary to fully describe the $CaCO_3 - NaCl - CO_2 - H_2O$ system which would be used to predict the solubility of calcite in NaCl rich waters. Data in almost all these systems is available at 25°C, and has been used to parameterize the HMW model. References to this data have been summarized in that article. Recent data for the carbonate system have been reported by Thurmond and Millero (1982) and Millero and Thurmond (1983). Higher temperature data references have been collected in Chapter 4, Mϕller (in press), Pabalan and Pitzer (in press), and Greenberg and Mϕller (in preparation). The importance of including as complete a data base as possible in the parameterization is illustrated in Figure 5 taken from Mϕller (in press). For systems which contain very slightly soluble minerals such as $CaSO_4$, solubility data (e.g. Marshall, et al. 1964; Blount and Dickson, 1973) may provide the only means of parameterizing a model. Parameters in the model may be very sensitive to data in a particular direction in composition space. For example, for the system illustrated by Figure 5, the predicted solubility of $CaSO_4$ (anhydrite) as a function of Na_2SO_4 is very sensitive to ion association. The reliability of a model for $CaSO_4$ solubility will be greatly improved by inclusion of solubility in the $CaSO_4$–Na_2SO_4–H_2O system. This explains the very large discrepancies between the ion pairing model predictions of solubility using the EQ 3/6 program (Wolery, 1983) and the data in Figure 5. Other strictly ion pairing models give similar results (Harvie and Weare, 1980). Recent versions of EQ 3/6 include the 25°C HMW model.

The most important limitation to the widespread application of the modeling approach we use is the large number of parameters required to model a complex mixture. The proliferation of parameters in the Pitzer approach stems from the necessity to describe ternary interactions (θ and ψ, Chapter 4). Data to evaluate these parameters is also more difficult to obtain. However, we have shown (see Fig. 5) that it is necessary to describe ternary mixing in order to reproduce solubility data with the accuracy required to interpret natural processes. In the Pitzer approach, all ternary parameters are constants, and therefore it is difficult to see how their the number may be reduced. Other models based on ion association introduce roughly the same number of parameters through the extra ion species added to describe non-ideality (Harvie and Weare, 1980). As yet, it has not been shown that these models can successfully predict solubility in mixed systems at concentrations greater than 0.5 m.

While we recognize that the procedure we are describing for parameterization is complicated, we believe it to be the most efficient way to build models of the accuracy necessary to make reliable predictions about natural systems. When all the parameters have been evaluated, the model predicts mineral solubilities with high accuracy in very complicated systems. Figure 6 and Table 4 compare model predictions to measured solubilities in the quinary seawater system. The agreement of the model phase diagram with the measurement is close to experimental error (see also Fig. 3 and Table 2). Similar results have been summarized in HMW for the seawater plus carbonate system. With such extensive validation with laboratory data, the model can be applied with confidence to field settings.

In the development of our models, great emphasis is placed on high accuracy. As illustrated in the next section, this accuracy is necessary in order to provide useful predictions in complex settings. In view of this requirement, it is important to assess the accuracy of other available models. Results of a study of this kind have been reported by Kerrisk (1981). Harvie and Weare (1980) compared the results of two ion pairing models, WATEQF (Truesdell and Jones, 1974; Plummer et al., 1976) and SOLMNEQ

Table 3. Data used to parameterize the $CaCO_3$–NaCl–CO_3–H_2O system

Species pairs	Data
Binary Interaction Parameters	
H^+, Cl^-; Na^+, Cl^-; Ca^{2+}, Cl^-; Na^+, OH^-; Na^+, HCO_3^-; Na^+, CO_3^{2-}	e.m.f. and osmotic coefficient data.
Ternary Interaction Parameters	
CO_2, Na^+; CO_2, Cl^-; CO_2, HCO_3^-; CO_2, CO_3^{2-}	Solubility data as a function of composition.
H^+, Na^+; Na^+, Ca^{2+}; Cl^-, OH^-; HCO_3^-, Cl^-; CO_3^{2-}, Cl^-	e.m.f, osmotic coefficient and solubility data.
Speciation Constants	
$H^+ + OH^- = H_2O$ (K_w); $H_2O + CO_2(aq) = HCO_3^- + H^+$ (K_1); $HCO_3^- = H^+ + CO_3^{2-}$ (K_2)	e.m.f. and osmotic coefficient data.
$Ca^{2+} + HCO_3^- = CaHCO_3^+$; $Ca^{2+} + CO_3^{2-} = CaCO_3^\circ$; $CaHCO_3^+, Cl^-$	e.m.f. and ternary solubility data.

Table 4. Invariant points for the quinary system Na–K–Mg–Cl–SO_4–H_2O at 25°C. Experimental values are in parenthesis (Harvie and Weare, 1980).

Point	m_{Na}	m_K	m_{Mg}	m_{SO_4}	Solid Phases
M	2.62 (2.69)	1.63 (1.58)	2.08 (1.97)	.84 (.78)	Halite + Sylvite + Glaserite + Schoenite
N	2.52 (2.41)	1.59 (1.51)	2.16 (2.19)	.84 (.78)	Halite + Leonite + Sylvite + Schoenite
P	1.16 (1.31)	1.01 (1.10)	3.40 (3.17)	.86 (.79)	Halite + Sylvite + Leonite + Kainite
Q	.48 (.49)	.57 (.64)	4.21 (4.05)	.32 (.29)	Halite + Sylvite + Carnallite + Kainite
R	.30 (.26)	.22 (.20)	4.75 (4.81)	.40 (.35)	Halite + Kieserite + Carnallite + Kainite
S	5.20 (5.32)	1.04 (.90)	.95 (.89)	1.31 (1.24)	Halite + Thenardite + Glaserite + Bloedite
T	3.08 (3.21)	1.31 (1.30)	2.00 (1.85)	1.14 (1.06)	Halite + Glaserite + Bloedite + Schoenite
U	2.49 (2.96)	1.18 (1.22)	2.40 (2.01)	1.11 (1.05)	Halite + Bloedite + Schoenite + Leonite
V	1.43 (1.25)	.85 (.72)	3.31 (3.46)	1.14 (1.10)	Halite + Epsomite + Bloedite + Leonite
W	1.16 (1.18)	.85 (.75)	3.51 (3.54)	1.03 (1.04)	Halite + Epsomite + Leonite + Kainite
X	.76 (.66)	.50 (.68)	3.95 (4.01)	.81 (.79)	Halite + Kainite + Hexahydrite + Epsomite
Y	.40 (.33)	.23 (.32)	4.57 (4.63)	.59 (.45)	Halite + Kainite + Hexahydrite + Kieserite
Z	.09 (.07)	.02 (.02)	5.74 (5.83)	.06 (.05)	Halite + Bischofite + Kieserite + Carnallite

Table 5. Hydrolysis constants for aluminum hydroxide species and the solubility product of gibbsite.

Species	$\ln K_H^1$		
	May *et. al.* (1979)	Baes and Mesmer (1976)	EQ3
$Al(OH)^{2+}$	-11.49	-11.44	-10.91
$Al(OH)_2^+$	-23.33	-21.41	-
$Al(OH)_3(aq)$	-	-34.54	-
$Al(OH)_4^-$	-51.02	-52.96	-51.02
$Al_2(OH)_2^{4+}$	-	-17.73	-
$Al_3(OH)_4^{5+}$	-	-32.10	-
$Al_{13}O_4(OH)_{24}^{7+}$	-	-227.33	-
	$\ln K_{sol}^2$		
$Al(OH)_3(s)$	18.675	19.572	18.328

[1] Reactions are of the form: $Al^{3+} + H_2O = Al(OH)^{2+} + H^+$
[2] The reaction is: $Al(OH)_3(s) + 3\,H^+ = Al^{3+} + 3\,H_2O$

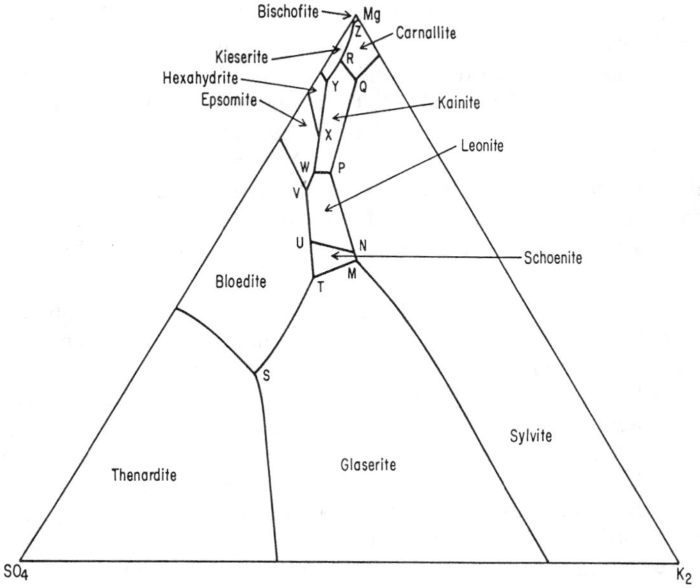

Figure 6. The calculated Na–K–Mg–Cl–SO_4–H_2O Janecke projection at 25°C (Harvie and Weare, 1980). All mineral zones denoted are also saturated with respect to halite.

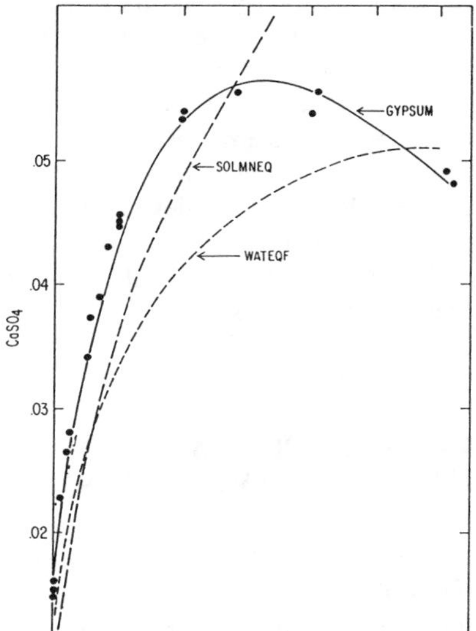

Figure 7. The calculated solubility of gypsum in NaCl aqueous solutions (Harvie and Weare, 1980) compared to experiment at 25°C. The results of two extended Debye-Hückel ion pairing models are also denoted (dashed curves).

(Kharaka and Barnes, 1973) to solubility data in the $CaSO_4-NaCl-H_2O$ system (see Fig. 7). A similar calculation using the EQ 3/6 program has been reported by Mϕller and is included in Figure 5 (a more recent version of the EQ 3/6 code includes the HMW model). As documented by Kerrisk (1981) for a wide variety of systems and a range of conditions, the results of Figures 5 and 7 are representative of the accuracy of all ion association models. Typically models such as EQ 3/6 and those represented in Figure 7 are expected to be more accurate in the important NaCl dominated systems. However, the basic problem with all these models for predictions at high concentration and for complex mixtures is the lack of recognition of the specific behavior of different electrolytes.

The inaccuracy of available models, however, is not limited to high concentration. As emphasized above before extrapolation to high concentration the chemistry of the system at low concentration must be carefully worked out from low concentration measurements. Even for some of the most common species in natural water the low concentration speciation chemistry has not been well described. Probably the most important system is aluminum. Large uncertainties in the experimental data for this system are a result of sluggish kinetics and the necessity of working at low concentrations because of the very low solubility of minerals containing aluminum. This has led to large discrepancies between the speciation schemes used in various models. In Table 5, the speciation constants for Al are taken from three different references. The constants from Baes and Mesmer (1976) represent selected values from a variety of experimental measurements including emf measurements and solubility measurements. The constants from May et al. (1979) were evaluated from solubility data. The EQ 3/6 constants were taken from the program. The very large differences in speciation schemes (e.g. different models don't even include the same set of species) is a result of the level of the understanding of the Al system. These calculations were made for 25°C. The comparison can only be made at this temperature because the necessary data to evaluate the speciation in the Al system at high temperature is not even available. This system has been reviewed in detail recently by Apps (1987). In presenting these results, I do not want to discourage the use of these models. However it is important to be aware of the large uncertainties in the chemistry of some very important systems. Even inaccurate models may provide insights into possible geochemical mechanisms. But the tremendous need for new data in order to refine models must be emphasized. I believe that given the necessary data, we have shown that models of the accuracy needed to describe natural systems can be based on existing phenomenologies. In the next section I will discuss the use of these models to provide insight into geochemical mechanisms.

OVERVIEW OF THE APPLICATION OF MODELS TO NATURAL ENVIRONMENTS

A model is most valuable when it can be used to distinguish between or provide constraints on possible geochemical mechanisms in field settings which are difficult to interpret on the basis of intuition or experience alone. The accuracy required of a model to provide such a capability is often not appreciated. It may vary from setting to setting and according to the question asked. For example, if the possible mineral assemblages that may be in equilibrium with the formation water are separated by large differences in free energy, a relatively inaccurate model may be able to provide a prediction of mineral paragenesis. On the other hand an accurate model will be necessary if the total amounts of minerals formed or the solution composition is important, or if there are a number of possible phase assemblages which have very similar free energies. In these cases, it is important to distinguish between the qualitative agreement of the model with field observations and the use of a model to make prediction or to test hypotheses about the mechanisms of formation. Even relatively inaccurate models may correctly predict the presence of a portion of the minerals in a particular assemblage, by for example, predicting supersaturation. However, it is a much more demanding test to predict equilibrium of the formation water (no supersaturation or undersaturation) with respect to the entire suite of minerals in the formation.

The time scales of the processes of interest must also be considered. Two time scales that are important are that of fluid composition change due to external control and the time scale of mineral formation or dissolution. For example, in an estuarine or salt lake environment the external time scale may be defined by the rate of evaporation (concentration) of the lake or estuary water. On the other hand, in diagenetic problems in deep sediments, the external time scale may be the convective flow of reacting fluids. Flow rates may differ by orders of magnitude for different settings. The rates of mineral formation and dissolution may also differ by many orders of magnitude. Laboratory rates for the formation of most common evaporite minerals (e.g. $CaSO_4$, NaCl) are of the order of days whereas feldspar dissolution experiments may take years to equilibrate. The problem is made even more complicated by the difficulty of extrapolating laboratory rate data to field settings.

The rates of the change in external controls versus the formation rates of the mineral in the system affects the state of equilibration in the formation. If the mineral reacts very quickly as in the case of evaporites (e.g. halite, calcite) the system should be in equilibrium with respect to the solid phases provided the external controls are varying on a slow time scale. Such a situation might be found in an estuarine environment in which the mechanism of change is slow evaporation. The field measurements on solar evaporation processes we will discuss in the following will show this to be the case for the minerals gypsum and halite. The situation is more complicated for calcite which also equilibrates in the laboratory fairly quickly. Unfortunately, for many environments, the great uncertainties in both the time scale of change of external control and the rates of mineral formation are so poorly known that it is difficult to estimate the state of equilibration.

This is an important constraint on the application of models. If a system may be assumed to be in equilibrium, thermodynamic models may be applied directly. On the other hand, if processes can be identified that hold the system out of equilibrium these processes must be treated with irreversible models. The uncertainty of parameters in models of non equilibrium properties i.e. permeabilities, diffusivities, reactions rates, etc., is much worse than the uncertainty in equilibrium properties, precluding quantitative modeling. Models of irreversible processes may provide important insights into the effects of disequilibrium, but much new data is necessary to define the transport and rate parameters.

Another concern that must be addressed when models are applied to natural systems is the quality of field data available. In the best situation, both accurate water chemistries and formation mineralogies are available. This relatively rare situation allows a much more detailed picture of the depositional process to be obtained. One of the most complete studies is represented by the work of Brantley et al. (1984) who studied the formation of modern evaporites in an estuarine environment. A brief discussion of this work is included here.

An accurate understanding of the geochemical controls that are operative in a particular setting is of extreme importance to a modeling program. Various geochemical controls such as fluid transfer between units in the formation and backreaction between the precipitated minerals and the evolving brine may profoundly influence the evolution of a geochemical process. In general it is not possible to predict these effects from theoretical models. Modeling studies will be most successful when field observations are combined with modeling efforts. An example of the importance of geochemical controls can be found in the application of modeling studies to the Searles Lake evaporite (Felmy and Weare, 1986). This work will also be discussed.

APPLICATION TO PERMIAN AND MIOCENE EVAPORITES IN THE SEAWATER SYSTEM

Van't Hoff and his coworkers (1905, 1909) carried out an extensive experimental program designed to investigate phase relationships in the marine evaporite system $Na-K-Mg-Ca-SO_4-Cl-H_2O$ with the objective of using this data to interpret the processes of formation and transformation in ancient evaporite deposits. Their prediction of the minerals produced from seawater evaporation were in qualitative agreement with mineral sequences observed in the German Permian Zechstein evaporite deposit. However, Van't Hoff knew that significant problems remained with the interpretation. Because of the limited dimensionality of phase diagram analysis, Van't Hoff and his coworkers based their predictions primarily on the evaporation path of the Ca free seawater system. While Ca levels in the solution phase remain very low during seawater evaporation because of the precipitation of insoluble Ca minerals, the neglect of Ca in an evaporation may not be justified because precipitated minerals may react with the evolving brine to form new minerals (backreaction).

In order to clarify the role of Ca and the effects of backreactions, Eugster et al. (1980) and Harvie et al. (1980) used the HMW model with carbonate removed to simulate the evaporation of waters in the seawater system. The most important results for seawater evaporation are summarized by Figure 8 and Figure 9 which is taken from their paper. Two evaporation pathways are plotted in Figure 8. The dashed line represents the evaporation path of seawater assuming complete equilibration between the precipitated minerals and the evolving brine (backreaction allowed). The solid line represents the equilibrium pathway when precipitated Ca minerals are not allowed to backreact with the evaporating brine (Ca minerals fractionated). This might be the situation for a laterally zoned evaporite such as the estuarine environment discussed in the next section. The path of the evaporation process with all Ca minerals fractionated, when plotted on a Janëke projection, is essentially the same as for the Ca free equilibrium system studied by Van't Hoff (Holser, 1979; Braitsch, 1971). In Figure 8, the two very different evaporation paths demonstrate that reactions between the solid phases and the evaporating solution play an important role in determining the evolution of the brine. In Figure 9, the mineral masses and solute concentrations are plotted for the case of complete equilibration (dashed line Fig. 8). The reactions responsible for the difference in pathways in Figure 8 have been discussed by Eugster et al. (1980). The important reactions are documented in Figure 9 where we see that as the system is concentrated slightly before halite saturation,

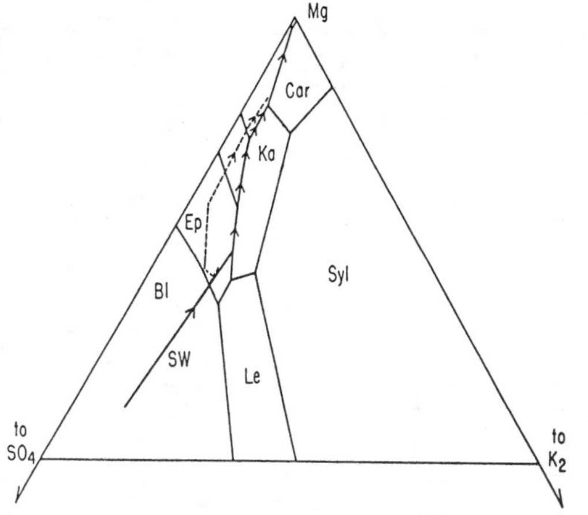

Figure 8. Seawater evaporation paths projected onto Figure 6. Dashed line: complete equilibrium between solution phase and mineral phases. Solid line: reaction between precipitated Ca minerals and solution phase prohibited.

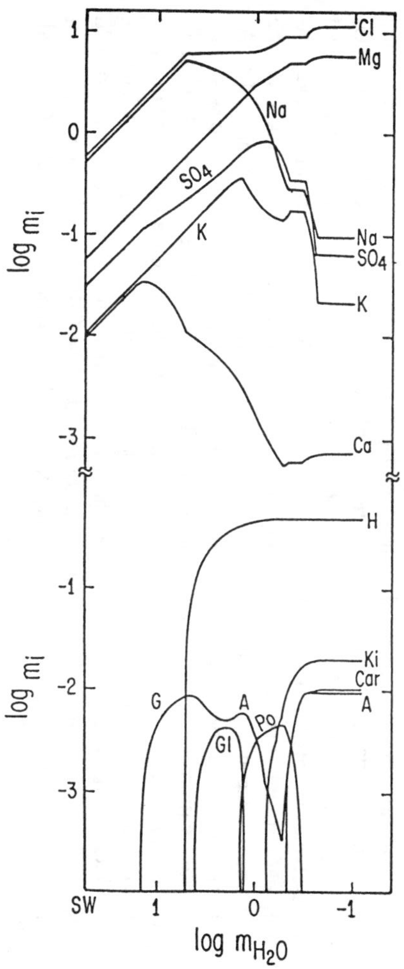

Figure 9. Equilibrium evaporation of 1 kg. seawater at 25°C. The upper half shows the molalities of the six major solutes as a function of evaporation, starting at 55.5 moles H_2O at the left. The lower half indicates the minerals present and their amounts (abbrevations are given in Table 6).

anhydrite ($CaSO_4$) replaces gypsum ($CaSO_4 \bullet 2H_2O$) which is precipitated very early in the evaporation. As the evaporation continues, anhydrite is partially replaced by glauberite ($Na_2Ca(SO_4)_2$). Glauberite is in turn replaced by polyhalite ($K_2MgCa_2(SO_4)_4 \bullet 2H_2O$) and anhydrite is further reduced. These replacements change the composition of the brine so that the dashed pathway in Figure 8 is followed and kainite ($KMgClSO_4 \bullet 3H_2O$) does not supersaturate. As the evaporation continues, polyhalite becomes unstable and redissolves. At this point, the equilibrium evaporation joins the Ca fractionated evaporation path in the Janëke projection and the evaporation continues to point z on the diagram. These are the main qualitative differences between the two reaction paths. As pointed out by Harvie et al. (1980), the equilibrium reaction path agrees quite well with field observations in the Zechstein. One of the significant difficulties of the work of Van't Hoff and his students was the appearance of primary kainite in the Ca free phase diagram derived evaporation sequence (the same as given by the solid line in Fig. 8). In Table 6, the mineral sequence produced by both pathways from Figure 8 are compared to field observations. Note that primary kainite is not found in the field. The Ca free pathway of Van't Hoft would be the same as the fractionated pathway (column 3 of Table 6) with all Ca minerals removed from the column.

Because Ca is present in the system, glauberite and polyhalite are also precipitated in the fractionated pathway (as well as bloedite). However, because $CaSO_4$ has been removed from the system as fractionated anhydrite, not enough of these minerals are present to move the evaporation path as plotted in Figure 8 (solid line) away from the Ca free pathway. The sequence calculated from the completely equilibrated path (dashed line Fig. 8) not only accounts for the absence of primary kainite but for the pseudomorphous replacement of gypsum and anhydrite by glauberite and polyhalite and the replacement of glauberite by polyhalite (Harvie et al., 1980). Other authors have attributed these replacements to thermal and solution metamorphism. Here, they result from backreactions in equilibrium evaporation.

These calculations illustrate the importance of including all the species in a calculation of mineral evolution. They also demonstrate an important effect of a geochemical control, in this case the chemical transfer between previously deposited minerals and the remaining concentrated brines. Major problems remain with this interpretation of the formation of the Zechstein II deposit. The important contribution of the model was to provide the true equilibrium sequence. Hydrological models will still have to explain how the equilibrium result can agree so closely with the observed sequence (Harvie et al., 1980).

A problem with the ancient evaporite study (Zechstein II) was the lack of water samples to further define the depositional process. Hardie and Eugster (1970) have pointed out the strong correlation between water chemistry and mineral formation. Recently McCaffrey et al. (in press) have reported research which promises to bring a new level of detail to the study of ancient depositional environments. In this work, the composition of fluid inclusions in late Miocene halites from a DSDP core at site 227 in the Red Sea (Whitmarsh et al., 1974), was analyzed. The molarity of Na vs Mg is plotted in Figure 10. Also plotted (solid line) is the HMW prediction of the behavior of Na concentration during equilibrium seawater evaporation. The sharp feature corresponds to the precipitation of halite as predicted by Hardie and Eugster (1970). The close agreement between the model and the inclusion measurements places some control on the composition of Miocene seawater and the depositional process. Similar studies of other components promise to provide a much more detailed view of ancient evaporite formation.

Table 6. Comparison of the Zechstein II sequence with that predicted for modern seawater evaporation at 25°C [a]. Abbrevations: G, gypsum; A, anhydrite; H, halite; Gl, glauberite; Po, polyhalite; Bl, bloedite; Ep, epsomite; Hx, hexadydrite; Ki, kieserite; Ka, kainite; Car, carnallite; and Bi, bischofite

Complete Equilibrium[b]	Observed Zechstein II	Fractionation[c]
G		G
A	A	A
A+H	A+H	A+H
Gl+A+H	Gl+A+H	Gl+H
Po+A+H	Po+A+H	Po+H
		Bl+Po+H
Ep+Po+A+H		Ep+Po+H
		Ka+Ep+H
Hx+Po+A+H		Hx+Ka+H
Ki+Po+A+H	Ki+Po+A+H	Ki+Ka+H
Car+Ki+Po+A+H	Car+Ki+Po+A+H	Car+Ki+H
Car+Ki+A+H	Car+Ki+A+H	
Bi+Car+Ki+A+H		Bi+Car+Ki+H+A

a. Some invariant points have been eliminated for brevity.
b. All minerals allowed to react with the evaporating brine.
c. Precipitated minerals not allowed to react with the evaporating brine.

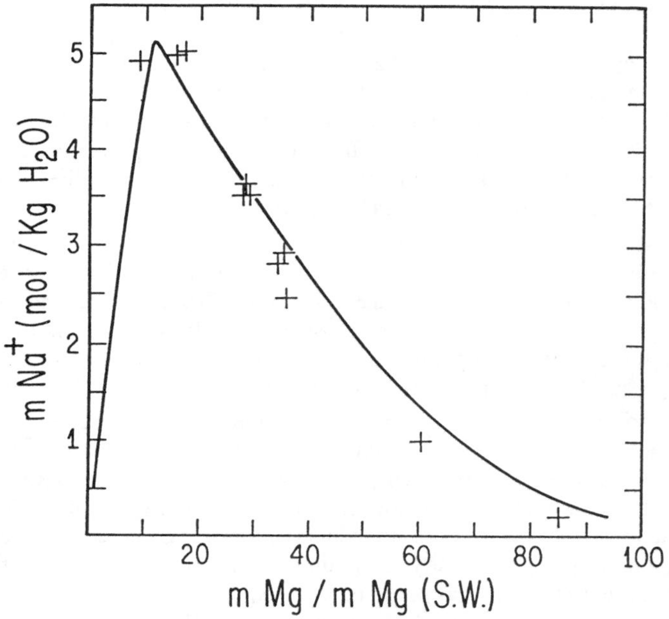

Figure 10. Na concentrations in Miocene halite fluid inclusions (+) plotted against the ratio of Mg concentration in the inclusion to Mg concentration in seawater (McCaffrey et al., in press). Solid line is the prediction of the HMW model for seawater evaporation.

APPLICATION TO RECENT AND PRESENT DAY EVAPORATION PROCESSES

Brantley et al. (1984) used the HMW model to study the formation of marine evaporites in a modern natural seawater evaporation process. In this study, water samples were taken from the Bocana de Virrila, a relict estuary on the cost of northwestern Peru in the Sechura desert. Samples were taken at intervals along the estuary and correlated with the minerals found in the sediment. Because there is essentially no water input into the estuary apart from the ocean, evaporation causes a continual increase in concentration as a function of distance from the mouth.

In the study of Brantley et al. (1984), the estuary was modeled as a laterally zoned evaporite basin assuming a fractionation process whereby minerals precipitating from the brine were not allowed to backreact. However, in this estuary concentration levels do not become high enough so that the qualitative effects of backreaction on the mineral sequence would be observed (as above). The HMW model was used to interpret the field data. The sequence of minerals predicted to form followed the expected sequence calcite $\rightarrow$ gypsum $\rightarrow$ halite. This is an example where at least the mineral sequence could be predicted from a considerably less accurate model. However, water samples were taken at intervals along the estuary. The composition of these samples can be predicted by the model, providing a much more detailed view of the depositional process. For example, Hardie and Eugster (1970) have discussed the sharp change in behavior of solution composition vs. extent of evaporation at the point minerals precipitate in an equilibrium concentration process. Such changes were observed in the field measurements of Brantley et al. (1984) and are illustrated in Figure 11 for the concentration of Ca. In the estuary Mg was observed to show conservative behavior and therefore was used as a measure of evaporation. The prediction of the theoretical model is given as the solid line. The general agreement between field observations and calculations is encouraging. The sharp break in the curve at Mg concentration of $\approx$0.21 m signals the precipitation of gypsum. This point closely coincided with the actual presences of this mineral in the sediments. These calculations show that for this system, the natural process of gypsum formation closely follows the equilibrium evaporation path. In other words, rate processes are not important for this mineral in this setting. Similar observations show that halite deposition also appears to be very close to equilibrium (as in Fig. 10).

The above examples demonstrate the close equilibration in natural environments for gypsum and halite. The situation may not be as straightforward for other minerals. In Figure 12, the alkalinity of the brines in the Bocana de Virrila (pluses) and the results of the laboratory evaporations of seawater of Lazar et al. (1983) (circles and triangles) are plotted against the degree of evaporation. Also plotted are the results of model calculations assuming equilibrium with calcite and atm CO_2 (P_{CO_2} = 0.000333) both with and without borate contributions to the alkalinity (note the large contributions of the borate species to the alkalinity after calcite precipitation). In this case, we find that there are significant problems with the model predictions even for the fairly simple mineral calcite, which equilibrates rapidly in the laboratory (not quite as fast as gypsum and halite). Preliminary calculations and new pH measurements (B. Lazar, private communication) indicate that the solution phase chemistry is in equilibrium. However, the solution may be out of equilibrium with both the gas phase (CO_2 disequilibrium) and with the solid phase. The fact that the evaporation results of Lazar et al. (1983) and Brantley et al. (1984) agree so closely suggest that there may be a control mechanism which operates the same

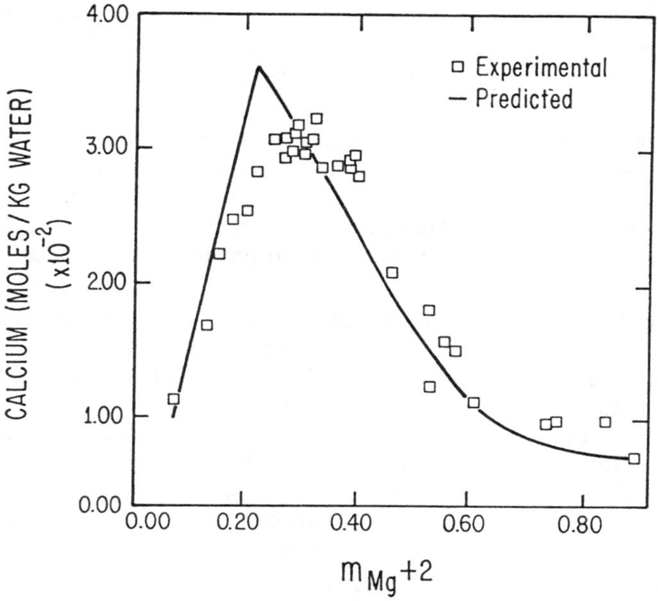

Figure 11. Concentrations of the Ca component of evaporated seawater taken from the Bocana de Virrila plotted as a function of degree of evaporation (Mg concentration). Field data are plotted as squares and the HMW model predictions are plotted as solid lines.

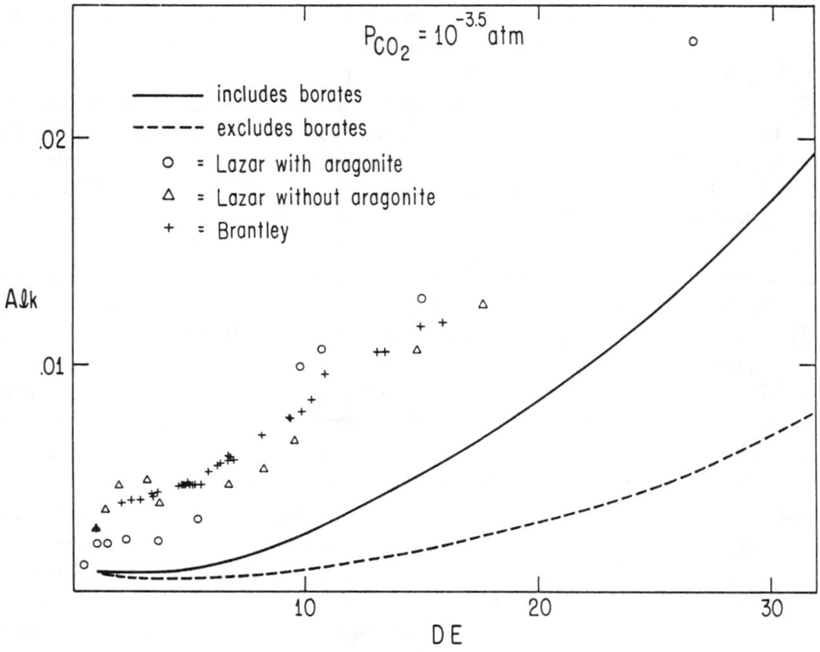

Figure 12. Alkalinity of evaporated seawater plotted versus degree of evaporation (DE). Pluses: field data of Brantley et al. (1984). Circles and triangles: laboratory data of Lazar et al. (1983). Solid line: Felmy and Weare (1986) model including borate contributions to the alkalinity. Dashed line: HMW model which does not include borate species.

way in a variety of settings. Disequilibrium with the atmospheric CO_2 has also been noted by Monnin and Schott (1984) in their modeling studies of carbonate depositional processes in Lake Magadi, Kenya and had already been pointed out by Hardie and Eugster (1970).

The above examples emphasized the seawater evaporite system. The chemistry of the brines in this system is relatively easily described. Other important systems such as aluminum and boron do not show such simple behavior. As shown in Table 5, the chemistry of Al is so poorly understood that quantitative studies are not possible. However for the borate system, there appears to be enough data to provide a detailed description at least at 25°C (see Fig. 2).

Felmy and Weare (1986) applied their model to the natural borate deposits of Searles Lake. Searles lake (Smith, 1979) is a dry salt pan about 100 km^2 in area. It was the third in a chain of as many as six lakes formed during pluvial periods of the Pleistocene. The proposed source of water for the lake was the Owens River which drains the east side of the Sierra Nevada. The deposits in Searles lake are composed of alternating layers of mud and evaporite salts supposedly corresponding to alternating periods of lake filling and dessication. The subsurface stratigraphic units are summarized in Smith (1979). The work of Felmy and Weare (1986) concentrated on the prediction of the mineralogy of the upper salt and its associated parting mud layer.

The mineralogy and water composition from the upper salt provide a stringent field test of the accuracy of the borate model of Felmy and Weare (1986). Water compositions from this unit are summarized in Table 7. If these concentrations are used in the model to calculate the free energy of the various possible minerals, the free energy values for the minerals that are present should match closely the values calculated for the appropriate combination of ions in the solution phase. That is, the calculation should be close to the measured solubility products. Results are usually summarized in terms of the saturation index, the log of the ratio of the activity product over K_{sp} for a particular mineral (Felmy and Weare, 1986). If this index is 0.0 the mineral is just saturated and in equilibrium. In Table 8 I have summarized the results of the saturation indices for the Searles lake application. With the exception of aphthitalite (Smith avg.), all minerals present are within 0.025 pK units of the expected equilibrium result of 0.0. This indicates both that the model correctly summarizes the thermodynamics of the system and that the minerals in the formation are in equilibrium with the formation water. Note that many minerals which are not found in the formation have saturation indices of approximately 0.1. Accuracy of better than 0.1 pk unit is necessary to describe this setting. The results of Table 8 provide an important field validation of the thermodynamic model. The results are even more encouraging when the highly complex nature of borate chemistry is recognized. All the information used to parameterize the model comes form laboratory data. Frequently when model calculations don't agree with field observations, special properties of natural phases are invoked (e.g. amorphous surface phases, rate control, etc.) to remove the discrepancy. In this case no such mechanisms were necessary.

The high reliability of the thermodynamic model encouraged us to apply it to studies of the formation of the Searles evaporite deposit. The mineralogy of the upper salt is summarized in the work of Smith (1979). Using these average values, we investigated whether the observed mineral assemblage could result form a straightforward evaporation of present day Owen's river water. The first attempt at evaporation allowed complete equilibration between the brine and the previously precipitated minerals. The results of the calculations are summarized in Table 9 (column 1), and are very different

Table 7. Published average brine compositions for the Upper Salt (Searles Lake).*

Component	Concentration (m) Teeple (1929)	Gale (1938)	Smith (1979)**
Na	7.54	7.48	7.43
K	1.01	.98	.78
total CO_3	.71	.69	.64
SO_4	.75	.76	.73
Cl	5.41	5.32	5.25
total boron	.47	.46	.46

*recalculated from wt% to molality
**average from analysis taken between 1935-1950.

Table 8. Calculated mineral saturation indices for interstitial brine in the Upper Salt (Searles Lake).

Mineral	S.I. (Teeple avg.)	S.I. (Gale avg.)	S.I. (Smith avg.)	Found in upper salt
Aphthitalite	-.0158	-.0210	-.0788	yes
Borax	.0215	.0145	.0246	yes
Boric Acid	-1.012	-1.030	-.9941	no
Burkeite	.0145	.0104	-.0047	yes
Halite	.0218	.0115	.0018	yes
Mirabilite	-.1817	-.1759	-.1684	no
Nahcolite	-.0669	-.0992	-.0832	no
Natron	-.2561	-.2543	-.2682	no
Sodium Metaborate	-.6460	-.6378	-.6548	no
Sylvite	-.2012	-.2139	-.2734	no
Teepleite	-.1627	-.1650	-.1807	no
Thenardite	.0042	.0015	-.0065	yes
Thermonatrite	-.2414	-.2472	-.2751	no
Trona	.0236	.0077	-.0016	yes

Table 9. Calculated mineral assemblages resulting from the evaporation of Owens River water. Ap: Aphthitalite; B: Borax; Bu: Burkeite; C: Calcite; D: Dolomite; G: Gaylussite; H: Halite; Ne: Nesquehonite; N: Northupite; P: Pirssonite; T: Trona; Th: Thenardite.

1.Calculated[1]	2. Searles Lake[2]	3. Calculated[3]
C+D	C+D[4]	C+Ne[5]
P	G P	P
P+B+Th	B T+N	B B+T+N
P+B+Th+Ap P+B+Th+Ap+H	B+T+Bu+Ap+H+Th[5]	B+T+Bu+N B+T+Bu+Ap+N B+T+Bu+Ap+H

[1] All minerals allowed to react with previous precipitates.
[2]Considers only Parting Mud and Upper Salt
[3] The HCO_3^-/SO_4^{2-} is adjusted in the Owens River water. $CaCO_3$ is not allowed to backreact to form pirssonite. and dolomite and magnesite are suppressed.
[4]Aragonite present but metastable
[5]Thenardite stable at higher P_{CO_2} (>.0006 atm). Hanksite present but not primary

from the mineralogy actually present (column 2). In particular, the calculations showed a large amount of pirssonite ($Na_2Ca(CO_3)_2{\bullet}2H_2O$) and no trona ($Na_3H(CO_3)_2{\bullet}2H_2O$) in the upper salt, whereas the actual formation is characterized by large amounts of trona and little pirssonite. If we consider the transformation

$$CaCO_3 \text{ (calcite)} + Na_2CO_3 + 2H_2O \rightarrow Na_2Ca(CO_3)_2{\bullet}2H_2O \text{ (pirssonite)} \quad , \quad (10)$$

we can see that the straight forward evaporation produces pirssonite at the expense of previously precipitated calcite. There is a backreaction effect (see above), producing a result which does not agree with field observations. A constraint must be placed on the model in order that the backreaction of aqueous Na_2CO_3 with precipitated $CaCO_3$ is limited.

In the deposit, most of the calcite is found in the clay layer below the upper salt. The presence of this relatively impermeable clay layer provides a possible mechanism for preventing the chemical communication between layers (Eugster and Smith, 1965). Further evidence for this control can be found by carrying out the calculation now preventing backreaction with calcite but otherwise allowing the system to equilibrate. In this case, the early sequence is in good agreement but the later sequence does not show the presense of northupite ($Na_3MgCl(CO_3)_2$). Northupite will only precipitate if the Mg level in the brine gets high enough. However, early formation of dolomite in our sequence removes enough Mg so that northupite never forms. We take this as evidence that primary dolomite probably did not form. Further transformation to dolomite during later stages of evaporation was prohibited by the impermeable mud layer. When dolomite is removed from the model, northupite forms. At this point, the only important discrepancy between the model predictions and the field is the appearance of burkeite ($Na_6CO_3(SO_4)_2$) before trona in the sequence. These two double salts are similar (both contain Na_2CO_3). Increasing the ratio of HCO_3 to SO_4 will decrease the stability of trona, causing it to appear before burkeite. With a minor adjustment of the ratio of these two species we can obtain good agreement with the field measurements (Table 9 column 3). I would like to emphasize that the dessication process represented here is a very simplified model of what may be a very complex natural process. The results are encouraging for the application of models but further changes and refinements will occur with an improved understanding of the field.

I believe that this application of a well defined chemical model to the complicated Searles lake evaporite has several important general implications for the application of models to natural systems. First, Table 8 shows that models can accurately predict the chemistry even in very complicated settings. Furthermore, when expected to be in chemical equilibrium, the solubilities of observed minerals were in close agreement with the model predictions from laboratory measurements. The level of accuracy required to separate between minerals present and undersaturated minerals ($\approx$ 0.02 pk units) may provide a guide to the accuracy of models of complex settings. I consider it to be an important priority to obtain such information from a variety of geological settings to assess the required accuracy of equilibrium model predictions. Secondly, (and perhaps most important for assessing the feasibility of using models to reliably predict geochemical processes) certain geochemical controls which could not be predicted from the model (e.g. late dolomitization and limited communication between units in the formation) were necessary to bring the model into good agreement with the field measurements. These controls affected only the application of the model and in no way involved the parameterization of the model. In some sense, the requirement of these additions to the model limits the application of models to poorly characterized systems. On the other hand, in this example, the required controls could be directly related to field observations. In this sense, the model has provided important information as to what might have happened in the formation process.

ACKNOWLEDGMENTS

The continuing collaboration with my colleagues Nancy Mϕller, Jerry Greenberg, Charles Harvie, and Andrew Felmy is greatfully acknowledged. Throughout this work the close association with Professors Ken Pitzer and Hans Eugster has been a pleasure. This research was sponsored in part by the Department of Energy Grant DE-AC-85SF-15522 and the National Science Foundation OCE Grant 85-07902.

APPENDIX

The parameters for the HMW model are collected in Tables A1-A4. The equations of Pitzer, (Chapter 4), when used with these parameters, should accurately reproduce solubilities in the Na–K–Mg–Ca–H–Cl–SO_4–OH–HCO_3–CO_3–CO_2–H_2O system to high ionic strength at 25°C. These parameters also serve as a guide to relative magnitude which we consider to be acceptable for parameters in the Pitzer model (see text).

Table A-1: Single electrolyte solution parameter values

Cation	Anion	$\beta_{ca}^{(0)}$	$\beta_{ca}^{(1)}$	$\beta_{ca}^{(2)}$	C_{ca}^{ϕ}
Na	Cl	.0765	.2644	-	.00127
Na	SO_4	.01958	1.113	-	.00497
Na	HSO_4	.0454	.398	-	-
Na	OH	.0864	.253	-	.0044
Na	HCO_3	.0277	.0411	-	-
Na	CO_3	.0399	1.389	-	.0044
K	Cl	.04835	.2122	-	-.00084
K	SO_4	.04995	.7793	-	-
K	HSO_4	-.0003	.1735	-	-
K	OH	.1298	.320	-	.0041
K	HCO_3	.0296	-.013	-	-.008
K	CO_3	.1488	1.43	-	-.0015
Ca	Cl	.3159	1.614	-	-.00034
Ca	SO_4	.20	3.1973	-54.24	-
Ca	HSO_4	.2145	2.53	-	-
Ca	OH	-.1747	-.2303	-5.72	-
Ca	HCO_3	.4	2.977	-	-
Ca	CO_3	-	-	-	-
Mg	Cl	.35235	1.6815	-	.00519
Mg	SO_4	.2210	3.343	-37.23	.025
Mg	HSO_4	.4746	1.729	-	-
Mg	OH	-	-	-	-
Mg	HCO_3	.329	.6072	-	-
Mg	CO_3	-	-	-	-
MgOH	Cl	-.10	1.658	-	-
MgOH	SO_4	-	-	-	-
MgOH	HSO_4	-	-	-	-
MgOH	OH	-	-	-	-
MgOH	HCO_3	-	-	-	-
MgOH	CO_3	-	-	-	-
H	Cl	.1775	.2945	-	.0008
H	SO_4	.0298	-	-	.0438
H	HSO_4	.2065	.5556	-	-
H	OH	-	-	-	-
H	HCO_3	-	-	-	-
H	CO_3	-	-	-	-

Table A-2: Common-ion two electrolyte parameter values.

c	c′	$\theta_{cc'}$	$\psi_{cc'Cl}$	$\psi_{cc'SO_4}$	$\psi_{cc'HSO_4}$	$\psi_{cc'OH_3}$	$\psi_{cc'HCO_3}$	$\psi_{cc'CO_3}$
Na	K	-.012	-.0018	-.010	-	-	-.003	.003
Na	Ca	.07	-.007	-.055	-	-	-	-
Na	Mg	.07	-.012	-.015	-	-	-	-
Na	MgOH	-	-	-	-	-	-	-
Na	H	.036	-.004	-	-.0129	-	-	-
K	Ca	.032	-.025	-	-	-	-	-
K	Mg	0.	-.022	-.048	-	-	-	-
K	MgOH	-	-	-	-	-	-	-
K	H	.005	-.011	.197	-.0265	-	-	-
Ca	Mg	.007	-.012	.024	-	-	-	-
Ca	MgOH	-	-	-	-	-	-	-
Ca	H	.092	-.015	-	-	-	-	-
Mg	MgOH	-	.028	-	-	-	-	-
Mg	H	.10	-.011	-	-.0178	-	-	-
MgOH	H	-	-	-	-	-	-	-
a	**a′**	$\theta_{aa'}$	$\psi_{aa'Na}$	$\psi_{aa'K}$	$\psi_{aa'Ca}$	$\psi_{aa'Mg}$	$\psi_{aa'MgOH}$	$\psi_{aa'H}$
Cl	SO_4	.02	.0014	-	-.018	-.004	-	-
Cl	HSO_4	-.006	-.006	-	-	-	-	.013
Cl	OH	-.050	-.006	-.006	-.025	-	-	-
Cl	HCO_3	.03	-.015	-	-	-.096	-	-
Cl	CO_3	-.02	.0085	.004	-	-	-	-
SO_4	HSO_4	-	-.0094	-.0677	-	-.0425	-	-
SO_4	OH	-.013	-.009	-.050	-	-	-	-
SO_4	HCO_3	.01	-.005	-	-	-.161	-	-
SO_4	CO_3	.02	-.005	-.009	-	-	-	-
HSO_4	OH	-	-	-	-	-	-	-
HSO_4	HCO_3	-	-	-	-	-	-	-
HSO_4	CO_3	-	-	-	-	-	-	-
OH	HCO_3	-	-	-	-	-	-	-
OH	CO_3	.10	-.017	-.01	-	-	-	-
HCO_3	CO_3	-.04	.002	.012	-	-	-	-

Table A-3: Neutral ion parameter values

i	$\lambda_{CO_2,i}$	$\lambda_{CaCO_3,i}$	$\lambda_{MgCO_3,i}$
H	0.0	-	-
Na	.100	-	-
K	.051	-	-
Ca	.183	-	-
Mg	.183	-	-
MgOH	-	-	-
Cl	-.005	-	-
SO_4	.097	-	-
HSO_4	-.003	-	-
OH	-	-	-
HCO_3	-	-	-
CO_3	-	-	-

Table A-4: Values for the standard chemical potentials of the aqueous solution species and minerals.

Species or Mineral	Chemical Formula	μ^o/RT
Water	H_2O	-95.6635
Sodium Ion	Na^+	-105.651
Potassium Ion	K^+	-113.957
Calcium Ion	Ca^{+2}	-223.30
Magnesium Ion	Mg^{+2}	-183.468
Magnesium Hydroxide Ion	$MgOH^+$	-251.94
Hydrogen Ion	H^+	0.
Chloride Ion	Cl^-	-52.955
Sulfate Ion	SO_4^{-2}	-300.386
Bisulfate Ion	HSO_4^-	-304.942
Hydroxide Ion	OH^-	-63.435
Bicarbonate Ion	HCO_3^-	-236.751
Carbonate Ion	CO_3^{-2}	-212.944
Aq. Calcium Carbonate	$CaCO_3^0$	-443.5
Aq. Magnesium Carbonate	$MgCO_3^0$	-403.155
Aq. Carbon Dioxide	CO_2^0	-155.68
Carbon Dioxide Gas	CO_2 (gas)	-159.082
Anhydrite	$CaSO_4$	-533.73
Aphthitalite (Glaserite)	$NaK_3(SO_4)_2$	-1057.05
Antarcticite	$CaCl_2 \bullet 6H_2O$	-893.65
Aragonite	$CaCO_3$	-455.17
Arcanite	K_2SO_4	-532.39
Bischofite	$MgCl_2 \bullet 6H_2O$	-853.1
Bloedite	$Na_2Mg(SO_4)_2 \bullet 4H_2O$	-1383.6
Brucite	$Mg(OH)_2$	-335.4
Burkeite	$Na_6CO_3(SO_4)_2$	-1449.4
Calcite	$CaCO_3$	-455.6
Calcium Chloride Tetrahydrate	$CaCl_2 \bullet 4H_2O$	-698.7
Calcium Oxychloride A	$Ca_4Cl_2(OH)_6 \bullet 13H_2O$	-2658.45
Calcium Oxychloride B	$Ca_2Cl_2(OH)_2 \bullet H_2O$	-778.41
Carnallite	$KMgCl_3 \bullet 6H_2O$	-1020.3
Dolomite	$CaMg(CO_3)_2$	-871.99
Epsonmite	$MgSO_4 \bullet 7H_2O$	-1157.83
Gaylussite	$CaNa_2(CO_3)_2 \bullet 5H_2O$	-1360.5
Glauberite	$Na_2Ca(SO_4)_2$	-1047.45
Gypsum	$CaSO_4 \bullet 2H_2O$	-725.56
Halite	NaCl	-154.99
Hexahydrite	$MgSO_4 \bullet 6H_2O$	-1061.60
Kainite	$KMgClSO_4 \bullet 3H_2O$	-938.20
Kalicinite	$KHCO_3$	-350.06
Kieserite	$MgSO_4 \bullet H_2O$	-579.80
Labile Salt	$Na_4Ca(SO_4)_3 \bullet 2H_2O$	-1751.45
Leonite	$K_2Mg(SO_4)_2 \bullet 4H_2O$	-1403.97
Magnesite	$MgCO_3$	-414.45
Magnesium Oxychloride	$Mg_2Cl(OH)_3 \bullet 4H_2O$	-1029.6
Mercallite	$KHSO_4$	-417.57
Mirabilite	$Na_2SO_4 \bullet 10H_2O$	-1471.15
Misenite	$K_8H_6(SO_4)_7$	-3039.24
Nahcolite	$NaHCO_3$	-343.33
Natron	$Na_2CO_3 \bullet 10H_2O$	-1382.78
Nesquehonite	$MgCO_3 \bullet 3H_2O$	-695.3
Picromerite (Schoenite)	$K_2Mg(SO_4)_2 \bullet 6H_2O$	-1596.1
Pirssonite	$Na_2Ca(CO_3)_2 \bullet 2H_2O$	-1073.1
Polyhalite	$K_2MgCa_2(SO_4)_4 \bullet 2H_2O$	-2282.5
Portlandite	$Ca(OH)_2$	-362.12
Potassium Carbonate	$K_2CO_3 \bullet 3/2H_2O$	-577.37
Potassium Sesquicarbonate	$K_8H_4(CO_3)_6 \bullet 3H_2O$	-2555.4
Potassium Sodium Carbonate	$KNaCO_3 \bullet 6H_2O$	-1006.8
Potassium Trona	$K_2NaH(CO_3)_2 \bullet 2H_2O$	-971.74
Sesquipotassium Sulfate	$K_3H(SO_4)_2$	-950.8
Sesquisodium Sulfate	$Na_3H(SO_4)_2$	-919.6
Sodium Carbonate Heptahydrate	$Na_2CO_3 \bullet 7H_2O$	-1094.95
Sylvite	KCl	-164.84
Syngenite	$K_2Ca(SO_4)_2 \bullet H_2O$	-1164.8
Tachyhydrite	$Mg_2CaCl_6 \bullet 12H_2O$	-2015.9
Thenardite	Na_2SO_4	-512.35
Thermonatrite	$Na_2CO_3 \bullet H_2O$	-518.8
Trona	$Na_3H(CO_3)_2 \bullet 2H_2O$	-960.38

REFERENCES

Archer, D.G., and Wood, R.H. (1985) Chemical equilibrium model applied to aqueous magnesium sulfate solutions. J. Solution Chem. **141**, 757-780.

Apps, J.A., Neil, J.M. and Jun, C.H. (1986) Thermochemical properties of gibbsite, boehmite, diaspore and aluminate ion between 0 and 350°C. NUREG Report, LBL 2148L.

Baes, C.F. and Mesmer, R.E. (1976) The hydrolysis of cations. Wiley-Interscience **489** John Wiley and Sons, New York.

Blount, C.W. (1977) Barite solubility and thermodynamic quantities up to 300°C and 1400 bars. Amer. Mineral. **62**, 942-957.

Blount, C.W. and Dickson, F.W. (1973) Gypsum-anhydrite equilibria in systems $CaSO_4-H_2O$ and $CaSO_4-NaCl-H_2O$. Amer. Mineral. **58**, 323-331.

Braitsch, O. (1971) Salt Deposits: Their Origin and Composition. Springer-Verlag, London.

Brantley, S.L., Crerar, D.A. , Mϕller, N.E., and Weare, J. H. (1984) Geochemistry of a modern marine evaporite: Bocana de Virrila, Peru. J. Sed. Petrol. **54** 477-462.

Bromley, L.A. (1973) Thermodynamic properties of strong electrolytes in aqueous solutions. AICHE *B10*, 313.

Dawson, B.S.W., Irish, D. E. and Toogood, G.E. (1986) Vibrational spectral studies of solutions at elevated temperature and pressure. 8. A raman spectral study of ammonium hydrogen sulfate solutions and the $HSO_4^- - SO_4^2$ equilibrium. J. Phys. Chem. **90**, 334.

Eugster, H.P., Harvie, C.E., and Weare, J.H. (1980) Mineral equilibria in a six-component seawater system, $Na-K-Mg-Ca-Cl-SO_4-H_2O$ at 25°C. Geochim. Cosmochim. Acta **44**, 1335.

Eugster, H. P. and Smith, G. I. (1965) Mineral equilibria in the Searles Lake evaporites, California. J. Petrol. **6**, 473.

Felmy, A. and Weare, J.H. (1986) The prediction of borate mineral equilibria in natural waters: Application to Searles Lake, California. Geochim. Cosmochim. Acta **50** 2771-2783.

Gale, W. A. (1938) Chemistry of the Trona Process from the standpoint of the Phase Rule. Indust. Eng. Chem. 30, (**8**), 867-871.

Garrels, R.M. and Thompson, M.E. (1962) A chemical model for seawater at 25°C and one atmosphere total pressure. Amer. J. Sci. **260**, 57-66.

Gromova, E.T. (1960) Zh. Neorgan. Khim. 5 **11**, 2575 (see Stephens and Stephens (1963)).

Hardie, L.A. and Eugster, H.P. (1970) The Evolution of Closed-Basin Brines. Mineral. Soc. Amer. Spec. Paper **3**, 273-290.

Harvie, C.E., Greenberg, J.P., and Weare, J.H. (1987) A chemical equilibrium algorithm for highly nonideal multiphase systems: Free energy minimization. Geochim. Cosmochim. Acta **51**, 1045-1057.

Harvie, C.E., Mϕller, N., and Weare, J.H. (1984) The prediction of mineral solubilities in natural waters: The $Na-K-Mg-Ca-H-Cl-SO_4-OH-HCO_3-H_2O$ system to high ionic strengths at 25°C. Geochim. Cosmochim. Acta **48**, 723-751.

Harvie, C.E. and Weare, J.H. (1980) The prediction of mineral solubilities in natural waters: the $Na-K-Mg-Ca-Cl-SO_4-H_2O$ system from zero to high concentration at 25°C. Geochim. Cosmochim. Acta **44**, 981-997.

Harvie, C.E., Weare, J.H., Hardie, L.A. and Eugster, H.P. (1980) Evaporation of seawater: calculated mineral sequences. Science **208**, 498-500.

Helgeson, H.C., Brown, T.H., Nigrini, A. and Jones, T.A. (1970) Calculation of mass transfer in geochemical processes involving aqueous solutions. Geochim. Cosmochim. Acta **34**, 569-592.

Hershey, J. P., Fernandez, M., Milne, P. J., and Millero, F. J. (1985) The ionization of boric acid in NaCl, Na-Ca-Cl and Na-Mg-Cl solutions at 25°C. Geochim. Cosmochim. Acta **50**, 143-148.

Holmes, H.F. and Mesmer, R.E. (1983) Thermodynamic properties of aqueous solutions at elevated temperatures II. NaCl + KCl mixtures. J. Phys. Chem. **87**, 1242-1255.

Holmes, H.F. and Mesmer, R.E. (1986) Thermodynamics of aqueous solutions of the alkai metal sulfates. J. Solution Chem. **15**, 495-518.

Holmes, H.F., Baes, C.F., and Mesmer, R.E. (1978) Isopiestic studies of aqueous solutions at elevated temperatures: 1. KCl, $CaCl_2$, and $MgCl_2$. J. Chem. Thermodyn. **19**, 983-996.

Holser, W.T. (1979) Mineralogy of evaporites. In R. G. Burns, Ed., Marine Minerals, Reviews in Mineralogy, **6**, 235-295 LithoCrafters, Inc., Chelsea, Michigan

Irish, D.E. , Dawson, B., Pursel, R., Stolberg, L., and Toogood, G.E. (1984) The application of raman spectroscopy to the study of ionic complexes in solutions. Paper 199, IUPAC Conf. on Chemical Thermodynamics, McMaster Univ., Hamilton, Ontario.

Kharaka, Y. K. and Barnes, I. (1973) SOLMNEQ: Solution-Mineral Equilibrium Computations. U.S. Geol. Surv., Menlo Park, California.

Kerrisk, J.K. (1981) Chemical equilibrium calculations for aqueous geothermal brines. Los Alamos Science Lab, Report LA-8851-MS Los Alamos, N.M.

Lazar, B., Starinsky, A., Katz, A., and Sass, E. (1983) The carbonate system in hypersaline solutions: Alkalinity and $CaCO_3$ solubility of evaporated seawater. Limnol. Oceanogr. 28, 978-986.

Linke, W.F. (1965) Solubilities of inorganic and metal organic compounds.. American Chemical Society (4th ed.).

Marshall, W.L., Slusher, R., and Jones, E.V. (1964) Solubility and thermodynamic relationships for $CaSO_4$ in $NaCl-H_2O$ solutions from 40°C to 200°C, O to 4 molal NCl. J. Chem. Eng. Data **9**, 187-191.

May, H.M., Helmke, P.A., and Jackson, M.L. (1979) Gibbsite solubility and thermodynamic properties of hydroxy-aluminum ions in aqueous solution at 25°C. Geochim. Cosmochim. Acta **43**, 861-868.

McCaffrey, M. A., Lazar, B., and Holland, H. D. (1987) J. Sed. Pet., in press

Meissner, H.P. and Kusik, C.L. (1972) Activity coefficients of strong electrolytes in aqueous solutions. Ind. Eng. Chem. Process. Des. Dev. **11**, 128.

Millero, F. J. (1979) Effect of temperature and pressure on activity coefficients. In R. Pytckowicz, Ed., Activity coefficients in electrolyte solutions II, 63-152, CRC Press

Millero, J. M.. Connaughton, L. M., Vinokurova, F., and Chetirkin, P. V. (1985) PVT Properties of Concentrated Aqueous Electrolytes. III. Volume Changes for Mixing the Major Sea Salts at I = 1.0 and 3.0 at 25°C. J. Sol. Chem. **14**, 837-851.

Millero, F. J., and Thurmond, V. (1983) The ionization of carbonic acid in Na-Mg-Cl solutions at 25°C. J. Sol. Chem. **12**, 401.

Mϕller, N. (1987) A chemical equilibrium model for the $CaSO_4-NaCl-CaCl_2-H_2O$ system, to high temperature and concentration. Geochim. Cosmochim. Acta, in press

Monnin, C. (1987) Densities and apparent molal volumes of $MgCl_2$ and $CaCl_2$ solutions, J. Sol. Chem., in press

Monnin, C. and Schott, J. (1984) The determination of the solubility products of sodium carbonate minerals and an application to trona deposition in Lake Magadi (Kenya). Geochim. Cosmochim. Acta **48**,

571-581.

Nordstrom, K., Plummer, L.N., Wigley, T.M.L., Wolery, T.J., Ball, J.W., Jenne, E.A., Bassett, R.L., Crerar, D.A., Florence, T.M., Fritz, B., Hoffman, M., Holdren, G.R. Jr., Lafon, G.M., Mattigold, S.V., McDuff, R.E., Morel, F., Reddy, M.M., Sposito, G. and Thrailkill, J., (1979) A comparison of computerized chemical models for equilibrium calculations in aqueous systems Chemical Modeling in Aqueous systems. E.A. Jenne, Ed., Amer. Chem. Soc. Symp. Ser. **93**, 857-892.

Owen, B.B. and King, E.J. (1943) The effect of sodium chloride upon the ionization of boric acid at various temperatures. J. Amer. Chem. Soc. **65**, 1612-1620.

Pabalan, R. T. and Pitzer, K. S. (1987) Thermodynamics of concentrated electrolyte mixtures and the prediction of mineral solubilities to high temperatures for mixtures in the system $Na-K-Mg-Cl-SO_4-OH-H_2O$. Geochim. Cosmochim. Acta **51** , in press.

Patterson, C. S., Busey, R. H. and Mesmer, R. E. (1984) Second Ionization of Carbonic Acid in NaCl Media to 250°C. **13** 647-661.

Patterson, C. S., Slocum, G. H., Busey, R. H. and Mesmer, R. E. (1982) Carbonate equilibria in hydrothermal systems: first ionization of carbonic acid in NaCl media to 300°C. Geochim. Cosmochim. Acta **46** 1653-1663.

Platford, R.F. (1968) Osmotic and activity coefficients of some simple borates in aqueous solution at 25°. Can. J. Chem. **47**, 2271-2273.

Peiper, J.C. and Pitzer, K.S. (1982) Thermodynamics of aqueous carbonate solutions including mixtures of sodium carbonate, bicarbonate, and chloride. J. Chem. Thermodyn. **7**, 613-638.

Plummer, L. N. and Busenberg, E. (1982) The solubilities of calcite, aragonite, and vaterite in CO_2-H_2O solutions between 0 and 90°C, and an evaluation of the aqueous model for the system $CaCO_3-CO_2-H_2O$ Geochim. Cosmochim. Acta **46**, 1011-1040

Plummer, L. N., Jones, B. F., and Truesdell, A. H. (1976) WATEQF - A Fortran IV Version of WATEQ, A Computer Program for Calculating Chemical Equilibrium of Natural Waters, U.S. Geol. Surv. Water-Resour. Invest. 76-13.

Smith, G.I. (1979) Subsurface stratigraphy and geochemistry of late Quaternary evaporites, Searles Lake, California. U.S. Geol. Surv. Prof. Paper 1043, 130p.

Spencer, R.J., Eugster, H., and Jones, B.F. (1985) Geochemistry of Great Salt Lake, Utah II: Pleistocene-Holocene evolution. Geochim. Cosmochim. Acta **49**, 739-747.

Stephens, H. and Stephens, T. (1963) Solubilities of inorganic and organic compounds. Macmillan, New York, NY.

Teeple, J.E. (1929) The industrial development of Searles Lake brines with equilibrium data. In Amer. Chem. Soc. Monograph Ser. 49, The Chemical Catalog Company, Inc..

Thurmond, V. and Millero, F. J. (1982) Ionization of carbonic acid in sodium chloride solutions at 25°C. J. Sol. Chem. **11**, 447.

Truesdell, A. H. and Jones, B. F. (1969) Ion association in natural brines. Chem. Geol. **4** 51-62

Van't Hoff, J. H. Zur Bildung der ozeanischen Salzlagerstätten Vieweg, Braunschweig, 1905 (first part) and 1909 (second part)

Whitmarsh, R.B., Weser, O.E., Ross, D.A., et al. (1974) Initial Reports of the Deep Sea Drilling Project., Site 227. Vol. 23, Washington, D.C., U.S. Gov't Printing Office, 601-676.

Wolery, T. J. (1983) EQ3NR A computer program for geochemical aqueous speciation-solubility calculations: user's guide and documentation. Lawrence Livermore Lab. Report UCRL-53414.

Chapter 6

Dimitri A. Sverjensky

CALCULATION OF THE THERMODYNAMIC PROPERTIES OF AQUEOUS SPECIES AND THE SOLUBILITIES OF MINERALS IN SUPERCRITICAL ELECTROLYTE SOLUTIONS

INTRODUCTION

Despite numerous experimental studies the solubilities of most rock-forming minerals as functions of temperature, pressure and fluid composition at supercritical temperatures and pressures are poorly known (see reviews by Eugster and Baumgartner, 1987; Eugster, 1981, 1986; Helgeson et al., 1981; and Walther, 1986a). This problem hampers severely our understanding of hydrothermal, metamorphic and magmatic processes involving supercritical aqueous electrolyte solutions in the Earth's crust. Because of the complexities of these processes, which involve so many minerals and fluids of widely varying and evolving compositions, prospects are remote for an experimental resolution of this problem unless experiments are accompanied by comprehensive theoretical calculations of mineral solubilities at supercritical conditions. Recent advances in aqueous solution chemistry (Shock and Helgeson, 1987; Sverjensky et al., 1987; Tanger and Helgeson, 1987), together with experimental investigation of the thermodynamic behavior of aqueous electrolytes (Gates, 1985; Hovey and Tremaine, 1986; Tremaine et al., 1986; Wood et al., 1983; and others), have led to equations of state and correlation algorithms that can be used to calculate the thermodynamic properties of aqueous ions and complexes in supercritical electrolyte solutions. These calculations permit prediction of the solubilities of the rock-forming minerals in supercritical electrolyte solutions. The purpose of the present chapter is to provide a summary of how to make these predictions.

COMPUTATIONAL STRATEGY FOR MINERAL SOLUBILITY CALCULATIONS

Calculation of mineral solubilities in supercritical aqueous electrolyte solutions can be carried out with the aid of speciation models for the aqueous phase which take into account explicitly the formation of aqueous complexes (e.g. Garrels and Christ, 1965; Garrels and Thompson, 1962; Helgeson, 1964, 1968, 1969, 1970; Helgeson et al., 1981; Nordstrom et al., 1979; Wolery, 1983). In the equilibrium constant formulation of this approach (Nordstrom et al., 1979) four principle sources of information are required: hydrolysis constants (or solubility products) of minerals, dissociation constants of aqueous complexes, activity coefficients of all the solute species, and the activity of the solvent. This information permits simultaneous solution of sets of non-linear equations derived from mass action, mass balance, and charge balance constraints.

As an example, consider the solubility of galena in the system PbS-NaCl-H_2S-H_2SO_4-H_2O. Even in this relatively simple system, a comprehensive solubility calculation requires consideration of the law of mass action for the following equilibria:

$$PbS + H^+ = Pb^{2+} + HS^- \tag{1}$$

$$PbCl^{2+} = Pb^{2+} + Cl^- \tag{2}$$

$$PbCl_2 = Pb^{2+} + 2Cl^- \quad (3)$$

$$PbCl_3^- = Pb^{2+} + 3Cl^- \quad (4)$$

$$PbCl_4^{2-} = Pb2+ + 4Cl^- \quad (5)$$

$$Pb(OH)^+ = Pb^{2+} + OH^- \quad (6)$$

$$Pb(OH)_2 = Pb^{2+} + 2OH^- \quad (7)$$

$$Pb(OH)_3^- = Pb^{2+} + 3OH^- \quad (8)$$

$$PbSO_4^\circ = Pb^{2+} + SO_4^{2-} \quad (9)$$

$$NaCl = Na^+ + Cl^- \quad (10)$$

$$Na(OH) = Na^+ + OH^- \quad (11)$$

$$HCl = H^+ + Cl^- \quad (12)$$

$$H_2O = H^+ + OH^- \quad (13)$$

$$H_2S = H+ + HS^- \quad (14)$$

$$HSO_4^- = H^+ + SO_4^{2-} \quad (15)$$

$$H_2S + O_{2(g)} = 2H^+ + SO_4^{2-} \quad (16)$$

$$H_2S + 0.5O_{2(g)} = H_2O + 0.5S_{2(g)} \quad (17)$$

The mass action expressions corresponding to these seventeen equilibria define relationships between the thermodynamic activities of PbS, $O_{2(g)}$, $S_{2(g)}$, H_2O and the twenty aqueous species in Equations (1)-(17). Consequently, for given activities of PbS, $O_{2(g)}$, $S_{2(g)}$ and H_2O there are twenty unknown activities related by seventeen mass action equations. By further specifying the total dissolved molalities of sodium ($m_{t,Na}$) and chloride ($m_{t,Cl}$) given by

$$m_{t,Na} = m_{Na^+} + m_{NaCl^\circ} + m_{Na(OH)^\circ} \quad (18)$$

and

$$m_{t,Cl} = m_{Cl^-} + m_{NaCl^\circ} + m_{HCl^\circ} + m_{PbCl^+}$$

$$+ 2m_{PbCl_2^\circ} + 3m_{PbCl_3^-} + 4m_{PbCl_4^{2-}} , \quad (19)$$

together with the charge balance equation

$$m_{Na^+} + m_{H^+} + 2m_{Pb^{2+}} + m_{Pb(OH)^+} =$$

$$m_{Cl^-} + m_{PbCl_3^-} + 2m_{PbCl_4^{2-}} + m_{HSO_4^-} + 2m_{SO_4^{2-}} + m_{HS^-} + m_{OH^-} \quad (20)$$

we obtain a set of twenty equations involving twenty unknowns. Alternate specifications of the problem might: (a) replace Equation (17) and the given value of $a_{S_2(g)}$ by specifying the total dissolved sulfur in the fluid ($m_{t,S}$) according to

$$m_{t,S} = m_{H_2S} + m_{HS^-} + m_{SO_4^{2-}} + m_{PbSO_4^\circ} + m_{HSO_4^-} \quad (21)$$

or (b) replace the charge balance equation (Eq. 20) by specifying the pH. In either case, simultaneous solution of the appropriate set of equations permits calculation of the total dissolved lead concentration of the fluid in equilibrium with galena ($m_{t,Pb}$) from

$$m_{t,Pb} = m_{Pb^{2+}} + m_{PbCl^+} + m_{PbCl_2} + m_{PbCl_3^-} +$$

$$m_{PbCl_4^{2-}} + m_{Pb(OH)^+} + m_{Pb(OH)_2} + m_{Pb(OH)_3^-}$$

$$+ m_{PbSO_4^\circ} \quad . \quad (22)$$

Extension of the above example to include other minerals would involve enlarging the chemical system under consideration to incorporate additional components such as FeO, Al_2O_3, and SiO_2. For example, the activities of $O_{2(g)}$ and $S_{2(g)}$ can be specified by including in the system the equilibrium assemblage, magnetite + pyrrhotite + pyrite. These activities are defined by the law of mass action for

$$FeS + 0.5S_2 = FeS_2 \quad (23)$$

and

$$3FeS + 2O_2 = Fe_3O_4 + 1.5S_2 \quad . \quad (24)$$

Further incorporation of the assemblage albite + paragonite + quartz specifies implicitly the a_{Na^+}/a_{H^+} ratio (Hemley, 1959; Montoya and Hemley, 1975) as a consequence of the law of the mass action for

$$1.5NaAlSi_3O_8 + H^+ = 0.5NaAl_3Si_3O_{10}(OH)_2 + Na^+ + 3SiO_2 \quad . \quad (25)$$

The solubility of galena can then be calculated from a combination of the law of mass action expressions for Equilibria (1)-(17) and (23-(25), and Equations (18)-(22), together with additional mass action constraints imposed by equilibria between Fe and Al-bearing aqueous species. Although the speciation of these two elements in alkali halide solutions is not well established (see Anderson et al., 1987; Chou and Eugster 1977; Crerar and Barnes, 1974; Crerar et al., 1978; Eugster, 1986; Whitney et al., 1985; Wood and Crerar, 1985), a number of iron and aluminum chloride complexes may form, for which we can write:

$$FeCl^+ = Fe^{2+} + Cl^- \quad (26)$$

$$FeCl_2 = Fe^{2+} + 2Cl^- \quad (27)$$

$$FeCl_3^- = Fe^{2+} + 3Cl^- \quad (28)$$

$$FeCl_4^{2-} = Fe^{2+} + 4Cl^- \quad (29)$$

$$AlCl^{2+} = Al^{3+} + Cl^- \quad (30)$$

Additional complexes such as $AlCl_2^+$, $AlCl_3^\circ$, $AlCl_4^-$, and $NaAlCl_4^\circ$ might also be present in appreciable concentrations. Equations (26)-(30) require modification of Equation (19), the chloride mass balance equation, to include the species $FeCl^+$, $FeCl_2^\circ$, $FeCl_3^-$, $FeCl_4^{2-}$, and $AlCl^{2+}$. If Equation (20), the charge balance equation is to be used, it must also be modified to include the additional charged species. Depending on the relative values of the equilibrium constants corresponding to Equilibria (26)-(30) compared to those for Equilibria (2)-(5), the Fe- and Al-bearing complexes may compete effectively with the Pb-bearing complexes for the chloride ligand (cf. Hemley et al., 1986). Consequently, the solubility of galena in the presence of the assemblage magnetite + pyrrhotite + pyrite + albite + paragonite + quartz should differ significantly from that in the presence of magnetite + pyrrhotite + pyrite, or from the solubility of galena in the absence of these mineral assemblages. This should be true even if the temperature, pressure, a_{O_2}, a_{S_2}, and m_{t,Cl^-} were identical for each of the three cases. It is thus crucial to consider all of the minerals present in a natural system, as well as all relevant aqueous species in the speciation model of the supercritical aqueous solution. This, in turn, requires values for all of the equilibrium constants over a wide range of pressures and temperatures.

Calculation of equilibrium constants requires calculation of the the apparent standard molal Gibbs free energies of formation (ΔG°_ϕ) of the ϕth species in each chemical reaction (e.g. Eq. 1-17) according to

$$\Delta G^\circ_{\phi;P,T} = \Delta G^\circ_{f,\phi;P_r,T_r} + (G^\circ_{\phi;P,T} - G^\circ_{\phi;P_r,T_r}) \;, \tag{31}$$

where the first term on the right ($\Delta G^\circ_{f,\phi;P_r,T_r}$) refers to the standard molal Gibbs free energy of formation of the ϕth species from its elements at the reference pressure (P_r) and temperature (T_r), and the term in parentheses on the right-hand side of Equation (31) refers to the change in the standard molal Gibbs free energy of the ϕth species associated with the change from P_r and T_r to P and T. The latter can be computed by evaluating specific expressions of the equation

$$G^\circ_{\phi;P,T} - G^\circ_{\phi;P_r,T_r} = - S^\circ_{\phi;P_r,T_r}(T-T_r) + \int_{P_r,T}^{P_r,T} (C^\circ_{P_r;\phi})dT$$

$$-T \int_{P_r,T_r}^{P_r,T} (C^\circ_{P_r;\phi})d\ln T + \int_{P_r,T}^{P,T} (V^\circ_{\phi;T})dP \;, \tag{32}$$

where $S^\circ_{\phi;P_r,T_r}$, $C^\circ_{P_r,\phi}$, and $V^\circ_{\phi,T}$ refer to the standard molal entropy, heat capacity and volume of the ϕth species at the subscripted pressures and temperatures. It is shown below that Equations (31) and (32) can be evaluated over wide ranges of pressure and temperature for minerals, gases and aqueous species, which provides a comprehensive basis for prediction of thermodynamic equilibrium constants. Let us first consider the calculation of equilibrium constants for mineral hydrolysis.

Comprehensive prediction of the hydrolysis constants of 120 sulfide, sulfate, carbonate, oxide, and silicate minerals to 600°C and 5 kbar were made by Bowers et al. (1984) based on evaluation of Equations (31) and (32) using equations of state and thermodynamic data for minerals, gases, and aqueous species taken from Helgeson et al. (1978), Helgeson and Kirkham (1974a), Helgeson et al. (1981) and Helgeson (1985a). Recent revision of the equations of state for aqueous species (Tanger and Helgeson, 1987), together with the development of correlation algorithms (Shock, 1987; Shock and Helgeson, 1987; Shock et al., in prep.) enables extension of such predictions to 1000°C and 5 kbar. Before carrying out these calculations let us consider, in turn, specific expressions of Equation (32) for minerals, gases, and aqueous species.

Standard molal Gibbs free energies of minerals

The standard state for minerals adopted in the present study refers to the pure mineral at the pressure and temperature of interest. Calculations of the apparent standard molal Gibbs free energy of formation of a mineral designated by α ($\Delta G^\circ_{\alpha;P,T}$) using Equations (31) and (32) can be carried out using values of $\Delta G^\circ_{f,\alpha;P_r,Tr}$ and $S^\circ_{\alpha;P_r,T_r}$, the standard molal free energy of formation of the mineral and the standard molal entropy of the mineral α, respectively, at P_r and T_r, together with expressions for the temperature dependence of $C^\circ_{P_r;\alpha}$ and $V^\circ_{T;\alpha}$.

The standard molal heat capacities of minerals can be calculated as a function of temperature at 1 bar from power functions (Maier and Kelley, 1932; Haas and Fisher, 1976; Berman and Brown, 1985). In the present study, the heat capacities of minerals are evaluated from the Maier-Kelley equation

$$C^\circ_{Pr,\alpha} = a_\alpha + b_\alpha T - c_\alpha T^{-2} \tag{33}$$

where a_α, b_α, and c_α represent coefficients obtained by fitting experimentally determined heat capacities, or by estimation from hypothetical equilibria involving phases with chemical compositions and structures as similar to the αth mineral as possible (Helgeson et al., 1978). With the exception of quartz, the standard molal volumes of most rock-forming minerals in the Earth's crust can be closely approximated by

$$V^\circ_{\alpha;P,T} = V^\circ_{\alpha;P_r,T_r} \tag{34}$$

(Helgeson et al., 1978; Walther and Helgeson, 1977). Taking account of Equations (31)-(34) results in the following expression for the apparent standard free energy of formation of the αth mineral at P,T:

$$\Delta G^\circ_{\alpha;P,T} = \Delta G^\circ_{f,\alpha;P_r,T_r} - S^\circ_{\alpha;P_r,T_r}(T-T_r)$$

$$+ a_\alpha(T-T_r-T\ln(T/T_r))$$

$$+ \frac{(c_\alpha - b_\alpha TT_r^2)(T-T_r)^2}{2TT_r^2}$$

$$+ V^\circ_{\alpha;P_r,T_r}(P-P_r) \quad , \tag{35}$$

which can be evaluated numerically using data summarized in Tables 8 and 9 in Helgeson et al. (1978).

Standard molal Gibbs free energies of gases

The standard state for gases adopted in the present study refers to the hypothetical perfect gas at 1 bar and the temperature of interest. As a consequence the fugacities of gases are equal to their thermodynamic activities at all temperatures and pressures. Because the apparent standard free energy of formation of the ith gas ($\Delta G^\circ_{i;P,T}$) is independent of pressure it can be calculated from a combination of appropriate statements of Equations (31)-(33), which results in

$$\Delta G_{i;P,T} = \Delta G^\circ_{f,i;P_r,T_r} - S^\circ_{i;P_r,T_r}(T-T_r)$$

$$+ a_i(T-T_r - T\ln(T/T_r))$$

$$+ \frac{(c_i-b_iTT_r^2)(T-T_r)^2}{2TT_r^2} \quad , \tag{36}$$

where a_i, b_i, c_i are the Maier-Kelley coefficients of the gas, and $S^\circ_{i;P_r,T_r}$ and $\Delta G^\circ_{f,i;P_r,T_r}$ refer to the standard entropy of the gas and the standard free energy of formation of the gas, respectively, at P_r and T_r.

Standard molal Gibbs free energies of aqueous species

The standard state for the jth aqueous species adopted in the present study refers to the hypothetical one molal solution consistent with Henry's Law. Computation of the apparent standard molal Gibbs free energy of formation of the jth aqueous species at supercritical pressures and temperatures ($\Delta\bar{G}^\circ_{j;P,T}$) using Equations (31) and (32) requires values of the standard partial molal Gibbs free energy of formation of the jth species at P_r and T_r ($\Delta\bar{G}^\circ_{j;P_r,T_r}$) and the standard partial molal entropy of the jth species at P_r and T_r ($\bar{S}^\circ_{j;P_r,T_r}$), and equations for the temperature dependency of the standard partial molal heat capacity of the jth species at P_r($\bar{C}^\circ_{P_r,j}$) and the pressure dependency of the standard partial molal volume of the jth species at T ($\bar{V}^\circ_j$). Although the temperature dependency of the standard partial molal heat capacities of aqueous ions and electrolytes can be calculated accurately with equations containing power functions of temperature (e.g. Phutela and Pitzer, 1986; Phutela et al., 1987), the equations cannot usefully be extrapolated to temperatures outside the range covered by experimental data. In contrast, equations of state for the temperature and pressure dependencies of the standard partial molal heat capacities and volumes of aqueous species are useful for extrapolation and contain a small number of coefficients that can be estimated in the absence of experimental measurements at high temperatures and pressures (e.g. Helgeson and Kirkham, 1974a,b; 1976; Helgeson et al., 1981; Shock and Helgeson, 1987; Tanger and Helgeson, 1987). The equations of state revised by Tanger and Helgeson can be written as

$$\overline{C}^{\circ}_{P_r,j} = c_{1,j} + \frac{c_{2,j}}{(T-\theta)^2} + \omega_j TX + 2TY\left(\frac{\partial \omega_j}{\partial T}\right)_P - T\left(\frac{1}{\varepsilon} - 1\right)\left(\frac{\partial^2 \omega_j}{\partial T^2}\right)_P \tag{37}$$

and

$$\overline{V}^{\circ}_j = a_{1,j} + \frac{a_{2,j}}{P+\psi} + \frac{a_{3,j}}{T-\theta} + \frac{a_{4,j}}{(P+\psi)(T-\theta)} - \omega_j Q + \left(\frac{1}{\varepsilon} - 1\right)\left(\frac{\partial \omega}{\partial P}\right)_T \quad , \tag{38}$$

where $c_{1,j}$, $c_{2,j}$, $a_{1,j}$, $a_{2,j}$, $a_{3,j}$, $a_{4,j}$ represent temperature/pressure independent coefficients characteristic of the jth aqueous species, and θ and ψ refer to solvent parameters with values of 228 K and 2600 bars, respectively. Y, Q, and X (Helgeson and Kirkham, 1974a) refer to the partial derivatives of the dialectric constant of H_2O (ε) given by

$$Y = \frac{1}{\varepsilon^2}\left(\frac{\partial \varepsilon}{\partial T}\right)_P \tag{39}$$

$$Q = \frac{1}{\varepsilon^2}\left(\frac{\partial \varepsilon}{\partial P}\right)_T \tag{40}$$

$$X = \frac{1}{\varepsilon^2}\left(\frac{\partial^2 \varepsilon}{\partial T^2}\right)_P - \frac{2}{\varepsilon}\left(\frac{\partial \varepsilon}{\partial T}\right)_P^2 \tag{41}$$

and ω_j denotes the conventional Born coefficient of the j aqueous species.

For the jth aqueous ion, the conventional Born coefficient, ω_j can be expressed as

$$\omega_j = \omega_j^{abs.} - Z_j(0.5387) \tag{42}$$

where

$$\omega_j^{abs.} = \frac{\eta Z_j^2}{r_{e,j}} \tag{43}$$

and $\eta = 1.66027 \times 10^5$ Å.cal.mole^{-1}, Z_j stands for the charge on the ion, and $r_{e,j}$ denotes the effective electrostatic radius of the ion, given by (Shock, 1987; Tanger and Helgeson, 1987):

$$r_{e,j} = r_{x,j} + |Z_j|(k_Z+g) \quad . \tag{44}$$

Table 1. Standard molal quantities and equation of state parameters used to predict the properties of the aqueous species in the reactions corresponding to the curves depicted in Figures 1-4.

Species	$\Delta\overline{G}^\circ_f$	$\overline{S}^\circ$	$\overline{V}^\circ$	$\overline{C}^\circ_p$	c_1	c_2	a_1	a_2	a_3	a_4	ω
H^+	0.0	0.0	0.0	0.0	0.0	0.0	0.0	0.0	0.0	0.0	0.0
Na^+	-62,591	13.96	-1.11	9.06	18.18	-2.98	1.8390	-2.2850	-3.2560	-2.726	0.3306
HS^-	2860	16.3	20.65	-22.17	3.42	-6.27	5.0778	4.9799	3.0144	-2.9849	1.4410
OH^-	-37,604	-2.6	-4.18	-32.79	4.15	-10.39	1.7759	0.0738	-1.8275	-2.7821	1.7246
Cl^-	-31379	13.56	17.79	-29.44	-4.40	-5.71	4.0320	4.8010	5.5630	-2.847	1.4560
Pb^{2+}	-5710	4.2	-15.6	-12.7	8.66	-5.62	-0.0051	-9.3957	12.8788	-2.3906	1.0788
$PbCl^+$	-39046	28.0	3.0	4.9	10.22	-2.04	2.2188	-3.1615	8.7041	-2.6483	0.1281
$PbCl^\circ_2$	-71190	47.0	21.7	2.0	3.91	-2.63	4.6088	3.5381	4.2176	-2.9253	-0.3718
$PbCl^-_3$	-102147	55.0	40.3	11.0	19.95	-0.79	7.5487	11.793	-1.3012	-3.2659	0.7961
$PbCl^{2-}_4$	-133254	59.0	59.0	9.6	33.15	-1.08	10.6222	20.3950	-7.0707	-3.6221	2.3187
$NaCl^\circ$	-92910	28.0	24.0	8.5	10.80	-1.30	5.0363	4.7365	3.4154	-2.9748	-0.038

Table 2. Solvent parameters used to predict the curves in Figures 1, 2 and 4-6. Units of ΔG°_f are cal. mole^{-1}. Units of the function g(P,T) are Å.

	T(°C)	ΔG°_f	ε	-100g
1 kbar	300	-62564	24.91	0.0364
	350	-64012	20.08	0.2216
	400	-65540	16.01	0.7145
	450	-67146	12.48	1.9007
	500	-68831	9.45	4.2621
	550	-70592	7.03	7.8765
	600	-72426	5.34	12.676
	650	-74323	4.24	18.255
	700	-76273	3.52	24.161
2 kbar	300	-62061	28.05	0.0
	350	-63477	23.32	0.0
	400	-64964	19.55	0.0
	450	-66518	16.45	0.2440
	500	-68136	13.86	0.6382
	550	-69815	11.69	1.3157
	600	-71550	9.87	2.2077
	650	-73339	8.36	3.2561
	700	-75178	7.14	4.4382
	750	-77062	6.16	5.7592
	800	-78988	5.37	7.2268
	850	-80953	4.74	8.8323
	900	-82953	4.24	10.545
3 kbar	300	-61588	30.43	0.0
	350	-62980	25.57	0.0
	400	-64441	21.77	0.0
	450	-65966	18.71	0.0
	500	-67550	16.20	0.0
	550	-69190	14.09	0.1236
	600	-70882	12.31	0.3747
	650	-72624	10.79	0.7228
	700	-74412	9.51	1.1412
	750	-76243	8.42	1.6117
	800	-78114	7.49	2.1271
	850	-80023	6.71	2.6881
	900	-81968	6.04	3.2974
4 kbar	300	-61134	32.42	0.0
	350	-62510	27.36	0.0
	400	-63951	23.45	0.0
	450	-65453	20.35	0.0
	500	-67013	17.82	0.0
	550	-68627	15.72	0.0
	600	-70291	13.95	0.0
	650	-72002	12.44	0.0107
	700	-73758	11.14	0.0514
	750	-75555	10.01	0.1242
	800	-77392	9.04	0.2231
	850	-79265	8.20	0.3423
	900	-81173	7.47	0.4776

In Equation (44), $r_{x,j}$ refers to the crystallographic radius of the jth ion, k_Z is a constant equal to 0.94 Å for cations and 0.0 Å for anions, and g represents a solvent function of pressure and temperature. Values of g(P,T) have been obtained by regression of experimental heat capacities and volumes of the electrolyte NaCl to 450°C and 2 kbar (Tanger and Helgeson, 1987). Values of g(P,T) at higher temperatures and pressures have been retrieved from experimental dissociation constants for the complex NaCl° (Shock et al., in prep.; see Table 2) and permit evaluualuation of Equations (42)-(44) to 1000°C and 5 kbar.

Combination of Equations (37) and (38) with appropriate statements of Equations (31) and (32) leads to

$$\Delta G^{\circ}_{j;P,T} = \Delta\overline{G}^{\circ}_{f,j;P,T} - \overline{S}^{\circ}_{j;P_rT_r}(T-T_r)$$

$$- c_{1,j}\left[T\ln\left(\frac{T}{T_r}\right) - T + T_r\right] + a_{1,j}(P-P_r)$$

$$+ a_{2,j} \ln \left(\frac{P + \psi}{P_r + \psi}\right)$$

$$- c_{2,j} \left[\left(\frac{1}{T-\theta} - \frac{1}{T_r-\theta}\right)\left(\frac{\theta-T}{\theta}\right) - \frac{T}{\theta^2} \ln \left(\frac{T_r(T-\theta)}{T(T_r-\theta}\right)\right]$$

$$+ \left(\frac{1}{T-\theta}\right)\left[a_{3,j}(P-P_r) + a_{4,j}\ln \left(\frac{P + \Psi}{P_r+\Psi}\right)\right]$$

$$+ \omega_{j;P,T}\left(\frac{1}{\varepsilon} - 1\right) - \omega_{j;P_r,T_r}\left(\frac{1}{\varepsilon_{P_r,T_r}} - 1\right)$$

$$+ \omega_{j;P_r,T_r} Y_{P_r,T_r}(T-T_r) \quad . \tag{45}$$

Values of the equation of state coefficients in Equation (45) have been generated by regression of experimentally determined standard molal volumes and heat capacities of a wide array of electrolytes (Shock, 1987; Shock and Helgeson, 1987; and Tanger and Helgeson, 1987; some typical sets of equation of state coeficients are listed in Table 1). In addition, Shock (1987) and Shock and Helgeson (1987) have demonstrated that the equation of state coefficients determined by regression of electrolyte data define a set of simple linear correlations with values of the standard molal entropies, volumes and heat capacities referring to 25°C and 1 bar. These correlations afford estimation of equation of state coefficients for any aqueous species. Estimated equation of state coefficients have now been generated for over 200 monatomic and polyatomic ions and neutral species (Shock, 1987; Shock and Helgeson, 1987).

Equation (45) enables computation of the temperature and pressure dependencies of the standard molal Gibbs free energies of any aqueous species. Values of $\varepsilon(P,T)$ for this purpose to 500°C and 5 kbar are tabulated in Helgeson and Kirkham (1974a). By using the Kirkwood equation (Pitzer, 1983), values of $\varepsilon(P,T)$ can be computed to 1000°C and 5 kbar (Table 2).

In summary, the apparent standard molal Gibbs free energies of minerals, gases and aqueous species can be calculated using Equations (35), (36) and (45), respectively. This information, together with values of the standard molal Gibbs free energy of H_2O (Haar et al., 1984; see Table 2), permits calculation of the standard Gibbs free energies, and hence the equilibrium constants, corresponding to the hydrolysis of the rock-forming minerals to 1000°C and 5 kbar. Such calculations are considerably facilitated by the computer program SUPCRT which is currently available (Shock, Oelkers, Tanger and Helgeson, 1987, pers. comm.).

Examples of the application of the equations, data and information summarized above are depicted in Figures 1 and 2 which contain curves representing equilibrium constants for reactions (1) and (25) calculated with the aid of the program SUPCRT from apparent standard molal Gibbs free energies predicted with Equations (35) and (45), and the parameters and coefficients given in Tables 1 and 2. Equilibrium constants such as those represented by the curves in Figures 1 and 2 constitute one of the four sources of information required to calculate the solubilities of minerals in supercritical aqueous electrolyte solutions (see above). Values of the dissociation constants of all possible aqueous complexes are also required. Computation of these will be discussed next.

DISSOCIATION CONSTANTS OF AQUEOUS SPECIES

The overall dissociation of the complex $(ML_y)^z$ referring to the dissociation of the complex into its constituent cation (M^{z-yl}) and y ligands (L^l), where z and l refer to the charges on the cation and ligand, respectively, can be represented by

$$(ML_y)^z = M^{z-yl} + yL^l \quad . \tag{46}$$

The overall dissociation constant β_y is given by

$$\beta_y = \frac{a_{M^{z-yl}}(a_{L^l})^y}{(a_{ML_y^z})} \quad . \tag{47}$$

Most attempts to predict values of β_y at high temperatures or pressures rely on extrapolating experimental measurements referring to 25°C and 1 bar using estimated or measured values of the standard entropies or enthalpies of dissociation together with simplifying assumptions about the heat capacities of reaction. One such method is the isocoulombic approach, according to which the standard heat capacities of reaction are assumed to be zero for reactions in which aqueous species with the same charge occur on each side of the reaction of interest (Gurney, 1936, 1938, 1953; Lindsay, 1980). Another method (Helgeson, 1967, 1969; Arnorsson et al., 1982, 1983; Smith et al., 1986), proceeds by assuming that the ratio of the nonelectrostatic standard heat capacity of reaction to the overall standard heat capacity of reaction is a constant to 150-200°C at low pressures. In the supercritical region, empirical methods have been proposed for extrapolation of equilibrium dissociation

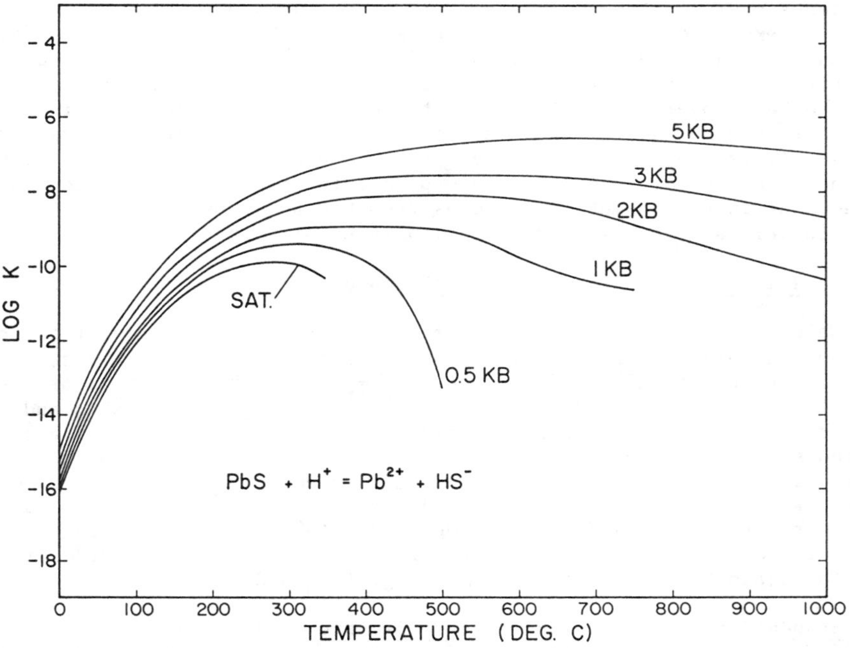

Figure 1. Predicted values of the hydrolysis constant of galena (Equilibrium 1, see text).

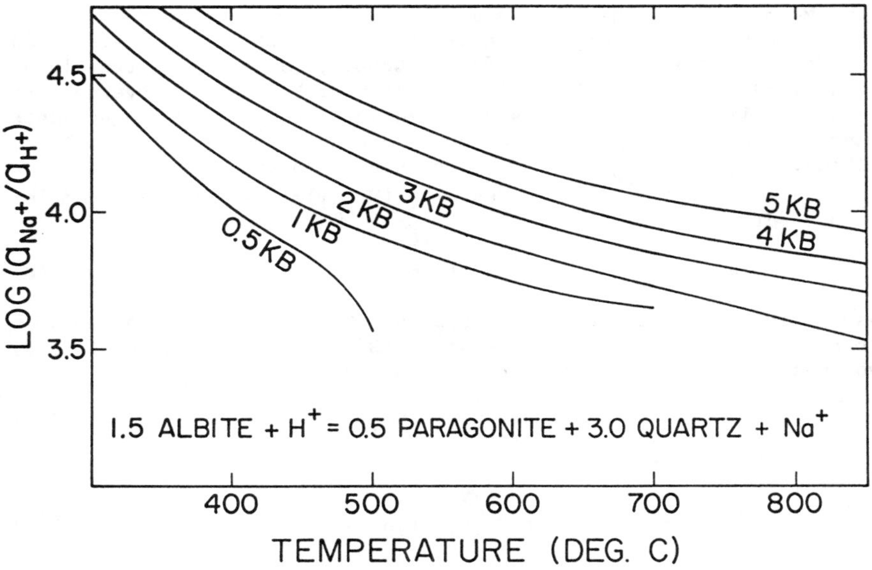

Figure 2. Predicted values of the log a_{Na^+}/a_{H^+} defined by the assemblage albite + paragonite + quartz (Equilibrium 25, see text).

constants over limited regions of P/T space based on empirical correlations of experimental dissociation constants with solvent density (e.g. Marshall, 1968, 1969, 1970, 1972a, 1972b; Franck, 1956, 1981; see also Chapter 10). None of the above methods enable prediction of dissociation contants over a wide range of temperatures and pressures from subcritical through supercritical conditions without experimental calibration. It is the purpose of the present section to describe recent advances that permit comprehensive prediction of dissociation constants of aqueous inorganic metal complexes from 0°C and 1 bar to 1000°C and 5 kbar (Sverjensky et al., 1987; Sverjensky et al., in prep.).

Standard molal Gibbs free energies of aqueous complexes

Calculation of the standard molal Gibbs free energy of the qth aqueous complex using Equation (45) requires values of the standard molal entropy at 25°C and 1 bar ($\bar{S}^\circ_{q;P_r,T_r}$) and the equation of state coefficients represented by $c_{1,q}$, $c_{2,q}$, $a_{1,q}$, $a_{2,q}$, $a_{3,q}$, $a_{4,q}$ and ω_q. By taking advantage of the correlations between equation of state coefficients and the standard molal entropies, heat capacities and volumes of aqueous ions, neutral species and complexes refered to above, equation of state coefficients for a wide variety of complexes can be estimated from the standard molal properties ($\bar{S}^\circ_q$, $\bar{C}^\circ_{P,q}$, $\bar{V}^\circ_q$) at 25°C and 1 bar (see below). Before reviewing this procedure, let us consider first how to obtain the requisite values of $\bar{S}^\circ_q$, $\bar{C}^\circ_{P,q}$ and $\bar{V}^\circ_q$.

Standard molal entropies, heat capacities and volumes of complexes at 25°C and 1 bar. Despite the fact that numerous experimentally derived values of $\bar{S}^\circ_q$, $\bar{C}^\circ_{P,q}$ and $\bar{V}^\circ_q$ have been reported in the literature (summarized by Akitt, 1980; Asano and le Noble, 1978; Baes and Mesmer, 1976, 1981; Hamann, 1974; Hogfeldt, 1982; Seward, 1981; Sillen and Martell, 1964, 1971; Smith and Martell, 1976; Wagman et al., 1982), most of the data are subject to such large uncertainties (e.g. discussion in Baes and Mesmer, 1981) that they are not useful for an equation of state approach. Previous attempts to estimate values of $\bar{S}^\circ_q$ have also been subject to large uncertainties (e.g. Cobble, 1953a,b; George, 1959; Helgeson, 1969; Nancollas, 1960, 1961). This problem can be overcome by employing the linear correlations depicted in Figures 3A-H to estimate stepwise entropies, heat capacities and volumes of association. These quantities, $\Delta S^\circ_{r,y}$, $\Delta Cp^\circ_{r,y}$ and $\Delta V^\circ_{r,y}$, correspond to the stepwise addition of a monovalent ligand L^- to the species $(ML_{y-1})^{Z+1}$ forming the complex $(ML_y)^Z$ according to

$$ML_{y-1}{}^{Z+1} + L^- = ML_y{}^Z \quad . \tag{48}$$

Values of the estimated stepwise entopies, heat capacities and volumes permit calculation of overall entropies (ΔS°_r), heat capacities ($\Delta C^\circ_{P,r}$) and volumes (ΔV°_r) of association corresponding to the reaction

$$M^{Z+y} + yL^- = (ML_y)^Z \quad , \tag{49}$$

where

$$\Delta S^\circ_r = \sum_{n=1}^{y} \Delta S_{r,n} \quad , \tag{50}$$

$$\Delta C^{\circ}_{P_r} = \sum_{n=1}^{y} \Delta C_{P_{r,n}} \quad , \tag{51}$$

and

$$\Delta V^{\circ}_r = \sum_{n=1}^{y} \Delta V_{r,n} \quad . \tag{52}$$

It follows that the required values of $\bar{S}^{\circ}_q$, $\bar{C}^{\circ}_{P,q}$ and $\bar{V}^{\circ}_q$ can be calculated from

$$\bar{S}^{\circ}_q = \Delta S^{\circ}_r + \bar{S}^{\circ}_c + y(\bar{S}^{\circ}_l) \quad , \tag{53}$$

$$\bar{C}^{\circ}_{P,q} = \Delta C^{\circ}_{P_r} + \bar{C}^{\circ}_{P,c} + y(\bar{C}^{\circ}_{P,l}) \quad , \tag{54}$$

and

$$\bar{V}^{\circ}_q = \Delta V^{\circ}_r + \bar{V}^{\circ}_c + y(\bar{V}^{\circ}_l) \quad , \tag{55}$$

where the subscripts c and l represent the cation (M^{Z+y}) and the ligand (L^-), respectively.

Equations (50)-(55) can be evaluated using equations for the slopes and intercepts of the linear correlations in Figures 3A-H. The latter are consistent with the following equations:

For stepwise entropies of association (Figs. 3A - D),

$$\Delta S^{\circ}_{r,y} = (a\bar{S}^{\circ abs}_l + a^l)\bar{S}^{\circ abs}_m + b\bar{S}^{\circ abs}_l + b^l \tag{56}$$

where $\bar{S}^{\circ abs}_l$ and $\bar{S}^{\circ abs}_m$ stand for the absolute standard molal entropies of the ligand (L^-) and the species $(ML_{y-1})^{Z+1}$ in Reaction (48) at 25°C and 1 bar. The coefficients a, a^l, b and b^l can be expressed as linear functions of the charge (Z), where Z = -3, -2, -1, 0, 1, 2, or 3, according to

$$a = (0.03420)Z - 0.01030 \tag{57}$$

$$a^l = (-0.641)Z + 0.1708 \tag{58}$$

$$b = (1.323)Z - 0.200 \tag{59}$$

$$b^l = (-9.08)Z + 0.5 \quad . \tag{60}$$

For stepwise heat capacities of association (Figs. 3E and F),

$$\Delta C^{\circ}_{P_{r,y}} = \left[(1.85)(Z+1) - 0.2\right]\bar{C}^{\circ abs}_{P_m} - (69.1)(Z+1) + 38.0 \quad , \tag{61}$$

where $\bar{C}^{\circ abs}_{P_m}$ refers to the absolute standard molal heat capacity of the species represented by $(ML_{y-1})^{Z+1}$ in reaction (48), Z = -1, 0, 1, 2 or 3, and the ligand in Reaction (48) is F^-, Cl^-, Br^-, I^-, or OH^-. For stepwise volumes of association, the study by Akitt (1980) is consistent with

$$\Delta V^{\circ}_{r,1} = \Delta V^{\circ}_{r,2} = \ . \ . \ . = \Delta V^{\circ}_{r,y-1} = \Delta V^{\circ}_{r,y} \tag{62}$$

which implies that

$$\Delta V^{\circ}_r = y(\Delta V^{\circ}_{r,y}) \tag{63}$$

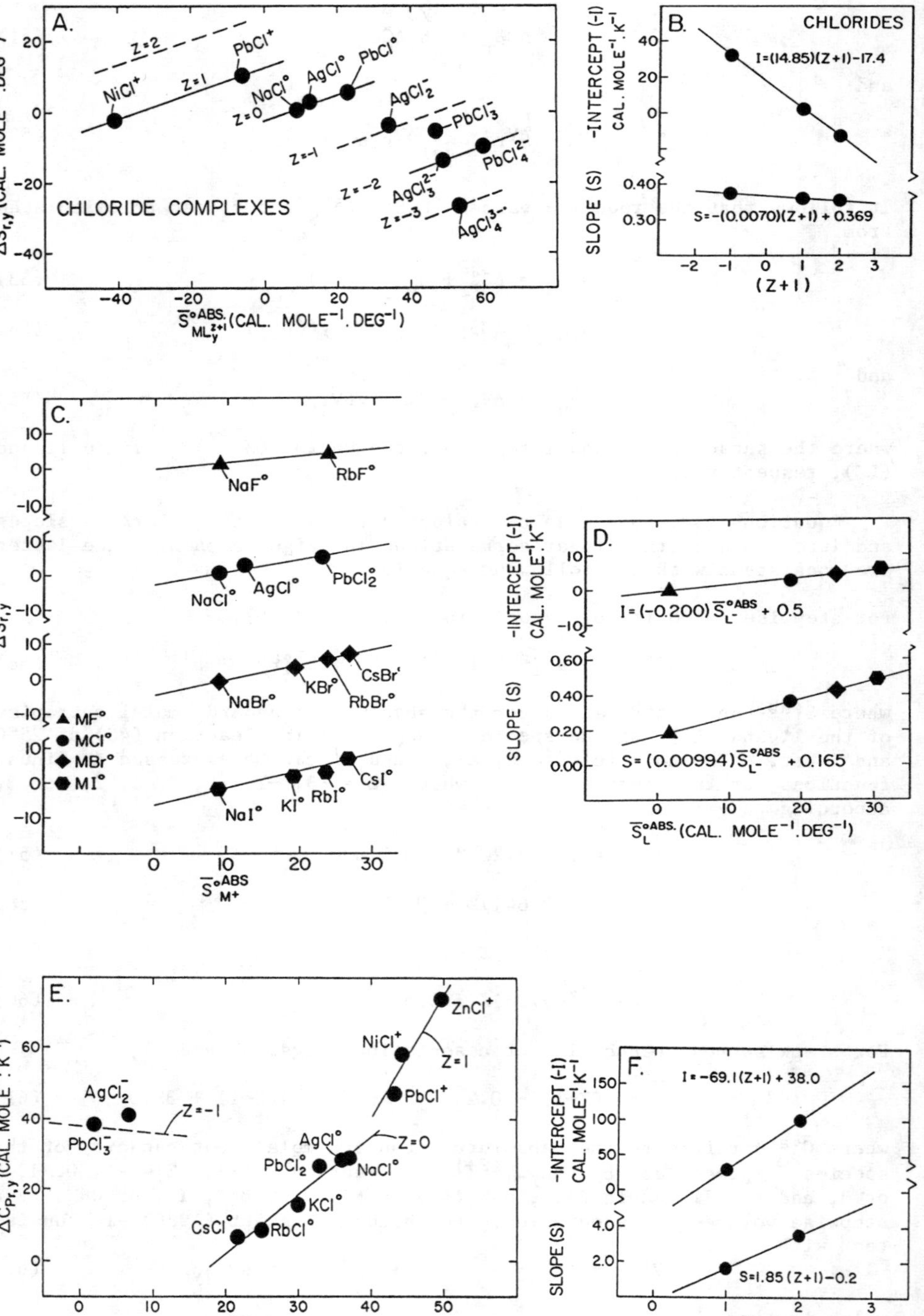

A.
CHLORIDE COMPLEXES
Z=2
Z=1
Z=0
Z=-1
Z=-2
Z=-3
NiCl+
PbCl+
NaCl°
AgCl°
PbCl°
AgCl2-
PbCl3-
AgCl3 2-
PbCl4 2-
AgCl4 3-
ΔS°r,y (CAL. MOLE-1.DEG-1)
S°ABS. ML_y Z+1 (CAL. MOLE-1. DEG-1)
B.
CHLORIDES
I=(14.85)(Z+1)-17.4
S=-(0.0070)(Z+1)+0.369
-INTERCEPT (-I) CAL. MOLE-1.K-1
SLOPE (S)
(Z+1)
C.
NaF°
RbF°
NaCl°
AgCl°
PbCl2°
NaBr°
KBr°
RbBr°
CsBr°
NaI°
KI°
RbI°
CsI°
MF°
MCl°
MBr°
MI°
ΔS°r,y
S°ABS M+
D.
I=(-0.200) S°ABS L- + 0.5
S=(0.00994) S°ABS L- + 0.165
-INTERCEPT (-I) CAL. MOLE-1.K-1
SLOPE (S)
S°ABS. L (CAL. MOLE-1. DEG-1)
E.
ZnCl+
NiCl+
PbCl+
Z=1
AgCl2-
PbCl3-
Z=-1
AgCl°
Z=0
PbCl2°
NaCl°
KCl°
CsCl°
RbCl°
ΔCPr,y (CAL. MOLE-1.K-1)
C°P(ML_y-1)Z+1
F.
I=-69.1(Z+1)+38.0
S=1.85(Z+1)-0.2
-INTERCEPT (-I) CAL. MOLE-1.K-1
SLOPE (S)
(Z+1)

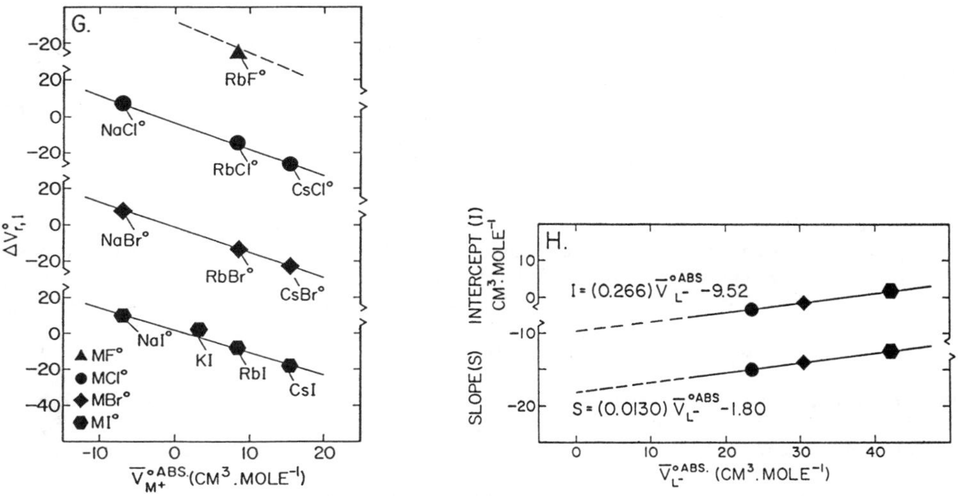

Figure 3A-H. Correlation of the standard stepwise entropies, heat capacities and volumes of aqueous metal complexes with the absolute standard molal properties of the constituent cations and ligands.

Combination of Equation (62) with equations representing the linear correlations in Figures 3G and H results in the equation

$$\Delta V^{\circ}_{r,y} = \frac{\lambda}{1-\lambda Z}\left(\bar{V}^{\circ\,abs}_{c}\right) + \frac{\lambda Z}{1-\lambda Z}\left(\bar{V}^{\circ\,abs}_{l}\right) + \mu \ , \tag{64}$$

where $\bar{V}^{\circ abs}_{c}$ and $\bar{V}^{\circ abs}_{l}$ represent the absolute standard molal volumes of the cth cation and lth anion in the complex $(ML_y)^Z$ (Z = -3, -2, -1, 0, 1, 2, or 3) and λ and μ represent coefficients that are consistent with

$$\lambda = (0.0130)\ \bar{V}^{\circ abs}_{l} - 1.80 \tag{65}$$

and

$$\mu = (0.266)\ \bar{V}^{\circ abs}_{l} - 9.52 \ . \tag{66}$$

Equating (49)-(66) provide the only general method for calculating the standard molal entropies, heat capacities and volumes of individual aqueous complexes at 25°C and 1 bar (Sverjensky et al., 1987). These results permit estimation of equation of state coefficients for aqueous complexes (reviewed next), which enables comprehensive prediction of the standard molal free energies at high temperatures and pressures for many complexes of geochemical interest.

Equation of state coefficients for aqueous complexes. The correlations between equation of state coefficients and the standard molal entropies, heat capacities and volumes of ions, neutral species, and complexes developed by Shock (1987) permits estimation of equation of state coefficients for any complex according to the following procedure:

(1) The conventional Born solvation coefficient of the mth charged complex, represented by ω_m, is calculated from Equations (42) and (43) using values of $r_{e,m}$ obtained from the equation

$$r_{e,m} = \frac{Z_m^2(\eta Y - 100)}{\bar{S}^\circ_m - 71.5 \left| Z_m \right|} \quad . \tag{67}$$

In Equation (67), Z_m represents the charge on the complex, $\bar{S}^\circ_m$ represents the standard molal entropy of the complex, Y represents a Born function (Helgeson and Kirkham, 1974a) which is equal to -5.81×10^{-5} K^{-1} at 25°C and 1 bar. For the nth neutral complex, values of ω_n can be calculated from

$$\omega_n = s\left[(1514.3874)\bar{S}^\circ_n + (0.38 \times 10^5)\right] - \left[0.04 \times 10^5\right] , \tag{68}$$

where $\bar{S}^\circ_n$ represents the standard molal entropy of the nth complex, and s is equal to 0 for polar, linear complexes (e.g. NaCl°, CuBr°, LiF°, etc.) and 1 for polar, non-linear complexes (e.g. $ZnCl_2^\circ$, $FeCl_3^\circ$, etc.).

(2) From values of the Born coefficient (ω_q) for the qth complex (charged or neutral), the solvation volume ($\Delta\bar{V}^\circ_{s,q}$) and heat capacity ($\Delta\bar{C}p^\circ_{s,q}$) referring to 25°C and 1 bar (where g = 0.0) can be calculated from

$$\Delta\bar{V}^\circ_{s,q} = -\omega_q Q \tag{69}$$

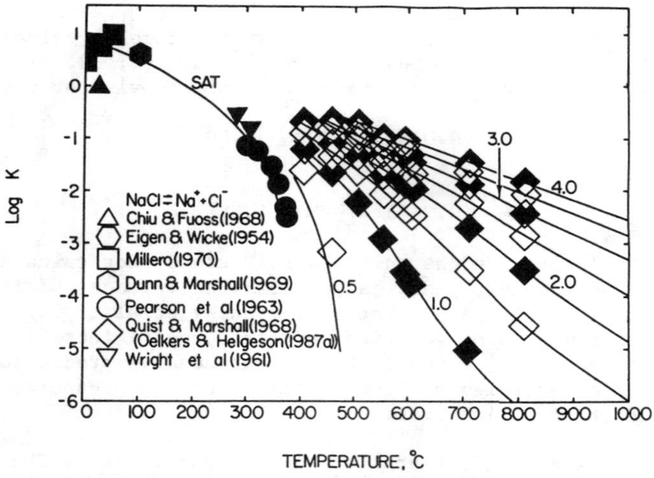

Figure 4. The dissociation constant of NaCl° calculated with the parameters in Tables 1 and 2.

and $$\Delta\bar{C}^{\circ}_{P_{s,q}} = \omega_q TX \quad , \tag{70}$$

where Q and X represent Born functions (Helgeson and Kirkham, 1974a) with values of 5.903×10^{-7} $(°K)^{-2}$ and -3.090×10^{-7} bar^{-1}, respectively. In turn, this permits calculation of the non-solvation volume $(\Delta\bar{V}^{\circ}_{n,q})$ from

$$\Delta\bar{V}^{\circ}_{n,q} = \bar{V}^{\circ}_{q} - \Delta\bar{V}^{\circ}_{s,q} \quad . \tag{71}$$

(3) From the non-solvation volume computed above, values of $a_{1,q}$, $a_{2,q}$ and $a_{4,q}$ can be calculated using the equations

$$a_{1,q} = (1.3684 \times 10^{-2})\ \Delta\bar{V}^{\circ}_{n,q} + 0.1765 \tag{72}$$

$$\sigma_q = (1.11)\Delta\bar{V}^{\circ}_{n,q} + 1.8 \tag{73}$$

$$a_{2,q} = (\sigma_q - a_{1,q})\ (P + \Psi) \tag{74}$$

$$a_{4,q} = -(4.134)\ a_2 - 27790 \tag{75}$$

where σ_q is a pressure-dependent, temperature independent parameter (Tanger and Helgeson, 1987).

(4) From the equation of state coefficients obtained above and the volume V_q at 25°C and 1 bar, the coefficient $a_{3,q}$ can be calculated using Equation (38).

(5) The equation of state coefficient $c_{2,q}$ can be estimated using the equation

$$c_{2,q} = (0.2037)\bar{C}^{\circ}_{P,q} - 3.0346 \tag{76}$$

Finally, the coefficient $c_{1,q}$ can be obtained because at 25°C and 1 bar the temperature derivatives of ω_q (Eq. 37) are zero (Tanger and Helgeson, 1987), which permits calculation of $c_{1,q}$ from Equation (37) and values of $c_{2,q}$ and $\bar{C}^{\circ}_{P,q}$. This completes the estimation procedure, resulting in a set of equation of state coefficients to be used in Equation (45) for comprehensive calculation of apparent standard molal Gibbs free energies of aqueous complexes as functions of temperature and pressure. In combination with calculated values of the apparent standard molal Gibbs free energies of the ions constituting the complexes, the above results enable calculation of the equilibrium dissociation constants of aqueous complexes to 1000°C and 5 kbar for the first time.

A few of the many possible examples of such calculations are given in Figures 4 and 5A-D. The solid curves in the figures represent equilibrium constants calculated using the equation of state parameters summarized in Table 1. It can be seen in Figures 4 and 5 that all the computed dissociation contants at supercritical temperatures exhibit strong decreases with decreasing pressure and increasing temperature. Combination of the equilibrium constants in Figure 1 with those referring to the dissociation equilibrium in Figure 5D yields the curves plotted in Figure 6. The latter has the opposite dependence on pressure to that exhibited in Figure 1, which suggests that the solubility of galena in saline solutions in which $PbCl_4^{2-}$ is the dominant lead-bearing species will vary inversely with pressure (see also Helgeson and Lichtner, 1987). Experimental measurements of galena solubility (Hemley et al., 1986) are consistent with this observation.

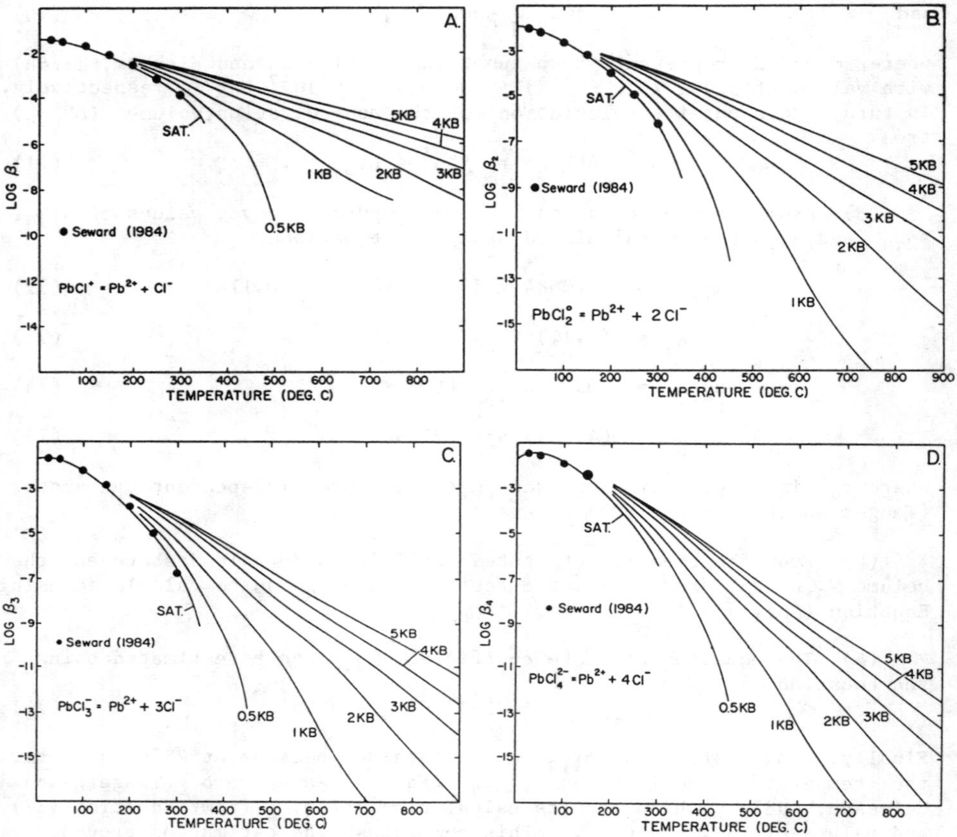

Figures 5A–D. Dissociation constants $\log\beta_n$ (n = 1 to 4) for lead chlorides calculated with the parameters in Tables 1 and 2.

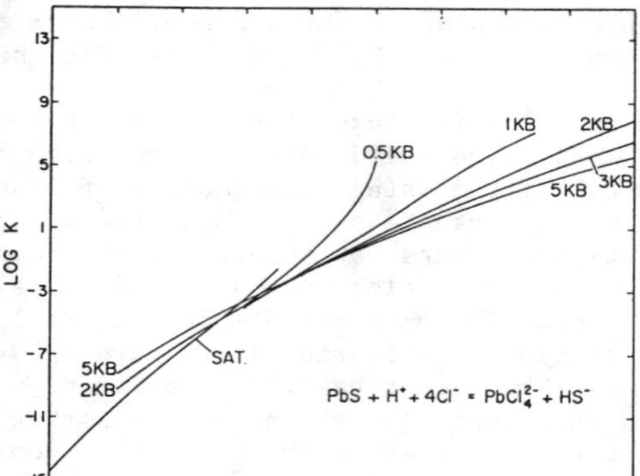

Figure 6. Combined hydrolysis constant of galena and the fourth dissociation constant (β_4). Compare with Figure 1.

With equilibrium constants such as those represented by the curves in Figures 4 and 5, the second of the four sources of information required to calculate the solubilities of minerals in supercritical aqueous electrolyte solutions (see above) is now available. The final two sources of information are discussed next.

ACTIVITY COEFFICIENTS OF SOLUTE SPECIES AND THE ACTIVITY OF THE SOLVENT

Although considerable attention has been paid to the calculation of activity coefficients of aqueous species at subcritical temperatures and pressures (e.g. Helgeson, 1969; Pabalan and Pitzer, 1987; Pitzer, 1973, 1975, 1977, 1979; Wood et al., 1984; and summaries in Harned and Owen, 1958; Helgeson et al., 1981; Lewis and Randall, 1961; and Chapter 4), few attempts have been made to calculate activity coefficients of inorganic ions in aqueous electrolyte solutions at supercritical temperatures and pressures (Helgeson et al., 1981). In the present chapter we will briefly review a geochemical approximation developed by Helgeson et al. (1981).

Solute Species

Charged Species. The activity coefficient of the jth aqueous ion or charged complex (γ_j) in NaCl solutions can be approximated by

$$\log \gamma_j = -\frac{A_\gamma Z_j^2 \bar{I}^{1/2}}{1+\mathring{a}B_\gamma \bar{I}^{1/2}} + \Gamma_\gamma + [\omega_j^{abs} b_{NaCl} + b_{Na^+Cl^-} - 0.19(|Z_j|-1)]\,\bar{I} \quad , \qquad (77)$$

where A_γ and B_γ are the Debye-Hückel solvent functions (Helgeson and Kirkham, 1974b; McKenzie and Helgeson, 1984), $\bar{I}$ refers to the "true" ionic strength of the solution defined by

$$\bar{I} = 0.5\Sigma m_j Z_j^2 \qquad (78)$$

where m_j is the molality of the jth aqueous species with charge Z_j, and $\mathring{a}$ refers to the ion-size parameter for the solution (3.72Å for NaCl dominated solutions, Helgeson et al., 1981). The term Γ_γ provides for conversion of the rational activity coefficient to the molal scale according to

$$\Gamma_\gamma = -\log(1 + 0.0180153m^*) \ , \qquad (79)$$

where m^* stands for the sum of the molalities of all solute species. The extended term parameters b_{NaCl} and $b_{Na^+Cl^-}$ refer to the dependence of solvation on ionic strength and short-range interactions between cations and anions, respectively.

Evaluation of an equation of state representation of the extended term parameters in Equation (77) to 550°C and 5 kbar (Helgeson et al., 1981) indicates that b_{NaCl} and $b_{Na^+Cl^-}$ have opposite signs at supercritical temperatures and pressures. In fact, the relative magnitudes of these two parameters and their trends with temperature and pressure are such that for $|Z_j| = 1$ and typical values of ω_j^{abs} (Helgeson and Kirkham,

1976), the term linear in $\bar{I}$ in Equation (77) is of the order of $(0.0 \pm 0.2)\bar{I}$ at pressures and temperatures within a large region of supercritical P-T space. This region extends to about 500°C at 1 kbar. However, at pressures greater than or equal to 2 kbar, it probably extends to to temperatures of about 750°C. Consequently, if $\bar{I} \leq 1.0$, the extended term parameters in Equation (77) probably contribute less than ± 0.2 to the overall log γ_j value. Under the same conditions, the term $\Gamma\gamma$ contributes even less to log γ_j. In contrast, the first term on the right-hand side of Equation (77), the Debye-Hückel contribution, is substantial (with an absolute magnitude of about 0.5-1.0 for $\bar{I}$ less than or equal to 1.0). Therefore, the Debye-Hückel contribution dominates under the conditions described above. As a consequence, close approximation of activity coefficients of aqueous species can be made from the Debye-Hückel equation alone (see also Wilson, 1986).

Neutral Species. Experimental evidence indicates that the activity coefficient of the nth neutral species (γ_n) is proportional to the total molality of the kth added electrolyte (m_k^n) according to

$$\log\gamma_n = b_{\gamma,n}\bar{I} , \tag{80}$$

where $b_{\gamma,n}$ is a coefficient characteristic of the neutral species and the electrolyte k (e.g. Helgeson et al., 1981; Marshall, 1980; summary in Lewis and Randall, 1961; Reardon and Langmuir, 1976). Theoretical calculations predict a limiting law containing a similar linear dependence on concentration (Lewis and Randall, 1961).

For supercritical temperatures and pressures, the traditional approach is to set $b_{\gamma,n} \simeq 0$ (e.g. Helgeson et al., 1981). This approximation is consistent with data for the solubility of quartz in supercritical aqueous fluids containing up to 1 molal KCl at 2 kbar and 500 - 700°C (Anderson et al., 1987), which can be computed from the solubilities of quartz in pure water. However, recent considerations of conductance data (Oelkers and Helgeson, 1986) suggest that at high temperatures and low pressures, $b_{\gamma,n}$ may be significantly greater than zero.

Activity of the solvent

The activity of the solvent (a_w) is related to the osmotic coefficient of an electrolyte solution (Φ) by

$$\log a_w = - \frac{m^* \Phi}{127.84} . \tag{81}$$

Evaluation of the dependence of ϕ on $\bar{I}$ at supercritical temperatures and pressures using equations and data summarized in Helgeson et al. (1981) indicates that for values of $\bar{I} \leq 1.0$ we can closely approximate the activity of H_2O in NaCl solutions (Helgeson, 1985b) by writing

$$a_{H_2O} = 1.0 . \tag{82}$$

COMPUTATIONAL APPROACH

With the information summarized above for prediction of the hydrolysis constants of minerals, the dissociation constants of aqueous species, and the activity coefficients of solute species, it is possible to solve sets of mass action, mass balance, and charge balance equations to predict aqueous speciation and the solubilities of minerals to temperatures and pressures of approximately 1000°C and 5 kbar. Numerous approaches to the solution of the governing sets of nonlinear equations and their application to aqueous speciation/mineral solubility calculations have been described (e.g. Helgeson, 1964, 1979; Crerar, 1975; Nordstrom et al., 1979; Wolery, 1983).

The most general code for the calculation of aqueous speciation and the solubilities of the rock-forming minerals at subcritical temperatures and pressures is EQ3NR (Wolery, 1983; Wolery et al., 1984). EQ3NR uses a Newton-Raphson algorithm for simultaneously solving the governing equations for mass action, mass balance, charge balance and ionic strength. A variety of techniques to aid convergence are employed, including the use of logarithmic iteration variables, under-relaxation techniques, automatic generation of starting estimates, optimization of starting estimates, and user-specified and automatic basic-switching (Wolery, 1983). The code is sufficiently general that it is virtually independent of the data base, so that adding or deleting species from the chemical model can be accomplished by changing the database alone. Similarily, changing the total range of temperatures and pressures that can be considered involves little more than changes in the database. All chemical species are referred to on the input file by name rather than index number and a large number of options for specifying the input of the specific problem of interest are available. For example, the concentration of an aqueous species can be input as a total dissolved concentration, or a thermodynamic activity, or by referring either to a mass action equation corresponding to a heterogeneous equilibrium or to the charge balance equation. Consequently, in the present study EQ3NR and its database have been modified to deal with the supercritical range of pressures and temperatures of interest, which enables preservation of all of the above features of the code and its generality. Calculation of aqueous speciation and mineral solubilities at supercritical temperatures and pressures for many multicomponent systems of geochemical interest can now be carried out for the first time. Some examples are discussed below.

ILLUSTRATIVE EXAMPLES

Aqueous speciation of lead and chloride in supercritical chloride-bearing fluids.

Aqueous speciation of lead and chloride in NaCl and HCl solutions fluids containing 0.1 and 1.0 molal total chloride from 400-750°C, 1-4 kbar, and pH 3-6 are depicted in Figures 7-10. The chemical speciation model from which the curves in Figures 6-9 were derived includes $PbCl_y^{2-y}$ (y = 1 to 4), $Pb(OH)^+$, NaCl°, HCl°, Na(OH)°, Cl^-, OH^-, H^+, Na^+ and Pb^{2+}. The curves shown in the Figures were generated from the equations and data summarized above (Tables 1 and 2) using mass action, mass balance, and charge balance equations like those in Equations (18)-(22).

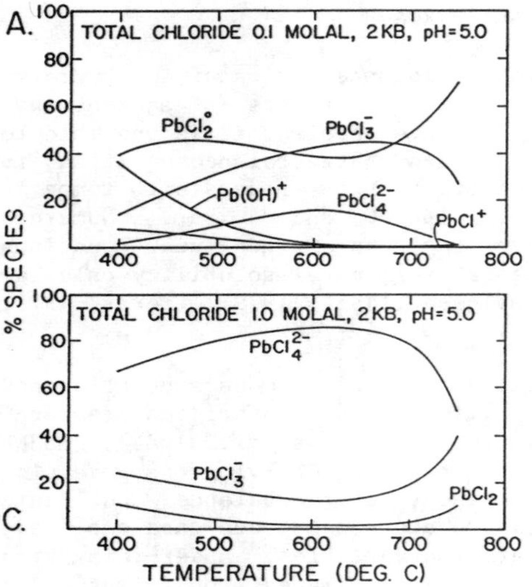

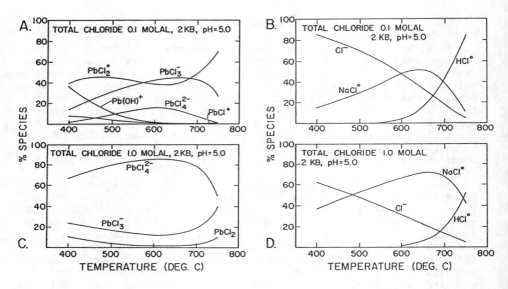

Figures 7A-D. Predicted speciation as a function of temperature at 2 kb and pH 5.0 for lead and chloride in NaCl fluids containing 0.1 and 1.0 molal total chloride and trace amounts of lead.

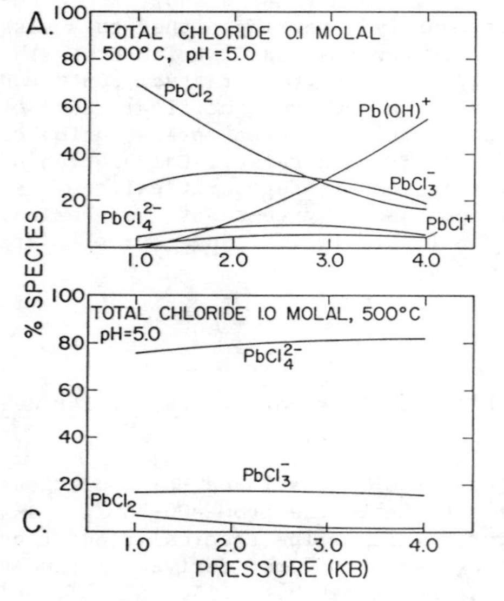

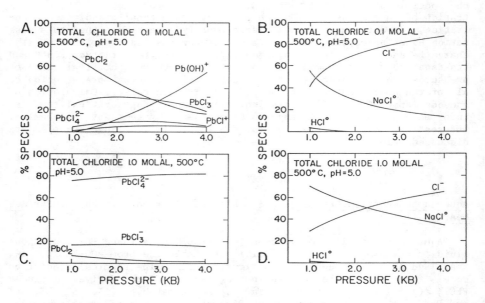

Figures 8A-D. Predicted speciation as a function of pressure at 500°C and pH 5.0 for lead and chloride in NaCl fluids containing 0.1 and 1.0 molal chlorideand trace amounts of lead.

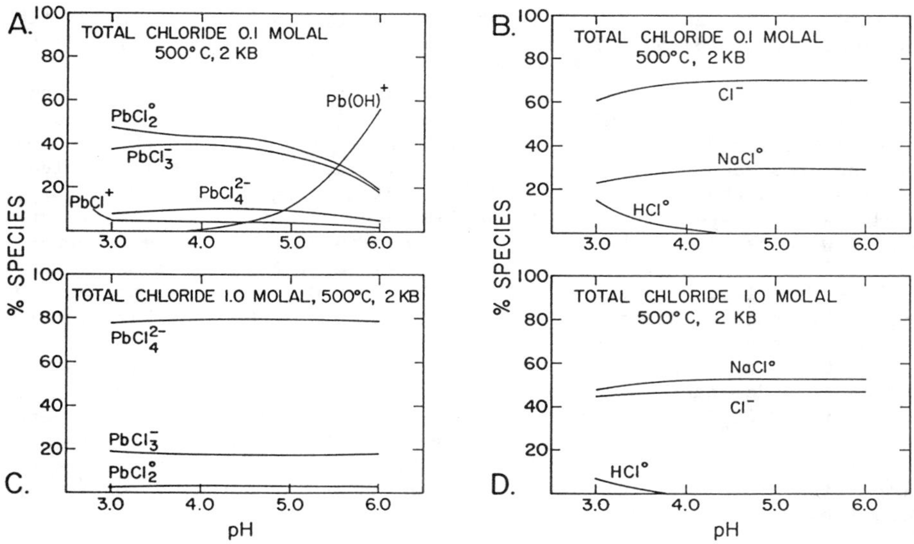

Figures 9A-D. Predicted speciation as a function of pH at 2 kb and 500°C for lead chloride complexes in NaCl fluids containing 0.1 and 1.0 molal chloride and trace amounts of lead.

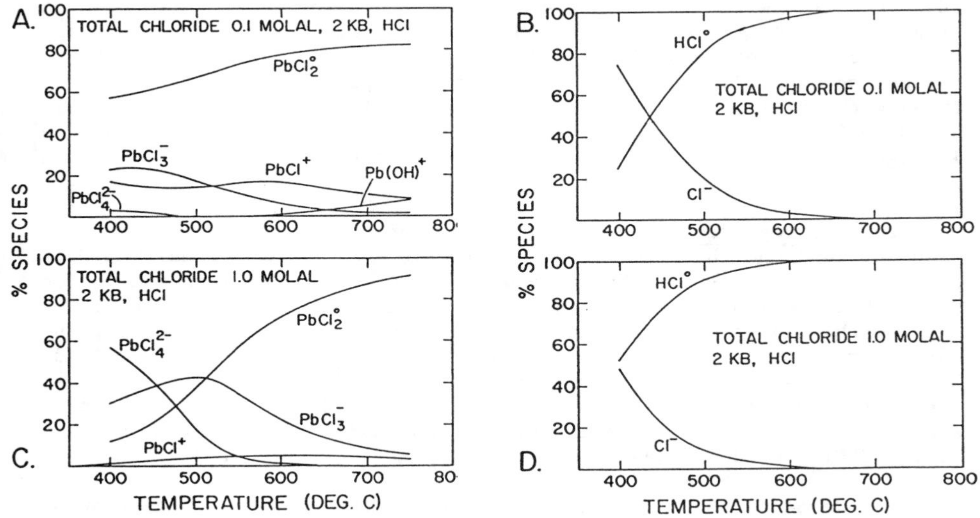

Figures 10A-D. Predicted speciation as a function of temperature at 2 kb for lead and chloride in HCl fluids containing 0.1 and 1.0 molal chloride. The pH is fixed by charge balance. Compare with Figures 7A-D.

Activity coefficients for all solute species (γ_j) were calculated with Equations (77), (78) and (79) with b_k, b_{il} and $b^j_{\gamma,n}$ set to zero.

Temperature dependence. The temperature dependence of the aqueous speciation of lead and chloride from 400-750°C at 2 kb and pH 5.0 is shown in Figures 7A-D. It can be seen in these figures that the predominant species depend strongly on both temperature and total dissolved chloride. In the 0.1 molal chloride fluids (Figs. 7A and B) $PbCl_2$, $Pb(OH)^+$ and Cl^- are predicted to predominate at the lowest temperature (400°C) whereas $PbCl^\circ_2$, $PbCl^-_3$ and HCl° predominate at the highest temperature (750°C). At intermediate temperatures (e.g. 600°C), combinations of these species, and others (e.g. NaCl°, $PbCl_4^{2-}$) are all present in appreciable concentrations. In contrast to this rather complex distribution of species, the speciation of the 1.0 molal chloride fluids (Figs. 7C and D) is much simpler. Below 700°C, where HCl° is a minor species, the dominant species are $PbCl_4^{2-}$, Cl^-, and NaCl°. In fact, the the species $PbCl_4^{2-}$ is the dominant Pb-bearing species between 400° and 700°C. It only decreases in importance at about 700-750°C where the molality of chloride ion has dropped to low values because of its association with H^+ to form HCl° (Fig. 7D), which promotes the formation of $PbCl_3^-$ (and $PbCl_2^\circ$) according to

$$PbCl_4^{2-} + H^+ \rightarrow PbCl_3^- + HCl^\circ \quad (83)$$

and

$$PbCl_4^{2-} + 2H^+ \rightarrow PbCl_2^\circ + 2HCl^\circ \quad . \quad (84)$$

However, even at 750°C, $PbCl_4^{2-}$ still constitutes 50% of the total dissolved lead. The dominance of $PbCl_4^{2-}$ over the range 400°-700°C at 2 kbar and pH 5.0 in NaCl solutions with total chloride of 1.0 molal (Fig. 7C), compared with its relative importance in the low chloride fluids (Fig. 7A), can be attributed not only to the relative magnitudes of the dissociation constants (Figs. 5A-D), but also to the fact that β_4 depends on $(a_{Cl^-})^4$. Consequently, it can be expected that for NaCl solutions with total chloride greater than 1.0 molal at 400-700°C, 2 kbar and pH 5.0, the dominance of $PbCl_4^{2-}$ would be even more pronounced than in Figure 7C. The effects of choosing different pressures and pH values are considered next.

Pressure dependence. The pressure dependence of the aqueous speciation of lead and chloride from 1.0 to 4.0 kbar at 500°C and pH 5.0 is shown in Figures 8A-D. The behaviour of the predicted lead speciation in the 0.1 and 1.0 molal chloride solutions is strikingly different. For the low chloride solution, the speciation is a strong function of pressure (Fig. 8A). At the lowest pressure, $PbCl_2^\circ$ is the dominant lead-bearing species, however, at the highest pressures $Pb(OH)^+$ predominates. At intermediate pressures, $PbCl_2^\circ$, $PbCl_3^-$, and $Pb(OH)^+$ are all present in approximately equal concentrations. In contrast, for the high chloride fluid, $PbCl_4^{2-}$ predominates from 1.0 to 4.0 kbar (Fig. 8C).

Dependence on pH. The pH dependence of the aqueous speciation of lead and chloride in solutions with pH values of 3.0 to 6.0 at 500°C and 2 kbar is shown in Figures 9A-D. In the low chloride fluid (Figs. 9A and B) the lead speciation at or below pH 5.0 is dominated by $PbCl_2$ and $PbCl_3^-$ (as noted above), however, by pH 6.0, $Pb(OH)^+$ predominates over the lead chloride complexes. In the high chloride fluid (Figs. 9C and D), $PbCl_4^{2-}$ is again the dominant species (cf. Figs. 7C, 8C, and 9C), almost independent of the pH.

In summary, the above speciation diagrams indicate that between 400°-700°C, 1-4 kbar, and pH 3.0-6.0, NaCl solutions with low total chloride contents (e.g. 0.1 molal or less) have lead and chloride speciation that are strong functions of temperature, pressure, and pH. Under these conditions $PbCl_2^\circ$ and $PbCl_3^-$ are the most abundant lead chloride complexes. However, in NaCl solutions with high total chloride contents (e.g. 1.0 molal or greater), $PbCl_4^{2-}$ is predicted to be the dominant lead-bearing species over virtually the whole range of temperature, pressure and pH. Consequently, the calculations described above strongly suggest that in natural hydrothermal or metamorphic fluids containing at least 1.0 molal chloride, $PbCl_4^{2-}$ is the dominant lead-bearing species. Preliminary calculations referring to subcritical temperatures suggest the same conclusion.

Dependence on HCl°. The role of HCl° in promoting the stabilities of lead chloride complexes with small ligand numbers can be seen by considering the aqueous speciation of Pb in HCl-fluids (e.g. Figs. 10A-D). At temperatures above about 550°C at 2 kbar, $PbCl_2^\circ$ is predicted to be the dominant Pb-bearing complex in HCl solutions with total chloride contents from 0.1 to 1.0 molal (Figs. 10A and C). Under these conditions HCl° accounts for 90% or more of the total Cl^- (Figs. 10B and D). The same situation prevails at subcritical temperatures and low pressures, which is consistent with the experimental results of Seward (1984), shown in Figure 5, who found no evidence of $PbCl_4^{2-}$ in HCl solutions at temperatures greater than about 150°C. The results of the present study indicate that experimental studies of metal speciation in HCl-fluids are particularly appropriate for the determination of metal chloride complexes with low ligand numbers, whereas studies of metal speciation in mixed NaCl-HCl fluids are more appropriate to the determination of complexes with the highest ligand numbers. This is consistent with available subcritical and supercritical mineral solubility studies (e.g. Boctor et al., 1980; Chou and Eugster, 1976, 1977; Eugster, 1986; Frantz and Eugster, 1973; Frantz and Popp, 1979; Whitney et al., 1985; Wilson, 1986).

Solubility of galena in supercritical chloride-bearing solutions

Figures 11A and B depict the predicted dependence of the solubility of galena on temperature from 400° to 750°C at pressures from 1.0 to 4.0 kbar for solutions containing 0.1 and 1.0 molal chloride respectively. The chemical speciation model used to generate the curves in Figures 11A and B includes the same aqueous species considered in Figures 7-10 above. However, the log (a_{Na^+}/a_{H^+}) ratio in Figures 11A and B is controlled by the albite + paragonite + quartz equilibrium (Eq. 25, and Fig. 2), and the log f_{O_2} and log f_{S_2} are fixed by pyrrhotite + pyrite + magnetite equilibria according to Equations (23) and (24). Neither iron nor aluminum complexes in the aqueous phase were considered in the calc-

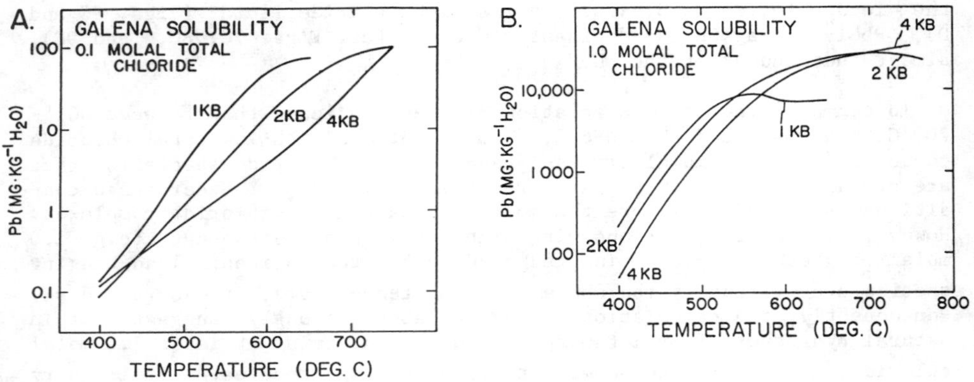

Figures 11A-B. Predicted solubility of galena as a function of temperature and pressure in 0.1 and 1.0 molal chloride solutions with log (a_{Na^+}/a_{H^+}) defined by albite + paragonite + quartz and f_{O_2}, f_{S_2} defined by pyrrhotite + pyrite + magnetite.

ulations responsible for Figures 11A and B because of the large uncertainties attending prediction of the speciation of these elements (see above). Preliminary calculations suggest that inclusion of iron and aluminum chloride complexes in the speciation model (Eqs. 26-30) used to calculate galena solubilities results in competition between iron, aluminum, sodium, and lead for the chloride ligand, which results in a significant decrease in the computed galena solubilities compared to those in Figures 11A and B. This result is consistent with the experimental observations made by Hemley et al. (1986). Consequently, the solubilities shown in Figures 11A and B are maximum values.

Examination of the curves in Figure 11A reveals that the solubilities of galena in the low chloride solution increase strongly with decreasing pressure and increasing temperature between 450° and 650°C. Under these conditions, $PbCl_2^\circ$ and $PbCl_3^-$ are the dominant Pb-bearing species in the solution and Cl^- is far more abundant than HCl°. Consequently, the solubility of galena is controlled by equilibria such as

$$PbS + 2H^+ + 2Cl^- = PbCl_2^\circ + H_2S \quad (85)$$

and

$$PbS + 2H^+ 3Cl^- = PbCl_3^- + H_2S \ . \quad (86)$$

However, at temperatures greater than 600°C, the solubilities are predicted to maximize with temperature at 1 kbar. At 2 kbar there is a trend towards a maximum, which disappears by 4 kbar. These progressive shifts with pressure are a consequence of the progressively changing ratio of HCl°/Cl^-. HCl° progressively complexes more of the Cl^- at the highest temperatures and lowest pressures. Under the latter conditions

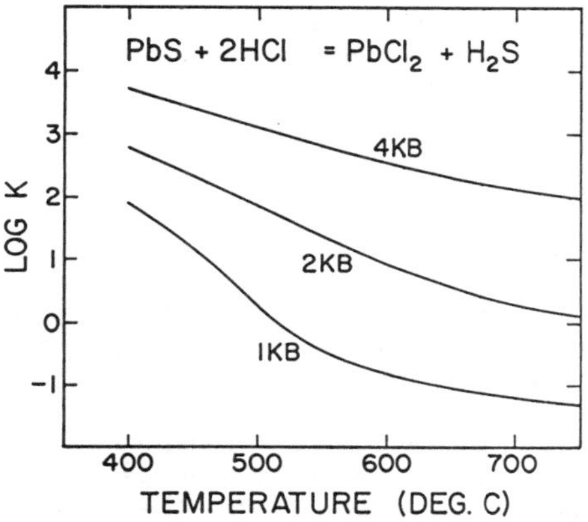

Figure 12. Calculated equilibrium constant corresponding to the dissolution of galena in terms of neutral aqueous species (Equilibrium 87 in the text).

the solubility of galena is controlled by the equilibrium

$$PbS + 2HCl^\circ = PbCl_2^\circ + H_2S \quad . \tag{87}$$

The same equilibrium controls the solubility of galena in the high chloride solution (Fig. 11B) above 550°C at 1 kbar, resulting in a change of the pressure dependence of the solubility. It can be seen in Figure 11B that below 550°C, the predicted solubilities increase with decreasing pressure (as in Fig. 11B) consistent with Figure 6 and the experimental measurements of Hemley et al. (1986). Above 550°C, the change of the pressure dependence of the solubility is a consequence of competition between the equilibria shown in Figs. 6 and 12, which have opposing pressure dependencies. The predicted sinusoidal shape of the 1 kbar solubility curve above 550°C is a consequence of the thermodynamic properties of the aqueous species in Equation (87), all of which are neutral species. This behaviour has been noted experimentally in solubility studies of other systems involving predominantly neutral aqueous species (e.g. Ragnarsdottir and Walther, 1985; Walther, 1986b; and Walther and Helgeson, 1977).

CONCLUDING REMARKS

The equations, data, and predictions described above demonstrate that considerable progress has been made in recent years toward predicting aqueous speciation and mineral solubilities in supercritical hydrothermal and metamorphic fluids in the Earth's crust. The examples of predicted supercritical aqueous speciation and mineral solubility discussed above represent just a few of the many calculations that are now possible. The application of this approach to geologic and geochemical problems requires consideration of all possible chemical components, and their speciation, in the supercritical fluids of interest. The latter can only be achieved by carrying out comprehensive aqueous speciation calculations involving prediction of the thermodynamic properties of these aqueous species. Calculations such as these can reveal the interdependencies of the compositions of coexisting equilibrium mineral assemblages and aqueous fluids. If the results of the calculations are integrated with the results of field investigations and experimental studies of mineral solubilities referring to elevated temperatures and pressures, which is now possible, a quantitative understanding of complex geochemical processes involving supercritical aqueous fluids in the Earth's crust will be attained.

ACKNOWLEDGMENTS

The research reported in this chaper was supported in part by NSF Grants EAR 84-12210 and EAR 84-19418. During the course of the research I have had the pleasure of a remarkably stimulating and fruitful collaboration with my friends and colleagues H.C. Helgeson and E.L. Shock. I am particularly grateful for their reviews and comments, and for permission to use their results in advance of publication. I would also like to acknowledge comments from and discussions with L. Baumgartner, H.P. Eugster and E. Oelkers, all of whom have helped to improve the chapter significantly. Finally, I would like to thank Kate Francis, Carolyn Spangler, and Maxine Mote for typing many drafts of this chapter.

REFERENCES

Akitt, J.W. (1980) Limiting single-ion molar volumes. Intrinsic volume as a function of the solvent parameters. J. Chem. Soc. Faraday Tr. I, 76, 2259-2284.

Anderson, G.M., Pascal, M.L., and Rao, J. (1987) Aluminum speciation in metamorphic fluids. NATO Advanced Study Institute Publication.

Arnórsson, S., Sigurdsson, S. and Svavarsson, H. (1982) The chemistry of geothermal waters in Iceland. I. Calculation of aqueous speciation from 0° to 370°C. Geochim. Cosmochim. Acta 46, 1513-1532.

__________ Gunnlaugsson, E. and Svavarsson, H. (1983) The chemistry of geothermal waters in Iceland. II. Mineral equilibria and independent variables controlling water compositions. Geochim. Cosmochim. Acta 47, 547-566.

Asano, T., and le Noble, W.J. (1978) Activation and reaction volumes in solution. Chem. Reviews, 78, 407-489.

Baes, C.F., Jr., and Mesmer, R.E. (1976) The Hydrolysis of Cations. New York, John Wiley & Sons, 489 p.

__________ (1981) The thermodynamics of cation hydrolysis. Amer. J. Sci. 281, 935-962.

Berman, R.G., and Brown, T.H. (1985) Heat capacity of minerals in the system Na_2O-K_2O-CaO-MgO-FeO-Fe_2O_3-Al_2O_3-SiO_2-TiO_2-H_2O-CO_2: representation, estimation, and high temperature extrapolation. Contrib. Mineral. Petrol. 89, 168-183.

Boctor, N.Z., Popp, R.K., and Frantz, J.D. (1980) Mineral solution equilibria IV. Solubilities and the thermodynamic properties of $FeCl_2^o$ in the system Fe_2O_3-H_2-H_2O-HCl. Geochim. Cosmochim. Acta 44, 1509-1518.

Bowers, T.S., Jackson, K.J., and Helgeson, H.E. (1984) Equilibrium activity diagrams for coexisting minerals and aqueous solutions at pressures and temperatures to 5 kb and 600°C. Springer-Verlog, New York, 397p.

Chou, I-M., and Eugster, H.P. (1976) Hydrothermal acid-base buffers: Fugacity control and dissociation constants of HBr and HI. Contrib. Mineral. Petrol. 56, 77-100.

__________ (1977) Solubility of magnetite in supercritical chloride solutions. Amer. J. Sci. 277, 1296-1314.

Cobble, J.W. (1953a) Empirical considerations of entropy. I. The entropies of the oxy-anions and related species. J. Chem. Physics 21, 1443-1446.

__________ (1953b) Empirical considerations of entropy. II. The entropies of inorganic complex ions. J. Chem. Physics 21, 1446-1450.

Crerar, D.A. (1975) A method for computing multicomponent equilibria based on equilibrium constants. Geochim. Cosmochim. Acta 39, 1375-1384.

__________ and Barnes, H.L. (1976) Ore solution chemistry V. Solubilities of chalcopyrite and chalcocite assemblages in hydrothermal solution at 200° to 350°C. Econ. Geol. 71, 772-794.

__________ Susak, N.J., Borcsik, M., and Schwartz, S. (1978) Solubility of the buffer assemblage + pyrrhotite + magnetite in NaCl solutions from 200° to 350°C. Geochim. Cosmochim. Acta 42, 1427-1437.

Eugster, H.P. (1981) Metamorphic solutions and reactions. In Wickman, F. and Rickard, D., eds., Proceedings, Nobel Symposium on the chemistry and geochemistry of solutions at high temperatures and pressures. Physics and Chemistry of the Earth 13/14, 461-507. Pergamon Press, New York.

__________ (1986) Minerals in hot water. Amer. Mineral. 71, 655-673.

Franck, E.U. (1956) Hochverdichteter Wasserdampf I. Elektrolytische Leitfähigkeit in $KCl-H_2O$-Lösungen bis 750°C. Zeits. Physik. Chemie, Neue Folge, 8, 92-106.

__________ (1981) Survey of selected non-thermodynamic properties and chemical phenomena of fluids and fluid mixtures. In D. Rickard and F. Wickman, eds., Proceedings, Nobel Symposium on the chemistry and geochemistry of solutions at high temperatures and pressures. Physics and Chemistry of the Earth, 13/14, 65-88. Pergamon Press, New York.

Eugster, H.P. and Baumgartner, L. (1987) Mineral solubilities and speciation in metamorphic fluids. Chapter 10, this volume.

Frantz, J.D. and Eugster, H.P. (1973) Acid-base buffers. Use of Ag + AgCl in the experimental control of solution equilibria at elevated pressures and temperatures. Amer. J. Sci. 273, 268-286.

__________ and Popp, R.K. (1979) Mineral solution equilibria - I. An experimental study of complexing and thermodynamic properties of aqueous $MgCl_2$ in the system $MgO-SiO_2-H_2O-HCl$. Geochim. Cosmochim. Acta 43, 1223-1239.

Garrels, R.M., and Christ, C.L. (1965) Solutions, Minerals and Equilibria. Freeman, Cooper and Co., San Francisco. 650 p.

__________ and Thompson, M.E. (1962) A chemical model for seawater at 25°C and one atmosphere total pressure. Amer. J. Sci. 260, 57-66.

Gates, J.A. (1985) Thermodynamics of aqueous electrolyte solutions at high temperatures and pressures: Ph.D. thesis, Univ. of Delaware, Newark, Delaware, 340 p.

George, J.H.B. (1959) Entropies of association of ions in aqueous solution. J. Amer. Chem. Soc. 81, 5530.

Gurney, R.W. (1936) Ions in solution. Cambridge Univ. Press, Cambridge. 206 p.

__________ (1938) J. Chem. Physics 6, 499.

__________ (1953) Ionic Processes in Solution. Dover Publ. Co., New York.

Haar, L., Gallagher, J.S. and Kell, G.S. (1984) NBS/NRC Steam Tables: Thermodynamic and Transport Properties and Computer Programs for Vapor and Liquid States of Water in SI Units. Hemisphere Publ., 318 pp.

Haas, J.L. Jr. and Fisher, J.R. (1976) Simultaneous evaluation and correlation of thermodynamic data. Amer. J. Sci. 276, 525-545.

Hamann, S.D. (1974) Electrolyte solutions at high pressure. In Bockris, J.O'M., ed., Modern Aspects of Electrochemistry, 9, 47-128, Plenum Press, New York.

Harned, H.S., and Owen, B.B. (1958) The physical chemistry of electrolytic solutions. Reinhold Book Corp., New York, 803 p.

Helgeson, H.C. (1964) Complexing and hydrothermal ore deposits. Pergamon Press, New York, 128 p.

__________ (1967) Thermodynamics of complex dissociation in aqueous solution at elevated temperatures. J. Phys. Chem. 71, 3121-3136.

__________ (1968) Evaluation of irreversible reactions in geochemical processes involving minerals and aqueous solutions. I. Thermodynamic relations. Geochim. Cosmochim. Acta 32, 853-877.

__________ (1969) Thermodynamics of hydrothermal systems at elevated temperatures and pressures. Amer. J. Sci. 267, 729-804.

__________ (1970) A chemical and thermodynamic model of ore deposition in hydrothermal systems. In Mineral. Soc. Amer. Spec. Paper 3, 155-186.

__________ (1979) Mass transfer among minerals and hydrothermal solutions. In Geochemistry of Hydrothermal Ore Deposits, K.L. Barnes, ed., 2nd edn., 568-610. John Wiley and Sons, New York, 747 p.

__________ (1985a) Errata II. Thermodynamics of minerals, reactions, and aqueous solutions at high pressures and temperatures. Amer. J. Sci. 285, 845-855.

__________ (1985b) Some thermodynamic aspects of geochemistry. Pure and Applied Chem. 57, 31-44.

__________ Delaney, J.M., Nesbitt, H.W., and Bird, D.K. (1978) Summary and critique of the thermodynamic properties of rock-forming minerals. Amer. J. Sci. 278-A, 1-229.

__________ and Kirkham, D.H. (1974a) Theoretical prediction of the thermodynamic behavior of aqueous electrolytes at high pressures and temperatures. I. Summary of the thermodynamic/electrostatic properties of the solvent. Amer. J. Sci. 274, 1089-1198.

__________ and Kirkham, D.H. (1974b) Theoretical prediction of the thermodynamic behavior of aqueous electrolytes at high pressures and temperatures. II. Debye-Hückel parameters for activity coefficients and relative partial molal properties. Amer. J. Sci. 274, 1199-1261.

__________ and Kirkham, D.H. (1976) Theoretical prediction of the thermodynamic properties of aqueous electrolytes at high pressures and temperatures. III. Equation of state for aqueous species at infinite dilution. Amer. J. Sci. 276, 97-240.

__________ Kirkham, D.H. and Flowers, G.C. (1981) Theoretical prediction of the thermodynamic behavior of aqueous electrolytes at high pressures and temperatures. IV. Calculation of activity coefficients, osmotic coefficients, and apparent molal and standard and relative partial molal properties to 5 kb and 600°C. Amer. J. Sci. 281, 1241-1516.

__________ and Lichtner, P.C. (1987) Fluid flow and mineral reactions at high temperatures and pressures. J. Geol. Soc. London (in press).

Hemley, J.J. (1959) Some mineralogical equilibria in the system K_2O-Al_2O_3-SiO_2-H_2O. Amer. J. Sci. 257, 241-270.

__________ Cygan, G.L., and d'Angelo, W.M. (1986) Effect of pressure on ore-mineral solubilities under hydrothermal conditions. Geology 14, 377-379.

Hogfeldt, E. (1984) Stability constants of metal-ion complexes. Part A. Inorganic ligands, IUPAC Chem. Data Ser. 22: Pergamon Press, Oxford.

Hovey, J.K. and Tremaine, P.R. (1986) Thermodynamics of aqueous aluminum: Standard partial molal heat capacities of Al^{3+} from 10 to 55°C. Geochim. Chosmochim. Acta 50, 453-459.

Lewis, G.N. and Randall, M. (1961) Thermodynamics, 2d ed., revised by Pitzer, K.S., and Brewer, L.: McGraw-Hill, New York, 723p.

Lindsay, W.T., Jr. (1980) Estimation of concentration quotients for ionic equilibria in high temperature water: the model substance approach. Proc. 41st Int'l. Water Conference, Pittsburgh, PA., 284-294.

Maier, C.G. and Kelley, K.K. (1932) An equation for the representation of high temperature heat content data. J. Amer. Chem. Soc. 54, 3243-3246.

Marshall, W.L. (1968) Conductances and equilibria of aqueous electrolytes over extreme ranges of temperature and pressure. Rev. Pure Appl. Chem. 181, 167-186.

__________ (1969) Correlations in aqueous electrolyte behavior to high temperatures and pressures. Rev. Chem. Progress 30, 61-84.

__________ (1970) Complete equilibrium constants, electrolyte equilibria, and reaction rates. J. Phys. Chem. 74, 346-355.

__________ (1972a) Predictions of the geochemical behavior of aqueous electrolytes at high temperatures and pressures. Chem. Geology 59-68.

__________ (1972b) A further description of complete equilibrium constants. J. Phys. Chem. 76, 720-731.

__________ (1980) Amorphous silica solubilities. III. Activity coefficient relations and predictions of solubility behaviour in salt solutions, 0-350°C. Geochim. Cosmochim. Acta 44.

McKenzie, W.F. and Helgeson, H.C. (1984) Estimation of the dielectric constant of H_2O and the thermodynamic properties of aqueous species to 900°C at 2 kb. Geochim. Cosmochim. Acta 48, 2167-2178.

Montoya, J.W. and Hemley, J.J. (1975) Activity relations and stabilities in alkali feldspar and mica alteration reactions. Econ. Geol. 70, 577-582.

Nancollas, G.H. (1960) Quart. Rev. 14, 402.

__________ (1961) The entropies of association of ions. J. Amer. Chem. Soc., 83 755.

Nordstrom, D.K., Plummer, L.N., Wigley, T.M., Wolery, T.J., Ball, J.W., Jenne, E.A., Bassett, R.L., Gerar, D.A., Florence, T.M., Fritz, B., Hoffman, M., Holdren, G.R., Lafon, R.M., Mattigod, S.V., McDuff, R. E., Morel, F., Reddy, M.M., Sposito, G., Thrailkill, J. (1979) A comparison of computerized chemical models for equilibrium calculations in aqueous systems. In Jenne, E.A., ed., Chemical Modeling in Aqueous Systems. Amer. Chem. Soc. Symp. Series 93, Washington, D.C., 857-892.

Oelkers, E.H. and Helgeson, H.C. (1986) Are hydrothermal solutions really associated at supercritical pressures and temperatures? Geol. Soc. Amer. Abstr. with Progr. 18, 709.

Pabalan, R.T. and Pitzer, K.S. (1987) Thermodynamics of NaOH(aq) in hydrothermal solutions. Geochim. Cosmochim. Acta 51, 829-838.

Phutela, R.C. and Pitzer, K.S. (1986) Heat capacity and other thermodynamic properties of aqueous magnesium sulfte to 473 K. J. Phys. Chem. 90, 895-901.

Phutela, R.C., Pitzer, K.S. and Saluja, P.P.S. (1987) Thermodynamics of aqueous magnesium chloride, calcium chloride and strontium chloride at elevated temperatures. J. Chem. Eng. Data 32, 76-80.

Pitzer, K. S. (1973) Thermodynamics of electrolytes. I. Theoretical bais and general equations. J. Phys. Chem. 77, 268-277.

__________ (1975) Thermodynamics of electrolytes. V. Effects of higher-order electrostatic terms. J. Solution Chem. 4, 249-265.

__________ (1977) Electrolyte theory - Improvements since Debye and Huckel. Accounts of Chem. Research 10, 371-377.

__________ (1979) Theory: Ion interaction approach. In Pytkowicz, R.M., ed., Activity Coefficients in Electrolyte Solutions. CRC Press, Boca Raton, Florida, 157-208.

__________ (1983) Dielectric constant of water at very high temperature and pressure. Proc. Nat'l. Acad. Sci., U.S.A. 80, 4575-4576.

Ragnarsdottir, K.V. and Walther, J.V. (1985) Experimental determination of corundum solubilities in pure water between 400-700°C and 1-3 kbar. Geochim. Cosmochim. Acta 49, 2109-2115.

Reardon, E.J., and Langmuir, D. (1976) Activity coefficients of $MgCO_3^{\circ}$ and $CaCO_3^{\circ}$ ion pairs as a function of ionic strength. Geochim. Cosmochim. Acta 40, 549-554.

Seward, T.M. (1981) Metal complex formation in aqueous solutions at elevated temperatures and pressures. In Wickman, F. and Rickard, D., eds., Proceedings, Nobel Symposium on the Chemistry and Geochemistry of Solutions at High Temperatures and Pressures. Phys. Chem. Earth 13/14, 113-132. Pergamon Press, New York.

__________ (1984) The formation of lead (II) chloride complexes to 300°C: a spectrophotometric study. Geochim. Cosmochim. Acta 48, 121-134.

Shock, E.L. (1987) Standard molal properties of ionic species and inorganic acids, dissolved gases and organic molecules in hydrothermal solutions. Unpubl. Ph.D. Thesis, Univ. California, Berkeley.

Shock, E.L., and Helgeson, H.C. (1987) Calculation of the thermodynamic and transport properties of aqueous species at high pressures and temperatures: Correlation algorithms for ionic species and equation of state predictions to 5 kb and 1000°C. Geochim. Cosmochim. Acta (submitted).

Sillen, L.G. and Martell, A.E. (1964) Stability Constants of Metal-Ion Complexes. Chem. Soc., London, 754 p.

__________ (1971) Stability Constants of Metal-Ion Complexes: Supplement No. 1: Chem. Soc., London, 865 p.

Smith, R.E. and Martell, A.E. (1976) Critical Stability Constants, V. 4: Inorganic Complexes. Plenum Press, New York, 256 p.

Smith, R.W., Popp, C.J. and Norman, D.E. (1986) The dissociation of oxy-acids at elevated temperatures. Geochim. Cosmochim. Acta 50, 137-142.

Smith-Magowan, D. and Wood, R.H. (1981) Heat capacity of aqueous sodium chloride from 320 to 600 K measured with a new flow calorimeter. J. Chem. Thermodynamics 13, 1047-1073.

Sverjensky, D.A., Shock, E.L., and Helgeson, H.C. (1987) Thermodynamics of aqueous metal complexes at elevated temperatures and pressures. Geol. Soc. Amer. Abstr.with Progr. 19, (submitted).

Tanger, J.C., and Helgeson, H.C. (1987) Calculation of the thermodynamic and transport properties of aqueous species at high pressures and temperatures: Revised equation of state for the standard partial molal properties of ions and electrolytes. Amer. J. Sci. (in press).

Tremaine, P.R., Sway, K. and Barbero, J.A. (1986) The apparent molar heat capacity of aqueous hydrochloric acid from 10° to 140°C. J. Sol'n. Chem. 15, 1-22.

Wagman, D.D., Evans, W.H., Parker, V.B., Schumm, R.H., Halow, I., Bailey, S.M., Churney, K.L. and Nuttall, R.L. (1982) The NBS tables of chemical thermodynamic properties. Selected values of inorganic and C_1 and C_2 organic substances in SI units. J. Phys. Chem. Ref. Data 11, supplement 2, 392.

Walther, J.V. (1986a) Mineral solubilities in supercritical H_2O solutions. Pure and Applied Chem. 58, 1585-1698.

__________ (1986) Experimental determination of portlandite and brucite solubilities in supercritical H_2O. Geochim. Coschim. Acta 50, 733-733-740.

__________ and Helgeson, H.C. (1977) Calculation of the thermodynamic properties of aqueous silica and the solubility of quartz and its polymorphs at high pressures and temperatures. Amer. J. Sci. 277, 1315-1351.

Whitney, J.A., Hemley, J.J. and Simon, F.O. (1985) The concentration of iron in chloride solutions equilibirated with synthetic granitic compositions: The sulfur-free system. Econ. Geol. 80, 444-460.

Wilson, G.A. (1986) Cassiterite solubility and metal chloride speciation in supercritical fluids. Unpubl. Ph.D. Thesis, The Johns Hopkins University, Baltimore, Maryland.

Wolery, T.J. (1983) EQ3NR: A computer program for geochemical aqueous speciation-solubility calculations, user's guide and documentation. UCRL-53414. Lawrence Livermore Lab., Univ. Calif., 191 p.

Wolery, T.J., Sherwood, D.J., Jackson, K.J., Delaney, J.M. and Puigdomenech, I. (1984) EQ3/6: Status and applications, UCRL-91884. Lawrence Livermore Lab., Univ. Calif., 12 p.

Wood, R.H., Smith-Magowan, D., Pitzer, K.S. and Rogers, P.S.Z. (1983) Comparison of experimental values of $\bar{V}_2^\circ$ $\bar{C}_{p_2}^\circ$, and $\bar{C}_{v_2}^\circ$ for aqueous NaCl with predictions using the Born equation at temperatures from 300 to 573.15 K at 17.7 MPa. J. Phys. Chem. 87, 3297-3300.

Wood, S.A. and Crerar, D.A. (1985) A numerical method for obtaining multiple linear regression parameters with physically realistic signs and magnitudes: Applications to the determination of equilibrium constants from solubility data. Geochim. Cosmochim. Acta 49,165-172.

__________ Crerar, D.A., Brantley, S.L. and Borcsik, M. (1984) Mean molal stoichiometric activity coefficients of alkali halides and related electrolytes in hydrothermal solutions. Amer. J. Sci. 284, 668-705.

Chapter 7

John R. Holloway

IGNEOUS FLUIDS

INTRODUCTION

The purpose of this chapter is to provide a chemical and thermodynamic description of fluids which could exist in the igneous environment. The igneous environment will be taken as occurring at temperatures above about 700°C and any pressure above atmospheric. There is obviously a considerable region of overlap with high-grade metamorphic conditions in this definition. Before going further, it is useful to define the term fluid as used in this section.

Fluid: A phase which consists of H_2O and other volatile molecules of low density such as CO_2, CH_4, etc. at supercritical temperatures. At pressures below the critical pressure of H_2O (about 218 bars, see Fig. 1) the fluid will have a very low density and the properties of a gas or vapor (steam if H_2O). At higher pressures, equivalent to the middle crust or deeper, fluids will have densities close to liquid water at room temperature (Fig. 1). At these densities, the fluid phase properties differ significantly from those of gas or vapor. Nevertheless, many authors use gas and vapor synonymously with fluid. Regardless of terminology, this phase is distinct from silicate melts which have densities greater than 2 grams/cm³.

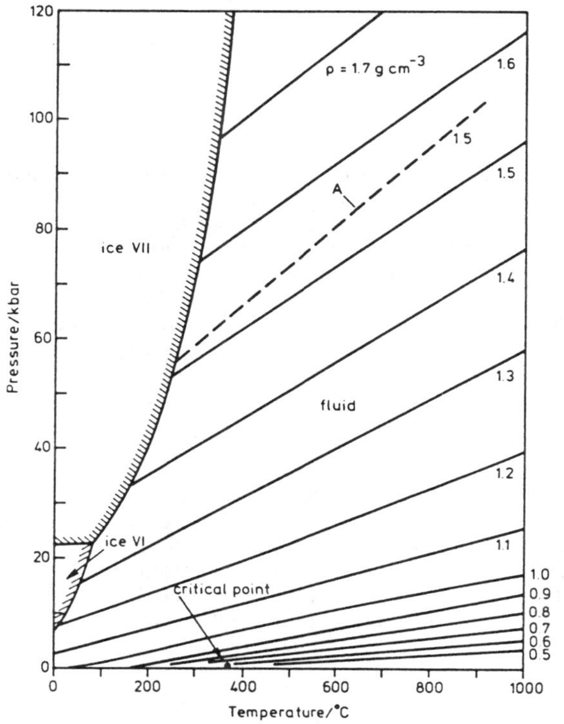

Figure 1. Phase diagram for H_2O showing isochores in density in grams/cm³.

Why study igneous fluids? What can they tell us about igneous rocks and processes? Any good petrologist knows by now that H_2O (and CO_2 at upper mantle conditions) can have profound effects on the physical properties, composition, and crystallization history of magmas. In cases where igneous fluids exist, they can be used to define activities of oxygen, H_2O, CO_2, H_2 , and carbon, if enough is known about the fluid. The composition of igneous fluids strongly influences the solubility, and hence transport, of major and trace element solutes. Fluids can have major mechanical effects such as in crack propagation or in explosive volcanism. Finally, interaction between fluids and magmas has had a major effect on the evolution of atmospheres on the terrestrial planets.

The aim of this chapter is to develop a means of calculating activities and fugacities of molecular species in fluids and to use that ability to place limits on the composition of fluids in the igneous environment. The discussion will begin with a characterization of properties of molecular species which affect their thermodynamic properties. The nature of equations of state is then considered, followed by a description of the simple, and modified, Redlich-Kwong equation and other geologically useful equations. The nature of aqueous fluids at igneous temperatures is discussed next and it is concluded that it is justifiable to treat igneous fluids as molecular mixtures at crustal and upper mantle conditions. Finally some techniques for calculating fluid compositions in the C-O-H system are presented and their extension to systems containing S, Cl and N is discussed.

PROPERTIES OF MOLECULAR SPECIES IN FLUIDS

In this section we will consider those properties of molecules such as their size, shape and electric charge distribution which effect the way a given kind of molecule interacts in a fluid. These characteristics describe molecules in the same way that crystal structure characterizes minerals, with the result that molecules can be arranged in classes with predicable characteristics.

Size and shape

The molecular species considered here are either spherical single atoms such as Ar, spherically symmetrical polyatomic molecules such as methane, or polyatomic molecules of lower symmetry. The shapes of the molecules considered here are well known, having been determined by a variety of spectroscopic techniques. The size of molecules is ambiguous in that outer electron density has no obvious boundary in quantum mechanical treatments. However, there is a growing body of evidence that molecules in fluids at normal densities behave as if they were hard spheres with well defined sizes (Chandler et al.,1983). For spherically symmetrical molecules this hard sphere size agrees closely with sizes determined by x-ray diffraction studies of crystalline molecular solids.

Attractive forces between molecules

The forces arising between uncharged molecules are all some form of electrostatic attraction caused by asymmetrical distributions of electrons. These attractive forces are responsible for condensation of gases to liquids, and in the case of molecular crystals (such as argon), for the crystallization of liquids. Extended discussions of these forces are given by Hirschfelder et al. (1964), Prausnitz (1969), Maitland et al. (1981), Prausnitz et al. (1986), and Rigby et al. (1986). The following abbreviated discussion is based on those sources.

We commonly refer to intermolecular forces, but in fact derive expressions for interactions between molecules in terms of potential energy, Γ. The relation between force, F, and potential energy is

$$F = -\,d\Gamma\ /\ dr,$$

where r is the intermolecular separation.

Permanent dipole-permanent dipole forces. These are the easiest to visualize. They

Table 1. Parameters for molecular species

Molecule	Polarizibility α ($cm^3 \cdot 10^{-25}$)	Dipole moment μ (Debye)	Quadrupole moment Θ ($10^{-40} C\, m^3$)	R-K a (10^{-6}bar $cm^6 K^{1/2}$)	R-K b (cm^3)
^4He	2.0	0	0	1.05	11.1
Ne	3.9	0	0	1.51	12.0
Ar	16.3	0	0	16.9	22.3
Kr	24.6	0	0	34.1	27.4
Xe	40.0	0	0	22.3	35.8
H_2	7.9	0	2.1	1.81	15.34
N_2	17.6	0	-4.9	15.6	26.8
O_2	15.7	0	-1.3	17.4	22.1
Cl_2	46.1	0	10.8	13.6	39.0
CO	19.5	0.11	-8.2	17.2	27.4
HF	24.6	1.93	7.8	78++	12.8++
HCl	26.3	1.07	12.3	67.4	28.1
CO_2	26.5	0	-14.9	64.7	29.7
SO_2	37.2	1.59	-16.2,12.7,3.4*	133.1++	37.4++
H_2O	14.8	1.86	8.7,-8.3,-0.4*	88++	14.6++
D_2O				144	21.46
H_2S				89	30.1
HCN		3.0	10.9	244	61
NH_3	22.6	1.47	-7.7	86.8	25.9
CH_4	26.0	0	0	32.0	29.7

*Three values of Θ for xx, yy, and zz moments.
+ $a = 2948\ T_c^{2.5}/P_c$, $b = 7.197\ T_c/P_c$. He, Ne, and H_2 values calculated using effective critical constants from Prausnitz (1969, p. 162).
++From Holloway (1981)

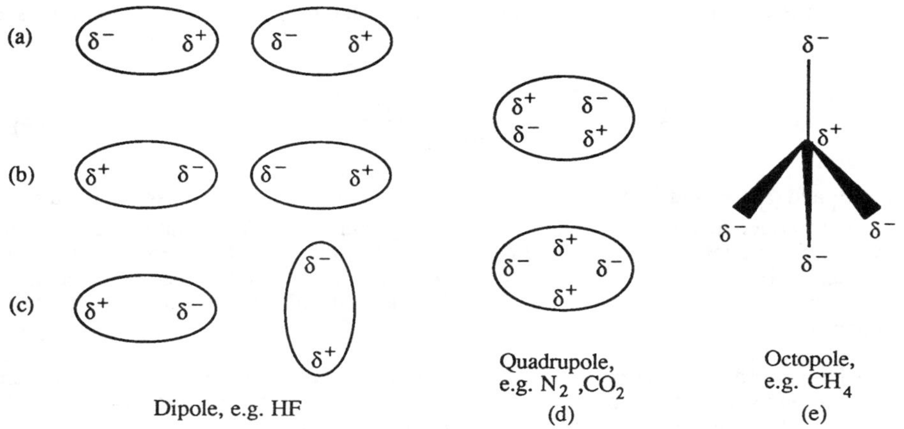

Figure 2. Electric multipoles. (a-c) show relations between dipole moments using HF as an example. (a) shows moments aligned so as to maximize the attractive force. (b) shows moments aligned to maximize repulsive force. (c) shows alignment that results in zero net force. (d) shows examples of quadrupoles, and (e) an octopole.

arise between molecules which have permanent dipole moments (Table 1). Consider the HF molecules shown in Figure 2(a). Within the molecule the electrons are not symmetrically distributed about the positively charged nuclei because of the very high electronegativity of F compared to H. This results in a net negative charge on the F end and a net positive charge on the H end of the molecule. The dipole moment within the molecule is given by the magnitude of the charge difference and the distance between the charges. If two HF molecules are aligned as shown in Figure 2(a) such that the positive end of one molecule is directly facing the negative end of the other, the attractive electrostatic force will have the maximum possible value for a given intermolecular separation, *r*. Changing the orientation of the dipoles to that shown is Figure 2(b) results in the maximum possible repulsive force and the orientation shown in Figure 2(c) results in a zero force. Any intermediate orientation will have an intermediate force. The attractive forces will tend to cause a preferential alignment of dipoles, while thermal motions will tend to cause random orientations (a perfectly random arrangement of a large number of molecules would result in a zero net force). It has been shown that the average potential energy for dipoles can be given as:

$$\Gamma_{\mu\mu} = \frac{-\ 2\ \mu_i^2\ \mu_j^2}{3\ r^6\ k\ T\ C^2} \tag{1}$$

where μ_i is the dipole moment of molecule i, k is the Boltzmann constant, T the absolute temperature, and C is a geometric constant. For fluids composed of a single species (pure fluids), i = j so $\mu_i = \mu_j$. In this case the potential energy depends of the fourth power of the dipole moment, so small variations in the moment from one kind of molecule to another will cause large changes is the contribution of this term to the overall intermolecular force. Note that the potential energy is inversely proportional to T, will go to zero at infinite T, but will decrease by only 1/3 from 700° to 1500°C.

Dispersion forces. These are also known as instantaneous dipole-induced dipole forces or London forces for F. London who discovered them. In classical terms they can be thought of as originating between (1) a molecule in which an instantaneous shift in its electron cloud has resulted in a momentary dipole and (2) a nearby molecule in which the dipole in the first molecule causes a perturbation in the electron cloud of the second molecule. The creation of the spontaneous dipole and its inductive effect on the neighboring molecule occurs on a very rapid time scale compared to molecular velocities. For two spherically symmetrical molecules the potential energy due to dispersion forces is given by:

$$\Gamma_{dis} = -\frac{3}{2}\ \frac{\alpha_i \alpha_j}{r^6} \left(\frac{h\nu_{oi}\, h\nu_{oj}}{h\nu_{oi} + h\nu_{oj}} \right) \tag{2}$$

where α_i and α_j are polarizibilities, h is Planck's constant and ν_{oi} and ν_{oj} are characteristic electronic frequencies for molecules i and j in the ground state. The relation between the characteristic frequency and refractive index gave rise to the use of the term dispersion in this context. Note the lack of a temperature dependence in this equation because there is no way to cause differences in orientation for spherical molecules and so thermal vibrations can have no effect.

Higher order permanent moments. Linear molecules such as CO_2 have permanent dipole moments because the internal C-O dipoles exactly cancel each other due to symmetry. However, the two dipoles result in four points of charge concentration in the molecule which form a permanent quadrupole moment. The potential energy expression is:

$$\Gamma_{\Theta\Theta} = \frac{-14\ \Theta_i^2\ \Theta_j^2}{5\,r^{10}\ k\ T\ C^2} \tag{3}$$

where Θ is the quadrupole moment. The potential energy resulting from these moments depends on the orientation of the molecules and hence is dependent on the reciprocal of absolute T as in the case of permanent dipoles. Note however that the dependence on intermolecular separation, r varies as the tenth power for quadrupoles compared to the sixth power for dipole and dispersion potentials. Thus quadrupole forces are important only at very close intermolecular distances equivalent to very high densities. There is an interaction between dipole and quadrupole moments which causes potential energy to vary as $1/r^8$. Higher order multipoles exist such as the octopole in methane. Examples of charge distribution in quadrupolar and octopolar molecules are shown in Figure 2(d) and (e). The dispersion potential for higher multipoles also has higher order, $1/r^8$ and $1/r^{10}$ terms.

Potential energy relations

The total potential energy between two molecules has historically been represented by separating the attractive and repulsive components:

$$\Gamma_{total} = \Gamma_{repulsive} + \Gamma_{attractive}.$$

The attractive term is usually represented as being proportional to $1/r^6$ which is justified by the dependence on r seen in Equations (1) and (2) for permanent dipole and dispersion forces (although not discussed here, the same dependence is shown by permanent dipole-induced dipole forces). Note that quadrupole and higher multipole moments are not adequately represented by a $1/r^6$ term. The repulsive term has been represented by terms proportional to $1/r^{12}$, $\exp(-r)$ and by a hard sphere in which $\Gamma_{repulsive}$ goes to infinity at a given value of r. Potential energy functions are often characterized by ε, the depth of the potential well, and σ, the distance at which the potential function crosses the zero energy line as shown in Figure 3.

Relative importance of attractive forces

In a pure fluid the total attractive force will be made up of the sum of dispersion, dipolar, quadrupolar and dipole-induced dipole forces (if dipoles exist). The contribution of each of these will depend on the relative magnitudes of polarizibility, dipole moment and quadrupole moment in different ways. Potentials from dipoles vary as the fourth power of the moment while potentials due to dispersion vary as the square of the polarizibility. A comparison of the relative contribution is shown in Table 2. Note the differences between H_2O and HCl. These differences can be attributed to the almost factor of two difference in dipole moment raised to the fourth power, and the similar difference in polarizibility squared (see Table 1). Note that monatomic noble molecules have only dispersion forces. Single bond energies are shown for comparison and it can be seen that the intermolecular forces are very small by comparison.

From the above discussion it can be seen that the attractive potential energy in general consists of a T independent part and a part which is proportional to 1/T and that both parts can further be divided into those which have $1/r^6$, $1/r^8$, and $1/r^{10}$ dependencies. Thus a general form of the attractive potential energy term would look like:

Table 2. Interaction energies (kJ/mol)

Molecule	r_m(nm)	Intermolecular attractive terms Γ_{dipole}	$\Gamma_{induction}$	$\Gamma_{dispersion}$	SBE*
Ar	0.37			-1.1	
Xe	0.43			-1.9	
CO	0.40	-4.x10^{-5}	-8.x10^{-4}	-1.3	343
HCl	0.42	-0.2	-0.07	-1.8	431
H_2O	0.30	-16.1	-0.9	-5.3	464
NH_3	0.29	-6.2	-0.9	-12.9	389

From Rigby et al. (1986)
*SBE is the single bond energy.

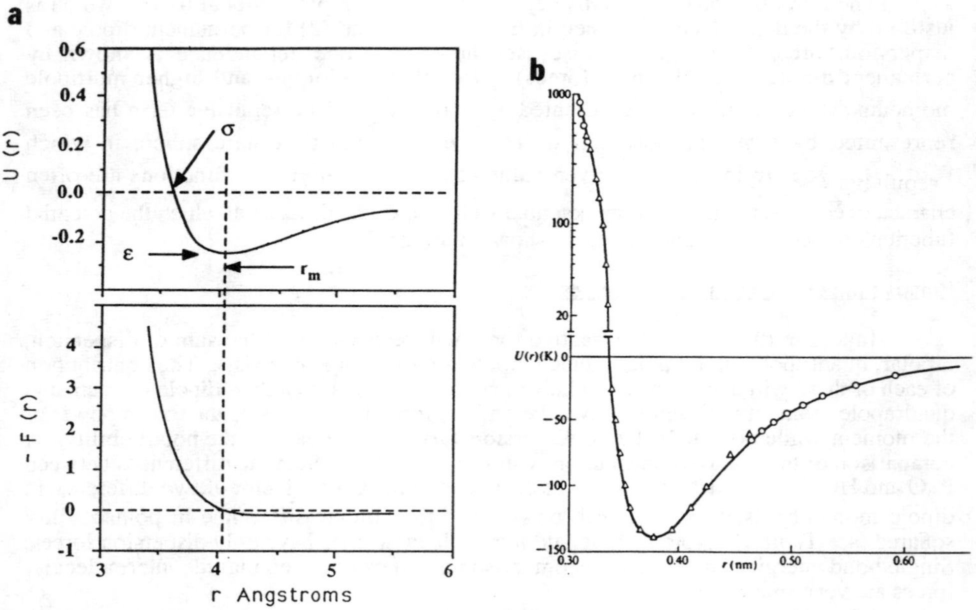

Figure 3. (a) Schematic potential energy and force plots. Note the positions of σ, ε, and r_m. (b) Potential energy function for a pair of argon molecules. From Rigby et al. (1986).

$$\Gamma_{ij} = c_1 \circ \left(\frac{c_2}{r^6} + \frac{c_3}{r^8} + \frac{c_4}{r^{10}}\right) + \frac{c_5}{T} \bullet \left(\frac{c_6}{r^6} + \frac{c_7}{r^8} \frac{c_8}{r^{10}}\right)$$

Great progress has been made in recent years in theoretical calculations of the total potential energy of pair-wise interactions of molecules (Rigby et al., 1986). Wilson (1986) gives a very good introduction to the numerical methods. It seems likely that there will be major advances in the theory of igneous fluids by making use of these approaches. For the present time however we must use the available equations of state which do not really make full use of these theoretical advances.

EQUATIONS OF STATE

Equations of state come in a bewildering variety of forms and complexities, partly because different types of equation do different things, and partly because for a time getting the universal equation of state was the holy Grail of one branch of physical chemistry. For our purposes we want an equation which can represent volume as a function of pressure and temperature. Such an equation directly provides V (or density) of a fluid and can be integrated to yield G, H, S, thermal expansion and compressibility. We also want to be able to extrapolate our equation to P, T regions well beyond the range of available P-V-T data. In some cases there are essentially no data available and so we want to be able to estimate parameters for our equation of state using corresponding state theory. In addition to these requirements the equation must be sufficiently accurate to be used in thermodynamic calculations of fluid - crystal - liquid equilibria. Finally we need to be able to predict thermodynamic properties in multispecies fluid mixtures.

This is a formidable list of requirements and none of the available equations satisfies all of them. In searching for an appropriate equation it is important to make a distinction between two broad categories, empirical equations which are not closely related to any theoretical model, and theoretically based equations. The empirical equations are used to precisely represent a set of experimental data and to interpolate within those data. Such equations have many adjustable parameters and cannot be extrapolated beyond the data base with any confidence. Theoretically based equations may have many or few adjustable parameters. A number of equations exist which are semi-theoretical in that they are based on a theoretical model but include some adjustable parameters with arbitrarily chosen forms. Because of limitations in available data, even theoretically based equations may not extrapolate well if they include many adjustable parameters. Currently available solutions to the problem are (a) use empirical equations for specific species for which data exists over the range of interest, (b) use a simple, theoretically based equation equation which can be extrapolated with confidence, and (c) use a corresponding state form of an empirical equation based on data covering a wide P-T range.

Two-parameter equations

For the reasons noted above, much effort has gone into very simple equations involving only two adjustable parameters for pure fluids (e.g. Peng and Robinson, 1976). These equations are all based on the van der Waals equation:

$$P = \frac{R\,T}{V - b} - a\,(V) \qquad (4)$$

This equation adds two correction terms to the ideal gas equation, the *b* parameter which accounts for the volume occupied by the molecules in a fluid, and the *a* parameter which accounts for the attractive potential between molecules. The *b* parameter has been used as a measure of molecular size and is approximately equivalent to the cube of the hard sphere radius. Note that P goes to infinity as V approaches *b*. The *a* parameter is a function of volume. In the van der Waals equation the functional relation is $a(V) = a / V^2$. Note that for dispersion forces the potential is given as C / r^6 and that r^6 is approximately equal to V^2, thus rationalizing the form of the van der Waals equation. However, the van der Waals equation is not successful at predicting fluid behavior over wide ranges of P and T using constant values of *a* and *b* even in the case of simple molecules such as argon. Its form is such that it requires isochores (lines of constant volume) to be straight lines ($P = C_1T - C_2$) but when fit to H_2O which does have straight line isochores above 2 kbar, the *a* and *b* parameters are found to be functions of density. Redlich and Kwong (1949) modified the *a*(V) term in the van der Waals equation in such a way that very good fits were obtained to experimental P-V-T data. The Redlich-Kwong equation is

$$P = \frac{RT}{V - b} - \frac{a}{\sqrt{T}\,(V^2 + V \cdot b)} \tag{5}$$

The difference between the *a*(V) functions between van der Waals and Redlich-Kwong is such that the potential energy term in the latter is a function of T as well as V, that is r^6 is now proportional to $T^{1/2}(V^2+bV)$. Redlich and Kwong (1949) stated that they had no theoretical justification for this modification, and there does not seem to have been a justification proposed in the intervening years. Indeed other formulations have been proposed (e.g. Peng and Robinson, 1976) but they either reduce to two parameter forms similar to the Redlich-Kwong equation or introduce other adjustable parameters. (It should be noted that changing the form of the attractive term causes the *b* parameter, as well as the *a* parameter, to have a different values depending on the equation used.) Because the Redlich-Kwong equation provides relatively good fits to non-polar molecules it must be assumed that the $T^{1/2}(V^2+bV)$ term provides a better approximation to the intermolecular separation than does the original van der Waals term. As such the $T^{1/2}$ is not considered to indicate a temperature dependence of the *a* parameter. Thus the original Redlich-Kwong formulation would not be expected to provide good fits to dipolar or quadrupolar fluids such as H_2O and CO_2. From equation (1) the temperature dependence of the *a* parameter is expected to vary as 1/T. In practice the *a* parameter is fit to experimental data (de Santis et al., 1974) and the resulting fit is better using a quadratic equation for the variation of a with T. This is the modified Redlich-Kwong equation proposed by de Santis et al. (1974). At high T equation (1) predicts that dipole effects should become small compared to dispersion, so the *a* parameter should assymtotically approach a limiting value (denoted a_o) at high T. The value of a_o can be obtained from solubility studies of simple gases in water (de Santis et al., 1974). We can thus extrapolate the *a*(T) function with confidence by using a_o as a limiting value. Some currently available Redlich-Kwong parameters are listed in Table 1.

Corresponding states

Corresponding state approaches were pioneered by van der Waals who showed that normalizing P, T and V for gases by dividing by their values at the critical point (Pc, Tc, Vc) resulted in nearly identical numerical behavior. This approach gives good results for simple, non-polar molecules and has been used extensively in the form of charts (Hougen and Watson, 1947) and to estimate Redlich-Kwong parameters. Corresponding states can be rationalized on a microscopic scale by normalizing the potential function Γ with ε and the intermolecular separation with σ (see Fig. 3). On the microscopic scale, all molecules whose potential energy as the same functional form when expressed in those normalized variables (Γ/ε and r/σ) would obey the law of corresponding states.

A recent application of corresponding states to fluids at very high P and T has been made by Saxena and Fei (1987). They normalized P-V-T data for O_2, H_2, CH_4, Ar, N_2 and CO_2 obtained in shock experiments to reduced variables and then regressed the reduced data to obtain temperature dependent coefficients of an empirical equation having the form of the virial equation on state. They also fit shock data for H_2O using their corresponding state equation, but found it necessary to significantly modify the temperature dependent coefficients to obtain reasonable fits to the H_2O data. In fitting the very high P, T shock data, they incorporated data from static experiments at much lower P and T to force their equation to mesh with existing equations and data in the low pressure range. This approach provides a very useful constraint on the general behavior of these fluids in the high pressure limit. The reduced variable data base used by Saxena and Fei (1987) is shown in Figure 4. By using corresponding states, Saxena and Fei were able to multiply the available data considerably, but the assumption that the above six species obey the law of corresponding states is implicit in their regression and so it is not surprising that their equation does a good job of reproducing the original data. They point out the poor fit for H_2 given by their equation and the critical P and T for H_2. One reason for this is that H_2 has such a low mass that it shows quantum mechanical behavior and so its critical values must be adjusted to give good corresponding states fits (Prausnitz et al., 1969). When the adjusted values are used in the Saxena and Fei equation the fit to the shock data for H_2 is much improved.

Empirical equations

Comparison of volumes of CO_2 calculated using the Saxena and Fei equation with those calculated by the two parameter modified Redlich-Kwong equation (de Santis et al., 1974) reveals a fundamental limitation in the two parameter Redlich-Kwong which is that a choice of parameters that provide a good fit at pressures up to about 10-20 kbar do not accurately fit very high pressure (and hence high density) data. This problem can be solved by making the *b* parameter a function of density or pressure as done by Kerrick and Jakobs (1981) and Bottinga and Richet (1981).

For very precise representations of high quality P-V-T data the best choice is some form of empirical equation with many adjustable parameters. Because the parameters are fit by least-squares regression techniques, usually to polynomial functions, and are usually highly correlated, the equations cannot be extrapolated beyond the range of available data. They often do not even fit the data near the limit of the range very well, and so should be used with caution.

Treatment of mixtures

Mixing rules have been discussed at length by de Santis et al. (1974) and Prausnitz et al. (1986). There is now a considerable amount of data available for binary mixtures containing H_2O, but only at temperatures up to about 400°C and pressures of 3 kbars (H_2, Seward and Franck, 1981; O_2, Japas and Franck 1985b; N_2, Japas and Franck,1985a; CH_4, see summary of data in Holloway, 1984). The proposed mixing rules are able to reproduce the two phase curve with fair accuracy, but there are not nearly enough data at higher temperature to be certain that the rules really describe fluid mixtures. This is especially true of H_2O - CO_2 mixtures because of the complications arising due to the bicarbonate ion pair (de Santis et al., 1974). For this reason mixing rules should be used with great caution.

THE NATURE OF IGNEOUS FLUIDS

The purpose of this section is to consider the chemical properties of possible igneous fluids with particular emphasis on aqueous fluids containing dissolved silicates, chlorides and relatively 'inert' volatiles such as CO_2.

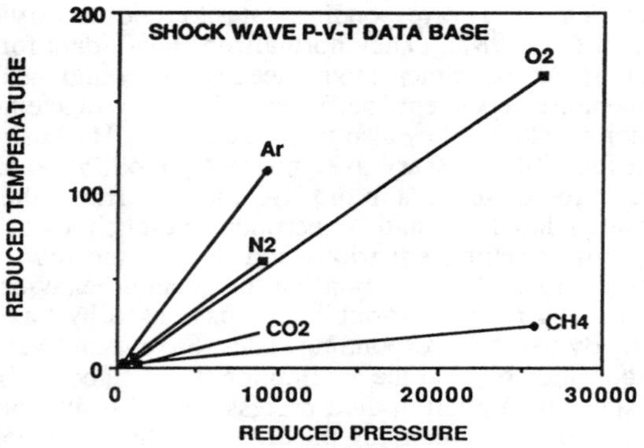

Figure 4. Data base used by Saxena and Fei (1987) for their corresponding state equation. For comparison the reduced P and T values for CH_4, CO_2, and H_2 equivalent to 2000°C and 100 kbar are 2155, 10.5; 1353, 6.6; and 45.9, 4888, respectively.

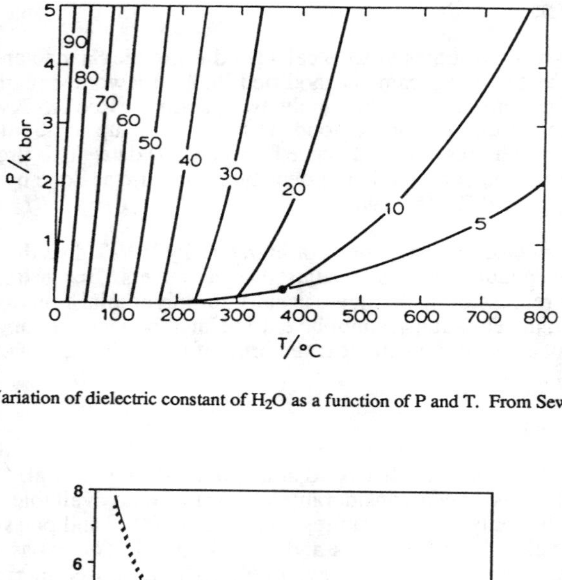

Figure 5. Variation of dielectric constant of H_2O as a function of P and T. From Seward (1981).

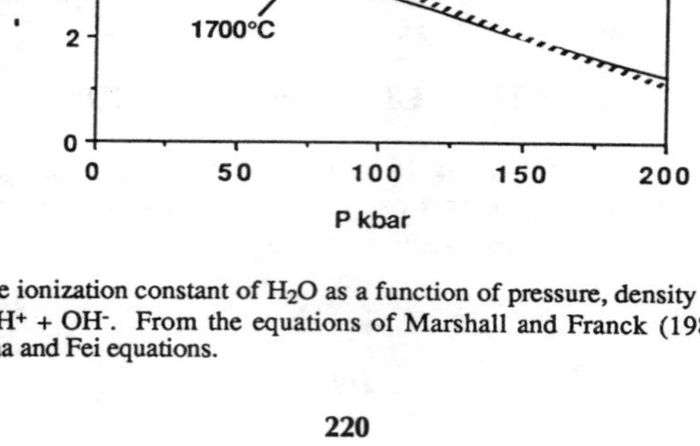

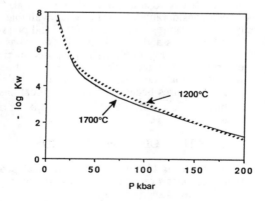

Figure 6. Variation in the ionization constant of H_2O as a function of pressure, density and temperature. K_w is for the reaction $H_2O = H^+ + OH^-$. From the equations of Marshall and Franck (1981) with H_2O densities calculated from the Saxena and Fei equations.

Properties of H_2O

Chemical properties of H_2O are dominated by its high dipole moment, its shape, and hydrogen bonding. There is evidence that, at least at one kbar, the effects of hydrogen bonding are not important, nor is tetrahedral ordering, at temperatures above 500°C (Gorbaty and Demianets, 1983). The main effect then is that of the dipole moment which is the cause of the very high dielectric constant for H_2O at liquid water densities.

The dielectric constant of a fluid is a measure of the fluid's ability to separate ions and hence has a large effect on the degree of ionization in a fluid. The dielectric constant is a function of the number of dipoles in a given volume and the extent to which the dipoles are lined up. The density of H_2O determines the number of dipoles per unit volume, and thermal motion reduces the dipole alignment. Consequently dielectric constants are directly proportional to fluid density and inversely proportional to T. The variation of dielectric constant with P and T is shown in Figure 5. Note the large decrease in the range 600° to 800°C. As will be shown below, these low dielectric constants will greatly decrease the degree of dissociation of acids, bases and salts.

The ionization of H_2O increases dramatically with P and, to a lessor extent, with T as shown in Figure 6. H_2O becomes highly ionized at very high temperatures and densities, and this has been observed in shock experiments (Mitchell and Nellis, 1982) which suggest extensive ionization at pressures above 200 kbar. This large change in one of the basic properties of H_2O should make one very suspicious of extrapolations of any kind concerning aqueous fluids into that pressure region.

Dissolved solutes

This discussion is restricted to the effect that solute material will have on the properties of the major volatile species in the fluid, a treatment of the solubility of silicate material or of partitioning of elements between fluids and melts or crystals is beyond the scope of the chapter.

Silica. Except for alkali chlorides, silica is usually the most abundant solute component in aqueous igneous fluids (Anderson and Burnham, 1983). In the crustal pressure range where it has been well studied, silica dissolves in a hydration reaction:

$$SiO_2 + n\,H_2O = Si(OH)_4 \cdot n'\,H_2O$$

In which n has values of about 4 (Crerar and Anderson, 1971). Anderson and Burnham (1983) show that the amount of dissolved silica in equilibrium with albite is about the same as in the case of quartz. It is probable that the other feldspars will dissolve to about the same extent. An 8 wt % quartz solubility translates to 2.8 mole % using a value of $n = 4$. For most mineral assemblages the solubility of silicates in aqueous fluids at upper mantle conditions is lower than this at pressures up to 25 kbar (Schneider and Eggler, 1984). The addition of volatile dilutents such as CO_2 will drastically decrease the solubility of species which form aqueous complexes with high hydration numbers such as silica. For a hydration number of 4, the solubility will decrease as the fourth power of the H_2O activity. Thus we can conclude that for most crustal and uppermost mantle conditions, the amount of dissolved silicate material will be too small to have a significant effect on igneous fluids other than that of simple dilution. This situation may change drastically at pressures in the 35-100 kbar range, but no data exist on fluid-rock interactions in that region.

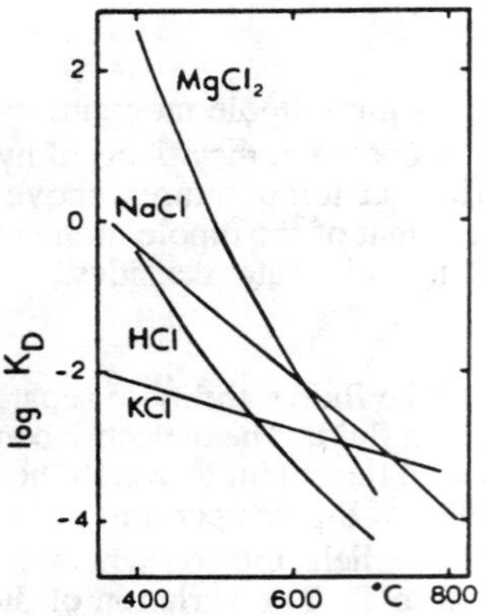

Figure 7. Ionization constants for some 'strong' electrolytes. Strong bases also follow the trends. From Eugster (1981).

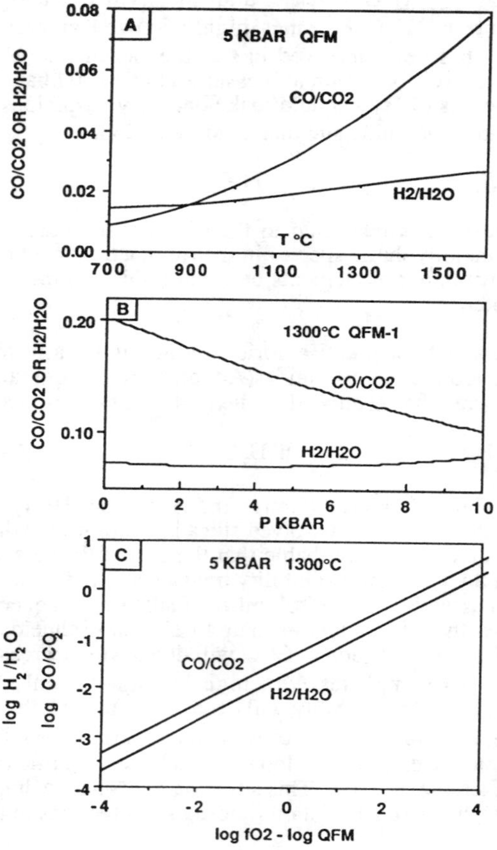

Figure 8. The effect of $f(O_2)$, T, and P on CO/CO_2 and H_2/H_2O ratios in fluids. Calculated for C-O or H-O systems, respectively.

Alkali chlorides. Alkali chlorides are present in significant amounts in many metamorphic environments and may be present in igneous environments. Their abundance is determined by the chlorine abundance in the system. At temperatures up to about 400°C acids, bases and salts are highly ionized. But as noted above, the degree of ionization of strong electrolytes is highly correlated with the dielectric constant of H_2O and this decreases with increasing T at constant P. Equilibrium constants for ionization follow this trend as shown in Figure 7. The low values of the ionization constants at T > 700°C indicate that for significant concentrations of alkali or alkaline earth chlorides (> about 1 molal, or 2 mole %) the species will be more than 95% associated and can thus be treated as highly polar, neutral molecules. As such their effect on H_2O activity in the fluid should be less than that of ions because neutral molecules will not perturb the local structure of the fluid by formation of hydration shells. Dilution of aqueous fluids with non-dipolar molecules also lowers the dielectric constant and this causes further decrease in the ionization constant of strong electrolytes (Hartmann and Franck, 1969). Thus alkali and alkaline earth chlorides will be even more strongly associated (into neutral molecules) in CO_2 -bearing aqueous fluids.

The nature of igneous fluids in the crust and upper mantle

From the above arguments it is probable that most fluids which exist in the crust and uppermost mantle at igneous temperatures will contain dissolved silicate material in amounts < 5 mole %. If the fluids contain significant amounts of alkali or alkaline earth chlorides then those species will be present as associated molecules. The dissolved silicates will lower the activity of volatile species by simple dilution. The presence of alkali chlorides will increase H_2O activity by relatively large amounts at the low T end of the igneous range, but that effect should be much smaller at upper mantle T.

It is not safe to extend these predictions to pressures much above the 35 kbar range and especially above the 100 kbar range due to the unknown effects of the ionization of H_2O and the possible increase in the dielectric constant of H_2O at the high densities of those pressures.

EQUILIBRIA IN C-O-H SYSTEMS

H_2O, CO_2, CO, H_2, and CH_4 are the most abundant species in fluids in the igneous pressure range under the oxygen fugacity conditions thought to exist in the crust and mantle (Holloway, 1981 and references therein). All of these species fall in the C-O-H three component system and that system thus provides a very good model for calculating the composition of igneous fluids. In the igneous temperature range the only phases found in this system are supercritical fluid and graphite (or diamond). We are interested in the molecular and atomic composition of the fluid as a function of P, T, and oxygen fugacity $f(O_2)$, and whether or not the fluid is saturated in graphite or diamond.

Free energy relations

It is useful to single out some simple reactions which have a strong effect on the fluid composition. There is a strong dependence of the fluid composition on f_{O_2} which is seen from the following:

$$CO + 1/2\, O_2 = CO_2 \tag{6}$$

$$K = f_{CO_2} / (f_{CO} \cdot f_{O_2}^{\ 1/2}) = \phi_{CO_2} \cdot X_{CO_2} / (\phi_{CO} \cdot X_{CO} \cdot f_{O_2}^{\ 1/2})$$

$$X_{CO_2} / X_{CO} = K' \cdot f_{O_2}^{\ 1/2}, \quad \text{where } K' = K \cdot (\phi_{CO} / \phi_{CO_2}).$$

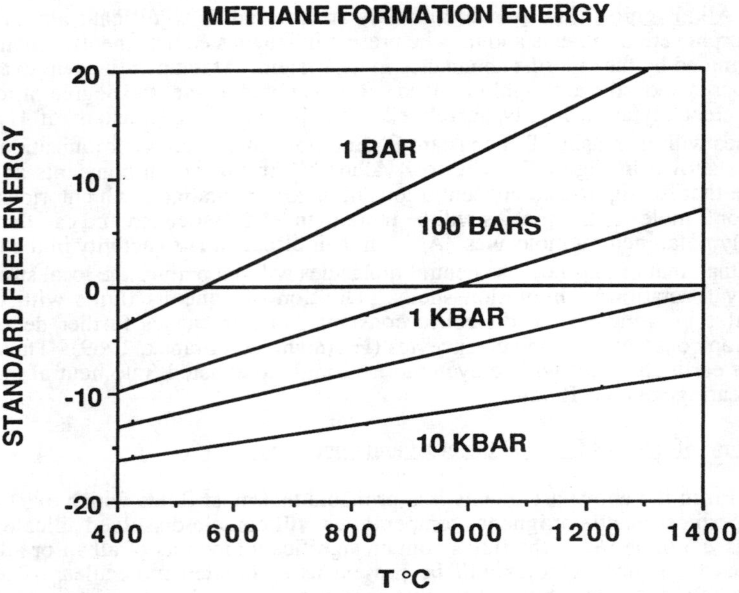

Figure 9. The effect of T and P on the free energy of Reaction (III) for the formation of methane.

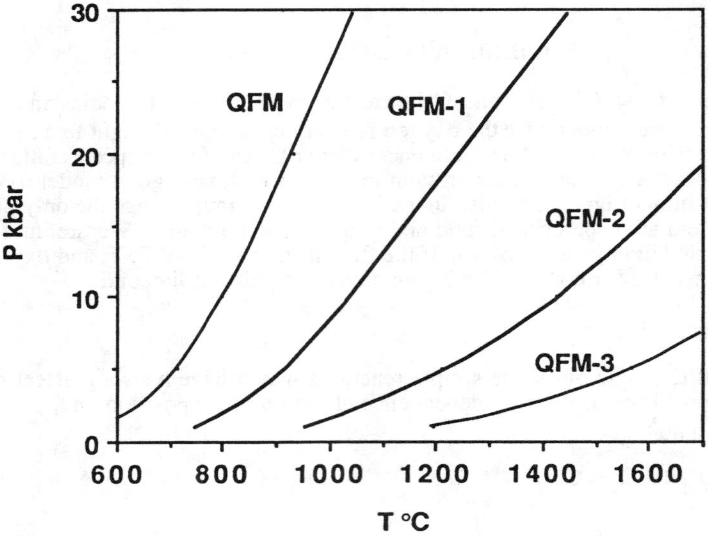

Figure 10. The graphite saturation surface in the C-O system. This system provides a high T limit for the existence of graphite in nature at a given $f(O_2)$. The $f(O_2)$ is indicated in log10 units relative to the QFM buffer, i.e. QFM-2 is two orders of magnitude lower than QFM. Graphite is stable to the left of the lines and unstable to the right.

ϕ_i is the fugacity coefficient for species i and is a function of P, T, and fluid composition. Rearrangement of the equilibrium constant expression for H_2O in the same way yields the following:

$$H_2 + 1/2\ O_2 = H_2O \qquad \text{(IV)}$$

$$X_{H_2O} / X_{H_2} = K' \cdot f_{O_2}^{\ 1/2}$$

The variations of CO/CO_2 and H_2/H_2O with T and $f(O_2)$ are shown in Figure 8 (a) and (c). Reactions (6) and (IV) are each pressure dependent because the the ratio of moles of fluid on the left is 1.5 times that on the right. However, because H_2 has a much smaller molar volume than the other species, reaction (6) is much more sensitive to pressure than is reaction (IV). The effect of pressure on the CO/CO_2 and H_2/H_2O ratios is also shown in Figure 8 (b).

Under oxidizing conditions, C-O-H fluids consist mainly of CO_2 and H_2O with small amounts of H_2. Under very reducing conditions CH_4 is an important species depending on P and T. The stability of CH_4 is governed by the reaction:

$$C + 2\ H_2 = CH_4 \qquad \text{(III)}$$

The free energy change for this reaction under standard conditions is strongly T dependent and becomes positive at about 650°C as shown in Figure 9. This reaction is strongly pressure dependent but the pressure dependence itself is also pressure dependent due to the large differences in compressibility between H_2 and CH_4. The change in free energy with P is also shown in Figure 9. These individual reactions have a strong influence on the overall chemistry of fluids in the C-O-H system.

Methods of equilibrium calculation

Two methods of calculating equilibria in this system are commonly used.

Minimization of total free energy is the most general kind of calculation. It has several advantages: Any number of species may be considered providing their standard free energy of formation and fugacity at P and T are known. No assumptions about phase assemblage need be made. Given a system bulk composition the relative proportions of phases is a result of the calculation. A disadvantages of this method is that it is more difficult to visualize the reactions producing the results. A minor inconvenience is that it requires much more computer time compared to the other method. A good summary of the free energy minimization technique is given by van Zeggeren and Storey (1970). Recent modifications are used by Ghiorso (1985) and Harvie et al. (1987).

Equilibrium constants and mass balance is the second method of calculation. This method requires that enough equilibrium constant equations are used so that, when combined with the equation for mass balance, there are an equal number of equations and unknowns. The advantages of this method are that the overall equilibrium is seen as a set of individual reactions. The major disadvantage is that a great deal of algebra must be done to add new components to the system or even to test to determine if a given species (C_2H_2, for instance) is present in significant amounts.

Before considering the details of the calculation it is useful to do a phase rule analysis of the C-O and C-O-H systems. The C-O system contains two components, C and O. In cases when the fluid is saturated in graphite there are two phases, graphite and fluid. So, f = c - ø + 2 = 2, if P and T are independently variable, and f = 0, if P and T are each specified. Thus at constant P and T the presence of graphite and fluid completely determines the composition of the fluid, including its $f(O_2)$. The C-O system provides an upper temperature limit on the stability of graphite in the Earth for a given value of $f(O_2)$ as shown in Figure 10. For example, at 10 kbar and 800°C graphite is unstable at $f(O_2)$ values higher than QFM.

The C-O-H system contains three components, but in a graphite saturated system there are still only two phases so f = 3 if neither P or T are fixed and f = 1 if both are fixed. This means that graphite saturated fluids at fixed P and T in the C-O-H system are univariant and one additional constraint is needed to specify (or calculate) the fluid composition. The variable chosen is commonly $f(O_2)$, but any other fugacity or fluid composition parameter, $X(H_2O)$ for instance, could be chosen.

Solution for the C-O-H system. This solution follows French (1966) and Eugster and Skippen (1967) except that it allows for read fluids and real mixtures. Holloway and Reese (1974) used a free energy minimization calculation to show that at temperatures >600°C only the five species CO_2, CO, CH_4, H_2, and H_2O were present in amounts greater that 0.1 mole %, so only these species need be considered. The presence of graphite or diamond fixes the activity of carbon (a_C) at one. In order to accurately define a_C accurately it is necessary to add a pressure correction to the standard free energy at one bar, and to use the free energy of diamond when in its stability field. As noted above, we must specify one variable in addition to P, T, and the condition of graphite (or diamond) saturation. The following derivation is developed for calculations based on a specified $f(O_2)$. Four reactions can be written:

$$C + O_2 = CO_2 \qquad \text{(I)}$$
$$C + 1/2\,O_2 = CO \qquad \text{(II)}$$
$$C + 2\,H_2 = CH_4 \qquad \text{(III)}$$
$$H_2 + 1/2\,O_2 = H_2O \qquad \text{(IV)}$$

The equilibrium constant expressions for these reactions are:

$$K_I = f_{CO_2} / (a_C \cdot f_{O_2})$$
$$K_{II} = f_{CO} / (a_C \cdot f_{O_2}{}^{1/2})$$
$$K_{III} = f_{CH_4} / (a_C \cdot f_{H_2})$$
$$K_{IV} = f_{H_2O} / (f_{H_2} \cdot f_{O_2}{}^{1/2})$$

The equilibrium constants are calculated from the standard free energies of formation (JANAF, 1986) at the temperature of interest. The standard state used here is the pure gas at one bar and T for the fluid species, and crystalline graphite at one bar and T. It is convenient to fit log K_i to a function of T. An additional constraint is that the sum of the mole fractions of the fluid species equal one:

$$1 = X_{CO_2} + X_{CO} + X_{CH_4} + X_{H_2} + X_{H_2O} + X_{O_2}. \qquad (7)$$

In practice $X(O_2)$ is so small in geologic environments that it can be omitted from the mass balance equation. The equilibrium constant equations can be rearranged as follows:

$$f_{CO_2} = K_I \cdot a_c \cdot f_{O_2} \quad (8)$$

$$f_{CO} = K_{II} \cdot a_c \cdot f_{O_2}^{1/2} \quad (9)$$

$$f_{CH_4} = K_{III} \cdot a_c \cdot f_{H_2}^2 \quad (10)$$

$$f_{H_2O} = K_{IV} \cdot f_{H_2} \cdot f_{O_2} \quad (11)$$

Defining the fugacity coefficient as (Redlich and Kwong, 1949):

$$\phi_i = f_i / (X_i \cdot P_T) \quad (12)$$

the mole fractions can be expressed as $X_i = f_i / (\phi_i \cdot P_T)$ and substituted into Equation (7) to yield:

$$P_T = \{(f_{CO_2}/\phi_{CO_2}) + (f_{CO}/\phi_{CO}) + (f_{CH_4}/\phi_{CH_4}) + (f_{H_2}/\phi_{H_2}) + (f_{H_2O}/\phi_{H_2O})\} \quad (13)$$

substituting Equations (8) through (11) into (13) and doing a lot of rearranging yields:

$$0 = (K_{III} \cdot \phi_{H_2}{}^2/\phi_{CH_4}) \cdot X_{H_2}{}^2 + (1 + [K_{IV} \cdot \phi_{H_2} \cdot f_{O_2}{}^{1/2}]/\phi_{H_2O}) \cdot X_{H_2}$$

$$+ ([K_I \cdot f_{O_2}/\phi_{CO_2}] + [K_{II} \cdot f_{O_2}{}^{1/2}/\phi_{CO}] - 1) \quad (14)$$

This equation is quadratic in $X(H_2)$ and so can be solved using the quadratic equation. After obtaining $X(H_2)$ equation (12) may be rearranged to calculate $f(H_2)$ which can be substituted into Equations (III) and (IV) to obtain $f(CH_4)$ and $f(H_2O)$ and these can be used with (12) to obtain $X(CH_4)$ and $X(H_2O)$. Note that defining $f(O_2)$, a_C, P and T directly fixes $f(CO_2)$ and $f(CO)$ from Equations (I) and (II). At this stage it is important to test for the stability of graphite at the chosen P, T, and $f(O_2)$ by summing the mole fractions. If graphite is stable the mole fractions will sum to one within the round off error of the computer. If the sum exceeds unity graphite is not stable at the conditions chosen. If the mole fractions do sum to unity we have a solution in the form of a set of mole fractions of CO_2, CO, CH_4, H_2, and H_2O coexisting with graphite at the P, T, and $f(O_2)$ we have specified. The atomic composition of the fluid may be calculated as follows:

$$C = X_{CH_4} + X_{CO} + X_{CO_2}$$

$$O = 2 \cdot X_{CO_2} + X_{CO} + X_{H_2O}$$

$$H = 4 \cdot X_{CH_4} + 2 \cdot X_{H_2} + X_{H_2O}$$

The C, O, and H values need to be normalized to a mole fraction of one, and then the composition of the fluid may be plotted on a C-O-H triangular diagram as shown in Figure 11 (a). In practice, the position of the graphite saturation surface is plotted on triangular diagrams such as Figure 11(a) by calculating a number of fluid compositions in equilibrium with graphite, plotting them and drawing a curve which represents the graphite saturation boundary for the given conditions.

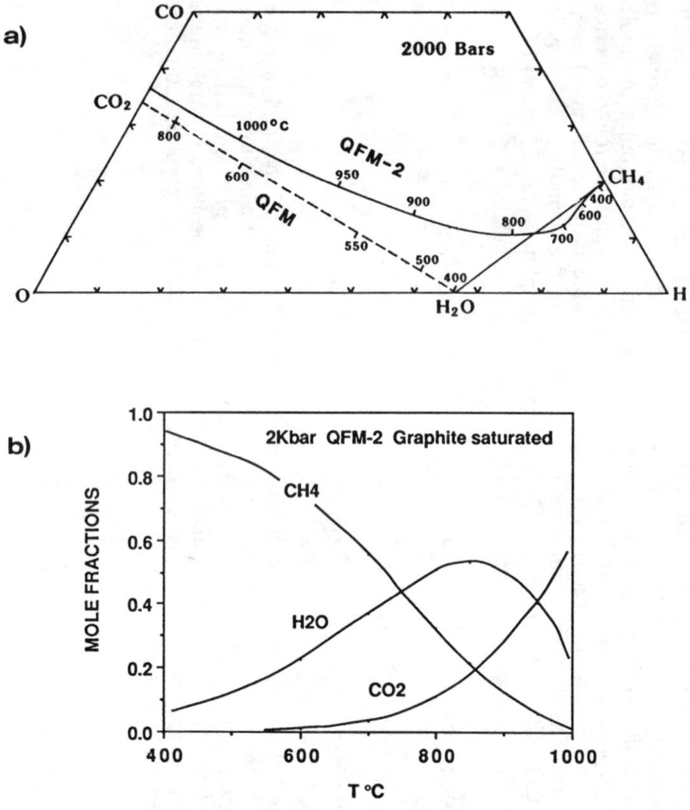

Figure 11 (a). The atomic composition of fluids coexisting with graphite at a total pressure of 2000 bars. The solid line is for an $f(O_2)$ two log units lower than defined by the quartz-fayalite-magnetite buffer (QFM-2), and the dashed line shows fluid compositions having an $f(O_2)$ equal to QFM. The small numbers along the lines indicate temperature in °C. (b). Variation in the amount of molecular species in the fluids shown in (a) at QFM-2.

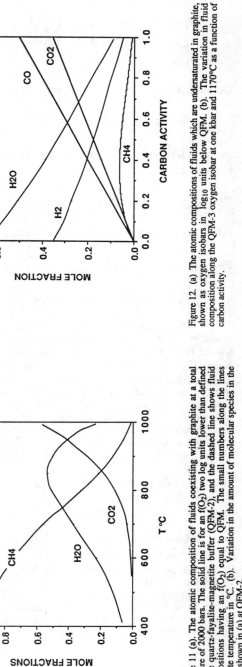

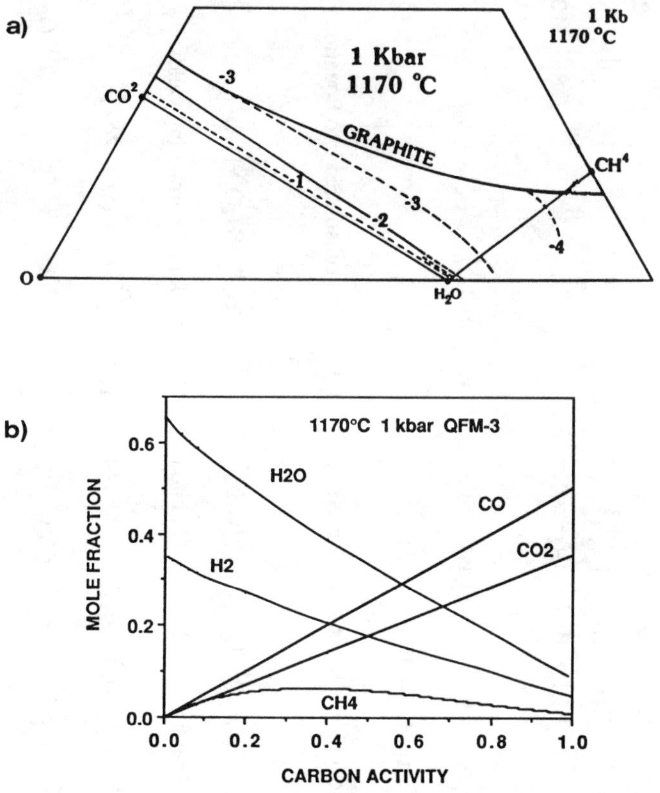

Figure 12. (a) The atomic compositions of fluids which are undersaturated in graphite, shown as oxygen isobars in $\log_{10}$ units below QFM. (b). The variation in fluid composition along the QFM-3 oxygen isobar at one kbar and 1170°C as a function of carbon activity.

Graphite undersaturated fluid calculations. Equations (8) through (10) include a term for carbon activity (a_C) which is set to unity if graphite saturation is specified. The activity of carbon in fluids undersaturated in graphite will be <1 and calculations may be done along lines of constant $f(O_2)$ by setting a_C equal to values between zero and one, combining a_C with K_I, K_{II}, and K_{III} to form $K'_I = K^\circ a_C$, etc., and substituting into equations (8), (9), and (14). An example of such calculations is shown in Figure 12 (a) which illustrates the position of oxygen isobars in the region below the graphite saturation surface. The variation of fluid composition along the QFM-3 oxygen isobar is shown in Figure 12 (b) as a function of a_C.

Representation of results. There are other ways of representing results of this kind of calculation besides the ones illustrated in Figures 11 and 12. Frost (1979) represents the graphite saturation surface on the type of diagram shown in Figure 13. These diagrams have also been used by Green et al. (1987). Ohmoto and Kerrick (1977) contour log $f(O_2)$ - T diagrams in fluid composition (e.g. $X(H_2O)$) which is a good way of comparing other $f(O_2)$ sensitive reactions. Another useful representation is to contour P-T diagrams in fluid composition for a given $f(O_2)$ as is shown in Figure 14.

ADDITION OF OTHER COMPONENTS

Several other elements form volatile molecular species which are important in some environments. Of those elements sulfur is probably the most important. Eugster and Skippen (1967) showed how sulfur could be included in the calculation scheme described above. The only two significant sulfur-bearing species are H_2S and SO_2.

The fugacities of these species can be defined by the following reactions:

$$H_2 + 1/2\ S_2 = H_2S, \qquad (V)$$

$$O_2 + 1/2\ S_2 = SO_2. \qquad (VI)$$

The equilibrium constant expressions for these reactions can be rearranged to give:

$$f_{H_2S} = K_V \cdot f_{H_2} \cdot f_{S_2}^{1/2}, \qquad (15)$$

$$f_{SO_2} = K_{VI} \cdot f_{O_2} \cdot f_{S_2}^{1/2}. \qquad (16)$$

These relations can then be incorporated into Equation (14) and the equation solved providing one more restriction is added to the system by fixing another compositional variable. Specifying $f(S_2)$ is appears to be the best choice because it is possible to place some limits on its variation in igneous environments. Sulfur fugacity is defined in several regions of the Fe-S system (Barton and Skinner, 1967). Pyrrhotite of a given composition determines $f(S_2)$ at a given T (the effect of P is small and much less than the uncertainty in pyrrhotite compositions, so will be ignored) so we can use pyrrhotite to specify $f(S_2)$ as a function of T. The $f(S_2)$ equivalent to 47.5 atom percent Fe in FeS is a convenient reference, and is given by:

$$\log_{10} f_{S_2}\ (47.5\%\ \text{Fe in FeS}) = 7.3 - 1.04 \cdot 10^4 / T(K) \qquad (17)$$

Fluid compositions in the C-O-H-S system at fixed P, T, $f(O_2)$ and a_C are shown in Figure 15. Note the cross-over in the proportions of SO_2 and H_2S at an intermediate value of $f(O_2)$.

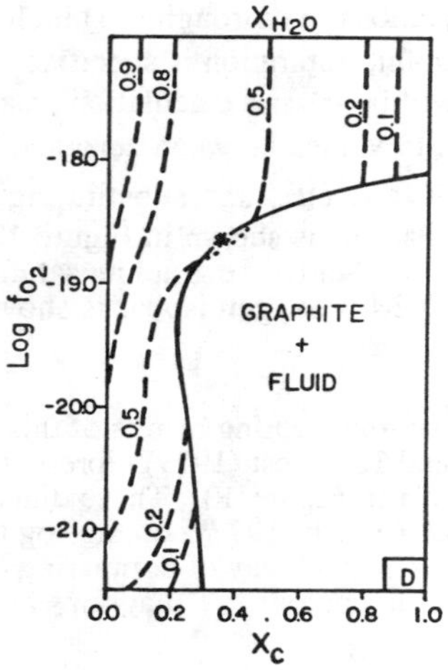

Figure 13. An alternative representation of graphite-fluid equilibria. X_C is calculated as $n_C/(n_C+n_{H_2})$. From Frost (1979).

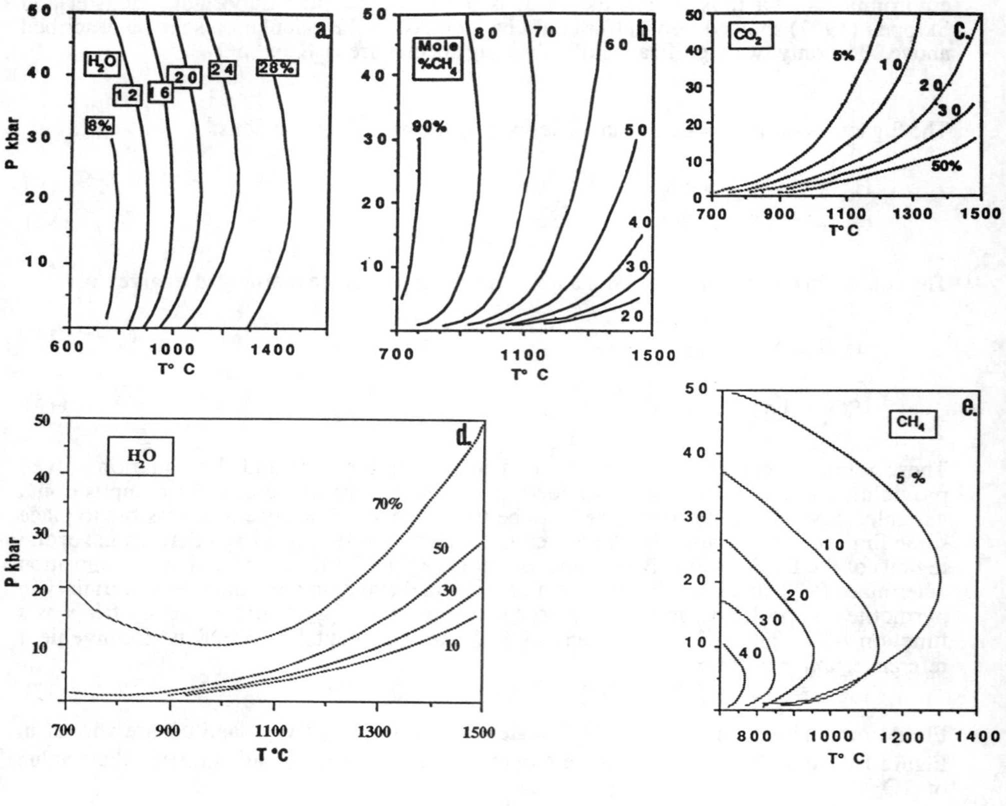

Figure 14. Fluid compositions in equilibrium with graphite. Mole fraction isopleths contoured in P-T projection. (a) and (b) are H_2O and CH_4 at $f(O_2)$ equal to the iron-wustite buffer. (c), (d) and (e) are CO_2, H_2O and CH_4 at $f(O_2)$ two log units below QFM.

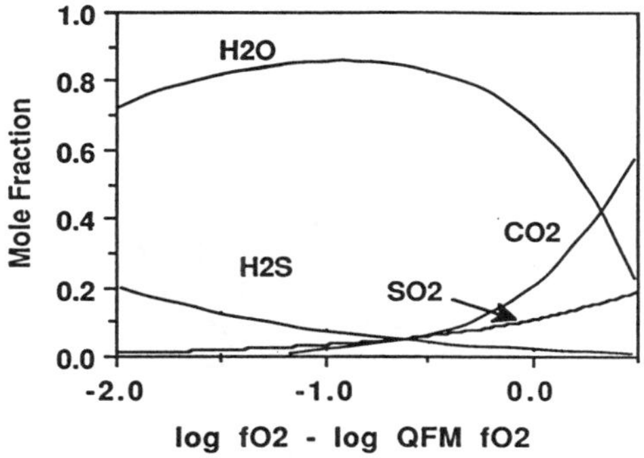

Figure 15. Variation in fluid compositions as a function of $f(O_2)$ at 10 kbar, 850°C with $f(S_2)$ defined by 47.5% Fe in pyrrhotite and carbon activity set at 0.1.

Cl, F, and N

Three other elements which form possibly important species in igneous fluids are Cl, F, and N. Addition of even one of these to the C-O-H-S system results in very complicated algebraic expressions and so it is preferable to use total free energy minimization to calculate equilibrium fluid compositions. This approach has the added benefit that we do not have to assume a particular set of important species, rather we can test for the importance of any species for which free energy of formation data is available. In testing for important species it is sufficient to assume an ideal gas approximation for uncommon species because in so doing the free energy of the species at high temperature will be lower (more negative) than it would be if a real gas fugacity were used, and this will result in an over-estimation of the species abundance. If the calculated concentration is insignificant even with the ideal gas approximation then it is safe to ignore the species in further calculations. If however the calculated abundance is significant, then the fugacity should be estimated using a corresponding states equation and critical constants (such as found in Kudchadker et al, 1968; and Mathews, 1972). Perhaps the most difficult task in doing calculations with species of these additional elements is in estimating the value of any additional parameters in the natural environment. One good possibility is to estimate fluorine or chlorine fugacities from their concentrations in hydrous minerals and use those as constraints in the calculations.

Modelling fluid/melt systems

Our current state of knowledge of volatiles in melts allows models to be made of H_2O and CO_2 evolution in ascending silicic magmas under crustal conditions (Holloway, 1976). Much more elaborate treatments are now possible by including the crystal/melt models of Burnham and Nekvasil (1986) and Nekvasil and Burnham (1987) with the fluid/melt models. The same type of fluid/melt modelling could be done in mafic and even ultramafic systems if we had detailed activity-composition data for CO_2 in those melts at high pressures.

IGNEOUS FLUID CALCULATIONS IN THE FUTURE

At the present time we have a fairly good data bases and equations of state for pure fluids of petrologic importance. The data base for aqueous fluid mixtures has been increasing, but it is currently limited to sub-igneous temperatures and relatively low pressures. The theoretical basis for treating mistures of polar and non-polar molecules is not well developed but is improving. More carefully designed experiments are needed, but the time is ripe for fluid dynamic calculations of aqueous mixtures for conditions found in the crust and mantle. Coupled with advances in this area should be the use of free energy minimization techniques which incorporate data for igneous minerals and fluids in a self-consistent set.

ACKNOWLEDGMENTS

This work was supported largely by NSF grant EAR 8617128. Most of the writting was done while on leave at the Division of Geological and Planetary Sciences, California Institute of Technology. I thank the inhabitants of the third floor of The Charles Arms Laboratory for providing a stimulating environment. I have greatly benefited from discussions covering the topics presented here with David B. Joyce.

REFERENCES

Anderson, G.M., and Burnham, C.W. (1965) The solubility of quartz in supercritical water. Amer. J. Sci. 263, 494-511.

_____ and Burnham, C.W. (1983) Feldspar solubility and the transport of aluminum under metamorphic conditions. Amer. J. Sci. 283-A, 283-297.

Barton, P.B. and Skinner, B.J. (1967) Sulfide mineral stabilities. In. Geochemistry of Hydrothermal Ore Deposits. H.L. Barnes, ed., Holt Rinehart, New York, p. 236-333.

Burnham, C.W. and Nekvasil, H. (1986) Equilibrium properties of granite pegmatite magmas. Amer. Mineral. 51, 239-263.

Chandler, D., Weeks, J.D., and Andersen, H.C. (1983) Van der Waals' picture of liquids, solids and phase transformations. Science 220, 787-794.

Crerar, D.A., and Anderson, G.M. (1971) Solubility and solvation reactions of quartz in dilute hydrothermal solutions. Chem. Geol. 8, 107-122.

de Santis, R., Breedvelt, G.J.F., and Prausnitz, J.M. (1974) Thermodynamic properties of aqueous gas mixtures at advanced pressures. Ind. Eng. Chem. Process Des. Develop. 13, 374-377.

Eugster, H.P. (1981) Metamorphic solutions and reactions. In: Physics and Chemistry of the Earth. D.T. Richard and F.E. Wickman, eds., vol. 13 and 14, pp. 461-507.

_____ and Skippen, G.B. (1967) Igneous and metamorphic equilibria involving gas equilibria. In: Researches in Geochemistry. P.H. Abelson, ed., vol. 2, J. Wiley, New York, p. 492-520.

French, B.M. (1966) Some geological implications of equilibrium between graphite and a C-H-O gas phase at high temperatures and pressures. Rev. Geophys. 4, 223-253.

Frost, B.R. (1979) Mineral equilibria involving mixed-volatiles in a C-O-H fluid phase: The stabilities of graphite and siderite. Amer. J. Sci. 279, 1033-1059.

Gorbaty, Yu.E., and Demianets, Yu.N. (1983) The pair-correlation functions of water at a pressure of 1000 bar in the temperature range 25-500°C. Chem. Phys. Lett. 100, 450-454.

Hartmann, D., and Franck, E.U. (1969) Elektrische Leitfahigkeit wassriger Losungen bei hohen Temperaturen und Drucken III. Ber. Bunsenges. Phys. Chem. 73, 308-314.

Hirschfelder, J.O., Curtis, C.F., and Bird, R.B. (1954) Molecular theory of gases and liquids. J. Wiley, New York.

Holloway, J.R. Fluids in the evolution of granitic magmas: Consequences of finite CO_2 solubility. Geol. Soc. Amer. Bull. 87, 1513-1518.

______ (1977) Fugacity and activity of molecular species in supercritical fluids. In: Thermodynamics in Geology. D.G. Fraser, ed., D. Reidel, Dordrecht, Holland, p. 161-181.

______ (1981) Volatile interactions in magmas. In: Thermodynamics of Minerals and Melts. R.C. Newton, A. Navrotsky, and B.J. Wood, eds., Adv. Physical Geochemistry 1, p. 273-293.

______ (1984) Graphite-CH_4-H_2O-CO_2 equilibria at low grade metamorphic conditions. Geology 12, 455-458.

______ (1986) Volatile solubilities in magmas: Transport of volatiles from mantles to planet surfaces. J. Geophys. Res. 91, D505-D508.

______ and Reese, R.L. (1974) The generation of N_2-CO_2-H_2O fluids for use in hydrothermal experimentation. I. Experimental method and equilibrium calculations in the C-O-H-N system. Amer. Mineral. 59. 589-597.

Japas, M.L., and Franck, E.U. (1985a) High pressure phase equilibria and PVT-data of the water-nitrogen system to 673 K and 250 MPa. Ber. Bunsenges. Phys. Chem. 89, 793-800.

______ and ______ (1985b) High pressure phase equilibria and PVT-data of the water-oxygen system including air-water to 673 K and 250 MPa. Ber. Bensenges. Phys. Chem. 89, 1268-1275.

Kerrick, D.M., and Jacobs, G.K. (1981) A modified Redlich-Kwong equation for H_2O, CO_2, and H_2O-mixtures at elevated pressures and temperatures. Amer. J. Sci. 281, 735-767.

Maitland, G.C., Rigby, M., Smith, E.B., and Wakeham, W.A. (1981) Intermolecular Forces--Their Determination. Clarendon Press, Oxford.

Marshall, W.L., and Franck, E.U. (1981) Ion product of water substance 0-1000°C, 1-10,000 bars, new international formulation and background. J. Phys. Chem. Ref. Data 10, 297-304.

Mitchell, A.C., and Nellis, W.J. (1982) Equation of state and electrical conductivity of water and ammonia shocked to the 100 GPa (1 Mbar) pressure range. J. Phys. Chem. 76, 6273-6281.

Nekvasil, H. and Burnham, C.W. (1987) The calculated individual effects of pressure and water content on phase equilibria in the granite system. In: Magmatic Processes: Physicochemical Principles. B.O. Mysen, ed. Spec. Pub. 1, The Geochemical Society, p. 433-445.

Ohmoto, J., and Kerrick, D. (1977) Devolatilization equilibria in graphitic systems. Amer. J. Sci. 277, 1013-1044.

Peng, D.-Yu., and Robinson, D.B. (1976) Two-constant equation of state. Ind. Eng. Chem. Fundam. 15, 59-63.

Prausnitz, J.M. (1969) Molecular Thermodynamics of Fluid Phase Equilibria. Prentice-Hall, New York, 523 pp.

______, Lichtenthaler, R.M., and de Azvedo, E.G. (1986) Molecular Thermodynamics of Fluid Phase Equilibria. Prentice-Hall, New York, 600 pp.

Redlich, O., and Kwong, J.N.S. (1949) On the thermodynamics of solutions. Chem. Rev. 44, 233-244.

Rigby, M., Smith, E.B., Wakeham, W.A., and Maitland, G.C. (1986) The Forces Between Molecules. Oxford Univ. Press, New York, 232 pp.

Saxena, S.K., and Fei, Y. (1987) High pressure and high temperature fluid fugacities. Geochim. Cosmochim. Acta 51, 783-791.

Schneider, M.E., and Eggler, D.H. (1986) Fluids in equilibrium with peridotite minerals: Implications for mantle metasomatism. Geochim. Cosmochim. Acta 50, 711-724.

Seward, T.M. (1981) Metal complex formation in aqueous solutions at elevat.ed temperatures and pressures. In: Physics and Chemistry of the Earth, D.T. Richard and F.E. Wickman, eds., vol. 13 and 14, p. 113-132.

Wilson, S. (1986) Chemistry by Computer. Plenum Press, New York, 233 pp.

Chapter 8

George H. Brimhall and David A. Crerar

ORE FLUIDS: MAGMATIC TO SUPERGENE

INTRODUCTION

A diverse spectrum of fluids exists in the Earth which is responsible for the transport and enrichment of ore-forming constituents including both metals and essential non-metallic species such as water, sulfur and halogens. Through their mobility and peculiar solvent characteristics, silicate magmas, aqueous fluids, and gasses may all, under particular circumstances, contribute to the mass and heat transfer involved in ore deposition. Such transport fluids, be they magmatic, gaseous, or aqueous, not only interact with wall rocks encountered along paths of migration by assimilation, thermal exchange, or hydrochemical reaction, but mutually interact themselves by phase separation as with boiling or exsolution of salt– and metal-rich magmatic water from a melt. Such release of solute-rich aqueous fluids promotes intense physical and chemical interaction of solids, aqueous fluids, and gasses and, because of enhancement of chemical mobility by complexing, results in selective metasomatic effects unsurpassed elsewhere in the Earth's crust.

In our treatment here, we have chosen to emphasize the continuum in nature in chemical and physical ore-forming processes and to develop relationships to those of common petrogenesis. Consequently, rather than writing separate chapters, we have combined our efforts into a single chapter, subdivided in three parts, beginning with the origin of ore-forming magmas and aqueous fluids in the context of crustal processes. We then develop the systematics in the physical chemistry of hydrothermal ore fluids, and end with the formation of primary ore deposits and their chemical weathering products. Metasomatic behavior is stressed.

In part, the high level of metasomatism is due to the chemical fractionation effects peculiar to fluid/fluid, fluid/rock, and fluid/solute interactions which efficiently extract ore elements from large source regions and quantitatively concentrate them in relatively small physical domains which are preserved for geological periods of time. It is the potential for such repetitive chemical focussing in the vicinity of available sources of thermal energy that drives ore-forming systems to extreme values of reaction progress and fluid-dominance. The attainment of end stages of chemical fractionation separates ores from the more common products of petrogenesis. The Earth as a thermally active planet surrounded by an oxygenated atmosphere provides ample opportunities for fractionation processes to begin internally; these ultimately are taken to completion by migrating fluids at shallow depths at or near the interface of the lithosphere and atmosphere, where chemical gradients are steep and disequilibrium common.

Many characteristics of these highly metasomatic systems such as rock permeability, porosity, redox state, and solution acidity may reach extreme values. In fact, it is now becoming obvious that constructive, that is mutually enhancing interrelationships, between such coupled factors as overall volume changes due to chemical reaction and rock permeability may be critical to attainment of ore-grade metasomatism. One reason why much more common crustal assemblages as in metamorphic rocks, indeed once considered isochemical products, have a more limited compositional spectrum than ores is that the heat and mass transport processes are themselves exceedingly more restricted in terms of time-integrated fluid and heat fluxes, solute load, permeability, and reactivity than in ore-forming systems, or at least less intense near the final stages of thermal or chemical dissipation. These ordinary crustal systems may in fact be self-limited while ore-forming systems may be runaway in comparison. Nevertheless, in many other respects the principles are identical to those governing igneous, metamorphic, and sedimentary processes within the Earth, but differ mainly in extent.

It is only now becoming possible to address rigorously such complex and interrelated physiochemical phenomena. First-principle forward modeling has largely been limited to consideration of either the chemical evolution or the fluid mechanics involved in ore transport and deposition, but not both simultaneously. Here we use the term "modeling" in a general sense, meaning any conceptualization of process based upon fundamental analytical considerations. This includes formulation of hypotheses from field, experimental, and theoretical perspectives. Given the complexity of ore deposits, recent modeling studies for the most part have had a limited scope, focussing on explaining central phenomena contributing to enrichment of metals at certain stages of the thermal, mechanical, or chemical evolution of specific types of ore deposits. While modeling has proceeded through such gross simplification, the results are nevertheless illuminating. Still, a challenging frontier remains in understanding fluid flow with chemical reaction from a theoretical, experimental, or field perspective. It is likely that insights from all three approaches will be necessary to meet the ultimate demands of the rich problems remaining in understanding ore fluids.

In this chapter we take a broad sweep through some of the processes responsible for deposition of ore and mineral deposits. We have out of necessity limited our treatment here considerably. Reviews of many relevant topics omitted here may be found in Brimhall (1987a), including magmatic ore deposits (Irvine et al., 1983), a topic neglected here altogether but one in which elegant thermo-mechanical-chemical modeling techniques have been developed. Instead, the focus here is on deposition from aqueous solutions, ranging from high-temperature fluids associated with felsic magmas, through hydrothermal systems at intermediate temperatures, down to near-surface oxidative weathering processes. We limit our analysis in this way so as to present the most complete, overall treatment of a single class of ore deposits, rather than err on the side of superficiality with a broader treatment. As a common thread, you might regard this as the local history of water-related deposits associated with a particular crystallizing pluton, from melting giving rise to a magma, its intrusion, primary mineralization, to final supergene or residual enrichment of ore metals. Metals, rather than being simply used as tracers to infer the nature of geochemical processes, are themselves the primary focus.

PART I. THE GENERATION OF MAGMAS AND ORE FLUIDS

PRE-METALLOGENIC HISTORY OF MAGMAS

It is important to keep in mind in the scenario which follows that metal enrichment does not simply begin with the emplacement of a parent pluton in the upper levels of the crust. Such localization of ore-grade concentrations of metallic compounds represents only the most recent metasomatic stage in a 4.5 billion-year history of selective enrichment of useful primordial elements in the form of ore deposits contained in preserved supracrustal environments (Hutchinson, 1981; Sawkins, 1984; Meyer, 1985). These supracrustal environments have themselves evolved considerably over geological time from the earliest Archean granite-submarine greenstone terranes through a variety of developing continental cratonic settings, up to the present plate tectonic environments at active subduction zones, continent-continent collision zones, passive continental margins, and submarine rift settings.

Ores occurring in these localized enriched states have as progenitors still earlier transport processes affecting large regions within the Earth. Within the first few hundred million years of Earth history, core/mantle interaction depleted the mantle in moderately siderophile transition metals (Fe, Co, Ni, Ag, Mo, and W) and highly siderophile metals (Au, and Pt-group metals) (Ringwood, 1979). The subsequent growth of the continental crust from the primitive mantle over a period of geological time spanning more than 3 Ga enriched the crust in incompatible rock-forming elements and ore metals (Brimhall, 1987b).

In this context ores are the outgrowths of efficient crustal fractionation processes beginning with the extraction of crustal components from the primitive mantle. Derivation and ascent of basaltic liquids enriched in incompatible elements by crystal/liquid partitioning results in formation of the oceanic crust. The relatively small mass of the crust in relation to the mantle yields maximum enrichment factors of approximately 100 which are limited essentially by mass balance constraints. For example, consider a highly incompatible element such as Cs which is thoroughly fractionated into the magmas giving rise to oceanic crust which ultimately is reprocessed through subduction to make continental crust. Figure 1 outlines bulk continental fractionation factors from the primitive mantle from Brimhall (1987b) using estimates of the composition of the primitive mantle by Wanke et al. (1984) and continental crust by Taylor and McLennan (1985). Most ore metals are enriched in the continental crust relative to the mantle by factors which vary systematically with Period and Group in the Periodic Table, as are all the incompatible elements, indicating a clear commonality of process and governing thermodynamic control. It is clear then that crust-forming processes, specifically magma generation in the mantle is the first order control on fractionation not only of rock-forming elements, but ore metals as well. Thus the conditions under which melting occurs in the mantle become the starting point for understanding fractionation of ore metals during crustal evolution. Subsequent remelting of continental crust gives rise to highly differentiated igneous products, which extends the spectrum of ore metal enrichment to still greater extremes. Many of the fractionation processes which ultimately produce near-surface ore deposits begin with the large-scale tectonic activity of the upper mantle and crust.

GENERATION OF MAGMAS AND PLUTONS AT SUBDUCTION ZONES

It has been known for a long time now that andesitic volcanoes and porphyry copper/molybdenum ore bodies tend to occur at a relatively uniform distance from oceanic trenches and subduction zones (e.g., Sillitoe, 1972; Turcotte and Schubert, 1973). The Circum-Pacific andesitic volcanoes, for example, form the well-known "ring of fire." Figures 2 and 3 outline a commonly accepted geochemical model for this feature. We will see that water plays an all-important role at every step. Our discussion is drawn largely from a paper by Burnham (1979) which you should consult for complete details.

Oceanic zones

Oceanic crust descending in a subduction zone carries with it some water in interstitial pores and vesicles, as well as the hydroxyl component in clays and chlorites. At pressures and temperatures represented approximately by point A on Figure 2, a subducting slab should have metamorphosed to the amphibolite facies with 2.0 to 3.0 wt % water bound in amphiboles. This descends further along the average P-T conditions for subduction zones illustrated as A→B on this figure. The solidus (beginning of melting) curves labeled S(a_W = 1.0 to 0.1) show the great effect water has on the melting temperature of amphibolite: Increasing water activity from 0.1 to 1.0 (saturation) decreases the melting temperature by roughly 500°C above 10 kbar. If this rock contains no pore fluid, it will not begin to melt until it intersects the "fluid-absent" hornblende solidus (Hb-S) at B. Here hornblende reacts to produce garnet peridotite and a silicate melt of water activity ≥ 0.3 (which you can read from the solidus curve for amphibolite). For this water activity, the melt must contain ≥ 6.4 wt % H_2O; this can be achieved by melting roughly 20% of an amphibolite containing 1.5 wt % H_2O.

This melange of solid + melt might continue on down the subduction zone for several more kilometers, perhaps to point C, melting further at greater depths. Somewhere over the narrow interval between B and C the melt will begin to rise (C→D) into hotter overlying lithosphere. Because it contains more water, it will melt or assimilate the low-melting fraction of overlying rocks and will crystallize more refractory minerals (pyroxene, olivine, garnet), becoming more felsic as it rises. Burnham (1979) estimates that by the time this syntectic melt rises to point D, it will have assimilated an additional 80% of overlying lithosphere, contain about 3% water and be roughly dioritic (andesitic) in composition. Since the initial melting at point C occurs within a very narrow depth range of 75-80 km, this overall process produces hydrous calc-alkaline magmas at a uniform distance from the oceanic trench. These magmas contain sufficient water to produce explosive island arc

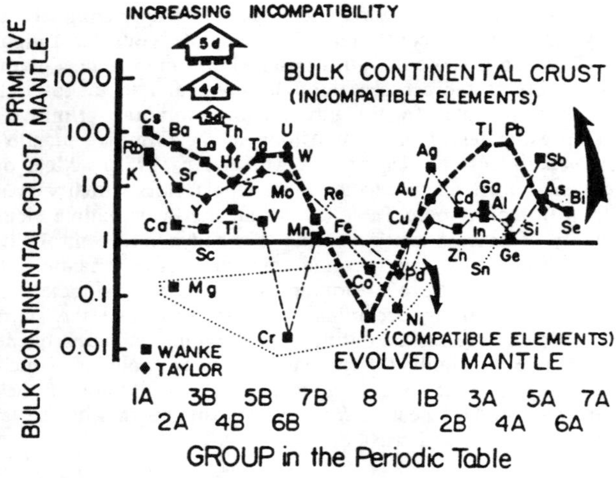

Figure 1. Bulk continental crust fractionation curves relative to the primitive mantle. Most metals are enriched in the crust by factors varying systematically with Group and Period in the Periodic Table. From Brimhall (1987b).

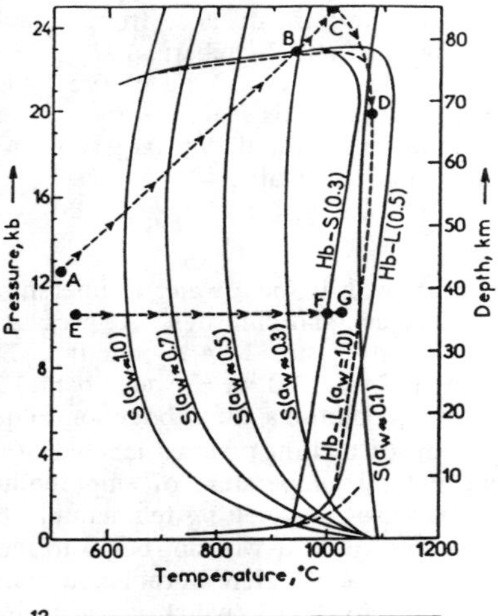

Figure 2. Melting relations for an amphibolite of olivine tholeiite composition. Curves marked S are the beginning of melting (solidus) for the water activities shown. Hb-S(0.3) is the beginning of melting for amphibolite for a mole fraction 0.3 of water in the melt. Hb-L(0.5) is the maximum stability of hornblende at a mole fraction 0.5 of water in the melt. Hb-L(a_w = 1.0) is the maximum stability of water-saturated hornblende. From Burnham (1979).

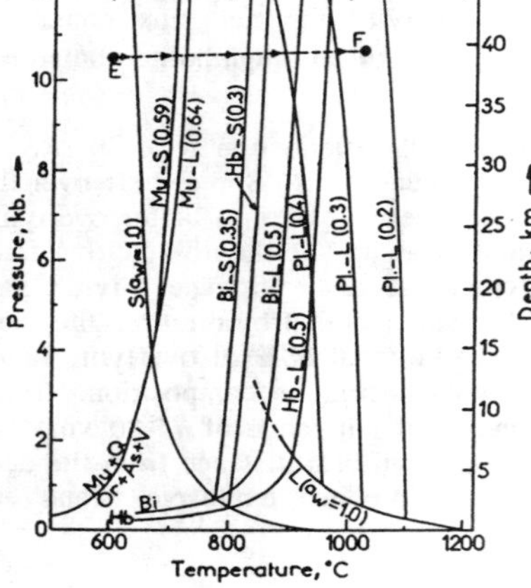

Figure 3. Melting relations for average hornblende-biotite-granodiorite composition. The curve Mu + Q = Or + As + V is the upper stability of muscovite + quartz. S(a_w = 1.0) and L(a_w = 1.0) are the water-saturated solidus and liquidus (plagioclase), respectively. Curves Mu-S, Bi-S and Hb-S are the fluid-absent solidi for assemblages containing muscovite, biotite and hornblende, respectively. The curves Mu-L, Bi-L, Hb-L and Pl-L are approximate thermal stabilities (liquidi) of muscovite, biotite, hornblende and plagioclase, respectively. Numbers in parentheses on each curve are the mole fraction of water in the melt. From Burnham (1979).

volcanism and associated hydrothermal deposits, as we shall see below.

Continental zones

The above process occurs under oceanic lithosphere. By contrast, the path E-F-G on Figure 2 represents intrusion of mafic magma at 1200°C into amphibolites underlying continents. Here the primary role of mantle-derived mafic magmas is simply to provide the heat to partially melt and assimilate deep crustal rocks. A mafic, water-poor magma intruded at 1200°C into lower amphibolitic crust at 40 km and 600°C can raise the temperature of equal proportions of magma plus (crustal) amphibolite above 1025°C (point G)—just above the solidus (F) for dry amphibolite. This can produce an anatectic melt (of roughly 50% the mass of the intruding magma) containing 3% H_2O, with the composition of a quartz diorite. Once again, the melt will rise while olivine, pyroxene and calcic plagioclase crystallize and sink, and H_2O-rich amphibolite melts at the top of the magma chamber.

Figure 3 illustrates what happens if the same mafic magma (at 1200°C) rises into a felsic granodioritic gneiss in deep continental crust (containing quartz, muscovite, biotite, plagioclase, orthoclase and amphibole). Starting at the same P-T conditions (point E), melt will appear at 710°C (muscovite solidus) containing 0.59 wt % water. If this melt rises without significant cooling, it could reach depths as shallow as 10-15 km before crystallizing (at the intersection with the solidus for water saturation, $a_W = 1.0$). Because the melt is water-saturated ($a_W = 1.0$) when it crystallizes, it could produce pegmatites (where large silicate crystals appear to have grown into immiscible aqueous bubbles in the silicate liquid). This is our first example of an aqueous ore fluid of magmatic origin. Pegmatites are frequently mined for elements which are incompatible in crystallizing silicates and are left behind to crystallize out in residual aqueous phases: examples are the rare earths (lanthanides), the actinides, and other anomalies such as tantalum minerals, boron in tourmaline, beryllium in beryl, and lithium-rich micas and pyroxenes.

If heating continues in Figure 3 on to point F, biotite, hornblende and plagioclase will melt in succession. This magma could reach a depth of less than 2.0 km before crystallizing (if it rises without significant cooling); for example, a magma at the intermediate temperature of 820°C will have melted muscovite, then biotite, and if it rises adiabatically it will not intersect the solidus (S for $a_W = 1.0$) until about 2 km; note that at this point it becomes water saturated ($a_W = 1.0$). Thus, granitic to dioritic magmas containing 3.0-9.0 wt % H_2O can be generated by intrusion of mafic mantle melts into lower crustal rocks. If these felsic magmas rise without significant cooling, they will not crystallize (intersect the solidus) until depths of several km, at which point they may become water saturated.

Figure 4 provides a closer look at the specific conditions which release water from a cooling felsic magma. This comes from the experimental phase equilibrium studies of Whitney (1975), and the figure represents a quartz monzonite system containing 3 wt % total water. Notice that at pressures greater than P2, aqueous vapor (V) and silicate liquid can coexist only over the very narrow temperature range of the field labeled Pl+Af+Q+L+V. Between P2 and P1 vapor and liquid coexist over a slightly wider field (including Pl+Af+L+V). However, below P1, aqueous vapor and silicate liquid coexist over the entire range of temperatures from approximately 720°C on up beyond 1200°C; under these conditions a felsic magma can be expected to become saturated in water and to exsolve or give off water as it crystallizes. A monzonitic magma containing 3 wt % water rising from depth will begin to exsolve copious water at P1 = 1 kbar, at a depth of about 3.5 km. The same thing will happen to a similar magma containing 4 wt % water at 1.3 kbar or about 4.5 km.

We emphasize that this evolution of water only occurs at fairly shallow depth, within the upper few kilometers of the Earth's crust. Also, most of the water originally contained in the magma must be given off as the system cools and crystallizes, because most felsic rocks contain <1 wt % water (as hydroxyl in micas and amphiboles). This means that an absolutely enormous quantity of water can be given off by a cooling magma. For example, a felsic magma of 1 km^3 volume containing 3 wt % water would exsolve approximately 10^8 metric tons of water as it solidified at shallow depth.

CLASSES OF ORE-FORMING PLUTONS

A critical question to ask about the aqueous magmatic fluids released from crystallizing plutons is whether or not there is significant diversity in their composition. There are certainly distinctive groups of base metal ores related to felsic magmas, indicating the existence of some major differences in overall processes between families of ore deposits. This may be attributable to essential differences in plutons and the ore-forming fluids specific to them.

In general, base metal ores related to plutons fall into three types. First, porphyry copper deposits related to granodiorites or diorites may have either molybdenum or gold as byproduct metals, depending upon whether the local tectonic setting is predominantly continental or oceanic, the former being relatively evolved chemically, the latter, primitive (Kesler, 1973; Hollister, 1978). Secondly, porphyry molybdenum deposits, often accompanied by a suite of highly lithophile trace metals W, Sn, Be, Zn, and Li, are related to highly differentiated silicic rhyolitic magmas of totally continental derivation (White et al., 1981, Barton, 1987; Christiansen and Lee, 1986). Thirdly, tin or tungsten deposits often form in association with true granites (Lehmann, 1982; Taylor, 1979).

These three types of mineralization are fundamentally distinct, raising the question whether the differences are due to the respective inheritance of metal suites or to other factors subsequent to formation of the magma body, such as composition of exsolved magmatic aqueous fluids and gasses. The latter possibility may depend on earlier processes affecting magmas on their ascent, such as assimilation or fractional crystallization. It is therefore essential to recognize and accurately interpret the salient differences between parent magmas for the three main types of base metal ore deposits related to plutons. It is logical to begin the analysis of this complex problem by examining the characteristics of magmatic source rocks as these attributes may constitute the primary difference, not only in terms of metal endowment, but physiochemical effects which are peculiar to pluton types.

Magmatic source rocks

Burnham (1981) outlined a number of physiochemical factors relating hydrothermal mineralization to intrinsic characteristics of parent magmas. These factors included water content, temperature or heat content, metal content, chlorine content, sulfur content, and oxidation state. He argued that each of these constraints is imposed differently on porphyry magmas by each of three magmatic source rocks assumed to exist in the source regions of partial melting. The critical source rocks are interpreted to be (1) hornblende-bearing mafic rocks, (2) biotite-bearing rocks of intermediate composition, and (3) muscovite-bearing metasedimentary rocks.

Utility of biotite mineral chemistry. Further support for Burnham's interpretation has come from mineral chemistry studies, particularly those of biotite, a hydrous mineral which reflects the relative fugacities of HF, HCl, O_2, H_2 and water during crystallization (Wones and Eugster, 1965; Munoz and Swenson, 1981; Munoz, 1984; Brimhall et al., 1985; Ague and Brimhall, 1987, 1988a,b; Brimhall and Ague, 1988), thereby contributing simultaneously to understanding several of the factors expressed by Burnham. The prime utility of biotite in this context is that it enters into equilibrium relationships with other common rock-forming minerals such as alkali feldspar and magnetite and thus buffers O_2 fugacity (Wones and Eugster, 1965), and hence oxidation state, while simultaneously reflecting the concentration of halogen species and water (Munoz, 1984). As a complex solid solution crystallizing from both magmas and high-temperature hydrothermal fluids, the substitution of F and Cl for OH which surround Mg and Fe in close octahedral coordination, makes biotite ideally-suited to serve this useful purpose in monitoring intensive variables in magmas and exsolved aqueous fluids from the magmatic stage to early hydrothermal processes

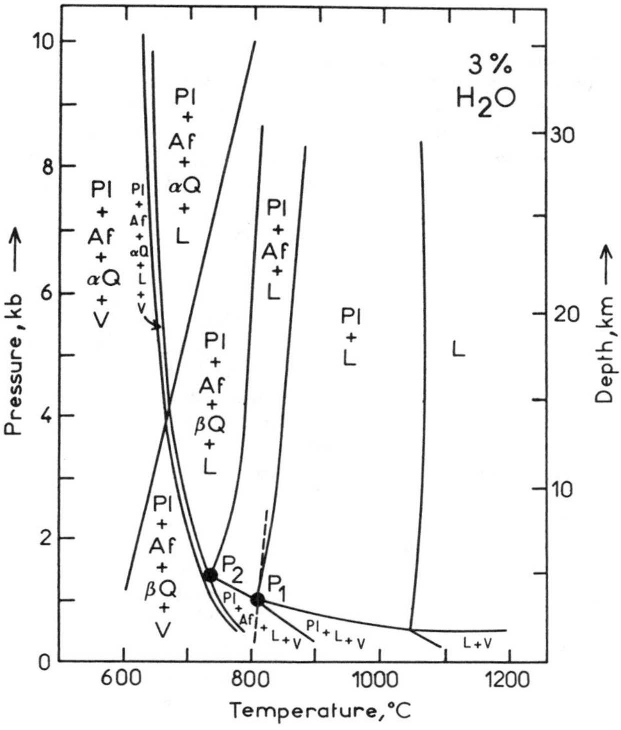

Figure 4. Phase diagram for synthetic quartz monzonite composition with magmatic water content of 3 wt %. The symbols Pl, Af, Q, L and V refer to plagioclase, alkali feldspar, quartz, liquid silicate, and water vapor, respectively. From Whitney (1975).

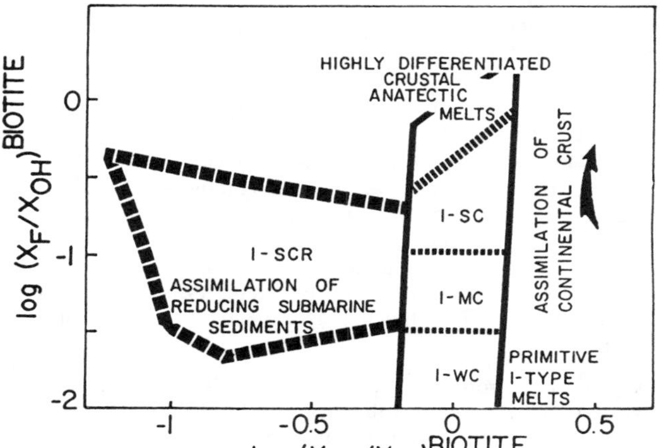

Figure 5. Summary of igneous biotite compositions within the batholiths of California. Highly differentiated crustal anatectic melts are from Colorado. From Brimhall and Ague (1988). Classification scheme of plutons uses biotite composition. I-scr type plutons are strongly contaminated and reduced through the assimilation of graphitic metasediments. The I-wc to I- sc trends are normal plutonic types from primitive oceanic crust-derived melts to strongly contaminated magmas derived from the pre-Cambrian crust or highly contaminated with continental crust.

responsible for the initial release of ore metals from the parent magma.

Using the X_{Mg} and X_F of alteration hydrothermal biotite in this capacity, Munoz and Swenson (1981) proved that there were in fact substantial differences in the HF/HCl fugacity between porphyry copper and porphyry molybdenum deposits pointing out, for the first time, quantitative thermodynamic differences between the geochemical evolution of such magmatic-hydrothermal systems at the hydrothermal stage. By considering biotite compositions more broadly in terms of X_F/X_{OH} and X_{Mg}/X_{Fe} it was demonstrated by Brimhall et al. (1985) that igneous biotite compositions, in distinction to hydrothermal products, are grouped into three main classes which each correlate with a distinct type of base metal ore deposit.

Hornblende, biotite, and muscovite. Both porphyry molybdenum deposits and porphyry Cu deposits have igneous biotites with nearly equal mole fractions of Mg and Fe, but the former have a much higher X_F/X_{OH} in biotite. In contrast, tungsten and tin deposits have magmatic biotites with much lower Mg contents reflecting relatively low oxygen fugacity. This correspondence of ore metal type and biotite composition serve to illuminate the source rocks involved in the genesis of ore-forming magmas. F-rich biotites probably reflect the melting of biotite-bearing gneisses in the pre-Cambrian craton of North America which gives rise to porphyry molybdenum deposits derived exclusively from highly fractionated radiogenic continental materials. Melting of mafic amphibolites with a correspondingly lower F content may form the parent magmas of porphyry copper deposits. The ferruginous nature of biotites in tungsten and tin deposits indicates relatively reducing conditions which have been ascribed to anatexis of pelitic metasedimentary rocks, often containing graphite. Burnham and Ohmoto (1980) have demonstrated that the redox lines of descent of these magma types are distinct, and accompanied by differences in the concentration of carbon and sulfur species.

A single orogenic belt in which major ore deposits occur can contain plutons of several types differing markedly in their biotite mineral chemistry. Field relationships are necessary to interpret the significance of the differences. Granitic batholiths of the Circum-Pacific region have received intense study and serve a useful purpose in illustrating the fundamental differences between pluton types and the controlling geological factors.

Classification by redox state and biotite halogen composition

Figure 5 shows the spectrum of biotite compositions from the granitic batholiths of California which range in composition from quartz diorites on the western oceanic side to true granites on the eastern continental side where plutons have intruded great thicknesses of pre-Cambrian continental crust. The primary variable in biotite composition used in this classification (Brimhall et al., 1985; Ague and Brimhall, 1987, 1988a,b) is $\log(X_F/X_{OH})$ which increases from a value of –2.0 in biotite from primitive mafic rocks (quartz diorites) to a high of 0.0 for the most felsic granites, with a range of a factor of 100 variation in X_F/X_{OH}, from 0.01 to 1.0. In reference to the inferred amount of interaction or derivation from the pre- Cambrian craton, the variations in these I-type plutons have been referred to as I-wc (weakly contaminated), I-mc (moderately contaminated), and I-sc (strongly contaminated). I-wc, I-mc, and I-sc pluton types define elongate belts parallel to the edge of the pre-Cambrian craton of North America (Ague and Brimhall, 1987). Well within the craton, for example in Colorado, highly differentiated crustal anatectic melts contain biotites with the highest F content studied.

The ratio $\log(X_{Mg}/X_{Fe})$ in biotite in I-wc, I-mc, and I-sc plutons remains essentially constant at a value near zero. However, a distinct type of pluton, I-scr (strongly contaminated and reduced) contains biotite with much lower ratios. Such plutons in California and elsewhere, for example Japan, occur only where plutons intrude graphite-bearing pelitic metasedimentary wall rocks. These special plutons are generally peraluminous, bearing muscovite and, rarely, garnet, and often contain ilmenite as the main Fe-Ti oxide in contrast to magnetite and ilmenite in the I-wc to I-sc series. The distribution of I-scr types is con-

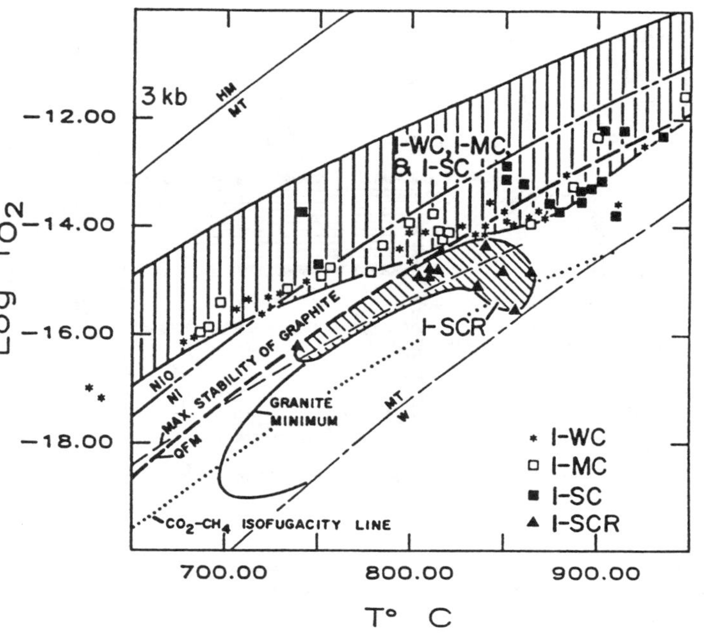

Figure 6. Calculated variation cf oxygen fugacity with temperature for I-wc to I-sc plutons in comparison to I-scr types. Based upon the ilmenite- granite buffer equilibria of Ague and Brimhall (1988b).

trolled solely by the nature of pre-batholitic wall rocks.

Classification of granitic plutons by intensive variables

Use of biotite mineral chemistry to subdivide distinct types of granitic plutons provides a simple means of extending their characterization to include ranges in intensive variables of importance to interpreting controls on mineralization. Through equilibrium rock-forming mineral assemblages and experimental activity-composition relations for biotite (Munoz, 1984), Ague and Brimhall (1988b) transform observed igneous biotite compositions in terms of X_{Mg}/X_{Fe} and X_F/X_{OH} into calculated values of oxygen fugacity and fugacity ratio of HF to H_2O. Thus a direct comparison between plutons may be made based on fugacities of HF/H_2O and O_2, intensive variables of thermodynamic significance.

Oxygen fugacity. Granitic plutonic rocks, in contrast to their extrusive equivalents which cool rapidly upon air quenching after eruption, have Fe-Ti oxide compositions which are affected by sub-solidus re-equilibration during slow cooling. Plutonic magnetite is generally pure Fe_3O_4, having lost its Ti content through diffusion during cooling. However, ilmenite is affected much less severely than magnetite by re-equilibration, and may be used to estimate oxygen fugacity of plutons. This is accomplished by modifying the granitic rock oxygen buffer equilibria of Wones and Eugster (1965) to use the hematite (Fe_2O_3) component of ilmenite instead of magnetite (Fe_3O_4). The equilibrium from (Ague and Brimhall, 1988b) is:

$$\text{annite} + 0.75\,O_{2(gas)} = \text{alkali feldspar} + 1.5\ \text{hematite} + \text{water}$$

Compositional data on biotites and ilmenites from plutons of the batholiths of California provide both oxygen fugacity and temperature estimates shown in Figure 6 from Ague and Brimhall (1988b). I-scr plutons crystallize at much lower oxygen fugacities than I-wc, I-mc, or I-sc plutons. The latter three types generally cluster along the Ni-NiO buffer,

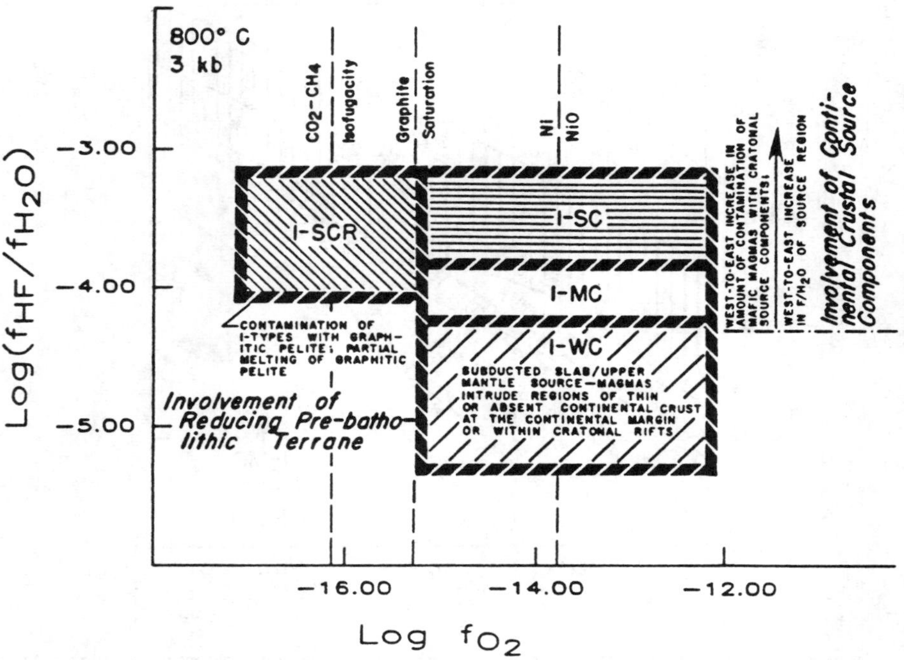

Figure 7. Calculated gas fugacities assuming a total pressure of 3 kbar and 800°C. From Ague and Brimhall (1988b).

below magnetite-hematite and above quartz-fayalite-magnetite. I-scr plutons have oxygen fugacities between quartz-fayalite-magnetite and the isofugacity curve of CO_2-CH_4, falling generally in the realm of the granite melting minimum in the presence of graphite as calculated by Ohmoto and Kerrick (1977); this supports the conclusions of Burnham and Ohmoto (1980) that some S-type (sediment-derived) granites cool along the latter gas buffer curve. I-scr type plutons are distinct from I-wc, I-mc, and I-sc types in terms of oxygen fugacity trends, and crystallize at much lower values, probably in equilibrium with graphite. It is likely that at least some of the I-scr plutons are in fact derived by partial melting of graphitic and pelitic wall rocks which gives rise to the peraluminous character of these plutons.

HF/H_2O fugacity. Using activity-composition relations of Munoz (1984), biotite compositions expressed in terms of X_F/X_{OH} may be interpreted as fugacity ratios of HF/H_2O. In Figure 7 we present these calculations (from Ague and Brimhall, 1988b) at crystallization conditions of 3 kilobars and 800°C. The separation of pluton classes in terms of fugacities of HF/H_2O and O_2 is clear. The oxygen fugacity trend of the I-wc to I-sc plutons along the Ni-NiO buffer is similar to common temperature-oxygen fugacity curves derived for volcanic rocks using magnetite-ulvospinel and ilmenite-hematite Fe-Ti oxides (Carmichael et al., 1974). It is clear then that with the exception of the reduced I-scr plutons, the I-wc to I-sc types do have extrusive equivalents. The I-scr plutons are distinct from the main igneous trends.

Correlation of ores with plutonic classes

In Figure 8 we show the compositions of biotites from mineralized plutons. Igneous and hydrothermal biotites from the three ore types Cu, Mo, and W are depicted (Brimhall and Ague, 1988). Included are data for porphyry copper deposits: Santa Rita, New Mexico (Jacobs and Parry, 1979) and Butte, Montana, the Henderson Colorado porphyry molybdenum deposit (Gunow et al., 1980), and two tungsten deposits (Pine Creek and Strawberry). For comparison, the biotite data from the unmineralized plutons of the California batholiths, and the Magnetite and Ilmenite Series of Japan (Ishihara et al., 1979) using data from

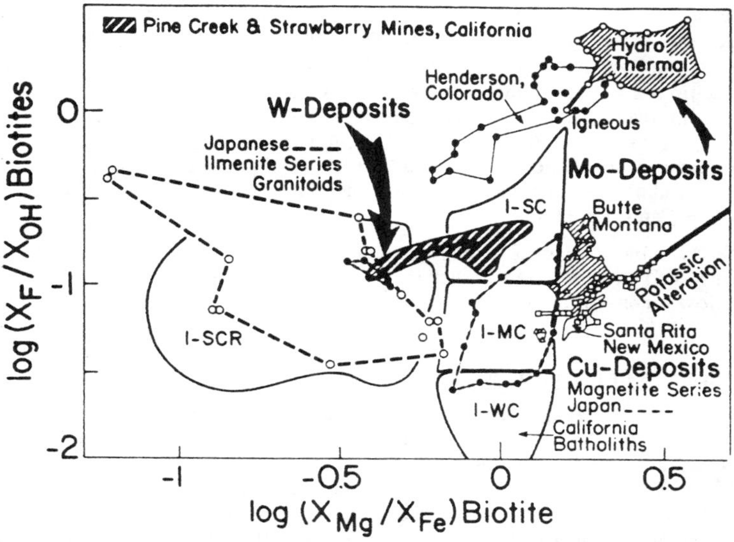

Figure 8. Comparison of hydrothermal biotite compositions with igneous biotites. Hydrothermal biotites form arrays with a positive slope extending away from the composition of the local igneous biotite composition. Hydrothermal biotites are shown for three types of base metal ore deposits related to porphyries; W, Mo, and Cu. The Japanese Ilmenite and Magnetite Series plutons are shown in comparison to the California batholiths. From Brimhall and Ague (1988).

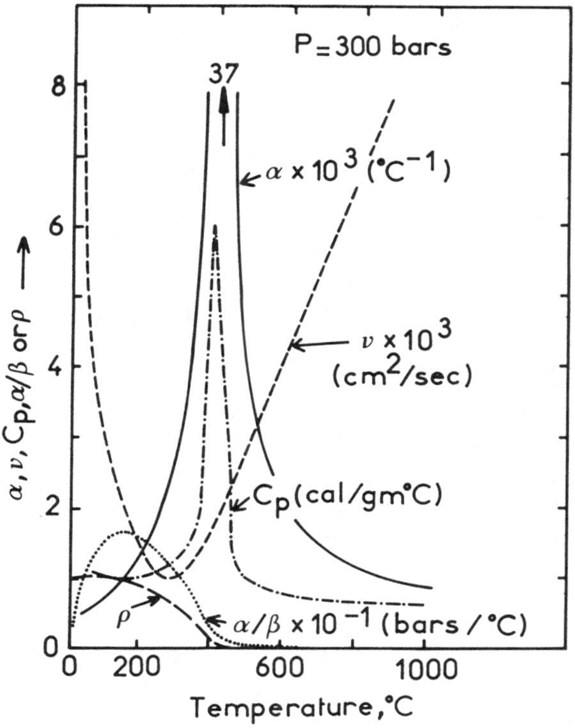

Figure 9. Physical properties of pure water at 300 bars versus temperature. The symbols α, β, ρ, C_p, and ν refer to coefficients of thermal expansion and compressibility, density, heat capacity and viscosity, respectively. From Norton (1984).

Czamanske et al. (1981) are shown. It is clear that assuming similar crystallization conditions, Cu, Mo, and W deposits are characterized by distinctive igneous biotite compositions, and hence by distinct values of oxygen and HF/H_2O fugacity. Both Mo and Cu systems are in general much more oxidizing than W deposits, a conclusion consistent with recent experimental studies indicating inverse dependence of scheelite and cassiterite solubilities with oxygen fugacity (Wilson and Eugster, 1984; Haselton and D'Angelo (1986). While having generally the same redox state, Mo systems are more fluorinated than Cu deposits, consistent with the radiogenic character of the parent rhyolitic intrusives derived from biotite gneiss of the pre-Cambrian craton.

Hydrothermal biotites form arrays extending away from the compositions of local igneous biotites, and in general are richer in magnesium and fluorine. Interpretation of this effect will follow a consideration of the composition of magmatic water and its release from plutons.

Physical implications of magmatic water

Energy release. The range of water concentrations of felsic magmas as estimated by Burnham (1979) is 2.5 to 6.5 wt % with a median close to 3.0%. The lower limit is necessary to produce biotite or hornblende, and the upper limit corresponds to saturation at 2.1 kbar or 8 km depth. As water exsolves from a melt, it undergoes a considerable change in volume. For example the partial molar volume of water in a silicate (albite) melt as measured by Burnham and Davis (1971) is 22 cm^3 $mole^{-1}$ at 800°C, 1kbar, while the molar volume of pure water at the same T and P is 78 cm^3 $mole^{-1}$ (Burnham et al., 1969); under these conditions, water expands 3.5 times as it exsolves from a magma. The total change in volume of the reaction H_2O-saturated melt $\rightarrow$ crystals + vapor generates tremendous P-V mechanical energy, on the order of 10^{16} J km^{-3} of magma. This is the approximate explosive energy of a 10 megaton bomb for each km^3 of magma -- for details, see Burnham (1979, 1985). The most spectacular consequence is explosive volcanism, such as the Mount St. Helens blast of May 18, 1980 which was equivalent to a 400 megaton explosion (Decker and Decker, 1981). More important for ore emplacement, less cataclysmic release can cause intense fracturing, brecciation, void volumes, and extensive comminution of the solidifying magma. Ore deposits (such as porphyries) associated with large plutons are commonly mineralized at scales ranging from large veins and breccia zones down to microscopic dissemination in finely comminuted host rock. Explosive volcanism is rare relative to this latter type of fracture release because the tensile strength of typical igneous wall rocks is only on the order of 100 bars (corresponding roughly to 1000 J kg^{-1}).

Hydrothermal convection. The intrusion of any heat source such as a magma into fluid-rich crustal rocks will generate convection cells, irrespective of whether magmatic fluid is released or not. The energetic release of magmatic water will intensify convection by increasing permeability and acting as an upward-moving, high-temperature plume. Some of the properties of pure water that guide its behavior in such systems are summarized in Figure 9, from Norton and Knight (1977) and Norton (1984). This is particularly interesting because it shows a near coincidence of conditions favorable for convection in the vicinity of 300°-400°C: maximum heat capacity and minimum density (promotes buoyant force and heat-carrying ability); minimum viscosity; and a maximum ratio of thermal expansivity to compressibility (develops maximum pressure per unit increase in temperature). This demonstrates that a hydrothermal fluid exerts considerable control over its own P-V-T conditions in a convecting system. It also suggests that the temperature range 300°-400°C provides optimal convective behavior, and might partly explain why temperatures measured in the most active ocean ridge vents often fall within this range.

There is certainly no dearth of hydrologic models for hydrothermal convection around magmatic systems (see, for example, Cathles, 1977; Norton and Knight, 1977; Norton, 1978; Henley and McNabb, 1978; and Norton, 1984, on convection around plutons; and Fehn, 1986; and Fehn and Cathles, 1986, on multiple convection cells at oceanic spreading centers). These models all have in common the entrainment of groundwaters within the convection cell, and some do not consider magmatic waters at all. Other models include provision for boiling or vapor formation, and we illustrate one of these by Cathles (1977) in Figure 10. This represents the behavior of a pluton 0.75 km in half-width, 2.25 km high, intruded at 700°C to a depth 2.75 km below the surface assuming a uniform permeability of 0.25 millidarcies. Free flow of water is permitted at the surface (as hot springs and as groundwater recharge). Initial discharge of magmatic water is not considered, so the model is conservative. The diagram shows a cap of vapor above the pluton which reaches a maximum for these conditions after 5000 years. This is formed by boiling if the fluid is sufficiently NaCl-rich to remain subcritical. This vapor phase rises and disappears after 10,000 years for these conditions and convection dies out in less than 100,000 years. The convection cell includes considerable entrained groundwater and total flow would be 250 kg water/cm^2 through the top of the pluton in the 10^4 year lifetime of this system. Cathles shows by simple mass balance that this flow is sufficient to leach 0.05 wt % Cu from the intrusive and concentrate it in an ore shell 1 km deep and 200 m thick of grade 0.43 wt % Cu, provided solubility exceeds 1000 ppm Cu, and suitable deposition mechanisms can be invoked. We will see below that these conditions are not unreasonable, particularly if initial magmatic water is included in the picture.

Active hydrothermal convection has now been observed at many sites along oceanic spreading centers, as illustrated, for example, by upwelling vents such as the celebrated 21°N "Black Smokers." Recently, downward circulation of hot, chloride– and metal-rich brines has been observed in crater lakes of active volcanoes, providing a potential for ore generation within volcanic stockworks (Brantley et al., 1987).

Lifetimes of hydrothermal systems

The general lifetime of hydrothermal ore-forming systems is not well known because radiometric dating techniques are insufficiently precise to date the beginning and end of deposition. Skinner (1979) suggests that 10^6 years is an upper limit for porphyry coppers, simply because some known deposits in the Pacific are about that old. He notes that seafloor hydrothermal deposits of the Kuroko and Cyprus type probably formed in several thousand years because they contain little sedimentary detritus. Observation of active seafloor vents seems to indicate that these change almost daily, but that large systems such as the Guaymas basin and Red Sea brines may persist for periods of 10^2 to 10^3 years at least. Theoretical models of convection also indicate relatively rapid formation ($<10^4$ years) as we have just seen.

SOURCES AND GENERAL COMPOSITIONS OF HYDROTHERMAL SOLUTIONS

Sources of water

The question of source has been debated by economic geologists for over a century. In fact, Skinner (1979) managed to find comments on the subject by Agricola (who favored meteoric water) and Descartes (a magmatist of sorts).

Aside from magmatic and entrained groundwater, it is also certain that seawater convects at oceanic spreading centers, perhaps to depths on the order of 7–10 km (the thickness of the oceanic crust). There is also considerable evidence in metamorphosed terrains suggesting that hydrothermal waters could have been produced by metamorphic dehydration reactions (Henley et al., 1976; Fyfe and Kerrich, 1984; Henley, 1985).

The greatest success in differentiating magmatic and meteoric water sources has come from oxygen and hydrogen isotope analyses. Figure 11 is a now quite familiar compilation of data for different ore deposits by Taylor (1979) showing the fields of meteoric and primary magmatic water. A complication with these diagrams is that many groundwaters of meteoric origin react with rock-forming minerals to become enriched in ^{18}O, which moves them to the right of the meteoric line. However, you can see from this diagram that some deposits such as the Kuroko ores must have formed predominantly from heated seawater;

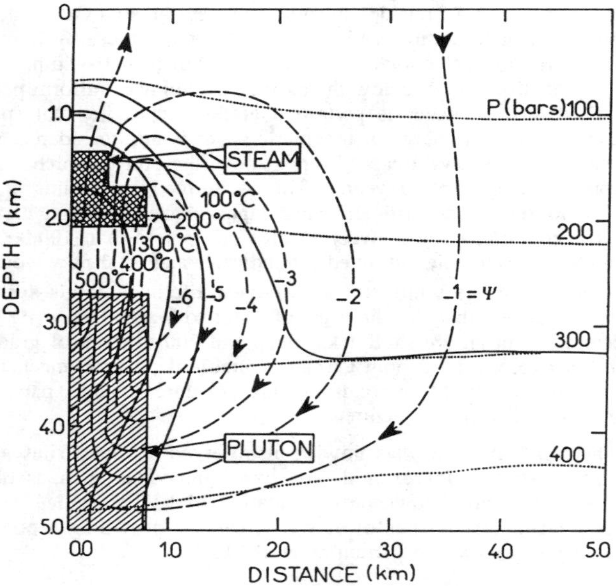

Figure 10. Convective behavior of pure water around a pluton 0.75 km in half-width, 2.25 km high, intruded at 700°C to a depth 2.75 km beneath the surface, assuming uniform permeability of 0.25 millidarcies, and calculated for a period 5000 years after intrusion. Free flow of water is permitted in and out of the surface. The shaded area above the pluton represents a region of vapor-like fluid, and would be generated by boiling in systems containing >10 wt % total dissolved salts. Temperature and hydrostatic pressure are contoured, and streamlines are given in units of cm^2/sec flow. From Cathles (1977).

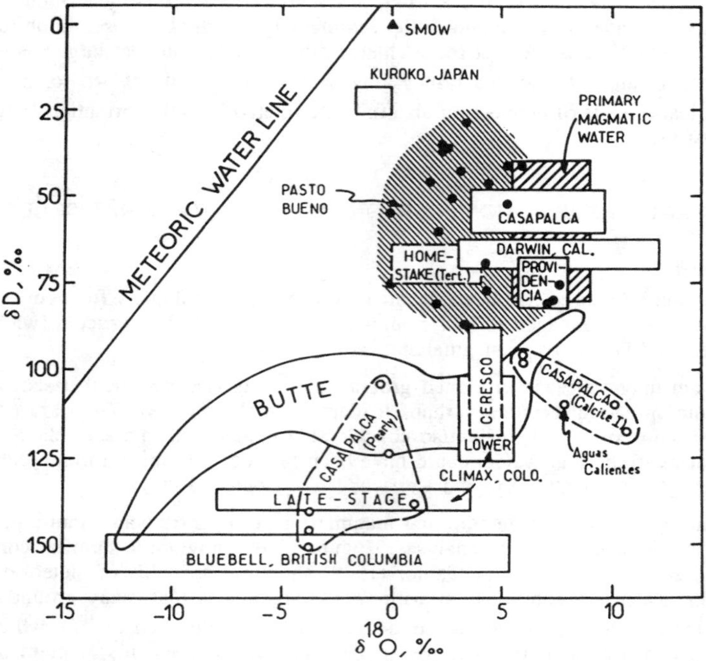

Figure 11. Hydrogen versus oxygen isotope plot showing observed values of hydrothermal fluids in a variety of ore deposits. SMOW is standard mean ocean water, and the fields of meteoric and magmatic water are delineated. Compilation by Taylor (1979).

Bluebell appears to have formed from a meteoric source; other deposits such as Homestake and Providencia were predominantly magmatic; and a third class started as predominantly magmatic and evolved to late-stage hydrothermal waters of meteoric origin (Casapalca, Climax, Butte).

Because of this variety of possible sources it is inappropriate to associate the term "hydrothermal solution" with waters of any one origin. Instead, take the word in its literal sense to mean any natural hot water.

Composition

The compositions of ore-forming hydrothermal solutions vary widely, but all can be thought of as brines with total dissolved solids ranging from roughly 1 to 50 wt %. The major components are generally Na, K, Ca and Cl, with lesser concentrations of Mg, Br, SO_4, H_2S, CO_2 and possibly NH_3 at concentrations frequently exceeding 1000 ppm. Ore-forming metals may be present in concentrations ranging from <1 ppm to >1000 ppm. The pH generally falls within 1-2 units of neutral; more acidic conditions predominate for many ore deposits, while many active geothermal systems are slightly basic. Barnes (1979) suggests that the most common, idealized ore-forming solution might have the following properties: 1m Cl^-, 0.1m carbonate, <0.1m total dissolved sulfur (H_2S or HSO_4^-), 0.01m NH_4^+ and pH 1 unit acid from neutral. It is commonly presumed that such solutions must contain at least 1 to 10 ppm of a dissolved metal to deposit potential ore. Very high concentrations (>10,000 ppm) of metals such as Fe, Cu, Mn and Zn have occasionally been observed in fluid inclusions (Roedder, 1979; Kwak et al., 1986), usually in more saline and/or higher temperature samples.

The observed compositions of hydrothermal solutions have been compiled by: Skinner (1979) and Barnes (1979) (hydrothermal, ore-forming); Ellis and Mahon (1977), Ellis (1979), Weissberg et al. (1979), Henley et al. (1984) and Henley (1985) (geothermal, typically sub-economic); and Von Damm et al. (1985 a,b) (hydrothermal vents at oceanic spreading centers).

COMPOSITION OF MAGMATIC WATER

As a hydrothermal solution exsolves from a cooling magma it will take with it any components that preferentially concentrate in aqueous rather than silicate fluids. This includes most highly volatile species, all electrolytic salts and many transition (and other) metals. Experimental data so far indicate that this is an extremely important fractionation process, capable of transporting many ore-forming components.

Water solubility in silicate melts

The primary volatile components of late-stage magmas include H_2O, H_2S, HCl, HF, CO_2, CH_4, SO_2 and H_2. Preliminary experimental results quoted by Burnham (1979) suggest that the solubilities of H_2O, H_2S and HCl in granitic melts are comparable and high. These should all be expected to fractionate strongly into an exsolving aqueous phase, because they are similar, strongly polar small molecules, fundamentally different from the large, polymerized structural units of silicate melts.

The first, obvious question is whether water is really significantly soluble in silicate melts. If so, then presumably other volatile, polar molecules important to ore-formation such as HCl and H_2S may behave similarly. In fact, water is extremely soluble in magmas covering the entire composition range from basaltic to pegmatitic as shown in Figure 12 from Burnham (1975). The larger plot expresses water solubility in these different rock types normalized against the molecular weight of albite (according to rules described by Burnham). The surprising feature here is that a silicate melt can actually contain more moles of water than silicate since 14 wt % H_2O (maximum measured solubility on the insert) is 70 mole % relative to albite. Note the strong pressure dependence; as pressure drops, water solubility falls causing exsolution at near-surface pressures as already observed in Figures 2 to 4.

Partitioning of ore components between magmas and exsolving water

Fractionation of different components between felsic melts and exsolving aqueous fluids has been studied by Kilinc (1969), Kilinc and Burnham (1972), Holland (1972), Flynn and Burnham (1978), Webster and Holloway (1980), Carron and Lagache (1980), Manning (1981), and Candela and Holland (1984, 1986).

Chloride and sulfur. The early experimental work by Kilinc (1969) and Kilinc and Burnham (1972) showed that chloride partitions very strongly from felsic silicate melts into coexisting aqueous fluids. They observed molal partitioning coefficients ($m_{Cl,aq}/m_{Cl,melt}$) of 43, 85, and 13 at 2, 6 and 8 kbar, respectively. Unpublished results by Kilinc (1969) show a similar fractionation of H_2S. If the granodioritic gneiss of our previous example (Fig. 3) with 3.0 wt % H_2O contains a conservative 0.1 wt % Cl, the first-formed aqueous fluid (0.6 kbar, 2 km depth, 90% melt) would contain 4.6 wt % Cl; at 8 km depth (2.1 kbar) exsolution of water begins only when the magma is 53% crystallized, and first-formed waters here should contain 7.5 wt % Cl (Burnham, 1979).

The quenched pH of these and similar experiments by Holland (1972) were very low with median values about pH=2. This is presumably due to strong fractionation of HCl from melt to aqueous fluid. This fractionation has been interpreted by Eugster (1985, 1986) as hydrolysis of NaCl dissolved in the melt:

$$2NaCl_{melt} + H_2O_{fluid} \rightarrow HCl_{fluid} + Na_2O_{melt} \ .$$

As we shall see below, HCl and other electrolytic components are highly associated as neutral species (HCl°) at magmatic temperatures, and under these conditions, the solutions are probably not far from neutral pH. However, as temperature falls HCl and similar components ionize, becoming strong acids. Thus, chloride partitioning into the aqueous fluid provides a complex-forming ligand for metal transport as well as acid potential. We will see that this acidity increases metal solubility at higher temperatures, and as temperature falls it causes wall rock alteration and concomitant metal precipitation. The partitioning of sulfur is important for similar reasons: it too becomes acidic at lower temperatures, can serve as a possible complex-forming ligand, and is obviously necessary for deposition of metallic sulfides (and sulfates such as barite and anhydrite).

Cations and metals. Experiments by Holland (1972) showed that partitioning of both Na and K from melt to aqueous vapor varies linearly with the total Cl concentration of the vapor. The molal partitioning coefficients for melts of granitic composition are: $m_{Na,aq}/m_{Na,melt} = 0.46\ m_{Cl,aq}$, and $m_{K,aq}/m_{K,melt} = 0.34\ m_{Cl,aq}$. A similar result was obtained by Candela and Holland (1984) for monovalent Cu, for which $m_{Cu,aq}/m_{Cu,melt} = 9.21\ m_{Cl,aq}$.

The divalent cations Ca, Mg, Zn and Mn also partition into the aqueous phase with Mn and Zn being most strongly concentrated (Holland, 1972). For these four cations, the partitioning coefficients increase more rapidly than the first power of Cl concentration: $m_{i,aq}/m_{i,melt} = k\ (m_{Cl,aq})^x$ (for $x > 1$).

Finally, Candela and Holland (1984) showed that Mo partitioning is independent of Cl: $m_{Mo,aq}/m_{Mo,melt} = 2.5$. The chloride dependence is necessary to maintain an electrical charge balance between chloride (the predominant anion) and all cations. At these magmatic temperatures the predominant aqueous species are associated, neutral molecules such as $CuCl^\circ$, $NaCl^\circ$, $CaCl^\circ_2$ and $ZnCl^\circ_2$. The 1:1 proportionality between univalent cations and Cl follows from these stoichiometries, and the 1:2 relations between divalent cations and Cl, from equilibria such as

$$CaCl_{2(aq)} + 2Na^+_{(melt)} = Ca^{2+}_{(melt)} + 2NaCl_{(aq)}$$

where $NaCl_{(aq)}$ varies directly with $Cl_{(aq)}$. Equilibrium constants for such reactions derived from the partitioning coefficients fit the observed data relatively well, suggesting that these

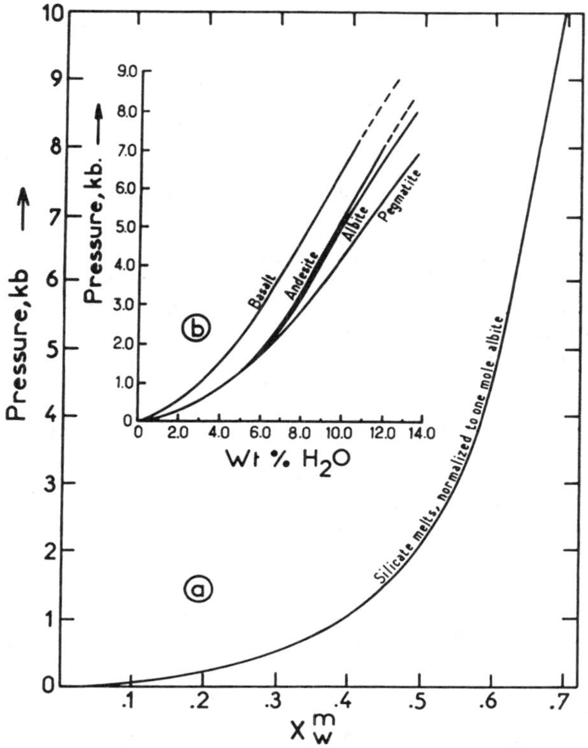

Figure 12. Observed solubility of water in silicate melts including basalt, andesite, Li-pegmatite and albite. The insert plots wt % solubility, and the larger figure gives the mole fraction of water in all four melt compositions normalized against albite composition. From Burnham (1979).

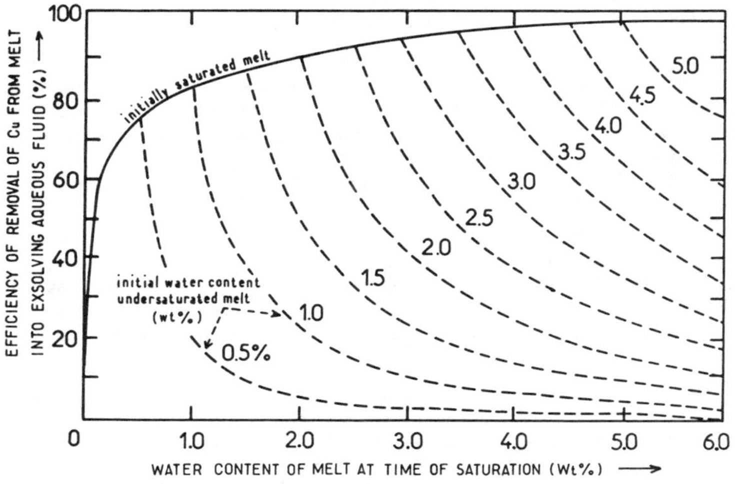

Figure 13. The percent efficiency of removal of Cu from a felsic silicate melt into an exsolving aqueous fluid as a function of both wt % water in the melt at saturation and of the initial wt % water in the melt. Calculated for a solid/melt Cu partition coefficient of 0.3 and an initial concentration ratio (chloride/water) in the melt of 0.1. From Candela and Holland (1986).

kinds of exchange reactions control the distribution of cations between the aqueous and silicate phases. Mo, which is Cl-independent, may be a special case which does not form appreciable chloride compounds , but instead partitions as mixed H_2O-OH^- complexes such as molybdic acid, $MoO_2(OH)_2$.

The partitioning coefficient for Cl, coupled with those for Zn, Mo, Cu and Mn, can be used to calculate the metal concentrations of hydrothermal solutions exsolving from felsic magmas as a function of initial Cl, H_2O and metal concentrations. This depends on the water content of the melt at the time of water saturation, and on the degree of crystallization of the melt. A representative calculation by Candela is included here as Figure 13, and many of the possible variations are discussed by Holland (1972) and Candela and Holland (1986). All metals studied so far can be very efficiently extracted from a melt into an exsolving aqueous phase, provided water concentrations at the time of melt saturation are sufficiently high. In Figure 13, for example, the fractionation of Cu into exsolving water is nearly complete (95%) for the 3 wt % water concentrations used in our representative felsic magmas above. Based on his own experimental work, Burnham (1979) calculated that an aqueous vapor coexisting with a granodioritic magma containing magnetite and 0.1 wt % Cl at 900°C and 1 kbar would have the following composition: roughly 1m total Cl, 0.40m Na, 0.23m K, 0.1m total dissolved Fe, 0.09m H (as H+) and 0.02m Ca. The very high concentration of iron has been borne out by more recent solubility studies mentioned below. In addition, the exsolving HCl-charged vapor phase will leach metals from the cooling, solid carapace of the pluton and from adjacent wall rocks.

Magmatic to hydrothermal transition: the biotite sensor

The phase separation accompanying the transition from magmatic to early hydrothermal processes affects considerable fractionation of ore metals and dissolved salts manifested in precipitation of abundant ore sulfides and the presence of solute-rich fluid inclusions in quartz in biotitic igneous- hydrothermal breccias occurring near the tops of plutons or dikes. Since biotite crystallizes both from magmas and as a reaction product in the potassic alteration assemblage, it serves a critical function of monitoring the earliest stages of hydrothermal fluid circulation. Hydrothermal biotite is easily recognized petrographically on the basis of its shreddy habit and occurrence in veinlets, and in their alteration halos. Fortunately hydrothermal biotites form in all three ore deposit types (Cu, Mo, and W-Sn) and from its composition and the associated mineral assemblages the nature of the magmatic to hydrothermal transition can be compared between systems. See Brimhall and Ague (1988) for more details on the thermodynamic arguments which follow.

Compositions of hydrothermal biotites. Figure 14 presents a generalized version of Figure 8, showing the compositions of hydrothermal biotites accompanying mineralization. In all three cases (Cu, Mo, and W deposits) biotite compositions form regular linear data arrays extending away from the specific composition of igneous biotite in the parent pluton, I-scr for W, I-mc for Cu, and crustal anatectic melt for Mo. The slopes of these arrays are positive, near a value of 3/2 shown with a solid bar in Figure 8, such that systematic F and Mg enrichment occur simultaneously during potassic alteration. One possible explanation of this positive slope is simply that it is due to the Fe-F avoidance principle (Munoz, 1984). Ague and Brimhall (1988b) have shown that this effect would produce a slope of only 0.6 in contrast to the observed 3/2 and have concluded that other processes are responsible for the Mg and F enrichment observed. Fe-OH and Mg-Fe exchange equilibria for chemical components in biotite in equilibrium with aqueous ions in solution can be used effectively for this purpose:

$$KMg_3AlSi_3O_{10}(OH)_2 + 2F^- + 3Fe^{2+} = KFe_3AlSi_3O_{10}(F_2) + 2OH^- + 3Mg^{2+} \quad . \qquad (1)$$

This expression is one of two possible mineral-solution equilibria which defines the exchange of F, OH, Mg, and Fe end-member components in biotite. The alternative set of exchange components is Mg-F and Fe-OH which we will return to later. The equilibrium constant for this reaction is:

$$K(p,t) = \frac{a^{biotite}_{KFe_3AlSi_3O_{10}(F_2)} \; a^{2\,fluid}_{(OH)^-} \; a^{3\,fluid}_{Mg^{2+}}}{a^{biotite}_{KMg_3AlSi_3O_{10}(OH)_2} \; a^{2\,fluid}_{F^-} \; a^{3\,fluid}_{Fe^{2+}}} \tag{2}$$

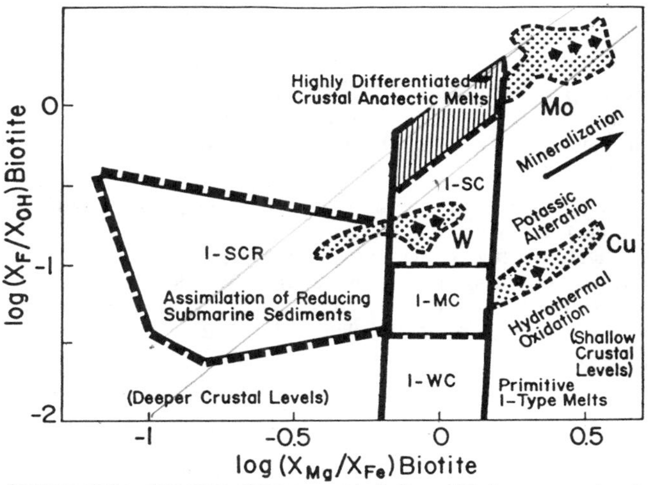

Figure 14. Summary of the relationships of igneous and hydrothermal biotites near ore deposits. From Brimhall and Ague (1988).

Assumption of ideal site mixing in biotite gives a useful simplification. For example:

$$a^{biotite}_{KFe_3AlSi_3O_{10}(F)_2} = X^3_{Fe} * X^2_F \quad , \tag{3}$$

$$a^{biotite}_{KMg_3AlSi_3O_{10}(OH)_2} = X^3_{Mg} * X^2_{OH} \quad . \tag{4}$$

The dependence of F and OH on Mg and Fe may be then derived after substituting the assumed ideal mixing activity-composition relations in (3) and (4), taking logarithms, and rearranging:

$$\log (X_F/X_{OH})^{biotite} = 1.5 \log (X_{Mg}/X_{Fe})^{biotite} + \log (a_{F^-}/a_{OH^-}) - 1.5 \log (a_{Mg^{2+}}/a_{Fe^{2+}})^{fluid} + 0.5 \log K \quad . \tag{5}$$

Therefore, at constant pressure and temperature and F^-/OH^- and Mg^{2+}/Fe^{2+} ion activity ratios, log (X_F/X_{OH}) in biotite should increase linearly with log (X_{Mg}/X_{Fe}) at a rate of 1.5. Written more formally:

$$\left(\frac{\partial \log (X_F/X_{OH})^{biotite}}{\partial \log (X_{Mg}/X_{Fe})^{biotite}} \right)_{fluid\ composition,p,t} = 1.5 \quad . \tag{6}$$

The alternative set of exchange components gives an analogous expression to (5) but with a slope (6) of negative 3/2. Since the data on natural biotites clearly indicate a positive correlation, the alternative set can be disregarded. Given that the observed slope is about 3/2, it is likely that hydrothermal biotite precipitates essentially under these constraints: essentially constant pressure, temperature, and the aqueous activity ratios indicated. What then causes the exchange reaction to proceed and cause the compositional variation which is characteristic of hydrothermal biotite in Cu, Mo, and W deposits?

Given the oxygen and halogen fugacities inferred from biotite and related mineral assemblages, the three magmatic-hydrothermal systems evolve from different starting points but apparently undergo processes held in common. In all three cases, the biotites formed are more magnesian than their igneous counterparts, and therefore it is likely that oxidation accompanied the hydrothermal process. This conclusion is supported by the granite oxygen fugacity buffer of Wones and Eugster (1965):

$$\text{annite} + 0.5\,O_2 = \text{magnetite} + \text{alkali feldspar} + \text{water} \quad . \tag{7}$$

Using ideal site mixing in biotites again gives an expression for oxygen fugacity:

$$\log f_{O_2} = -6\log(1 - X_{Mg}) - 4\log X_{OH} + 2\log f_{H_2O} - 2\log K \quad . \tag{8}$$

Oxygen fugacity is expected to increase with X_{Mg} as with the activity of water. Consequently, oxidation could be the result of water saturation in the parent magma, although this explanation seems unlikely since the effect may be limited to melts containing only low iron contents (Candela, 1986).

Early high temperature hydrothermal oxidation. A more likely possibility is that oxidation is driven by changes in the solutes contained in magmatic aqueous fluids. Of these, perhaps the most likely to affect oxidation are the acidic components such as HCl, which at high temperatures are quite associated but ionize rapidly with decreasing temperature. With the quantitative partitioning of such acid constituents into aqueous fluids, their collective ionization could dramatically lower the fluid pH. Writing the granite oxygen fugacity buffer again in terms of H^+ and OH^- instead of water provides a way to relate oxidation to ionization.

$$\text{annite} + 0.5O_2 = \text{magnetite} + \text{alkali feldspar} + H^+ + OH^- \quad . \tag{9}$$

With the H^+ activity increasing by acid ionization and charge balance being maintained by increased chloride ion activity, the granite buffer reaction could be driven to the left, and in so doing, increase oxygen fugacity and modify biotite compositions producing the magnesium enrichment arrays present in all three types of base metal deposits. The simultaneous enrichment of fluorine may be due to maintenance of the biotite exchange component equilibrium (1) and the proportionality required (5) and (6).

Hydrothermal biotites and the compositional exchange path, probably due to late-stage magmatic and early-stage hydrothermal oxidation, are rare in major batholiths such as the Sierra Nevada, and so far have only been recognized in and near major ore deposits, the Pine Creek and Strawberry Mine tungsten skarn deposits (Brimhall and Ague, 1988). The technique of using biotite as a sensor for detecting mineralizing aqueous fluids is therefore specific to the oxidative events responsible for major deposition of ore metals (tungsten, copper and molybdenum), and therefore offers a means to differentiate between mineralized and unmineralized plutons.

Relative importance of magmatic and meteoric waters

Much of the above discussion supports the century-old theory that cooling plutons can exsolve hydrothermal brines which carry sufficient dissolved metal and sulfur to produce economically attractive deposits. The high acid potential and alkali chloride content of these solutions is also consistent with observed alteration of neighboring wall rocks and the high-salt fluid inclusions commonly associated with pluton-related ores. This is sufficient evidence to terminate the old debate on the importance of magmatic waters in ore deposition: exsolved magmatic waters have the full potential to produce ores given the appropriate physical environment for deposition and preservation.

At the same time, high-temperature aqueous solutions, no matter what their origin, are capable of leaching metals and other components from wall rocks as well as magmas. Thus, meteoric groundwaters entrained in a hydrothermal convection system also become potential ore-forming fluids; in some cases, such as the Mississippi Valley-type Pb-Zn deposits, they appear to be the sole aqueous component since no igneous intrusions occur anywhere nearby. In other cases, heated and recirculated seawater seems to be the ore– forming fluid (Kuroko, oceanic spreading centers), although here a small magmatic component cannot be entirely dismissed. Many ore bodies appear to have formed from multiple sources, perhaps starting with predominantly magmatic waters, and finishing with predominantly meteoric input (as apparent in Fig. 11). Quite clearly, there are many possible sources of ore-forming solutions, all important to varying degrees in different environments.

The historic confusion and long-standing debate over the relative ore- forming potential of hydrothermal fluids from different sources is based more on geological than chemical arguments. In the field it is obvious that different ores were produced by different geological processes: porphyry coppers quite clearly formed by a different process and from a different source than Mississippi Valley-type Pb-Zn ores; these are different again from the Cyprus-type umbers or stratiform sulfides. Nevertheless, the chemical properties of water are the same, no matter what the origin. Aqueous solutions of different composition are governed by the same chemical principles. From the chemist's point of view, ore-forming aqueous solutions appear much less diverse. In the following sections we discuss some of these underlying chemical controls.

PART II. PHYSICAL CHEMISTRY OF HYDROTHERMAL ORE FLUIDS

SOLVENT-SOLUTE CONTROLS ON ORE SOLUTIONS

Water is, of course, the most abundant liquid in the Earth's crust and governs many chemical processes from surface to upper mantle. It seems ironic that water is also a very anomalous solvent, unlike most other liquids in many respects. For example, it has been called "the universal solvent" and "one of the most corrosive substances known" (Franks, 1972, p. 20). These peculiarities arise from the structure of the water molecule itself. The following discussion is taken largely from the series on the chemistry of water edited by Franks (1982, and preceding volumes). For further references, see Eisenberg and Kauzmann (1969), Horne (1969, 1972) and Neilson and Enderby (1986).

The water molecule

Figure 15 illustrates (very schematically) the electron structure of a gaseous water molecule. This has a distorted tetrahedral shape (the perfect tetrahedral angle being 109°28') with two sp^3 hybrid sigma bonds between O and H. Recall that the electron configuration of O is $1s^2\ 2s^2\ 2p^4$ (Fig. 15A). In the H_2O molecule the H-H repulsion increases the H-O-H angle from 90° to 92°. Shrinking of the two orbitals opposite the H atoms ($+p_y$, $-p_x$) and other interactions further increases this angle to 104°40' (nearly tetrahedral). The $2s^2$ and $2p_z^2$ electrons are not bonded to hydrogen and are left behind as two "lone-pairs" in approximate tetrahedral positions (Fig. 15B).

Water structure, hydrogen bonding and polarity

These lone-pair electrons cause much of the anomalous behavior of water. For example, they interact with electron-deficient species (such as cations). Perhaps the most important interaction is electrostatic bonding with hydrogen atoms in other water molecules. In the structure of ice, for example, all lone-pairs are bound to hydrogens in neighboring water molecules, and the lattice consists of a puckered, rigid network of $O\text{-}H_4^-$ tetrahedra analogous to silicate structures (Fig. 15C). In liquid water, hydrogen bonds are formed and broken, with a strength as high as 1/10 that of the sigma H-O molecular bond in the individual water molecules themselves. This imparts a structure to water consisting of random "flickering clusters" of hydrogen-bonded molecules (the Frank model), bending hydrogen bonds (Pople's model) and/or "iceberg" structures where individual molecules fill the interstices of broken-down ice structures (the Bernal and Fowler model). These models and their variations are reviewed by Horne (1969). In other words, liquid water retains some of the tetrahedral structure characteristic of ice through lone-pair interaction and hydrogen bonding. Hydrogen bonding accounts for many of the anomalous properties of water, many of which have tremendous geological implications: high heats of fusion and vaporization and high heat capacity (regulating effect on climate); high surface tension (capillary movement in soils, sediments and living organisms); high melting and boiling points (without

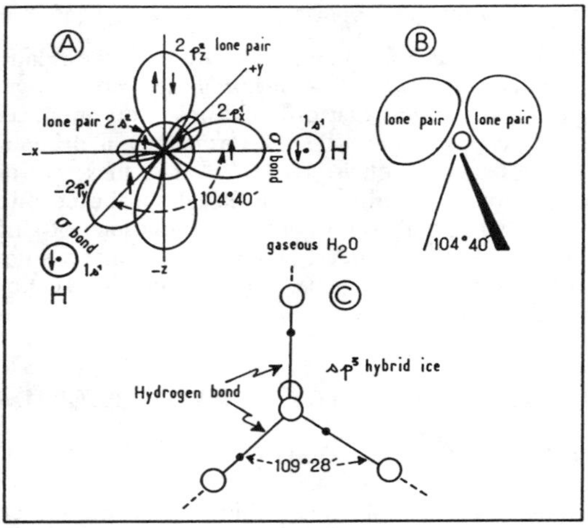

Figure 15. Schematic structure of the water molecule. (A) Electron orbitals just prior to formation of gaseous water showing 2s and 2p lone pairs and distortion of the 2p orbitals. (B) Gaseous water molecule showing near-tetrahedral H-O-H angle and two lone-pair distributions. (C) Tetrahedral structure of water in ice showing H-bonds. The hydrogens are constantly moving and, in general, no two hydrogens are at the same distance from any oxygen at one instant.

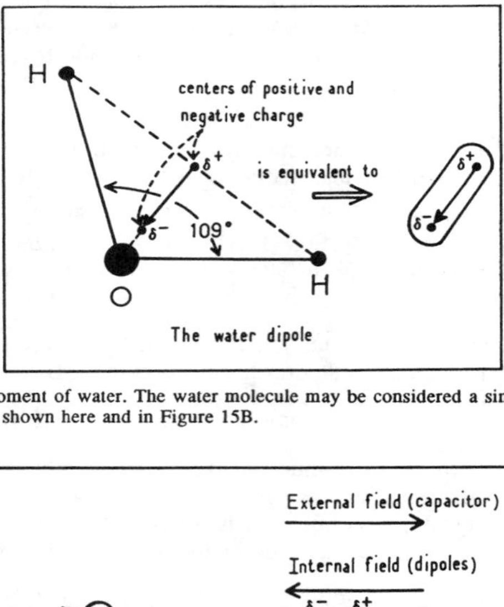

Figure 16. The dipole moment of water. The water molecule may be considered a simple electrical dipole because of the charge distribution shown here and in Figure 15B.

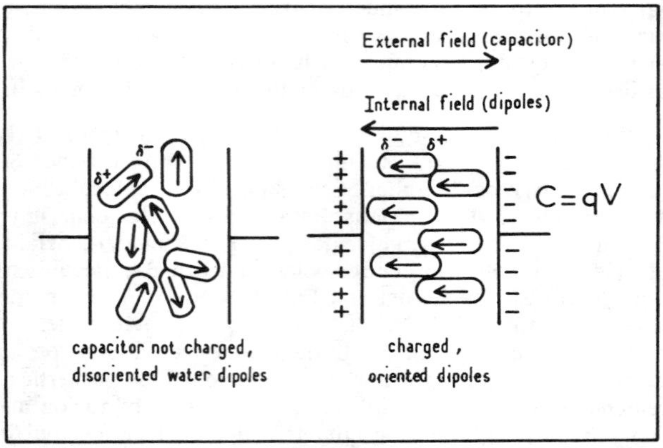

Figure 17. The dielectric constant of water, illustrated by the orientation of water molecules in an electrical field imposed between two capacitor plates. Oriented water dipoles create an internal field which opposes the external field of the capacitor and reduces net voltage across the plates. The dielectric constant of water is defined as the voltage across the plates when separated by a vacuum divided by the voltage when separated by water.

hydrogen bonding, water would be a gas at ambient temperature like NH_3, and Earth would have a hot, aqueous atmosphere); a local density maximum at 4°C (ice floats and insulates underlying water, but if it sank like solid NH_3, most regions of the Earth's oceans would freeze solid). An equally important property attributable to the same lone-pair structure is that water is quite polar. By this we mean that free water molecules (in liquid or gas) have negative and positive directions, or behave like electrical dipoles, as illustrated in Figure 16. The dipole moment of free gaseous water is 1.83×10^{-18} e.s.u., and this increases when associated with other molecules as in liquid water. By comparison, the dipole moments of other molecules, in units of 10^{-18} e.s.u. are: CO_2 = 0 (linear O-C-O molecule); CCl_4 = 0 (perfectly tetrahedral, sp^3 covalent molecule, no lone-pairs); NH_3 = 1.3 (distorted tetrahedron, sp^3, one lone-pair); H_2S = 1.10 (similar to water, two lone-pairs, but $3s^2$ $3p^4$ configuration); HCl = 1.08 (ionic); CsCl = 10.42 (extremely ionic).

Dielectric constant of water

Because of this high dipole moment, water molecules align themselves in an electrical field, as illustrated in Figure 17. Think of these as the two plates of a capacitor at voltage V with stored charge q. As shown, the alignment of water dipoles sets up an opposing field, so that the net potential across the plates decreases. In fact, the individual water molecules become even more polar in an electrical field, and their effect is accentuated by hydrogen bonding. For the same stored charge on the capacitor plates, you would find that if you started with a vacuum between the plates and then added water, the voltage across the plates would drop by a factor of 78.47 at 25°C. Recalling that capacitance is $C = q/V$, we define the dielectric constant of the medium between the plates relative to a vacuum as $C_{diel}/C_{vac} = V_{vac}/V_{diel}$. At 25°C and 1 bar, the dielectric constant of water (78.47) is extremely high, exceeded by only a few geologically unlikely liquids (such as pure hydrocyanic acid). For comparison, the dielectric constant of CCl_4 is 2.2, pure HCl is 4.6, H_2S is 9.1 (at −78°C) and pure NH_3 is 16.9.

Solvating power of water

Coulomb's law. There are several main reasons why water is an unusually good solvent for ionic (electrolytic) compounds such as NaCl. These all relate to the two lone-pair electrons and polarity of the water molecule itself; one of the most important effects stems from the dielectric constant of water. First, recall that Coulomb's law for the electric force between two charges in a medium of dielectric constant D is

$$F = \frac{q_1 q_2}{D r^2} \quad , \tag{10}$$

where r is the separation between the charges. The higher the dielectric constant, the lower the force of attraction between anions and cations. Thus, tightly bonded ionic crystals, such as NaCl break apart and ionize to Na^+ and Cl^- in water. By contrast, they remain associated in non-polar solvents such as CCl_4 and have much lower solubilities.

Hydration. A second, related effect, is that water dipoles align themselves around charged ions as shown in Figure 18, forming hydration shells. The physical picture here is called Gurney's co-sphere model of ionic hydration (see Gurney, 1962; Franks, 1973). In the primary, innermost hydration layer, called zone I, water molecules are relatively fixed (not free to rotate), are more compact than in normal water, and therefore have lower specific volume and entropy. In the next shell, zone II, water molecules are disoriented, pulled one way by the field of zone I and another by the flickering field of outside bulk water; thus zone II water has higher volume and entropy (disorder) than pure water. We should emphasize that the hydrated species itself is dynamic and that all water molecules are constantly moving and exchanging with a half-life as low as pure water structures (10^{-10}–10^{-11} sec.). In general, cations tend to be more highly hydrated than anions of similar charge and size, because the positive region of the water dipole is more dispersed than the negative (as you can see in Figure 16). Small, highly charged (Z) cations such as Li^+,

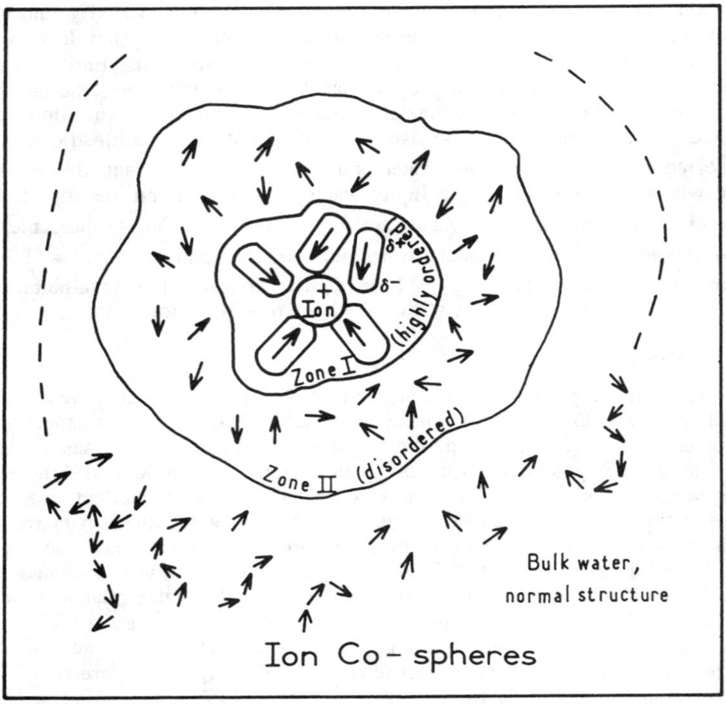

Figure 18. Gurney's co-sphere model of ion hydration. Zone I is an inner region of water dipoles relatively fixed about the ion. Zone II is intermediate and more disordered than bulk water. Water outside this zone is shown with a flickering cluster structure caused by H-bonding.

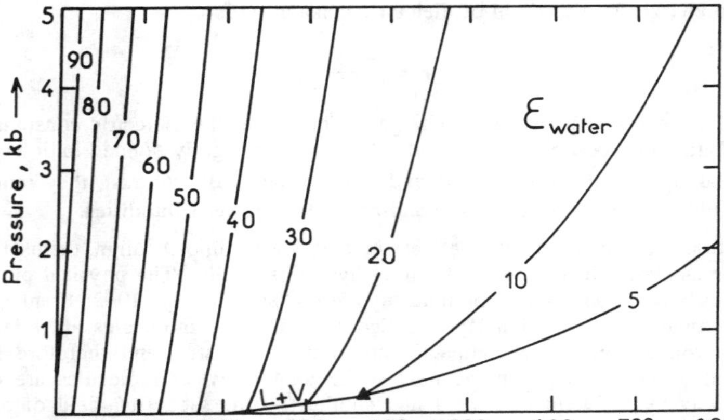

Figure 19. The dielectric constant of water as a function of temperature and pressure. The triangular spot marks the critical point of water. Data sources are cited by Seward (1981) and Eugster (1986).

Na^+, H_3O^+, Ca^{2+}, Ba^{2+}, Mg^{2+} (high Z/r) tend to be most highly hydrated, have a predominant zone I, and are called electrostrictive structure-making ions. These tend to decrease net solution entropy (increase order), to lower total density and to increase viscosity. In contrast, larger cations with low charge and some anions tend to have the opposite effect and are called structure breakers (K^+, NH^{3+}, Rb^+, Cs^+, Cl^-, Br^-, NO_3^-, BrO_3^-, IO_3^-, ClO_4^-).

These kinds of qualitative models need to be applied with care, but do account for observations such as the low viscosity of K^+, Rb^+ and Cs^+ salt solutions (lower than pure water) and the low electrophoretic mobility (movement in electrical fields) of Li^+. With the transition metals, spectroscopic studies mentioned below indicate that the primary hydration shell is bound in specific geometries, forming true molecular entities or complex ions.

The two combined properties of hydration and dielectric constant help explain the extraordinary solvating power water has for ionic compounds, at least at ambient T and P. Dissolved ions are effectively shielded or insulated from each other by their hydration shells and by the high dielectric constant of the solvent between each hydrated complex.

Solvation energies. As might be expected, there have been a great many attempts to quantify these effects, many summarized again in the volumes edited by Francks (1982 and preceding), and by Helgeson et al. (1981). Two of the oldest and more successful approaches have also received much use in geochemistry. The first is the Debye-Hückel equation for activity coefficients, which predicts the non-ideal (electrostatic) free energy of interaction based on Coulomb's law. The second approach is the Born Equation (1920) for the free energy change associated with removing an ion of radius r_i and charge Z_ie from a vacuum and placing it in a solvent of dielectric constant D:

$$G = -N_A \frac{(Z_ie)^2}{2r_i}\left(1 - \frac{1}{D}\right) . \tag{11}$$

Here N_A is Avogadro's number, so the equation applies per mole of ions. This comes directly from Coulomb's law (10), and despite its overly simplistic picture, is in reasonable agreement with experimental observation at 25°C (Bockris and Reddy, 1970, p. 69); growing evidence indicates it works even better at higher temperatures (e.g., Tremaine et al., 1986, and references therein). It is an important component of the model used by Helgeson et al. (1981) for electrolytic solutes in hydrothermal solutions. Note that since the dielectric constant of any medium is always greater than that of a vacuum, D>1 and the Born free energy (11) will always be negative; this also predicts that the free energy of ionic solvation should become more negative for smaller, more highly charged ions in solvents of higher dielectric constant. For singly charged ions of radius 1.5 Angstroms, Equation (11) predicts a free energy of solvation of roughly -100 kcal $mole^{-1}$ at 25°C. These large negative free energies help explain the stability of ions in aqueous solutions even without taking into account the additional stabilizing effect of hydration shells.

Effects of temperature and pressure on the dielectric constant

The chemical behavior and solvating ability of water changes enormously with temperature and pressure. This is perhaps best illustrated by the changes in a fundamental property such as the dielectric constant shown in Figure 19. Notice that at liquid-vapor equilibrium the dielectric constant falls by almost an order of magnitude from O°C to 374°C, the critical point of pure water. Under supercritical conditions the dielectric constant is 25 or less; at low temperatures it is high and almost independent of P, and at high temperatures, it increases with P.

Temperature. The effect of temperature can be explained, to a good approximation, by the Kirkwood equation, derived and discussed in detail by Bockris and Reddy (1970, p. 152):

$$\frac{(D - 1)(2D + 1)}{9D} = \frac{4\pi n}{3}\left[\alpha + \frac{\mu^2(1 + g\cos\gamma)^2}{3kT}\right] . \tag{12}$$

This includes the effects of molecular clusters (bound by hydrogen bonds) orienting in an electric field. Here α is a measure of the degree to which an electric field induces a dipole in any one molecule; g is the number of nearest-neighbor water molecules linked with a central water molecule as a cluster; cos γ is the average of the cosines between the dipole moment of the central molecule and those of its neighbors in the same cluster. This illustrates several important effects. First, the dielectric constant should decrease at higher temperatures simply from the inverse relationship in (12). Increasing temperature at constant P will also lower the quantities g and cos γ by breaking hydrogen bonds and disorienting individual molecules, causing a further decrease in D. This is expected intuitively because at higher temperatures molecular vibrations (including rotations and translations) increase, and the molecule is less capable of aligning itself in an applied electric field. Equation (12) also illustrates the importance of hydrogen bonding which increases the size of orientable clusters, g, and of cos γ. As an example, the dipole moments of liquid H_2O and SO_2 are similar, at 1.83 and 1.67×10^{-18} e.s.u., respectively. However, because water is hydrogen bonded and SO_2 is not, their dielectric constants at 25°C are 78.5 and 12.35.

Pressure. The effect of pressure on the dielectric constant of water may be predicted by analogy with Equation (12). At higher pressures individual molecules squeeze closer together, increasing interactions such as hydrogen bonding; this increases the dielectric constant. The detailed effects of pressure have been summarized by Millero (1971), Helgeson and Kirkham (1974, 1976), Seward (1981) and Eugster (1986). Of primary importance is the electrostrictive volume decrease due to ion hydration and collapse of water structure in the hydration shell (zone I). Again, this can be modeled very simplistically by Coulomb's law (10) for electrostatic interaction of water dipoles and charged ions. Differentiating the Born Equation (11) with respect to pressure leads to the expression

$$\overline{V}_{ion(electrostriction)} = \frac{Z^2e^2}{2Dr}\left(\frac{\partial \ln D}{\partial P}\right)_T \quad . \tag{13}$$

Rather surprisingly, this was derived before the Born equation by Drude and Nernst (1894). This gives the theoretical contraction of solvent of dielectric constant D, around a sphere of radius r with total charge Ze. This is a conservative estimate because it does not include specific solvent collapse and structuring in the primary hydration shell. For a simple ion of unit charge, Equation (13) predicts a volume decrease of roughly 10 cm^3 $mole^{-1}$ due simply to the electrostatic constriction of water around the ion (Seward, 1981, p. 119). For further details, see Hamann (1981).

The effect of this electrostrictive volume decrease around ions is to increase the tendency toward ionization at higher pressures. We guessed a similar result above by analogy with Equation (12); higher pressure should raise the solvent dielectric constant, thereby increasing ionization.

We have purposely chosen a very simple picture of electrostatic interactions in the above discussion, with Equations (11) through (13) following directly from Coulomb's law (10). Given what we have also said about water structure and solute hydration along with some of the effects to be considered below, the Coulombic view of hydrothermal solutions seems almost absurdly unrealistic. However, this simple electrostatic picture works to a first approximation for salt solutions that are not too concentrated, and it is an essential part of other more rigorous models. For example, the Debye-Hückel equation for activity coefficients of dissolved ionic species assumes Coulombic interactions. The more detailed expressions for activity coefficients derived by Pitzer and his colleagues (e.g., Pitzer, 1979; Pitzer and Weare, this volume) add a virial equation for non-ideal interaction to the basic Debye-Hückel equation. The Born Equation (11) and Debye-Hückel equation are the two primary components of the model proposed by Helgeson et al. (1981) for activity and osmotic coefficients of hydrothermal solutions (see also Sverjensky, this volume). The fact that Coulombic models work at all attests to the importance of electrostatic interactions between ions and water dipoles. For present purposes these models provide a simple and intuitively satisfying way of thinking about systems that in reality must be extraordinarily complicated.

Effects of temperature and pressure on ionization

With this kind of reasoning we can predict that electrostatic interactions should increase at lower pressures and higher temperatures where the dielectric constant of water is reduced. This means that we should expect considerably increased ion-association for such conditions along with a tendency towards formation of electrically neutral or low-charge ion pairs and complex ions. Highly charged species such as $CuCl_4^{3-}$ are unstable at these lower dielectric constants, and electrically neutral complexes such as $CuCl^\circ$ and $PbCl^\circ_2$ predominate.

Similarly, association reactions such as

$$Na^+ + Cl^- \rightarrow NaCl^\circ \quad (14)$$

will increase, and compounds that we think of as strongly ionized at room temperature may become almost entirely associated at high T and low P. Compounds such as HCl which are strong acids (highly ionized) at ambient P and T, become weak to moderate acids at high T and low P. As an example, the dissociation constants of $NaCl^\circ$ and HCl° are plotted in Figure 20 as a function of P and T. At constant P, these decrease exponentially with T; at constant T, they increase with P, just as predicted above. Similarly, Figure 21 shows the temperature variation of dissociation constants for several geologically important acids at vapor-saturated pressure; all become more associated at higher T (including H_2O, which rises above 11.5 above 350°C). At the lower pressures of these vapor-saturated conditions, the effect of temperature predominates the change in dielectric constant and ionization, and pressure effects are negligible (see Fig. 19). By contrast, at very high pressures ionization may become extreme: it has been estimated, for example that water itself almost completely ionizes to OH^- and H_3O^+ at pressures in excess of 200 kbar and 800°-1000°C (Franck, 1981; Hamann, 1981). Under these conditions water would resemble a fused salt or ionic fluid, with interesting implications for behavior in the mantle.

Other effects of pressure and temperature on water-solute interactions

While changes in the solvent dielectric constant are extremely important, other factors also control changes in the chemical behavior of hydrothermal solutions with temperature and pressure.

Molecular vibration. At higher temperatures, molecular vibrations increase, making molecules such as the $NaCl^\circ$ species more likely to dissociate, and reversing the trend of Reaction (14). The two competing effects at higher T of decreasing dielectric constant (lowers ionization) and increasing vibration (increases ionization) account for the maxima commonly observed in the dissociation constants of ionic compounds as you can observe in Figure 21 (this was discussed at length 20 years ago by Helgeson, 1967).

Ligand field stabilization. With transition metals dissolved as aqueous complex ions, an additional stabilizing energy is produced by splitting of electronic d-orbitals, as illustrated in Figure 22. This is termed ligand-field stabilization energy (LFSE) and is proportional to $1/r^5$, the average separation between anions and cation in a complex ion (Dunn et al., 1965, p. 12). As we shall see below, this is a significant control on the behavior of transition metal ions in hydrothermal solutions, affecting the relative stabilities of the different metal complexes across each transition row. However, as predicted by the r^{-5} proportionality, the effect becomes less important at higher T and lower P. This is the P-T region where ion association is most intense because the dielectric constant is lower; the effect of LFSE might therefore be partly masked by important electrostatic, ionic interactions under the high-T, lower-P conditions typical of shallow ore-fluids at or near magmatic temperatures. While some preliminary work has been done on this problem for ore-forming conditions, much remains to be learned in this potentially quite important area (see discussions by Buback, 1981; Susak and Crerar, 1985; and Buback et al., 1987).

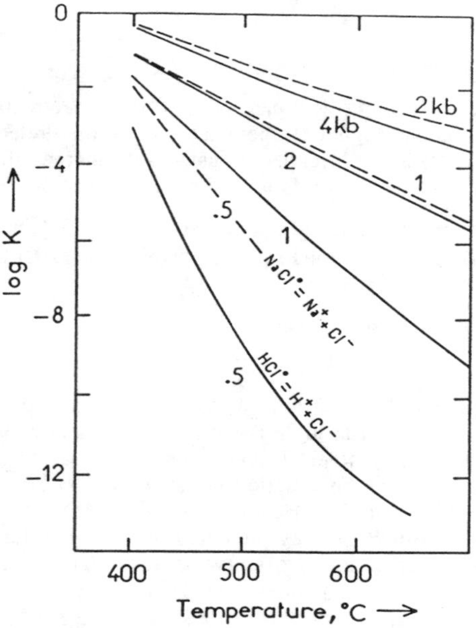

Figure 20. Dissociation constants of HCl and NaCl as a function of temperature and pressure. From Eugster (1986) after measurements by Quist and Marshall (1968) for NaCl, and Frantz and Marshall (1984) for HCl.

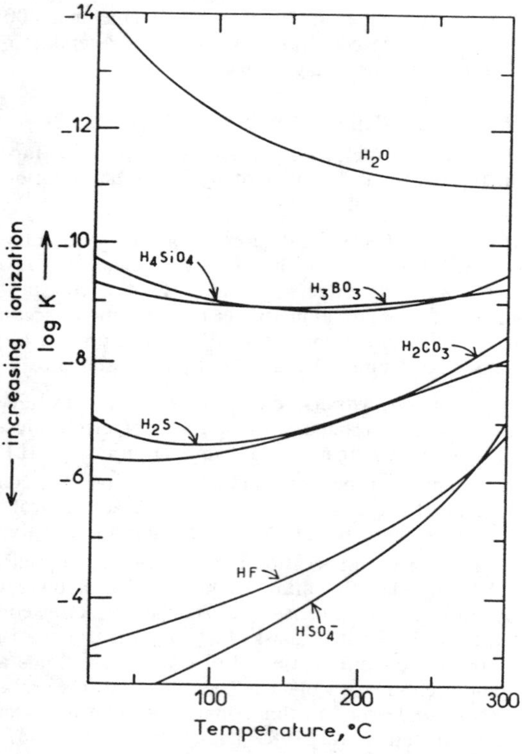

Figure 21. Variation of selected acid ionization constants with temperature at saturated water vapor pressures. Note tendency towards a maximum at intermediate temperatures. From Ellis and Mahon (1977).

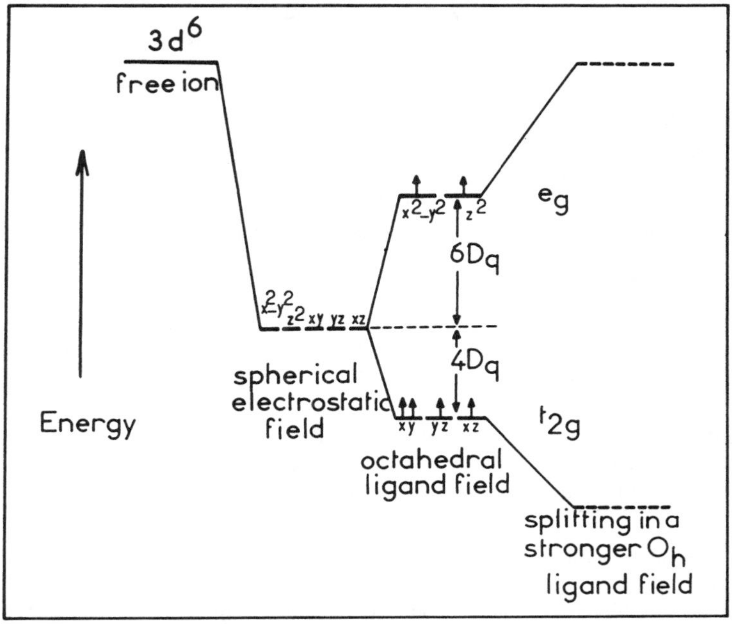

Figure 22. Splitting of 3d orbitals into two separate energies (e_g and t_{2g}) in an applied ligand field (such as occurs in transition metal complex ions). Note that 10Dq (the difference between the two new energy levels) depends strongly on the strength of the ligand field (hence on ligand type and metal-ligand bond length), and can change dramatically as shown to the right. Illustrated here with a high-spin d_6 ion such as Fe^{2+}. From Crerar et al. (1985).

Pressure-induced electron spin-pairing. Another related phenomenon is the possibility of an electronic transition from high- to low-spin state at elevated pressures (and with ligands which impose strong electrical fields). Figure 22 is drawn for the normal, high-spin case in which all electrons fill orbitals in accordance with Hund's rules. Under high pressures (or field strengths) it is possible to raise the upper e_g energy level illustrated here so far that electrons spin-pair and fill the lower level first, giving a low-spin configuration. The LFSE of high-spin Fe^{2+} shown here is $(2 \times 6Dq) - (4 \times (4Dq)) = -4Dq$; for the low-spin case, all six electrons would occupy the lower t_{2g} orbital with a total LFSE of $-6 \times 4Dq = -24Dq$. The quantity Dq is a measure of the field strength and is on the order of 5 kcal mole^{-1} (or higher at higher pressures, Crerar et al., 1985). Hence low-spin LFSE could be -120 kcal mole^{-1} or higher as opposed to high-spin which is only about -20 kcal mole^{-1} for this ion. Therefore, transition to the low-spin state at higher pressures is potentially very important, capable of stabilizing complexes and increasing total solubility of complex-forming minerals. The low- to high-spin transition is also accompanied by a significant decrease in partial molal volume (Seward, 1981). Unfortunately, once again, there have been no experimental studies of geologically interesting systems under appropriate conditions. Several geochemical laboratories have the experimental capability at the present time, so we may hope for direct measurement of this effect in the near future.

TRANSITION METAL COMPLEX IONS

It is now recognized that complex ions are responsible for transport of transition metals in hydrothermal solutions. Aqueous transition-metal complex ions can be regarded as well-defined molecular entities having a specific geometry and coordination number and co-ordinate bonding described by ligand-field/molecular-orbital theory (Figgis , 1966; Huheey, 1978; Crerar et al., 1985). The chemical controls on metal complex behavior in natural systems have been summarized in several recent reviews (Barnes, 1979; Seward, 1981; Crerar et al., 1985; Eugster, 1986), and we will only touch the more important points here.

Geologically important ligands

The metal in a complex ion is bound to, or coordinated by, ligands which serve as electron donors to the molecule. The potential ligands of most importance geologically have been discussed at length by Barnes (1979). These include: Cl^-, OH^-, HS^- and, of course, H_2O (probably the four most important); other probable ligands include organic acids, NH_3, F^-, S_x^{2-}, $S_2O_3^{2-}$ and HCO_3^-. Spectroscopic studies show that the transition metals are usually coordinated to water so that the formula for an octahedral one-chloro Cu(I) complex, for example, should really be written as $Cu(H_2O)_5Cl^\circ$. Available ligands are always competing with water for coordination sites about transition metal ions in aqueous solutions, and as we have seen above, water is a strong complexer, forming stable hydration shells.

Recent studies show that the speciation of natural hydrothermal systems may be much more diverse and complicated than was generally believed a decade or more ago. Based on a comprehensive solubility study of ten ore-forming minerals in hydrothermal solutions, for example, Scott Wood has recently concluded,

> "Our data underscore the highly complicated nature of ore-forming solutions. No single complex or species can be expected to predominate for any metal over reasonable ranges of solution composition and temperature, many different ligands may be significant in any given solution, different metals are likely to be transported by quite different mechanisms..., and mixed-ligand and perhaps also polynuclear species can be expected. In the mid-temperature range (approximately 200°-400°C) the most complicated solution chemistry and speciation is to be expected. At lower temperatures

Table 1. Classification of geological metals and ligands‡

Hard Acids	*Borderline Acids*
H^+, Li^+, Na^+, K^+, Rb^+, Cs^+	Fe^{2+}, Co^{2+}, Ni^{2+}, Cu^{2+}, Zn^{2+}
Ca^{2+}, Mg^{2+}, Ba^{2+}, Ti^{4+}, Sn^{4+}	Sn^{2+}, Pb^{2+}, Sb^{3+}, Bi^{3+}, SO_2
MoO^{3+}, WO^{4+}, Fe^{3+}, Al^{3+}, CO_2	
Soft Acids	
Cu^+, Ag^+, Au^+, Cd^{2+}, Hg^+, Hg^{2+}, M° (metal atoms and bulk metals)	
Hard Bases	
NH_3, H_2O, OH^-, CO_3^{2-}, NO_3^-, PO_4^{3-}, SO_4^{2-}, F^-, Cl^-	
Borderline Bases	*Soft Bases*
Br^-	CN^-, CO, H_2S, HS^-, I^-

‡According to relative hardness. Condensed from Huheey (1978).

Relative hardness of common metal ions and ligands*

$F^- > Cl^- > Br^- > I^-$	$Zn^{2+} > Pb^{2+}$
$Cu^+ > Ag^+ > Au^+$	$H^+ > Li^+ > Na^+ > K^+ > Rb^+ > Cs^+$
$Zn^{2+} > Cd^{2+} > Hg^{2+}$	$As^{3+} > Sb^{3+} = Bi^{3+}$

*Hardness decreases toward the right.

major electrolyte components are generally more dissociated and ion-pairing is less important. However, true inner sphere complexes remain important for some metals. At higher temperatures, there appears to be a simplification of metal complexes and a preponderance of species with neutral charge; the major electrolyte components (NaCl, KCl, $CaCl_2$) become essentially fully associated as ion pairs." (Wood et al., 1987).

Chemical controls

The ionic behavior of all transition metals is governed by d-orbital chemistry. There are four main chemical controls (Crerar et al., 1985): (1) Bonding changes from predominantly ionic to more covalent across each transition row from d^0 to d^{10} ions. (2) There is a general increase in electrostatic interactions with anions across each row (because ionic potential Z/r increases and ionic radius r decreases from Ti^{2+} to Cu^{2+} somewhat like the lanthanide contraction). Hence, complexes formed with a common ligand increase in stability from left to right across each row. (3) The splitting of d-orbitals mentioned above (Fig. 22) adds an additional LFSE (stabilization energy) to complexes formed by cations with configurations other than d^0, d^5 and d^{10}. (4) A relativistic effect dramatically increases covalency down each column of the heavier transition metals. Combined with rule #1, this means that the heavier d^{10} cations such as Au^+ and Hg^{2+} should form the strongest covalent complexes.

Hard-soft behavior

The interaction between specific metal ions and ligands (coordinating species, such as Cl^- or H_2O) can be regarded as acid-base reactions, with metal and ligand acting as electron acceptor and donor, respectively. In predicting which metals form complexes with which ligands, it is very helpful to use the hard-soft classification of Pearson (1963) and others (Ahrland, 1968, 1973; summary by Huheey, 1978). Class-A (or hard), metals and ligands are generally small, highly charged, and are only slightly polarizable. Class-B (or soft) species are large, relatively low in charge and highly polarizable. Hard species behave ionically, and soft species more covalently. The important point for present purposes is that, given competition between several ligands and metals, soft metals bind preferentially to soft ligands and hard metals bind with hard ligands. Table 1, from Crerar et al. (1985), classifies geologically important metals and ligands as hard, soft or borderline. Using this table we might guess, for example, that soft ligands such as HS^- might form relatively strong, predominantly covalent complexes with Hg, Au, Ag, Cu and Sb (all soft), weaker complexes with Pb and Zn (borderline), extremely weak complexes with Fe and Sn, and probably do not complex at all with W and Mo (hard). Borderline ligands such as Cl^- fall between the cracks and should form relatively stable complexes with most transition metals (except for the d^0 ions Sc^{3+}, Ti^{4+}, etc., which tend to form very weak chloride ion pairs). Finally, we should not expect stable complexes with mixed ligands if the ligands differ considerably in hardness; as an example, HS^- is much softer than Cl^- or OH^- and we should not expect mixed metal-Cl^--HS^- or M-OH^--HS^- species to be geologically important.

All transition metals should display increased type-A (hard acid) behavior at higher temperatures (and lower pressures) where electrostatic interactions increase as noted above. Thus complexes with intermediate or hard ligands such as Cl^- and OH^- should become more stable at higher T. This prediction is borne out by the observed increase in hydroxy complexing (Khodakovskiy and Yelkin, 1975; Baes and Mesmer, 1981), and the increased stability of chloro complexes, with temperature (Crerar et al., 1978; Barnes, 1979; Seward, 1981, 1984; Ruaya and Seward, 1986).

Electronegativity, LFSE and ionic potential. The effect of these three important variables on transition metal behavior is summarized in Figures 23 and 24. Try to picture these two diagrams as a single three-dimensional plot with axes Z/r, electronegativity and LFSE, since all three variables apply simultaneously. Ionic potential, Z/r, is a measure of the relative strength of electrostatic interaction and increases from left to right in each transition row (as in rule #2, above). Electronegativity gives the relative degree of ionic or covalent bonding. LFSE is determined by the degree of splitting of metal d-orbitals in the presence of ligands (Fig. 22) and stabilizes complexes with configurations other than d^0, d^5 or d^{10}.

It is possible to outline four general regions on Figure 23 where different types of complex ions predominate. As we have mentioned above, bisulfide (HS^-) will preferentially complex the most electronegative, lower Z/r metals. We would expect this to be particularly important for Au^+ which, as you can see from this figure, is anomalously covalent; it is thought that Au and its neighbors Pt, Ag, Pb, Bi and Sb are so electronegative because of an interesting relativistic property summarized by Crerar et al. (1985) (the velocities of the inner s and p electrons of these metals approach the speed of light, their orbitals contract and more effectively shield the outer d and f orbitals). In fact, strong $Au(HS)_2^-$ complexes are known to exist in hydrothermal systems (Seward, 1973). Based on limited available data for gold chloride systems, Seward (1983) estimated that $Au(HS)_2^-$ complexes should be four orders of magnitude more concentrated than $AuCl_2^-$ complexes at near-neutral pH, total Cl = 1.0m and total S (reduced) = 0.05 m. Somewhat less dramatic differences might be expected with Ag+ and Hg^{2+} based on Figure 23. However, we should emphasize that at higher temperatures, strong bisulfide complexing is expected only with these most covalent metals. In general, bisulfide complexing should be less important for all other metals of intermediate or lower electronegativity; aside from its soft base behavior, the HS^- species predominates relative to $H_2S°$ (a weak ligand) only at basic pH at temperatures above 25°C (e.g., Crerar and Barnes, 1976). Instead, for most transition metals, Figure 23 shows that chloride and/or hydroxy complexes predominate. These borderline to hard bases preferentially bond to metal cations of intermediate ionic potential and electronegativity. Hence the most important transition metal species in many geological environments are probably OH^- and Cl^- complexes. For the extreme case of metals with very high ionic potentials, strong oxyanions such as molybdic and tungstic acids (H_2MoO_4, H_2WO_4) are expected instead of chloro or even aquo complexes; here metal-oxygen bond strengths exceed that of the H-O bond in water, and oxygen is effectively stripped from the water molecule itself.

Figure 24 is the parallel diagram for LFSE versus ionic potential, and we can broadly delineate two fields for ions that either commonly or rarely form hydrothermal ore deposits. This diagram seems paradoxical at first sight because it indicates that metals with the highest LSFE's (which should form the most stable aqueous complexes) are least likely to form hydrothermal ores. In fact, as noted by Crerar et al. (1985) only four of the ten metals in the first transition row (from Sc to Zn) typically form large hydrothermal deposits: Mn, Fe, Cu and Zn. These four metals have the lowest (or zero) LFSE's of the first transition row. The probable explanation is that the remaining six metals have even greater LFSE's in minerals or magmas and are not as easily leached by hydrothermal solutions in the first place.

Of the 30 transition metals only 9 or 10 commonly form sizable hydrothermal deposits (Mn, Fe, Cu, Zn, Mo, Ag, W, Au, Hg and occasionally Co), and this in no way correlates with average crustal abundance. This interesting problem needs further research. Much can probably be learned about the metals that do form ores from the chemistry of those that do not. Figure 24 suggests that LFSE is one of the more promising chemical properties on which to focus initial attention. A possible relationship between LFSE and porphyry copper mineralization has already been recognized by Feiss (1978); he showed that economic mineralization correlates with the $Al_2O_3/(K_2O + Na_2O + CaO)$ ratio of plutons in the American southwest. High Al/alkali magmas have more liquid octahedral structures (with high LFSE's); crystallizing plutons comprising such magmas might be able to retain metals longer, perhaps to the point of water saturation. Susak and Crerar (1982,1985) have suggested that the coordination and structure of metal ion complexes might help control deposition; there is some preliminary evidence that large deposits correlate with conditions that produce complexes of tetrahedral (or lower) coordination—see Figure 25 and discussion

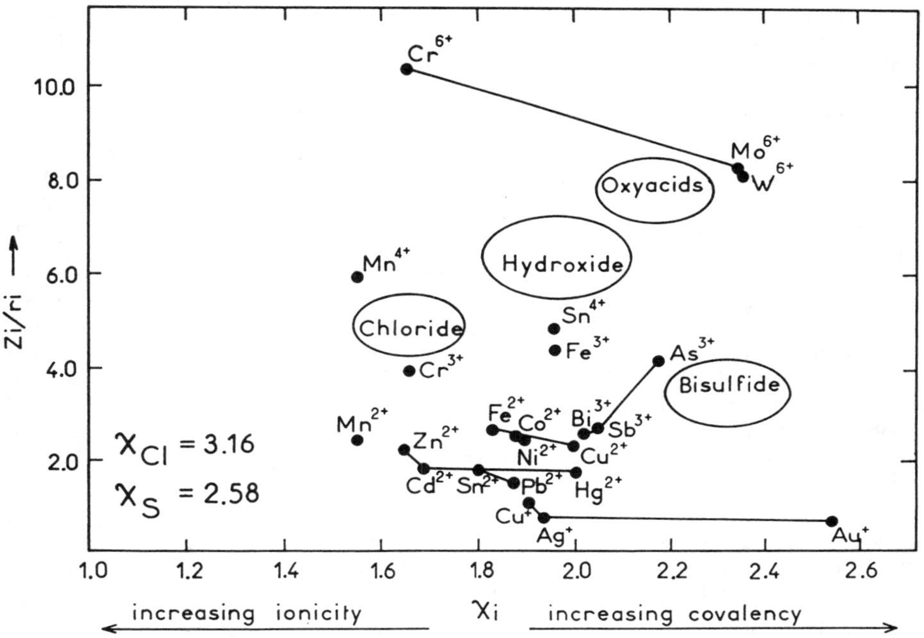

Figure 23. Plot of ionic potential (Z/r) versus Pauling electronegativity for selected ions. Crystal radii and electronegativities from Huheey (1978). Ions in selected columns of the periodic table are connected by solid lines, as are ions of the first transition series. Note that the ions group into four overlapping, general fields forming oxyacids, hydroxide, chloride and bisulfide complexes. From Crerar et al. (1985).

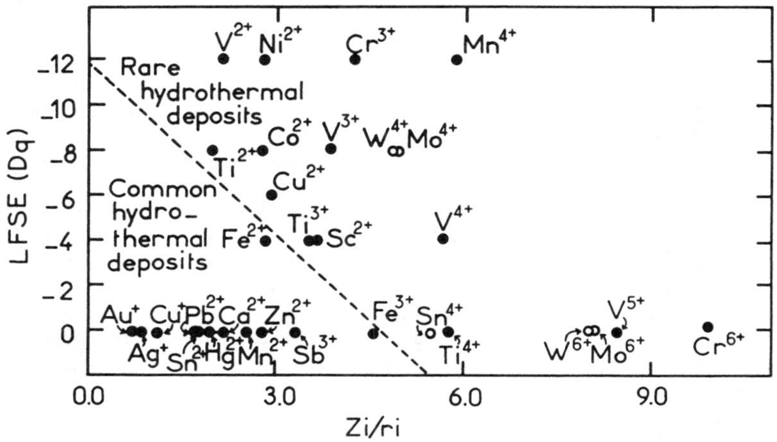

Figure 24. Ligand field stabilization energy (LFSE) versus ionic potential (Z/r) for selected metals. The valences shown are common in geological systems. A rough discrimination is drawn between metals that either commonly or rarely form large hydrothermal ore deposits. The tetravalent Sn, Mo and W ions are exceptions since they are heavy and more electronegative; on a three-dimensional plot including electronegativity, LFSE and Z/r, these three metals fall in the field of hydrothermal ores. From Crerar et al. (1985).

below. Since metals with higher LFSE tend to form complexes of octahedral (and possibly higher) coordination, this is another potential clue to the inverse relation between LFSE and ore deposition indicated on Figure 24.

Why solubilities increase with temperature

There is a general tendency for the solubilities of most substances in water to increase at higher temperature. This is a matter of everyday experience, dissolving sugar or salt in hot water for example. It is true that there are some minerals that show an anomalous retrograde solubility (become less soluble in hot water), at least up to certain limiting temperatures; this includes carbonates (calcite, dolomite, strontianite, witherite) and some sulfates (anhydrite, celestite) (see Holland and Malinin, 1979). However, these minerals are exceptions. Magma-water element partitioning experiments summarized above suggest that some metals might reach concentrations on the order of 1 wt % in exsolving water, and high-temperature fluid inclusion data from mineral deposits occasionally indicates similar high metal content (Roedder, 1979; Kwak et al., 1986). In most ore mineral solubility studies, solubility increases at higher temperatures. For example, experiments by Whitney et al. (1985) and Hemley et al. (1986) with magnetite, and sulfides in NaCl-H_2O-quartz monzonite systems above 500°C gave metal concentrations (Fe, Mn and Zn) on the order of 1 wt %.

This is really a very complicated question, with different explanations for different minerals. Many variables such as pH, oxygen and sulfur fugacity and even solution density change with T, and these all influence solubility. However, we will single out a few general effects which are directly related to temperature from our preceding discussion.

First, molecular vibration increases at higher T, increasing the probability for molecular dissociation; this partly explained the maxima in ionization constants observed in Figure 21. For most metal oxides, sulfides and silicates, this effect should increase the solubility product (see for example, mineral solubility products tabulated as a function of temperature by Helgeson, 1969).

We have also observed that at higher T there is a general increase in electrostatic interactions because of the decreased dielectric constant of water. We noted that transition metals all show increased type-A (hard acid) behavior at higher T. Now the relative stability of any species is determined by its Gibbs free energy (more stable compounds have more negative free energies). This, in turn, is a balance between enthalpy and entropy

$$\Delta G = \Delta H - T\Delta S \quad , \tag{15}$$

with entropy becoming increasingly important at higher T because of the $T\Delta S$ term. As summarized by Ahrland (1968) and Seward (1981), soft-soft interactions are characterized by exothermic (negative) enthalpies and low entropies (possibly even negative, which indicates increased molecular order). Thus the free energy of these species is dominated by the large negative enthalpy of predominantly covalent bonds. The low to negative entropy is less significant, and arises from the decrease in the total number of particles (through complex formation) and ordering of ligands around metal ions. Disruption of water structure (which would cause an increase in entropy) is minimal because these type-B complexes do not display strong electrostatic interactions and are not heavily solvated by water dipoles. The energy required to displace water dipoles from the hydration shell by a coordinating ligand such as Cl^- must be less than that gained by the metal-ligand bond (with the end result being an exothermic process).

At higher T where the dielectric constant falls, the situation reverses and ion-solvent interactions become increasingly electrostatic. Type A (hard) interactions are characterized by large positive enthalpies and entropies of complex formation (Ahrland, 1968). Recall that from the Born Equation (11) and the related Drude-Nernst Equation (13) we now expect a major effect from the electrostatic attraction between metal cations and water dipoles as well as other coordinating ligands. This would occur largely in zone I of the co-sphere model in Figure 18, with concomitant disorder in zone II. The electrostriction of water around the metal ion and its displacement by coordinating ligands causes a large,

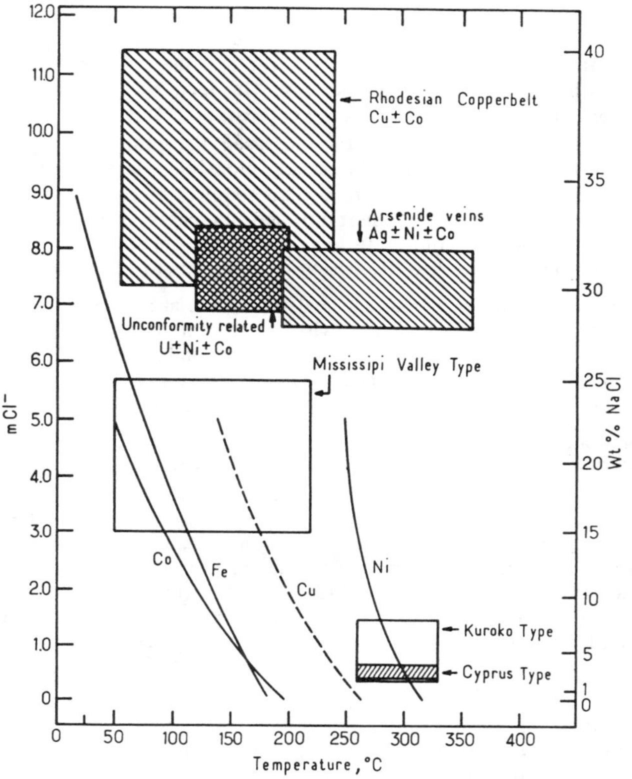

Figure 25. First appearance of tetrahedral complex ions for aqueous chloro complexes of four metals. Octahedral complexes predominate to the left, and tetrahedral complexes begin to the right of each line, with tetrahedral coordination eventually predominating at high T and/or Cl concentration. Data for Co(II) and Fe(II) from Susak and Crerar (1985) and Vogel et al. (1987), respectively. Lines for Cu(II) and Ni(II) are estimated by Susak and Crerar (1985). Environments of deposition for various types of ore deposits tend to occur in the tetrahedral region for each metal. Data on depositional fields are summarized by Susak and Crerar (1985).

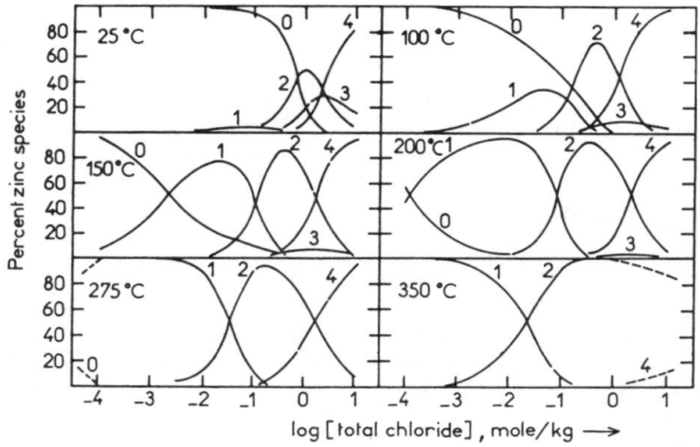

Figure 26. Distribution of aqueous chloro-zinc(II) complexes at different temperatures and total chloride concentrations, calculated from measured stability constants. The number of chloride ligands attached to each Zn nucleus is indicated on each curve (e.g., "1" refers to the $ZnCl^+$ complex). Notice that the one- and two-chloro species predominate at high T, and that speciation is much more diverse at lower T. From Ruaya and Seward (1986).

positive (endothermic) enthalpy. In other words, energy has to be put into the system to create this kind of inner sphere ordering and volume constriction (Nancollas, 1970). According to (15) the free energy would be positive and the complex unstable unless the entropy term overwhelms this positive enthalpy. In fact, the entropy term usually is sufficiently large, and dominates the free energy expression. Thus ligation by an anion such as Cl^- displaces waters in co-sphere I, causing increased disruption in co-sphere II (Fig. 18) and perhaps also in bulk solvent (which is already more disordered because of the higher temperature and decreased hydrogen bonding). This requires more energy than is gained by the metal-ligand bond (hence is endothermic) and increases disorder (positive entropy change). These changes in speciation with temperature are apparent in Figure 26 for aqueous chloro-Zn(II) complexes—see also discussion of complex stoichiometries below.

Of course, covalent molecular bonding is also possible at higher T (this is indicated by the Raman peaks of Figure 27, for example, which represent strong, predominantly covalent vibrations). The complex itself is a polar entity though, and at high T (and low dielectric constant), electrostatic interactions also occur. The net result is a complex doubly stabilized by both true molecular bonding and electrostatic attraction at high T.

These effects are summarized in Table 2. This shows the free energy, enthalpy and entropy of formation of the three neutral complexes $AgCl^\circ$, $PbCl^\circ_2$ and $ZnCl^\circ_2$, all of which predominate at 300°C-350°C at vapor-saturated pressures, from data of Seward (1981, 1984) and Ruaya and Seward (1986). These properties were derived from the observed variation of complex stability constants with T, and are not as reliable as quantities obtained directly by calorimetry, but the trends are significant. In all cases, the enthalpies and entropies of reaction increase at higher T; the free energy decreases, signifying greater stability; and the overall formation constants increase with temperature. This, coupled with increased mineral solubility products, results in higher solubility at higher temperature.

At high temperatures, water dissolves most minerals, whereas at low T it is a better solvent of ionic salts. As pointed out by Eugster (1986), low-T water tends to dissolve more covalent structures such as silicates only at low or high pH, whereas at high T most silicates dissolve readily and at roughly the same rate (Wood and Walther, 1983; Walther and Wood, 1986).

Note that large positive enthalpies are required for increased stability at higher T as predicted by the van't Hoff equation:

$$\left(\frac{\partial \ln K}{\partial T}\right)_P = \frac{\Delta H^\circ_r}{RT^2} \quad . \tag{16}$$

Species such as the $ZnCl_3^-$ which die out at higher T have exothermic (negative) enthalpies of formation. This illustrates the importance of electrostatic interactions at elevated T, since they produce large positive enthalpies.

RECENT EXAMPLES AND APPLICATIONS

It is no longer possible in a single article to adequately review all research involving metal-bearing hydrothermal solutions. Instead we will illustrate some of the chemical controls described above with recent experimental results. The examples are chosen to display some of the enormous diversity of the field as well as the many different experimental approaches now being used.

Molecular structures of complex ions

Aqueous transition metal complexes have clearly defined structures that can be determined by spectroscopic techniques. The two most successful approaches to date have been laser Raman spectroscopy and UV/Visible/Near-IR optical absorption spectroscopy. Raman

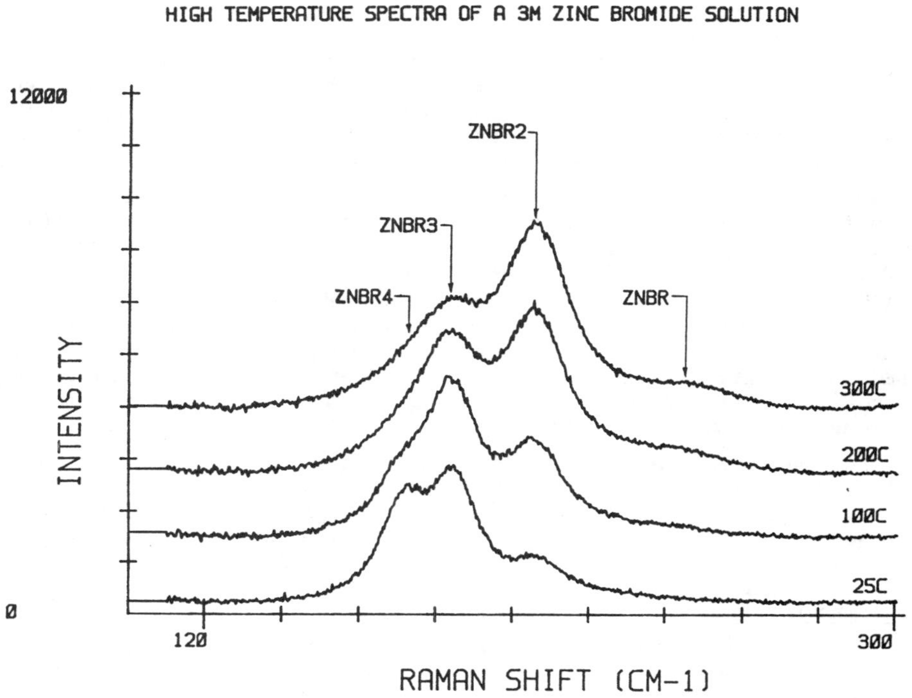

Figure 27. Raman spectra of aqueous zinc(II) bromide complex ions as a function of temperature. Peaks represent the totally symmetric vibration of each species. Notice that the dibromo complex predominates at high T, while speciation is more diverse at lower T. Spectra by Mary Yang, Princeton University.

Table 2. Enthalpy, entropy, Gibbs free energy and overall formation constants for the $AgCl^o$, $PbCl_2^o$ and $ZnCl_2^o$ aqueous complex ions from 25° to 350° C. Data from Seward (1976, 1984, 1986).

	$Ag^+ + Cl^- = AgCl^o$				$Pb^{2+} + 2Cl^- = PbCl_2^o$				$Zn^{2+} + 2Cl^- = ZnCl_2^o$			
T° C	log K	ΔG_r^o kJ mole^{-1}	ΔH_r^o kJ mole^{-1}	ΔS_r^o J K^{-1}mole^{-1}	log K	ΔG_r^o kJ mole^{-1}	ΔH_r^o kJ mole^{-1}	ΔS_r^o J K^{-1}mole^{-1}	log K	ΔG_r^o kJ mole^{-1}	ΔH_r^o kJ mole^{-1}	ΔS_r^o J K^{-1}mole
25	3.27	−18.66	−12.5	20	1.95	−11.14	14.76	87	0.620	−3.54	30.3	112
50	3.10	−19.17	−12.5	20	2.16	−13.37	16.72	93	1.04	−6.46	33.0	121
100	2.88	−20.57	−5.9	39	2.61	−18.65	26.5	121	1.89	−13.5	48.5	170
150	2.88	−23.33	0	55	3.18	−25.75	47.4	168	2.90	−23.5	77.4	240
200	2.87	−26.00	2.0	59	3.95	−35.8	75.7	236	4.12	−37.7	119	334
250	3.07	−30.74	36.6	130	4.96	−49.7	119.2	320	5.70	−57.1	176	449
300	3.52	−38.62	73.6	200	6.23	−68.4	178.1	430	7.52	−82.5	245	570
350	4.21	−50.22	121	280	—	—	—	—	9.59	−114.4	328	710

spectroscopy measures molecular vibrations and optical absorption spectra result from electronic transitions between molecular orbitals. The peak energies and shapes of both kinds of spectra are controlled by the geometry, bond strengths, and ligation numbers of the complex ion, so these techniques probe the molecule itself. The method requires sealing windows (quartz, sapphire or diamond) into high-pressure-temperature autoclaves, which is not entirely straightforward; some of the ingenious methods devised by different laboratories to do this have been summarized by Buback (1981) and Buback et al. (1987).

Experimental measurements to date indicate that most transition metal complexes are distorted tetrahedral or octahedral molecules, with the metal at the center surrounded by four or six coordinating ligands (e.g., Susak and Crerar, 1984). The higher coordination numbers are favored by higher LFSE's. Recent Raman studies by Mary Yang at Princeton University indicate that Zn(II)-chloro complexes are probably linear, as predicted by Crerar et al. (1985) for this and other d^{10} ions (see discussion of Figure 27, below). Coordination numbers as high as 8 or even 12 might be possible at high pressures, although there is no experimental evidence for this as yet. The octahedral $\rightarrow$ tetrahedral transition occurs at higher T and chloride activity as indicated in Figure 25 (which includes recent data for Fe(II)-chloro complexes by Vogel et al., 1987).

As noted above, Susak and Crerar (1982, 1985) have suggested that large ore deposits might correlate with conditions that promote tetrahedral (or even lower) coordination; this is evident on Figure 25. Recall that Figure 24 shows an apparent inverse relation between LFSE and ore deposition. Electronic spectroscopy gives both the LFSE and geometry of metal complexes directly, and is therefore ideally suited for work on such questions.

Stoichiometries

The formulas or stoichiometries of complex ions can be determined in many different ways. Many of these methods give the stability constants for the predominant complexes at the same time. The most common approach is statistical analysis of mineral solubility data as a function of ligand concentration (for different techniques, see Crerar et al., 1978; Barnes, 1981; Wood and Crerar, 1985; Eugster, 1986; Ruaya and Seward, 1986). Spectroscopic techniques have also been used successfully for both purposes at hydrothermal conditions (references above, and Irish and Brooker, 1976; Seward, 1984).

An example of the changing speciation of Zn-chloro complexes with temperature at saturated vapor pressures is given in Figure 26. This comes from recent solubility studies by Ruaya and Seward (1986) which determined complex stoichiometries and stability constants simultaneously. The most important point to notice here is the relative simplification of species at higher temperatures. At 25°C all species from Zn^{2+} to $ZnCl_4^{2-}$ occur together at significant concentrations; at higher temperatures only the $ZnCl^+$ and $ZnCl^\circ_2$ complexes are important, and $ZnCl^\circ_2$ predominates at moderate chloride concentrations. This trend toward neutral-charge at higher T was predicted above from the decrease in dielectric constant of water with temperature.

A similar trend is evident in the Raman spectra of Zn-bromo complexes shown in Figure 27, measured recently by Mary Yang. This is a spectroscopist's view of the totally symmetric vibration of these tetrahedral complexes at different temperatures and the same total bromide concentration. Once again, the neutral $ZnBr_2$ species predominates at the highest temperature while higher ligation numbers are significant at and below 200°C. The $ZnBr^+$ complex is apparently linear because there is no evidence for associated waters elsewhere in the spectrum as there is with $ZnBr_2(H_2O)^\circ_2$ and $ZnBr_3(H_2O)^-$.

Stability constants

Stability constants of metal complexes in hydrothermal systems can be determined by many techniques, as outlined in the references cited in the previous section. Compilations and references to complex stability constants at elevated temperatures are given by Baes and Mesmer (1976, 1981), Barnes (1979, 1981), Frantz et al. (1981), Seward (1981, 1984), Eugster (1986), and Wood et al. (1987).

There are two fundamentally different ways of measuring stability constants for metal-complexing reactions, each with its pros and cons. We will illustrate these using Fe(II) and the mineral pyrite as an example.

First, you can consider the aqueous species alone, and determine overall formation constants for ionic reactions such as:

$$Fe^{2+} + nCl^- = FeCl_n^{2-n} \;;\; K_n = \frac{(FeCl_n^{2-n})}{(Fe^{2+})(Cl^-)^n} \quad (n = 1,2,...,6) \;. \tag{17}$$

In this and the following equations, parentheses denote activities (and thus implicitly include activity coefficient corrections). There is one Expression (17) for each different complex, depending on the value of n. The measurement is usually accomplished by spectrophotometric or electrochemical techniques which are sensitive to the concentrations of individual aqueous species. There are innumerable examples at 25°C (see Smith and Martell, 1976), and one outstanding study at hydrothermal conditions by Seward (1984) who determined Pb(II)-chloro stability constants from UV charge-transfer spectra. These methods should give the most accurate and precise measurements of stability constants for ionic reactions.

However, if stability constants such as (17) are to be used to calculate mineral solubilities, additional information is required. To calculate the solubility of pyrite at given temperature, Cl^- activity, pH, and oxygen and sulfur fugacities, for example, we would also need equilibrium constants for the following reactions:

$$FeS_2 + 2H^+ + H_2O = Fe^{2+} + 2H_2S + 1/2O_2 \;;\; K = \frac{f_{O_2}^{1/2}(H_2S)^2(Fe^{2+})}{(H^+)^2} \;, \tag{18}$$

and

$$H_2S + 1/2O_2 = H_2O + 1/2S_2 \;;\; K = \frac{f_{S_2}^{1/2}}{f_{O_2}^{1/2}(H_2S)} \;. \tag{19}$$

Finally, Reactions (17) + 2 × (19) + (18) give

$$FeS_2 + 2H^+ + nCl^- + 1/2O_2 = FeCl_n^{2-n} + S_2 + H_2O \;;\; K_n = \frac{f_{S_2}(FeCl_n^{2-n})}{f_{O_2}^{1/2}(H^+)^2(Cl^-)^n} \;. \tag{20}$$

This is the kind of reaction geologists need: the solubility of a mineral is expressed in terms of accessible variables such as pH, Cl^- activity, and oxygen and sulfur fugacities. Unfortunately, it is two steps removed (Reactions 18 and 19) from the stability constant (17). There are two alternatives. We could devise some means of measuring or estimating equilibrium constants for Reactions (18) and (19), which would then allow us to calculate the constant for Reaction (20) from $K_{20} = K_{17} \times 2K_{19} \times K_{18}$; the cumulative errors here could be daunting. A more direct approach would be to measure the equilibrium constant for Reaction (20) by determining the solubility of pyrite as a function of T, P, pH, Cl^- activity, and sulfur and oxygen fugacities.

Such experiments are difficult but possible, and this is the second common approach (see Barnes, 1981, for an overview). Here the most difficult problem is controlling or measuring all necessary variables required to evaluate the equilibrium constant: for reactions such as (20) these variables are pH, oxygen and sulfur fugacity, chloride ion activity, and metal ion activities. Since one equilibrium constant K_n applies to each complex $FeCl_n^{2-n}$, the individual stability constants can be found by rewriting (20) as the sum of all metal ion concentrations:

$$\sum Fe = \frac{f_{O_2}^{1/2}}{f_{S_2}}(H^+)^2 \sum_{i=1}^{n} \frac{K_i(Cl^-)^i}{\gamma_i} \quad . \tag{21}$$

The right side is divided here by the activity coefficient γ_i of each metal complex to convert from metal activities to concentrations. Expression (21) is fitted as a regression equation to total Fe versus Cl^- activity data, which gives the stability constants K_n for each complex as regression parameters (Crerar et al., 1978; Wood and Crerar, 1985). This same expression may also be used to calculate solubilities for given T, P, Cl^- activity, pH, and oxygen and sulfur fugacities, once the equilibrium constants are evaluated. Note that once again, this requires knowledge of activity coefficients for all species in Expression (21); this is a serious problem and is discussed below.

We mentioned that there are pros and cons for both general methods of obtaining stability constants for reactions such as (20) or (17). The positive aspect of this second method is obvious—a mineral solubility is obtained directly in terms of geologically useful or reasonable variables. On the negative side, the experiments require control or measurement of many variables and thus cumulative errors can be large. In theory, it should be possible to use such experiments to obtain equilibrium constants for the purely ionic reactions such as (17); for example, if the stability constants K_3 and K_2 are measured for the 3- and 2-chloro forms of Reaction (20) then the constant for the reaction

$$FeCl^\circ_2 + Cl^- = FeCl_3^-$$

is obtained from K_3/K_2 (equivalent to subtracting reaction 20 with n = 2 from the same reaction for n = 3). This is equivalent to (17) as determined by spectroscopic or electrochemical methods. However, the equilibrium constant obtained by the solubility approach is likely to be less accurate for the reasons just given. In other words, solubility experiments are better suited to measuring full mineral dissolution reactions such as (20); the other approaches are best for determining stepwise stability constants such as (17).

Ore zoning

It has been recognized for over a century that there is a rough zoning of metals within many hydrothermal ore deposits, particularly those associated with large igneous intrusions. This is a subject of great controversy with absolutely no common agreement, but a very general, simplified sequence of metals from source to periphery might be Mo→Fe→Ni→Sn→Au→Cu→Zn-Pb→Sb-Hg (e.g., Barnes, 1975). You should not expect to find this exact sequence in any one deposit since it is an average drawn from many different places and types of ores. Ore zoning is much more than a simple effect of temperature. Susak and Crerar (1982), for example, have recently emphasized that it depends on many variables such as degree of saturation, T, P, pH, sulfur and oxygen fugacities, mineral stoichiometries, crystal structures and the molecular and thermodynamic properties of aqueous metal complexes.

This is a wonderful chemical problem, of obvious interest to the mining community as well. However, it is not likely to be answered definitively until much more is known about mineral solubility and deposition reactions. At present it is difficult to compare data for different minerals because of possible systematic errors involved in the various experimental approaches; there are even some differences in what the various experimental methods actually measure. One possible approach is to run solubility experiments containing many compatible minerals. Here, most systematic errors should apply to all minerals simultaneously, and the dissolution behavior of different minerals can be compared more easily. In a recent experiment of this type, Wood et al. (1987) measured simultaneous solubilities of the minerals pyrite, pyrrhotite, magnetite, sphalerite, galena, gold, stibnite, bismuthinite, argentite and molybdenite in hydrothermal NaCl solutions to 350°C at controlled oxygen and sulfur fugacities and fixed CO_2 partial pressure. Relative solubilities followed the order

Sb>Fe>Zn>Pb>Ag,Mo>Au in chloride-free solutions, and Fe>Sb, Zn>Pb>Au>Ag, Mo, Bi in concentrated chloride solutions. This is in rough agreement with the zoning sequence suggested by Barnes, above. Major differences in chemical behavior were apparent in this data set. Fe, Zn , Pb, Au and Ag formed chloride or hydroxy-chloride complexes in 0.5 to 5.0 m NaCl solutions; in contrast, Mo formed an oxyacid (tentatively, molybdic acid), and Sb appeared to dissolve as an neutral hydroxy complex. In chloride-free solutions, Au and Ag apparently formed bisulfide complexes; Fe formed the simple aquo complex; Zn, Sb and Bi formed hydroxy species; and Pb formed a carbonate. Much of this behavior could be predicted from the chemical controls outlined above. This diversity suggests that natural ore-forming solutions may be extremely complicated, with no single type of complex predominating over a wide range of conditions. The application to questions such as ore zoning will require much more work, methodically defining these species and their dissolution-precipitation reactions.

Metal-organic complexing

The importance of natural compounds in dissolving and transporting metals has been clearly demonstrated for surface- and ground-waters (see, for example, Reuter and Perdue, 1977; Jackson et al., 1978). Many natural humic and fulvic acids can chelate (form multiple bonds) metals and have stability constants on the order of 10^6 for metal complex formation at 25°C (Sohn and Hughes, 1981). By contrast, monochloro complex stability constants for many transition metals at 25°C are on the order of 1 to 100 (Smith and Martell, 1976). Organic matter is commonly associated with sedimentary and lower-temperature hydrothermal ore deposits (e.g., Nissenbaum and Swaine, 1976; Macqueen and Powell, 1983). Coupled with the strong metal-complexing ability, this has led to suggestions that the acids may participate in metal transport in hydrothermal systems (Barnes, 1979; Giordano and Barnes, 1981; Giordano, 1985). There are three interrelated problems here: what is the concentration of organic ligands in hydrothermal solutions? To what temperature do these ligands persist without degrading? What is their chelating capability at higher temperatures?

None of these questions can be answered very satisfactorily at the present time, but each is the topic of considerable current research in several laboratories. Data from Willey et al., (1975) and Carothers and Kharaka (1980) summarized by Giordano (1985) suggests that concentrations of potential organic ligands in sedimentary basin brines may range from < 1 ppm to several thousand ppm. Experimental work on organic degradation kinetics in aqueous systems shows that aliphatic and aromatic compounds with metal-binding carboxy and phenolic –OH functional groups can persist to temperatures between 100° and 200°C (Kharaka et al., 1983; Drummond and Palmer, 1986; Boles et al., 1987). Electrochemical measurements in progress at Princeton by Remy Hennet (1987) show Pb-organic stability constants ranging from log K = 2.3 (acetate, 85°C) to 7.8 (dipicolinate, 90°C). Given sufficient concentrations, such ligands could be significant metal transporting agents. However, this requires that the concentrations of other competing cations are not too high. Preliminary calculations by Giordano (1985) and Hennet (1987), for example, show that Na, Ca and Mg may bind organic ligands in solutions believed typical of the Mississippi Valley-type Pb-Zn ore-forming environment; if so, there would not be sufficient free organic ligands left to complex significant concentrations of Pb or Zn. At present, the question is not resolved. It appears likely that metal-organic interactions will be important in organic-rich environments, contributing to diagenetic processes, but that very unusual conditions will be required for organic deposition of metallic ores (high organic concentrations, low reduced sulfide and low alkali and akaline earth concentrations).

Activity coefficients

Now we come to one of the most perplexing barriers to understanding ore-forming solutions. Activity coefficients are essential in deriving thermodynamic data from experimental measurements (as with Eqns. (17) and (21), above). They are also necessary in calculating solution behavior from known stability constants and thermodynamic data. All species, whether ionic or neutral require this correction. Unfortunately, it is not certain what the correction should actually be in many cases.

Major salts. Experimental information is most complete for the major salt components of hydrothermal solutions, NaCl, KCl and $CaCl_2$. Much of this is reviewed by Wood et al. (1984). There are considerable data on mixed salts so that it is now possible to describe some multi-component salt solutions of geological interest to temperatures of 300° to 350°C. For this purpose, you should use the Pitzer (1979) formalism which is also described in this book (Pitzer and Weare, this volume). With this approach data on two salt systems can be used to predict activity coefficients in much more complex mixtures containing the same salt components.

Activity coefficients of several common salts at 350°C are compared in Figure 28.

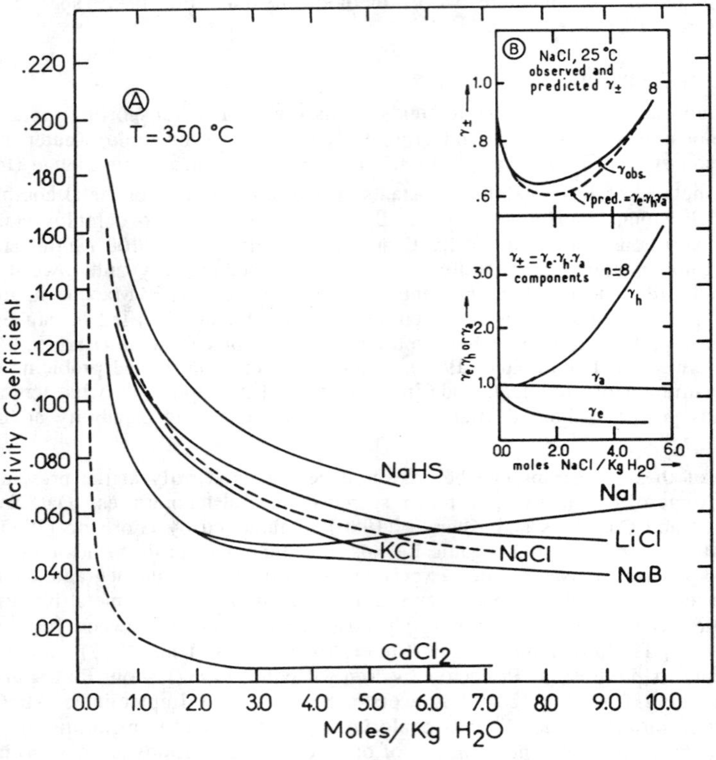

Figure 28. Mean molal stoichiometric activity coefficients for selected salts at 350°C versus salt concentration. Note that coefficients of all 1:1 salts are similar but that $CaCl_2$ is much lower. Activity coefficients of all salts at this temperature are roughly independent of concentration for concentrations above 2 m. Data from Wood et al. (1984) and Crerar et al. (1985).

At this higher temperature, the correction is considerable; the activity coefficient remains roughly the same at total salt concentrations exceeding 1 molal; and the activity coefficients of many of the 1:1 salts are similar. Lindsay (1980) suggested that at 300° to 350°C the structure of water becomes so thermally disrupted that all ions begin to act as structure-makers and will have similar activity coefficients. This approximation appears to work reasonably well for the alkali halides, but Wood et al. (1984) showed that it fails with other halides such as LiI. As should be expected from the charge difference alone, 2:1 electrolytes such as $CaCl_2$ show quite different behavior, and the activity coefficient for $CaCl_2$ at these elevated temperatures is on the order of 10^{-3}. These low values can be attributed in part to formation of the $CaCl^+$ ion pair; there have now been several problems fitting the Pitzer equations to data for NaCl-$CaCl_2$ mixtures both at 25°C (Ananthaswamy and Atkinson,

1982) and at elevated temperatures (Brantley, 1986) which could be attributed to such species. This is a problem for people working with hydrothermal solutions since NaCl and $CaCl_2$ are often predominant components. Resolution of this difficulty could require inclusion of Ca-Cl speciation in the fitting procedure.

Three main non-ideal effects. The non-ideal effects which contribute to activity coefficients can be divided into three broad categories based on a simple model described by Crerar (1973) and summarized by Wood et al. (1984). These include ion association (γ_a, short-range solute interactions), ion hydration (γ_h, short-range ion-solvent interactions) and electrostatic effects (γ_e, long-range solute interactions). As shown in the insert, Part B, of Figure 28, each effect can be represented by an activity coefficient γ_e, γ_h, and γ_a; the multiple $\gamma_e \cdot \gamma_h \cdot \gamma_a$ of these three coefficients gives the overall stoichiometric activity coefficient $\gamma_\pm$. Agreement between predicted and measured activity coefficients is surprisingly good using hydration numbers derived by Marshall (see Marshall, 1972, and references therein).

Because the physical model involved in our approach is too simplistic, we do not recommend it for predicting activity coefficients; however, it is a very helpful way of visualizing non-ideal processes in electrolyte solutions. Ion association and electrostatic interaction both contribute activity coefficients less than unity, while hydration causes large positive deviations, particularly at higher ionic strengths. At lower temperatures activity coefficients for many salts reach values as high as 10-30, reflecting low electrostatic interaction and ion-association, and high hydration. At temperatures above 300°C, activities fall below 0.1-0.001 because electrostatic interactions and association have increased significantly.

Minor components in concentrated solutions. The behavior of minor components in concentrated salt solutions is a major, unresolved problem. All components, whether dilute or concentrated, can have significant activity coefficients; this includes minor species such as dissolved metal complexes as well as the major dissolved salts like NaCl. At present there is no really adequate model for these minor species under the full range of ore-forming conditions.

For highly dilute solutions (less than 0.01m ionic strength), the Debye-Hückel (D-H) equation can be used. This gives γ_e, and the effects of association and hydration are assumed insignificant by comparison. This might be appropriate with a low temperature, relatively pure ground- or surface-water, for example. At high T and relatively low P where the dielectric constant is minimized, ion association becomes most important and neutral species such as NaCl° or CuCl° predominate. Here the ionic strength ($1/2\ \Sigma\ m_i Z_i^2$) may become sufficiently low that the D-H equation can be used again, provided that your calculation includes all relevant association reactions (i.e., calculate ionic strength and activity coefficients, use them to calculate association again and iterate to convergence). The D-H equation has been recommended by Eugster (1986), for example, for high-T, supercritical ore fluids. This might work to a first approximation, but does not take into account the effect of hydration (which may be significant at low P and low dielectric constant). It is also not clear how to treat neutral species under these (or any other) conditions, since the D-H equation applies only to charged ions. Common recommendations are to treat neutral species such as NaCl° as ideal and to assign them unit activity coefficients (e.g., Helgeson and Kirkham, 1981, p. 1478), or to treat them the same as CO_2 dissolved in a solution of the same ionic strength (Helgeson, 1969). Neither approach seems satisfactory, since intermolecular interactions, hydration and even electrostatic effects (with polar molecules) can be expected to vary with different species. At high ionic strengths, activity coefficients of neutral species might rise above unity simply because the other ionic species in solution are hydrated.

One point of related interest comes from a study of brucite dehydration by Barnes and Ernst (1963). They showed that NaOH exists predominantly as a neutral ion pair above 400°C in hydrothermal solutions, and that it mixes approximately ideally with water above 500°C. This means that both water and NaOH° obey the Lewis fugacity rule, $f_i = X_i f°_i$

(where f_i is fugacity of i in solution, f°_i is the fugacity of pure i at the same P and T, and X_i is its mole fraction; see Nordstrum and Munoz, 1985, p. 148) This is convenient for calculating activities of volatile components (like water), but does not help particularly with $NaOH^\circ$ unless you know its fugacity at T and P. Also, components may mix ideally and obey the Lewis fugacity rule without being ideal themselves: in the Barnes and Ernst experiments, water fugacity differed from water pressure, hence neither H_2O nor NaOH were ideal (which would require $f_i = P_i$).

At any rate it appears that the simplest activity corrections may apply at the two temperature extremes: dilute low-T solutions, and supercritical, high-T, lower-P fluids. The truly difficult problem lies between these two limits; unfortunately these are precisely the conditions expected for most ore fluids. Here, as we have already seen, speciation is most complicated, and ionic strengths may still be quite high.

The approach most frequently used by geochemists over the past several decades for calculating activities of minor components in concentrated salt solutions was suggested by Helgeson (1969); this in turn was an outgrowth of earlier work by chemists such as Scatchard and Harned summarized by Pitzer and Brewer (in Lewis and Randall, 1961, pp. 326, 578 and Appendix 4). A deviation function "B-dot" was defined as the difference between observed and predicted activity coefficients for an electrolyte such as NaCl:

$$B^\bullet(\bar{I}) = \frac{\log \gamma_{m_\pm}^{obs} + A\,|Z_+Z_-|\,\sqrt{\bar{I}}/(1 + \mathring{a}\,B\sqrt{\bar{I}})}{\bar{I}} = \log \gamma_{m_\pm}^{obs.} - \log \gamma_{m_\pm}^{D-H} \quad . \qquad (22)$$

Here the second term in the numerator is the D-H equation, A and B are D-H constants at any T and P and å is an adjustable parameter specific to each solute (representing the "distance of closest approach of two ions"). In Helgeson's treatment, the true (association-corrected) ionic strength $\bar{I}$ is used. According to our simple model above, association is therefore accounted for in (22) and electrostatic interactions are partly included by the D-H equation. $B^\bullet$ should therefore mostly represent hydration corrections and other left-over non-ideal interactions not included in our model (solvent electrostriction and structural changes, etc.).

Now (22) can be rearranged to give a guess at the activity coefficient of a minor species (such as several ppm $FeCl^{2+}$) in a strong salt solution (perhaps 2m NaCl):

$$\log \gamma_i^{minor\ species} = \frac{-A\,|Z_i|^2\sqrt{\bar{I}}}{1 + \mathring{a}_i B\sqrt{\bar{I}}} + B^\bullet(\bar{I})\cdot\bar{I} \quad . \qquad (23)$$

Here, γ_i is the activity coefficient for the minor component, Z_i and å are for the minor component (not NaCl), $\bar{I}$ is calculated from the total salt concentration (and known association constants and activity coefficients), and $B^\bullet(\bar{I})$ for NaCl over a range of P and T is tabulated by Helgeson (1969, and 1981, pp. 1345 and 1457). This presumes that the deviations from the D-H equation are the same for minor species as for NaCl, and to a rough approximation this might be so (for example, if removal of free water from the system by hydration of NaCl is predominant, it would have roughly the same effect on the activity coefficients of minor species as on NaCl). The problem is that $FeCl^+$ is a very different entity from Na^+ and Cl^- ions and will undergo different interactions, each capable of changing its activity coefficient.

More recently, Helgeson et al. (1981) have revised Equation (23) by splitting the $B^\bullet$ term into two parts:

$$\log \gamma_i = \frac{-AZ_i^2\sqrt{\bar{I}}}{1 + r_{e,i}B\sqrt{\bar{I}}} + \Gamma + \left[\omega_i b_{NaCl} + b_{Na^+Cl^-} - 0.19(|Z_i| - 1)\right]\bar{I} \quad . \qquad (24)$$

The first term here is the D-H equation again; Γ is a small correction to change from mole fraction to molar concentration (Helgeson et al., 1981, p. 1322); $\omega_i b_{NaCl}$ is a hydration parameter derived from the Born Equation (11); and all the rest accounts for remaining non-ideal short-range interactions (and any other effects). The Born parameter $\omega_j = 1.66027 \times 10^5 Z_j^2/r_{e,j}$, and values of the Born ionic radius $r_{e,j}$ are tabulated by Helgeson et al. (1981, p. 1304). Values of b_{NaCl} and $b_{Na^+Cl^-}$ for NaCl to 500°C and 5 kbar are tabulated on p. 1477 of the same article. The D-H parameters A and B are given by Helgeson and Kirkham (1974, pp. 1202 and 1256).

This equation is more flexible than (23) but similar concerns still apply. As before, the only parameter specific to a minor component (one other than NaCl) in (24) is the radius $r_{e,j}$. Equation (24) is a major contribution and the best method available for estimating activities of minor components in the absence of direct experimental data. Unfortunately, until such data appear we are left with considerable uncertainty in modeling multicomponent hydrothermal solutions.

Mineral solubility calculations from thermodynamic data

Geologists interested in the formation of a specific ore deposit commonly need to calculate the mineral solubilities for the presumed conditions at the time of ore deposition. This gives the saturation compositions of ore components in the solution and often provides insight into deposition processes. This is also important for engineers working with geothermal reservoirs. There are many examples of such calculations in the literature, and the following list is selected to cover some of the variations and complications that can arise: see for example, Crerar and Barnes, (1976), Barton et al. (1977), Crerar et al. (1978), Barnes (1979), Frantz et al. (1981), Henley et al. (1984), Seward (1976, 1983, 1984), Henley (1985) and Ruaya and Seward (1986).

There are two fundamentally different ways of working through a solubility calculation based on the available thermodynamic data. First, if you are lucky, you will find an accurate equilibrium constant for a mineral dissolution reaction such as (20) above. In this case you will need estimates for T, P, and for pertinent variables which appear in the equilibrium constant (in this case, pH, oxygen and sulfur fugacity and total chloride). Finally, you will have to estimate activity coefficients for each species. Using an equation such as (21), which sums all the Reactions (20) for each of the different complexes (or values of n), you then have an estimate of total metal solubility. NaCl activity has been measured to 350°C, and if concentrations of other components are insignificant by comparison, then you know the chloride activity coefficients (tabulated by Wood et al., 1984).

The second approach is more difficult, and we will illustrate it with calculations being completed currently by G. M. Anderson (Barrett and Anderson, 1987, and personal communication). In this case data for a mineral solubility reaction such as (20) are not available. However, there do exist stepwise formation constants for aqueous complexes such as (17), and data for the "solubility product" or equilibrium constant for the solubility Reaction (18) forming the simple ion.

Anderson wanted to know the solubility of galena and sphalerite to 300°C. We start with Seward's (1984) overall formation constants for PbS and Ruaya and Seward's (1986) analogous constants for ZnS to 300°C:

$$M^{2+} + nCl^- = MCl_n^{2-n} \;;\; \beta_n = \frac{\left(MCl_n^{2-n}\right)}{(M^{2+})(Cl^-)^n} \quad . \tag{25}$$

Here parentheses denote activities (so activity coefficients are included), and M stands for either Pb or Zn. There is a separate formation constant (25) for each value of n (e.g., $ZnCl^+$, $ZnCl^\circ_2$,...,$ZnCl_n^{2-n}$). Next, we obtain equilibrium constants from the compilation of Bowers et al. (1984) for the hydrolysis reactions

$$MS + 2H^+ = M^{2+} + H_2S_{(aq)} \;;\; K_{MS} = \frac{(M^{2+})(H_2S)}{(H^+)^2} \quad . \tag{26}$$

Adding Reactions (26) and (25) gives an expression for the total metal concentration:

$$\sum M = M^{2+} + \sum_n MCl_n^{2-n} = a_{M^{2+}}/\gamma_{M^{2+}} + \sum_n a_{MCl_n}/\gamma_{MCl_n} \quad , \tag{27}$$

where we have divided the activities on the right hand side by the appropriate activity coefficients to obtain concentrations. Substituting (25) and (26) for (M^{2+}) and (MCl_n^{2-n}) into (27) gives

$$\sum M = \frac{K_{MS}(H^+)^2}{\gamma_{M^{2+}}(H_2S)} \left[1 + \sum_n \frac{\beta_n (Cl^-)^n \gamma_{M^{2+}}}{\gamma_{MCl_n}} \right] \quad . \tag{28}$$

This is our desired expression for total concentration of Pb and Zn as a function of dissolved H_2S, Cl^- activity and pH. This is also where the real problems start, since we now need activity coefficients for all the quantities in parentheses and for all the metal species. We outline the calculation here since it is so crucial.

The first problem is to calculate the true ionic strength. If we choose a total 3m NaCl concentration, then true (association-corrected) ionic strength can be calculated from the known dissociation constant of NaCl° (Helgeson et al., 1981, pp. 1427 and 1428) and the measured activity coefficients of NaCl (Wood et al., 1984). The true ionic strength is required by Equation (24) and sums the true concentrations of all ions, corrected for association:

$$\bar{I} = 1/2 \sum m_i Z_i^2 \quad (= m_{Na} \text{ for NaCl}) \tag{29}$$

This is different from the stoichiometric ionic strength which presumes complete dissociation of all NaCl:

$$I = m_{NaCl}^{tot} \tag{30}$$

There are both rigorous (and more tedious) and approximate (simpler) ways to calculate true ionic strength, and we will illustrate the tedious way first.

The equilibrium constant for the association Reaction (14) of NaCl° is

$$K = \frac{m_{NaCl^\circ}}{m_{Na^+} m_{Cl^-}} \frac{\gamma_o}{\gamma_+ \gamma_-} = \frac{[m_{NaCl}^{tot} - m_{Na^+}]}{m_{Na^+}^2} \frac{\gamma_o}{\gamma_+ \gamma_-} \quad . \tag{31}$$

Solving this for m_{Na} gives

$$m_{Na} = \bar{I} = \frac{-1 + \sqrt{1 + 4 K m_{NaCl}^{tot} \dfrac{\gamma_o}{\gamma_+ \gamma_-}}}{2K \dfrac{\gamma_o}{\gamma_+ \gamma_-}} \tag{32}$$

This can be solved initially by setting all activity coefficients equal to 1.0 (or by using the measured stoichiometric coefficients at T, P and total NaCl concentration). With the new value for $\bar{I}$ calculate new activity coefficients from Equation (24). Start the procedure over again, using the new activity coefficients to calculate a better true ionic strength from (32), and iterate to convergence. The difference between true and stoichiometric ionic strength is considerable at higher T; such calculations show, for example, that at 300° a 1m NaCl solution is 41% associated and a 3m solution is 57% associated.

There are two, simpler but approximate ways to calculate true ionic strength. The first was suggested by Helgeson (1981, p. 162): approximate the true ionic strength by calculating stoichiometric ionic strength up to 300°C and 3m NaCl, since there is little effect on calculated activity coefficients below this limit. The second assumes the ratio of the true to stoichiometric ionic strength stays fairly constant over small ranges of concentration and uses the following equation (Helgeson, 1981, p. 162):

$$I = I + \frac{\gamma_{\pm}^2 \bar{I}^2}{K_n \gamma_n} . \qquad (33)$$

If you want to calculate the true ionic strength of a 3m NaCl solution, substitute 3 for $\bar{I}$ to give I. Then the desired true ionic strength is given by scaling total molality according to the ratio $\bar{I}/I \times 3$ = desired true ionic strength. This can also be solved by iterative interval-halving on $\bar{I}$ for an exact solution.

With the true ionic strength estimated, Equation (24) can now be used to calculate activity coefficients of individual species for use in Equation (28). For this you will need the ionic radii $r_{e,j}$. These are available for everything but the $PbCl_n^{2-n}$ complexes for this particular example. However, these too can be estimated from correlation plots between $r_{e,j}$ and ionic entropy given by Helgeson et al. (1981, pp. 1302 and 1303). In this case, the necessary entropies of Pb and Zn chloro complexes can be taken from Seward's (1984) and Ruaya and Seward's (1986) data, respectively.

How accurate are calculated solubilities?

At this point the solubilities of galena and sphalerite can be calculated from Equation (28). Calculated results are shown in Figure 29 where they are compared with actual measured solubilities by Barrett and Anderson. There is good agreement between observed and predicted solubilities at 25°C, but calculated solubilities are roughly an order of magnitude higher than observed values at 80°C. Both the theoretical calculations and the experimental measurements are arguably state-of-the-art. Hence this probably represents the minimum difference or error we can expect between observed and calculated properties at the present time. There are many sources of potential error, including any of the equilibrium constants used in the calculation as well as errors involved in using Equation (24) for the activity coefficients of minor species in concentrated NaCl solutions.

Estimating chemical conditions in mineral deposits

As we have seen, solubility expressions such as (21) and (28) require knowledge of variables such as pH, oxygen and sulfur fugacities, T, P, chloride concentration and so on. With mineral deposits the original solution has long since disappeared (with the important exception of fluid inclusions), but many of the chemical properties at the time of deposition can be estimated from the minerals themselves. Fluid inclusions can provide important information on many of these parameters, and are the most commonly used method of estimating temperature and chloride concentration (see reviews by Roedder, 1979; Hollister and Crawford, 1981; and references therein).

Another common approach is to use coexisting minerals to provide estimates of chemical conditions at the time they were formed. Here you must be careful to show that equilibrium was attained for each assemblage used. Of course, different minerals may have appeared at different times so textures must be studied quite carefully. We will use Figure 30, a phase diagram by Crerar and Barnes (1976), to illustrate the method. Referring to the stability fields in this figure you might argue that many porphyry type deposits contain bornite, pyrite, chalcopyrite, but no graphite or native sulfur. Often calcite, barite and/or anhydrite appear to have been mobile (depositing and dissolving), and the ores frequently form where K-feldspar has altered to muscovite (sericite zone). This is a simplification because these minerals probably did not deposit simultaneously. However, it does suggest some reasonable chemical limits for an ore-forming fluid: This would be somewhere near the stippled center of the diagram, constraining pH (slightly acid), oxygen fugacity (about 10^{-37}), sulfur fugacity (average 10^{-11}) and so on. We have contoured pyrite solubilities as ppm Fe in this region of the diagram; according to Reaction (18) if we decreased total dissolved sulfur to 10^{-2} m, iron concentrations would increase 100 times, which gives quite

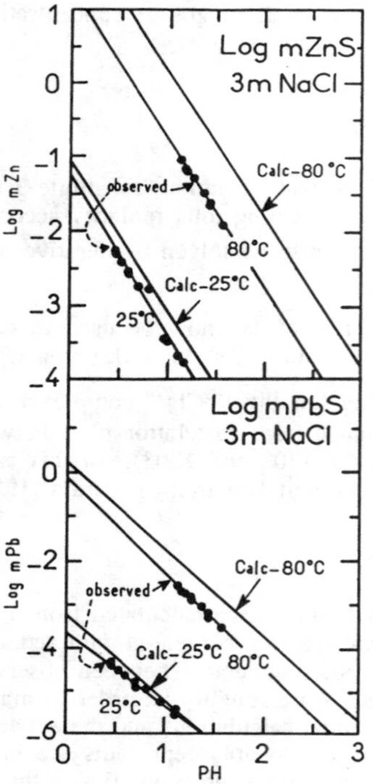

Figure 29. Comparison of calculated and measured ZnS and PbS solubilities in 3 m NaCl solutions at H_2S-saturated pressures as a function of pH and temperature. From T.J. Barrett and G.M. Anderson (personal communication).

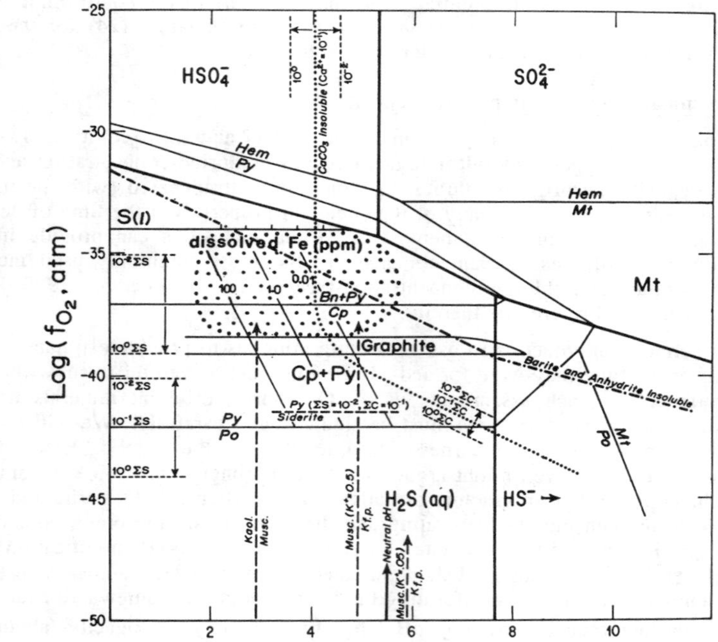

Figure 30. Oxygen fugacity versus pH stability fields of Cu-Fe-S-O minerals, plus calcite, barite, anhydrite, graphite, sericite (muscovite), and aqueous sulfur species at 250°C. Drawn for total S = 0.1 m; Ba^{2+} = 10^{-3} m; total carbon = 0.1 m; K^+ = 0.5 m; Ca^{2+} = 0.1 m. The stippled region is the "most probable ore fluid" for porphyry deposits discussed in the text. Pyrite solubility is contoured as ppm Fe within this region based on data in Crerar et al. (1978). Pyrite solubility increases 100-fold for a 10-fold decrease in total dissolved sulfur. Revised from Crerar and Barnes (1976).

respectable solubilities of 1 to 1000 ppm Fe in the stippled field. One of the best examples of this kind of reasoning is the study by Barton et al. (1977) of the Creede ore deposit. A similar approach is often used with active geothermal reservoirs where deep hydrothermal waters can often be sampled directly; in such cases solution chemistry can sometimes be described in considerable detail (see Ellis and Mahon, 1977, p. 101, and Ellis, 1979, for examples).

The variables chosen as axes for these diagrams should depend on what you know about the system or wish to describe. For example, if the deposit contains many sulfide and oxide minerals, oxygen and sulfur fugacity would be useful coordinates. If some information is available on dissolved components, activity ratios of cations can be very helpful. Details on the calculation of such diagrams are given by Holland (1959), Barnes and Kullerud (1961), Garrels and Christ (1965, Ch. 10), Barton and Skinner (1979) and Stumm and Morgan (1981, Ch. 9). An extensive compilation of activity diagrams to 5 kbar and 600°C, together with thermodynamic data used in their calculation, has been provided by Bowers et al. (1984).

A third, important class of techniques for estimating the physical and chemical parameters of ore deposition requires isotopic analyses. Since this could be (and is) the subject of several books, we will only recommend several comprehensive reviews for the details. First, isotope ratios between coexisting minerals can be used to estimate deposition temperatures (provided the minerals equilibrated). The most useful isotope geothermometers have proven to be $^{18}O/^{16}O$, summarized by Taylor (1979), and $^{34}S/^{32}S$ (Ohmoto and Rye, 1979). With sulfur and $^{13}C/^{12}C$ isotopes it is often also possible to trace the evolution of pH and oxygen fugacity of ore-forming fluids throughout the period of active ore deposition; this was discovered by Hiroshi Ohmoto (1972). In conjunction with the other methods outlined in this section and information on mineral solubilities, the isotope record often provides remarkably detailed information on the chemical history of ore deposits (for some examples, see Rye and Ohmoto, 1974).

PART III. FORMATION OF PRIMARY AND SECONDARY ORE DEPOSITS

PRIMARY ORE DEPOSITION

Up to this point, we have been primarily concerned with generation and transport of ore components in hydrothermal solutions. We now consider the reverse problem, that of precipitating ore minerals from solution. There are many possible precipitation processes, many of them interrelated. The list includes: decreasing solubility by decreasing temperature; decreasing solubility by raising pH; decreasing ligand concentration (by dilution with groundwater); boiling; reaction with reduced sulfur; and changing oxygen fugacity.

Initial acidity

Starting with a high-T vapor phase which has just exsolved from a magma, we have seen that it is likely to be charged with ore components, as well as acidic components such as HCl°. Most species should be electrically neutral and associated at these temperatures. The exsolving solution will also contain dissolved sulfur, which you will recall partitions about as strongly as Cl from magma into vapor. The initial oxidation state of this exsolved sulfur depends on the oxygen fugacity at magmatic temperatures; for all natural mineral buffers (such as hematite-magnetite, $2Fe_3O_4 + 1/2\ O_2 = 3Fe_2O_3$) oxygen fugacity increases exponentially with temperature (see tabulation by Huebner, 1971). The oxygen fugacity of rhyolitic and dacitic lavas summarized by Carmichael et al. (1974, p. 330) fall between the QFM and HM buffers with values on the order of 10^{-11}-10^{-15} bars. For a similar argument, see Henley et al. (1984, p. 156). Figure 30 provides a dramatic contrast with the oxygen fugacity of roughly 10^{-34}-10^{-37} bars estimated for a porphyry-type ore environment at 250°C. Under these magmatic conditions, much of the exsolved sulfur may be oxidized to SO_2. Burnham (1979), for example, estimates that about 50-90% of the exsolving sulfur will be oxidized by reactions such as

$$SH^-_{melt} + 5\,OH^-_{melt} = 3\,O^{2-}_{melt} + SO_{2,vapor} + 3\,H_{2,vapor} \quad . \tag{34}$$

This SO_2 has considerable acid potential too (like undissociated HCl°), and as temperature falls it should hydrolyze, producing H_2S and H_2SO_4 in a 1:3 ratio by the Reaction (Holland and Malinin, 1979):

$$4\,SO_2 + 4\,H_2O \rightarrow H_2S + 3\,H_2SO_4 \quad . \tag{35}$$

The ionization constant of HCl° also increases exponentially at lower temperatures (Fig. 20) so considerable acidity can be generated by both sulfur and chlorine as our exsolving solution cools.

Sulfur

Reaction (35) has immediate implications for ore deposition: Reduced sulfur will precipitate metal sulfides. The sulfate is the probable source of barite and particularly anhydrite which can be quite abundant in higher-T alteration zones (notice that the stability lines for both minerals pass through the center of our "most probable" porphyry ore fluid zone on Fig. 30). Eugster (1985, p. 20) summarizes isotopic evidence that "most or at least some of the sulfur in Sn-W deposits" associated with granites came from the original magma. The country rock itself is the only other possible source of sulfur. This could be derived in part from evaporitic sulfates (e.g., the carbonate-hosted Sn-W Dachang ores of SE China, Eugster, 1985; sulfide deposits in the Red Sea brines, Shanks and Bischoff, 1977), or from disseminated sulfide and sulfate minerals in adjacent wall rocks. Thermal or bacterial degradation of organic components is often invoked as the source of reduced sulfur for lower temperature sediment-hosted deposits such as the Mississippi Valley-type Pb-Zn ores (see review by Anderson, 1975). H_2S-charged brines and gas pockets are relatively common in deep sedimentary basins; mixing with metalliferous chloride brines could produce the sulfide deposits characteristic of the rims of large sedimentary basins (Jackson and Beales, 1967). There is no doubt about the sulfide-rich rims, but whether or not this mixing process really happens has been a topic of great debate and considerable interest for many years (for a critical review read Sverjensky, 1986).

pH and alteration reactions

Almost all ore mineral precipitation reactions such as

$$FeCl_n^{2-n} + 2\,H_2S + 1/2\,O_2 \rightarrow FeS_2 + 2\,H^+ + H_2O + nCl^- \tag{36}$$

not only generate acidity (H^+), but are actually driven to the left by acidity. (Notice on Fig. 30 that pyrite solubility increases by two orders of magnitude for each unit decrease in pH). This is certainly a complicating factor since we have just observed that ore solutions should become increasingly acidic as they cool. Reaction (35) generates three times as much sulfuric acid as H_2S (and even H_2S is a weak acid), and the ionization of HCl° with decreasing T compounds the problem. We clearly need some means of titrating out the acidity of cooling hydrothermal solutions if ore minerals are to precipitate at all.

This problem was resolved with the work of Julian Hemley (1959; Montoya and Hemley, 1975) and many others since (summary by Rose and Burt, 1979). The minerals of the host rocks themselves react with these corrosive ore fluids. Many of these wall-rock alteration reactions consume acidity with the minerals acting as proton sinks or Brønsted-Lowry bases. This can produce an enormously complex assemblage of changing alteration zones around and within large pluton-related deposits. For example, in the typical porphyry copper deposit, a central potassic zone of biotite-orthoclase alteration grades outward through quartz-sericite to sericite-kaolinite and finally to propylitic chlorite-epidote-calcite zones; primary copper sulfides are commonly most abundant in the sericitic zone near the potassic core (Lowell and Guilbert, 1970; Crerar and Barnes, 1976). Silicate alteration is typified by the orthoclase-sericite-kaolinite reactions:

$$3\,KAlSi_3O_8 + 2\,H^+ = KAl_2AlSi_3O_{10}(OH)_2 + 2K^+ + 6SiO_2 \;;\; K' + \frac{(K^+)^2}{(H^+)^2} \quad , \tag{37}$$
orthoclase — sericite

$$2\,KAl_2AlSi_3O_{10}(OH)_2 + 2\,H^+ + 3\,H_2O = 3\,Al_2Si_2O_5(OH)_4 + 2K^+ \;;\; K'' + \frac{(K^+)^2}{(H^+)^2} \quad . \tag{38}$$
sericite — kaolinite

Equilibrium constants for these two reactions have been measured by Montoya and Hemley (1975) both in terms of total m_{KCl}/m_{HCl} and the activity ratios (37) and (38) (both are necessary since the degree of ionization of KCl° and HCl° increases at lower temperatures). Given an estimate for K concentration and temperature, the pH of the ore fluid can be calculated from these constants. The vertical arrows on Figure 30 delimiting the fields of orthoclase, sericite and kaolinite alteration were derived this way. Equilibrium constants for many alteration reactions have been tabulated by Frantz et al. (1981) and Bowers et al. (1984).

Alteration processes such as (37) and (38) continue down to the lowest ore-forming temperatures and play an important role in supergene, freshwater, and marine environments. At lower temperatures these are usually thought of as weathering reactions, and are discussed in detail in the low-temperature aqueous literature (e.g., Stumm and Morgan, 1981, Ch. 9; Morel, 1983, Ch. 5).

Ore mineral deposition in an alteration environment probably proceeds by two mechanisms, both of which can produce replacement textures. First, the wall rock minerals themselves might react with ore fluids, raising pH and thereby driving precipitation reactions such as (36). The second possibility is that the mineral precipitates in response to some other stimulus; with Reaction (36) this might be decreasing T or chloride activity, or increasing oxygen fugacity or H_2S. Sulfide precipitation releases protons (2 moles of H^+ for every mole of pyrite deposited), which then react with host rock minerals. For a precipitation reaction such as

$$ZnCl°_2 + H_2S \rightarrow ZnS + 2\,HCl° \;; \tag{39}$$

the total reaction in carbonate host rocks might be:

$$ZnCl°_2 + CaCO_3 + H_2S \rightarrow ZnS + CaCl°_2 + H_2CO°_3 \quad , \tag{40}$$
calcite — sphalerite

or, in silicates:

$$ZnCl°_2 + 3KAlSi_3O_8 + H_2S \rightarrow ZnS + KAl_3Si_3O_{10}(OH)_2 + 6\,SiO_2 + 2\,KCl° \quad . \tag{41}$$
orthoclase — sphalerite — muscovite

The field record indicates that these are very important, common causes of ore mineral deposition. Sulfide/silicate or sulfide/carbonate replacement textures are very common; large hydrothermal deposits related to igneous plutons are almost invariably associated with extensive wall rock alteration.

pH buffer capacity. This is an important control on the potential for corrosion or mineral deposition which has not received much attention to date. pH buffer capacity is defined as the increment of acid or base that causes unit change in pH. A hydrothermal solution with a high buffer capacity will resist change in pH during wall rock alteration and mineral precipitation reactions; it will be highly corrosive and will show less tendency to precipitate metallic minerals. Crerar et al. (1985) have calculated the pH buffer capacities of some idealized ore fluids: these vary by 5 orders of magnitude, from simple quartz-saturated NaCl solutions (low buffering) to high-T seawater and basic NaCl-NaHS solution (highest). Many calculated buffer capacities also increase exponentially with temperature.

The geological ramifications are considerable. Weakly buffered systems which react with wall rocks will become basic more quickly and will rapidly deposit metals via reactions such as (39). If the same solutions do not react with wall rocks, the precipitation of a trace of metal (same reaction) will lower pH and metals will stay in solution, so the alteration reactions are necessary for ore deposition. Our calculations suggest that the NaCl-rich ore fluids associated with felsic plutons are weakly buffered and that they rapidly precipitate copper, molybdenum and other sulfide ores in response to extensive wall rock alteration reactions at moderate to high temperatures. As temperature falls, the buffer capacity of these same solutions becomes even less and most metals are probably dumped before the solutions become too cool to react quickly with wall rocks. In contrast, fluids in carbonate terranes are highly buffered by the H2CO3-HCO3-conjugate pair; this helps explain how Mississippi Valley-type ore fluids can travel great distances without precipitating much metal; it also suggests that in this case metals may be precipitated by something other than a pH change (for example, meeting H_2S brines). Finally, seawater heated above 200°C has a very high buffer capacity, and also becomes acidic at elevated temperatures (Bischoff and Seyfried, 1978), both of which make it very corrosive; this is why circulating hydrothermal systems at oceanic spreading centers are so effective at leaching metals up and out of the oceanic crust. This is now recognized as a major control on the chemistry of the world ocean and a likely cause of many massive sulfide deposits (see Von Damm et al., 1985a,b; Bowers et al., 1985 and references therein).

Boiling

Extremely high salt concentrations (roughly 40 wt %) and variable liquid/vapor ratios are not uncommon in fluid inclusions (e.g., Roedder, 1979) and suggest the solution was boiling at the time of entrapment. There are many known geothermal reservoirs which discharge steam and boil at depth (Larderello, The Geysers, Matsukawa, and others described by Ellis and Mahon, 1977, p. 52). Hydrothermal fluids appear to have boiled in specific deposits ranging from porphyry copper-molybdenum and tin-tungsten ores to silver-gold and polymetallic vein deposits (summarized by Drummond and Ohmoto, 1985).

Physical models for convective flow around plutons, such as Figure 10 indicate boiling regions (see Cathles, 1977; Henley and McNabb, 1978). The "vapor" region shown in Figure 10 above the pluton is actually generated under supercritical conditions for pure water (so is not true boiling). However, addition of only 10 wt % salinity extends the boiling region (critical curve) so that boiling occurs for the same conditions. The computed lifetime of these vapor-dominated regions is less than 5,000-10,000 years, so if the computations are physically realistic, boiling is geologically short-lived.

The qualitative model for ore deposition in boiling systems is clearly outlined by Barton et al. (1977) for the Creede polymetallic vein deposits (see also Slack, 1980):

> "The ores were deposited from a freely convecting hydrothermal system that probably was initially charged by meteoric solutions, although the salts, metals, and sulfur may well have been derived from deeper sources. The circulating solutions deposited gangue and ore minerals near the top of the convecting cell in a hypogene enrichment process that extracted metals and sulfur from whatever sources were available at depth and swept them toward the surface. Boiling, with the loss of acid components (H_2S and CO_2) which recondensed in the cooler overlying rocks, led to the formation of an intensely altered sericitic capping above the ore. Precipitation of the ore is attributed to cooling and perhaps to a slight pH rise complementary to the loss of acid constituents through boiling."

The possible relation between boiling and ore deposition has now been modeled quantitatively by Drummond and Ohmoto (1985). In their work the essential features of the Creede model remain unchanged, except that the effect of boiling on pH is even greater than previously envisioned. Because natural hydrothermal solutions are complicated, multicomponent systems, the detailed behavior is highly dependent on solution composition. The effects are most pronounced for open, saline systems at lower temperatures with initially high CO_2/H^+ and CO_2/SO_4 ratios. The predominant volatile components partition from the liquid into a boiling vapor phase in the order $H_2>CH_4>CO_2>H_2S>SO_2$. While the computed results vary considerably with initial composition, the computation shown in Figure 31 is not atypical. Here you can see that the first 5% boiling (5% liquid converted to vapor) raises pH roughly 1 unit, which in turn precipitates most dissolved metals. The effect is highly temperature-dependent, and all changes are maximized at about 300°C. Drummond's model shows that typical boiling hydrothermal solutions should lose most of their volatile components to the steam phase and precipitate most of their metals by the time the volumes of vapor and liquid are equal; this occurs at 1, 2, 5 and 10% boiling at 200°, 250°, 300° and 350°C, respectively. Obviously, boiling can be a powerful deposition mechanism, for those circumstances where it occurs. It remains to be demonstrated how common this is geologically, since not all systems are sufficiently hot and shallow, and/or sufficiently saline, to boil, and when this does occur, hydrodynamic models indicate boiling is relatively short-lived.

Remaining deposition controls

The remaining controls seem rather obvious at first sight, but even here there are potential complications, suggesting enough interesting research directions to keep us all occupied for years. It is unreasonable to think that any one deposition mechanism predominates, and in fact, it is likely that some or all are important to varying degrees in different environments.

Temperature. This seems the most obvious control since the solubilities of most ore minerals decrease at lower temperatures. Based on measured solubilities of chalcopyrite, pyrite and bornite, the concentration of $CuCl^{\circ}$ should drop two orders of magnitude from 350° to 250°C in the center of the "average" ore fluid zone of Figure 30, and Fe concentrations should fall roughly an order of magnitude. For this reason, Crerar and Barnes (1976) suggested that the general, large-scale deposition of porphyry coppers might be attributable to decreasing temperature, with wall rock alteration and increasing pH being a secondary cause of local, highly disseminated precipitation. That idea lasted one year, until Larry Cathles (1977) argued that porphyry copper ore shells are not very thick (averaging about 200 m thickness by 1 km depth) and that the temperature gradients across such zones are probably too small (or too short-lived) in circulating systems to be the major cause of deposition. His preferred mechanism was boiling, but we have just pointed out some possible objections to that theory, too.

Dilution. The argument here is that dilution by relatively pure groundwaters entrained in a more saline, convecting hydrothermal system decreases ligand concentrations, driving precipitation reactions such as (36) to the right. A pronounced decrease in the chloride concentrations of fluid inclusions at greater distances from intrusive source rocks is commonly observed (see references in Crerar and Barnes, 1976), but this does not necessarily imply a cause-and-effect relation with ore deposition. If the pluton was the source of chloride then such gradients are expected. Dilution also reduces metal concentration (and degree of saturation) as well as that of other species such as H_2S; if a 1:1 complex such as $FeCl^{\circ}$ predominates there should be no net result. If complexes with higher ligation numbers predominate, then precipitation might be possible, and becomes more probable if the entrained solutions also decrease temperature or raise pH.

Oxygen fugacity. The oxygen and sulfur fugacities of typical mineral buffers increase exponentially with temperature (Eugster and Wones, 1962; Huebner, 1971; Barton and Skinner, 1979). In our discussion of initial acidity above, we observed that oxygen fugacity drops by about 20 orders of magnitude from magmatic temperatures to 250°C in porphyry systems. Many mineral dissolution reactions such as (36) are dependent on oxygen (or hydrogen) fugacity, so it seems reasonable to expect a profound effect on solubilities. Oxygen fugacity has opposing effects on different minerals as illustrated by magnetite and pyrite dissolution:

$$\underset{\text{pyrite}}{FeS_2} + 2H^+ + 1/2\,O_2 = Fe^{2+} + S_2 + H_2O \ ; \qquad (42)$$

$$1/3\ \underset{\text{magnetite}}{Fe_3O_4} + 2H^+ = Fe^{2+} + H_2O + 1/6\,O_2 \ . \qquad (43)$$

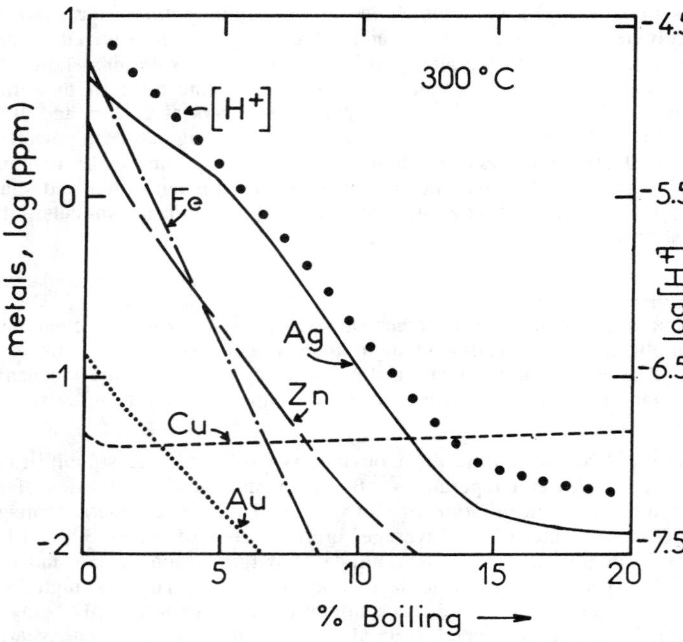

Figure 31. Change in metal and proton concentrations as a function of boiling for an open hydrothermal system at 300°C. The % boiling axis represents the wt % liquid converted to steam. Drawn for total carbonate = 3 m, and total sulfate = 3 × 10^{-9} m. Solubility data are summarized by Drummond and Ohmoto (1985). Diagram from Drummond and Ohmoto (1985).

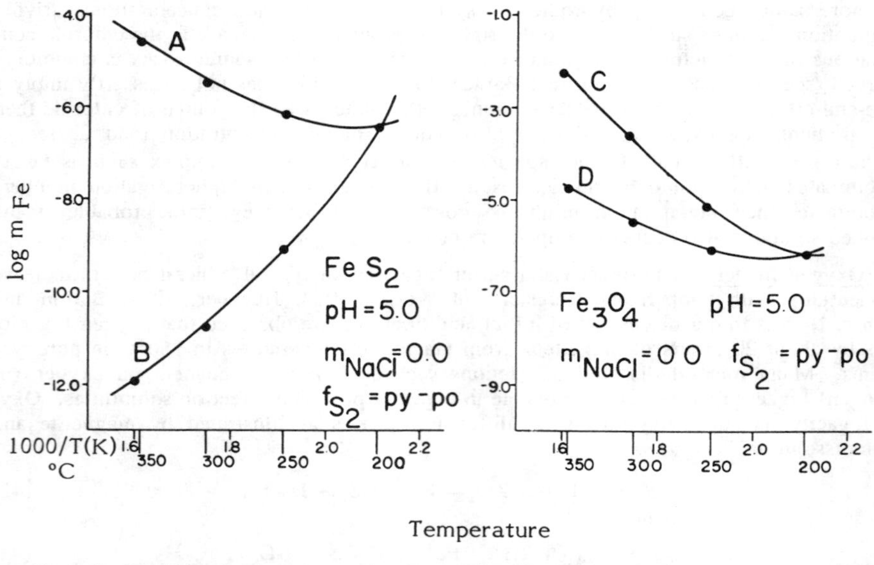

Figure 32. Solubilities of pyrite and magnetite versus 1/T for different oxygen fugacities in pure water. (A) Pyrite solubility when oxygen fugacity is buffered by pyrite + pyrrhotite + magnetite. (B) Pyrite solubility at constant log $f(O_2)$ = −43.9. (C) Magnetite solubility at constant log $f(O_2)$ = −43.9. (D) Magnetite solubility when oxygen fugacity is buffered by pyrite + pyrrhotite + magnetite. Solubility data from Crerar et al. (1978) and original diagram from Crerar et al. (1985).

The combined effects of oxygen fugacity and temperature on solubilities of both minerals are illustrated in Figure 32 (from solubility studies by Crerar et al., 1978). Curves B and C are for constant oxygen fugacity while A and D represent the fugacity change of the pyrite + pyrrhotite + magnetite buffer with temperature. The oxygen fugacity of this buffer increases roughly 13 orders of magnitude between from 200° to 350°C. Despite the reverse sense of oxygen in Reaction (43), magnetite solubility increases with temperature for both cases (the equilibrium constant for this reaction increases sufficiently with temperature to outweigh the opposing effect of higher oxygen fugacities). However, the solubilities for curve D (increasing oxygen fugacity) fall to 1% of that for C (constant oxygen) by 350°C. For pyrite, the effects are reversed and more pronounced, with concentration increasing 100-fold to 350°C for A (increasing oxygen) and decreasing enormously at constant oxygen fugacity. In general, the solubility of minerals such as gold, pyrite and molybdenite (for which solubility increases with oxygen fugacity) should increase dramatically with temperature. This helps explain the sharp rise in solubility of minerals such as gold with temperature (Fyfe and Hemley, 1973; Seward, 1983) and is consistent with the common occurrence of metals such as Au, Mo and Cu in higher-T deposits. Minerals such as cassiterite (SnO_2) and magnetite which show the reverse trend (since dissolved species are more reduced than the minerals) pose a different problem; here solubilities can decrease at higher oxygen fugacities (hence at higher T if the same mineral assemblage controls oxygen fugacity at all temperatures). See, for example, the data for cassiterite (SnO_2) solubility summarized by Eugster (1986), and the study of magnetite solubility under supercritical conditions by Chou and Eugster (1977).

Natural processes responsible for sulfide precipitation are complex, and depending upon specific circumstances, an array of mechanisms such as boiling or variation of sulfide solubilities with temperature may operate. However, in the absence of fluid inclusion data indicating boiling or strong temperature gradients, it is likely that alteration reactions are necessary to explain the observed metal concentrations within ore zones. Genetic and spatial associations of hydrothermally altered wall rocks and fracture- controlled sulfides are common (Meyer et al.,1968; Brimhall, 1977) indicating the importance of irreversible reactions between ore-forming solutions and surrounding wall rocks to the final deposition of sulfides. Circulation of hydrothermal fluids through previously-mineralized wall rocks occurs frequently in nature and is the norm not the exception.

Multi-stage mineralization and ore metal remobilization

Field studies of hydrothermal mineralization in porphyry systems have shown that primary mineralization, that is ore which is unoxidized by chemical weathering, is generally composed of superimposed networks of veins and veinlets which control alteration patterns, either in halos around individual veins or pervasive zones in regions of high fracture density. The fracture networks and corresponding alteration is formed during single or multiple hydrothermal events, each of which can be attributed to the intrusion, convective fluid circulation, and cooling of a specific parent pluton (Meyer et al., 1968; Lowell and Guilbert, 1970; Gustafson and Hunt, 1975; Brimhall, 1977 and 1979).

Ore metal remobilization versus introduction. Certain systems, particularly porphyry molybdenum deposits often have multiple intrusions, each having localized ore grade mineralization (Wallace et al., 1968) near the apex of each successive intrusion as at Climax, Colorado. Field and petrographic evidence strongly suggest that each intrusive released its own ore-forming fluid. In contrast, copper rich porphyry systems seldom have as many mineralized plutons as in the molybdenum systems. In fact in copper systems driven by only a single thermal event, the nature of mineralization and alteration may change drastically during the cooling history with the incursion of meteoric water. The superposition of phyllic (sericite-bearing, biotite destructive) alteration on earlier, higher temperature potassic alteration assemblage (biotite, muscovite, and alkali feldspar) is due to this late-stage modification of the ore– forming hydrothermal fluid (Taylor 1979).

Ore metals are clearly first emplaced locally in a district from saline magmatic aqueous fluids through circulation in the highest temperature vein networks and biotitic igneous breccias which control potassic alteration. With cooling and dilution by meteoric water, the ore-forming aqueous fluids become highly acid and circulate more extensively than before. Intense hydrolysis reactions pervade the wall rocks as well as the earlier, sulfide- bearing potassium silicate altered proto– ore (protore). At this stage, the source or sources of metals is difficult to ascertain. The basic question is whether or not ore metals in veins within phyllic alteration halos were derived by dissolution of earlier sulfides related to potassic alteration mineral assemblages, or whether they were simply scavenged by circulation of late hydrothermal fluids through ordinary unmineralized wall rock. The former mechanism is referred to as ore metal remobilization, the latter as scavenging. Both processes actually involve remobilization, but the former mechanism refers to extraction of metals from a previously mineralized state, rather than from ordinary wall rocks. In actuality, both mechanisms may occur, and given the enormous size of hydrothermal convection cells, the possible source regions for metals is not only gigantic but potentially quite varied as well. Resolution of this question is important as it relates to two of the most fundamental aspects of understanding ore fluids: the source of metals and their behavior during superposition of hydrothermal circulation systems.

Relationships of wall rock alteration to mineralization. Assessment of the effects of superposition of young hydrothermal events upon pre-existing protores can only be addressed from a practical standpoint within available exposures in mines. The deep reaches of convective systems are therefore rarely accessible as they are low grade, and hence are of little interest to mine operators. This leaves only three possibilities for addressing the issue of remobilization: (1) petrologic and geochemical analysis of the few optimal mining exposures where source regions of metals have been recognized, (2) theoretical analysis using irreversible chemical thermodynamics to model multi-stage mineralization, and (3) experiment. In all cases, the essential question is the net effect that superimposed alteration reactions have on protore sulfide, silicate, and oxide mineral assemblages. There is much evidence to suggest that hydrothermal remobilization and redistribution is common, at least for copper. Molybdenum, tungsten, and tin have received much less attention is this respect. The varied array of copper sulfides (chalcopyrite, bornite, chalcocite, and digenite) which are known as hydrothermal products and the thermochemical conditions of their stability may contribute to the relative mobility and complex history of copper in contrast to molybdenum and tungsten.

Hypogene leaching. Lacy and Hosmer (1956) recognized evidence suggesting the leaching of metals during hypogene mineralization at Cerro de Pasco, Peru. This important discovery opened the way for pursuing the behavior of metals during multiple hydrothermal events and for improving understanding of the space-time patterns, relationships of mineralization to alteration, plutonism, and temporal evolution of fluid composition. Continued pursuit of this problem has been possible in the extensive exposures in the Butte Mining District of Montana where multi-stage mineralization was recognized (Meyer et al., 1968). Here a large fissure vein system with phyllic and argillic alteration, the Main Stage, was superimposed upon a pre-Main Stage ore, a fracture-controlled disseminated chalcopyrite-pyrite-magnetite porphyry copper deposit with potassic alteration containing secondary biotite. The late veins are particularly well-developed as mineralization was syntectonic with regional compressive stress during Laramide thrusting and folding (Woodward, 1986), producing high fracture permeability and intimate fluid/rock interaction. Using the concept of reaction progress (De Donder, 1928, 1936; Prigogine, 1955; Helgeson, 1968) for monitoring the net extent of Main Stage alteration and mineralization effects on the pre-Main Stage protore, Brimhall (1977) developed a lithologic method for evaluating the transfer of metals between an early type of mineralization and a later ore-forming fluid with subsequent precipitation of sulfides.

Chalcopyrite abundance, determined from quantitative modal analysis on heavy mineral separates, can be used to monitor the amount of Main Stage reaction progress. Figure 33 shows the progressive leaching of chalcopyrite, destruction of pre-Main Stage mineral assemblage (biotite, magnetite, hematite, orthoclase), and the precipitation of Main Stage ore minerals (enargite, covellite, digenite, chalcocite, quartz, and pyrite. From this lithologic analysis, it is clear that phyllic (sericitic) alteration can leach copper from

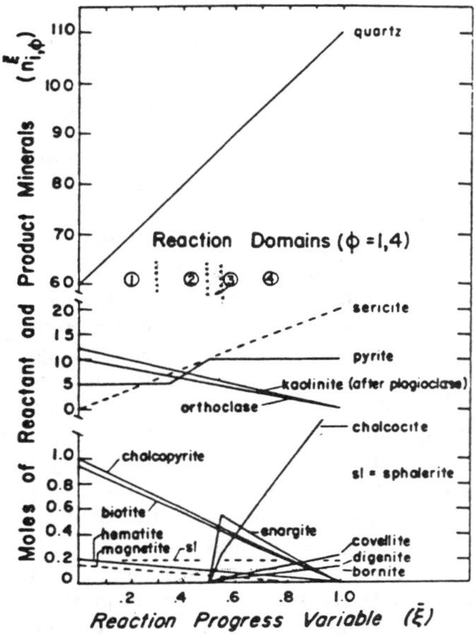

Figure 33. Generalized molar variation patterns based on lithological analysis. Four reaction domains are recognized as shown with abrupt changes in the slope of piecewise continuous molar curves. Slopes of molar variation curves with reaction progress (based on chalcopyrite abundance) are equal to stoichiometric reaction coefficients. From Brimhall (1979).

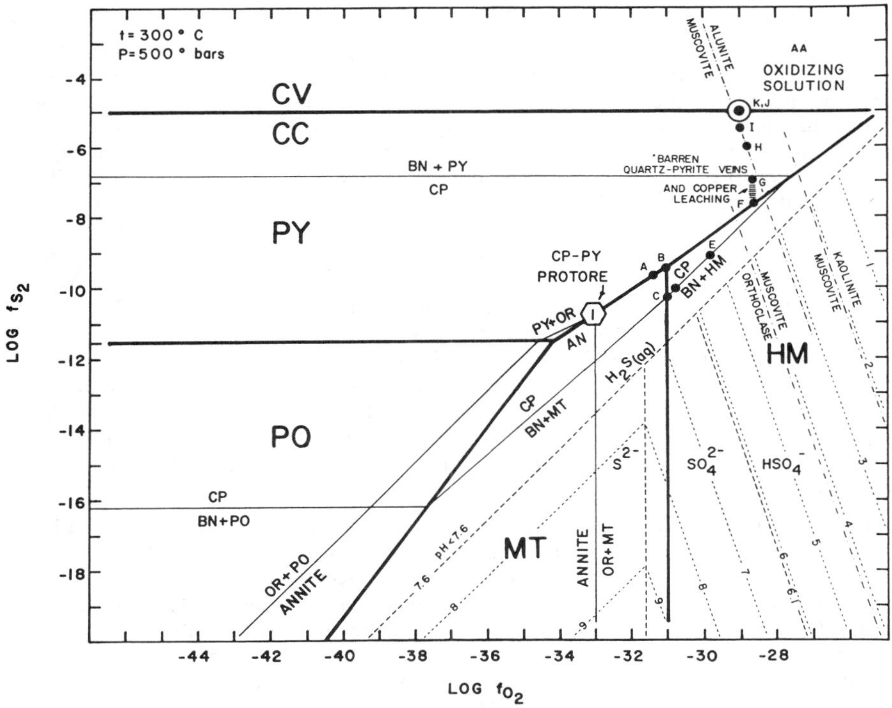

Figure 34. Isothermal isobaric fugacity diagram showing the stability fields of covellite (CV), chalcocite (CC), pyrite (PY), pyrrhotite (PO), magnetite (MT), and hematite (HM). Chalcopyrite field is surrounded by bornite plus an additional sulfide. Annite stability field is surrounded by orthoclase plus sulfides and oxides. Position of potassium-silicate protore at Butte, Montana is given at 1. Position of advanced argillic alteration assemblage and Main Stage oxidizing fluid is at the intersection of CV-CC phase boundary and that of alunite- muscovite. From Brimhall (1980) and Brimhall and Ghiorso (1983).

chalcopyrite-bearing potassic alteration assemblages. Hence, the pre-Main Stage mineralization serves as a protore for leaching copper which is remobilized and reprecipitated during Main Stage hydrothermal mineralization.

Thermodynamic modeling of hypogene oxidation and sulfidation: effects of magmatic volatiles on hydrothermal fluids and protores. Thermodynamic modeling of the copper mass transfer process described is possible using standard thermochemical data available in SUPCRIT (Helgeson et al., 1978) in conjunction with numerical modeling programs to simulate the irreversible reaction chemistry (Wolery, 1979). First, phase equilibria are examined which are relevant to the mineral assemblages present in the potassic alteration protore. Second, the end product Main Stage mineral assemblage is characterized in terms of its equilibrium fluid composition at conditions typical of vein formation (300°C and 1 kbar). The protore mineral assemblage chalcopyrite, pyrite, biotite magnetite is the starting point for the Main Stage reaction path which leads ultimately to the mineral assemblage chalcocite-covellite-muscovite-alunite as seen in Figure 33, an oxygen-sulfur fugacity diagram (Holland, 1959, 1965; Meyer and Hemley, 1967; Brimhall, 1979, 1980). While the pre-Main Stage assemblage formed at much higher temperatures, near 650 °C, Main Stage reactions proceeded at a lower temperatures, near 300°C, based on fluid inclusion homogenation temperatures. The Main Stage fluid, in the areas of most intense alteration, is in equilibrium with the advanced argillic assemblage (alunite, muscovite, kaolinite, quartz), often with covellite and chalcocite (Hemley and Jones, 1964; Meyer and Hemley, 1967; Hemley et al., 1969; 1980). It is clear in Figure 34, that this high f_{S_2} fluid composition is at considerable disequilibrium with respect to the chalcopyrite-pyrite-biotite-magnetite protore. The highly acid fluids of the Main Stage were probably generated by degassing of late volatile-rich magmas, volatile condensation, ionization, and subsequent hydrothermal metasomatic reactions between protore, magmatic gasses (SO_2 and H_2S), and ground water (Brimhall and Ghiorso, 1983; Head et al., 1987).

Destruction of wall rock buffer control: the role of biotite. The isothermal calculation of the reaction path is shown in Figure 34 extending from the protore containing chalcopyrite along the pyrite- magnetite phase boundary, along magnetite-hematite, then along chalcopyrite- hematite, finally leaving the chalcopyrite field and arriving ultimately at the covellite-chalcocite boundary in the presence of alunite-muscovite. Upon leaving the chalcocite field, an interval of the reaction path is attained over which the fluid is not in equilibrium with any copper-bearing sulfide. The copper released from the dissolution of chalcopyrite is all contained in the fluid phase. This is thermodynamic confirmation of copper leaching from the protore. To understand the phase equilibria involved it is useful to introduce the $a_{H_2S_{(aq)}}$ and $a_{Fe^{2+}}/a_{Cu^+}$ variables. Figure 35 presents phase diagrams at different a_{H_2S}, and shows how the biotite stability field is limited to log a_{H_2S} values less than -0.5. The composition of the protore is shown in Figure 35d which, given the sulfide-oxide-silicate assemblage present, buffers the log f_{O_2} at a value of -33. In contrast, the Main Stage fluid shown in Figure 35f, has a log f_{O_2} at a much higher value, -29. The role of biotite in buffering the oxygen fugacity, and tending to keep it at a low value in equilibrium with the protore is evident in Figure 35d. Figure 36 shows the dramatic effect of H_2S fugacity on the equilibria, especially the Fe to Cu activity ratio in the fluid. This figure is a plot of the triple points shown in Figure 35.

Combining Figures 35 and 36 in Figure 37 conveniently shows the region occupied by biotite, a domain characterized by a high $a_{Fe^{2+}}/a_{Cu^+}$ and low f_{H_2S} and f_{O_2}. The pyrite-chalcopyrite-orthoclase-biotite protore assemblage plots directly on the upper and outer phase boundary of biotite. At higher values of f_{O_2}, magnetite is stable instead of biotite. The computed Main Stage reaction path departs this protore composition only when biotite is destroyed by hydrolysis reactions in which muscovite (sericite) is produced. At this point, the buffering effect of the protore assemblage to maintain an equilibrium fluid composition is eliminated, and the fluid composition is free to evolve towards the Main Stage fluid, a composition rich in aqueous H_2S and Cu (AA for advanced argillic assemblage). Over this path leading up to the advanced argillic assemblage, the fluid becomes progressively more oxidized and sulfidized. This assemblage is attained in nature only in the

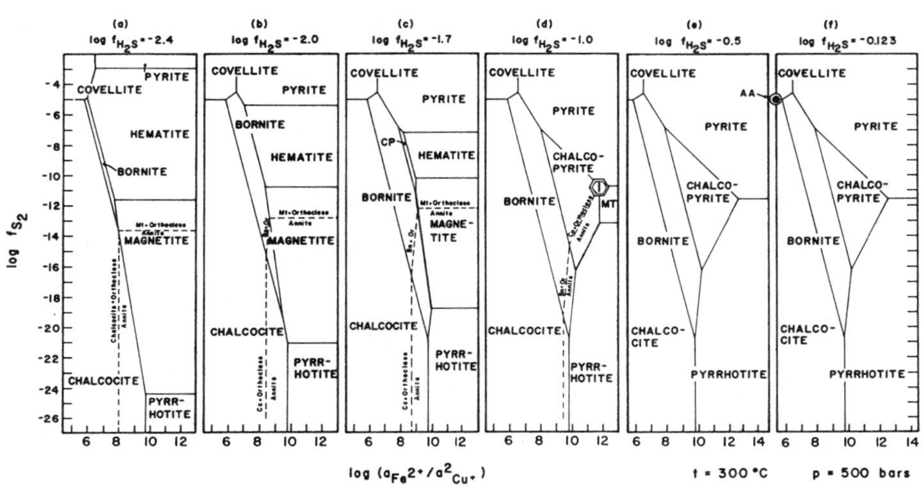

Figure 35. Oxygen fugacity-iron/copper activity diagrams at 300°C and 500 bars. Figures 35a through f are cross sections of Figure 36 at different H_2S fugacities where the triple points of Figure 35 are shown. The composition of the potassic-alteration protore assemblage is shown in relation to the composition of the Main Stage fluid. The annite (biotite) stability field occurs in the lower right of each figure except at highest H_2S fugacities where it is unstable. From Brimhall (1980).

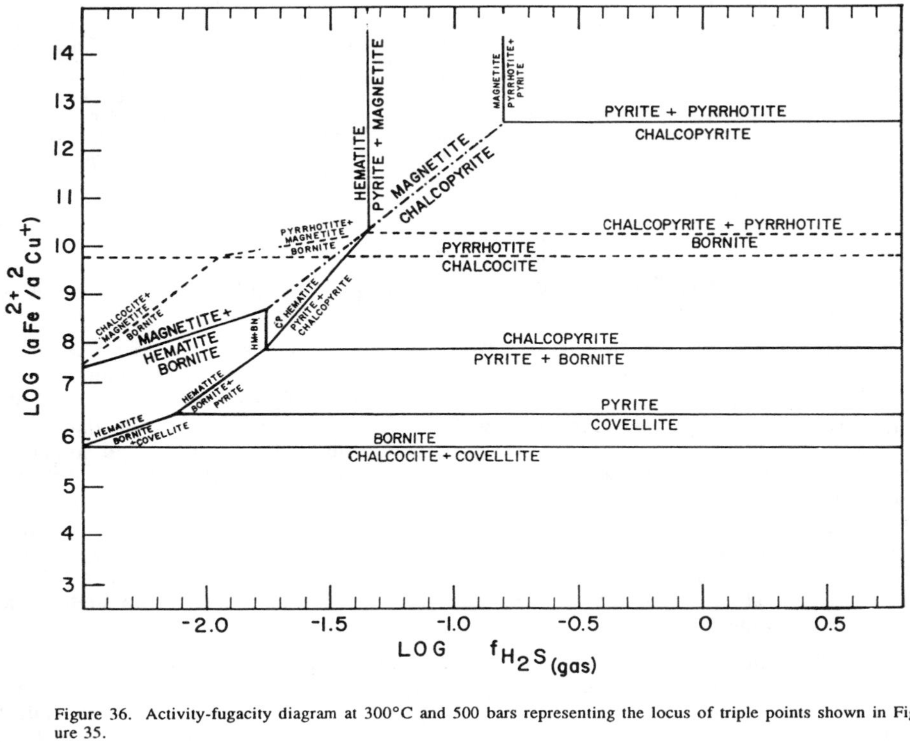

Figure 36. Activity-fugacity diagram at 300°C and 500 bars representing the locus of triple points shown in Figure 35.

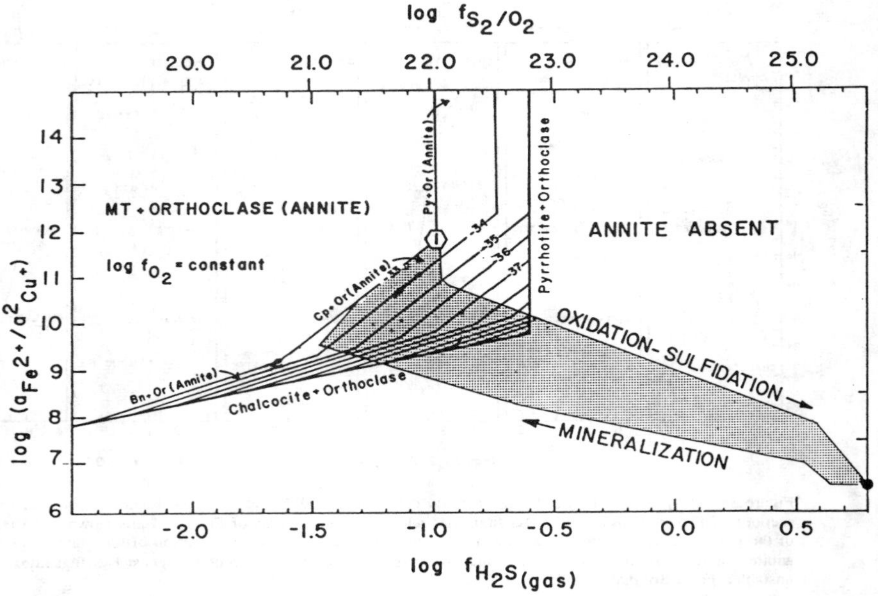

Figure 37. Calculated reaction loop for growth of Main Stage hydrothermal veins from Potassic alteration protore which plots at point 1 on the phase boundary of annite-chalcopyrite-orthoclase-pyrite. Biotite (annite component) is restricted to the upper left portion of the diagram at high aqueous Fe/Cu ratios and low activity of H_2S. Oxygen fugacity contours are shown. Above a log fugacity of oxygen of −33, annite is unstable, yielding to magnetite, orthoclase, and water. Biotite imposes strong buffering effects on reacting fluids, and until it is destroyed, fluid composition is largely fixed. Upon alteration of biotite by sericitization, the reaction path proceeds to the advanced argillic composition (alunite, muscovite, kaolinite) (AA), as magmatic volatiles are released and ionize reacting with the protore. With cessation of magmatic gas contamination of the hydrothermal fluid, reaction occurs between the advanced argillic fluid and the protore along the mineralization path. In contrast to the oxidation-sulfidation path leading the the AA fluid over which copper is leached, the mineralization path precipitates ore sulfides. These are the Main Stage Veins. Their copper has been remobilized from the older disseminated chalcopyrite protore. From Brimhall (1980).

regions of most intense fluid circulation, at the veinward edge of alteration halos, or in regions of highest fracture density within or near the parent pluton.

There are two parts to the calculated reaction path. The first is due to the development of the Main Stage fluid, the hypogene oxidation-sulfidation path leading to the advanced argillic assemblage (Brimhall and Ghiorso, 1983). This process has modeled the degassing of SO_2-H_2S-rich felsic magmas, with interaction with ground water in equilibrium with the protore. The magmatic volatiles disproportionate to yield sulfuric acid solutions. Over this leg, copper-bearing sulfides are only briefly thermodynamically stable as the fluid composition has such a high f_{O_2} that pyrite is the only stable sulfide see Figure 38 (and Fig. 35e). During this excursion, copper is leached from the protore and enriched in the Main Stage fluid. Finally, during reaction between this advanced argillic fluid and the protore, the stability of copper sulfides is once again attained (Brimhall, 1980). This is the Main Stage mineralization leg of the reaction path (Figs. 37 and 38) over which high grade copper sulfide veins have formed.

Feedback of chemical reaction and fluid flow: fluid dominated threshold states and the importance of the advanced argillic alteration mineral assemblage. The theoretical mass transport calculations modeling redistributive phenomena of vein formation help isolate some of the critical attributes necessary for ore metal remobilization to occur. What

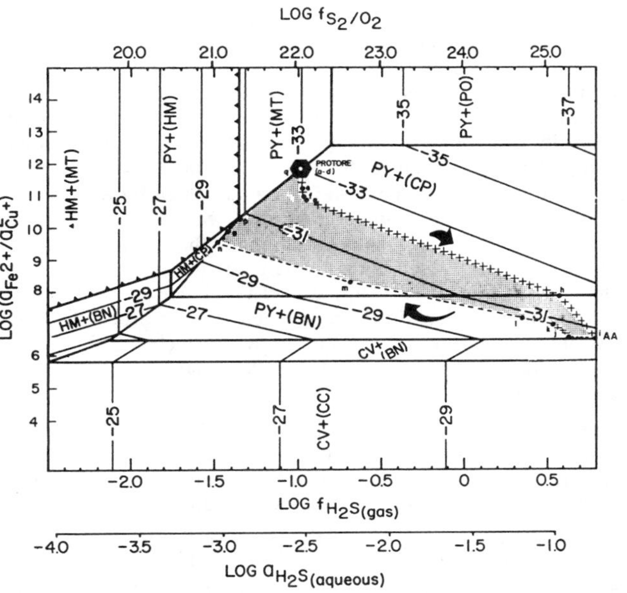

Figure 38. Same coordinate system as Figure 37, but this figure is contoured on log oxygen fugacity. Phase boundaries are shown as surfaces contoured on log oxygen fugacity with the mineral above the surface (high) fugacity phase shown without parentheses and the subjacent phase (low oxygen fugacity) phase in parentheses. For example, PY+ (CP) means pyrite exists at a higher oxygen fugacity than chalcopyrite as shown in Figure 35. The reaction loop of Figure 37 is shown. This path is shown with a plus sign when it is up in the pyrite (PY) field and as a minus sign when it occurs along a phase boundary with a copper-bearing sulfide. The oxidation-sulfidation path leaches copper, the mineralization path fixes copper. From Brimhall (1980). Points along reaction path labeled (a-q) are also shown in Figure 39 for comparison.

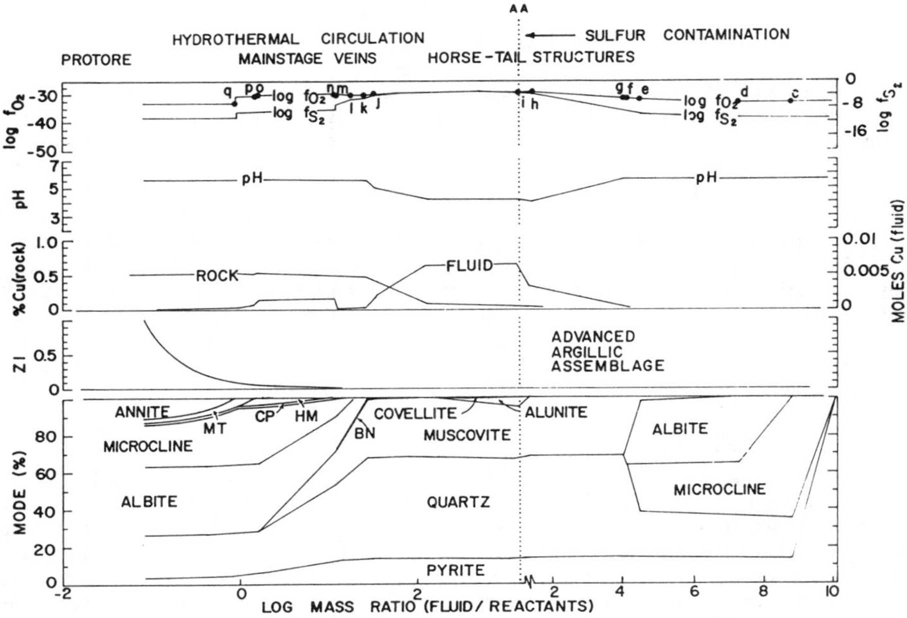

Figure 39. Complete reaction cycle at 300°C and 500 bars. Read from left to right the sulfur contamination path followed by the the hydrothermal mineralization phase with the advanced argillic (AA) zone in between. The AA represents extreme values in aqueous copper, pH, and molar volume change of reaction. Negative volume change may mean that porosity or permeability are enhanced at the AA stage. From Brimhall and Ghiorso (1983).

emerges from the simulation is the fact that the advanced argillic alteration mineral assemblage is a unique hydrochemical state in the chemical evolution of sulfur-rich hydrothermal systems. Not only are the oxidation and sulfidation states extremely high, so high in fact that copper-bearing protore host minerals become unstable releasing copper to the ore fluid, but a structural threshold is attained as well. Figure 39 shows that during the development of the advanced argillic alteration assemblage a negative calculated volume change (products minus reactants) of the alteration reaction occurs. This state corresponds also to maxima in f_{O_2}, f_{S_2}, and aqueous copper content and is accompanied by a minima in solution pH. With the indicated negative volume change, creation of void space or enhancement of fracture permeability is likely. Once this state is attained, then it is probable that continued reaction with wall rock mineral assemblages will tend to neutralize the advanced argillic fluid, and ultimately reprecipitate ore sulfides. The enhanced porosity or permeability increases the likelihood that such hydrothermal reactions can proceed to completion. The advanced argillic assemblage can be viewed then as a physiochemical threshold state, which if attained, can have major ore-forming consequences, particularly metal redistribution in the porphyry copper environment.

Consistent with these conclusions based on chemical thermodynamic modeling, are more recent calculations combining fluid flow with chemical reaction (Lichtner, 1985; Helgeson and Lichtner, 1987) which indicate that the decrease in fluid pressure accompanying upward flow of acid solutions favors dissolution of sulfides and precipitation of quartz.

<u>Epithermal Systems: manifestations of deep porphyry mineralization?</u> There is mounting evidence suggesting that the alteration- and ore mineral assemblages in acid-sulfate precious metal deposits and in some active geothermal systems within calc-alkaline stratovolcanoes may be the upper manifestations of deeper base metal mineralization (Wallace, 1979; Henley and Ellis, 1983; Brimhall and Ghiorso, 1983). However the question remains as to whether acid-sulfate epithermal precious metal deposits are typically located above porphyry copper deposits, whether they are near surface, low temperature equivalents, or whether they are shallow deposits formed from a reworked porphyry copper deposit at depth (Healds et al., 1987).

SECONDARY ORE DEPOSITION

Upon erosion of the overlying volcanic or plutonic edifice, base metal ores related to porphyritic intrusives are exposed to reactive surficial fluids, both ground water and air. All primary ores, of porphyry type or otherwise, are subject to intense chemical weathering in the near surface environment under conditions far different than at depth. Reaction products formed during imposition of surficial conditions are referred to as secondary ores, in contrast to all the older products of multi-stage processes which enriched ore metals at greater depths, generally under the influence of hydrothermal fluids. Not all ore-grade weathering products have hydrothermal ores as a precursor. In fact, under certain conditions, weathering produces minable ores from rather ordinary rocks containing ore metals at only crustal abundance levels, for example the formation of bauxites from granites. Analysis of these processes will be included here as they are part of a family of related secondary transport processes and help to illustrate the full range of possibilities of hydrochemical behavior.

<u>Atmosphere-dominated states</u>

To a large extent, given the uniquely oxygen-rich composition of the earth's atmosphere, surficial oxidation is a ubiquitous factor in hydrochemical modification of sulfide ores and exposed rocks in general. So intense are the oxidation and hydrolysis reactions which proceed to minimize the state of disequilibrium between the atmosphere and rocks formed in the subsurface, that weathered material may often retain little resemblance to its protolith. Ore-forming constituents such as sulfides, formed at depth, may contribute additionally to the intensity and extent of weathering reactions by producing natural sulfuric acid which destroys the capacity of rock mineral assemblages to buffer oxygen fugacity, and hence resist modification by imposed fluids. Ultimately, atmospheric gasses dominate the pore spaces of weathered rocks, a condition which may be maintained to the depth of the ground water table. We will see that the transition from unsaturated to saturated conditions is an interface between geochemical systems dominated by the Earth's atmosphere, and a subsurface domain controlled by the primary mineral assemblages, protected from oxidation by the presence of water in pore spaces which precludes oxidizing atmospheric gasses.

Chemical weathering is responsible for the final enrichment of ore metals in large low-grade ore deposits which in many cases, for example copper, is necessary for the mineralization to be considered minable. Because of the diversity of primary ore types and the zoning within a single deposit, chemical effects due to weathering are varied and depend upon the level of exposure and position within a primary district zoning pattern.

Some enrichment mechanisms involve the dissolution of ore minerals with subsequent transport and ultimate redeposition of ore metals in an enriched state. Other processes enrich ore metals without their migration simply by virtue of removing other more mobile species. In all cases, weathering results in strong physical and chemical modification of primary ores. Consequently, in order to interpret elemental mobility patterns, it is necessary to consider both physical properties as well as chemical composition in addressing secondary enrichment. The relationships between rock density, chemical composition, porosity, and deformation are described by constitutive mass balance models given in Brimhall et al. (1985) and Brimhall and Dietrich (1987). Mass balance models describe the conservation of mass of an element of interest and provide a practical means of using chemical elements as geochemical tracers for interpreting mass transport mechanisms. Mass balances provide the ultimate limits on enrichment processes.

Constitutive mass balance models and simplified chemical controls

Four major classes of surficial transport behavior account for the secondary enrichment of ore deposits: (1) residual, (2) supergene, (3) hypogene, and (4) perfectly mobile or continuously leached. These differ primarily as to whether or not an ore metal of interest is a mobile chemical species, and if so, whether its migration is upwards or downward, and is reprecipitated or not. In the case of mobile elements, the concern for this sense of transport has to do with identification of the source region of the metal of interest, and hence the direction from which metals have come: "supergene" literally from above, "hypogene" from below. This directionality in transport of mobile metals is also of significance because of a need to understand the contrasting physiochemical conditions may which at first release metals from the source, mobilize them into ground water, and ultimately, reprecipitate them in an enriched state. Considering the broad range of metals in primary ores, all four types of transport behavior often occur within the same secondary hydrochemical system. The specific behavior taken by an element depends upon the mineralogy and pore structure of the protolith, mineral stability and kinetics of a particular system in relation to advective flow and gas diffusion rates of ground water and ground gasses respectively. The behavior of a given element can vary markedly from system to system, in one being immobile and residually enriched, to nearly complete removal as a mobile species.

Residual enrichment. If an ore metal of interest is present in a rock before weathering, and is geochemically immobile, it may become enriched simply by removal of perfectly mobile elements from the rock by migrating ground water. This passive enrichment is referred to as residual, and results for the most part by a rock becoming more porous, as unstable minerals are attacked and dissolved. The mass balance model describing conservation of mass of an immobile species during residual enrichment is simply that the total mass of an element contained within a representative elementary volume after weathering is equal to the mass of that element before weathering. The mass of such an element is given by the product of the volume of the system of interest, V, bulk rock density, ρ, and concentration, C, before weathering in the protolith state, p, and after weathering, w (44):

$$V_p \rho_p C_{i,p} = V_w \rho_w C_{i,w} \quad . \tag{44}$$

This may be reduced from a three dimensional volume to one dimension, B. The possible effects of compaction or expansion during weathering can be accounted for by defining strain, ε_i, as the change in length in the vertical direction divided by the initial length (45)

$$\varepsilon_{i,w} = \frac{B_{i,w} - B_{i,P}}{B_{i,P}} \quad . \tag{45}$$

With this definition of deformational strain, and reduction to one dimension where V terms reduce to one dimension, B, (44) may be solved in non-dimensional form for an enrichment factor (46), the ratio of the concentration of an immobile element in the weathered state of a rock to that of the protolith.

$$\frac{C_{i,w}}{C_{i,p}} = \frac{B_{i,p}}{B_{i,w}} \frac{\rho_p}{\rho_w} = \frac{1}{\varepsilon_{i,w} + 1} \frac{\rho_p}{\rho_w} \quad . \tag{46}$$

This expression offers a simple means of interpreting the enrichment factor of an element in terms of changes in physical properties of a rock after weathering. In Figure 40 the residual enrichment mass balance model is shown diagrammatically, and in Figure 41 a set of data is presented on nickel enrichment in weathered ultramafic protoliths (Brimhall and Dietrich, 1987). Closed system residual enrichment of nickel is evident as a harzburgite protore is converted into laterite saprolite and pisolitic reddish soil, all without demonstrable deformation, a slope of one for $\varepsilon = 0.0$. It is clear that from (46) simple residual enrichment is governed primarily by the magnitude of the change in rock density, and secondarily by collapse, that is negative strain.

Figure 42 explains the observed paragenesis in typical nickeliferous laterite profiles in terms of the equilibrium solubilities of minerals (Golightly, 1981) given as moles per liter of cationic species. In this figure ideal or stoichiometric solubility as a function of solution pH, is consistent with the vertical weathering profile with the least soluble mineral, goethite at the top, and the most soluble mineral, forsterite at the bottom. The most soluble mineral is followed upwards by pyroxenes, serpentines, chlorite, talc, nepouite (nickel serpentine), kerolite (nickel talc), kaolinite, gibbsite, and finally goethite on top. The pH of surface water in laterites has been measured at values near 5, while at depth the fluids are alkaline, with a pH of 8.5. The nickel originally contained in forsterite at a concentration of generally 0.20 wt.%, is ultimately retained in the weathering product, and enriched in accord with the reduction in density.

The chemical weathering path by which this dramatic density reduction occurs and causes residual enrichment is shown in Figure 43, extending from a non-porous protolith with an initial bulk density of 3 and an average grain density of 3.1 to saprolites with bulk densities of 0.6, grain densities of 2.5 and porosities of 0.75 (75%). Porosity is calculated from bulk and grain density from the relationship: porosity equals one minus bulk density divided by grain density.

Supergene enrichment. The second type of transport process which causes metal enrichment during weathering is supergene enrichment. This is the dominant type of secondary enrichment leading to ore-grade deposits as the supergene enrichment factors are typically several times those of maximum residual enrichment. The difference between the mobility of elements which are enriched by supergene and hypogene processes and perfectly mobile elements, is that in the former two cases, once mobilized, elements are at least partially reprecipitated. The efficiency of reprecipitation is addressed in (Brimhall et al., 1985). In supergene enrichment ore metals are introduced to a zone of enrichment from above where leaching has released them to migrating meteoric water.

Figure 44 shows that in copper systems, leaching of copper from primary sulfides (typically chalcopyrite and bornite) occurs, followed by reprecipitation of copper as secondary chalcocite and covellite. Mg, Ca, and Na are leached without reprecipitation, behaving as perfectly mobile elements. In deposits other than porphyry copper deposits, for example gold-rich laterites developed over primary mineralization in greenstone terranes, Au, Ce, Cr, Mo and Fe behave as supergene elements with Nb, Sn, Th, V, W, Zr, Al, and Ti being

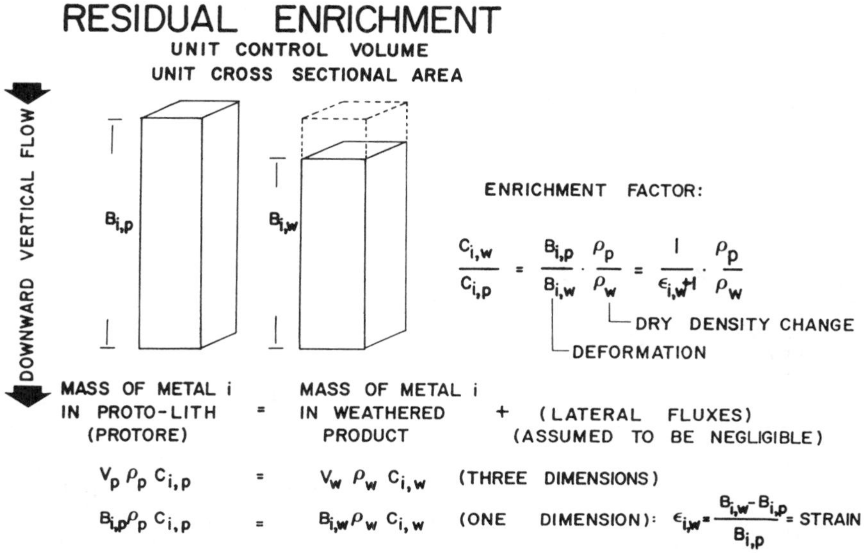

Figure 40. Mass conservation during residual enrichment, showing how three-dimensional volume changes may be represented in one dimension using strain. From Brimhall and Dietrich (1987).

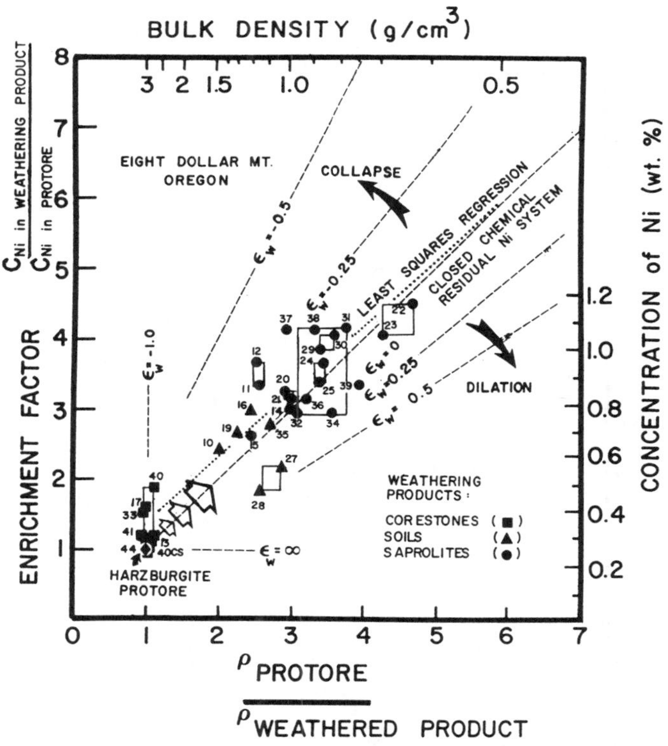

Figure 41. Data array for Eight Dollar Mountain Ni laterite in southwestern Oregon. The boxes represent data for two samples taken at the same depth in the weathering profile. Weathering path is shown with arrows extending away from the ultrmafic protore through saprolites and soils. Position of data array with respect to lines of constant strain (collapse or expansion indicate that there was no systematic deformation during weathering. From Brimhall and Dietrich (1987).

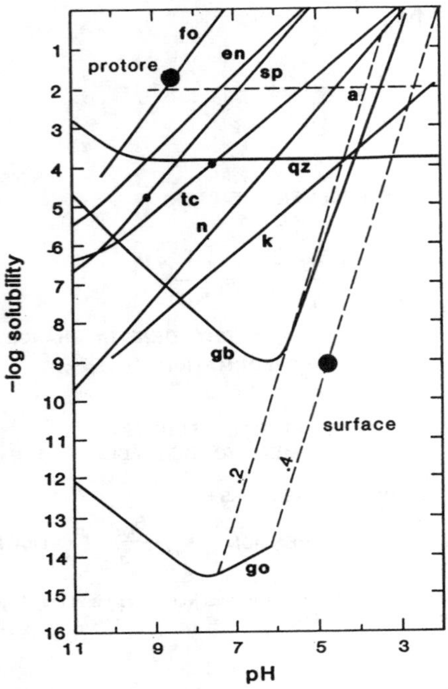

Figure 42. The pH dependence of the congruent solubility of minerals in laterite development. S is the calculated equilibrium concentration in moles per liter of Mg, Si, Fe, Al, or Ni as appropriate. Abbreviations are: fo=forsterite, en=enstatitie, sp=serpentine, tc=talc, n=nickel serpentine (nepouite), k=nickel talc (kerolite), qz=quartz, a=amorphous silica, gb=gibbsite, go=goethite. pH at the surface is about 4.5 and 8.5 at depth. From Golightly (1981).

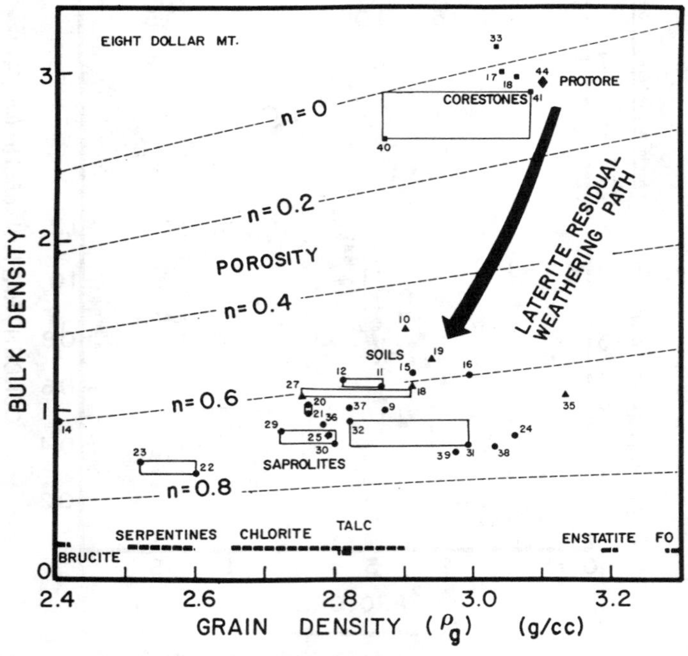

Figure 43. Variation in bulk density as a function of grain density showing the weathering path from the protore for the Eight Eight Dollar Mountain deposit, Oregon (Fig. 41). From Brimhall and Dietrich (1987).

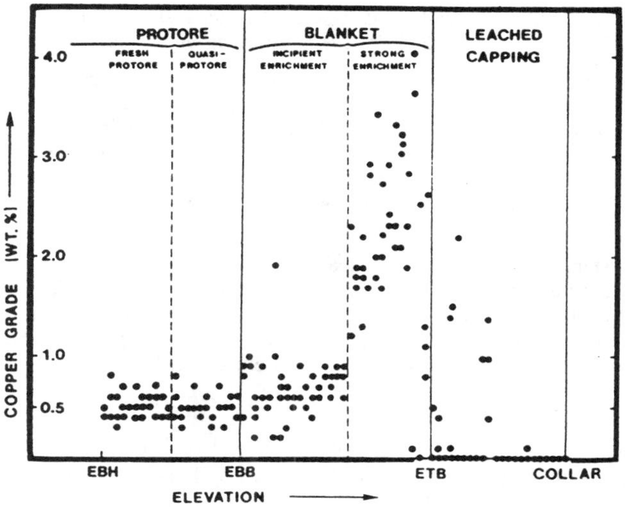

Figure 44. Supergene copper grade profile for the La Escondida porphyry copper deposit, northern Chile. From Brimhall et al. (1985).

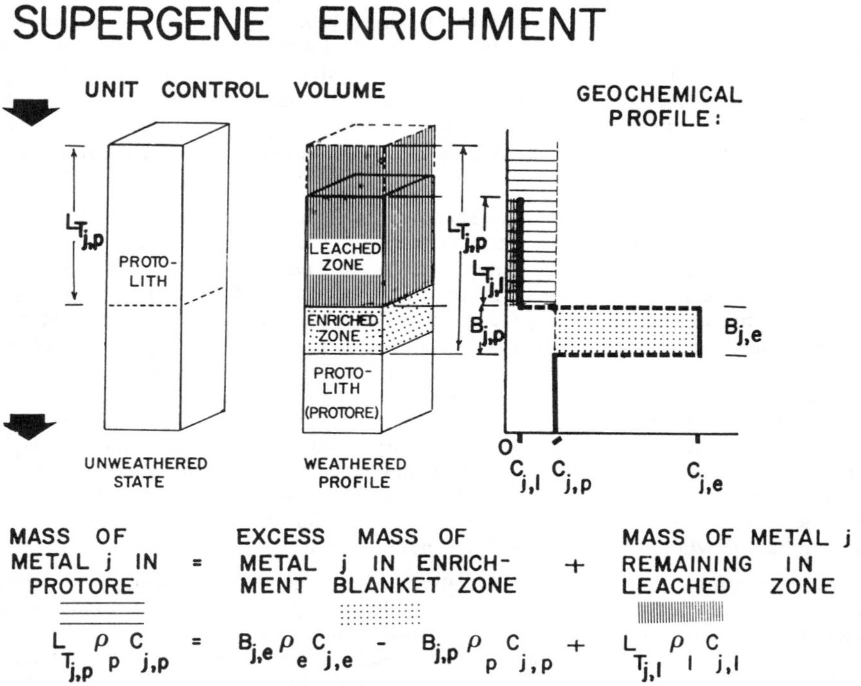

Figure 45. Mass balance model for supergene enrichment depicting the meaning of terms in the algebraic model. Mass of metal within a control volume is given by a product of column height, bulk density, and metal concentration. From Brimhall and Dietrich (1987).

immobile and enriched only residually (Davy and El-Ansary, 1986). Ba, Co, La, Mn, Rb, Sr, Ta, Y, Zn, Ca, K, Mg, and Na are all leached, behaving as perfectly mobile elements.

Local secondary gold enrichment, increase in gold fineness, and formation of gold nuggets are currently of major importance in mineral exploration, and have been ascribed to gold complexing by transient sulfur-bearing ligands such as thiosulfate and bisulfide complexes (Webster and Mann, 1984; Webster, 1986; Stoffregen, 1986) in regions where chloride-rich ground water is often lacking, or alternatively by chloride complexes where such fluids are known (Mann, 1984; Webster and Mann, 1984). Gold is clearly transported by supergene processes.

Supergene systems are therefore chemically differentiated into two related subsystems positioned sequentially along a fluid flow path, a source region and a sink, where reprecipitation occurs. Obviously, the two subsystems are related through rigid mass conservation constraints of a mobile element, j. Equation (47) gives the supergene enrichment factor in non-dimensional form equivalent to (46) for residual enrichment.

$$\frac{C_{j,e}}{C_{j,p}} = \frac{(L_{T_{j,p}} + B_{j,p})}{B_{j,p}} \frac{1}{(\varepsilon_{i,e} + 1)} \frac{\rho_p}{\rho_e} - \frac{L_{T_{j,p}}}{B_{j,p}} \frac{C_{j,l}}{C_{j,p}} \frac{(\varepsilon_{i,l} + 1)}{(\varepsilon_{i,e} + 1)} \frac{\rho_l}{\rho_e} \quad . \tag{47}$$

These terms are illustrated in Figure 45. In (47) $L_{T_{j,p}}$ is the total thickness of the zone of leaching of supergene element, j. $B_{j,p}$ is the thickness of the zone of enrichment of element,j. The strain terms for immobile element i undergoing simple residual enrichment, are $\varepsilon_{i,e}$ for the zone of enrichment and $\varepsilon_{i,l}$ for the leached zone. The density terms are rp, re, and rl for the protore, enriched and leached zones respectively. The concentration terms for supergene element, j, are $C_{j,e}$, $C_{j,p}$, and $C_{j,l}$ for the enriched zone, protolith and leached zone respectively. Note that (47) reduces down to simple residual enrichment in the limit when the thickness of the leached zone, $L_{T_{j,p}}$ approaches zero. Residual and supergene enrichment are therefore two extreme cases of related processes. The prime difference is the focussing effect of removing elements from a leached zone and concentrating them in a zone of enrichment. This volumetric effect is portrayed in the first term in (47) $(L_{T_{j,p}} + B_{j,p})/B_{j,p}$, which enhances the enrichment due to change in density (rp/re) as leached zones are typically several times as thick as enriched zones. Collapse, negative strain, tends to increase the enrichment factor as well, as does complete leaching in the leached zone as $C_{j,l}$ approaches zero.

In (47), the only unknown is $L_{T_{j,p}}$. This may be found by substituting in appropriate values of the other variables, concentration, density, and strain. Given the difference between the calculated total leached column height $L_{T,p}$ and the thickness of the present leached zone, the amount of erosion can be calculated, and with age dates of secondary minerals, the erosion rate determined (Alpers and Brimhall, 1988a,b). Also pre-weathering paleo-topography can be reconstructed from addition of $L_{T_{j,p}}$ terms for weathering columns to the elevation of the top of secondary enrichment. This provides a means to evaluate sub-surface horizontal fluid fluxes in terms of hydrological gradients and regional fracture patterns. This use of mobile supergene elements in appropriate mass balance models affords a means to relate near surface geochemistry to surficial processes and geomorphic evolution. Thus there is clearly a complex interplay between heterogeneous mineral-solution-gas reactions affecting chemical transport and weathering and the ground water table. These effects are amenable to thermodynamic modeling, and illustrate the necessity of treating fluid flow along with chemical transport in order to approach reality.

<u>Thermodynamic and fluid flow modeling of supergene enrichment.</u> The division between the leached and enriched zones is generally the top of the capillary fringe, a zone of saturation above the ground water table in which capillary forces between grains induce tension saturation. Above the capillary fringe the voids are filled with air making oxidation a dominant process. From the top of the capillary fringe downward, ground water fills the voids precluding air except as a relatively minor dissolved constituent. The top of the capillary fringe separates two quite distinct environments, and hence represents a dramatic gradient in chemical composition.

The oxidation of pyrite within the unsaturated zone is a process of great importance in supergene enrichment since hydrogen ion is generated by pyrite destruction. The rate at which acid generation occurs is largely controlled by the amount of oxygen available. The movement of O_2 gas then is of critical importance, especially its replenishment in the unsaturated zone. Diffusion of oxygen through air-filled pore spaces is considered to be the dominant process of oxygen interchange between the atmosphere and the porous rock undergoing weathering. Troeh et al. (1982) developed Fick's first law type equations for diffusive transport in porous materials (48).

$$Q = D_0 \frac{V_v(1 - S_{sat})}{V_T} A \frac{\Delta C}{\Delta L} \tag{48}$$

Here Q is mass per second of a gas diffusing across cross sectional area A. $\Delta C/\Delta L$ is the oxygen concentration gradient. V_v is the void volume, and S_{sat} is the saturation state, and D_0 is the maximum diffusional rate of oxygen through air. The saturation term reduces the oxygen flux. Therefore, above the capillary fringe, oxygen diffusion is much more rapid than below. This appears to be the rate limiting step in pyrite oxidation and supergene enrichment of copper. At 25°C, the computed logarithms of the oxygen fugacities in the leached, enriched, and protore portions of a porphyry copper deposit are −38, −63, and −71 showing that the effect of oxygen diffusion is dramatic. It is this change in redox state which causes elements to be leached under oxidizing conditions, transported, and finally reprecipitated in reducing environments.

By incorporating (48) describing oxygen diffusion, Cunningham (1984) has theoretically modeled the irreversible reaction chemistry of supergene copper enrichment of a pyrite-chalcopyrite-magnetite-biotite-alkali feldspar protore along with fluid flow in a simplified integrated finite difference model system consisting from top to bottom of a leached zone, enriched zone and protore—see Figure 46. The following preliminary results are offered as an illustration of a technique of approaching an important ore-forming porous media flow problem and the necessity of subdividing it into related unsaturated and saturated zones.

The computer programs used were EQ3/EQ6 (Wolery, 1979) for reaction chemistry and speciation of the aqueous fluid, and TRUST (Narasimhan et al., 1977) for fluid flow through the unsaturated zone into the capillary fringe, and into the saturated zone. EQ3/EQ6 solves a series of differential equations describing mass balance, mass action, and ionic strength. The algorithm is based on the work of Helgeson (1968). Kinetic data for minerals used are as follows: rates of silicates are from Wood and Walther (1983), dissolution rate of pyrite is from Wiersma and Rimstidt (1984), and chalcopyrite dissolution rate from Braithwaite (1976). Program TRUST uses the integrated finite differences method to solve for ground water flow conditions, specifically mass conservation of water, the hydraulic potential gradient, and fluid fluxes.

Computed chemical reaction paths, considering both fluid flow and chemical reaction are shown in Figures 47, 48, and 49. In this modeling an explicit coupling of fluid flow and chemical reaction has been accomplished by alternating between fluid flow calculations, fluid mixing, and chemical reaction over short time steps. No feedback between fluid flow and such chemical variables as volume change of net chemical reactions is considered here. The starting fluid composition is rain water, which reacts with the minerals contained in the potassium-silicate protore in which a steady, non- transient flow regime exists divided into unsaturated and saturated zones. The effluent fluid from the base of each system enters the subjacent system, and mixes with the contained fluids. Reaction progress in EQ3/EQ6 is related to real time necessary for kinetics and fluid flow through the total surface area of exposed mineral grains.

The equilibrium composition of the copper enrichment blanket and protore are distinct, the former along the chalcocite-covellite phase boundary in the alunite field, the latter in the pyrite-chalcopyrite field near equilibrium with biotite (Fig. 47), consistent with petrographic observation. The relatively reducing conditions prevailing in the enrichment blanket and protore are evident in Figure 47 in contrast to those in the leached zone. The low sulfur

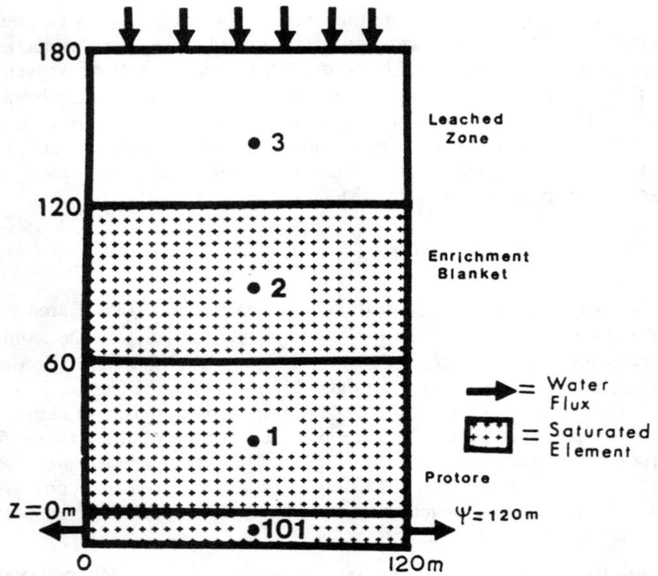

Figure 46. Idealized three-cell computation framework for steady state flow hydrological modeling of supergene enrichment. Cell 3 is unsaturated, and cells 1 and 2 are saturated. The top of the capillary fringe is taken to be the interface of cells 2 and 3. Depth below the surface is shown in meters. From Cunningham (1984).

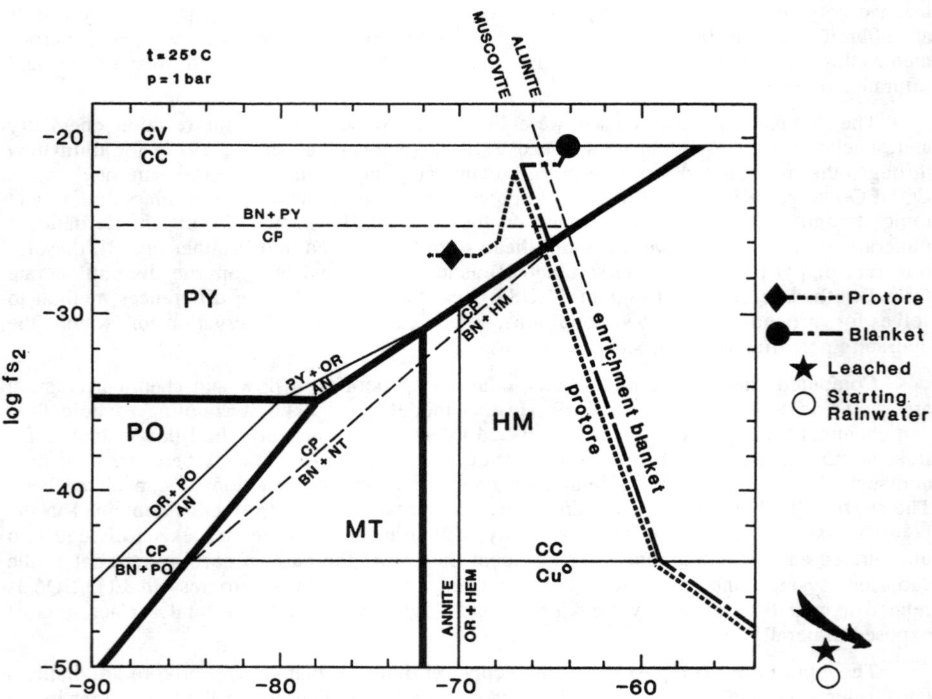

Figure 47. Computed chemical reaction paths at 25°C and 1 bar for supergene enrichment. Shown are the paths for the enrichment blanket and the protore. The starting rainwater and leached zone compositions are off-scale. Notice that the enrichment blanket reaches the chalcocite-covellite phase boundary. From Cunningham (1984).

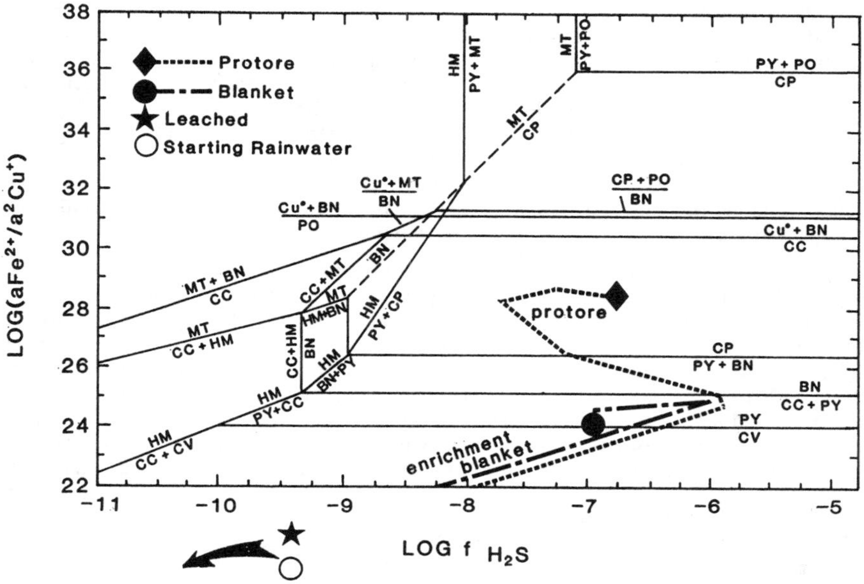

Figure 48. Computed chemical reaction paths at 25°C and 1 bar for supergene enrichment showing enrichment blanket and protore paths. Compare with the vein forming hydrothermal paths in Figures 37 and 38. From Cunningham (1984). Supergene and hydrothermal hypogene oxidation produce very distinct paths.

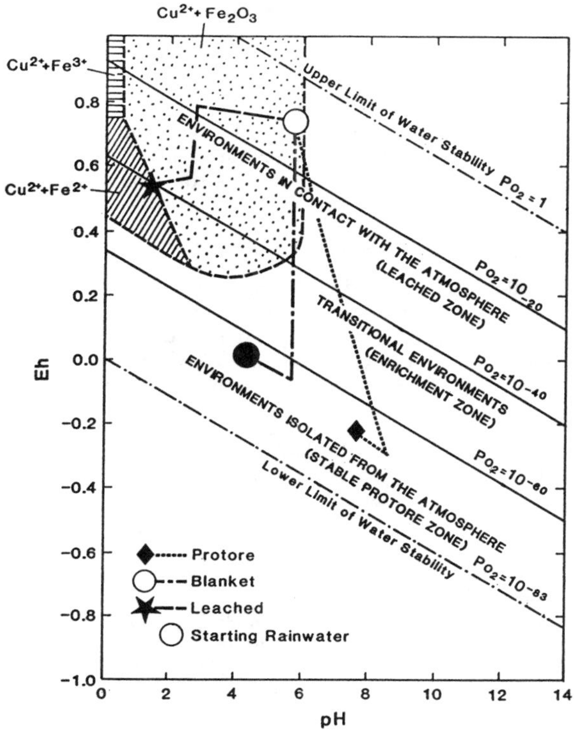

Figure 49. Computed Eh-pH diagram for supergene leaching and enrichment. The main reason for copper leaching is illustrated here as the leached zone path remains in the aqueous copper field at high Eh values. From Cunningham (1984).

fugacity in the leached zone is due to the quantitative oxidation of sulfide to sulfate which is flushed down into the enrichment blanket. The striking disequilibrium between the mineral assemblages contained in the leached zone and enrichment blanket is illustrated in Figure 48, as is the similarity of fluid composition between the enrichment blanket and the protore. The reaction paths are plotted finally on an Eh-pH diagram (Fig. 49) which confirms the similarity of enriched zone and protore in terms of relatively reducing conditions (similar oxygen partial pressures) in environments removed from the atmosphere, and the contrast with the leached zone and surficial fluid dominated by oxygen-rich gasses. This contrast is the fundamental reason for the release of copper and its reprecipitation after a short excursion by downward aqueous fluid advection.

Hypogene enrichment by ferrolysis. An analogous type of transport to supergene behavior is known for iron and perhaps manganese, differing only in that certain metals migrate upward instead of downward, and move not by advection but by chemical diffusion (Webster and Mann, 1984). The driving forces and metal concentration gradients are related to Eh and pH gradients created between the water table and ground surface during oxidation. This process has been called ferrolysis (Brinkman, 1977), a mechanism involving oxidation and hydrolysis of iron (Mann, 1984). It is not widely considered as an iron or manganese ore- forming processes in itself, but is mainly of interest because of its effects on gold solubility and supergene deposition (Mann, 1984, 1985; Webster and Mann, 1984; Mann and Ollier, 1985). Ferrolysis is however, widespread, and is likely to be important in the formation of ferruginous nodules and ferricretes in laterites and bauxites (Grubb, 1970, 1971; Sadleir and Gilkes, 1976; Valeton, 1972) in general involving an evolution from iron-rich mottles in a kaolinite matrix to Al-rich hematite, to Al-rich goethite nodules to pisolites following a cortification processes (Tardy and Nahon, 1985).

According to Mann (1984) in the case of a laterite profile iron oxidation probably occurs in two discrete steps, the first at the weathering bedrock front (49).

$$FeS_2 + 2\,H_2O + 7\,O_2 = 2\,Fe^{2+} + 4\,SO_4^{2-} + 4\,H^+ \quad . \qquad (49)$$

The second step occurs at or near the water table at a higher oxygen fugacity (50).

$$2\,Fe^{2+} + 3\,H_2O + 0.5\,O_2 = 2\,FeOOH + 4\,H^+ \quad . \qquad (50)$$

The applicability of (50) is observed by use of water samples of active weathering profiles in Western Australia (Mann, 1984). The path of oxidation and hydrolysis of iron is reconstructed on a pH-Eh diagram Figure 50 in comparison to a theoretically derived path (51) from this study assuming unit activity coefficients of dissolved species.

$$Fe^{2+} + 2\,H_2O = FeOOH + 3\,H^+ + e^- \quad . \qquad (51)$$

The measured and calculated ferrolysis paths have precisely the same slope (one electron for each iron oxidized from ferrous to ferric), but are coincident for an aqueous iron concentration of 200 milligrams per liter, the highest measured concentration. The postulated conditions and direction of iron migration by diffusion after its release at the weathering front is shown in Figure 51. There is an increase in Eh and decrease in pH up towards the water table.

The hypothesis relating gold transport to ferrolysis is illustrated in Figure 51. Dissolution of gold and silver from a primary deposit exposed to weathering undergoes dissolution by (52) at or near the water table where Cl^-, H^+, and O_2 are abundant, and there is little or no ferrous iron.

$$4\,Au + 16\,Cl^- + 3\,O_2 + 12\,H^+ = 4\,AuCl_4^- + 16\,H_2O \quad . \qquad (52)$$

Gold chloride complexes migrate, and precipitate pure elemental gold (53). On the other hand, goethite forms by hydrolysis in the mottled zone after its upward ascent and near-surface enrichment of iron—see Figure 52. The gold precipitation is thought to be due to reduction of $AuCl_4^-$ with ferrous iron (Mann, 1984).

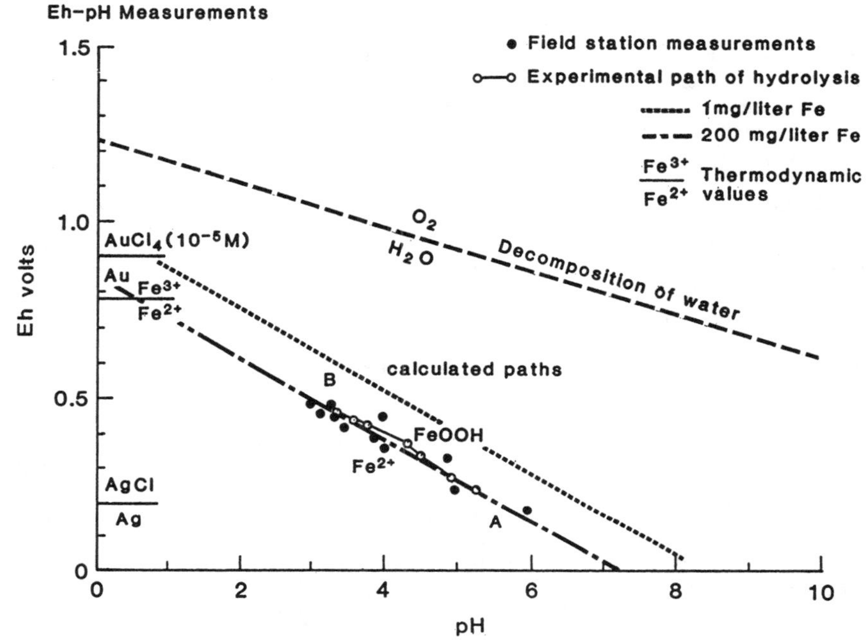

Figure 50. Eh-Ph diagram showing experimental data on weathering laterites where ferrolysis is occurring. From Mann (1984). Computed path is shown for comparison using 200 milligrams dissolved iron per liter.

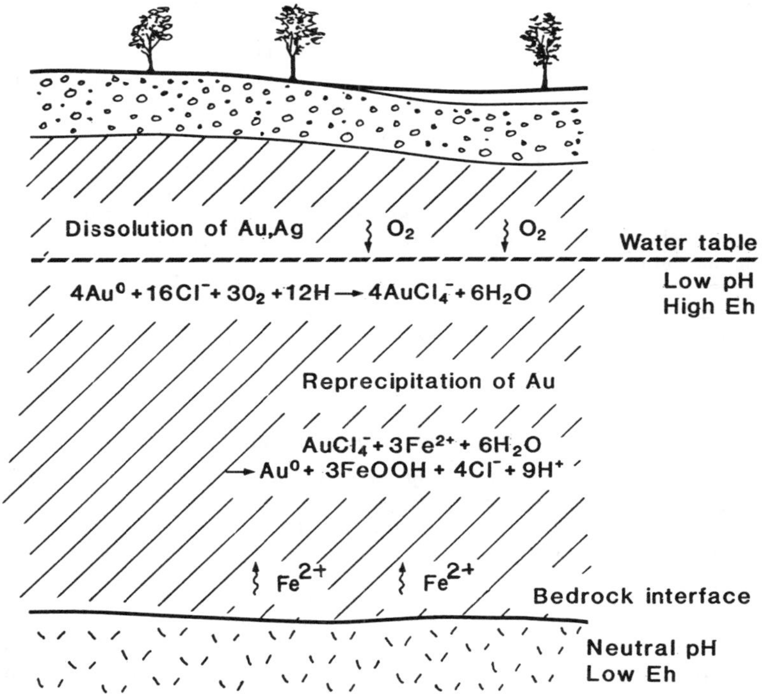

Figure 51. Chemical model for the redistribution of gold from a primary lode system by weathering in a lateritic profile. From Mann (1984).

$$AuCl_4^- + 3\,Fe^{2+} + 6\,H_2O = Au + 3\,FeOOH + 4\,Cl^- + 9\,H^+ \quad . \qquad (53)$$

The mass balance model for hypogene chemical weathering is identical to supergene in form, but the direction of transport is inverted. Many of the same limiting factors are expected to control enrichment: thickness of the leached zone in relation to the thickness of the zone of enrichment, density ratios of the enriched and leached zones with respect to protore density, and concentration in the leached zone.

Internal factors

Weathering paths in physical properties. Primary permeability in a protolith before chemical weathering has been viewed as a controlling factor of water percolating through rocks (Samana, 1986), and thus a major internal factor of oxidative weathering. Measurements of permeability of weathered rocks are presently quite rare, but a mounting data base of porosity measurements exists (Brimhall and Dietrich, 1987) which suffices to illustrate the limitations of the role of permeability in surficial environments. The results are surprising.

Figure 53 shows that there is little correlation between the initial porosity of a protolith and the final porosity after intense weathering. Bulk density and average density serve a useful purpose in providing a comparative framework for viewing weathering paths in diverse protoliths on a uniform basis regardless of the bulk composition of the rock. Bulk density varies both with the density of constituent minerals and the abundance and size of voids, and thus responds to dissolution and precipitation reactions as well as deformation. Grain density is simply the average density of the constituent mineral assemblage and is therefore a mineralogic effect. The two densities combine to give porosity. In Figure 53, five protoliths are depicted: disseminated sulfide deposits, granites, ultramafic rocks, massive sulfide deposits, and granular material such as beach sand. These parent material are converted to leached capping, bauxite, laterites, gossans, and podzols respectively. For example, ultramafic rocks with very low porosity can be converted to laterites with porosities of almost 0.8 (80 %) while beach sands with an initial porosity of 0.35 (35 %) can weather to podzols with a porosity of only 0.6 (60 %). Many other factors play a role besides porosity, including the amount of deformation, particularly collapse due to weathering, which can be as high as 50%. Also, secondary precipitation in voids can reduce porosity.

Primary permeability. Rather than consider porosity and permeability as primary factors controlling and limiting chemical weathering, it may be more accurate to consider them as being, to a large extent, consequences of chemical weathering. The principal role of permeability in the context of thermodynamic modeling may be as an important aspect of initial conditions which are rapidly modified once hydrochemical processes are initiated. Such changes in permeability may ultimately be so large that they outweigh the initial differences between systems. Clearly then, a major importance of permeability is its relationship to the fluid flow regime within the full context of surface processes including rock composition and hydrological patterns, especially the interface between unsaturated and saturated conditions. The capillary rise of water under tension saturation in the capillary fringe varies inversely with average pore radius, making a hydrology, lithology, and pore structure complexly intertwined, even at the start of chemical weathering. The selection of one controlling factor such as permeability is misleading in this respect.

Available sulfur. Another factor of major importance in oxidative weathering systems such as supergene enrichment of primary copper ores is available sulfur. Here the major effect is in the production of sulfuric acid from the oxidation of pyrite. Once hydrogen ion is formed, hydrolysis reactions proceed which modify the protolith extensively. Depending upon the pyrite to copper sulfide ratio, oxidation may be total or incomplete (Samana, 1986). In the simplest sense, chalcopyrite has a sulfur to total metal ratio S/M of 1 which means it can be extensively oxidized. Chalcocite is sulfur deficient in this respect, S/M = 0.5, requiring 1 mole of sulfur from another source for it to be oxidized. Pyrite (S/M = 2) can provide sulfur for the oxidation of other sulfide minerals. These sulfur to metal ratios are the minimum required for total oxidation since the analysis does not include the

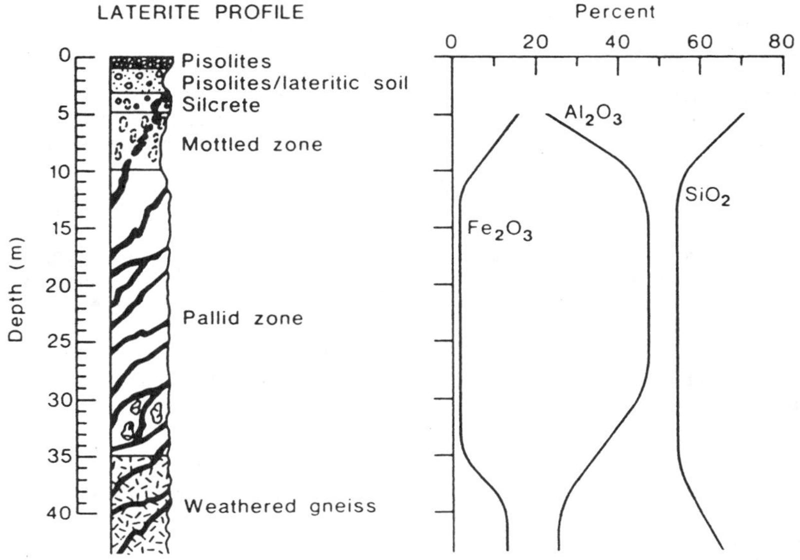

Figure 52. Chemical profiles showing iron, aluminum, and silica variation with depth. From Webster and Mann (1984).

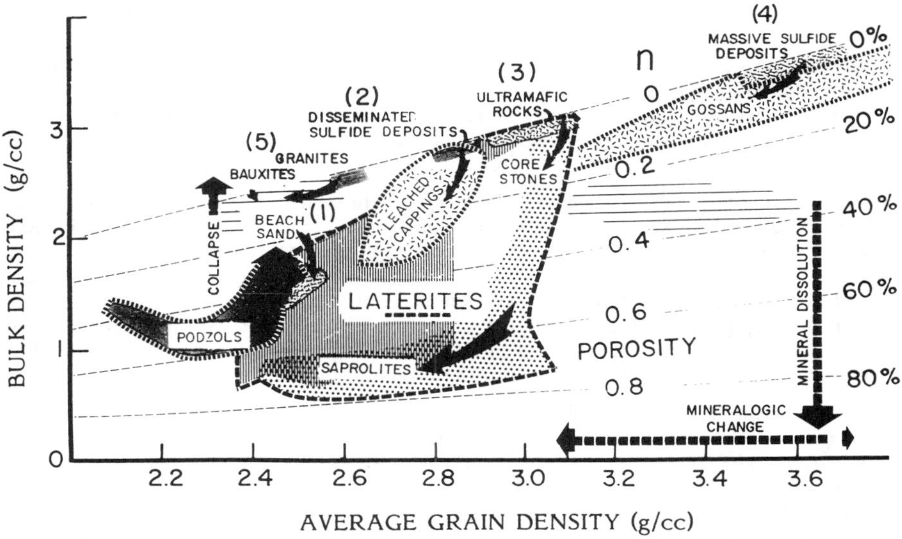

Figure 53. Comparison of physical weathering paths in terms of bulk density and average grain density for five parent materials: beach sand, porphyry copper ores, ultramafic rocks, massive sulfide deposits, and granites. These weather to podzols, leached cappin, laterites, gossans, and bauxites, respectively. From Brimhall and Dietrich (1987).

reactivity of gangue minerals. These are based primarily on the sulfide-oxide-silicate buffer assemblages, and the capacity of the rock to neutralize strong acids. It is thought that a pyrite to copper sulfide ratio of four is optimal for oxidation and secondary enrichment. Chalcocite enrichment blankets may be stable partly because of the relatively low pyrite to copper sulfide ratio, and the inhibition of oxidation-leaching processes. Perched enrichment blankets may result from lowering of the ground water table, and the relative slow rates of sulfide oxidation.

External factors

Geomorphic conditions are of major importance in surficial weathering, especially the evolution in time as changes have particularly marked effects on the ground water table and the rates of erosion in relation to rates of chemical weathering. In this respect, the single most important external factor seems to be climate, as it relates to rainfall, infiltration rates, the nature of runoff, temperature, and biota such as bacteria which catalyze pyrite oxidation. Tectonic factors contribute to the uplift rate which is complexly related to hydrology through fluvial processes.

Optimal conditions for secondary enrichment and preservation

Steady state versus transient flow effects. The downward cumulative enrichment which characterizes supergene transport is optimized in a transient fluid flow regime in which the ground water table descends exposing increasing column heights of sulfides to oxidizing conditions above the ground water table (Brimhall et al., 1985; Samana, 1986). Similar conclusions have been reached about the importance of water table recession during the onset of arid climates favoring deep zones of bleached clays and duricrusts (Butt, 1981). There is a distinct asymmetry in the effects produced by changes in the position of the ground water table. Descent enhances supergene enrichment. Ascent simply submerges sulfide enrichment blankets putting them in a temporary state of preservation through reduction as pores previously filled with air in the unsaturated zones are filled with water. Therefore, seasonal fluctuations only produce significant supergene transport during the descent phase.

The preservation of such secondarily enriched products over geological periods of time depends either upon chance events such as burial of leached capping beneath a capping conglomerate as at San Manuel-Kalamazoo, Arizona (Lowell and Guilbert, 1970) or some other effects. Enrichment blankets developed in porphyry copper deposits are so common that it seems therefore highly unlikely that such chance events are the main cause of preservation of these enriched states. It seems more likely that preservation is somehow intrinsically related to the fundamental processes of enrichment so that their preservation would be expected and not simply an aberration.

It is clear that optimal development of supergene enrichment processes is due to a balance of rates of ground water table descent, mechanical erosion, and sulfide oxidation (Alpers and Brimhall, 1988a,b). With excessively high erosion rates, leached capping is stripped away and ultimately enriched blanket material is removed. If the rate of ground water descent is much higher than the equivalent rate at which sulfides are oxidized, then leaching is incomplete, and the enrichment processes is inefficient. If the ground water table is static, then an enrichment system may become stagnant, and downward cumulative enrichment ceases, as in the case of ascent of the groundwater table.

Dynamic evolution of supergene metal transport systems has been demonstrated in the Atacama Desert of northern Chile (Alpers and Brimhall, 1988a,b). This chemical and hydrological evolution was quite likely in direct response to transient groundwater flow during Tertiary climatic desiccation with uplift of the Andes which produced an essentially monotonic decline of the ground water table. Related to this hydrologic trend are surficial processes which were similarly changing during the Tertiary. Long-term average rates of erosion decreased markedly from approximately 0.1 mm/year during primary hydrothermal hypogene mineralization in the La Escondida porphyry copper deposit. During supergene enrichment the rate slowed to 0.04 mm/year and subsequent to this enrichment, 0.009

mm/year. Therefore, optimal conditions for supergene enrichment were simultaneously accompanied by a major decrease in the rate of erosion, providing a simple mechanism for insuring the long-term preservation of the enrichment system. This coupling of surficial rates, migration of the ground water table, sulfide oxidation, and erosion is only one example of mutually enhancing effects in transport processes which operate with chemical, thermodynamic, and fluid mechanical aspects.

ACKNOWLEDGMENTS

First we thank Joan Bossart and Jan Dennie for their considerable patience and word processing skills in assembling this document. We thank Mary Yang, Remy Hennet, Gregor M. Anderson, Charles Alpers, Jay Ague, and Aric Cunningham for kindly letting us use their unpublished data. D.A.C. particularly thanks Jacques Schott and the Laboratoire de Géochimie, Université Paul Sabatier, Toulouse for the stimulating scientific and cultural environment in which this was written, and C.N.R.S., France, for financial support. L'inspiration gastronomique de cet article a été fournie gracieusement par Messieurs Lucien Vanel (Toulouse), Yves Thuries (Cordes sur Ciel), André et Arnaud Daguin (Auch), Alain Chapel (Mionnay) et beaucoup d'autres artistes. Financial support came from NSF grants EAR-8517254 to D.A.C. and EAR-8416790 to G.H B.

REFERENCES

Ague, J.J. and Brimhall, G.H (1987) Granites of the batholiths of California: Products of local assimilation and regional-scale crustal contamination. Geology 15, 63-66.

Ague, J.J. and Brimhall, G.H (1988a) Regional variations in bulk chemistry, mineralogy, and the compositions of mafic and accessory minerals in the batholiths of California. Geol. Soc. Amer. Bull. (submitted).

Ague, J.J. and Brimhall, G.H (1988b) Diverse regional controls on magmatic arc asymmetry and distribution of anomalous plutonic belts in the batholiths of California: Effects of assimilation, cratonal rifting, and depth of crystallization. Geol. Soc. Amer. Bull. (submitted).

Ahrland, S. (1968) Thermodynamics of complex formation between hard and soft acceptors and donors. Struct. Bonding 5, 118-149.

Ahrland, S. (1973) Thermodynamics of stepwise formation of metal-ion complexes in aqueous solution. Struct. Bonding 15, 167-188.

Alpers, C.N. and Brimhall, G.H. (1988a) Tertiary climatic dessication and erosion rates in the Atacama Desert, Northern Chile: Optimal Conditions for the Supergene Enrichment and Preservation of Ore Deposits. Submitted to Geol. Soc. Amer. Bull.

Alpers, C.N. and Brimhall, G.H (1988b) Dynamic evolution of supergene metal transport systems in response to transient groundwater flow: Results from La Escondida, Chile. Submitted to Econ. Geol.

Ananthaswany, J. and Atkinson, G. (1982) Thermodynamics of concentrated electrolyte mixtures. I. Activity coefficients in aqueous $NaCl-CaCl_2$ at 25°C. J. Solution Chem. 11, 509-527.

Anderson, G.M. (1975) Precipitation of Mississippi Valley-type ores. Econ. Geol. 70, 937-942.

Anderson, J.A. (1982) Characteristics of leached capping and techniques of appraisal. In: Advances in the Geology of Porphyry Copper Deposits, Southwestern North America. S.R. Titley, ed., University of Arizona Press, Tucson, p. 275-295.

Baes, C.F., Jr. and Mesmer, R.E. (1976) The Hydrolysis of Cations. Wiley-Interscience, New York, 489 pp.

Baes, C.F., Jr. and Mesmer, R.E. (1981) The thermodynamics of cation hydrolysis. Amer. J. Sci. 281, 935-962.

Barnes, H.L. (1975) Zoning of ore deposits: Types and causes. Trans. Royal Soc. Edinburgh 69, 295-310.

Barnes, H.L. (1979) Solubilities of ore minerals. In: Geochemistry of Hydrothermal Ore Deposits, 2nd ed. H.L. Barnes, ed., Wiley-Interscience, New York, p. 404-460.

Barnes, H.L. (1981) Measuring thermodynamically interpretable solubilities at high pressures and temperatures. In: Chemistry and Geochemistry of Solutions at High Temperatures and Pressures. D. Rickard and F. Wickman, eds., Physics and Chemistry of the Earth, v. 13/14, Pergamon Press, New York, p. 321-343.
Barnes, H.L. and Ernst, W.G. (1963) Ideality and ionization in hydrothermal fluids: The system $MgO-H_2O-NaOH$. Amer. J. Sci. 261, 129-150.
Barnes, H.L. and Kullerud, G. (1961) Equilibria in sulfur-containing aqueous solutions, in the system Fe-S-O, and their correlation during ore deposition. Econ. Geol. 56, 648-688.
Barrett, T.J. and Anderson, G.M. (1987) The solubility of sphalerite and galena in 1-5 M NaCl solution to 300°C. Geochim. Cosmochim. Acta, in press.
Barton, M.D. (1983) Metallogenesis. Reviews in Geophys. and Space Physics 21, 1407-1419.
Barton, M.D. (1987) Lithophile-element mineralization associated with Late Cretaceous two-mica granites in the Great Basin. Geology 15, 337-340.
Barton, P.B., Jr., Bethke, P.M. and Roedder, E. (1977) Environment of ore deposition in the Creede mining district, San Juan Mountains, Colorado: Part III. Progress toward interpretation of the chemistry of the ore-forming fluid for the OH vein. Econ. Geol. 72, 1-24.
Barton, P.B., Jr. and Skinner, B.J. (1979) Sulfide mineral stabilities. In: Geochemistry of Hydrothermal Ore Deposits, 2nd ed. H.L. Barnes, ed., Wiley-Interscience, New York, p. 278-403.
Bateman, A.M. (1950) Economic Mineral Deposits. John Wiley, New York, p. 245-287.
Bischoff, J.L. and Seyfried, W.E. (1978) Hydrothermal chemistry of seawater from 25°C to 350°C. Amer. J. Sci. 278, 838-860.
Blain, C.F. and Andrew, R.L. (1977) Sulfide weathering and the evaluation of gossans in mineral exploration. Mineral Sci. Eng. 9, 119-149.
Bockris, J.O'M. and Reddy, A.K.N. (1970) Modern Electrochemistry, v. 1. Plenum, New York, 622 pp.
Born, M. (1920) Volumen und Hydrationswärme der Ionen. Zeit. Physik. 1, 45-48.
Boles, J., Crerar, D., Grissom, G. and Key, T. (1987) Aqueous thermal degradation and diagenesis of naturally occurring aromatic acids. Geochim. Cosmochim. Acta, in press.
Bowers, T.S., Jackson, K.J. and Helgeson, H.C. (1984) Equilibrium Activity Diagrams. Springer-Verlag, New York, 397 pp.
Bowers, T.S., Von Damm, K.L. and Edmond, J.M. (1985) Chemical evolution of mid-ocean ridge hot springs. Geochim. Cosmochim. Acta 49, 2239-2252.
Braithwaite, J.W. (1976) Simulated deep solution mining of chalcopyrite and chalcocite. Unpublished thesis, University of Utah.
Brantley, S.L. (1986) The Chemistry and Thermodynamics of Natural Brines and the Kinetics of Dissolution-Precipitation Reactions of Quartz and Water. Ph.D. dissertation, Princeton University, p. 39-62.
Brantley, S.L., Rowe, G., Fernandez, J.F., Reynolds, J.R., and Borgia, A. (1987) Acid crater lake brine of Poás volcano, Costa Rica. Nature, in press.
Brimhall, G.H (1977) Early fracture-controlled disseminated mineralization at Butte, Montana. Econ. Geol. 72, 37-59.
Brimhall, G.H (1979) Lithologic determination of mass transfer mechanisms of multiple-stage porphyry copper mineralization at Butte, Montana: Vein formation by hypogene leaching and enrichment of potassium silicate protore. Econ. Geol. 74, 556-589.
Brimhall, G.H (1980) Deep hypogene oxidation of porphyry copper potassium-silicate protore at Butte, Montana: A theoretical evaluation of the copper remobilization hypothesis. Econ. Geol. 75, 384-407.
Brimhall, G.H, Jr. (1987a) Metallogenesis. Reviews in Geophys. and Space Physics 25, 1079-1088.
Brimhall, G.H (1987b) Preliminary fractionation patterns of ore metals through earth history. Chem. Geol. 64, 1-16.
Brimhall, G.H, Agee, C., and Stoffregen, R. (1985) The hydrothermal conversion of hornblende to biotite. Canadian Mineral. 23, 369-379.
Brimhall, G.H and Ague, J.J. (1988) Granite systems. In: Hydrothermal Processes—Applications to Ore Genesis. H.L. Barnes and H. Ohmoto, eds., Reidel Publishers, Dordrecht, Holland, 33 pp. (in press).
Brimhall, G.H, Alpers, C.N., and Cunningham, A.B. (1985) Analysis of supergene ore-forming processes and ground water solute transport processes using mass balance principles. Econ.

Geol. 80, 1227-1256.

Brimhall, G.H and Dietrich, W.E. (1987d) Constitutive mass balance relations between chemical composition, volume, density, porosity, and strain in metasomatic hydrochemical systems: Results on weathering and pedogenesis. Geochim. Cosmochim. Acta 51, 567-587.

Brimhall, G.H and Ghiorso, M.S. (1983) Origin and ore-forming consequences of the advanced argillic alteration process in hypogene environments by magmatic gas contamination of meteoric fluids. Econ. Geol. 78, 73-90.

Brinkman, R. (1977) Surface-water gley soils in Bangladesh: Genesis. Geoderma 17, 111-144.

Buback, M. (1981) Spectroscopic investigations of fluids. In: Chemistry and Geochemistry of Solutions at High Temperatures and Pressures. D. Rickard and F. Wickman, eds., Physics and Chemistry of the Earth, v. 13/14, Pergamon Press, New York, p. 345-360.

Buback, M., Crerar, D. and Vogel, L. (1987) Vibrational and electronic spectroscopy of hydrothermal systems. In: Hydrothermal Experimental Techniques. G.C. Ulmer and H.L. Barnes, eds., Wiley-Interscience, New York, Ch. 14, in press.

Burnham, C.W. (1975) Water and magmas; a mixing model. Geochim. Cosmochim. Acta 39, 1077-1084.

Burnham, C.W. (1979) Magmas and hydrothermal fluids. In: Geochemistry of Hydrothermal Ore Deposits, 2nd ed. H.L. Barnes, ed., Wiley-Interscience, New York, p. 71-136.

Burnham, C.W. (1981) Physiochemical constraints on porphyry mineralization. In: Relations of Tectonics to Ore Deposits in the Southern Cordillera. W.R. Dickenson and W.D. Payne, eds., Arizona Geol. Soc. Digest 14, p. 71-77.

Burnham, C.W. (1985) Energy release in subvolcanic environments: Implications for breccia formation. Econ. Geol. 80, 1515-1522.

Burnham, C.W. and Davis, N.F. (1971) The role of H_2O in silicate melts. I. P-V-T relations in the system $NaAlSi_3O_8$-H_2O. Amer. J. Sci. 270, 54-79.

Burnham, C.W., Holloway, J.R. and Davis, N.F. (1969) Thermodynamic Properties of Water to 1000°C and 10,000 bars. Geol. Soc. Amer. Special Paper 132, 96 pp.

Burnham, C.W. and Ohmoto, H. (1980) Late-stage processes of felsic magmatism. In: Granitic Magmatism and Related Mineralization. S. Ishihara and S. Takenouchi, eds., The Soc. of Mining Geologists of Japan, Tokyo, p. 1-12.

Butt, C.R.M. (1981) The nature and origin of the lateritic weathering mantle with particular reference to Western Australia. In: Geophysical Prospecting in Deeply-Weathered Terranes. J.E. Glober and D.F. Groves, eds., Univ. Western Australia Extension Service Pub. No. 66, p. 11-29.

Butt, C.R.M. and Nickel, E.H. (1981) Mineralogy and geochemistry of ore weathering of the disseminated nickel sulfide deposit at Mt. Keith, Western Australia. Econ. Geol. 76, 1736-1751.

Candela, P.A. and Holland, H.D. (1984) The partitioning of copper and molybdenum between silicate melts and aqueous fluids. Geochim. Cosmochim. Acta 48, 373-380.

Candela, P.A. and Holland, H.D. (1984) A mass transfer model for Cu and Mo in magmatic hydrothermal systems: The origin of porphyry-type ore deposits. Econ. Geol. 81, 1-19.

Candela, P.A. (1986) The evolution of vapor from silicate melts: Effect on oxygen fugacity. Geochim. Cosmochim. Acta 50, 1205-1211.

Carmichael, I.S.E., Turner, F.J. and Verhoogen, J. (1974) Igneous Petrology. McGraw-Hill, New York, 739 pp.

Carothers, W.W. and Kharaka, Y.K. (1980) Stable carbon isotopes of HCO_3 in oil-field waters—implications for the origin of CO_2. Geochim. Cosmochim. Acta 44, 323-332.

Carron, J.P. and Lagache, M. (1980) Etude experimentale du fractionnement des élements Rb, Cs, Sr et Ba entre feldspaths alcalins, solutions hydrothermales et liquides silicates dans le système Q.Ab.Or.H_2O à 2 kbar entre 700 et 800°C. Bull. Mineral. 703, 571-578.

Cathles, L.M. (1977) An analysis of the cooling of intrusives by ground-water convection which includes boiling. Econ. Geol. 72, 804-826.

Chou, I-Ming and Eugster, H.P. (1977) Solubility of magnetite in supercritical chloride solutions. Amer. J. Sci. 277, 1296-1314.

Christiansen, E.C. and Lee, D.E. (1986) Fluorine and chlorine in granitoids from the Basin and Range Province, western United States. Econ. Geol. 81, 1481-1494.

Crerar, D.A. (1973) The estimation of activity coefficients of electrolyte solutions and related thermodynamic models. Rpt., available from Dept. Geol. and Geophys. Sci. Library, Princeton

University, Princeton, NJ, 195 pp.
Crerar, D.A. and Barnes, H.L. (1976) Ore solution chemistry—V. Solubilities of chalcopyrite and chalcocite assemblages in hydrothermal solution at 200-350°C. Econ. Geol. 71, 722-794.
Crerar, D.A., Susak, N.J., Borcsik, M. and Schwartz, S. (1978) Solubility of the buffer assemblage pyrite + pyrrhotite + magnetite in NaCl solutions from 200 to 350°C. Geochim. Cosmochim. Acta 42, 1427-1437.
Crerar, D., Wood, S., Brantley, S. and Bocarsly, A. (1985) Chemical controls on solubility of ore-forming minerals in hydrothermal solutions. Canadian Mineral. 23, 333-352.
Cumberlidge, J.T. and Chase, M.C. (1968) Geology of the Nickel Mountain Mine, Riddle, Oregon. In: Ore Deposits of the United States, 1933-1967. J.D. Ridge, ed., Graton-Sales, New York, pp. 1650-1672.
Cunningham, A.B. (1984) Geologically constrained hydrologic and geochemical modeling of supergene weathering processes using physical rock parameters, geochemical profiles, and modal data. Unpublished M.S. thesis, University of California, Berkeley, 122 p.
Czamanske, G.K., Ishihara, S., and Atkin, S.A. (1981) Chemistry of rock-forming minerals of the Cretaceous-Paleocene batholith in southwestern Japan and implications for magma genesis. J. Geophys. Res. 86, 10431-10469.
Davy, R. and El-Ansary, M. (1986) Geochemical patterns in the laterite profile at the Boddington Gold Deposit, Western Australia. J. Geochem. Expl. 26, 119-144.
Decker, R. and Decker, B. (1981) The eruptions of Mount St. Helens. Sci. Amer. 244, 68-83.
De Donder, Th. (1928) L'Affinité. Gauthier-Villars, Paris.
De Donder, Th. and Rysselberghe, P.V. (1936) Affinity. Stanford University Press, Stanford, CA, 142 pp.
Drude, P. and Nernst, W. (1894) Über Elektrostriktion durch freie Ionen. Zeit. Phys. Chem. 15, 79-85.
Drummond, S.E. and Ohmoto, H. (1985) Chemical evolution and mineral deposition in boiling hydrothermal systems. Econ. Geol. 80, 126-147.
Drummond, S.E. and Palmer, D.A. (1986) Thermal decarboxylation of acetate—Part II. Boundary conditions for the role of acetate in the primary migration of natural gas and the transportation of metals in hydrothermal systems. Geochim. Cosmochim. Acta 50, 825-833.
Dunn, T.M., McClure, D.S. and Pearson, R.G. (1965) Some Aspects of Crystal Field Theory. Harper and Row, New York, 115 pp.
Eisenberg, D. and Kauzmann, W. (1969) The Structure and Properties of Water. Oxford Press, New York, 296 pp.
Ellis, A.J. (1979) Explored geothermal systems. In: Geochemistry of Hydrothermal Ore Deposits, 2nd ed. H.L. Barnes, ed., Wiley-Interscience, New York, p. 632-737.
Ellis, A.J. and Mahon, W.A.J. (1977) Chemistry and Geothermal Systems. Academic Press, New York, 392 pp.
Eugster, H.P. (1985) Granites and hydrothermal ore deposits: A geochemical framework. Mineral. Mag. 49, 7-23.
Eugster, H.P. (1986) Minerals in hot water. Amer. Mineral. 71, 655-673.
Eugster, H.P. and Wones, D.R. (1962) Stability relations of the ferruginous biotite, annite. J. Petrol. 3, 82-125.
Fehn, U. (1986) The evolution of low-temperature convection cells near spreading centers: A mechanism for the formation of the Galapagos mounds and similar manganese deposits. Econ. Geol. 81, 1396-1407.
Fehn, U. and Cathles, L.M. (1986) The influence of plate movement on the evolution of hydrothermal convection cells in the oceanic crust. Tectonophys. 125, 289-312.
Feiss, P.G. (1978) Magmatic sources of copper in porphyry copper deposits. Econ. Geol. 73, 397-404.
Figgis, B.N. (1966) Introduction to Ligand Fields. Wiley-Interscience, New York, 351 pp.
Flynn, R.T. and Burnham, C.W. (1978) An experimental determination of rare earth partition coefficients between a chloride-containing vapor phase and silicate melts. Geochim. Cosmochim. Acta 42, 685-701.
Franck, E.U. (1981) Survey of selected non-thermodynamic properties and chemical phenomena of fluids and fluid mixtures. In: Chemistry and Geochemistry of Solutions at High Temperatures and Pressures. D. Rickard and F. Wickman, eds., Physics and Chemistry of the Earth, v.

13/14, Pergamon Press, New York, p. 65-88.
Franks, F. (1972) Introduction—Water, the unique chemical. In: Water. A Comprehensive Treatise. F. Franks, ed., Plenum, New York, p. 1-20.
Franks, F. (1973) The solvent properties of water. In: Water. A Comprehensive Treatise, v. 2. F. Franks, ed., Plenum, New York, p. 1-54.
Franks, F., ed. (1982) Water. A Comprehensive Treatise, v. 7. Plenum, New York, 484 pp.
Frantz, J.D. and Marshall, W.L. (1984) Electrical conductances and ionization constants of salts, acids, and bases in supercritical aqueous fluids: I. Hydrochloric acid from 100° to 700°C and at pressures to 4000 bars. Amer. J. Sci. 284, 651-667.
Frantz, J.D., Popp, R.K. and Boctor, N.Z. (1981) Mineral-solution equilibria. V. Solubilities of rock-forming minerals in supercritical fluids. Geochim. Cosmochim. Acta 45, 69-77.
Fyfe, W.S. and Henley, R.W. (1973) Some thoughts on chemical transport processes with particular reference to gold. Minerals Sci. Eng. 5, 295-298.
Fyfe, W.S. and Kerrich, R. (1984) Gold: Natural concentration processes. In: Gold '82, the Geology, Geochemistry, and Genesis of Gold Deposits. R.P. Foster, ed., Geol. Soc. Zimbabwe Spec. Publ. 1, p. 99-128.
Garrels, R.M. and Christ, C.L. (1965) Solutions, Minerals and Equilibria. Harper and Row, New York, 450 pp.
Giordano, T.H. (1985) A preliminary evaluation of organic ligands and metal-organic complexing in Mississippi Valley-type ore solutions. Econ. Geol. 80, 96-106.
Giordano, T.H. and Barnes, H.L. (1981) Lead transport in Mississippi Valley-type ore solutions. Econ. Geol. 76, 2200-2211.
Golightly, J.P. (1979) Nickeliferous laterites: A general description. In: International Laterite Symposium. D.J.I. Evans and R.S. Shoemaker, eds., Soc. Mining Engineers, New York, pp. 3-23.
Golightly, J.P. (1981) Nickeliferous laterite deposits. In: Econ. Geol. 75th Anniv. Vol. (1905-1980). B.J. Skinner, ed., pp. 710-734.
Grubb, P.L.C. (1970) Mineralogy, geochemistry, and genesis of the bauxite deposits on the Gove and Mitchell Plateaus, northern Australia. Mineralia Deposita 5, 248-272.
Grubb, P.L.C. (1971) Genesis of the Weipa bauxite deposits, N.E. Australia. Mineralia Deposita 5, 265-274.
Grubb, P.L.C. (1979) Genesis of bauxite deposits in the lower Amazon basin and Guianas coastal plain. Econ. Geol. 74, 735-750.
Gunow, A.J., Ludington, S., and Munoz, J.L. (1980) Fluorine in micas from the Henderson molybdenite deposit, Colorado. Econ. Geol. 75, 1127-1137.
Gurney, R.W. (1962) Ionic Processes in Solution. Dover, New York, 275 pp.
Gustafson, L.B. and Hunt, J.P. (1975) The porphyry copper deposit at El Salvador, Chile. Econ. Geol. 70, 857-912.
Hamann, S.D. (1981) Properties of electrolyte solutions at high pressures and temperatures. In: Chemistry and Geochemistry of Solutions at High Temperatures and Pressures. D. Rickard and F. Wickman, eds., Physics and Chemistry of the Earth, v. 13/14, Pergamon Press, New York, p. 89-111.
Haselton, H.T., Jr. and D'Angelo, W.M. (1986) Tin and tungsten solubilities (500-700°C, 1 kbar) in the presence of a synthetic quartz monzonite [abs]. Amer. Geophys. Union, EOS 67, 388.
Heald, P., Foley, N.K. and Hayba, D.O. (1987) Comparative anatomy of volcanic-hosted epithermal deposits: Acid-sulfate and adularia-seriate types. Econ. Geol. 82, 1-26.
Helgeson, H.C. (1967) Thermodynamics of complex dissociation in aqueous solution at elevated temperatures. J. Phys. Chem. 71, 3121-3136.
Helgeson, H.C. (1968) Evaluation of irreversible reactions in geochemical processes involving minerals and aqueous solutions—I. Thermodynamic relations. Geochim. Cosmochim. Acta 32, 853-877.
Helgeson, H.C. (1969) Thermodynamics of hydrothermal systems at elevated temperatures and pressures. Amer. J. Sci. 267, 729-804.
Helgeson, H.C. (1981) Prediction of the thermodynamic properties of electrolytes at high temperatures and pressures. In: Chemistry and Geochemistry of Solutions at High Temperatures and Pressures. D. Rickard and F. Wickman, eds., Physics and Chemistry of the Earth, v. 13/14, Pergamon Press, New York, p. 133-177.

Helgeson, H.C., Delany, J.M., Nesbitt, H.W., and Bird, D.K. (1978) Summary and critique of the thermodynamic properties of rock-forming minerals. Amer. J. Sci. 278-A, 1-229.

Helgeson, H.C. and Kirkham, D.H. (1974) Theoretical prediction of the thermodynamic behavior of aqueous electrolytes at high pressures and temperatures: I. Summary of the thermodynamic/electrostatic properties of the solvent. Amer. J. Sci. 274, 1089-1198.

Helgeson, H.C. and Kirkham, D.H. (1974) Theoretical prediction of the thermodynamic behavior of aqueous electrolytes at high pressures and temperatures: II. Debye-Hückel parameters for activity coefficients and relative partial molal properties. Amer. J. Sci. 274, 1199-1261.

Helgeson, H.C. and Kirkham, D.H. (1976) Theoretical prediction of the thermodynamic properties of aqueous electrolytes at high pressures and temperatures: III. Equation of state for aqueous species at infinite dilution. Amer. J. Sci. 276, 97-240.

Helgeson, H.C., Kirkham, D.H. and Flowers, G.C. (1981) Theoretical prediction of the thermodynamic behavior of aqueous electrolytes at high pressures and temperatures: IV. Calculation of activity and osmotic coefficients and apparent molal and standard and relative partial molal properties to 600°C and 5 kb. Amer. J. Sci. 281, 1249-1493.

Helgeson, H.C. and Lichtner, P. (1987) Fluid flow and mineral reactions at high temperatures and pressures. J. Geol. Soc. London 144, 313-326.

Hemley, J.J. (1959) Some mineralogical equilibria in the system K_2O-Al_2O_3-SiO_2-H_2O. Amer. J. Sci. 257, 241-270.

Hemley, J.J., Cygan, G.L. and d'Angelo, W.M. (1986) Effect of pressure on ore mineral solubilities under hydrothermal conditions. Geol. 14, 377-379.

Hemley, J.J., Hostetler, P.B., Gude, A.J., and Mountjoy, W.T. (1969) Some stability relations of alunite. Econ. Geol. 64, 599-612.

Hemley, J.J. and Jones, W.R. (1964) Chemical aspects of hydrothermal alteration with emphasis on hydrogen metasomatism. Econ. Geol. 59, 538-569.

Hemley, J.J., Montoya, J.W., Marinenko, J.W. and Luce, R.W. (1980) Equilibria in the system Al_2O_3-SiO_2-H_2O and some general implications for alteration/mineralization processes. Econ. Geol. 75, 210-228.

Henley, R.W. (1985) The geothermal framework of epithermal deposits. In: Geology and Geochemistry of Epithermal Systems. B.R. Berger and P.M. Bethke, eds., Reviews in Econ. Geol. v. 2, Econ. Geol. Publ. Co., El Paso, Texas, p. 1-24.

Henley, R.W. and Brown, K.L. (1985) A practical guide to the thermodynamics of geothermal fluids and hydrothermal ore deposits. In: Geology and Geochemistry of Epithermal Systems. B.R. Berger and P.M. Bethke, eds., Reviews in Econ. Geol. v. 2, Econ. Geol. Publ. Co., El Paso, Texas, p. 25-44.

Henley, R.W. and Ellis, A.J. (1983) Geothermal systems ancient and modern: A geochemical review. Earth Sci. Rev. 19, 1-50.

Henley, R.W. and McNabb, A. (1978) Magmatic vapor plumes and ground-water interaction in porphyry copper emplacement. Econ. Geol. 73, 1-20.

Henley, R.W., Norris, R.J. and Paterson, C.J. (1976) Multistage ore genesis in the New Zealand geosyncline—A history of post-metamorphic lode emplacement. Mineralium Deposita 11, 180-196.

Henley, R.W., Truesdell, A.H., Barton, P.B., Jr. and Whitney, J.A. (1984) Fluid-Mineral Equilibria in Hydrothermal Systems. Reviews in Econ. Geol. v. 1, Econ. Geol. Publ. Co., El Paso, Texas, 267 pp.

Hennet, R. (1987) The effect of organic complexing and CO_2 partial pressure on metal transport in low-temperature hydrothermal systems. Ph.D. Dissertation, Princeton Univ., 308 pp.

Holland, H.D. (1959) Some applications of thermochemical data to problems of ore deposits. I. Stability relations among the oxides, sulfides, sulfates and carbonates of ore and gangue minerals. Econ. Geol. 54, 184-233.

Holland, H.D. (1965) Some applications of thermochemical data to problems of ore deposits. II. Mineral assemblages and the composition of ore-forming fluids. Econ. Geol. 60, 1101-1166.

Holland, H.D. (1972) Granites, solutions, and base metal deposits. Econ. Geol. 67, 281-301.

Holland, H.D. and Malinin, S.D. (1979) Solubility and occurrence of non-ore minerals. In: Geochemistry of Hydrothermal Ore Deposits, 2nd ed. H.L. Barnes, ed., Wiley-Interscience, New York, p. 461-508.

Hollister, V.F. (1978) Geology of the Porphyry Copper Deposits of the Western Hemisphere. American Institute of Mining, Metallurgy and Petroleum Engineers, New York, pp. 29-137.

Hollister, L.S. and Crawford, M.L., eds. (1981) Short Course in Fluid Inclusions: Applications to Petrology. Mineral. Assoc. Canada, Short Course Handbook, v. 6, Toronto, 304 pp.
Horne, R.H. (1969) Marine Chemistry: The Structure of Water and the Chemistry of the Hydrosphere. Wiley-Interscience, New York, 568 pp.
Horne, R.A., ed. (1972) Water and Aqueous Solutions. Wiley-Interscience, New York, 837 pp.
Huebner, J.S. (1971) Buffering techniques for hydrostatic systems at elevated pressures. In: Research Techniques for High Pressure and High Temperature. G.C. Ulmer, ed., Springer-Verlag, New York, p. 123-178.
Huheey, J.E. (1978) Inorganic Chemistry, 2nd ed. Harper and Row, New York, 889 pp.
Hutchinson, R.W. (1981) Mineral deposits as guides to supracrustal evolution. In: Evolution of the Earth. R.J. O'Conel and W.S. Fyfe, eds., Amer. Geophys. Union 5, p. 120-132.
Irish, D.E. and Brooker, M.H. (1976) Raman and infrared spectral studies of electrolytes. In: Advances in Infrared and Raman Spectroscopy, v. 2. R.J.H. Clark and R.E. Hester, eds., Heyden, London, p. 212-311.
Irvine, T.N., Keith, D.W., and Todd, S.G. (1983) The J-M platinum and palladium reef of the Stillwater Complex, Montana. II. Origin by double diffusive convection magma mixing and applications for the Bushveld Complex. Econ. Geol. 78, 1287-1334.
Ishihara, S. (1977) The magnetite-series and ilmenite-series rocks. Mining Geol. 27, 293-305.
Ishihara, S., Sawata, H., Arpornsuwan, S., Busaracombe, P., and Bungbrakearti, N. (1979) The magnetite-series and ilmenite-series granitoids and their bearing on tin mineralization, particularly of the Malay Peninsula region. Geol. Soc. Malaysia Bull. 11, 103-110.
Jackson, K.S., Jonasson, I.R. and Skippen, G.B. (1978) The nature of metals-sediment-water interactions in freshwater bodies, with emphasis on the role of organic matter. Earth Sci. Rev. 14, 97-146.
Jackson, S.A. and Beales, F.W. (1967) An aspect of sedimentary basin evolution: The concentration of Mississippi Valley-type ores during late stages of diagenesis. Bull. Can. Petroleum Geol. 15, 383-433.
Jacobs, D.C. and Parry, W.T. (1979) Geochemistry of biotite in the Santa Rita porphyry copper deposit, New Mexico. Econ. Geol. 74, 860-887.
Kesler, S.E. (1973) Copper, molybdenum, and gold abundances in porphyry copper deposits. Econ. Geol. 68, 106-112.
Kharaka, Y.K., Carothers, W.W. and Rosenbauer, R.J. (1983) Thermal decarboxylation of acetic acid: Implications for origin of natural gas. Geochim. Cosmochim. Acta 47, 397-402.
Khodakovskiy, I.L. and Yelkin, A. Ye. (1975) Measurement of the solubility of zincite in aqueous NaOH at 100, 150 and 200°C. Geochem. Int. 12, 127-133.
Kilinc, I.A. (1969) Experimental Metamorphism and Anatexis of Shales and Graywackes. Ph.D. dissertation, Pennsylvania State University, State College, PA.
Kilinc, I.A. and Burnham, C.W. (1972) Partitioning of chloride between a silicate melt and coexisting aqueous phase from 2 to 8 kilobars. Econ. Geol. 67, 231-235.
Kwak, T.A.P., Brown, W.M., Abeysinghe, P.B. and Tan, T.H. (1986) Fe solubilities in very saline hydrothermal fluids: Their relation to zoning in some ore deposits. Econ. Geol. 81, 447-465.
Lacy, W.C. and Hosmer, H.L. (1956) Hydrothermal leaching in central Peru. Econ. Geol. 51, 69-79.
Lehmann, B. (1982) Metallogeny of tin: Magmatic differentiation versus geochemical heritage. Econ. Geol. 77, 50-59.
Lewis, G.N. and Randall, M. (1961) Thermodynamics, 2nd ed. Revised by K.S. Pitzer and L. Brewer, McGraw-Hill, New York, 723 pp.
Lichtner, P.C. (1985) Continuum model for simultaneous chemical reactions and mass transport in hydrothermal system. Geochim. Cosmochim. Acta 49, 779-800.
Lindsay, W.T. (1980) Estimation of concentration quotients for ionic equilibria in high temperature water: The model substance approach. Proceedings, 41st Int'l. Water Conference, Pittsburgh, PA, p. 284-294.
Lowell, J.D. and Guilbert, J.M. (1970) Lateral and vertical zoning in porphyry ore deposits. Econ. Geol. 65, 373-408.
Macqueen, R.W. and Powell, T.G. (1983) Organic geochemistry of the Pine Point lead-zinc ore field and region, Northwest Territories, Canada. Econ. Geol. 78, 1-25.
Mann, A.W. (1984) Mobility of gold and silver in lateritic weathering profiles: Some observations from western Australia. Econ. Geol. 79, 38-49.

Mann, A.W. and Ollier, C.D. (1985) Chemical diffusion and ferricrete formation, Catena Supplement, 6, 152-157.

Manning, D.A.C. (1981) The effect of fluorine on liquidus phase relationships in the system Qz-Ab-Or with excess water at 1 kb. Contrib. Mineral. Petrol. 76, 206-215.

Marshall, W.L. (1972) A further description of complete equilibrium constants. J. Phys. Chem. 76, 720-731.

Meyer, C., and Hemley, J.J. (1967) Wall rock alteration. In: Geochemistry of Hydrothermal Ore Deposits. H.L. Barnes, ed., Holt, Rinehart and Winston, New York, p. 166-235.

Meyer, C. (1985) Ore metals through geological history. Science 227, 1421-1428.

Meyer, C., Shea, E., Goddard, C.C., Zeihen, L.G., Guilbert, J.M., Miller, R.N., McAleer, J.F., Brox, G.B., Ingersoll, R.G., Burns, G.J., and Wigal, T. (1968). In: Ore Deposits in the United States 1933/1967. J.D. Ridge, ed., American Institute of Mining, Metallurgy and Petroleum Engineers, New York, p. 1373-1416.

Millero, F.J. (1971) The partial molal volumes of electrolytes in aqueous solutions. In: Water and Aqueous Solutions. R.A. Horne, ed., Wiley, New York, p. 519-564.

Montoya, J.W. and Hemley, J.J. (1975) Activity relations and stabilities in alkali feldspar and mica alteration reactions. Econ. Geol. 70, 577-594.

Morel, F.M.M. (1983) Principles of Aquatic Chemistry. Wiley-Interscience, New York, 446 pp.

Munoz, J.L. (1984) F^-OH and Cl^-OH exchange in micas with applications to hydrothermal ore deposits: Micas. In: S.W. Bailey, ed., Reviews in Mineralogy, 13, 469-494.

Munoz, J.L. and Swenson, A. (1981) Chloride-hydroxl exchange in biotite: An estimation of relative HCl/HF activities in hydrothermal fluids. Econ. Geol. 766, 2212-2221.

Narasimhan, T.N., Witherspoon, P.A., and Edwards, E.A. (1978) Numerical model for saturated-unsaturated flow in deformable porous media, 2. The algorithm. Water Resources Research 14, 255-261.

Nancollas, G.H. (1970). The thermodynamics of metal complex and ion-pair formation. Coord. Chem. Rev. 5, 379-415.

Neilson, G.W. and Enderby, J.E., eds. (1986) Water and Aqueous Solutions. Adam Hilger, Bristol, England, 349 pp.

Nissenbaum, A. and Swaine, D.J. (1976) Organic matter-metal interactions in recent sediments: The role of humic substances. Geochim. Cosmochim. Acta 40, 809-816.

Nordstrom, D.K. and Munoz, J.L. (1985) Geochemical Thermodynamics. Benjamin/Cummings, Inc., Menlo Park, CA, 477 pp.

Norton, D. (1978) Sourcelines, sourceregions, and pathlines for fluids in hydrothermal systems related to cooling plutons. Econ. Geol. 73, 21-28.

Norton, D.L. (1984) Theory of hydrothermal systems. Ann. Rev. Earth Planet. Sci. 12, 155-177.

Norton, D. and Knight, J. (1977) Transport phenomena in hydrothermal systems: Cooling plutons. Amer. J. Sci. 277, 937-981.

Ohmoto, H. (1972) Systematics of sulfur and carbon isotopes in hydrothermal ore deposits. Econ. Geol. 67, 551-579.

Ohmoto, H., and Kerrick, D. (1977) Devolatilization equilibria in graphitic systems. Amer. J. Sci. 277, 1013-1044.

Ohmoto, H. and Rye, R.O. (1979) Isotopes of sulfur and carbon. In: Geochemistry of Hydrothermal Ore Deposits, 2nd ed. H.L. Barnes, ed., Wiley-Interscience, New York, p. 509-567.

Pearson, R.G. (1963) Hard and soft acids and bases. J. Amer. Chem. Soc. 85, 3533-3539.

Pitzer, K.S. (1979) Theory: Ion interaction approach. In: Activity Coefficients in Electrolyte Solutions, v. I. R.M. Pytkowicz, ed., CRC Press, Boca Raton, Florida, p. 157-208.

Pollard, P.J., Taylor, R.G., and Cuff, C. (1983) Metallogeny of tin: Magmatic differentiation versus geochemical heritage—A discussion. Econ. Geol. 78, 543-545.

Prigogine, I. (1955) Introductiion to Thermodynamics of Irreversible Processes. John Wiley, New York, 119 p.

Quist, A.S. and Marshall, W.L. (1968) Electrical conductances of aqueous sodium chloride solutions from 0 to 800° C and at pressures to 4000 bars. J. Phys. Chem. 72, 684-703.

Reuter, J.H. and Perdue, E.M. (1977) Importance of heavy metal-organic matter interactions in natural waters. Geochim. Cosmochim. Acta 41, 325-334.

Ringwood, A.E. (1979) Origin of the Earth and Moon. Springer, New York, 295 p.

Roedder, E. (1979) Fluid inclusions as samples of ore fluids. In: Geochemistry of Hydrothermal Ore Deposits, 2nd ed. H.L. Barnes, ed., Wiley-Interscience, New York, p. 684-737.

Rose, A.W. and Burt, D.M. (1979) Hydrothermal alteration. In: Geochemistry of Hydrothermal Ore Deposits, 2nd ed. H.L. Barnes, ed., Wiley-Interscience, New York, p. 173-235.

Ruaya, J.R. and Seward, T.M. (1986) The stability of chlorozinc (II) complexes in hydrothermal solutions up to 350°C. Geochim. Cosmochim. Acta 50, 651-661.

Rye, R.O. and Ohmoto, H. (1974) Sulfur and carbon isotopes and ore genesis: A review. Econ. Geol. 69, 826-842.

Sadleir,. S.B. and Gilkes, R.J. (1976) Bauxite in relation to parent material. J. Geol. Soc. Australia 23, 333-344.

Samana, J.C. (1986) Ore Fields and Continental Weathering. Van Nostrand-Reinhold, New York, 124-129.

Sangameshwar, S.R. and Barnes, H.L. (1983) Supergene processes in zinc-lead-silver sulfide ores in carbonates. Econ. Geol. 78, 1379-1397.

Sawkins, F.J. (1984) Metal Deposits in Relation to Plate Tectonics. Springer Verlag, Berlin, 330 pp.

Seward, T.M. (1973) Thio complexes of gold and the transport of gold in hydrothermal ore solutions. Geochim. Cosmochim. Acta 37, 379-399.

Seward, T.M. (1976) The stability of chloride complexes of silver in hydrothermal solutions up to 350°C. Geochim. Cosmochim. Acta 40, 1329-1341.

Seward, T.M. (1981) Metal complex formation in aqueous solutions at elevated temperatures and pressures. In: Chemistry and Geochemistry of Solutions at High Temperatures and Pressures. D. Rickard and F. Wickman, eds., Physics and Chemistry of the Earth, v. 13/14, Pergamon Press, New York, p. 113-132.

Seward, T.M. (1983) The transport and deposition of gold in hydrothermal systems. In: Gold '82: The Geology, Geochemistry and Genesis of Gold Deposits. R.P. Foster, ed., Geol. Soc. Zimbabwe, Spec. Publ. No. 1, p. 165-181.

Seward, T. (1984) The formation of lead (II) chloride complexes to 300°C: A spectrophotometric study. Geochim. Cosmochim. Acta 48, 121-134.

Shanks, W.C. and Bischoff, J.L. (1977) Ore transport and deposition in the Red Sea geothermal system: A geochemical model. Geochim. Cosmochim. Acta 41, 1507-1519.

Sillitoe, R.H. (1972) A plate tectonic model for the origin of porphyry copper deposits. Econ. Geol. 76, 184-197.

Skinner, B.J. (1979) The many origins of hydrothermal mineral deposits. In: Geochemistry of Hydrothermal Ore Deposits, 2nd ed. H.L. Barnes, ed., Wiley-Interscience, New York, p. 1-21.

Slack, J.F. (1980) Multistage vein ores of the Lake City district, western San Juan Mountains, Colorado. Econ. Geol. 75, 963-991.

Smith, R.M. and Martell, A.E. (1976) Critical Stability Constants, v. 4: Inorganic Complexes. Plenum, New York, 257 pp.

Sohn, M.L. and Hughes, M.C. (1981) Metal ion formation constants of some sedimentary humic acids with Zn (II), Cu (II), and Cd (II). Geochim. Cosmochim. Acta 45, 2393-2399.

Stoffregen, R. (1986) Observations on the behavior of gold during supergene oxidation at Summitville, Colorado, U.S.A., and implications for electron stability in the weathering environment. Appl. Geochem. 1, 549-558.

Stumm, W. and Morgan, J.J. (1981) Aquatic Chemistry, 2nd ed. Wiley-Interscience, New York, 780 pp.

Susak, N. and Crerar, D. (1982) Factors controlling mineral zoning in hydrothermal ore deposits. Econ. Geol. 77, 476-482.

Susak, N. and Crerar, D. (1985) Spectra and coordination changes of transition metals in hydrothermal solutions: Implications for ore genesis. Geochim. Cosmochim. Acta 49, 555-564.

Sverjensky, D.A. (1986) Genesis of Mississippi Valley-type lead-zinc deposits. Ann. Rev. Earth Planet. Sci. 14, 177-199.

Tardy, I. and Nahon, D. (1985) Geochemistry of laterites, stability of Al-goethite, Al-hematite, and Fe^{3+}-Kaolinite in bauxites and ferricretes: An approach to the mechanism of concretion formation. Amer. J. Sci. 285, 865-903.

Taylor, H.P. (1979) Oxygen and hydrogen isotope relationships in hydrothermal mineral deposits. In: Geochemistry of Hydrothermal Ore Deposits, 2nd ed. H.L. Barnes, ed., Wiley-Interscience, New York, p. 236-277.

Taylor, R.G. (1979) Geology of Tin Deposits. Elsevier, Amsterdam, 543 pp.

Taylor, S.R. and McClennan, S.M. (1985) The continental crust: Its composition and evolution. An Examination of the Geochemical Record Preserved in Sedimentary Rocks. Blackwell Scientific Publications, Oxford, 301 pp.

Titley, S.R. and Beane, R.E. (1981) Porphyry copper deposits. In: Econ. Geol. 75th Anniv. Vol., B.J. Skinner, ed., p. 214-269.

Titley, S.R., Thompson, R.C., Haynes, F.M., Manske, S.L., Robinson, L.C., and White, J.L. (1986) Evolution of fractures and alteration in the Sierrita-Espernanza hydrothermal system, Pima Country, Arizona. Econ. Geol. 81, p. 343-370.

Tremaine, P.R., Sway, K. and Barbero, J.A. (1986) The apparent molar heat capacity of aqueous hydrochloric acid from 10 to 140°C. J. Sol. Chem. 15, 1-22.

Troeh, G., Jabro, J.D., and Kirkham, D. (1982) Gaseous diffusion equations for porous materials. Geoderma 27, 239-253.

Turcotte, D.L. and Schubert, T. (1973) Frictional heating on the descending lithosphere. J. Geophys. Res. 78, 5876-5886.

Vogel, L.M., McClure, D.S. and Crerar, D.A. (1987) A spectroscopic study of iron (II) chloro complexes in LiCl-DCl-D_2O solutions. Inorg. Chem. 26, 308-313.

Von Damm, K.L., Edmond, J.M., Grant, B., Measures, C.I., Walden, B. and Weiss, R.F. (1985a) Chemistry of submarine hydrothermal solutions at 21° N, East Pacific Rise. Geochim. Cosmochim. Acta 49, 2197-2220.

Von Damm, K.L., Edmond, J.M., Measures, C.I. and Grant, B. (1985b) Chemistry of submarine hydrothermal solutions at Guaymas Basin, Gulf of California. Geochim. Cosmochim. Acta 49, 2221-2238.

Wallace, A.B. (1979) Possible signatures of buried porphyry-copper deposits in middle to late Tertiary volcanic rocks of western Nevada. In: Proceedings of the Fifth Quadrennial IAGOD Symposium. J.D. Ridge, ed., University of Nevada-Mackay School of Mines, Reno, Vol. 2, p. 69-76.

Wallace, S.R., Muncaster, N.K., Jonson, D.C., MacKenzie, W.B., Bookstrom, A.A., and Surface, V.E. (1968) Multiple intrusion and mineralization at Climax, Colorado. In: Ore Deposits in the United States 1933/1967. J.D. Ridge, ed., American Institute of Mining, Metallurgy and Petroleum Engineers, New York, p. 605-640.

Walther, J.V. and Wood, B.J. (1986) Mineral-fluid reaction rates. In: Fluid-rock interactions during metamorphism. J.V. Walther and B.J. Wood, eds., Advances in Physical Geochemistry, v. 5, Springer-Verlag, New York, p. 194-213.

Wanke, H.G., Dreibus, G.D., and Jagoutz, E. (1984) Mantle chemistry and accretion history of the earth. In: Archean History of the Earth. A. Kroner, ed., Springer-Verlag, Berlin, p. 1-22.

Webster, E.A. and Holloway, J.R. (1980) The partitioning of REE'S, Sc, Rb and Cs between a silicic melt and a Cl fluid. EOS Amer. Geophys. Union Trans. 61, 1152.

Webster, J.G. (1986) The solubility of gold and silver in the system Au-Ag-S-O_2-H_2O at 25°C and 1 atm. Geochim. Cosmochim. Acta 50, 1837-1845.

Webster, J.G. and Mann, A.W. (1984) The influence of climate, geomorphology and primary geology on the supergene migration of gold and silver, J. Geochem. Expl. 22, 21-42.

Weissberg, B.G., Browne, P.L. and Seward, T.M. (1979) Ore metals in active geothermal systems. In: Geochemistry of Hydrothermal Ore Deposits, 2nd ed. H.L. Barnes, ed., Wiley-Interscience, New York, p. 738-780.

White, W.H., Bookstrom, A.A., Kamilli, R.J., Ganster, M.W., Smith, R.P., Ranta, D.E., and Steininger, R.C. (1981) Character and origin of Climax-type molybdenum deposits. Econ. Geol. 75th Anniv. Vol. B.J. Skinner, ed., p. 270-316.

Whitney, J.A. (1975) Vapor generation in a quartz monzonite magma: A synthetic model with application to porphyry copper deposits. Econ. Geol. 70, 346-358.

Whitney, J.A., Hemley, J.J. and Simon, F.O. (1985) The concentration of iron in chloride solutions equilibrated with synthetic granitic compositions: The sulfur-free system. Econ. Geol. 80, 444-460.

Wiersma, E.L. and Rimstidt, J.D. (1984) Rates of reaction of pyrite and marcasite with ferric iron at pH: 2. Geochim. Cosmochim. Acta 48, 85-92.

Willey, L.M., Kharaka, Y.K., Presser, T.S., Rapp, J.B. and Barnes, I. (1975) Short chain aliphatic acid anions in oil field waters and their contribution to the measured alkalinity. Geochim.

Cosmochim. Acta 39, 1707-1711.

Wilson, G.A. and Eugster, H.P. (1984) Cassiterite solubility and tin-chloride speciation in supercritical solutions [abs]. Geol. Soc. Amer. Abstr. Program 16, 696.

Wolery, T.J. (1979) Calculation of chemical equilibrium between aqueous solution and minerals: The EQ3/6 soft-wares package. Univ. California Lawrence Livermore Lab. Bull. 526658, 31 p.

Wones, D.R. and Eugster, H.P. (1965) Stability of biotite: Experiment, theory and applications. Amer. Mineral. 50, 1228-1272.

Wood, S.A. and Crerar, D.A. (1985) A numerical method for obtaining multiple linear regression parameters with physically realistic signs and magnitudes: Applications to the determination of equilibrium constants from solubility data. Geochim. Cosmochim. Acta 49, 165-172.

Wood, S.A., Crerar, D.A. and Borcsik, M.P. (1987) Solubility of the assemblage pyrite-pyrrhotite-magnetite-sphalerite-galena-gold-stibnite-bismuthinite-argentite-molybdenite in H_2O-NaCl-CO_2 solutions from 200° to 350°C. Econ. Geol. 81, in press.

Wood, S., Crerar, D., Brantley, S. and Borcsik, M. (1984) Mean molal stoichiometric activity coefficients of alkali halides and related electrolytes in hydrothermal solutions. Amer. J. Sci. 284, 668-705.

Wood, B.J. and Walther, J.V. (1983) Rates of hydrothermal reactions. Science 222, 413-415.

Woodward, L.A. (1986) Tectonic origin of fractures for fissure vein emplacement in the Boulder batholith and adjacent rocks, Montana. Econ. Geol. 81, 1387-1395.

Chapter 9 John M. Ferry and Lukas Baumgartner

THERMODYNAMIC MODELS OF MOLECULAR FLUIDS AT THE ELEVATED PRESSURES AND TEMPERATURES OF CRUSTAL METAMORPHISM

INTRODUCTION

One goal of geochemistry and theoretical metamorphic petrology is prediction of mineral-fluid equilibria at elevated pressures and temperatures. Attainment of this goal requires a quantitative thermodynamic description (i.e., equation of state) of the fluid. In this chapter we (1) review the thermodynamics of molecular fluids, (2) summarize and evaluate equations of state for molecular fluids that are currently used in geochemistry and petrology, and (3) apply one of these equations of state to several problems of mineral-fluid equilibrium at pressure-temperature conditions typical of metamorphism.

The behavior of fluids at low pressures may be quantitatively described by the ideal-gas equation:

$$P\overline{V} = RT \tag{1}$$

(see Table 1 for all notation). For a fluid composed only of component i:

$$(\mu_i^\circ)^{P_2,T} - (\mu_i^\circ)^{P_1,T} = \int_{P_1}^{P_2} \left(\frac{\partial \mu_i^\circ}{\partial P}\right)_T dP = \int_{P_1}^{P_2} \overline{V}_i^\circ dP \quad , \tag{2}$$

where states (P_2,T) and (P_1,T) may be chosen arbitrarily. For ideal gas, therefore

$$(\mu_i^\circ)^{P_2,T} - (\mu_i^\circ)^{P_1,T} = RT\ln(P_2/P_1) \quad . \tag{3}$$

Although fluids do not behave according to the ideal-gas equation at metamorphic pressures and temperatures, Equations (1) and (3) are important because more accurate equations of state often preserve their form.

Fluids whose behavior deviate from that prescribed by Equation (1) are referred to as real fluids, and their behavior is described by the function f, fugacity. G.N. Lewis defined fugacity in terms of the chemical potential of component i in such a way as to preserve the formalism of Equation (3) for ideal gas:

$$(\mu_i)^{\text{state 2}} - (\mu_i)^{\text{state 1}} = RT\ln[(f_i)^{\text{state 2}}/(f_i)^{\text{state 1}}] \quad , \tag{4}$$

where states 1 and 2 are at the same temperature but may differ arbitrarily in pressure and/or fluid composition. Equation (4) shows that

Table 1. Notation.

P	pressure (bars, except where noted); 1 Kbar = 1000 bars
T	temperature (°K, except where noted)
R	universal gas constant (83.14 cm^3bar/°Kmole, except where noted)
n	total number of moles
$\underline{V}$	total volume (cm^3, $\underline{V} = n\overline{V}$)
$\overline{V}$	molar volume (cm^3/mole)
$\underline{U}$	total internal energy (cal)
$\overline{U}$	molar internal energy (cal/mole)
$\underline{S}$	total entropy (cal/°K)
$\overline{S}$	molar entropy (cal/°Kmole)
$A^{V,T}$	total Helmholtz free energy (cal)
$\underline{H}$	total enthalpy (cal)
$\overline{H}$	molar enthalpy (cal/mole)
$\overline{G}$	molar Gibbs free energy (cal/mole)
μ_i	chemical potential of component i (cal)
f_i	fugacity of component i
a_i	activity of component i
X_i	mole fraction of component i
γ_i	fugacity coefficient of component i
λ_i	activity coefficient of component i
Z	compressibility factor ($Z=PV/nRT$)
ρ	density ($1/\overline{V}$)
c_P	isobaric heat capacity (cal/°Kmole)
K	equilibrium constant
ν_i	stoichiometric coefficient for species i in reaction (positive for products; negative for reactants)
Γ_{ij}	potential energy associated with attraction between molecules i and j
a	empirical coefficient in Redlich-Kwong equation of state ($bar \cdot cm^6 °K^{\frac{1}{2}}/mole^2$)
b	empirical coefficient in Redlich-Kwong equation of state (cm^3/mole)
A	$A^2 = a/(R^2T^{2.5})$ in Redlich-Kwong equation of state
B	second virial coefficient in virial equation of state (cm^3/mole); empirical coefficient in modified version of Berthelot's Law (cm^3/mole); b/RT in Redlich-Kwong equation of state
C	third virial coefficient in virial equation of state ($cm^6/mole^2$)
D	fourth virial coefficient in virial equation of state ($cm^9/mole^3$)
y	$b/4\overline{V}$ in Redlich-Kwong equation of state
w	weight percent NaCl relative to NaCl+H_2O in fluid mixture

Subscript Notation.

i	refers to component i
j	refers to component j
mix	mixing
m	refers to fluid mixture
R	denotes reduced variable
C	refers to variable at critical point
s	refers to solid mineral phases

Superscript Notation.

°	refers to pure substance composed only of component indicated by subscript
*	refers to state T-P*-V* at sufficiently low pressure that fluid mixes ideally and behaves as an ideal gas. P* associated with 1 bar
P,T	variable at state (P,T)
V,T	variable at state (V,T)
obs	observed variable
calc	calculated variable

fugacity represents a way to quantitatively evaluate the change in chemical potential of a component in a one-component real fluid with pressure at constant temperature:

$$(\mu_i^\circ)^{P_2,T} - (\mu_i^\circ)^{P_1,T} = \int_{P_1}^{P_2} \left(\frac{\partial \mu_i^\circ}{\partial P}\right)_T dP = \int_{P_1}^{P_2} \overline{V}_i^\circ dP = RT\,ln[(f_i^\circ)^{P_2,T}/(f_i^\circ)^{P_1,T}]. \quad (5)$$

At very low pressure P* (e.g., 1 bar) fluids behave as ideal gases and

$$(f_i)^{P^*,T} \rightarrow X_i P^* \quad . \quad (6)$$

If we then associate state (1) with 1 bar and state (2) with any higher pressure P, Equation (4) or (5) becomes, for a fluid composed only of component i,

$$(\mu_i^\circ)^{P,T} - (\mu_i^\circ)^{1,T} = RT\,ln[(f_i^\circ)^{P,T}/1 \text{ bar}] \quad . \quad (7)$$

Fugacity in a one-component fluid may also be represented in terms of the fugacity coefficient, γ_i°, which is defined by

$$(f_i^\circ)^{P,T} = P\cdot(\gamma_i^\circ)^{P,T} \quad . \quad (8)$$

The fugacity coefficient is dimensionless, species-specific, and a function of both pressure and temperature.

Equation (4) applies not only to fluids composed of a single molecular species but also to fluid mixtures. For fluid mixtures state (1) in Equation (4) is associated with the fluid mixture at elevated pressure P and temperature T and state (2) with a fluid composed only of component i at the same temperature T but at sufficiently low P* (usually taken as 1 bar) that it behaves as an ideal gas. Thus

$$(\mu_i)^{P,T} - (\mu_i^\circ)^{1,T} = RT\,ln(f_i)^{P,T} = RT\,ln[P\cdot(\gamma_i)^{P,T}] \quad , \quad (9)$$

where $(f_i)^{P,T}$ and $(\gamma_i)^{P,T}$ now refer to the fugacity of component i in the fluid solution.

Because the activity of i (a_i) and the activity coefficient for i (λ_i) in the fluid mixture are defined by

$$(\mu_i)^{P,T} - (\mu_i^\circ)^{P,T} = RT\,ln a_i = RT\,ln(X_i\lambda_i) \quad , \quad (10)$$

there exists a simple relationship between the activity and fugacity of component i in the mixture and between the activity coefficient and fugacity coefficient for i:

$$(\mu_i)^{P,T} - (\mu_i^\circ)^{P,T} = [(\mu_i)^{P,T} - (\mu_i^\circ)^{1,T}] - [(\mu_i^\circ)^{P,T} - (\mu_i^\circ)^{1,T}] \tag{11}$$

$$RT\,ln\,a_i = RT\,ln(f_i)^{P,T} - RT\,ln(f_i^\circ)^{P,T} \tag{12}$$

$$a_i = (f_i)^{P,T}/(f_i^\circ)^{P,T} \tag{13}$$

$$\lambda_i = (\gamma_i)^{P,T}/X_i(\gamma_i^\circ)^{P,T} \quad . \tag{14}$$

Equations (1) and (3) show the formal relationship among the equation of state for an ideal one-component gas, pressure, and chemical potential. In the next sections we develop the corresponding relationships among equations of state, fugacity, and chemical potential for real fluids. Combining these relationships with Equations (13) and (14), in turn, leads to calculations of mineral-fluid equilibria for real fluid mixtures of geological interest.

THERMODYNAMICS OF FLUIDS WITH VARIABLES V AND T

Equations of state for real gases may be written in volume-explicit or pressure-explicit forms:

$$\overline{V} = f(P,T) \tag{15}$$

or

$$P = f'(\overline{V},T) \quad , \tag{16}$$

where functions f(P,T) and f'(V,T) are determined from experimental P-V-T data for fluids. P-V-T data commonly are better fit by pressure-explicit functions whose independent variables are V and T (Prausnitz, 1969). In order to relate equations of state like (16) to fugacity and activity, the fundamental thermodynamic functions for fluids must, in turn, be represented in forms with V and T as independent variables. The development of these equations follows, in part, that of Beattie (1955).

Internal Energy

For n moles fluid mixture

$$U^{V,T} = n\bar{U}^* + \int_{V^*}^{V} \left(\frac{\partial U}{\partial V}\right)_{T,n} dV \quad , \tag{17}$$

where superscript "*" refers to a T-V*-P* state at sufficiently low pressure that the fluid mixes ideally and behaves according to the ideal-gas Equation (1). Let P* be associated with 1 bar pressure. Because the fluid mixes ideally at T-V*-P*,

$$n\bar{U}^* = \sum_i n_i (\bar{U}_i^\circ)^* \tag{18}$$

$$dU = TdS - PdV = T\left(\frac{\partial S}{\partial T}\right)_{V,n} dT + T\left(\frac{\partial S}{\partial V}\right)_{T,n} dV - PdV = T\left(\frac{\partial S}{\partial T}\right)_{V,n} dT + T\left(\frac{\partial P}{\partial T}\right)_{V,n} dV - PdV. \tag{19}$$

Therefore

$$\left(\frac{\partial U}{\partial V}\right)_{T,n} = [T\left(\frac{\partial P}{\partial T}\right)_{V,n} - P] \quad , \tag{20}$$

and

$$U^{V,T} = \sum_i n_i (\bar{U}_i^\circ)^* + \int_{V^*}^{V} [T\left(\frac{\partial P}{\partial T}\right)_{V,n} - P] dV \quad . \tag{21}$$

Because at $P \leq P^*$ (and at $V \geq V^*$) the fluid obeys the ideal-gas equation,

$$\int_{V^*}^{\infty} [P - T\left(\frac{\partial P}{\partial T}\right)_{V,n}] dV = \int_{V^*}^{\infty} [P - (nRT/V)] dV = 0 \quad , \tag{22}$$

and the V* limit of integration may be replaced by ∞ :

$$U^{V,T} = \sum_i n_i (\bar{U}_i^\circ)^* + \int_{V}^{\infty} [P - T\left(\frac{\partial P}{\partial T}\right)_{V,n}] dV \quad . \tag{23}$$

Entropy

$$S^{V,T} = n\bar{S}^* + \int_{V^*}^{V} \left(\frac{\partial S}{\partial V}\right)_{T,n} dV = n\bar{S}^* + \int_{V^*}^{V} \left(\frac{\partial P}{\partial T}\right)_{V,n} dV = n\bar{S}^* - \int_{V}^{\infty} \left(\frac{\partial P}{\partial T}\right)_{V,n} dV + \int_{V^*}^{\infty} \left(\frac{\partial P}{\partial T}\right)_{V,n} dV \tag{24}$$

At state T-V*-P* the fluid mixes ideally and

$$n\bar{S}^* = \sum_i n_i(\bar{S}_i^\circ)^* + n\bar{S}^*_{mix} = \sum_i n_i(\bar{S}_i^\circ)^* - nR\sum_i X_i \ln X_i \tag{25}$$

Because fluid behaves as an ideal gas at $P\leq P^*$ (and $V\geq V^*$),

$$\int_{V^*}^{\infty} \left(\frac{\partial P}{\partial T}\right)_{V,n} dV = \int_{V^*}^{\infty} (nR/V)dV = \int_{V}^{\infty} (nR/V)dV - \int_{V}^{V^*} (nR/V)dV \tag{26}$$

$$\int_{V^*}^{\infty} \left(\frac{\partial P}{\partial T}\right)_{V,n} dV = \int_{V}^{\infty} (nR/V)dV + nR\ln V - nR\ln V^* \tag{27}$$

$$\int_{V^*}^{\infty} \left(\frac{\partial P}{\partial T}\right)_{V,n} dV = \int_{V}^{\infty} (nR/V)dV + nR\ln V - nR\ln(nRT) \quad . \tag{28}$$

(recall P* = 1 bar). Combining equations (24), (25), and (28),

$$S^{V,T} = \sum_i n_i(\bar{S}_i^\circ)^* + \int_{V}^{\infty} [(nR/V) - \left(\frac{\partial P}{\partial T}\right)_{V,n}]dV + nR\ln(V/nRT) - nR\sum_i X_i \ln X_i \ . \tag{29}$$

Because

$$n = \sum_i n_i = \sum_i nX_i \quad , \tag{30}$$

$$S^{V,T} = \sum_i n_i(\bar{S}_i^\circ)^* + \int_{V}^{\infty} [(\sum_i n_i R)/V - \left(\frac{\partial P}{\partial T}\right)_{V,n}]dV + \sum_i n_i R\ln(V/n_i RT) \quad . \tag{31}$$

Helmholtz free energy

$$A^{V,T} = U^{V,T} - TS^{V,T} = \sum_i n_i[(\overline{U}_i^\circ)^* - T(\overline{S}_i^\circ)^*] + \int_V^\infty [P - (\sum_i n_i RT)/V]dV - \sum_i n_i RT\,ln(V/n_i RT) \quad . \tag{32}$$

Chemical potential

$$(\mu_i)^{V,T} = \left(\frac{\partial A^{V,T}}{\partial n_i}\right)_{V,T,n_j \neq n_i} = (\overline{U}_i^\circ)^* - T(\overline{S}_i^\circ)^* + \int_V^\infty [\left(\frac{\partial P}{\partial n_i}\right)_{V,T,n_j} - RT/V]dV - RT\,ln(V/n_i RT) + RT \quad . \tag{33}$$

Because

$$H = U + PV \quad , \tag{34}$$

$$\overline{H}^* = \overline{U}^* + P^*\overline{V}^* = \overline{U}^* + RT \quad , \tag{35}$$

and

$$(\mu_i)^{V,T} = (\overline{H}_i^\circ)^* - T(\overline{S}_i^\circ)^* + \int_V^\infty [\left(\frac{\partial P}{\partial n_i}\right)_{T,V,n_j} - RT/V]dV - RT\,ln(V/n_i RT) \quad . \tag{36}$$

Fugacity and fugacity coefficient

$$(\mu_i)^{V,T} = (\mu_i^\circ)^{V^*,T} + RT\,ln(f_i)^{V,T} = (\overline{H}_i^\circ)^* - T(\overline{S}_i^\circ)^* + RT\,ln(f_i)^{V,T} \quad . \tag{37}$$

Comparing Equations (36) and (37),

$$RT\,ln(f_i)^{V,T} = \int_V^\infty [\left(\frac{\partial P}{\partial n_i}\right)_{T,V,n_j} - RT/V]dV - RT\,ln(V/n_i RT) \quad , \tag{38}$$

and

$$RT\,ln(\gamma_i)^{P,T} = RT\,ln[(f_i)^{V,T}/X_i P] = \int_V^\infty [\left(\frac{\partial P}{\partial n_i}\right)_{T,V,n_j} - RT/V]dV - RT\,ln Z \quad , \tag{39}$$

where Z is the compressibility factor

$$Z = PV/nRT \quad . \tag{40}$$

For a pure substance (one-component fluid) Equations (38) and (39) reduce to

$$RT\,ln(f_i^\circ)^{V,T} = \int_V^\infty [\left(\frac{\partial P}{\partial n}\right)_{T,V} - RT/V_i^\circ]dV - RT\,ln(V_i^\circ/nRT) \tag{41}$$

and

$$RT\,ln(\gamma_i^\circ)^{V,T} = \int_V^\infty [\left(\frac{\partial P}{\partial n}\right)_{T,V} - RT/V_i^\circ]dV - RT\,ln Z^\circ \quad . \tag{42}$$

EQUATIONS OF STATE

Equations (38), (39), (41), and (42) provide a method for obtaining fugacity coefficients from pressure-explicit equations of state. Numerous equations of state have been proposed over the years to describe the P-V-T properties of real fluids. We limit the following review to two that are used in geochemistry and petrology.

Virial equation

One-component fluids. The virial equation represents the compressibility factor, Z, as a power series in reciprocal molar volume:

$$Z = 1 + B/\overline{V} + C/\overline{V}^2 + D/\overline{V}^3 + \ldots = 1 + B\rho + C\rho^2 + D\rho^3 + \ldots \tag{43}$$

where $\rho = 1/\overline{V}$ is density, B is referred to as the second virial coefficient, C the third virial coefficient, etc. The coefficients B, C, D,... are functions of temperature and independent of pressure. Equation (43) therefore is a pressure-explicit equation of state:

$$P = RT/\overline{V} + BRT/\overline{V}^2 + CRT/\overline{V}^3 + DRT/\overline{V}^4 + \ldots \tag{44}$$

Experimental P-V-T data are usually adequately represented if the series is truncated after the third term in equation (43) (Prausnitz, 1969; Saxena and Fei, 1987a). Coefficients B and C may be evaluated from low-pressure P-V-T data as

$$B = \lim_{\rho \to 0}\left(\frac{\partial Z}{\partial \rho}\right)_T \tag{45}$$

and

$$C = \lim_{\rho \to 0}\left(\frac{1}{2}\right)\left(\frac{\partial^2 Z}{\partial \rho^2}\right)_T \quad . \tag{46}$$

Alternatively the virial equation may be expressed in volume-explicit form:

$$Z = 1 + B'P + C'P^2 + D'P^3 + \ldots \quad (47)$$

The equivalence of Equation (43) and (47) can be seen from the explicit relationships among the coefficients of the two equations:

$$B' = B/RT \quad (48)$$

$$C' = (C-B^2)/(RT)^2 \quad (49)$$

$$D' = (D-3BC+2B^3)/(RT)^3 \quad . \quad (50)$$

Both the volume-explicit and pressure-explicit equations of state may represent the P-V-T properties of a wide range of fluids with the same B, C, D, B', C', D',... coefficients if the equations are written in terms of reduced variables T_R, P_R, and V_R where $T_R = T/T_C$, $P_R = P/P_C$, and $V_R = V/V_C$ with T_C the critical temperature, P_C the critical pressure, and V_C the critical molar volume of the fluid. In reduced variables Equation (43) becomes

$$Z = 1 + B/\overline{V}_R + C/\overline{V}_R^2 + D/\overline{V}_R^3 + \ldots \quad (51)$$

and Equation (47) becomes

$$Z = 1 + B'P_R + C'P_R^2 + D'P_R^3 + \ldots \quad (52)$$

and B, C, D, B', C', D',... are universal functions of T_R.

Fluid mixtures. A powerful feature of the virial equation of state is that it applies equally well to fluid mixtures as to pure substances. For fluid mixtures B, C, D, etc. in Equation (43) refer to the fluid mixture. The relationships between these coefficients for the mixture and fluid composition are referred to as "mixing rules."

Citing statistical thermodynamic arguments, Prauznitz (1969) showed

$$B_m = \sum_{ij}\sum X_i X_j B_{ij} \qquad (B_{ij}=B_{ji}) \quad (53)$$

and

$$C_m = \sum_{ijk}\sum\sum X_i X_j X_k C_{ijk} \qquad (C_{iij} = C_{iji} = C_{jii}, \text{ etc.}) \quad . \quad (54)$$

Table 2. Models of Potential Energy between Molecules.

1. Ideal gas potential:

   $\Gamma(r) = 0$ for all r.

2. Hard-sphere potential:

   $\Gamma(r) = 0$ for $r>\sigma$
   $\Gamma(r) = \infty$ for $r\leq\sigma$

   σ = sum of radii of the hard-sphere molecules.

3. Sutherland potential:

   $\Gamma(r) = \infty$ for $r\leq\sigma$
   $\Gamma(r) = -D/r^6$ for $r>\sigma$

   D = species-specific constant.

4. Lennard-Jones' form of Mie's potential:

   $\Gamma(r) = 4\varepsilon[(\sigma/r)^{12} - (\sigma/r)^6]$

   ε = depth of energy well (see Fig. 1).

5. Square-well potential:

   $\Gamma(r) = \infty$ for $r\leq\sigma$
   $\Gamma(r) = -\varepsilon$ for $\sigma<r\leq R\sigma$
   $\Gamma(r) = 0$ for $r>R\sigma$

   R = reduced well width (see Fig. 1).

6. Exp-6 potential:

   $\Gamma(r) = \varepsilon/[1-(6/\gamma)]\{(6/\gamma)\exp[\gamma(1-r/r_m)] - (r_m/r)^6]\}$

   γ = adjustable parameter that determines steepness of potential well
   r_m = intermolecular separation at minimum potential energy (see Fig. 1).

7. Kihara potential:

   $\Gamma(r) = \infty$ for $r<2a$
   $\Gamma(r) = 4\varepsilon\{[(\sigma-2a)/(r-2a)]^{12} - [(\sigma-2a)/(r-2a)]^6\}$

   2a = sum of radii of hard sphere molecular core.

8. Stockmayer potential:

   $\Gamma(r) = 4\varepsilon[(\sigma/r)^{12} - (\sigma/r)^6](\mu^2/r^3)f_\theta(\theta_1, \theta_2, \theta_3)$

   μ = dipole moment
   f_θ relates potential energy to orientation of dipole molecules defined by angles θ_1, θ_2, θ_3.

The terms B_{ii} and C_{iii} are the second and third virial coefficients, respectively, for fluid composed only of component i (etc. for B_{jj} and C_{jjj}). The terms B_{ij}, C_{ijk}, etc. refer to interactions among unlike molecules in a fluid mixture. There are numerous statistical thermodynamic models for the B_{ij} and C_{ijk} terms (Prausnitz, 1969). In each case the virial coefficients are related to the potential energy between pairs of unlike molecules i, j, and k, Γ_{ij}, Γ_{ik}, and Γ_{jk}, by

$$B_{ij} = 2\pi N\int_0^\infty [1 - \exp(-\Gamma_{ij}/kT)]r^2dr \tag{55}$$

and

$$C_{ijk} = \frac{-8\pi^2N^2}{3}\int_0^\infty\int_0^\infty\int_{|r_{ij}-r_{jk}|}^{r_{ij}+r_{jk}} f_{ij}f_{ik}f_{jk}r_{ij}r_{ik}r_{jk}dr_{ij}dr_{ik}dr_{jk}$$

$$f_{ij} = \exp(-\Gamma_{ij}/kT) - 1 \text{ (etc. for } f_{ik}, f_{jk}) \quad , \tag{56}$$

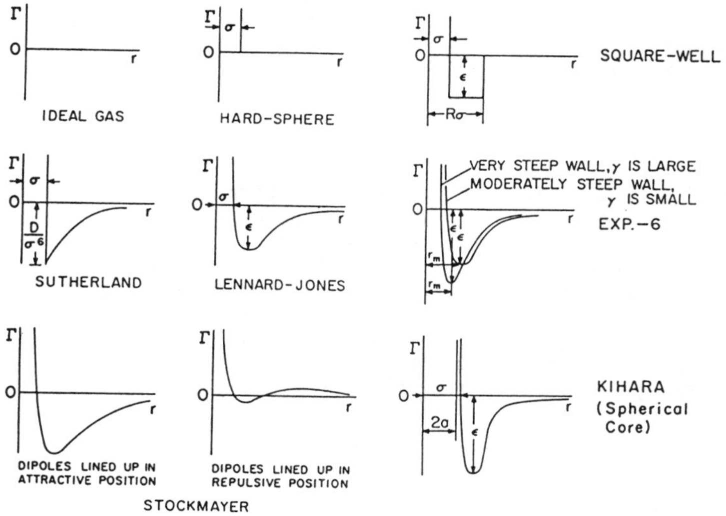

Figure 1. Models for the potential energy (Γ) of interaction between two molecules as a function of the separation of their centers (r). Mathematical formulations and symbols explained in Table 2. Each of the potential energy models allows calculation of virial coefficients B and C in Equations (53)-(56). From Prausntiz (1969).

where N is Avagadro's constant, k is Boltzmann's constant, and r is the separation between centers of molecules. Table 2 summarizes eight models of the potential energy, and Figure 1 qualitatively illustrates each model for Γ as a function of r. The various potential models are used to calculate virial coefficients B_{ij} and C_{ijk} through equations (55) and (56) and coefficients B_m and C_m through Equations (53) and (54).

Redlich-Kwong Equation

One-component fluids. Redlich and Kwong (1949) proposed an equation of state that retains the form of the ideal-gas equation and that satisfactorily represents P-V-T properties of fluids at metamorphic P-T conditions:

$$P = \frac{RT}{\overline{V}-b} - \frac{a}{T^{\frac{1}{2}}\overline{V}(\overline{V}+b)} \quad . \tag{57}$$

There is no rigorous theoretical basis for Equation (57); the best justification for its use is its empirical success in reproducing P-V-T data for a wide range of fluids over a wide range of P-T conditions (Prausnitz, 1969; Holloway, 1977).

Coefficient b refers to the volume occupied by the molecules themselves ("excluded volume"), and therefore is species-specific and independent of pressure and temperature (for geological processes). Coeffient a refers to attractive forces among the molecules and is species-specific, temperature-dependent (in some cases), and independent of pressure. Coefficient a usually is represented as:

$$a = a^\circ + a_1(T) \quad . \tag{58}$$

The term a° refers to intermolecular attraction due to dispersion forces among molecules in the fluid. There is no theoretical basis for treating a° as anything other than a species-specific constant independent of pressure and temperature (Prausnitz, 1969; Holloway, 1977). The function $a_1(T)$ refers to attractive forces among permanent dipole and permanent quadripole molecules in the fluid and/or to attractive chemical forces (e.g., hydrogen bonding, association, chemical reaction). For simple, non-polar fluids $a_1(T) = 0$. For other fluids there is no theoretical basis for treating $a_1(T)$ as anything other than species-specific and independent of pressure (Prausnitz, 1969; Holloway, 1977). The function $a_1(T)$ is usually approximated by a power series in temperature:

$$a_1(T) = L + MT + NT^2 + \ldots \quad , \tag{59}$$

with L, M, N, constants.

Numerical values for b, a°, L, M, N,... are normally obtained from regression of experimental P-V-T data. Alternatively, as with the virial equation, the Redlich-Kwong equation may be used with universal values of a and b if reduced variables are substituted into Equation (57). The resulting equation in reduced variables, however, is less accurate than an equation with parameters fit for a specific molecular species. Because at the critical point for any phase

$$\left(\frac{\partial P}{\partial \overline{V}}\right)_T = 0 \tag{60}$$

and

$$\left(\frac{\partial^2 P}{\partial \overline{V}^2}\right)_T = 0 \quad , \tag{61}$$

it follows that the Redlich-Kwong equation in reduced variables is equivalent to applying Equation (57) with a and b calculated from the following universal function of P_C and T_C:

$$a = 0.4278R^2T_C^{2.5}/P_C \tag{62}$$

$$b = 0.0867RT_C/P_C \quad . \tag{63}$$

Equations (62) and (63) may be derived by solving Equations (57), (60), and (61) for a, b, and V_C (= 3.8473b) at P_C and T_C. In the absence of experimental P-V-T data, Equations (62) and (63), combined with values of P_C and T_C, recently compiled by Hiza et al. (1982) and Mathews (1972), may be used with equation (57) to predict the P-V-T properties of real fluids.

Relationship between Redlich-Kwong and virial equations. Equations (45) and (46), along with corresponding equations for higher-order virial coefficients, are an explicit statement of the equivalence of the virial to the Redlich-Kwong (or any other pressure-explicit) equation of state. Applying equations (45) and (46), the P-V-T properties of fluids that follow the Redlich-Kwong equation, for example, may be represented by the virial equation (truncated after the third term) with

$$B = b - a/RT^{1.5} \tag{64}$$

and

$$C = b^2 + ab/RT^{1.5} \quad . \tag{65}$$

Fluid Mixtures. Equation (57) also applies to fluid mixtures when the a and b parameters are taken for the mixture (a_m and b_m).

Evaluation of b term for mixtures. - The b term in Equation (57) refers to the volume of fluid occupied by the molecules themselves. The mixing rule for the b term, therefore, traditionally is:

$$b_m = \sum_i X_i b_i \quad , \tag{66}$$

where b_i is the b term in an equation of state for fluid that is pure component i. The b_m term can be calculated from Equation (66) using the P-V-T properties of the individual fluid components.

Evaluation of a term for mixtures: general equation. - Following the justification for the mixing rule for the second virial coefficient, the mixing rule for the a term traditionally is:

$$a_m = \sum_i \sum_j X_i X_j a_{ij} \quad . \tag{67}$$

Evaluation of a term for mixtures of non-polar molecules. - The a term in Equation (57) refers to attractions among molecules in the fluid. Prausnitz (1969) and Holloway (1977) show that the potential energy, Γ_{ij}, of the interaction of different simple, non-polar molecules i and j in a mixture due to dispersion forces is the geometric mean of the potential energy of the same same interaction between like molecules, i.e.,

$$\Gamma_{ij} = (\Gamma_{ii}\Gamma_{jj})^{\frac{1}{2}} \quad . \tag{68}$$

This is the principal theoretical justification for estimating a_{ij} in mixtures of non-polar molecules from

$$a_{ij} = (a_i a_j)^{\frac{1}{2}} \quad . \tag{69}$$

Equation (69), however, is probably better justified from the empirical observation that it leads to adequate predictions of the P-V-T properties for many mixtures of non-polar molecules (e.g., Redlich and Kwong, 1949).

Evaluation of a term for mixtures of polar and non-polar molecules (but without both CO_2 and H_2O). - In mixtures of polar and non-polar molecules the only attraction between polar and non-polar molecules is due to dispersion forces. Because $a_1(T)$ in Equation (58) refers to intermolecular attractions due to forces other than dispersion, forces described by $a_1(T)$ do not affect interactions between polar and non-polar molecules. The mixing rule for the a_{ij} coefficient in mixtures of polar and non-polar molecules therefore is:

$$a_{ij} = (a_i^\circ a_j^\circ)^{\frac{1}{2}} \quad , \tag{70}$$

where i refers to a polar molecule and j a non-polar molecule.

Evaluation of a term for mixtures with both CO_2 and H_2O. - Evaluation of the $a_{CO_2-H_2O}$ term in Equation (67) for mixtures containing both CO_2 and H_2O is complicated by the tendency for H_2O and CO_2 to chemically react according to:

$$H_2O + CO_2 = H_2CO_3 \quad . \tag{71}$$

The effect of this reaction on the value of $a_{CO_2-H_2O}$ has been estimated by considering the effect of any general reaction

$$1 + 2 = 12 \tag{72}$$

on the a_{12} coefficient for fluid mixtures of non-polar molecules 1 and 2 Lambert et al., 1949; de Santis et al., 1974).

Consider pressure sufficiently low that a 1-2 fluid mixture obeys a simplified form of Berthelot's Law,

$$P = RT/(\overline{V}-B) \quad , \tag{73}$$

where B is an empirically-determined coefficient. Application of equation (45) shows that B in Equation (73) is identical to the second virial coefficient in an equivalent pressure-explicit virial equation of state state for the fluid. Let K be the equilibrium constant for reaction (72) expressed in terms of partial pressure:

$$K = \frac{X_{12}P}{(X_1)(X_2)P^2} \quad . \tag{74}$$

Consider one mole fluid defined as 6.02×10^{23} molecules 1 and 2 (i.e., before reaction and formation of 12). Let α be the fraction of 1 and 2 molecules that form 12.

$$n_{12} = \alpha/2 \tag{75}$$

$$n_1 + n_2 = 1 - \alpha \tag{76}$$

after reaction and

$$K = \frac{P\alpha/[2(1 - \alpha/2)]}{(X_1)(X_2)P^2} \quad . \tag{77}$$

For a small amount of reaction, $1 - \alpha/2 \sim 1$,

$$\alpha/2 = KP(X_1)(X_2) \tag{78}$$

and

$$n_T = 1 - KP(X_1)(X_2) \quad , \tag{79}$$

where n_T is the total number of molecules present after reaction. Let B refer to the empirical coefficient in equation (73) for a fictive 1-2 mixture in which Reaction (72) does not occur and B^{obs} refer to the observed empirical coefficient for a real 1-2 mixture in which the reaction has occurred. The difference between B and B^{obs} may be quantitatively estimated by noting that the effect of Reaction (73) is to reduce the total number of molecules present. From Equation (73)

$$PV/RT = n_T + n_T PB/RT \quad . \tag{80}$$

After reaction (Eqn. 79):

$$PV/RT = 1 + P\left[\frac{n_T B - RTK(X_1)(X_2)}{RT}\right] \quad . \tag{81}$$

Because α is small, $n_T \sim 1$ and

$$P\overline{V}/RT = 1 + P\left[\frac{n_T B - RTK(X_1)(X_2)}{RT}\right] \quad . \tag{82}$$

$$P\overline{V} = RT + P[B - RTK(X_1)(X_2)] \quad . \tag{83}$$

Because (from Eqn. 73)

$$P\overline{V} = RT + PB^{obs} \quad , \tag{84}$$

evidently

$$B^{obs} = B - RTK(X_1)(X_2) \quad . \tag{85}$$

Applying Equation (53)

$$B^{obs} = B_1(X_1)^2 + 2B_{12}^{obs}(X_1)(X_2) + B_2(X_2)^2 \quad . \tag{86}$$

For a fictive 1-2 mixture without reaction

$$B = B_1(X_1)^2 + 2B_{12}(X_1)(X_2) + B_2(X_2)^2 \quad , \tag{87}$$

and because of Equation (85)

$$2B_{12}^{obs}(X_1)(X_2) = 2B_{12}(X_1)(X_2) - RTK(X_1)(X_2) \quad ; \tag{88}$$

therefore

$$B_{12}^{obs} = B_{12} - \tfrac{1}{2}RTK \quad . \tag{89}$$

B_{12} in Equation (89) is the same as B_{12} (physical) in equation (8) of de Santis et al. (1974). Equation (89) shows that the observed B^{obs} coefficient incorporates effects of both physical interactions between 1 and 2 molecules (B_{12}) and chemical interactions ($-\tfrac{1}{2}RTK$).

Combining Equations (53), (64), (66), and (67) for the 1-2 mixture:

$$B_m^{obs} = B_1(X_1)^2 + 2B_{12}^{obs}(X_1)(X_2) + B_2(X_2)^2$$

$$= b_1X_1 + b_2X_2 - \frac{a_1(X_1)^2 + 2a_{12}(X_1)(X_2) + a_2(X_2)^2}{RT^{1.5}} \quad . \tag{90}$$

Comparing terms,

$$B_{12}^{obs} = (b_1X_1 + b_2X_2/[2(X_1)(X_2)] - a_{12}/(RT^{1.5}) \quad . \tag{91}$$

For a fictive 1-2 mixture in which Reaction (72) does not occur

$$B_{12} = \frac{b_1X_1 + b_2X_2}{2(X_1)(X_2)} - \frac{[(a_1^\circ)(a_2^\circ)]^{\frac{1}{2}}}{RT^{1.5}} \tag{92}$$

(Eqns. 53, 64, 66, 67, 69). Combining Equations (89), (91), and (92):

$$a_{12} = (a_1^\circ a_2^\circ)^{\frac{1}{2}} + \tfrac{1}{2}R^2T^{2.5}K \quad . \tag{93}$$

Although Equation (93) has been derived for non-polar molecules at conditions of low pressure and small degrees of reaction, it has been used by de Santis et al. (1974) to evaluate a in calculations of the P-V-T properties of CO_2-H_2O mixtures at pressures up to 3000 bars and by Holloway (1977), Flowers (1979), and Bowers and Helgeson (1983b), among others, in calculations for comparable or even higher pressures. Values of K in Equation (93) for Reaction (71) are given by de Santis et al. (1974):

$$\ln K = -11.071 + 5953/T - (2746 \cdot 10^3)/T^2 + (464.6 \cdot 10^6)/T^3 \quad , \tag{94}$$

where K is in reciprocal atmospheres. The coefficients in Equation (94) were obtained by a fit to solubility and P-V-T data in the range 25°-750°C and 1 - 1500 bars.

Equations (41) and (42) allow calculation of fugacity or fugacity coefficient for a one-component fluid for which there is an equation state (e.g., Eqns. 43, 57). Equations (38) and (39) allow calculation of the fugacity or fugacity coefficient for a component in a fluid mixture from equations of state when they are combined with the appropriate mixing rules (Eqns. 53-54; 66-67; 69-70; 93).

Fugacities from virial equation of state

For a one-component (i) fluid that obeys Equation (43) truncated after the third term:

$$\left(\frac{\partial P}{\partial n}\right)_{T,V} = RT/\overline{V} + 2nB_i RT/\overline{V}^2 + 3n^2C_i RT/\overline{V}^3 \quad . \tag{95}$$

Applying Equation (41)

$$ln(f_i^\circ)^{V,T} = 2B_i/\overline{V} + 3C_i/(2\overline{V}^2) - ln(\overline{V}/RT) \quad , \tag{96}$$

with $\overline{V}$ calculated at arbitrary pressure and temperature from Equation (43).

The fugacity of component i in a fluid solution that obeys Equation (43) may be calculated similarily using equation (38) and the mixing rules, Equations (53) and (54). For example, for a mixture of n_1 moles component 1 and n_2 moles component 2 and a virial Equation (43) truncated after the third term:

$$\left(\frac{\partial P}{\partial n_1}\right)_{T,V,n_2} = RT/\overline{V}_m + RT(2n_1B_1 + 2n_2B_{12})/\overline{V}_m^2 + RT(3n_1^2C_1 + 6n_1n_2C_{112} + 3n_2^2C_{122})/\overline{V}_m^3 \quad , \tag{97}$$

and

$$ln(f_1)^{V,T} = 2(X_1B_1 + X_2B_{12})/\overline{V}_m + 3(X_1^2C_1 + 2X_1X_2C_{112} + X_2^2C_{122})/(2\overline{V}_m^2) - ln(\overline{V}_m/RT) \quad , \tag{98}$$

with $\overline{V}_m$ calculated at arbitrary pressure and temperature from Equations (43), (53), and (54).

Fugacities from Redlich-Kwong equation of state

For a one-component fluid containing n moles component i that obeys Equation (57):

$$\left(\frac{\partial P}{\partial n}\right)_{T,V} = \frac{RT}{(V - nb_i)} + \frac{nb_iRT}{(V - nb_i)^2} - \frac{2a_in}{T^{\frac{1}{2}}V(V + nb_i)} + \frac{a_ib_in^2}{T^{\frac{1}{2}}V(V + nb_i)^2} \quad . \tag{99}$$

Applying Equation (41)

$$ln(f_i^\circ)^{V,T} = ln\left(\frac{\overline{V}}{\overline{V} - b_i}\right) + \frac{b_i}{\overline{V} - b_i} - \frac{a_i}{RT^{1.5}(\overline{V} + b_i)} + \left(\frac{a_i}{Rb_iT^{1.5}}\right) ln\left(\frac{\overline{V}}{\overline{V} + b_i}\right) - ln\left(\frac{\overline{V}}{RT}\right) \quad . \tag{100}$$

Equation (100) is often recast in terms of A_i and B_i, where

$$B_i \equiv b_i/(RT) \tag{101}$$

and

$$A_i^2 \equiv a_i/(R^2T^{2.5}) \tag{102}$$

(Redlich and Kwong, 1949; Holloway, 1977). Equation (100) then becomes

$$ln(f_i^\circ)^{V,T} = -ln(Z - B_iP) - (A_i^2/B_i)ln[(Z + B_iP)/Z] + Z + lnP - 1 \quad . \tag{103}$$

For n moles fluid mixtures containing n_i moles component i that obeys Equation (57):

$$\left(\frac{\partial P}{\partial n_i}\right)_{T,V,n_j \neq n_i} = \frac{RT}{V_m - nb_m} + \left[\frac{nRT}{(V_m - nb_m)^2}\right]\left(\frac{\partial nb_m}{\partial n_i}\right)_{T,V,n_j} - \frac{\left(\frac{\partial a_m n^2}{\partial n_i}\right)_{T,V,n_j}}{T^{\frac{1}{2}}V_m(V_m + nb_m)} + \left[\frac{a_m n^2}{T^{\frac{1}{2}}V_m(V_m + nb_m)^2}\right]\left(\frac{\partial nb_m}{\partial n_i}\right)_{T,V,n_j} \tag{104}$$

and

$$ln(f_i)^{V,T} = ln\left(\frac{\overline{V}_m}{\overline{V}_m - b_m}\right) + \frac{\left(\frac{\partial nb_m}{\partial n_i}\right)_{T,V,n_j}}{\overline{V}_m - b_m} - \frac{\left(\frac{\partial a_m n^2}{\partial n_i}\right)_{T,V,n_j}}{RT^{1.5}nb_m} ln\left(\frac{\overline{V}_m + b_m}{\overline{V}_m}\right) +$$

$$\left[\frac{a_m\left(\frac{\partial nb_m}{\partial n_i}\right)_{T,V,n_j}}{b_m^2RT^{1.5}}\right]\left[ln\left(\frac{\overline{V}_m + b_m}{\overline{V}_m}\right) - \left(\frac{b_m}{b_m + \overline{V}_m}\right)\right] - ln(\overline{V}_m/RT) \text{ , } \quad (105)$$

with a_m, b_m, and $\overline{V}_m$ calculated from Equation (57) and mixing rules, Equations (66)-(67), (69)-(70), and (93). The partial differentials may be evaluated using these mixing rules:

$$\left(\frac{\partial nb_m}{\partial n_i}\right)_{T,V,n_j} = b_i \quad (106)$$

$$\left(\frac{\partial a_m n^2}{\partial n_i}\right)_{T,V,n_j} = 2n(X_i a_i + \sum_{j \neq i} X_j a_{ij}) \quad , \quad (107)$$

and Equation (105) becomes

$$ln(f_i)^{V,T} = ln\left(\frac{\overline{V}_m}{\overline{V}_m - b_m}\right) + \frac{b_i}{\overline{V}_m - b_m} + \frac{2\sum_{j \neq i} X_j a_{ij}}{RT^{1.5}b_m} ln\left(\frac{\overline{V}_m + b_m}{\overline{V}_m}\right) +$$

$$\left(\frac{a_m b_i}{b_m^2RT^{1.5}}\right)\left[ln\left(\frac{\overline{V}_m + b_m}{\overline{V}_m}\right) - \left(\frac{b_m}{b_m + \overline{V}_m}\right)\right] - ln(\overline{V}_m/RT) \quad . \quad (108)$$

A note on the Lewis and Randall Rule

The Lewis and Randall Rule is commonly used by geochemists and petrologists to calculate the fugacity of a component in a fluid solution.

The Lewis and Randall Rule may be derived from Equations (39) and (42) and Amagat's Law:

$$\overline{V}_i = \overline{V}_i^\circ \quad . \tag{109}$$

From Equations (39) and (42),

$$RT\ln[(f_i)^{V,T}/(X_iP)] - RT\ln[(f_i^\circ)^{V,T}/P] =$$

$$\int_V^\infty \left[\left(\frac{\partial P_{mixture}}{\partial n_i}\right)_{T,V,n_{j\neq i}} - \frac{RT}{V}\right]dV - \int_V^\infty \left[\left(\frac{\partial P_{pure\ fluid}}{\partial n}\right)_{T,V} - \frac{RT}{V^\circ}\right]dV - RT\ln Z + RT\ln Z^\circ . \tag{110}$$

If the fluid mixture obeys Equation (109) (i.e., ideal mixing with respect to volume) for all components over the entire volume range between V and ∞ ,

$$\left(\frac{\partial P_{mixture}}{\partial n_i}\right)_{T,V,n_j} = \left(\frac{\partial P_{pure\ fluid}}{\partial n}\right)_{T,V} \tag{111}$$

and

$$RT\ln\left[\frac{(f_i)^{V,T}}{X_i(f_i^\circ)^{V,T}}\right] = \int_V^\infty \left(-RT/V + RT/V^\circ\right)dV - RT\ln(Z/Z^\circ) = 0 \tag{112}$$

or

$$(f_i)^{V,T} = X_i(f_i^\circ)^{V,T} \quad , \tag{113}$$

which is the Lewis and Randall Rule. Equations (109)-(113) emphasize that the Lewis and Randall Rule is only valid for fluid mixtures that obey Amagat's Law over the range of volumes involved in the integrals in Equation (110). Because of the significant deviation of fluid mixtures from Amagat's Law at moderate pressures, Prausnitz (1969) concluded that the Lewis and Randall Rule is of limited utility.

Table 3. Expressions for A(T), B(T), C(T), and D(T) in Virial-like Equations of State from Saxena and Fei (1987a, b).

A. Expressions for Ar, Xe, N_2, O_2, CO, CO_2, CH_4, and H_2 for equation (114).

$A = 2.0614 - 2.2351T_R^{-2} - 0.39411 \ln T_R$

$B = 5.5125\cdot10^{-2}T_R^{-1} + 3.9344\cdot10^{-2}T_R^{-2}$

$C = -1.8935\cdot10^{-6}T_R^{-1} - 1.1092\cdot10^{-5}T_R^{-2} - 2.1892\cdot10^{-5}T_R^{-3}$

$D = 5.0527\cdot10^{-11}T_R^{-1} - 6.3033\cdot10^{-21}T_R^{3}$

B. Expressions for H_2O for equation (114).

$A = 1.4937 - 1.8626T_R^{-2} + 0.80003T_R^{-3} - 0.3941 \ln T_R$

$B = 4.2410\cdot10^{-2}T_R^{-1} + 2.4097\cdot10^{-2}T_R^{-2} - 8.9634\cdot10^{-3}T_R^{-3}$

$C = -9.016\cdot10^{-7}T_R^{-1} - 6.1345\cdot10^{-5}T_R^{-2} + 2.2380\cdot10^{-5}T_R^{-3} + 5.2335\cdot10^{-7} \ln T_R$

$D = -7.6707\cdot10^{-9}T_R^{-1} + 4.1108\cdot10^{-8}T_R^{-2} - 1.4798\cdot10^{-8}T_R^{-3} - 6.3033\cdot10^{-21}T_R^{3}$

C. Expressions for N_2, O_2, CO, CO_2, CH_4, and H_2 for equation (115).

$A = 1 - 0.5917T_R^{-2}$

$B = 0.09122T_R^{-1}$

$C = 1.4164\cdot10^{-4}T_R^{-2} - 2.8349\cdot10^{-6} \ln T_R$

(P > 1 kbar; T > 400°K) and

$A = 1$

$B = 0.09827T_R^{-1} - 0.2709T_R^{-3}$

$C = 0.01472T_R^{-4} - 0.00103T_R^{-1.5}$

(P < 1 kbar; T > 400°K).

D. Expressions for H_2O for equation (116).

$A = -0.7025 + 1.16\cdot10^{-3}T + 99.6799T^{-1}$

$B = 0.2143T^{-1} - 3.1423\cdot10^{-14}T^{3}$

$C = -2.249\cdot10^{-6}T^{-1} - 0.1459T^{-3} + 2.1690\cdot10^{-15}T^{2}$

E. Expressions for H_2O for equation (117).

$B' = -2.20960T^{0.5} + 3.35460\cdot10^{-8}T^{3}$

$C' = 3.4569\cdot10^{-5}T^{2.5} + 64.9764 \ln T$

Table 4. Values for a° and b and Expressions for $a_{CO_2}(T)$ and $a_{H_2O}(T)$ in Holloway-Flowers Version of the Redlich-Kwong Equation of State (from Holloway, 1981).

	b	a°
CO_2	29.7	$46\cdot10^{6}$
CO	27.38	$16.98\cdot10^{6}$
CH_4	29.70	$31.59\cdot10^{6}$
H_2	15.15	$3.56\cdot10^{6}$
H_2O	14.6	$35.0\cdot10^{6}$
H_2S	20.0	$87.9\cdot10^{6}$
SO_2	39.4	$142.6\cdot10^{6}$
N_2	26.8	$15.382\cdot10^{6}$

$a_{CO_2}(T) = [73.03 - 0.0714T(°C) + 2.157\cdot10^{-5}T^2(°C)]\cdot10^5$

600°C ≤ T ≤ 1200°C:

$a_{H_2O}(T) = [166.8 - 0.19308T(°C) + 0.1864\cdot10^{-3}T^2(°C) - 0.71288\cdot10^{-7}T^3(°C)]\cdot10^5$

T < 600°C:

$a_{H_2O}(T) = [4.221\cdot10^3 - 3.1227\cdot10^1T(°C) + 8.7485\cdot10^{-2}T^2(°C) - 1.07295\cdot10^{-4}T^3(°C) + 4.86111\cdot10^{-8}T^4(°C)]\cdot10^5$

T > 1200°C:

$a_{H_2O}(T) = [140 - 0.050T(°C)]\cdot10^5$

a in units of $atm\cdot cm^6 °K^{\frac{1}{2}}/mole^2$

b in units of $cm^3/mole$

EQUATIONS OF STATE IN THE GEOCHEMICAL/PETROLOGICAL LITERATURE

Virial equations of state

Saxena and Fei (1987a) derived a volume-explicit equation of state in reduced variables similar to the virial equation from shock wave and conventional P-V-T data:

$$Z = A + BP_R + CP_R^2 + DP_R^3 \quad . \tag{114}$$

Equation (114) is strictly equivalent to the virial equation only when A = 1. Data for Ar, Xe, N_2, O_2, CO, CO_2, CH_4 and H_2 were fit by Equation (114) with temperature-dependent functions for A, B, C, and D (Table 3A). Data for H_2O are not well-represented by the same A, B, C, D in Table 3A. For H_2O, Saxena and Fei presented separate expressions for A, B, C, and D (Table 3B). Equation (114) may be used along with expressions in Tables 3A and 3B and tabulations of critical pressures and critical temperatures (e.g., Mathews, 1972; Hiza et al., 1982) to calculate the molar volume of a one-component fluid composed of any of the 9 species at arbitrary pressure and temperature. Because coefficients in Tables 3A and 3B incorporate data for P-T conditions up to 15,000K and 1 megabar, Saxena and Fei believe their equation of state is valid to at least 3000K and 1 megabar. They do not recommend, however, using Equation (114) and coefficients in Table 3B for P < 2 kbar nor coefficients in Table 3A for CO_2 for P < 5 kbar.

For crustal pressures and temperatures Saxena (1987b) derived a a simpler volume-explicit, virial-like equation of state in reduced variables from P-V-T data alone:

$$Z = A + BP_R + CP_R^2 \tag{115}$$

Data for H_2, O_2, CO, CH_4, N_2 and CO_2 were fit by Equation (115) with the temperature-dependent functions for A, B, and C in Table 3C. Data for H_2O are not well-represented at crustal pressures and temperatures by Equation (115) and expressions in Table 3C. For H_2O, Saxena and Fei (1987b) fit P-V-T data for P > 1 kbar to

$$Z = A + BP + CP^2 \quad , \tag{116}$$

with temperature-dependent expressions for A, B, and C in Table 3D. Data for H_2O at 100 < P < 1000 bars were fit to

$$Z = 1 + B'/\overline{V} + C'/\overline{V}^2 \quad , \tag{117}$$

with temperature-dependent expressions for B' and C' in Table 3E. Saxena and Fei (1987b) do not recommend using Equations (115) and (116) for T < 400K nor Equation (117) for T < 673K. They state no upper pressure or upper temperature limits for Equations (115)-(117).

Currently efforts are underway to extend the virial (or virial-like) equation of state to fluid mixtures of H_2, O_2, CO_2, H_2O, CO and CH_4 using the Kihara and Stockmayer models of intermolecular potential energy (Fei and Saxena, 1987).

Redlich-Kwong equations of state

Holloway-Flowers version. Holloway (1977) introduced the Redlich-Kwong equation of state to geochemists and petrologists. Flowers (1979) later noted that Holloway's treatment of the mixing rules violated the Gibbs-Duhem relationship. The correct treatment of the mixing rules, as shown by Flowers (1979) and Flowers and Helgeson (1983), are Equations (66)-(67), (69)-(70), (93), (106), and (107). Holloway (1977) recommended the following treatment of the a terms: (1) Equation (69) for mixtures of simple, non-polar molecules; (2) Equation (70) for mixtures of simple, non-polar molecules with CO_2 or H_2O (but not both); (3) Equation (93) for $a_{CO_2\text{-}H_2O}$ in mixtures containing both CO_2 and H_2O. Table 4 contains b and a° values from Holloway (1981) for fluid species of geological interest as well as expressions for $a_{CO_2}(T)$ and $a_{H_2O}(T)$. We note that the a_{H_2O} functions in Table 2 are not continuous in temperature. Calculated values of f_{H_2O} therefore will also not be continuous in temperature. The Holloway-Flowers version is the theoretically most justified form of the Redlich-Kwong equation used by geochemists because it treats b and a° independent of pressure and temperature and a as a function of temperature only. Holloway (1977, 1981) used the Redlich-Kwong equation to calculate fugacity and activity in a variety of geologically-relevant fluid mixtures at P-T conditions typical of metamorphism in the crust. Holloway (1981) stated no pressure or temperature limits to his equation of state.

Bottinga-Touret-Richet version. Touret and Bottinga (1979) presented an equation of state for CO_2 based on the Redlich-Kwong equation (57) with

$$a = a_1(r^3 - r^6) + a_2 \quad , \tag{118}$$

$$b = [\ln(\overline{V}/a_3) + b_1]/b_2 \quad , \tag{119}$$

where $r = a_3/\overline{V}$ and a_1, a_2, a_3, b_1, and b_2 are adjustable parameters (see Table 5). Bottinga and Richet (1981) used equations (57) and (118)-(119) to calculate density, Gibbs free energy, entropy, heat capacity, thermal expansion, and compressibility of CO_2 to 2100K and 50 kbar. They did not attempt to apply their equation of state to fluid mixtures. The Bottinga-Touret-Richet version of the Redlich-Kwong equation treats a and b as functions of volume which allowed a better fit of Equation (57) to P-V-T data for CO_2. There is no theoretical justification, however, for treating a and b as functions of volume. Bottinga and Richet (1981) state that their equation satisfactorily fits P-V-T data for gas-

Table 5. Expressions for a_1, a_2, a_3, b_1, and b_2 in Equations (118) and (119) from Touret and Bottinga (1979).

$a_1 = 6.566 \cdot 10^7 bar°K^{\frac{1}{2}}(cm^3mole^{-1})^2$

$a_2 = 7.276 \cdot 10^7 bar°K^{\frac{1}{2}}(cm^3mole^{-1})^2$

$a_3 = 37.3\ cm^3mole^{-1}$

For $\overline{V} \geq 180\ cm^3mole^{-1}$: $b_1 = 7.352629$
$b_2 = 0.241413\ mole \cdot cm^{-3}$

For $180 > \overline{V} > 47.22\ cm^3mole^{-1}$: $b_1 = 11.707864$
$b_2 = 0.363955\ mole \cdot cm^{-3}$

For $\overline{V} \leq 47.22\ cm^3mole^{-1}$: $b_1 = 1.857669$
$b_2 = 0.0637935\ mole \cdot cm^{-3}$

Table 6. Expressions for $a^\circ_{H_2O}(T,w)$, $a_{H_2O}(T,w)$, and $b_{H_2O}(w)$ in Bowers-Helgeson Version of the Redlich-Kwong Equation of State (from Bowers and Helgeson, 1983a).

$a^\circ_{H_2O} \cdot 10^{-6} = \exp(\Theta + w\Xi + w^2\Upsilon + w^3\Phi)$ where

$\Theta = 4.881243 + 0.1823047 \cdot 10^{-2}T - 0.1712269 \cdot 10^{-4}T^2 + 0.6479419 \cdot 10^{-8}T^3$

$\Xi = 0.02636494 - 0.536994 \cdot 10^{-3}T + 0.2687074 \cdot 10^{-5}T^2 - 0.4321741 \cdot 10^{-8}T^3$

$\Upsilon = 0.6802827 \cdot 10^{-2} - 0.948023 \cdot 10^{-4}T + 0.3770339 \cdot 10^{-6}T^2 - 0.5075318 \cdot 10^{-9}T^3$

$\Phi = 0.5235827 \cdot 10^{-4} - 0.3505272 \cdot 10^{-7}T$

$a_{H_2O} \cdot 10^{-6} = \Gamma + w\Lambda$ where

$\Gamma = 111.3057 + 50.70033\exp(-0.982646 \cdot 10^{-2}T)$

$\Lambda = -8.05658\exp(-0.982646 \cdot 10^{-2}T)$

$b_{H_2O} = \Psi + w\Omega$ where

$\Psi = 14.6$

$\Omega = -0.04420283$

a in units of $bar \cdot cm^6 °K^{\frac{1}{2}}/mole^2$

b in units of $cm^3/mole$

eous CO_2 over the P-T range -50° to 1000°C and 1 - 10,000 bars. They believe it may be applied in the temperature range 400 - 2100K and in the pressure range 1 - 50,000 bars.

Halbach-Chatterjee version. Halbach and Chatterjee (1982) derived an equation of state for H_2O based on the Redlich-Kwong Equation (57) with

$$a(T) = 1.616\cdot10^8 - 4.989\cdot10^4 T - 7.358\cdot10^9 T^{-1} \tag{120}$$

and

$$b(P) = \frac{1 + 3.4505\cdot10^{-4}P + 3.898\cdot10^{-9}P^2 - 2.7756\cdot10^{-15}P^3}{6.3944\cdot10^{-2} + 2.3776\cdot10^{-5}P + 4.5717\cdot10^{-10}P^2} \tag{121}$$

Although Halbach and Chatterjee (1982) achieved a better fit of P-V-T data to Equation (57) with a pressure-dependent b term, there is no theoretical justification for doing so. They did not extend their equation of state to fluid mixtures. Halbach and Chatterjee believe their equation may be applied over the P-T range 100° - 1000°C and up to 200 kbar.

Bowers-Helgeson version. Bowers and Helgeson (1983a) presented a Redlich-Kwong equation of state for H_2O-CO_2-NaCl mixtures. Rather than treat NaCl explicitly as a third component, they included it implicitly in the a_{H_2O} and b_{H_2O} terms by making them a function of w (weight precent NaCl relative to NaCl+H_2O in the mixture). Their preferred functions $a_{H_2O}(T,w)$ and $b_{H_2O}(w)$ are presented in Table 6. Bowers and Helgeson utilized the same values for b_{CO_2}, $a^{\circ}_{CO_2}$, and $a_{CO_2}(T)$ as Holloway (1977) (our Table 4). Fugacities and fugacity coefficients for CO_2 and H_2O in the ternary mixture were calculated from Equations (38) and (39) utilizing the mixing rules in Equations (66) - (67) and (93). Because of the length and complexity of the expressions for b and a in Table 6, the explicit expression for fugacity and fugacity coefficient were not published by Bowers and Helgeson (1983a) and we did not attempt to include them here. Bowers and Helgeson achieved a better fit of P-V-T data to Equation (57) by allowing $a^{\circ}_{H_2O}$ to be a function of temperature. Because of the significance of a°, there is no theoretical justification for taking $a^{\circ}_{H_2O}$ as anything other than a constant independent of pressure and temperature. In a companion paper, Bowers and Helgeson (1983b) used their equation of state to calculate numerous phase diagrams depicting equilibria involving minerals and CO_2-H_2O-NaCl fluid at P-T conditions typical of metamorphism. Bowers and Helgeson (1985) state that their equation is intended for applications in the range T = 350° - 600° and P > 500 bars and to fluids with w < 35%.

Kerrick-Jacobs version. Kerrick and Jacobs (1981) and Jacobs and Kerrick (1981a) used a modification of the Redlich-Kwong equation proposed by Carnahan and Starling (1969, 1972) to derive an equation of state for CO_2-H_2O-CH_4 mixtures. Carnahan and Starling replaced the repulsive term in the Redlich-Kwong Equation (57) (first term on righthand side) with $RT(1+y+y^2-y^3)/\overline{V}(1-y)^3$ where $y=b/4V$. Carnahan and

Starling (1969, 1972) demonstrate that the new repulsive term is justified theoretically and provides a better fit to P-V-T data than the conventional repulsive term $RT/(\overline{V}-b)$. Kerrick and Jacobs fit the modified Redlich-Kwong equation:

$$P = \frac{RT(1 + y + y^2 - y^3)}{\overline{V}(1 - y)^3} - \frac{a}{T^{\frac{1}{2}}\overline{V}(\overline{V} + b)} = \frac{RT}{\overline{V}}\left[1 + \frac{4y - 2y^2}{(1 - y)^3}\right] - \frac{a}{T^{\frac{1}{2}}\overline{V}(\overline{V} + b)} \tag{122}$$

to P-V-T data for CO_2, H_2O, and CH_4. Their preferred values for b are listed in Table 7A. The a terms were fit to a power series in reciprocal molar volume truncated after the third term:

$$a_i = c_i + d_i/\overline{V} + e_i/\overline{V}^2 \quad . \tag{123}$$

Their preferred functions of c_i, d_i and e_i for CO_2, H_2O, and CH_4 are listed in Table 7B. In contrast to the Holloway-Flowers or Bowers-Helgeson versions of the Redlich-Kwong equation, Kerrick and Jacobs used

Table 7. Expressions for b_i, c_i, d_i, and e_i in Kerrick-Jacobs Version of the Redlich-Kwong Equation of State (from Kerrick and Jacobs, 1981).

A. Values for b parameters.

$b_{CO_2} = 58$

$b_{H_2O} = 29$

$b_{CH_4} = 60$

B. Expressions for c, d, and e parameters.

$c_{H_2O} = (290.78 - 0.30276T + 1.4774\cdot10^{-4}T^2)\cdot10^6$

$d_{H_2O} = (-8374 + 19.437T - 8.148\cdot10^{-3}T^2)\cdot10^6$

$e_{H_2O} = (76600 - 133.9T + 0.1071T^2)\cdot10^6$

$c_{CO_2} = (28.31 + 0.10721T - 8.81\cdot10^{-6}T^2)\cdot10^6$

$d_{CO_2} = (9380 - 8.53T + 1.189\cdot10^{-3}T^2)\cdot10^6$

$e_{CO_2} = (-368654 + 715.9T + 0.1534T^2)\cdot10^6$

$c_{CH_4} = 13.403\cdot10^6 + 9.28\cdot10^4T + 2.7T^2$

$d_{CH_4} = 5.216\cdot10^9 - 6.8\cdot10^6T + 3.28\cdot10^3T^2$

$e_{CH_4} = -2.3322\cdot10^{11} + 6.738\cdot10^8T + 3.179\cdot10^5T^2$

b_i in units $cm^3/mole$

c_i in units $bar\cdot cm^6\,{}^\circ K^{\frac{1}{2}} mole^{-2}$

d_i in units $bar\cdot cm^9\,{}^\circ K^{\frac{1}{2}} mole^{-3}$

e_i in units $bar\cdot cm^{12}\,{}^\circ K^{\frac{1}{2}} mole^{-4}$

Equations (66), (67), and (69) as their mixing rules even for fluid mixtures containing both CO_2 and H_2O. Kerrick and Jacobs (1981) justify their choice of mixing rules by noting that the rules have theoretical justification only for equations of state with a firm theoretical foundation. As Prausnitz (1969) pointed out, mixing rules for empirical equations of state are largely arbitrary and are better justified by empirical tests. In the absence of empirical tests, relatively simple mixing rules are preferred over more complicated ones. Kerrick and Jacobs calculated fugacities and fugacity coefficients for pure CO_2, pure H_2O, and pure CH_4 fluids using Equations (122), (123), (41), and (42). For example, for n moles one-component fluid (considering $(\partial y/\partial n)_{T,V} = y/n$ and $\overline{V} = V/n$),

$$\left(\frac{\partial P}{\partial n}\right)_{T,V} = \frac{RT}{V} + \frac{RT}{V}\left[\frac{8y - 6y^2}{(1-y)^3} + \frac{12y - 6y^3}{(1-y)^4}\right] - \frac{2cn}{T^{\frac{1}{2}}V(V+bn)} + \frac{bcn^2}{T^{\frac{1}{2}}V(V+bn)^2} -$$

$$\frac{3dn^2}{T^{\frac{1}{2}}V^2(V+bn)} + \frac{bdn^3}{T^{\frac{1}{2}}V^2(V+bn)^2} - \frac{4en^3}{T^{\frac{1}{2}}V^3(V+bn)} + \frac{ebn^4}{T^{\frac{1}{2}}V^3(V+bn)^2} \quad , \tag{124}$$

and because $d\overline{V} = -dy/y$, application of Equation (42) leads to

$$ln(\gamma^\circ)^{V,T} = \frac{8y - 9y^2 + 3y^3}{(1-y)^3} - lnZ - \frac{c}{RT^{1.5}(\overline{V}+b)} - \frac{d}{RT^{1.5}\overline{V}(\overline{V}+b)} -$$

$$\frac{e}{RT^{1.5}\overline{V}^2(\overline{V}+b)} + \left(\frac{c}{RT^{1.5}b}\right) ln\left(\frac{\overline{V}}{\overline{V}+b}\right) - \frac{d}{RT^{1.5}b\overline{V}} +$$

$$\left(\frac{d}{RT^{1.5}b^2}\right) ln\left(\frac{\overline{V}+b}{\overline{V}}\right) - \frac{e}{RT^{1.5}2b\overline{V}^2} + \frac{e}{RT^{1.5}b^2\overline{V}} -$$

$$\left(\frac{e}{RT^{1.5}b^3}\right) ln\left(\frac{\overline{V}+b}{\overline{V}}\right) \quad , \tag{125}$$

with a, b, c, d, and e determined from equations in Table 7 and $\overline{V}$ calcucalculated from Equation (122) at the pressure and temperature of interest.

Similarly, for n moles CO_2 (component 1) - H_2O (component 2) mixture with composition X_1-X_2, from mixing rules, Equations (66), (67), and (69):

$$\left(\frac{\partial bn}{\partial n_1}\right)_{T,V,n_2} = b_1 \tag{126}$$

$$\left(\frac{\partial n^2c}{\partial n_1}\right)_{T,V,n_2} = 2c_1n_1 + 2c_{12}n_2 \tag{127}$$

$$\left(\frac{\partial n^2d}{\partial n_1}\right)_{T,V,n_2} = 2d_1n_1 + 2d_{12}n_2 \tag{128}$$

$$\left(\frac{\partial n^2e}{\partial n_1}\right)_{T,V,n_2} = 2e_1n_1 + 2e_{12}n_2 \tag{129}$$

and

$$\left(\frac{\partial P}{\partial n_1}\right)_{T,V,n_2} = \frac{RT}{\overline{V}_m} + \frac{RT}{\overline{V}_m}\left[\frac{(4b_my_m + 4b_1y_m - 2b_my_m^2 - 4b_1y_m^2)}{b_m(1 - y_m)^3} + \frac{12b_1y_m^2 - 6b_1y_m^3}{b_m(1 - y_m)^4}\right] -$$

$$\frac{2c_1n_1 + 2c_{12}n_2}{T^{\frac{1}{2}}V_m(V_m + nb_m)} + \frac{c_mn^2b_1}{T^{\frac{1}{2}}V_m(V_m + nb_m)^2} - \frac{n^2d_m + n(2d_1n_1 + 2d_{12}n_2)}{T^{\frac{1}{2}}V_m^2(V_m + nb_m)} +$$

$$\frac{d_mn^3b_1}{T^{\frac{1}{2}}V_m^2(V_m + nb_m)^2} - \frac{2e_mn^3 + n^2(2e_1n_1 + 2e_{12}n_2)}{T^{\frac{1}{2}}V_m^3(V_m + nb_m)} + \frac{e_mn^4b_1}{T^{\frac{1}{2}}V_m^3(V_m + nb_m)^2} \tag{130}$$

where subscript m refers to the mixture. Substitution of Equation (130) into Equation (39) and integration leads to:

$$\ln(\gamma_1)^{V,T} = \frac{4y_m - 3y_m^2}{(1 - y_m)^2} + \frac{b_1(4y_m - 2y_m^2)}{b_m(1 - y_m)^3} - \left(\frac{2c_1X_1 + 2X_2c_{12}}{RT^{1.5}b_m}\right)\ln\left(\frac{\overline{V}_m + b_m}{\overline{V}_m}\right) -$$

$$\frac{c_mb_1}{RT^{1.5}b_m(\overline{V}_m + b_m)} + \left(\frac{c_mb_1}{RT^{1.5}b_m^2}\right)\ln\left(\frac{\overline{V}_m + b_m}{\overline{V}_m}\right) - \frac{2d_1X_1 + 2X_2d_{12} + d_m}{RT^{1.5}b_m\overline{V}_m} +$$

$$\left(\frac{2d_1X_1 + 2X_2d_{12} + d_m}{RT^{1.5}b_m^2}\right)\ln\left(\frac{\overline{V}_m + b_m}{\overline{V}_m}\right) + \frac{b_1d_m}{RT^{1.5}\overline{V}_mb_m(\overline{V}_m + b_m)} +$$

$$\frac{2b_1d_m}{RT^{1.5}b_m^2(\overline{V}_m + b_m)} - \left(\frac{2b_1d_m}{RT^{1.5}b_m^3}\right)\ln\left(\frac{\overline{V}_m + b_m}{\overline{V}_m}\right) - \frac{2e_1X_1 + 2X_2e_{12} + 2e_m}{RT^{1.5}2b_m\overline{V}_m^2} +$$

$$\frac{2e_1X_1 + 2e_{12}X_2 + 2e_m}{RT^{1.5}b_m^2\overline{V}_m} - \left(\frac{2e_1X_1 + 2e_{12}X_2 + 2e_m}{RT^{1.5}b_m^3}\right)\ln\left(\frac{\overline{V}_m + b_m}{\overline{V}_m}\right) +$$

$$\frac{e_mb_1}{RT^{1.5}2b_m\overline{V}_m^2(\overline{V}_m + b_m)} - \frac{3e_mb_1}{RT^{1.5}2b_m^2\overline{V}_m(\overline{V}_m + b_m)} +$$

$$\left(\frac{3e_mb_1}{RT^{1.5}b_m^4}\right)\ln\left(\frac{\overline{V}_m + b_m}{\overline{V}_m}\right) - \frac{3e_mb_1}{RT^{1.5}b_m^3(\overline{V}_m + b_m)} - \ln Z_m\ , \tag{131}$$

with y_m and $\overline{V}_m$ determined from Equation (122) and expressions in Table 7 at the pressure, temperature, and composition of interest.

Kerrick and Jacobs achieved a better fit of P-V-T data to Equation (122) by allowing a to be a function of both volume and temperature which has no theoretical justification. In a companion paper Jacobs and Kerrick (1981b) used their equation of state to calculate numerous phase diagrams depicting devolatilization equilibria among minerals and CO_2-H_2O fluid at elevated pressures and temperatures. Jacobs and Kerrick (1981c) state that their equation may be applied to CO_2 - H_2O mixtures over the range 300° - 1050°C and 1 - 20,000 bars. Kerrick and Jacobs (1981), however, do not recommend application of their version of the Redlich-Kwong equation for $P \leq 500$ bars and T = 400° - 500°C.

MINERAL-FLUID EQUILIBRIA AND EVALUATION OF EQUATIONS OF STATE

In this section we show how experimentally-determined mineral-fluid equilibria may be used to critically evaluate several equations of state for fluid at the elevated P-T conditions of metamorphism. We restricted our attention to the Holloway-Flowers, Bowers-Helgeson, and Kerrick-Jacobs versions of the Redlich-Kwong equation as they apply to CO_2-H_2O mixtures because (1) they are currently the only equations of state that have been extended to fluid mixtures, (2) they are in wide use among geochemists and petrologists, and (3) there are independent experimental data on equilibrium between minerals and CO_2-H_2O fluids. As introduction to the evaluation, we briefly review the basic equation for mineral-fluid equilibrium and discuss the thermodynamic data used in our calculations.

Basic Equation for mineral-fluid equilibrium

If a stoichiometric reaction relationship can be written among components in minerals and fluid at equilibrium at pressure P and temperature T:

$$(\Delta\overline{G})^{P,T} = (\Delta\overline{G}°)^{1,298} + \int_{298}^{T} (\Delta c_p^\circ)dT - T\int_{298}^{T} (\Delta c_p^\circ/T)dT + \Delta\overline{V}_s^\circ P +$$

$$\sum_i \nu_{i,s} RT\,ln(a_{i,s})^{P,T} + \sum_j \nu_j RT\,ln(f_j)^{P,T} = 0 \quad , \tag{132}$$

where subscripts "i" and "s" refer to components i in the solid phases and subscript "j" refers to components j in the fluid solution. If values of $(\Delta\overline{G}°)^{1,298}$, Δc_p°, $\Delta\overline{V}_s^\circ$, and $a_{i,s}$ are known, then Equation (132) allows

$$\sum_j \nu_j RT\,ln(f_j)^{P,T} \tag{133}$$

to be calculated at any pressure and temperature of interest.

Equilibria and the thermodynamic data base

We evaluated the three equations of state using experimental data for four mineral equilibria (all with CO_2-H_2O fluid):

$$\underset{\text{calcite}}{CaCO_3} + \underset{\text{quartz}}{SiO_2} = \underset{\text{wollastonite}}{CaSiO_3} + CO_2 \tag{134}$$

$$\underset{\text{muscovite}}{KAl_3Si_3O_{10}(OH)_2} + \underset{\text{calcite}}{CaCO_3} + \underset{\text{quartz}}{2\,SiO_2} = \underset{\text{sanidine}}{KAlSi_3O_8} + \underset{\text{anorthite}}{CaAl_2Si_2O_8} + H_2O + CO_2 \tag{135}$$

$$\underset{\text{dolomite}}{3\,CaMg(CO_3)_2} + \underset{\text{quartz}}{4\,SiO_2} + H_2O = \underset{\text{talc}}{Mg_3Si_4O_{10}(OH)_2} + \underset{\text{calcite}}{3\,CaCO_3} + 3\,CO_2 \tag{136}$$

$$\underset{\text{zoisite}}{2\,Ca_2Al_3Si_3O_{12}(OH)} + CO_2 = \underset{\text{anorthite}}{3\,CaAl_2Si_2O_8} + \underset{\text{calcite}}{CaCO_3} + H_2O \quad . \tag{137}$$

Reactions (134)-(137) represent each of the four qualitatively different kinds of decarbonation/dehydration equilibria. Because the minerals in the experiments were pure substances, all $a_{i,s}$ terms in Equation (132) are one and

$$\sum_i \nu_{i,s} RT\,ln(a_{i,s})^{P,T} = 0 \quad . \tag{138}$$

We used Equation (132) along with c_p°, $\overline{V}_s^\circ$, and $(\overline{G}^\circ)^{1,298}$ data from Helgeson et al. (1978) to calculate $ln F_{calc}$ for Equilibria (134)-(137) at pressure-temperature conditions of interest where we define

$$ln F_{calc} \equiv \sum_j \nu_j\, ln(f_j^{calc})^{P,T} \quad . \tag{139}$$

Diagrams to evaluate equations of state

For each experimental determination of a P-T-X_{CO_2} state of equilibrium for mineral-fluid Equilibria (134)-(137), we may calculate, using either the Holloway-Flowers, Bowers-Helgeson, or Kerrick-Jacobs equation of state:

$$ln F_{obs} \equiv \sum_j \nu_j\, ln(f_j^{obs})^{P,T} \quad , \tag{140}$$

where f_j^{obs} is the fugacity of CO_2 or H_2O in the fluid solution. We define

$$\Delta ln F \equiv ln F_{obs} - ln F_{calc} \quad , \tag{141}$$

and $\Delta \ell n F$ is a test for the agreement between the equations of state and the experimental data. In Figures 2-5, we plot $\Delta \ell n F$ vs. X_{CO_2} for experimental P-T-X_{CO_2} determinations of Equilibria (134)-(137) to test for this agreement, to check whether any deviation from agreement shows a systematic relationship with fluid composition, and to evaluate mixing behavior.

In our calculations based on the Bowers-Helgeson version we set $a^{\circ}_{H_2O}$ constant and equal to the preferred value of Holloway (1977) (our Table 4). We then assigned the temperature-dependent portion of the expression for $a^{\circ}_{H_2O}$ in Table 6 to the K term in Equation (93). Our modification of the Bowers-Helgeson equation is prompted by the lack of theoretical justification for treating $a^{\circ}_{H_2O}$ as anything other than a constant independent of temperature. The modification has no effect on calculations for CO_2-H_2O mixtures but a small one for mixtures of H_2O with other molecular species.

<u>Dolomite-quartz-talc-calcite</u>. Figure 2 plots $\Delta \ell n F$ vs. X_{CO_2} for Equilibrium (136) from the experimental data of Metz and Puhan (1970, 1971). In Figures 2-5 panel A refers to the Holloway-Flowers version of the Redlich-Kwong equation, panel B to the Bowers-Helgeson version, and panel C to the Kerrick-Jacobs version. Each box represents one experimental determination of the P-T-X_{CO_2} conditions of equilibrium. The length of the horizontal sides of the boxes refer to the experimental uncertainty in determining equilibrium X_{CO_2}. If there were perfect agreement among experimental data, Helgeson et al.'s (1978) thermochemical data, and the equations of state, a horizontal line could be drawn in Figure 2 at $\Delta \ell n F = 0$ that passes through all boxes. If the $(\overline{G}^{\circ})^{1,298}$ terms are in error but there is perfect agreement between c°_p and $\overline{V}^{\circ}_s$ data, experimental data, and the equations of state, a horizontal line could still be drawn through all boxes but at $\Delta \ell n F \neq 0$. If $(\overline{G}^{\circ})^{1,298}$ and $\overline{V}^{\circ}_s$ data are in error but there is perfect agreement between c°_p data, experimental data, and the equations of state, a horizontal line could be drawn through subsets of isobaric boxes but again at $\Delta \ell n F \neq 0$. We believe that the c°_p data of Helgeson et al. (1978) and the experimental data are sufficiently accurate that the data in Figure 2 constitute a valid test of the equations of state for the fluid.

A horizontal line cannot be drawn through all boxes in any of Figures 2A-2C. Furthermore, no horizontal line can be drawn through any of the isobaric boxes in Figure 2. Apparently none of the versions of the Redlich-Kwong equation currently used in geochemistry and petrology is consistent with the experimental data of Metz and Puhan (1970, 1971). Because boxes in Figures 2B and 2C plot in a more nearly horizontal array, however, it appears that the Bowers-Helgeson and Kerrick-Jacobs versions are more consistent with the experimental data. The positive slope of the array of boxes in Figure 2 suggest that all three equations of state tend to exaggerate non-ideality of the fluid at low X_{CO_2} and underestimate the fluid's non-ideality at high X_{CO_2}.

<u>Calcite-quartz-wollastonite</u>. Figure 3 plots $\Delta \ell n F$ vs. X_{CO_2} for Equilibrium (134) from the experimental data of Greenwood (1967) and Harker and Tuttle (1956). Filled symbols refer to P-T-X_{CO_2} conditions at which calcite+quartz is stable; open symbols to P-T-X_{CO_2} conditions at which wollastonite is stable. Greenwood did not report uncertainties in experimentally-determined X_{CO_2}; the size of symbols in Figure 3 therefore has no significance. If there were perfect agreement between experimental data, c°_p and $\overline{V}^{\circ}_s$ data, and the equations of state, a horizontal line could be drawn that would perfectly separate open and filled symbols. No such horizontal line can be drawn in any of Figures 3A-3C.

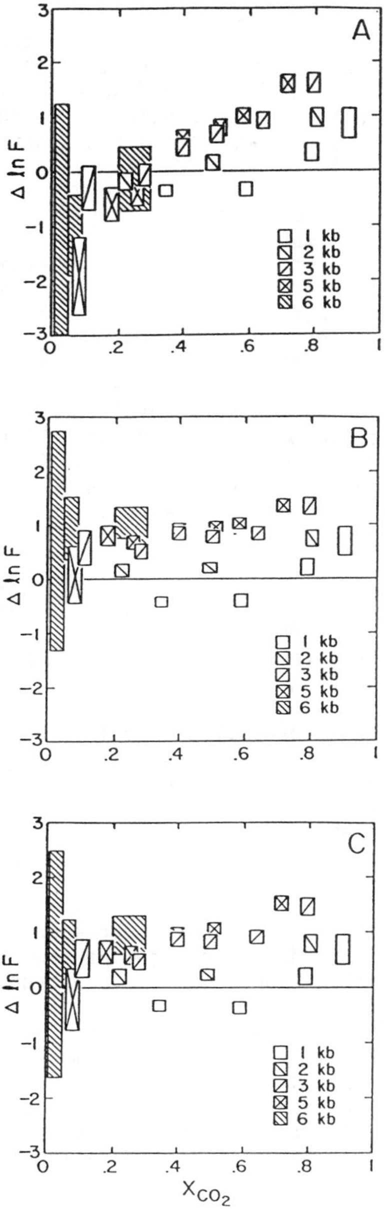

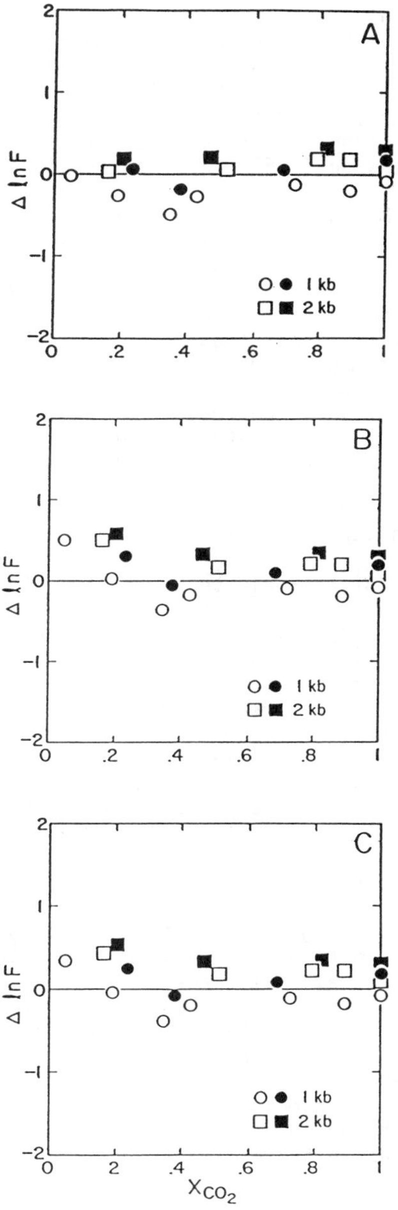

Figure 2 (left). Evaluation of modified Redlich-Kwong equations of state using experimental phase equilibrium data of Metz and Puhan (1970, 1971) for Reaction (136). Sizes of symbols reflect uncertainties in position of equilibrium in terms of plotted variables. A = Holloway-Flowers version; B = Bowers-Helgeson version; C = Kerrick-Jacobs version. See text for discussion.

Figure 3 (right). Evaluation of modified Redlich-Kwong equations of state using experimental phase equilibrium data of Greenwood (1967) and Harker and Tuttle (1956) for Reaction (134). Filled symbols: conditions at which calcite + quartz is stable; open symbols: conditions at which wollastonite is stable. Size of symbols has no significance. See text for discussion. Panels A-C have same significance as in Figure 2.

Furthermore, while a horizontal line in Figure 3A separates solid from open 2 kbar symbols, no horizontal line can separate solid from open 1 kbar symbols in Figure 3A nor can a horizontal line separate solid from open isobaric symbols in Figures 3B and 3C at either 1 or 2 kbar. Unless c_p° data are greatly in error, it therefore appears that none of the three equations of state are fully consistent with the experimental data of Greenwood (1967) and Harker and Tuttle (1956).

Muscovite-calcite-quartz-sanidine-anorthite. Figure 4 plots $\Delta ln F$ vs. X_{CO_2} for Equilibrium (135) from the experimental data of Hewitt (1973). Filled symbols refer to P-T-X conditions at which muscovite+calcite+quartz is stable; open symbols to conditions at which sanidine+anorthite is stable. Size of symbols reflect experimental uncertainty in temperature and X_{CO_2} for each measurement. Because Hewitt performed experiments over a wide range of X_{CO_2} only at 2 kbar, we only consider the 2 kbar data as a test of the equations of state. As in Figure 3, if there were perfect agreement among c_p° data, experimental data, and equations of state, a horizontal line could be could be drawn in Figure 4 that perfectly separates solid and open symbols for 2 kbar. Such a horizontal line can be drawn in each of Figures 4A-4C. The experimental 2 kbar data of Hewitt (1973) appear to be consistent with each version of the Redlich-Kwong equation of state.

Zoisite-calcite-anorthite. Figure 5 plots $\Delta ln F$ vs. X_{CO_2} for Equilibrium (137) from the experimental data of Allen and Fawcett (1982). Filled symbols refer to the stability of zoisite and open symbols to the stability of anorthite+calcite with their size reflecting uncertainty in temperature and X_{CO_2} for each measurement. All experiments were performed at 5 kbar pressure. If there were perfect agreement between c_p° data, experimental data, and equations of state, a horizontal line could be drawn in Figures 5A-5C that perfectly separates solid from open symbols. Because no such line can be drawn in any of Figures 5A-5C, the experimental data of Allen and Fawcett (1982) apparently disagree with each of the versions of the Redlich-Kwong equation of state.

Discussion

Although none of the equations of state are in perfect agreement with all experimental data that we selected, the deviation from agreement is not large. All three versions of the Redlich-Kwong equation are probably satisfactory for many calculations of mineral-fluid equilibria of interest in metamorphic petrology. Figure 2 suggests that the Bowers-Helgeson and Kerrick-Jacobs versions are slightly better models for CO_2-H_2O fluids than the Holloway-Flowers version. Because of its easier use in numerical calculations, we prefer the Bowers-Helgeson version (with our modification of $a_{H_2O}^\circ$) over the Kerrick-Jacobs version although we hasten to emphasize that the two versions appear to be indistinguishable in their accuracy.

Comparison of Figures 2-5 shows that hydration/decarbonation or carbonation/dehydration equilibria of the form

$$\text{Reactant Minerals} + H_2O = \text{Product Minerals} + CO_2 \qquad (142)$$

are more sensitive tests of equations of state of the fluid than either decarbonation or dehydration/decarbonation equilibria. This is because $\Delta\bar{S}$ for Equilibria (142) are smaller per mole volatiles involved, and

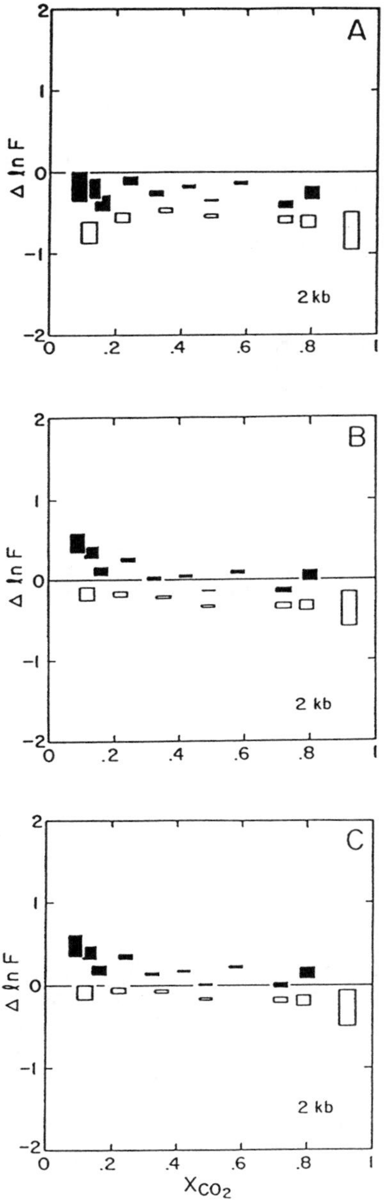

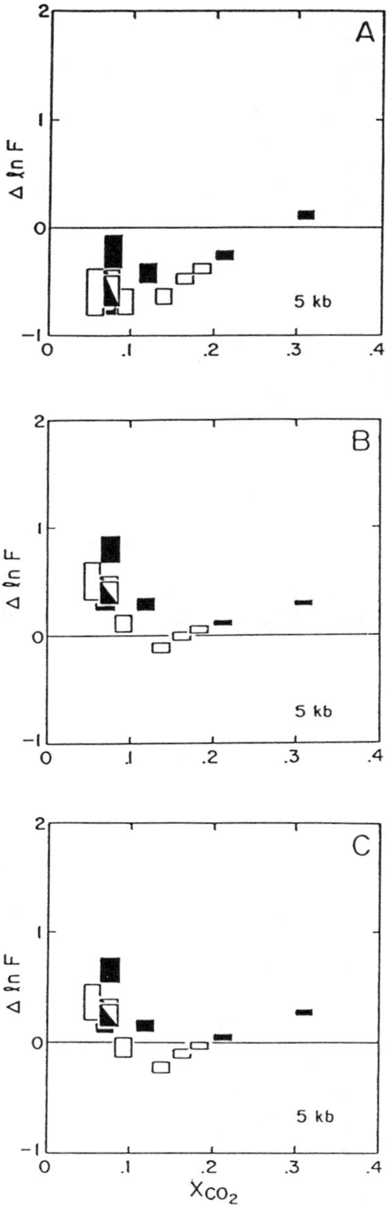

Figure 4 (left). Evaluation of modified Redlich-Kwong equations of state using experimental phase equilibrium data of Hewitt (1973) for Reaction (135). Filled symbols: conditions at which muscovite + calcite + quartz is stable; open symbols: conditions at which anorthite + sanidine is stable. Size of symbols reflects uncertainties in plotted variables. See text for discussion. Panels A-C have same significance as in Figure 2.

Figure 5 (right). Evaluation of modified Redlich-Kwong equations of state using experimental phase equilibrium data of Allen and Fawcett (1982) for Reaction (137). Filled symbols: conditions at which zoisite is stable; open symbols: conditions at which anorthite + calcite is stable; half-filled symbols: conditions at which zoisite + anorthite + calcite is stable. Size of symbols reflects uncertainties in plotted variables. See text for discussion. Panels A-C have same significance as in Figure 2.

hence the equilibria are relatively more sensitive to the mixing properties of the fluid. Clearly experimental data for reactions like (142), especially at low temperatures where non-ideality of the fluid is more pronounced, will provide the best evaluations of equations of state for CO_2-H_2O mixtures. We believe it is also clear that a significantly more accurate version of the Redlich-Kwong equation for CO_2-H_2O mixtures will probably not be developed until more P-V-T data on CO_2-H_2O mixtures at elevated pressures and temperatures become available.

APPLICATIONS OF THE REDLICH-KWONG EQUATION TO TERNARY AND HIGHER-ORDER FLUID SOLUTIONS

In this section we briefly illustrate some applications of the Redlich-Kwong equation to calculations of mineral-fluid equilibria. These calculations are not an exhaustive treatment but rather are intended to show quantitatively the effects of non-ideal mixing and to encourage similar calculations more closely tailored to particular petrologic problems.

The system C-O-H

We used the Redlich-Kwong equation to calculate the composition of C-O-H fluid in equilibrium with graphite at several P-T conditions typical of metamorphism. We considered the fluid to be a mixture of CO_2, H_2O, H_2, CH_4, CO, and O_2 with $(\Delta G°)^{1,T}$ data for

$$C + O_2 = CO_2 \tag{143}$$

$$CH_4 + 2O_2 = CO_2 + 2H_2O \tag{144}$$

$$CO + \tfrac{1}{2}O_2 = CO_2 \tag{145}$$

$$H_2 + \tfrac{1}{2}O_2 = H_2O \quad , \tag{146}$$

from Ohmoto and Kerrick (1977). The molar volume of graphite was taken from Helgeson et al. (1978). The a and b values for H_2O were those of Bowers and Helgeson (1983a) (our Table 6); values of a and b for the other C-O-H species were taken from Holloway (1981) (our Table 4). We used equations (66)-(67),(69)-(70), and (93) as mixing rules. For each choice of pressure and temperature we calculated the mole fraction of the species using equations of the form of (132) over the range of physically-accessible values of f_{O_2} along with the constraint:

$$\sum_i X_i = 1 \quad . \tag{147}$$

Results are presented in Figure 6 along with those obtained assuming ideal mixing of species in the fluid (Lewis and Randall Rule). The effect of increasing pressure, decreasing temperature, and increasing non-ideality of the fluid all act to shift the graphite-saturation surface towards the CO_2-H_2O and H_2O-CH_4 binary joins. Even at moderate pressures and temperatures, the graphite-saturation surface is almost

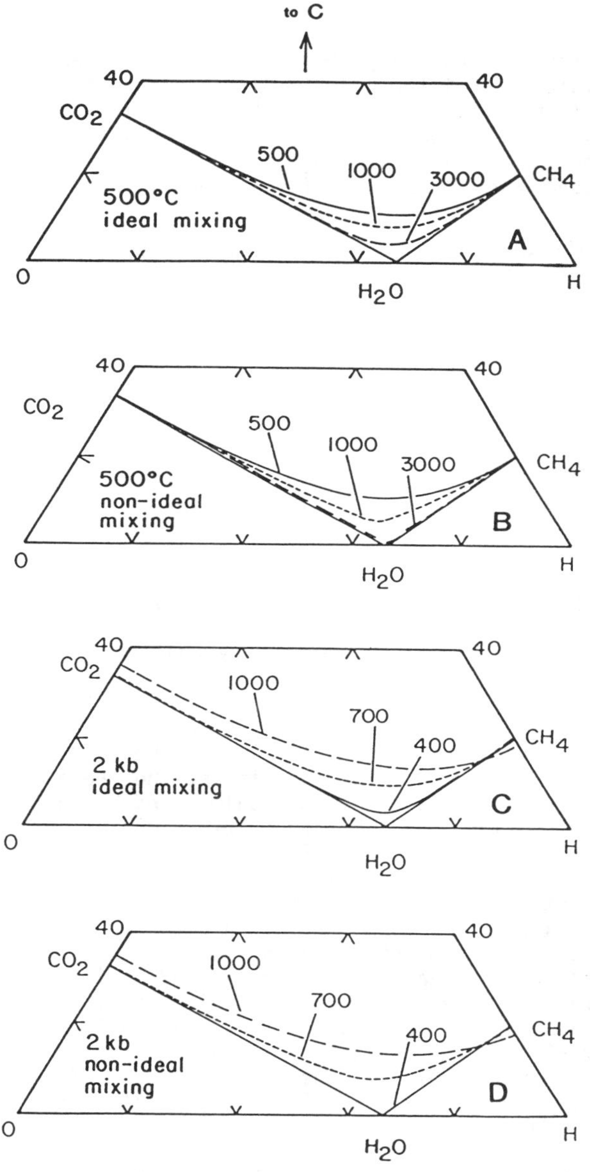

Figure 6. Calculated position of the graphite + fluid equilibrium in the system C-O-H assuming both ideal mixing (Lewis and Randall Rule) and non-ideal CO_2-H_2O mixing specified by the Bowers-Helgeson version of the Redlich-Kwong equation. Non-ideal mixing of other C-O-H species specified by the Holloway-Flowers version. A and B: position at 500, 1000, and 3000 bars pressure and 500°C. C and D: position at 400°, 700°, and 1000°C and 2000 bars. See text for further discussion.

coincident with the CO_2-H_2O and H_2O-CH_4 joins. While it is beyond the scope of this chapter to discuss in detail the petrologic significance of the near coincidence, we note that it has important implications to metamorphic fluid-rock interactions (e.g., Rumble and Hoering, 1986; Rumble et al., 1986).

The system C-O-H-S

We used the Redlich-Kwong equation to calculate the composition of C-O-H-S fluid in equilibrium with pyrite+pyrrhotite+graphite+biotite+K-feldspar at 1 and 3.5 kbars and 350°-600°C. The assemblage is common in metamorphosed graphitic sulfidic schists (Ferry, 1981). We considered the fluid to be a mixture of CO_2, H_2O, H_2, O_2, CH_4, CO, S_2, H_2S, SO_2, and COS, and we used Equilibria (143)-(146) along with

$$\tfrac{1}{2}S_2 + H_2O = H_2S + \tfrac{1}{2}O_2 \tag{148}$$

$$\tfrac{1}{2}S_2 + O_2 = SO_2 \tag{149}$$

$$\tfrac{1}{2}S_2 + CO = COS \tag{150}$$

$$2\ FeS_2 = 2\ FeS + S_2 \tag{151}$$

$$\underset{\text{biotite}}{KFe_3AlSi_3O_{10}(OH)_2} + 1.5S_2 = \underset{\text{K-feldspar}}{KAlSi_3O_8} + \underset{\text{pyrrhotite}}{3\ FeS} + 1.5O_2 + H_2O\ . \tag{152}$$

$(\Delta\bar{G}°)^{1,T}$ for Equilibria (148)-(150) were taken from Ohmoto and Kerrick (1977). $(\Delta\bar{G}°)^{1,T}$ for Equilibrium (151) was taken from Helgeson et al. (1978). Sulfur fugacity defined by pyrite+pyrrhotite and the activity of FeS in pyrrhotite were taken from Toulmin and Barton (1964). $(\Delta\bar{G}°)^{1,T}$ for Equilibrium (152) was derived from experimental data of Tso et al. (1979) and Equation (9) of Wones (1982) assuming ideal ionic mixing of Fe and Mg in synthetic biotite solid solutions. Molar volumes of minerals were taken from Helgeson et al. (1978). Coefficients a and b in the Redlich-Kwong equation for H_2O were calculated from the Bowers-Helgeson version (our Table 6); for other C-O-H-S species other than COS, a and b coefficients were taken from Holloway (1981) (our Table 4); a and b coefficients for COS were calculated from its critical temperature and pressure (Mathews, 1972) and Equations (62) and (63). We used Equations (66)-(67), (69)-(70), and (93) as the mixing rules. From mineral composition data in Ferry (1981) we assumed $a_{or,Kf} = X_{or,Kf} = 0.95$ and $a_{annite,Biotite} = (Fe^*/3)^3 \sim (0.15/3)^3$ where Fe* is the number of iron atoms per standard biotite formula unit in the mineral assemblage.

Results are presented in Figure 7. For comparison, we also calculated fluid compositions assuming ideal mixing of species in the fluid (Lewis and Randall Rule). The composition of C-O-H-S fluid is strongly temperature-dependent and, using our methods and assumptions, there are two possible compositions of fluid in equilibrium with the minerals at each temperature. At low temperature (T < ~400°C) fluids are either nearly

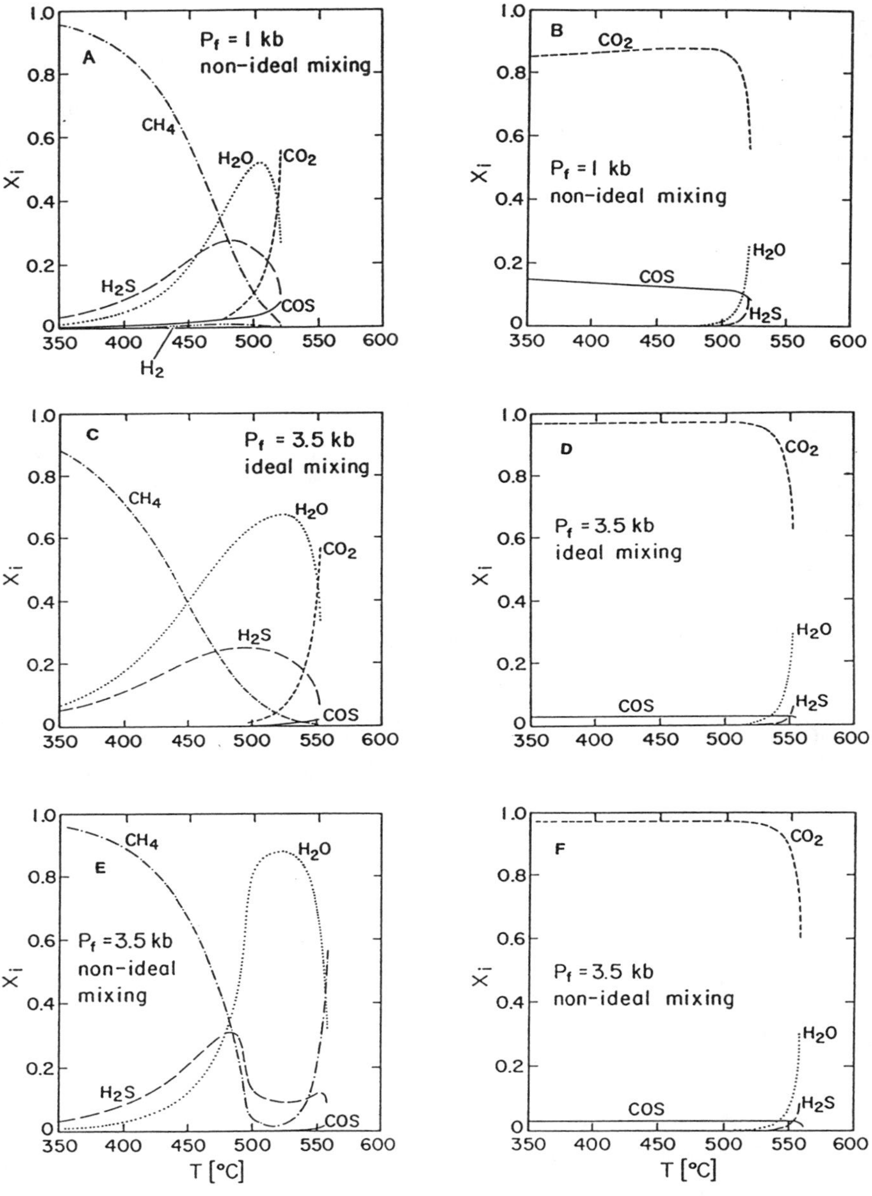

Figure 7. Mole fraction of species in C-O-H-S fluid in equilibrium with pyrite + pyrrhotite + graphite + biotite + K-feldspar at 1000 (panels A,B) and 3500 (panels C-F) bars pressure as a function of temperature in the range 350°-600°C. Results for ideal mixing (panels C and D) assume species mix according to the Lewis and Randall Rule. Results for non-ideal mixing (panels A, B, E, F) assume CO_2 and H_2O mix according to the Bowers-Helgeson version of the Redlich-Kwong equation and other species according to the Holloway-Flowers version. At each pressure and temperature, two equilibrium fluid compositions are possible: one relatively reduced (panels A, C, E) and one relatively oxidized (panels B, D, F). See text for further discussion.

pure CH_4 or CO_2-COS mixtures. At elevated temperatures (T > 500°C) fluids are either CO_2-COS or H_2O-CO_2-H_2S mixtures. The curves in Figure 7 terminate at either ~520°C or ~560°C because those temperatures are the upper stability limit for the assemblage at 1 kbar and 3.5 kbar, respectively. Calculated fluid compositions are qualitatively the same regardless of the mixing model chosen for the fluid. In detail, however, estimations of fluid composition clearly depend on the equation of state adopted for the fluid solution.

The system C-O-H-S-N

As a final application of the Redlich-Kwong equation, we calculated the composition of C-O-H-S-N fluid in equilibrium with pyrite+pyrrhotite+graphite+biotite+K-feldspar using the same procedures described above and specifying

$$P_N^{tot} \equiv N/(N + O + C + H + S)P_{total} = 0.2P_{total} \quad . \tag{153}$$

Preliminary calculations indicated that N_2 and NH_3 are the only significant nitrogen-bearing fluid species, and we therefore assumed the fluid to be a mixture of CO_2, H_2, H_2O, O_2, CH_4, CO, S_2, H_2S, SO_2, COS, N_2 and NH_3. We obtained $(\Delta\bar{G}°)^{1,T}$ for

$$\tfrac{1}{2}N_2 + 1.5H_2 = NH_3 \tag{154}$$

from the JANAF Tables (1965) and estimated a and b parameters in Equation (57) for N_2 and NH_3 from their critical pressures and temperatures (Mathews, 1972) and Equations (62) and (63). Results are shown in Figure 8. The effect of N principally is to dilute the C-O-H-S species. Because the conditions of Figure 8 are quite reducing, we believe that N_2 is probably the only significant nitrogen-bearing fluid species under almost all crustal metamorphic conditions.

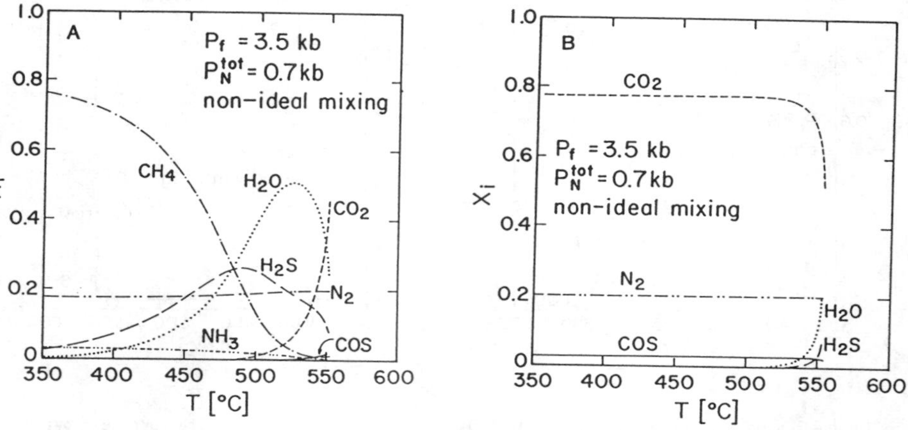

Figure 8. Mole fraction of species in C-O-H-S-N fluid in equilibrium with pyrite + pyrrhotite + graphite + biotite + K-feldspar at 3500 bars pressure and 350°-600°C. Sum of mole fraction of N-bearing species arbitrarily fixed at 0.2. CO_2 and H_2O assumed to mix according to the Bowers-Helgeson version of the Redlich-Kwong equation. All other species assumed mix according to the Holloway-Flowers version. At each pressure and temperature two equilibrium fluid compositions are possible: one relatively reduced (panel A) and one relatively oxidized (panel B). See text for further discussion.

ACKNOWLEDGEMENTS

We thank Terri Bowers for her comments on the chapter and Kate Francis for her patient typing of the manuscript. We gratefully acknowledge support from the U. S. National Science Foundation, Grant EAR-86-06864 (to J.M.F.) and from the Swiss National Science Foundation, Fonds zur Foerderung junger Wissenschaftler (to L.K.B.).

REFERENCES

Allen, J.M. and Fawcett, J.J. (1982) Zoisite-anorthite-calcite stability relations in H_2O-CO_2 fluids at 5000 bars: An experimental and SEM study. J. Petrol. 23, 215-239.

Beattie, J.A. (1955) Thermodynamic properties of real gases and mixtures of real gases. In: Rossini, F.D., ed., Thermodynamics and Physics of Matter. 1, 240-338. Princeton University Press, Princeton, NJ.

Bottinga, Y. and Richet, P. (1981) High pressure and temperature equation of state and calculation of the thermodynamic properties of gaseous carbon dioxide. Amer. J. Sci. 281, 615-660.

Bowers, T.S. and Helgeson, H.C. (1983a) Calculation of the thermodynamic and geochemical consequences of nonideal mixing in the system H_2O-CO_2-NaCl on phase relations in geologic systems:Equation of state for H_2O-CO_2-NaCl fluids at high pressures and temperatures. Geochim. Cosmochim. Acta 47, 1247-1275.

Bowers, T.S. and Helgeson, H.C. (1983b) Calculation of the thermodynamic and geochemical consequences of nonideal mixing in the system H_2O-CO_2-NaCl on phase relations in geologic systems: Metamorphic equilibria at high pressures and temperatures. Amer. Mineral. 68, 1059-1075.

Bowers, T.S. and Helgeson, H.C. (1985) Fortran programs for generating fluid inclusion isochores and fugacity coefficients for the system H_2O-CO_2-NaCl at high pressures and temperatures. Computers Geosciences 11, 203-213.

Carnahan, N.F. and Starling, K.E. (1969) Equation of state for non-attracting rigid spheres. J. Phys. Chem. 51, 635-636.

Carnahan, N.F. and Starling, K.E. (1972) Intermolecular repulsions and the equation of state for fluids. Amer. Inst. Chem. Eng. 18, 1184-1189.

de Santis, R., Breedveld, G.J.F. and Prausnitz, J.M. (1974) Thermodynamic properties of aqueous gas mixtures at advanced pressures. Ind. Eng. Chem. Process Des. Develop. 13, 374-377.

Ferry, J.M. (1981) Petrology of graphitic sulfide-rich schists from south-central Maine: An example of desulfidation during prograde regional metamorphism. Amer. Mineral. 66, 908-930.

Flowers, G.C. (1979) Correction of Holloway's (1977) adaptation of the modified Redlich-Kwong equation of state for calculation of the fugacities of molecular species in supercritical fluids of geologic interest. Contrib. Mineral. Petrol. 69, 315-318.

Flowers, G.C. and Helgeson, H.C. (1983) Equilibrium and mass transfer during progressive metamorphism of siliceous dolomites. Amer. J. Sci. 283, 230-286.

Greenwood, H.J. (1967) Wollastonite: Stability in H_2O-CO_2 mixtures and occurrence in a contact-metamorphic aureole near Salmo, British Columbia, Canada. Amer. Mineral. 52, 1669-1680.

Fei, Y. and Saxena, S.K. (1987) Fluid mixtures at crustal pressures and temperatures. Trans. Amer. Geophys. Union 68, 451.

Halbach, H. and Chatterjee, N.D. (1982) An empirical Redlich-Kwong-type equation of state for water to 1,000°C and 200 kbar. Contrib. Mineral. Petrol. 79, 337-345.

Harker, R.I. and Tuttle, O.F. (1955) Experimental data on the P_{CO_2}-T curve for the reaction: calcite + quartz = wollastonite + carbon dioxide. Amer. J. Sci. 254, 239-256.

Helgeson, H.C., Delany, J.M., Nesbitt, H.W. and Bird, D.K. (1978) Summary and critique of the thermodynamic properties of rock-forming minerals. Amer. J. Sci. 278A, 1-229.

Hewitt, D.A. (1973) Stability of the assemblage muscovite-calcite-quartz. Amer. Mineral. 58, 785-791.

Hiza, M.J., Kidnay, A.J. and Miller, R.C. (1982) Equilibrium Properties of Fluid Mixtures. 2. A Bibliography of Experimental Data on Selected Fluids. NSRDS Bibliographic Series. Plenum, New York.

Holloway, J.R. (1977) Fugacity and activity of molecular species in supercritical fluids. In: Fraser, D.G., ed., Thermodynamics in Geology. Reidel, Dordrecht, Holland, p. 161-181.

Holloway, J.R. (1981) Compositions and volumes of supercritical fluids in the Earth's crust. In: Hollister, L.S. and Crawford, M.L., eds., Fluid Inclusions: Applications to Petrology, p. 13-38, Mineral. Assoc. of Canada, Toronto,

Jacobs, G.K. and Kerrick, D.M. (1981a) Methane: An equation of state with application to the ternary system H_2O-CO_2-CH_4. Geochim. Cosmochim. Acta 45, 607-614.

Jacobs, G.K. and Kerrick, D.M. (1981b) Devolatilization equilibria in H_2O-CO_2 and H_2O-CO_2-NaCl fluids: An experimental and thermodynamic evaluation at elevated pressures and temperatures. Amer. Mineral. 66, 1135-1153.

Jacobs, G.K. and Kerrick, D.M. (1981c) APL and FORTRAN programs for a new equation of state for H_2O, CO_2, and their mixtures at supercritical conditions. Computers Geosciences 7, 131-143.

Joint Army-Navy-Air Force Thermochemical Tables (1965) U.S. Dept. Commerce P.B. 168370, Washington, D.C.

Kerrick, D.M. and Jacobs, G.K. (1981) A modified Redlich-Kwong equation for H_2O, CO_2, and H_2O-CO_2 mixtures at elevated pressures and temperatures. Amer. J. Sci. 281, 735-767.

Lambert, J.D., Roberts, G.A.H., Rowlinson, J.S. and Wilkinson, V.J. (1949) The second virial coefficient of organic vapours. Proc. Roy. Soc. London 196A, 113-125.

Mathews, J.F. (1972) The critical constants of inorganic substances. Phys. Rev. 72, 71-100.

Metz, P. and Puhan, D. (1970) Experimentelle untersuchung der metamorphose von kieselig dolomitischen sedimenten. I. Die gleichgewichtsdaten der reaktion 3 dolomit + 4 quartz + H_2O = talc + 3 calcit + 3 CO_2 fur die gesamtgasdrucke von 1000, 3000, und 5000 bar. Contrib. Mineral. Petrol. 26, 302-314.

Metz, P. and Puhan, D. (1971) Korrektur zur arbeit "Experimentelle untersuchung der metamorphose von kieselig dolomitischen sedimenten. I. Die gleichgewichtsdaten der reaktion 3 dolomit + 4 quartz + H_2O = talc + 3 calcit + 3 CO_2 fur die gesamtgasdrucke von 1000, 3000, und 5000 bar." Contrib. Mineral. Petrol. 31, 169-170.

Ohmoto, H. and Kerrick, D.M. (1977) Devolatilization equilibria in graphitic systems. Amer. J. Sci. 277, 1013-1044.

Prausnitz, J.M. (1969) Molecular Thermodynamics of Fluid-Phase Equilibria. Prentice-Hall, Englewood Cliffs, New Jersey.

Redlich, O. and Kwong, J.N.S. (1949) On the thermodynamics of solutions. V. An equation of state. Fugacities of gaseous solutions. Chem.

Rev. 44, 233-244.
Rumble, D. and Hoering, T.C. (1986) Carbon isotope geochemistry of graphite vein deposits from New Hampshire, U.S.A. Geochim. Cosmochim. Acta 50, 1239-1247.
Rumble, D., Duke, E.F. and Hoering, T.C. (1986) Hydrothermal graphite in New Hampshire: Evidence of carbon mobility during regional metamorphism. Geology 14, 452-455.
Saxena, S.K. and Fei, Y. (1987a) High pressure and high temperature fluid fugacities. Geochim. Cosmochim. Acta 51, 783-792.
Saxena, S. K. and Fei, Y. (1987b) Fluids at crustal pressures and temperatures. I. Pure species. Contrib. Mineral. Petrol. 95, 370-375.
Toulmin, P. and Barton, P.B., Jr. (1964) A thermodynamic study of pyrite and pyrrhotite. Geochim. Cosmochim. Acta 28, 641-671.
Touret, J. and Bottinga, Y. (1979) Equation d'etat pour le CO_2; Application aux inclusions carboniques. Bull. Mineral. 102, 577-583.
Tso, J.L., Gilbert, M.C. and Craig, J.R. (1979) Sulfidation of synthetic biotites. Amer. Mineral. 64, 304-316.
Wones, D.R. (1981) Mafic silicates as indicators of intensive variables in granitic magmas. Mining Geology 31, 191-212.

Chapter 10 Hans P. Eugster and Lukas Baumgartner

MINERAL SOLUBILITIES AND SPECIATION IN SUPERCRITICAL METAMORPHIC FLUIDS

INTRODUCTION

Most metamorphic fluids cannot be sampled directly or at least not at the conditions at which the associated mineral assemblages formed. Exceptions are geothermal fluids, but they are restricted to the low pressure-low temperature, subcritical L+V region (see for instance Henley et al., 1984). Two complementary approaches are available for deducing the compositions of fossil metamorphic fluids. They consist of treating fluid inclusions as samples of such fluids (for summaries see Hollister and Crawford, 1981; Roedder, 1984) and of using mineral solubility information extrapolated to the appropriate P-T conditions to infer the composition of the fluid in equilibrium with a particular mineral assemblage.

Fluid inclusions provide information on composition and physical conditions during entrapment, but the timing of this entrapment in relation to the metamorphic evolution is often difficult to establish. The chemical composition of fluid inclusions may be determined for molecular species such as those of the system C-O-H-S-N (see Chapter 9), whereas dissolved salts are more difficult to identify. Judging from fluid inclusion studies (Roedder, 1984; Hollister and Crawford, 1981), metamorphic fluids do not contain pure water, but rather chloride brines of low to very high concentrations with additional ligands, such as carbonate-bicarbonate, fluoride, borate.

On the other hand, solubilities are usually measured in a simple anionic medium and cation concentrations are reported. Less frequently an effort is made to identify solute species responsible for mineral dissolution and precipitation. In this contribution we will concentrate on such experimental information in the near to supercritical region.

Although the data available is limited and of variable quality, this information is critical for modelling mass transport, mineral dissolution and precipitation. A change in the nature of the dominant species can profoundly affect solubility behavior. Whereas some solubility aspects can be predicted from thermodynamic calculations (Helgeson et al., 1978, 1981; McKenzie and Helgeson, 1984) the formation constants of the neutral and charged complexes present in solution are more difficult to calculate (see Chapter 6) and experimental determinations are essential. Three measurement techniques can be used to provide the necessary information: electrical conductivity, optical spectroscopy and solubility. These data are reviewed here insofar as they pertain to major solute species to be expected in supercritical metamorphic fluids.

Thermodynamic constants derived from these laboratory measurements have considerable uncertainties attached to them, partly because of experimental difficulties and partly because of uncertainties in the constants used in the data reduction. Furthermore, measurements frequently are restricted to a narrow range of conditions and extrapolations beyond that range introduces large uncertainties. For instance, conductivity measurements are carried out mostly in very dilute solutions and hence often do not have the range of anion concentrations necessary to include complexes with higher ligand numbers. With spectroscopic techniques the limitation is with pressure, since they usually refer to L + V conditions. Through electrostriction dissociation leads to a volume decrease because of the denser packing of water dipoles around charged particles. Consequently, dissociation increases with increasing pressure. Solubility measurements are plagued by the difficulty of establishing equilibrium, by quench problems and by uncertainties associated with activity coefficient corrections.

In this chapter we present a semitheoretical equation of state which allows us to interpolate and extrapolate solubility and dissociation data from 400°C, 1 kbar to at least 5 kbar, 700°C and probably higher.

It is perhaps useful to point out that our understanding of the thermodynamics of supercritical electrolytes is roughly at the same stage now that thermodynamics of minerals was twenty-five years ago. Most of the work lies ahead of us.

EQUATION OF STATE FOR SOLUTES

The Gibbs free energy of a phase is given by

$$G^{P,T} = G^{P^\circ,T^\circ} - \int_{T^\circ}^{T} S\,dT + \int_{P^\circ}^{P} V\,dP \quad . \qquad (1)$$

Several workers have attempted to predict or express the free energy of solutes in supercritical conditions. Helgeson and coworkers (Helgeson and Kirkham, 1974a,b, 1976; Helgeson, 1981; Tanger and Helgeson, 1987) extracted partial molar properties of the solvent and of the ions, in order to calculate their free energies at P and T by integration of Equation (1). This approach has been extended to include complexes (see Chapter 6). Quist and Marshall (1968) and Marshall (1970, 1972) found that for a given T the logarithms of dissociation constants plotted as straight lines against the logarithm of the molal water concentration. Frantz et al. (1981) described the free energy difference between a metal chloride complex $MeCl_n$ and HCl, $[G_{MeCl_n} - nG_{HCl}]$, by the Van't Hoff equation.

The free energy change of a reaction between minerals and one mole of gas i can be expressed by

$$(\Delta G_r)_{P,T} = \Delta G^\circ + \int_{P^\circ}^{P} \Delta V_s\,dP + RT \ln f_i = 0 \quad , \qquad (2)$$

where standard states are pure solids at P and T and unit fugacity at 1 and T, and ΔV_s is the volume change of the solids. This equation was introduced by Eugster and Wones (1962) to calculate the effect of pressure on the oxygen fugacities of a mineral buffer and it is often written in the form

$$\log K = A + B/T + \frac{C(P-1)}{T} , \qquad (3)$$

where log K includes terms for the activities of solids and the fugacity of the gas. The constant A is related to the entropy change, B to the enthalpy change of the reaction, and C to the volume change of the solids during the reaction. A large number of solid-gas reactions calibrated experimentally have been expressed in this manner (see for instance Skippen, 1971; Kerrick, 1974).

Equation (3) assumes that the heat capacity, Cp, and the volume V_s are constant. We assume that this equation also describes the free energy of a bare, or unsolvated ion or ion complex. Thus the free energy of a solute at infinite dilution can be split up into

$$G_{solute} = G_{bare} + G_{solvation} . \qquad (4)$$

Following the arguments of Marshall (1970, 1972), we describe the free energy change due to solvation by $\nu RT \ln C_{water}$, where ν is related to the solvation number.

$$G^{P,T} = H^{P°,T°} - S(T-T°) + (P-P°)V + \nu RT \ln d = - RT \ln K . \qquad (5)$$

Here the density d of water at P and T, rather than the molar concentration C is used. Rearranging and dividing by -RT yields

$$\log K = A + \frac{B}{T} + \frac{CP}{T} + D \log d . \qquad (6)$$

We found that this simple equation adequately describes the available solubility and conductivity measurements of aqueous silica and the major rock forming metal chloride species. Despite the lack of rigor in its derivation, Equation (6) is a simple and valuable tool for interpolating and extrapolating solute free energies above 400°C, 1 kbar. In the vicinity of the critical point, the solvent properties change rapidly and an expression such as Equation (6) may not be adequate at lower temperatures and pressures, although the necessary tests have not yet been carried out.

In order to extract free energies of formation from the elements for metal chloride complexes for each solubility experiment, we used mineral data listed in Helgeson et al. (1978) and the equation of state for H_2O from Bowers and Helgeson (1983). The results were fit to Equation (6) using the Marquardt fit procedure (Bevington, 1969; Press et al., 1986). Unless otherwise noted, data were weighted according to their estimated uncertainties.

WATER AS A SOLVENT

The bulk of the mineral solubility measurements have been carried out in pure water or in H_2O-CO_2 mixtures. Holland and Malinin (1979) and Barnes (1979) summarized the data available for nonsilicate minerals such as oxides, carbonates and sulfides. In their reviews, Eugster (1981) and Walther (1986a) also included silicates.

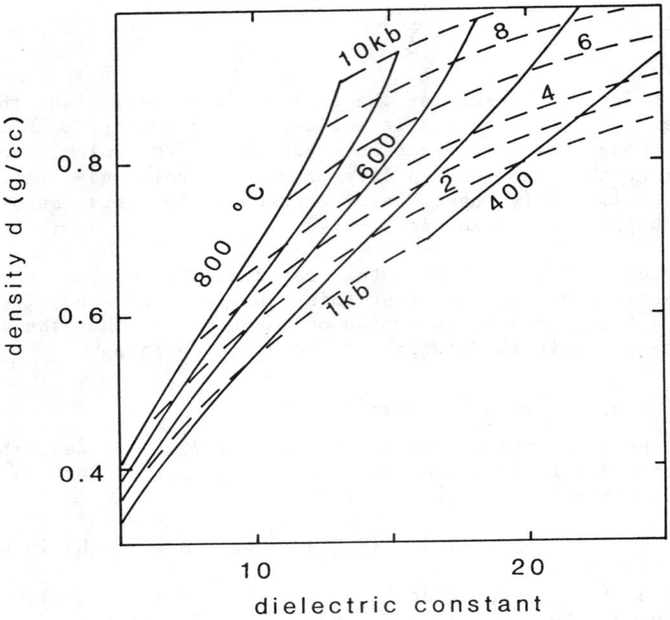

Figure 1. Dielectric constant ε as a function of the density of water (Pitzer, 1983).

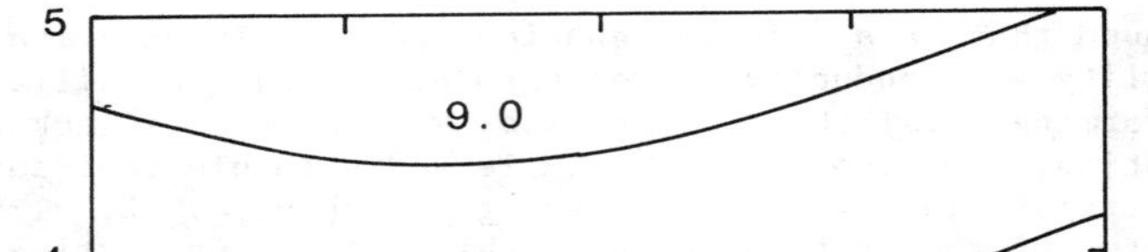

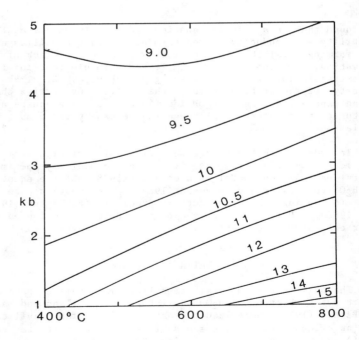

Figure 2. Ionic dissociation constant of water, plotted as -log K_8, as a function of P and T (Marshall and Franck, 1981).

The solvent

The properties which govern the efficacy of water as a solvent of minerals at elevated P and T are its density d, dielectric constant ε and ionic dissociation constant K_{H_2O}. Mineral dissolution in its simplest form is a hydration process and hence solubility generally increases monotonically with increasing density. Densities for pure water are well known from experiments (e.g. Burnham et al., 1969; Haar et al., 1984; Holloway, 1979, 1981), and are readily calculated from equations of state. In this chapter we will use Haar's equation of state (Haar et al., 1979). For typical values see Table 1.

The dielectric constant ε expresses the ability of the solvent to shield charged particles by solvation. According to the Kirkwood equation (Kirkwood, 1939; Bockris and Reddy, 1970), the magnitude of ε depends on the molecular polarizability α and dipole moment μ of the solvent and its density d.

The Kirkwood equation adequately describes the dielectric constant of a liquid containing associated dipoles:

$$\frac{(\varepsilon-1)(2\varepsilon+1)}{g\,\varepsilon} = \frac{4\pi N_o d}{3M} + \left(\alpha + \frac{\mu^2 g}{3kT}\right) , \quad (7)$$

where N_o is Avogadro's number, d the density of the liquid, M the molecular weight, k the Boltzmann constant and g the Kirkwood correlation factor. Only g varies with P and T. We will use Pitzer's (1983) formulation which expands g as

$$g = 1 + a_1 d + a_2 d^5 \left[(a_3/T)^{a_4} - 1\right] \quad (7a)$$

and the following constants

$$\alpha = 1.444 \cdot 10^{-24}; \; \mu = 1.84 \cdot 10^{-18}$$

$$a_1 = 2.68; \; a_2 = 6.69; \; a_3 = 565; \; a_4 = 0.3 \; .$$

The dependence of ε on d is shown in Figure 1, calculated from the Kirkwood equation of Pitzer (1983). ε decreases with increasing T and decreasing P.

The ionic dissociation of H_2O

$$H_2O = H^+ + OH^- \quad (8)$$

affects all dissociation equilibria and defines the point of acid-base neutrality as a function of P and T. Calculations of K_{H_2O} from conductivity measurements have been reviewed by Marshall and Franck (1981) and their recommended values for our P-T range can be fitted to Equation (6). Fit parameters are given in Table 2. Results are shown in Figure 2. Neutral pH can be as low as 4.5 and as high as 7.

Quartz

The most complete data set for mineral solubility under crustal conditions refers to SiO_2. Walther and Helgeson (1977) calculated

Table 1. Dielectric constant ε (Pitzer, 1983) and density d (Haar, et al., 1979) (in g/cm^3).

P [bar]	T [°C]	d g/cm^3	ε	P [bar]	T [°C]	d	ε
1000	450	0.614	12.49	2000	450	0.743	16.46
	500	0.528	9.45		500	0.691	13.87
	550	0.444	7.04		550	0.640	11.69
	600	0.374	5.34		600	0.580	9.87
	650	0.321	4.24		650	0.543	8.37
	700	0.282	3.52		700	0.500	7.14
	750	0.253	3.04		750	0.462	6.16
	800	0.231	2.69		800	0.429	5.37
3000	400	0.854	21.78	4000	400	0.900	23.46
	450	0.814	18.72		450	0.865	20.35
	500	0.773	16.20		500	0.830	17.83
	550	0.732	14.10		550	0.796	15.73
	600	0.693	12.31		600	0.761	13.95
	650	0.654	10.80		650	0.728	12.44
	700	0.618	9.51		700	0.696	11.14
	750	0.584	8.42		750	0.666	10.02
	800	0.553	7.49		800	0.637	9.05
5000	400	0.938	24.82				
	450	0.907	21.63				
	500	0.876	19.06				
	550	0.845	16.94				
	600	0.815	15.17				
	650	0.785	13.65				
	700	0.756	12.35				
	750	0.728	11.21				
	800	0.702	10.22				

Table 2. Fit coefficients for equation 6 for the molal dissociation reactions discussed in text.

	A	B	C	D
H_2O	-4.247	-2959.4	0.00928	13.493
HCl	-5.406	3874.9	—	13.93
NaCl	-1.276	1287.4	—	8.966
KCl	-2.594	535.6	—	2.414
$CaCl_2$	-3.21	2407.	—	9.6
$CaCl^+$	-5.05	3112.	—	15.5
$MgCl_2$	-2.9	2459.	—	11.2
$MgCl^+$	-4.80	3415.	—	18.2

Table 3. Fit coefficients for equation 6 for SiO_2^{aq} and metal chloride species as discussed in text.

	A	B	C	D
$-\frac{G_{SiO_2^{aq}}}{2.303RT}$	6.215	40'933.4	-.0857	1.539
$-\frac{G_{NaCl}-G_{HCl}}{2.303\ RT}$	-9.147	21'598.9	—	-1.719
$-\frac{G_{KCl}-G_{HCl}}{2.303\ RT}$	1.984	15'293.3	-.0312	2.289
$-\frac{G_{CaCl_2}-2G_{HCl}}{2.303\ RT}$	-7.415	32'834.9	—	2.849
$-\frac{G_{MgCl_2}-2G_{HCl}}{2.303\ RT}$	-5.913	26'022.	—	9.463
$-\frac{G_{FeCl_2}-2G_{HCl}}{2.303\ RT}$	-4.313	7'589.6	—	4.172

thermodynamic properties of aqueous silica from published data. In metamorphic fluids, aqueous silica will always be one of the most abundant solutes. Figure 3 shows concentrations to be expected in the presence of quartz as a function of fluid pressure and T. By measuring quartz solubility in H_2O-CO_2 mixtures, Walther and Orville (1983) were able to identify the dominant species. Plotting log a_{H_2O} in the mixture vs. log m_{SiO_2} produces a slope close to 4, defining the dissolution reaction as

$$SiO_{2(q)} + 4H_2O = Si(OH)_4 \cdot 2H_2O \quad . \qquad (9)$$

At lower mol fractions of H_2O there is an indication of dehydration to a species such as $Si(OH)_4 \cdot H_2O$ with a slope of 3. The issue remains unresolved because of the uncertainty in the activity coefficient of H_2O in the mixtures. Knowledge of the hydration number is crucial for solutions in which a_{H_2O} departs significantly from unity. This is readily apparent from Equation (9).

The existence of alkaline silicate complexes such as NaH_3SiO_4 and KH_3SiO_4 have been suggested by Seward (1974) and Anderson and Burnham (1967, 1983) for alkaline solutions to account for the increased solubility of quartz at high pH and elevated P and T. Walther (1986a) has expressed the complexes as $NaH_3SiO_4 \cdot H_2O$ and $KH_3SiO_4 \cdot H_2O$, but the precise stoichiometries have not been determined.

Dissolution of SiO_2 in pure water can be written as

$$SiO_{2(s)} = H_4SiO_{4(aq)} - 2\,H_2O \quad , \qquad (10)$$

assuming H_4SiO_4 as the only solute species. Walther and Helgeson (1977) reported free energy values for $SiO_{2(aq)}$ based on the reaction

$$SiO_{2(s)} = SiO_{2(aq)} \quad . \qquad (11)$$

These free energy values are equivalent to the free energy difference of $G_{H_4SiO_4} - 2G_{H_2O}$. Expressing these free energies in terms of

$$\log K = - \frac{G_{H_4SiO_4} - 2G_{H_2O}}{2.303RT} \qquad (12)$$

reveals that the four-term expression of Equation (6) reproduces the data within the limits of their accuracy. $G_{SiO_{2(aq)}}$ was calculated by Walther and Helgeson (1977) from published quartz solubility data for Reaction (11). Their data from 400°-550°C, 1-5 kbar are plotted in Figure 4 as a function of log d_{H_2O}, fitted to Equation (6). The fit parameters are listed in Table 3. The curvature of the isotherms clearly indicates the need for the pressure correction term C.

Corundum

Corundum is the key mineral for defining Al speciation, and such information is crucial for our understanding of the dissolution and precipitation of all aluminum silicates. According to Anderson et al. (1987), corundum dissolves in pure water by a reaction such as

$$0.5Al_2O_3 + (1{\cdot}5{+}n)H_2O = Al(OH)_3 \cdot nH_2O \quad . \qquad (13)$$

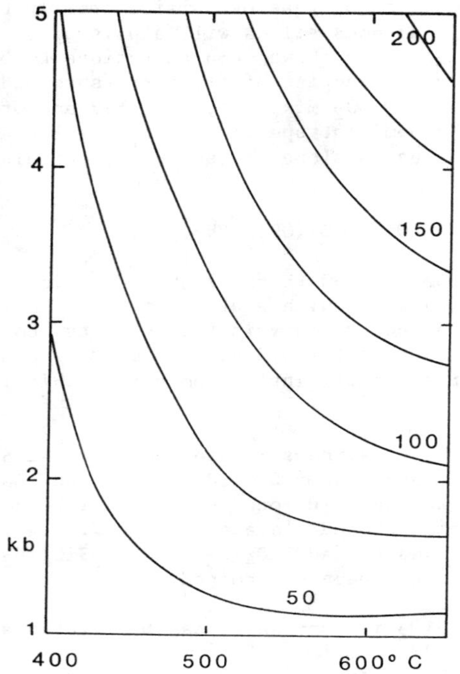

Figure 3. Concentration, in mmol/kg, of aqueous silica in equilibrium with quartz as a function of P and T. After Walther and Helgeson (1977).

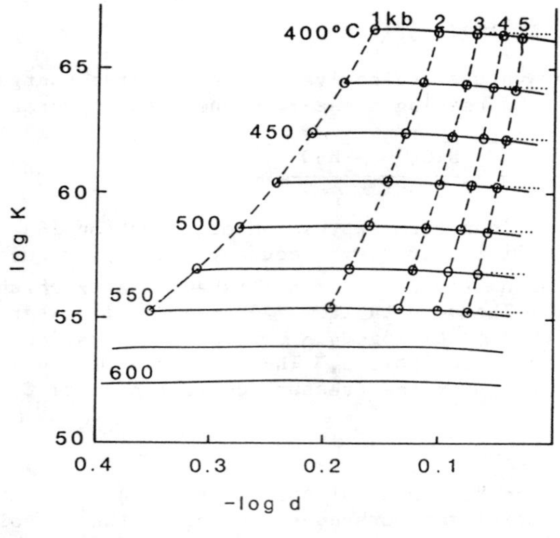

Figure 4. -G/2.303RT for aqueous silica (Walther and Helgeson, 1977) as a function of water density d. Isotherms (solid lines) and isobars (dashed lines) calculated from the fit parameters of Table 3. The dotted lines show the fit for C = 0.

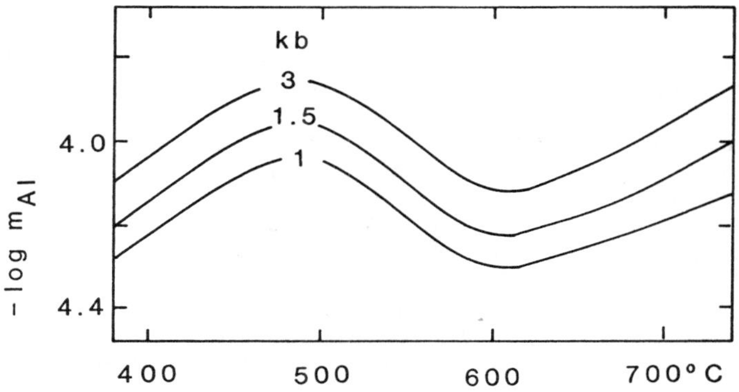

Figure 5. Corundum solubility after Ragnarsdottir and Walther (1985).

Becker et al. (1983) and Ragnarsdottir and Walther (1985) report Al-concentrations in solution. A strong pressure dependency at constant temperature is evident as it is for silica, and it is caused by the increase in the density of the solvent. Becker et al. (1983) did not explore the temperature effect, whereas Ragnarsdottir and Walther (1985) found little increase between 400° and 700°C and 2 kbar but report a maximum at 500° and a minimum at 600°C. These data are shown in Figure 5. Whether the extrema are real or not, the Al content of water in equilibrium with corundum under metamorphic conditions can be expected to range from a few to a few tens of ppm.

K-feldspar and muscovite

Because of the uncertainties with respect to Al speciation, the solubilities of feldspars are not well characterized. Anderson and Burnham (1983) have discussed published data on albite-H_2O and sanidine-H_2O and Anderson et al. (1987) presented data on KCl-H_2O fluids equilibrated with aluminum silicate assemblages, in which the activities of silica and alumina were buffered, ranging from quartz-K-feldspar-muscovite to muscovite-leucite-corundum. The principal species present in the supercritical fluid are inferred to be $Si(OH)_4$, $Al(OH)_3$, $KAl(OH)_4$, $NaAl(OH)_4$ and their hydrates as well as a (K,Na)-Al-Si-H_2O complex of unknown stoichiometry. The existence of an uncharged $KAl(OH)_4$ complex is deduced from the corundum-KOH solubility data of Pascal (1984) which show no dependence of Al concentration with pH, but a linear dependence with KOH and NaOH, with the ratio of $Al_{(total)}/(Na,K)_{total}$ increasing with temperature. Anderson et al. (1987) present speciation calculations, assuming unit activity coefficients for all neutral species and unit activity for H_2O. The data indicate that the bulk of the potassium and aluminum are present in a complex such as $KAl(OH)_4$, although the precise stoichiometry was not established. Anderson et al. (1987) also suggest the presence of an alkali-Al-Si complex in silica-rich solutions, but the evidence is even more tenuous, and they conclude that "... deducing the nature of solute species from a combination of sparse solubility data and ionization constants is a somewhat unsatisfactory exercise."

Albite and paragonite

The equivalent sodium system has been studied by Woodland and Walther (1987) using albite + paragonite + H_2O. Again, the activities of Al_2O_3 and SiO_2 are fixed by the solids and Al concentration in solutions between 400 and 500°C at 1 and 2 kbar are much too high to be accounted for by $Al(OH)°_3$ as the main solute. A neutral complex of the stoichiometry $NaAl(OH)°_4 \cdot x\ nSiO_2$ is suggested with n between 0 and 3. Using the procedures of Anderson et al. (1987), Woodland and Walther (1987) calculated the molality of $NaAl(OH)_4$ from aluminum concentrations in solution. Speciation calculations for albite + paragonite + quartz + H_2O show that the solutions are distinctly alkaline. This is no surprise, since $(OH)^-$ is the only anion considered. Because of the stability of the (Na,K)-Al complexes, Al solubility in these supercritical fluids is appreciable, enhancing the mobility of aluminum in metamorphic reactions. The presence of NaCl does not seem to increase the abundance of the alkali-alumina complexes.

Brucite and portlandite

Information on Ca and Mg speciation in supercritical H_2O fluids was provided by Walther (1986b) from solubility data of portlandite, $Ca(OH)_2$, and brucite, $Mg(OH)_2$. Solutions have calculated pH values of some 2 units more alkaline than neutral. Both solubilities decrease rather strongly with temperature, except for brucite between 450°-550°C at 2 kbar. Pressure dependence is more regular. The pH range is too narrow to permit speciation information to be extracted. The procedure Walther (1986b) adopted consists of choosing the most likely species, such as Ca^{2+} and Mg^{2+} and to calculate its free energies of formation from the solubility data. These results are then compared with values calculated for the ions by Helgeson and Kirkham (1976). If the fit is reasonable, this indicates that the correct ion was selected, if not, the choice was incorrect or another species is also present. For portlandite, Ca^{2+} gives an acceptable fit, except for temperatures above 500°C and the dissolution can be written as

$$Ca(OH)_2 + 2H^+ = Ca^{2+} + 2H_2O \quad . \qquad (14)$$

For Mg^{2+}, the misfit between the two equivalent data sets is as high as an order of magnitude. Although the calculated pH is roughly one unit less basic, $Mg(OH)^+$ is chosen as a likely candidate for the dominant solute with dissolution represented by

$$Mg(OH)_2 + H^+ = Mg(OH)^+ + H_2O \quad . \qquad (15)$$

Using the solubility data and making the assumption that $Mg(OH)^+$ is the only species present in solution, the equilibrium constant K_{15} can be calculated. The reliability of these results is unknown. The uncertainties probably are large because species identification was not possible and all activity coefficients have been assumed to be one.

Magnetite

No data are available for the solubility of magnetite in pure supercritical water. Based on the subcritical data of Tremaine and LeBlanc (1980), and the supercritical data of Chou and Eugster (1977) and Boctor et al. (1980) in H_2O-HCl mixtures, we can expect iron to be present as Fe^{2+} for reasonable crustal f_{O_2} values. Hence, with increasing pH the following sequence of species can be expected:

$$Fe^{2+} \rightarrow Fe(OH)^+ \rightarrow Fe(OH)_2 \rightarrow Fe(OH)_3^- \quad .$$

DISSOCIATION CONSTANTS OF CHLORIDE COMPLEXES FROM CONDUCTIVITY MEASUREMENTS

Numerous measurements have been published of metal-chloride dissociation constants based on conductivity data in dilute solutions. Frantz and Marshall (1982, 1984) have fitted the dissociation constant for HCl, the first and second dissociation constants for $CaCl_2$ and the second constant for $MgCl_2$ to an expression of the form

$$\log K = A + B/T + D \log d_{H_2O} \quad , \tag{16}$$

which is equivalent to equation (6) with C = 0.

We have similarly fitted the published dissociation data for NaCl, KCl and $MgCl_2$.

HCl

The most recent conductivity measurements of the dissociation of HCl,

$$HCl = H^+ + Cl^- \quad , \tag{17}$$

were reported by Frantz and Marshall (1984), using 0.002, 0.005 and 0.01 molal solutions from 100°-700°C and at pressures from 500 to 4000 bars. The data on the molar dissociation constant, K_{HCl} agrees well with earlier measurements of Franck (1956). Fit parameters for Equation (6) are listed in Table 2.

NaCl

Quist and Marshall (1968) have presented the most complete set of dissociation data for NaCl,

$$NaCl = Na^+ + Cl^- \quad , \tag{18}$$

obtained from conductivity measurements on dilute solutions (0.001-0.1 m) in the P-T range of interest. In their Figure 26 they demonstrated that $\log K_{NaCl}$ is a liner function of $\log C_{H_2O}$ at constant T, where C_{H_2O} is the molar concentration of water:

$$C_{H_2O} = d_{H_2O} \times 55.51 \quad . \tag{19}$$

Fitting their data to Equation (6) we obtain the constants in Table 2, which reproduces their data within experimental error (see Fig. 6).

KCl

Dissociation constants for KCl,

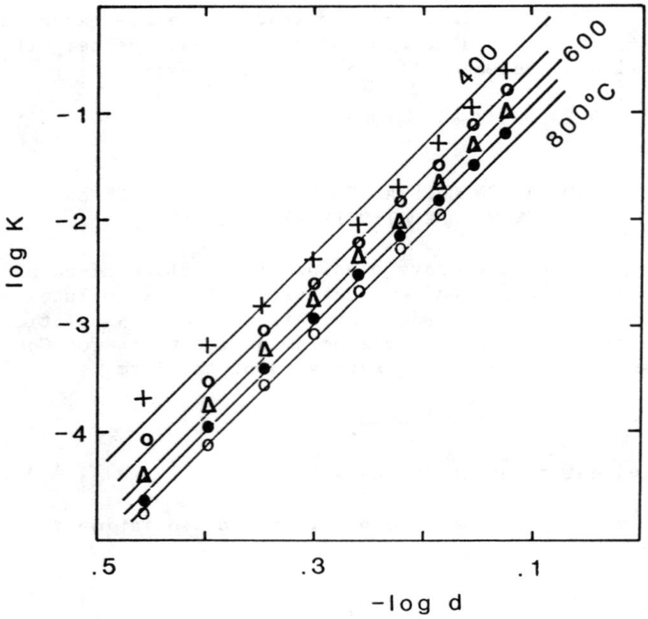

Figure 6. Dissociation constant for NaCl, fitted to Equation (6). Crosses, circles, triangles, dots and circles represent, respectively, data for 400°, 500°, 600°, 700° and 800°C.

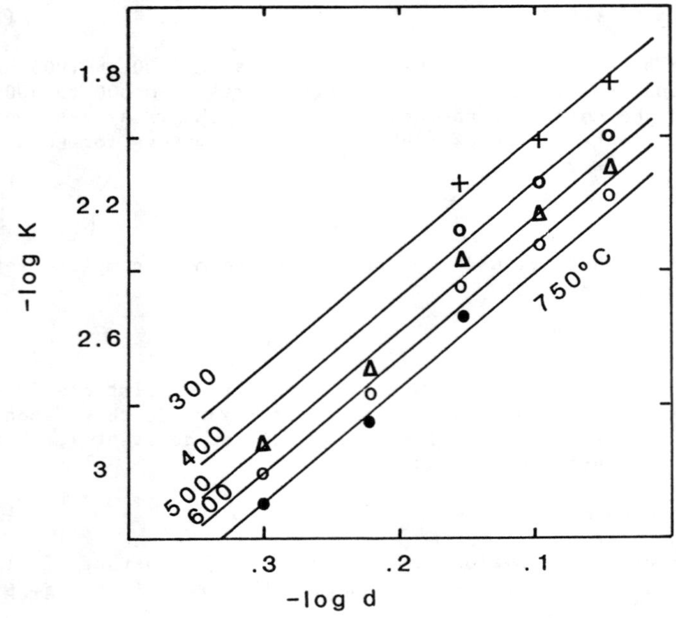

Figure 7. Dissociation constant for KCl fitted to Equation 6 with the data of Ritzert and Franck (1968).

$$KCl = K^+ + Cl^- \tag{20}$$

were reported by Ritzert and Franck (1968) to 750°C and 6 kbar in 0.001, 0.002 and 0.01 molal KCl solutions. Mangold and Franck (1968) extended the measurements to 1000°C and 12 kbar. Within experimental error (see Fig. 7), the data for this dissociation is represented by Equation (6) using the constants of Table 2.

$CaCl_2$

The first and second ionization constants for $CaCl_2$,

$$CaCl_2 = CaCl^+ + Cl^- , \tag{21}$$

$$CaCl^+ = Ca^+ + Cl^- , \tag{22}$$

were measured by Frantz and Marshall (1982) to 600°C and 4 kbar in 0.001 and 0.005 molal solutions, and we have listed their A, B and D constants in Table 2.

$MgCl_2$

Frantz and Marshall (1982) measured dissociation constants for $MgCl_2$,

$$MgCl_2 = MgCl^+ + Cl^- , \tag{23}$$

$$MgCl^+ = Mg^{2+} + Cl^- . \tag{24}$$

Because of brucite or periclase precipitation below 3000 bars and 600°C, they were able to calculate only three values for the first dissociation constant between 500° and 600°C at densities of 0.55, 0.6 and 0.65. These values were compatible with the assumption that the first dissociation constant for $MgCl_2$ has the same value as the equivalent constant for $CaCl_2$. We have refitted these data by combining them with the values reported by Wilson (1986) for 2 kbar and 600°, 500°, 475° and 450°C (see Fig. 8). These latter data were calculated from the solubility data of Frantz and Popp (1979) and Luce et al. (1985) on Mg-silicates in HCl solutions. Appropriate activity coefficient corrections were applied using Debye-Hückel constants, to bring them in line with the conductivity data from dilute solutions. The resulting K values in the form of Equation (6) are very preliminary and differ somewhat from log K_{MgCl_2} of Frantz and Marshall (1982). For log K_{MgCl^+} we accept their reported values (Table 2).

$FeCl_2$

No conductivity measurements have been reported for supercritical $FeCl_2$ solutions.

Dissociation constants for the chlorides listed in Table 2 are plotted in Figure 9.

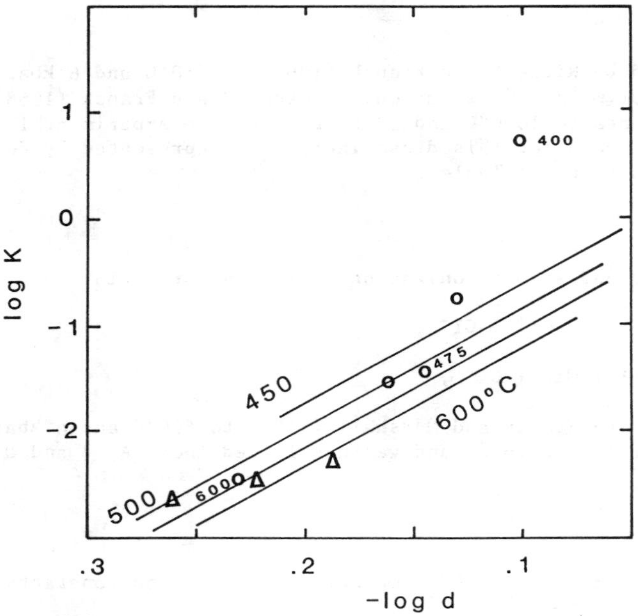

Figure 8. The first dissociation constant for $MgCl_2$, K_{23}, fitted to Equation 6, using the data of Frantz and Marshall (1982) (triangles) and Wilson (1986) (circles). The 400°C point was not used. Temperatures for points not located on an isotherm are listed.

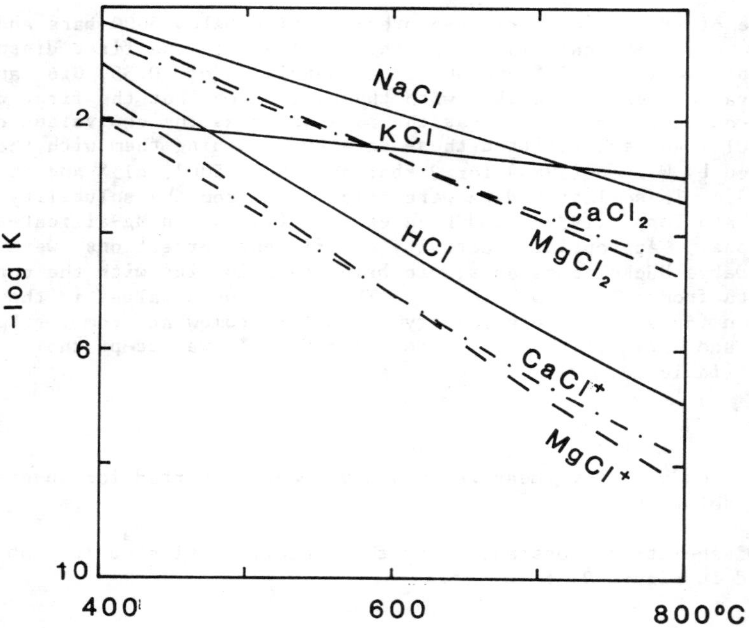

Figure 9. Comparison of the dissociation constants of NaCl, KCl, HCl, $MgCl_2$, $CaCl_2$, $MgCl^+$ and $CaCl^+$ from the data of Table 2 at 2 kbar. The KCl data clearly do not conform to the general trend.

MINERAL SOLUBILITIES IN SUPERCRITICAL H_2O-HCl MIXTURES

Background

The importance of low pH solutions for mineral dissolution and alteration was stressed by Hemley (1959) with reference to hydrothermal alteration associated with porphyry copper deposits. Reactions such as K-feldspar-muscovite-quartz,

$$3KAlSi_3O_8 + 2HCl = KAl_3Si_3O_{10}(OH)_2 + 6SiO_2 + 2KCl \,, \qquad (25)$$

were calibrated in the laboratory in terms of concentration ratios, assuming that Al is conserved. Solubilities of Na-silicates in supercritical HCl-H_2O solutions were studied by Hemley and Jones (1964), Popp and Frantz (1979), K-silicates by Shade (1974), Montoya and Hemley (1975), Gunter and Eugster (1980), Mg-silicates by Hemley et al. (1977a, b), Frantz and Popp (1979), Luce et al. (1985), Ca-silicates by Gunter and Eugster (1978), Popp and Frantz (1979), Luce et al. (1985), Fe-oxides by Chou and Eugster (1977), Boctor et al. (1980) and Mn silicates by Boctor (1985). In many of these studies, chloride ion concentrations were not varied sufficiently to allow information on speciation to be extracted. Without such information, solubility products for the minerals and free energies for the metal-chloride complexes cannot be calculated. However, by assuming the presence of a single species and by assigning a likely identity to that species, equilibrium constants and free energies for that solute can be calculated. This procedure was used for $CaCl_2$ (Gunter and Eugster, 1978) and $FeCl_2$ (Chou and Eugster, 1977), but it clearly is restricted to high temperature (fully associated species) or low temperature (fully dissociated) conditions.

Experimental methods

The experimental methods used to determine mineral solubilities and speciation of the chloride complexes in supercritical HCl-H_2O mixtures were reviewed recently by Eugster et al. (1987). Because of the low pH involved, most of these measurements are carried out in sealed noble metal capsules in rapid quench bombs. This implies analysis of solutions in the microliter range. The methods consist either of using the Ag-Ag-AgCl buffer (Frantz and Eugster, 1973), the hydrogen (Chou and Eugster, 1976) and HCl sensor (Popp and Frantz, 1979) or the solution composition combined with the mass balance and mass action equations (Montoya and Hemley, 1975; Wilson, 1986; Eugster et al., 1987). If a single species is dominant, the "variation of slope" method is successful (Frantz and Popp, 1979; Popp and Frantz, 1979; Boctor et al. 1980), which consists of plotting total metal concentrations against HCl° molalities. If two or more species are abundant, the results are equivocal and the method of Wilson and Eugster (1985) gives more reliable information.

Figure 10 shows the three experimental arrangements. The Ag-AgCl buffer combined with an oxygen buffer (Fig. 10a) controls the fugacity of associated HCl at P and T (Frantz and Eugster, 1973). Experimental calibrations have been presented by Frantz and Eugster (1973), Chou and Frantz (1977), Frantz and Popp (1979), Luce et al. (1985) and summarized in Eugster et al. (1987). In practice, because of the high values of f_{HCl} imposed, HM is the only feasible oxygen buffer.

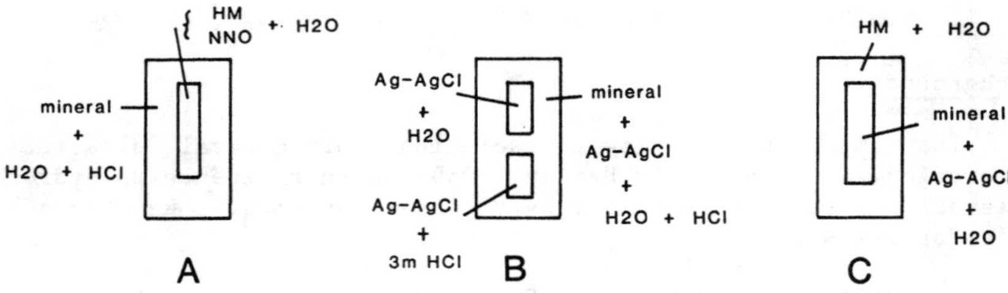

Figure 10. Experimental arrangements for solubility and speciation measurements in supercritical fluids. All inner tubes are sealed Pt, whereas outer tubes are sealed Au. (a) Wilson and Eugster (1984), Eugster and Wilson (1985), using an oxygen buffer in a sealed inner tube. (b) Frantz and Popp (1979), Popp and Frantz (1979), the "unbuffered method," "modified buffer" method or "HCl sensor" method (Boctor et al., 1980) which contains two hydrogen (HCl) sensors in sealed inner tubes. (c) Frantz and Eugster (1973), the Ag-AgCl buffer method which contains an oxygen buffer in the outer gold tube.

The Ag-AgCl buffer is not particularly useful for speciation studies, because it does not permit a wide variation in total chloride molality. In order to overcome this limitation, Frantz and Popp (1979) omitted the oxygen buffer (Fig. 10b). Instead, they added Ag + AgCl to the charge and placed two hydrogen sensors (Chou and Eugster, 1976) in a sealed Pt tube inside the charge system. After quench, a pH or chloride measurement is sufficient to calculate f_{H_2} imposed by the charge on the sensors and hence also f_{HCl} in the charge (for details of the calculations see Frantz and Popp, 1979; Eugster et al., 1987). Since f_{HCl} is not imposed, it can be varied by changing the composition of the charge fluid. This method is possible only for metals which have a single valence in solids and fluid.

Figure 10c illustrates the method of Wilson and Eugster (1984) which combines oxygen fugacity control with the ability to vary chloride concentrations widely. It does not use the Ag + AgCl assemblage, relying instead on mass balance equations. Quench is rapid and during analysis air-contamination is carefully excluded. Metal valence x at P and T is determined from the charge balance equation at 25°C:

$$x(Me) + (H^+) = (Cl^-) + (OH^-) \ . \qquad (26)$$

Omitting (OH^-) for acid solutions and assuming the presence of a single metal-chloride complex yields the charge balance at P and T,

$$(x-n)(MeCl_n^{x-n})_{P,T} + (H^+)_{P,T} = (Cl^-)_{P,T} \ , \qquad (27)$$

where n is the ligand number. Total metal and chloride concentrations are measured after quench:

$$(Me)_{tot} = (MeCl_n^{x-n})_{P,T} \ , \qquad (28)$$

$$(Cl)_{tot} = (Cl^-)_{P,T} + (HCl)_{P,T} + n(MeCl_n^{x-n})_{P,T} \ . \qquad (29)$$

The mass action relation for HCl is given by

$$\log K_{HCl} = \log (Cl^-)_{P,T} + \log (H^+)_{P,T} - \log (HCl)_{P,T} \ . \quad (30)$$

Equations (27)-(30) suffice to calculate the concentrations of (H^+), (Cl^-), (HCl) and $(MeCl_n^{x-n})$ at P and T. The ligand number n is not yet defined, except that it must be a small integer. Hence, equations (27)-(30) are solved for each data point by successively assigning a value of 0, 1, 2, 3, 4 ... to n, and the results are plotted in terms of the dissolution reaction. For MgO, for instance that reaction would be

$$MgO + 2H^+ + nCl^- = MgCl_n^{2-n} + H_2O$$

and

$$\log K_{31} = \log (MgCl_n^{2-n}) - 2 \log (H^+) - n \log (Cl^-) \ . \quad (31)$$

Plotting $[\log (MgCl_n^{2-n}) - 2 \log (H^+)]$ vs. $[n \log (Cl^-)]$ reveals which value of n produces the best fit to a straight line of slope n. Wilson and Eugster (1985) have extended this approach to include more than one chloride complex.

METAL CHLORIDE FREE ENERGIES AND SPECIATION

During the last twenty years an extensive data set has accumulated on metal-chloride free energies and on mineral solubilities in chloride fluids. We will briefly review information on the metal chlorides NaCl, KCl, $CaCl_2$, $MgCl_2$, $FeCl_2$, $MnCl_2$ and $NiCl_2$ obtained from supercritical mineral solubilities. This is the data base which we use to interpolate and extrapolate dissociation constants and solute free energies. An earlier summary was presented by Frantz et al. (1981).

KCl

Potassium silicate solubilities in supercritical HCl solutions have been reported by Shade (1974), Montoya and Hemley (1975), Lagache and Weisbrod (1977), Gunter and Eugster (1980), and Vidale (1983). Popp and Frantz (1980) derived an expression for the free energy difference between KCl° and HCl° from the solubility data of Montoya and Hemley (1975) and Shade (1974) for the assemblages K-feldspar-muscovite-quartz and muscovite-andalusite-quartz at 1 and 2 kbar.

Of the metal-chloride complexes, measurements for KCl have the widest range of H_2O densities, with the data of Shade (1974) extending to 7 kbar at 500° and 600°C. Because of the uncertainties in HCl free energies, we have followed the procedure of Frantz et al. (1981) in reporting only the difference in free energies between metal chlorides and HCl. This is permissible since in mineral dissolution and precipitation reactions, HCl always balances metal chloride complexes. These free energy differences were calculated from the data of Shade (1974), Montoya and Hemley (1975) and Gunter and Eugster (1980). All of the data used are in the P-T range of associated species and they were weighted equally. The calculated free energies can be combined with the dissociation constants listed in Table 2 for speciation calculations. -[G(KCl)

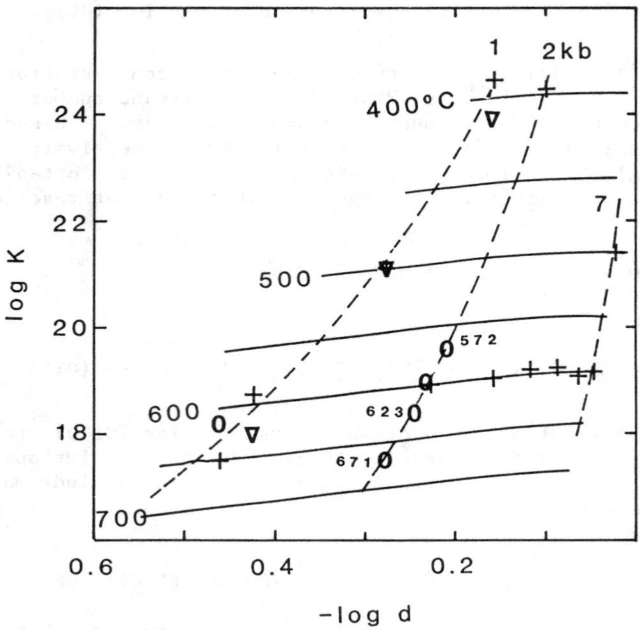

Figure 11. Log K = $-[G_{(KCl)}-G_{HCl})]/2.303RT$ as a function of the density of water from the data of Shade (1974; crosses), Montoya and Hemley (1975; triangles) and Gunter and Eugster (1980; circles). Temperatures (solid lines) and isobars (dashed lines) calculated from the constants of Table 3. Temperatures which do not fit isotherms are indicated.

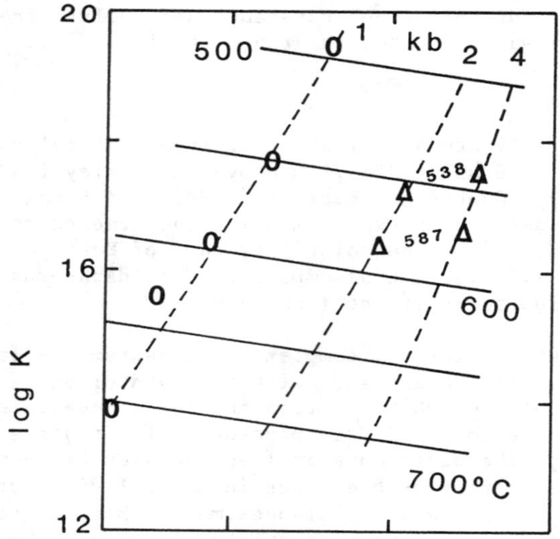

Figure 12. Log K for $(G_{NaCl}-G_{HCl})$ vs. density of water from the data of Popp and Frantz (1980; circles) and Vidale (1983; triangles). Isotherms and isobars from the constants of Table 3. Temperatures which do not fit isotherms are indicated.

$- G(HCl)]/(2.303 RT)$ has been plotted against H_2O density in Figure 11 using the data of Shade (1974), Montoya and Hemley (1975) and Gunter and Eugster (1980). Fit parameters for equation (6) are listed in Table 3. The curvature of the isotherms is described adequately by the linear pressure correction term.

NaCl

Sodium silicate solubilities in supercritical HCl solutions were reported by Montoya and Hemley (1975), Popp and Frantz (1980) and Vidale (1983). By combining their solubility data on albite + andalusite + quartz + NaCl with the dissociation data of Quist and Marshall (1968), Popp and Frantz (1979) were able to calculate NaCl° concentrations in their experimental range of 500-650°C, 1 kbar. This permitted them to evaluate the difference in free energy between NaCl° and HCl° at 1 kbar The results agree with the data by Montoya and Hemley (1975) on the assemblage albite + paragonite + quartz + NaCl. At 2 kbar the free energy difference was calculated by combining their free energy values for (KCl-HCl) with the exchange constants of Lagache and Weisbrod (1977) for the reaction

$$KAlSi_3O_8 + NaCl = NaAlSi_3O_8 + KCl \ . \quad (32)$$

Data which permit extraction of NaCl free energies are less complete, and they have been plotted in Figure 12. We have used the 1 kbar data of Popp and Frantz (1980) from 500°-700°C and have combined them with information extracted from Vidale (1983) at 538° and 587°C, 2 and 4 kbar, using mean values of all molalities. The parameters are listed in Table 3. Needless to say, this fit is very preliminary and additional measurements are badly needed at the high densities for a range of temperatures.

$MgCl_2$

Preliminary data on the dissolution of talc in HCl solutions were reported by Frantz and Eugster (1973) and a more complete treatment was published by Frantz and Popp (1979). Buffered as well as unbuffered runs were used in the latter study in order to extend the range of chloride concentrations and define speciation. Free energy differences between $MgCl_2$ and 2HCl were calculated for 1, 1.5 and 2 kbar assuming that $MgCl_2$ is the major solute above 550°C and Mg^{2+} at 400°C. The presence of $MgCl^+$ was not detected.

Luce et al. (1985) measured $MgCl_2/(HCl)^2$ ratios for the talc-quartz equilibrium, at 2 kbar and 500°-700°C. Similar measurements were carried out for tremolite-talc, diopside-tremolite, wollastonite-diopside, wollastonite-quartz and diopside-talc. By assuming that in their experiments $MgCl_2$ was the only abundant species and by combining their results with thermodynamic data for minerals and Mg^{2+} of Robie et al. (1979), Helgeson and Kirkham (1974, 1976), Luce et al. (1985) derived the total dissociation constant for $MgCl_2$, which is the product of the first and second dissociation constants. The differences between these results based on a mixture of phase equilibria and thermodynamic calculations and the conductivity data of Frantz and Marshall (1982) are within experimental error.

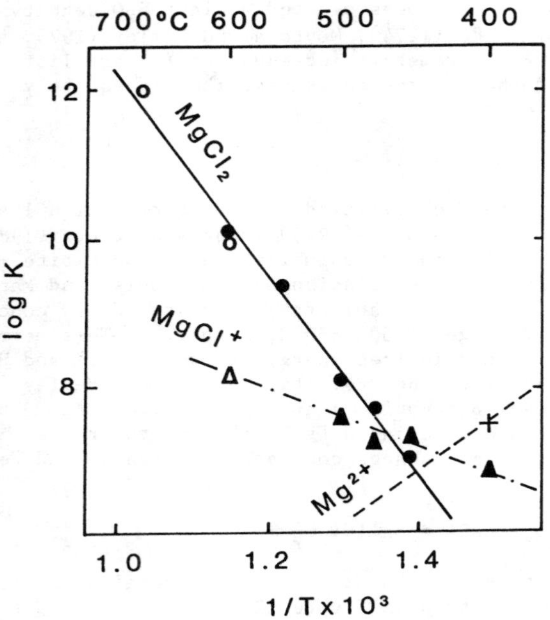

Figure 13. Speciation calculations for Mg chlorides at 2 kbar for talc + quartz from Wilson (1986). Open symbols data of Luce et al. (1985), filled symbols data of Frantz and Popp (1979).

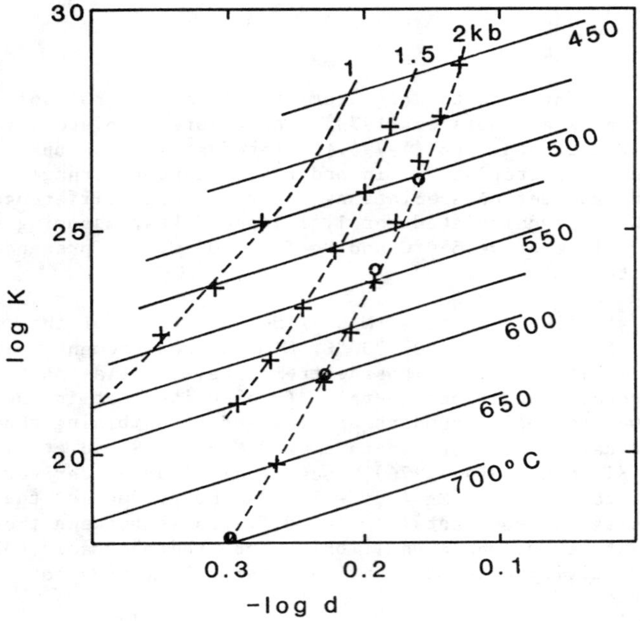

Figure 14. Log K for $[G_{MgCl_2}-2G_{HCl}]$ vs. density of water from the data of Frantz and Popp (1979; crosses) and Luce et al. (1985; circles).

$MgCl_2$ concentrations in solutions equilibrated with more comlex phlogopite-bearing mineral assemblages were reported by Vidale (1983).

Wilson (1986) has recalculated the solubility data of Frantz and Popp (1979) and Luce et al. (1985) using the procedure of Wilson and Eugster (1984) and has verified the presence of $MgCl^+$. His reinterpretation is shown in Figure 13.

We have calculated $MgCl_2$ free energies from the data of Frantz and Popp (1979) and Luce et al. (1985), shown in Figure 14 together with the fit. The data are restricted to 1, 1.5 and 2 kbar and do not involve high enough water densities to define C. However, the two data sets agree well. Fit parameters are listed in Table 3.

$CaCl_2$

The free energy of aqueous $CaCl_2$ was calculated by Gunter and Eugster (1978) from solubility data on wollastonite-quartz between 750° and 850°C. The measurements were carried out in the presence of Ag-AgCl and it was assumed that $CaCl_2$ was fully associated at these high temeratures. Popp and Frantz (1979) extended the study to temperatures of 425-600°C by eliminating the oxygen buffer and thus avoiding the problem of $CaCl_2$ saturation encountered by Gunter and Eugster (1978). Free energy differences of $CaCl_2$-2HCl were calculated by Popp and Frantz (1979) assuming $CaCl_2$ to be the dominant species. These data are compatible with the values of Gunter and Eugster (1978).

Luce et al. (1985) also measured reactions involving wollastonite, tremolite and diopside at temperatures between 500° and 700°C at 2 kbar. Combined dissociation constants were calculated for $CaCl_2$ using the same methods as for $MgCl_2$. The comparison of this combined constant with the conductivity data of Frantz and Marshall (1982) for $CaCl_2$ is less satisfactory with the difference somewhat larger than an order of magnitude. The source of that discrepancy is not known at this time.

Vidale (1983) also included $CaCl_2$ in her solutions with the calcium provided by the dissolution of plagioclase.

Our free energy calculations are shown in Figure 15. The 2 kbar free energy data extend from 450° to 840°C and agree well, but there seems to be some problem with the 1 kbar data of Popp and Frantz (1980), perhaps related to the short run durations. Nevertheless, giving all data equal weight produces a reasonable fit (Table 3).

$FeCl_2$

The free energy of $FeCl_2$ was calculated by Chou and Eugster (1977) from the magnetite dissolution data between 500° and 650°C at 2 kbar, buffered by HM and Ag-AgCl. Iron valence was found to be 2 and $FeCl_2$ was assumed to be the major solute. The apparent standard free energy can be expressed by

$$G^*_{(FeCl_2)}/T(K) = -382.49/T(K) + 0.141 \text{ kJ/deg mol} . \quad (33)$$

Boctor et al.(1980) carried out similar measurements on hematite between

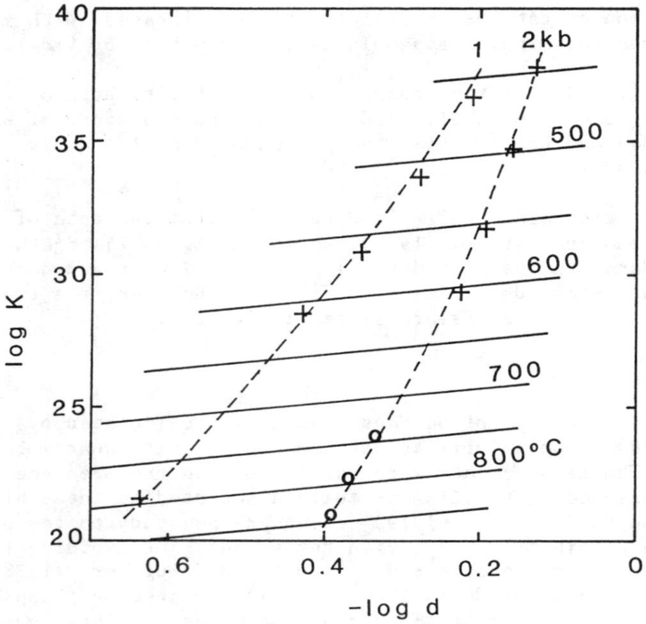

Figure 15. Log K for $[G_{CaCl_2}-2G_{HCl}]$ vs. density of water from the data of Popp and Frantz (19791; crosses) and Gunter and Eugster (1978; circles).

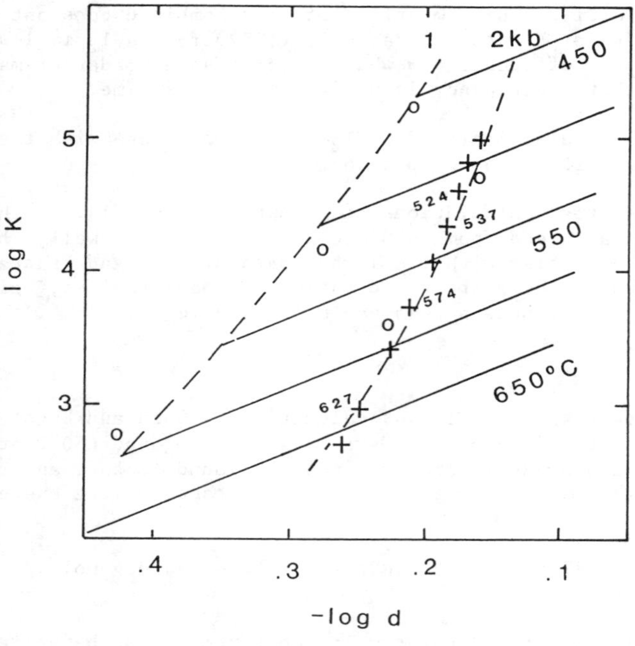

Figure 16. Log K for $[G_{FeCl_2}-2G_{HCl}]$ vs. density of water from the data of Chou and Eugster (1977; crosses) and Boctor et al. (1980; circles). Fit based on equal weighting shown in solid lines. Temperatures which do not fit isotherms are indicated.

400°-600°C, 1 and 2 kbar, using the "unbuffered" or "modified buffer" method (see Fig. 10). They confirmed $FeCl_2$ as the dominant species. The two data sets were found to be compatible.

We have combined both data sets and the results of our free energy calculations are shown in Figure 16 and the fit coefficients are listed in Table 3.

The data of Chou and Eugster (1977) on $FeCl_2$ are all at a high enough temperature so that dissociation does not present a problem. However of the data of Boctor et al. (1980) only the measurements above 450°C have been included. A considerably more extensive data set is necessary before $FeCl_2$ free energies can be considered defined.

$MnCl_2$ and $NiCl_2$

In a parallel study, Boctor (1985) determined the free energy of $MnCl_2$ from the reaction

$$MnSiO_3 + 2HCl = SiO_2 + MnCl_2 + H_2O \ . \qquad (34)$$

Mn was found to be bivalent and complexed with chloride as $MnCl_2$. Lin and Popp (1984) measured NiO in HCl solutions and calculated $NiCl_2$ free energies.

H_2O-CO_2 MIXTURES AS SOLVENTS

Metamorphic fluids frequently are mixtures of H_2O and CO_2 (see also Chapter 9), particularly in carbonate terranes (Skippen, 1974; Ferry, 1983). Unfortunately very few solubility studies have been carried out on carbonates at elevated P and T and most of these are restricted to calcite. Sharp and Kennedy (1965) and Fein and Walther (1987) measured calcite solubility in H_2O-CO_2 fluids to 620°C and 2 kbar pressure. At constant mol fraction of CO_2, calcite solubility decreases with increasing T and decreasing P. The solubility is highest in water-rich solutions, presumably because of solvation effects and it reaches a maximum at H_2O mol fractions of around 0.95 (see Fig. 17).

Fein and Walther (1987) evaluated speciation by comparing their results with those of Walther and Long (1986) on calcite solubilities in supercritical H_2O. The comparison yields the first dissociation constant of H_2CO_3 (see Fig. 18),

$$H_2CO_3 = H^+ + HCO_3^- \ , \qquad (35)$$

indicating that the reaction is very strongly displaced to the left. A comparison of the calcite data with the portlandite solubility data of Walther (1986b) eliminates Ca-OH species. Similarly, Ca-HCO_3 complexes are ruled out which leads to the conclusion that calcium ions are the only abundant calcium species. Since Ca concentrations are very low in pure H_2O-CO_2 mixtures, they can easily be balanced by $(OH)^-$ and $(HCO_3)^-$.

The addition of chloride will dramatically change this situation. In the neutral to acid pH range, Ca-concentrations can be expected to be much higher and complexing, especially with HCO_3^- will become a very real possibility. However, the necessary documentation is lacking.

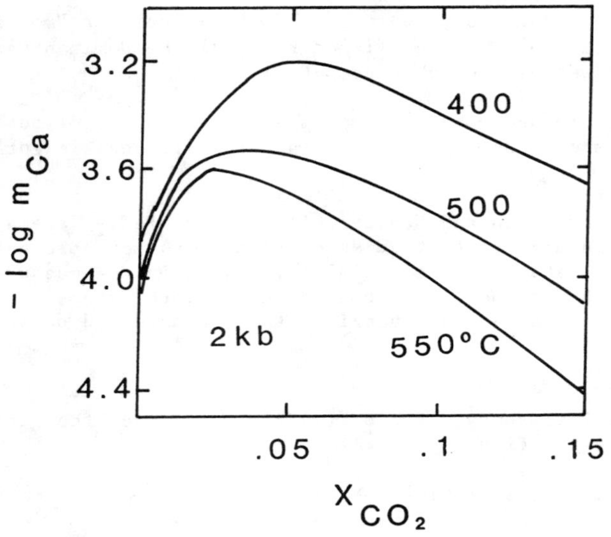

Figure 17. Solubility of calcite in H_2O-CO_2 mixtures in terms of total Ca molality and mol fraction of CO_2. Adapted from Fein and Walther (1987).

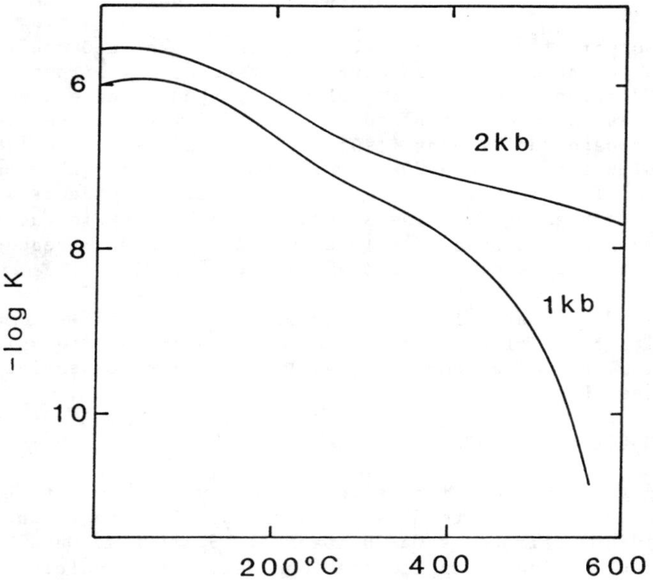

Figure 18. First dissociation constant of H_2CO_3, K_{35}, after Fein and Walther (1987).

ACTIVITY COEFFICIENTS

Activity coefficients in the supercritical region are now available for many rockforming minerals, common gases and gas mixtures, but not for solutes. Traditionally, uncharged solutes such as H_4SiO_4 or NaCl have been assumed to behave ideally and hence have been assigned activity coefficients of one. This has recently been challenged by Oelkers and Helgeson (1986), who stated that "the activity coefficients of neutral species increase dramatically with increasing temperature and decreasing pressure .... The degrees of association of 1:1 electrolyte solutions maximize at a stoichiometric ionic strength of $\sim$ 0.2 molal instead of increasing monotonically toward complete association at high concentration in response to $\gamma n = 1$." (p. 709). The basis for this statement has not yet been published. We shall continue to use unit activity coefficients for neutral species, implying that supercritical elecrolytes are largely associated. Activity coefficients for charged species can be calculated from the Debye-Hückel equation (Helgeson and Kirkham, 1974). This seems to be permissible even in saline fluids because ionic strengths in supercritical fluids are distinctly lower than they would be in the dissociated region. As in subcritical saline fluids, Pitzer coefficients (Pitzer, 1973; Harvie and Weare, 1980, Chapter 4) may offer the most practical approach.

A test of the assumptions involved in using the Debye-Hückel equation has been carried out by Wilson (1986) with respect to the first dissociation constant of $MgCl_2$,

$$MgCl_2 = MgCl^+ + Cl^- . \qquad (36)$$

The constant derived from the solubility data of Frantz and Popp (1979) and Luce et al. (1985) in concentrated chloride solutions shown in Figure 13 can be compared to the equilibrium constant determined in dilute solutions by conductivity measurements (Frantz and Marshall, 1984), where all activity coefficients should be close to unity. The difference between the two determinations should yield mean ion activity coefficients for P,T and ionic strength of the solubility experiment. These can then be compared with Debye-Hückel calculations. As mentioned earlier, the first dissocition constant for $MgCl_2$ was measured by Frantz and Marshall (1984) only at 500° and 600°C and densities of 0.55, 0.6 and 0.65 because of saturation problems. The comparison for those temperatures is shown in Figure 19 and the agreement between experimental and calculated coefficients is quite satisfactory. However, many more such tests are desirable.

SPECIATION CALCULATIONS

Methodology

The calculation of solute species distribution leads to a set of nonlinear equations. Several algorithms have been used, including free energy minimization and Newton-Raphson interation (e.g. Wolery, 1979; Wolery et al., 1984; van Zegeren and Storey, 1970; Harvie et al., 1987). For m solute species, m independent mass action equations can be written

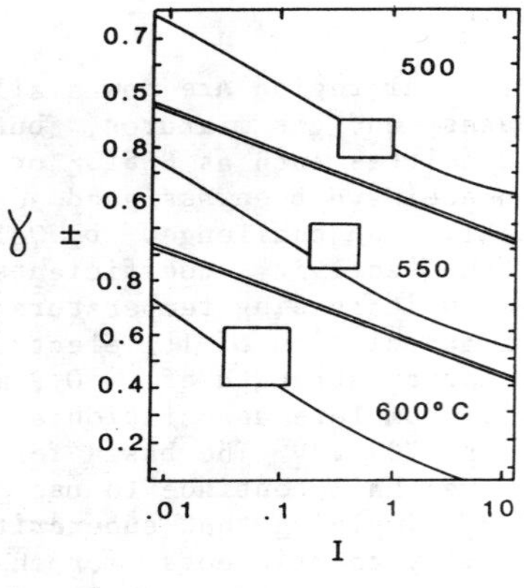

Figure 19. Mean ion activity coefficient for $MgCl_2$ as a function of ionic strength I and temperature at 2 kbar. Experimental determinations (see text) shown as rectangles, Debye-Hückel calculations as curves. From Wilson (1986).

Table 4. Dissolution of brucite in 1 molar chloride solution at T = 500°C, P = 2000 bar, $\log K^{diss}_{brucite} = 5.04$.

neutrality:	$m_{MgCl^+} + 2\,m_{Mg^{2+}} - m_{Cl^-} + m_{H^+} - m_{OH^-} = 0$
total chloride:	$2m_{MgCl_2} + m_{HCl} + m_{MgCl^+} + m_{Cl^-} - 2 = 0$
Brucite dissolution:	$\log m_{MgCl_2} - 2\log m_{HCl} - 5.04 = 0$
$MgCl_2$ dissociation:	$-\log m_{MgCl_2} + \log m_{MgCl^+} + \log m_{Cl^-} + 1.68 = 0$
$MgCl^+$ dissociation:	$-\log m_{MgCl^+} + \log m_{Mg^{2+}} + \log m_{Cl^-} + 3.62 = 0$
HCl dissociation:	$-\log m_{HCl} + \log m_{Cl^-} + \log m_{H^+} + 2.79 = 0$
H_2O dissociation:	$\log m_{H^+} + \log m_{OH^-} + 10.2 = 0$

Choosing $m_i^o = .01$ leads to the linear equation set:

	$MgCl_2$	HCl	$MgCl^+$	Mg^{2+}	Cl^-	H^+	OH^-				
Neutrality	0	0	1	2	-1	1	-1		m_{MgCl_2}		-.02
m^{tot}_{Cl}	2	1	1	0	1	0	0		m_{HCl}		0.95
Brucite	100	-200	0	0	0	0	0		m_{MgCl^+}		3.043
$MgCl_2$	-100	0	100	0	100	0	0	x	$m_{Mg^{2+}}$	=	0.4823
$MgCl^+$	0	0	-100	100	100	0	0		m_{Cl^-}		-1.3025
HCl	0	-100	0	0	100	100	0		m_{H^+}		-0.6277
H_2O	0	0	0	0	0	100	100		m_{OH^-}		-6.2149

Solving for Δm_i yields:

$$\Delta m_i^1 = \begin{pmatrix} .2744 \\ .12198 \\ .13857 \\ -.01510 \\ .14065 \\ -.02494 \\ -.03721 \end{pmatrix} \quad \text{and} \quad m_i^1 = \begin{pmatrix} .284 \\ .132 \\ .149 \\ -.0051 \\ .151 \\ -.0149 \\ -.0272 \end{pmatrix}$$

Repeating this procedure several times leads to the solution for the molar concentration of the species:

$$m_{MgCl_2} = 0.390;\; m_{MgCl^+} = 0.108;\; m_{Mg^{2+}} = 4.93\cdot10^{-4}$$

$$m_{HCl} = 1.88\cdot10^{-3};\; m_{Cl^-} = 0.109$$

$$m_{H^+} = 4.05\cdot10^{-5};\; m_{OH^-} = 1.50\cdot10^{-6}$$

$$\sum_i \nu_{ij} \log a_i - \log k_j = 0; \; j = 1 \ldots m \tag{37}$$

where ν_{ij} are stoichiometric coefficients, a_i is the activity of species i, k_j the equilibrium constants of reaction j, and i is summed over all fluid and solid species involved in the j th reaction. These mass action equations generally contain $\ell > m$ unknowns. In order to uniquely define them, a minimum of $\ell - m$ additional equations have to be found. The electrical neutrality condition is expressed as

$$\sum_i m_i z_i = 0 , \tag{38}$$

where m_i is the molarity of species i, z_i the charge of i. If additional constraints are needed, they are supplied either by specifying activities or by mass balance conditions, that is by fixing total molalities, e.g. m_{Cl}^{tot} present in the fluid:

$$\sum_i \nu_{ij} m_{ij} = m_{i,j}^{tot} . \tag{39}$$

Once ℓ equations of form (37), (38) and (39) are available, the task is reduced to finding the roots of ℓ nonlinear equations. A straight forward technique that normally converges satisfactorily is the Newton-Raphson iteration (for example Press et al., 1986; Wolery, 1980; Frantz et al., 1981). Beginning with a guess for the concentration of each species, the Newton-Raphson method successively improves the concentration values, following the gradient of the ℓ-dimensional hyperplanes in concentration space, defined by Equations (37) to (39).
Abbreviating this set of equations as

$$f_j = f_j(m_1, m_2, \ldots m_l) = 0, \; j = 1, \ldots \ell \; , \tag{40}$$

and inserting the initial guess for all m_i^o, a Taylor series is written of the form:

$$f_j(m_1 + \Delta m_1, m_2 + \Delta m_2 \ldots) = f_j(m_i{}^o) + \sum_i^{\ell} \frac{\partial f_j}{\partial m_i} \Delta m_i + \sum_{i=1}^{\ell} \frac{\partial^2 f_j}{\partial m_i 2} \Delta m_i^2 \; \ldots\ldots \tag{41}$$

Linearizing the Taylor series gives

$$\sum_{i=1}^{\ell} \frac{\partial f_j}{\partial m_i} \Delta m_i \cong f_j(m_i^o) \quad . \tag{42}$$

This linear set of equations is solved for Δm_i, and a new estimate m_i^{k+1} is calculated by

$$m_i^{k+1} = m_i^k + \Delta m_i \; , \tag{43}$$

and the procedure is repeated, until the desired accuracy is reached.

Table 5. Stable mineral asemblages of the five model ultrabasic rocks.

Bulk Chemistry	Mineral Zone 1	2	3	4
A	diopside-brucite-serpentine	------- diopside-forsterite-brucite -----		
B	diopside-brucite serpentine	diopside-serpentine-fosterite	- diopside-tremolite-fosterite	
C	diopside-tremolite-serpentine		tremolite-forsterite - serpentine	tremolite-forsterite-talc
D	------- tremolite-serpentine-talc --------			tremolite-forsterite-talc
E	-------------- tremolite-talc-quartz ----------------------			

As an example, speciation calculations for a fluid in equilibrium with brucite at 500°C, 2 kbar and 1 m total chloride are presented in Table 4, assuming activity coefficients to be equal to 1. The Newton-Raphson iteration can fail to converge. This is normally due either to ill conditioning of the matrix or to computer roundoff errors. Normalizing Equation (39) to the largest coefficient in each equation normally helps convergence.

Speciation in the system $MgO-CaO-SiO_2-H_2O-HCl$.

Combining the free energy data of the metal chloride complexes listed in Table 3 and the dissociation constants of Table 2 with free energy data for minerals permits calculation of mineral solubilities in terms of solute concentrations and speciation, provided information is available on anion abundances. Such calculations for supercritical fluids have been reported, among others, by Gunter and Eugster (1980), Eugster and Gunter (1981), and Frantz et al. (1981). However, at the present time lack of data on Al chloride species hampers solubility calculations for the all important Al-bearing silicates. We present here some typical results for minerals of the Al-free ultrabasic system $MgO-CaO-SiO_2-H_2O-HCl$.

Calculations have been carried out for five different bulk compositions, labelled A, B, C, D, and E (Fig. 20, Table 5) arranged in sequence of increasing silica contents. The concentrations of solute species in a fluid containing 0.1 molar total chloride were computed between 350°C and 650°C at 2 kbar total pressure. Four mineral zones are present within this temperature range (Evans and Trommsdorff, 1970), separated by the reactions

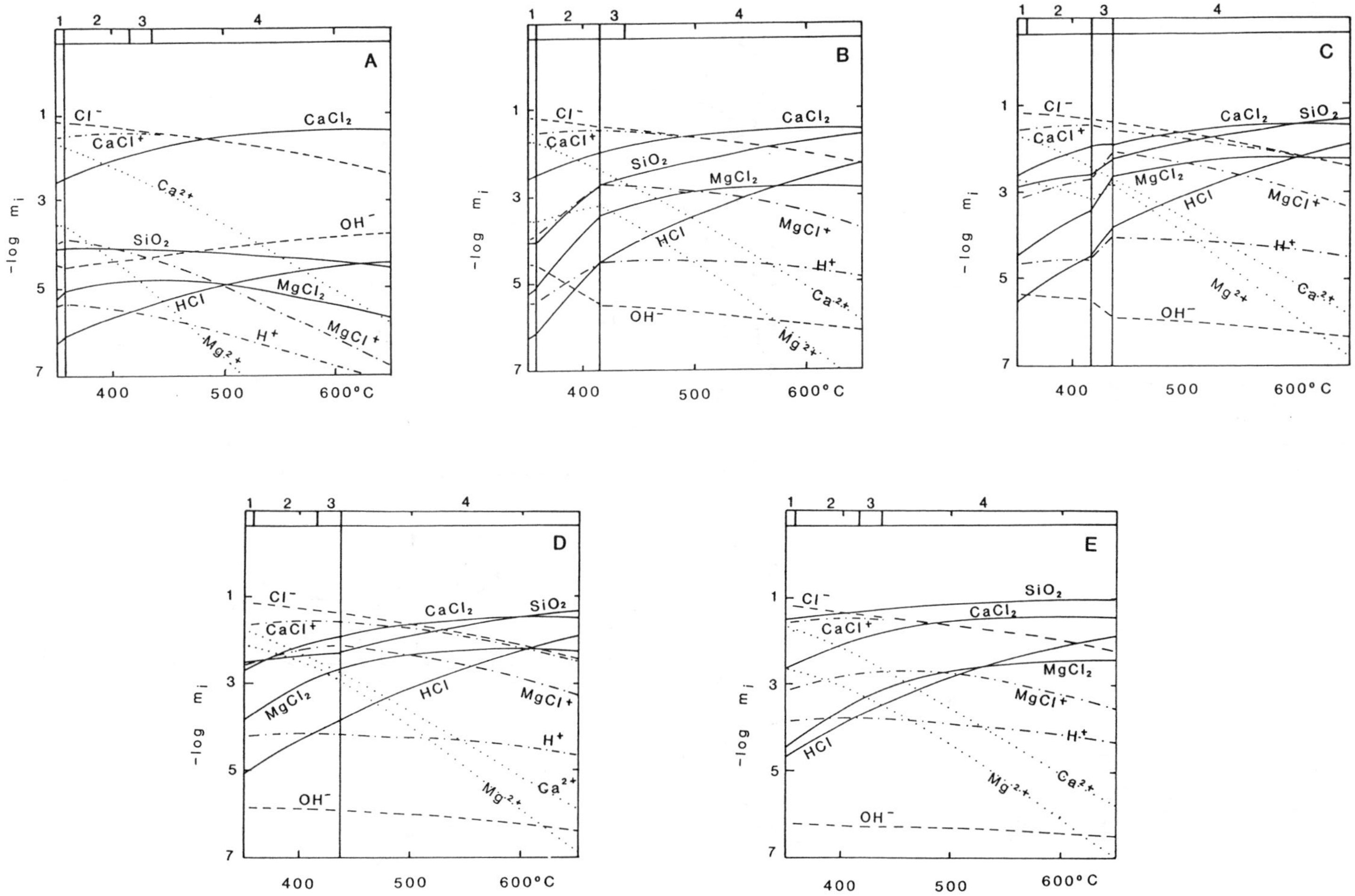

[Legends for Figures 20A,B,C,D,E are on the next page.]

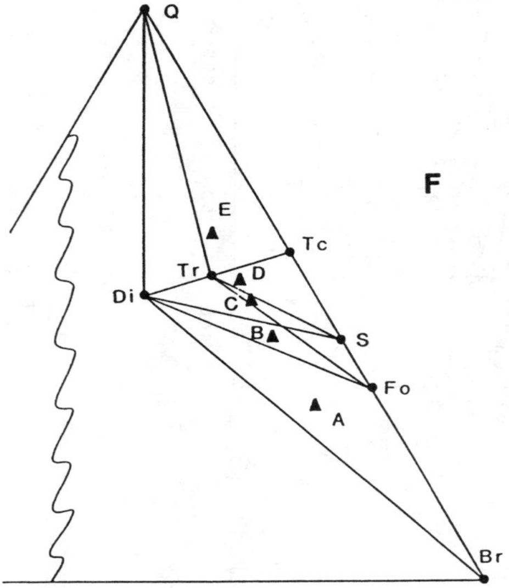

Figure 20. Speciation calculations for mineral assemblages involving quartz (Q), diopside (Di), tremolite (Tr), talc (Tc), serpentine (S), forsterite (Fo) and brucite (Br) at 2 kbar and total chloride of 0.1 mol with all activity coefficients set to unity. Bulk compositions for diagrams A, B, C, D and E are plotted in diagram F. Solid curves for neutral, dashed for anionic, dash-dot for singly charged cationic and dotted for doubly charged cationic species. Mineral assemblages 1, 2, 3 and 4 are listed in Table 5.

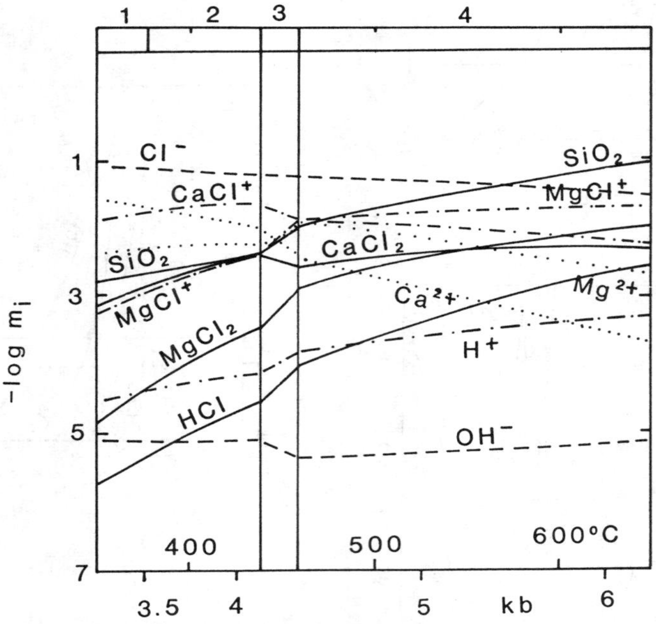

Figure 21. Speciation calculations for bulk composition D of Figure 20, but assuming a geothermal gradient of 0.1°C/bar.

serpentine + brucite --> forsterite + H_2O ,

diopside + serpentine --> forsterite + tremolite + H_2O ,

serpentine --> forsterite + talc + H_2O .

These mineral reactions are represented by distinct breaks in species concentration profiles. If fluid flow occurs across the mineral sequences, precipitation or dissolution is likely to occur near these breaks.

Composition A contains brucite, and pH of the solution varies between 5 and 7, that is from neutral to two units more basic than neutral. For compositions C to E, the pH is around 4, one unit more acid than neutral. pH decreases, as silica content in the rock increases. Whereas Ca-species concentrations are essentially pH independent, the Mg-species concentrations increase with decreasing pH from rock composition A to C, even though in the bulk composition Mg decreases. This demonstrates the importance of acid solutions in mobilizing metals (Eugster, 1985). Aqueous silica rises substantially from bulk composition A to quartz saturation in E.

The results presented in Figure 20 document that chloride-rich supercritical fluids equilibrated with Ca-and Mg-silicates are dominated by calcium-bearing solutes and, for SiO_2-rich bulk compositions, by silica. This conclusion conforms to the frequently cited observations of the chemistries of ridge crest hydrothermal fluids and seawater dominated geothermal brines (Edmond et al., 1982).

Figure 21 was calculated for bulk composition C using a typical geothermal gradient of 0.1°C/bar ($\sim$35°C/km). These calculations suggest that single charged metal chloride species such as $MgCl^+$, $CaCl^+$, are the dominant metal species over a wide range of metamorphic grades. This is due to the increased dissociation with increasing pressure. The fluids in equilibrium with Ca-Mg-silicates are Ca-dominated at low metamorphic grade, and Mg-dominated at higher grade.

SUMMARY AND CONCLUSIONS

The importance of fluids in crustal and mantle processes has been appreciated for well over a century (for an historic account see Eugster, 1986). Fluids are involved in the growth and dissolution of minerals and in the transport of material in response to gradients in composition, pressure and temperature. In order to model mass transport processes quantitatively, we need data on mineral solubilities as a function of P and T, as well as dissociation constants for the dominant solute complexes. This information can be obtained from conductivity, spectroscopic and solubility measurements, as well as from thermodynamic predictions such as those presented in Chapter 6. Laboratory measurements must include identification of the abundant solute species to make extraction of thermodynamic parameters possible. Most of the earlier solubility measurements in the near and supercritical region do not satisfy this requirement. During the last decade, however, a nearly complete data set has been assembled for major solutes in H_2O-HCl mixtures. The important speciation data for Al are now being collected.

For purposes of interpolation and extrapolation beyond 400°C and 1 kbar, these data can be fit to simple expressions which include entropy, enthalpy, volume and solvation terms. Solute free energy data can then be combined with mineral data to calculate solution compositions for a wide range of pressures, temperatures and mineral assemblages, essential information for modelling metamorphic, metasomatic and hydrothermal processes.

The supercritical data are restricted largely to major solutes in H_2O-HCl fluids, a logical choice because of the abundance of chloride in natural fluids. However, many metamorphic terranes contain CO_2-bearing fluids and the effect of CO_2 additions on solubility and speciation must be evaluated. Data on ore-forming elements such as Cu, Pb, Zn, Sn, Ag, Au and others are necessary to improve our understanding of hydrothermal ore deposits. A promising start has been made by Seward (1976, 1981, 1984) and Wilson (1986). Finally we need to bridge efficiently from the subcritical to the supercritical region. Although thermodynamic predictions (see Chapter 6) extend through the critical conditions, very few data sets are complete enough to calibrate the predictions and such experimental verification is essential.

In this chapter we have restricted ourselves to laboratory data obtained in the supercritical region and we have emphasized procedures for data reduction. We hope the results will encourage others to join us in further exploring this important area.

ACKNOWLEDGEMENTS

Unpublished manuscripts were generously provided by Greg Anderson, John Walther and Glenn Wilson and comments on the manuscript by Dimitri Sverjensky. Obviously, this treatment would not have been possible without the enthusiasm, ideas and efforts of former students John Frantz, Bill Gunter, I-Ming Chou and Glenn Wilson. Supported by NSF grant #EAR 8411050, and by the Fonds zur Foerderung junger Wissenschaftler of the Swiss National Science Foundation (Lukas Baumgartner).

REFERENCES

Anderson, G.M. and Burnham, C.W. (1967) Reaction of quartz and corundum with aqueous chloride and hydroxide solutions at high temperatures and pressures. Amer. J. Sci. 265, 12-27.

Anderson, G.M. and Burnham, C.W. (1983) Feldspar solubility and the transport of aluminum under metamorphic conditions. Amer. J. Sci. 283-A, 283-297.

Anderson, G.M., Pascal, M.L. and Rao, J. (1987) Aluminum Speciation in in Metamorphic Fluids. NATO ASI Series. H.C. Helgeson, ed. Reidel, Dordrecht, Holland.

Becker, K.H., Cemic, L. and Langer, K.E.O.E. (1983) Solubility of corundum in supercritical water. Geochim. Cosmochim. Acta 47, 1573-1578.

Barnes, H.L. (1979) Solubilities of ore minerals. In: Geochemistry of Hydrothermal Ore Deposits (2nd edition) H.L. Barnes, ed., John Wiley & Sons, New York, 404-460.

Bevington, P.R. (1969) Data Reduction and Error Analysis for the Physical Sciences. McGraw-Hill, New York.

Bockris, J.O'M. and Reddy, A.K.N. (1970) Modern Electrochemistry. Plenum Press, New York.

Boctor, N.Z. (1985) Rhodonite solubility and thermodynamic properties of aqueous $MnCl_2$ in the system $MnO-SiO_2-HCl-H_2O$. Geochim. Cosmochim. Acta, 49, 565-575.

Boctor, N.Z., Popp, R.K. and Frantz, J.D. (1980) Mineral solution equilibria IV. Solubilities and the thermodynamic properties of $FeCl_2^o$ in the system $Fe_2O_3-H_2-H_2O-HCl$. Geochim. Cosmochim. Acta, 44, 1509-1518.

Bowers, T.S. and Helgeson, H.C. (1983) Calculation of the thermodynamic and geochemical consequences of nonideal mixing in the system H_2O-CO_2-NaCl on phase relations in geologic systems: Equation of state for H_2O-CO_2-NaCl fluids at high pressures and temperatures. Geochim. Cosmochim. Acta 47, 1247-1275.

Burnham, C.W., Holloway, J.R., Davis, N.F. (1969) Thermodynamic properties of water to 1,000°C and 10,000 bars. Geol. Soc. Amer. Spec. Paper 132, 96 pp.

Chou, I-M and Eugster, H.P. (1976) Hydrothermal acid-base buffers: Fugacity control and dissociation constants of HBr and HI. Contrib. Mineral. and Petrol., 56, 77-100.

Chou, I-M and Eugster, H.P. (1977) Solubility of magnetite in supercritical chloride solutions. Amer. J. Sci. 277, 1296-1314.

Chou, I-M and Frantz, J.D. (1977) Recalibration of Ag + AgCl acid buffer at elevated pressures and temperatures. Amer. J. Sci. 277, 1067-1067-1072.

Edmond, J.M., von Damm, K.L., McDuff, R.E. and Measures, C.I. (1982) Chemistry of hot springs on the East Pacific Rise and their effluent dispersal. Nature 297, 187-191.

Eugster, H.P. (1981) Metamorphic solutions and reactions. In: Chemistry and Geochemistry of Solutions at High Temperature and Pressure, Physics and Chemistry of the Earth, 13/14, F.E. Wickman and D.T. Rickard, eds., Pergamon Press, New York, 461-507.

Eugster, H.P. (1985) Granites and hydrothermal ore deposits: A geochemical framework. Mineral. Mag. 49, 619-635.

Eugster, H.P. (1986) Minerals in hot water. Amer. Mineral. 71, 665-673.

Eugster, H.P., Chou, I-M. and Wilson, G.A. (1987) Mineral solubility and speciation in supercritical chloride fluids. In: Hydrothermal Experimental Techniques. G.C. Ulmer and H.L. Barnes, eds., John Wiley & Sons, New York (in press).

Eugster, H.P. and Gunter, W.D. (1981) The compositions of supercritical metamorphic solutions. Bull. Mineral. 104, 817-826.

Eugster, H.P. and Wones, D.R. (1962) Stability relations of the ferruginous biotite, annite. J. Petrol. 3, 82-125.

Evans, B.W. and Trommsdorff, V. (1970) Regional metamorphism of ultramafic rocks in the Central Alps: Paragenesis in the system $CaO-MgO-SiO_2-H_2O$. Schweiz. Mineral. Petrogr. Mitt. 50, 481-492.

Fein, J.B. and Walther, J.V. (1987) Calcite solubility in supercritical CO_2-H_2O fluids. Geochim. Cosmochim. Acta 51, 1665-1674.

Ferry, J.M. (1983) Regional metamorphism of the Vassalboro Formation, south-central Maine, USA: a case study of the role of fluid in metamorphic petrogenesis. J. Geol. Soc. 140, 551-576.

Franck, E.U. (1956) Hochverdichteter Wasserdampf I. Elektrolytische Leitfähigkeit in $KCl-H_2O$-Lösungen bis 750°C. Zeits. Physikal. Chemie, Neue Folge, 8, 92-106.

Frantz, J.D. and Eugster, H.P. (1973) Acid-base buffer: Use of Ag + AgCl in the experimental control of solution equilibria at elevated pressures and temperatures. Amer. J. Sci. 273, 268-286.

Frantz, J.D. and Marshall, W.L. (1982) Electrical conductances and ionization constants of calcium chloride and magnesium chloride in aqueous solutions at temperatures to 600°C and pressures to 4000 bars. Amer. J. Sci. 282, 1666-1693.

Frantz, J.D. and Marshall, W.L. (1984) Electrical conductances and ionization constants of salts, acids, and bases in supercritical aqueous fluids: I. Hydrochloric acid from 100° to 700°C and at pressures to 4000 bars. Amer. J. Sci. 284, 651-667.

Frantz, J.D. and Popp, R.K. (1979) Mineral-solution equilibria--I. An experimental study of complexing and thermodynamic properties of aqueous $MgCl_2$ in the system $MgO-SiO_2-H_2O-HCl$. Geochim. Cosmochim. Acta 43, 1223-1239.

Frantz, J.D., Popp, R.K. and Boctor, N.Z. (1981) Mineral-solution equilibria--V. Solubilities of rock-forming minerals in supercritical fluids. Geochim. Cosmochim. Acta 45, 69-77.

Gunter, W.D. and Eugster, H.P. (1978) Wollastonite solubility and free energy of supercritical aqueous $CaCl_2$. Contrib. Mineral. Petrol. 66, 271-281.

Gunter, W.D. and Eugster, H.P. (1980) Mica-feldspar equilibria in supercritical alkali chloride solutions. Contrib. Mineral. Petrol. 75, 235-250.

Haar, L., Gallagher, J.S. and Kell, G.S. (1979) Thermodynamic properties of fluid water. In: Contributions to the 9th Int. Conf. on Properties of Steam, Munich, Germany

Haar, L., Gallagher, J.S. and Kell, G.S. (1984) NBS/NRC Steam Tables. Thermodynamic and transport properties and computer programs for vapor and liquid states of water in SI units. Hemisphere Publishing Corp. McGraw Hill, New York.

Harvie, C.E. and Weare, J.H. (1980) The prediction of mineral solubilities in natural waters: the $Na-K-Mg-Ca-Cl-SO_4-H_2O$ system from zero to high concentration at 25°C. Geochim. Cosmochim. Acta 44, 981-997.

Harvie, , C.E., Greenberg,, J.P. and Weare, J.H. (1987) A chemical equilibrium algorithm for highly non-ideal multiphase systems: Free energy minimization. Geochim. Cosmochim. Acta 51, 1045-1058.

Helgeson, H.C. and Kirkham, D.H. (1974a) Theoretical prediction of the thermodynamic behavior of aqueous electrolytes at high pressures and temperatures: I. Summary of the thermodynamic/electrostatic properties of the solvent. Amer. J. Sci. 274, 1089-1198.

Helgeson, H.C. and Kirkham, D.H. (1974b) Theoretical prediction of the thermodynamic behavior of aqueous electrolytes at high pressures and temperatures: II. Debey-Hückel parameters for activity coefficients and relative partial molal properties. Amer. J. Sci. 274, 1199-1261.

Helgeson, H.C. and Kirkham, D.H. (1976) Theoretical prediction of the thermodynamic properties of aqueous electrolytes at high pressures and temperatures: III. Equation of state for aqueous species at infinite dilution. Amer. J. Sci. 276, 97-240.

Helgeson, H.C., Delany, J.M., Nesbitt, H.W. and Bird, D.K. (1978) Summary and critique of the thermodynamic properties of rock-forming minerals. Amer. J. Sci. 278-A, 1-229.

Helgeson, H.C., Kirkham, D.H. and Flowers, G.C. (1981) Theoretical prediction of the thermodynamic behavior of aqueous electrolytes at high pressures and temperatures. IV. Calculation of activity and osmotic coefficients and apparent molal and standard and relative molal properties to 600°C and 5 kb. Amer. J. Sci. 281, 1249-1493.

Hemley, J.J. (1959) Some mineralogical equilibria in the system K_2O-Al_2O_3-SiO_2-H_2O. Amer. J. Sci. 257, 241-270.

Hemley, J.J. and Jones, W.R. (1964) Chemical aspects of hydrothermal alteration with emphasis on hydrogen metasomatism. Econ. Geol. 59, 538-569.

Hemley, J.J., Montoya, J.W., Christ, C.L. and Hostetler, P.B. (1977a) Mineral equilibria in the MgO-SiO_2-H_2O system: Talc-chrysotile-forsterite-brucite stability relations. Amer. J. Sci. 277, 322-351.

Hemley, J.J., Montoya, J.W., Shaw, D.R. and Luce, R.W. (1977b) Mineral equilibria in the MgO-SiO_2-H_2O system. II. Talc-antigorite-forsterite-anthophyllite-enstatite stability relations and some geologic implications in the system. Amer. J. Sci. 277, 353-383.

Henley R.W., Truesdell, A.H., and Barton, P.B. (1984) Fluid-mineral equilibria in hydrothermal systems, Reviews in Econ. Geol. 1, 1-267.

Holland, H. D. and Malinin, S. D. (1979) The solubility and occurrence of non-ore minerals. In: Geochemistry of Hydrothermal Ore Deposits H. L. Barnes, ed., John Wiley & Sons, New York, 461-508.

Hollister, L.S. and Crawford, M.L., eds. (1981) Fluid Inclusions: Applications to Petrology. Mineral. Assoc. Canada Short Course Handbook 6, 304 p.

Holloway, J.R. (1979) Fugacity and activity of molecular species in supercritical fluids. In: Thermodynamics in Geology. D.G. Fraser, ed., Reidel, Dordrecht Holland, p. 161-181.

Holloway, J.R. (1981) Compositions and volumes of supercritical fluids in the Earth's crust. In: Fluid Inclusions: Applications to Petrology. L.S. Hollister, and M.L. Crawford, eds., Mineral. Assoc. Canada, p. 13-38.

Kerrick, D.M. (1974) Review of metamorphic mixed-volatile (H_2O-CO_2) equilibria. Amer. Mineral. 59, 729-762.

Kirkwood, J.G. (1939) The dielectric polarization of polar liquids. J. Chem. Phys. 7, 911-919.

Lagache, M. and Weisbrod, A. (1977) The system: Two alkali feldspars-KCl-$NaCl$-H_2O at moderate to high temperatures and low pressures. Contrib. Mineral. Petrol. 62, 77-101.

Lin, S. and Popp, R.K. (1984) Solubility and complexing of Ni in the system NiO-H_2O-HCl. Geochim. Cosmochim. Acta, 48, 2713-2722.

Luce, R.W., Cygan, G.L., Hemley, J.J. and D'Angelo, W.M. (1985) Some mineral stability relations in the system CaO-MgO-SiO_2-H_2-HCl. Geochim. Cosmochim. Acta, 49, 525-538.

Mangold, Von K. and Franck, E.U. (1969) Elektrische Leitfähigkeit wässriger Lösungen bei hohen Temperaturen und Drucken. II. Alkalichloride in Wasser bis 1000°C und 12 kbar. Ber. Bunsenges. 73, 21-27.

Marshall, W.L. (1970) Complete equilibrium constants, electrolyte equilibria, and reaction rates. J. Phys. Chem. 74, 346-355.

Marshall, W.L. (1972) A further description of complete equilibrium constants. J. Phys. Chem. 76, 720-731.

Marshall, W.L. and Franck, E.U. (1981) Ion product of water substance, 0-1000°C, 1-10,000 bars. New international formulation and its background. J. Physics and Chem. Reference Data 10, 295-303.

McKenzie, W.F. and Helgeson, H.C. (1984) Estimation of the dielectric constant of H_2O from experimental solubilities of quartz, and calculation of the thermodynamic properties of aqueous species to 900°C at 2 kb. Geochim. Cosmochim. Acta 48, 2167-2178.

Montoya, J.W. and Hemley, J.J. (1975) Activity relations and stabilities in alkali feldspar and mica alteration reactions. Econ. Geol. 70, 577-585.

Oelkers, E.H. and Helgeson, H.C. (1986) Are hydrothermal solutions really associated at supercritical pressures and tempertures? Geol. Soc. Amer. Abstracts with Programs 18, 709.

Pascal, M.L. (1984) Nature et propriétés des éspèces en solution dans le système K_2O-Na_2O-SiO_2-Al_2O_3-H_2O-HCl: contribution expérimentale. Thèse de doctorat d'état, l'Université Pierre et Marie Curie, Paris.

Pitzer, K.S. (1973) Thermodynamics of electrolytes I: Theoretical basis and general equations. J. Phys. Chem. 77, 268-277.

Pitzer, K.S. (1983) Dielectric constant of water at very high temperature and pressure. Proc. National Academy of Sciences USA, 80, 4575-4576.

Popp, R.K. and Frantz, J.D. (1979) Mineral solution equilibria II. An experimental study of mineral solubilities and the thermodynamic properties of aqueous $CaCl_2$ in the system CaO-SiO_2-H_2O-HCl. Geochim. Cosmochim. Acta 43, 1777-1790.

Popp, R.K. and Frantz, J.D. (1980) Mineral solution equilibria III. The system Na_2O-Al_2O_3-SiO_2-H_2O-HCl. Geochim. Cosmochim. Acta 44, 1029-1037.

Press, W.H., Flannery, B.P., Teukolsky, S.A., Vetterling, W.T. (1986) Numerical Recipes. Cambridge University Press, 818 pp.

Quist, A.S. and Marshall, W.L. (1968) Electrical conductances of aqueous sodium chloride solutions from 0 to 800°C and at pressures to 4000 bar. J. Phys. Chem. 72, 684-703.

Ragnarsdottir, K.V. and Walther, J.V. (1985) Experimental determination of corundum solubilities in pure water between 400-700°C and 1-3 kbar. Geochim. Cosmochim. Acta 49, 2109-2115.

Ritzert von, G. and Franck, E.U. (1968) Elektrische Leitfähigkeit wässriger Lösungen bei hohen Temperaturen und Drucken. I. KCl, $BaCl_2$, $Ba(OH)_2$ und $MgSO_4$ bis 750°C und 6 kbar. Ber. Bunsengesell. für Physik Chemie 72, 798-808.

Robie, R.A., Hemingway, B.S. and Fisher, J.R. (1979) Thermodynamic properties of minerals and related substances at 298.15 K and 1 bar (10^5) Pascals) pressure and at higher temperatures. U.S. Geol. Surv. Bull. 1452.

Roedder, E. (1984) Fluid Inclusions. Reviews in Mineralogy 12, 644 p.

Seward, T.M. (1974) Determination of the first ionization constant of silicic acid from quartz solubility in borate buffer solutions to 350°C. Geochim. Cosmochim. Acta 38, 1651-1664.

Seward, T.M. (1976) The stability of chloride complexes of silver in hydrothermal solutions up to 350°C. Geochim. Cosmochim. Acta 40, 1329-1341.

Seward, T.M. (1981) Metal complex formation in aqueous solutions at elevated temperatures and pressures. In: Chemistry and Geochemistry of Solutions at High Temperatures and Pressures, Physics and Chemistry of the Earth, 13 & 14, 113-132, Pergamon Press, New York.

Seward, T.M. (1984) The formation of Pb (II) chloride complexes to 300°C: a spectrophotometric study. Geochim. Cosmochim. Acta 48, 121-134.

Shade, J.W. (1974) Hydrolysis reactions in the SiO_2 excess portion of the system K_2O-Al_2O_3-SiO_2-H_2O in chloride fluids at magmatic conditions. Econ. Geol. 69, 218-228.

Sharp, W.E. and Kennedy, G.C. (1965) The system CaO-CO_2-H_2O in the two-phase region calcite and aqueous solution. J. Geol. 73, 391-403.

Skippen, G.B. (1971) Experimental data for reactions in siliceous marbles. J. Geol. 79, 451-481.
Skippen, G.B. (1974) An experimental model for low pressure metamorphism of siliceous dolomitic marble. Amer. J. Sci. 274, 487-509.
Tanger, J.C., IV, and Helgeson, H.C. (1987) Calculation of the thermodynamic and transport properties of aqueous species at high pressures and temperatures: I. Revised equation of state for the standard partial molal properties of ions and electrolytes. Amer. J. Sci., in press.
Tremaine, P.R. and LeBlanc, J.C. (1980) The solubility of magnetite and the hydrolysis and oxidation of Fe^{2+} in water to 300°C. Jour. Solution Chemistry 9, 415-442.
Van Zeggeren, F., Storey, S.H. (1970) The computation of chemical equilibrium. Cambridge University press, Cambridge. 176 pp.
Vidale, R. (1983) Pore solution compositions in a pelitic system at high temperatures, pressures and salinities. Amer. J. Sci. 283 A, 298-313.
Walther, J.V. (1986a) Mineral solubilities in supercritical H_2O solutions. Pure & Appl. Chem. 58, 1585-1698.
Walther, J.V. (1986b) Experimental determination of portlandite and brucite solubilities in supercritical H_2O. Geochim. Cosmochim. Acta 50, 733-739.
Walther, J.V. and Helgeson, H.C. (1977) Calculation of the thermodynamic properties of aqueous silica and solubility of quartz and its polymorphs at high pressures and temperatures. Amer. J. Sci. 277, 1315-1351.
Walther, J.V. and Long, M.I. (1986) Experimental determination of calcite solubilities in supercritical H_2O. Extended Abstracts, Fifth International Symposium on Water-Rock Interaction, Orkustofnun, Reykjavik, Iceland, 609-611.
Walther, J.V. and Orville, P.M. (1983) The extraction-quench technique for determination of the thermodynamic properties of solute complexes: Application to quartz solubility in fluid mixtures. Amer. Mineral. 68, 731-741.
Wilson G.A. (1986) Cassiterite solubility and metal chloride speciation in supercritical solutions. Ph.D. thesis, Johns Hopkins University, Baltimore, MD, 158 pp.
Wilson, G.A. and Eugster, H.P. (1984) Cassiterite solubility and tin-chloride speciation in supercritical solutions. Geol. Soc. Amer. Abstr. with Programs 16, 696.
Wolery, T. (1979) Calculation of chemical equilibria between aqueous solutions and minerals: the EQ3/6 software package. Lawrence Livermore National Lab. Livermore, CA URCL-52658, 191 pp.
Wolery, T.J., Sherwood, D.J., Jackson, K.J., Delaney, J.M. and Puigdomenech, I. (1984) EQ3/6: Status and applications. Lawrence Livermore National Lab., Livermore, CA, UCRL-91884, 12 pp.
Woodland, A.B. and Walther, J.V. (1987) Experimental determination of the solubility of the assemblage paragonite, albite, and quartz in supercritical H_2O. Geochim. Cosmochim. Acta (in press).

Chapter 11 Robert G. Berman and Thomas H. Brown

DEVELOPMENT OF MODELS FOR MULTICOMPONENT MELTS: ANALYSIS OF SYNTHETIC SYSTEMS

INTRODUCTION

During the past two decades geologists have begun to address the fundamental problem of formulating thermodynamic models for multicomponent melts which can be used to quantify and illuminate the diverse processes of igneous petrology and planetary geology. Thermodynamic description of the melt phase provides the means to compute stability relationships that can be used to constrain melting and differentiation processes, and to formulate a variety of thermobarometric techniques for routine use in geologic studies. In addition, it provides the basis for performing complete energy budget calculations necessary for exploring the plausibility and implications of partial melting, assimilation, and magma mixing processes. Another important role that a thermodynamic model fulfills is the systemization of and interpolation between the various experimental results pertaining to specific melt compositions. In igneous systems this role is particularly acute since the experimental coverage of the almost limitless compositional range of natural magmas will always be incomplete.

Many empirical models have recently been developed which attempt to meet the immediate demands for thermobarometric techniques that can be applied to magmatic systems (e.g. Langmuir and Weaver, 1982; Nielsen and Dungan, 1983; Glazner, 1984), but none of these models attempts to provide a general thermodynamic formulation of the melt phase. For this we are presently limited to the models of Ghiorso et al. (1983) and Burnham and Nekvasil (1986). The former covers a range of natural melt compositions and has been used to great advantage in quantitatively modelling a variety of igneous processes (Ghiorso and Carmichael, 1985; Nicholls et al., 1986). Its main shortcoming is its inability to account for phase relationships in simple binary and ternary systems, which leads one to question the significance of the empirical fit parameters and to recognize the danger in using it outside the P, T, X space in which it was calibrated. The model of Burnham and Nekvasil (1986) accounts for experimental data in simple and complex systems but has been calibrated solely in granitic systems. Moreover, the present formulation of this model assumes ideal mixing among all melt species, which restricts its applicability to systems without liquid immiscibility. In spite of these limitations, the fact that these two models have had considerable success in reproducing phase relationships for natural melt compositions should provide the necessary encouragement for earth scientists to continue with the development of improved thermodynamic models for magmatic systems.

In this chapter, we look in detail at the thermodynamics of synthetic melt systems which offer two important advantages over natural melt systems. First, phase relationships are known with reasonable precision and in considerable detail over a large portion of many synthetic systems making it possible to test rigorously the adequacy of thermodynamic models. Second, additional experimental measurements (e.g. activities, heats of mixing) are available with which to verify model predictions. We review the basic ingredients of a general thermodynamic model for melt systems with the following question in mind: What is required to construct a melt model that can accurately reproduce the variety of experimental data gathered in synthetic systems and also can be extended compositionally to describe natural melt systems? We stress at the outset that there are three aspects of this problem. The first is formulation of an adequate thermodynamic model. The second is development of a methodology for calibrating the model from the diverse set of experimental data which constrain not only the thermodynamic properties of melts but also, if liquidus relations are to be reproduced, those of minerals as well. The third is testing the consequences of the thermodynamic model, usually by prediction of stable and metastable phase relationships. The theoretical aspects of this problem are discussed in the first section of this chapter, after which the experimental data and methodology are reviewed. Lastly, we present a brief summary of some of the more encouraging and illustrative attempts to formulate thermodynamic models for multicomponent melts in synthetic systems.

THEORETICAL CONSIDERATIONS

The essential element of any thermodynamic model for multicomponent melts is the ability to calculate the Gibbs free energy (G) of the melt phase as a function of temperature, pressure and composition. All other thermodynamic properties can then be expressed as a function of the free energy and its derivatives with respect to the independent variables (T, P, and X). Although, we have elected to use the Gibbs free energy as the dependent variable to derive the thermodynamic relationships of the melt phase, other variables (i.e., enthalpy, entropy, etc.) can be used with equivalent results.

The variety of models that have been proposed to describe the free energy of silicate melts can be classified into what we will call *speciation* and *stoichiometric* approaches. In general, the speciation approach leads to a *complex solution* phase in which the number of chemical entities ('species') mixing in the solution is greater than the number of thermodynamic components (rank of the species composition matrix). The stoichiometric approach leads to a *simple solution* in which the number of chemical entities is equal to the number of components. For either type of solution, mixing may be ideal or nonideal. Before describing these two approaches to solution models, we briefly review some fundamental equations and definitions of equilibrium thermodynamics that apply in either type of solution model.

At a temperature, T in Kelvins, and pressure, P in bars, the molar Gibbs free energy of any phase can be expressed by

$$G = \sum_{i}^{ncs} x_i \mu_i \quad , \tag{1}$$

where ncs is the number of components or species, x_i is the mole fraction and μ_i the chemical potential of the ith component or species. The latter is defined by

$$\mu_i = \mu_i^\circ + RT \ln a_i \quad , \tag{2}$$

where R is the gas constant and a the activity. For any component or species, the standard state chemical potential, μ°, can be calculated from

$$\mu^\circ = \Delta_f H^{T_r,P_r} + \int_{T_r}^{T} Cp dT - T[S^{T_r,P_r} + \int_{T_r}^{T} \frac{Cp}{T} dT] + \int_{P_r}^{P} V dP \quad , \tag{3}$$

where $\Delta_f H^{T_r,P_r}$ and S^{T_r,P_r} are the standard state enthalpy of formation and third law entropy at the reference temperature (T_r = 298.15 K) and pressure (P_r = 1 bar), and Cp and V are the as yet unspecified heat capacity and volume functions of the mineral or melt species. The standard state that we will use throughout this chapter is that of unit activity for any pure mineral or melt species at any temperature and pressure. Note as well that the function given by Equation (3) is an apparent free energy of formation (Benson, 1968) which neglects the heat capacities and entropies of the elements. This is a convenient simplification since the element properties cancel when any equilibrium is considered.

The free energy of a solution (1) can be expanded in terms of Equation (2) and the definition of the activity coefficient ($a = \gamma x$):

$$G_{\text{solution}} = \sum_{i}^{ncs} x_i[\mu_i^\circ + RT \ln x_i + RT \ln \gamma_i] \quad . \tag{4}$$

The summation over the first two terms gives the mechanical mixing and ideal contributions to the total free energy of solution, while the latter term gives the excess free energy of solution:

$$G_{\text{excess}} = \sum_{i=1}^{ncs} x_i RT \ln \gamma_i \quad . \tag{5}$$

A great many theoretical and empirical formulations of G_{excess} as a function of composition have been proposed, and an excellent summary is given by Acree (1984). The interrelationships among some of these equations have been discussed by Grover (1977) and more recently by Tomiska (1986), while Fei et al. (1986) compare the ability of several formulations to represent enthalpy of mixing data for minerals. A safe conclusion to reach from these and other studies is that no one formulation is best suited for all applications. For the purposes of illustration in this paper, we will use a general polynomial expansion to represent the activity coefficients as a function of composition. The popular use in geologic studies of second and third degree polynomial expansions to approximate G_{excess} arises from the work of Thompson (1967). The polynomial expansions are simplified by application of the boundary conditions that G_{excess} approaches zero as each mole fraction approaches unity (Raoultian and Henrian behavior), and by recombination such that all terms are of the same degree. The coefficients of the recombined terms, W's, are called Margules parameters and are linear combinations of the original polynomial coefficients. In applying second and third degree polynomial expressions to G_{excess} there have been inconsistencies in the literature, as pointed out by Andersen and Lindsley (1981), with regard to subscript notation for the Margules parameters. In an effort to resolve such inconsistencies Berman and Brown (1984) presented a generalized equation which can readily be applied to solutions of any number of components or species and can incorporate any degree polynomial in the initial expansion:

$$G_{\text{excess}} = \sum_{i_1=1}^{ncs-1} \sum_{i_2=i_1}^{ncs} \cdots \sum_{\substack{i_p=i_{p-1} \\ i_p \neq i_1}}^{ncs} W_{i_1 i_2 \ldots i_p}(x_{i_1} x_{i_2} \ldots x_{i_p}) \quad , \tag{6}$$

where p is the degree of the polynomial. For example, a third degree polynomial applied to a ternary solution reduces Equation (6) to

$$G_{\text{excess}} = \sum_{i=1}^{2} \sum_{j=i}^{3} \sum_{\substack{k=j \\ k \neq i}}^{3} W_{ijk}(x_i x_j x_k) \quad , \tag{7}$$

which when expanded gives

$$\begin{aligned} G_{\text{excess}} = & \ W_{112}x_1x_1x_2 + W_{122}x_1x_2x_2 + W_{113}x_1x_1x_3 \\ + & \ W_{133}x_1x_3x_3 + W_{223}x_2x_2x_3 + W_{233}x_2x_3x_3 \\ + & \ W_{123}x_1x_2x_3 \quad . \end{aligned} \tag{8}$$

As pointed out by Barron(1986), this general formulation of G_{excess} simplifies the derivation of free energy derivatives and spinodal equations and reduces their form to single lines of text. The notation seems preferable as it is well suited to computer manipulation, and the subscripts of the Margules parameters can be extended to any degree polynomial. Furthermore, these subscripts serve as reminders of the powers of the mole fractions in each term (see Eqn. 8).

Equations (5) and (6) can be used to derive a general equation for the calculation of the activity coefficient of the mth component or species, using a polynomial of any degree (see Berman and Brown, 1984 for a complete derivation):

$$RT \ln \gamma_m = \sum_{i_1=1}^{nc} \sum_{i_2=i_1}^{nc} \cdots \sum_{\substack{i_p=i_p-1 \\ i_p \neq i_1}}^{nc} W_{i_1 i_2 \ldots i_p} [Q_m x_{i_1} x_{i_2} \ldots x_{i_p} / x_m + (1-p) x_{i_1} x_{i_2} \ldots x_{i_p}] \quad , \tag{9}$$

where Q_m is a term which sums how many of the i_1 to i_p indices are equal to m.

A temperature and pressure dependency for the excess free energy can be included by writing each of the above Margules parameters as:

$$W_{i_1 i_2 \ldots i_p} = W_{G_{i_1 i_2 \ldots i_p}} = W_{H_{i_1 i_2 \ldots i_p}} - T \cdot W_{S_{i_1 i_2 \ldots i_p}} + (P-1) W_{V_{i_1 i_2 \ldots i_p}} \quad . \tag{10}$$

This equation can also be expanded to include other terms such as excess heat capacity. The use of Equations (9) and (10) to express the composition, temperature, and pressure dependence of the excess free energy results in a great deal of flexibility in describing properties of nonideal solutions. In order that solution properties extrapolate reasonably, additional constraints can be placed on the temperature dependence of the excess free energy. For speciation models in which the mixing species reflect the actual species present, G_{excess} should approach zero at high temperature as the solution becomes ideal. For stoichiometric models and speciation models in which the mixing species do not reflect the actual species present, this same constraint cannot be applied. In such cases, Barron (1986) has suggested that the excess entropy be constrained to never exceed the configurational entropy in magnitude, thereby preventing the solution from becoming more nonideal with increasing temperature.

Speciation models

As with gas and aqueous phase equilibrium calculations, an optimal formulation would base the free energy function for the melt phase on the actual species and polymers present. With a speciation model, the molar Gibbs free energy of a melt consisting of *nsp* number of species is computed from:

$$G_{\text{melt}} = \sum_{i}^{\text{nsp}} x_i \mu_i \quad . \tag{11}$$

Under equilibrium conditions G_{melt} is at a minimum subject to the bulk compositional constraints:

$$X_j = \sum_i^{\text{nsp}} \nu_{ij} x_i \quad , \tag{12}$$

where X_j is the mole fraction of the jth component, and ν_{ij} is the stoichiometric amount of the jth component in the ith species.

Because the number of species in a real system is greater than the number of thermodynamic components, mathematical techniques both for calibrating speciation models and for recalculating phase relations are necessarily more complex. For any melt composition, iterative techniques are needed to determine the equilibrium distribution of melt species and consequently their thermodynamic properties as well (see discussion of Bottinga et al., 1981, pp. 224-229). For any reasonably complicated melt, the iterative computation of chemical equilibria is accomplished by solving a set of nonlinear equations or by utilizing free energy minimization techniques (van Zeggeren and Storey, 1970; Erikson, 1975; Ghiorso, 1985; deCapitani and Brown, 1987).

One of the advantages of a speciation model over stoichiometric formulations is that a greater amount of the solution's energy is accounted for in the mechanical and configurational mixing terms, and therefore mixing between species is generally closer to ideal. For example, Engi (1983) showed that a speciation model for quaternary spinel solid solutions results in much smaller deviations from ideality than the classical reciprocal salt solution (Wood and Nicholls, 1978). Nevertheless, mixing between some of the species in silicate melt systems must necessarily be nonideal in order to account for the widespread occurrence of liquid immiscibility especially in silica-rich portions of synthetic systems (Grieg, 1927). Moreover, the preliminary studies of Bottinga and Richet (1978) and Bottinga et al. (1981) suggest that, in order to reproduce liquidus relations, nonideal mixing parameters may be necessary between many species if the number of species is kept small (see 'Applications' below).

The development of the speciation approach has been hampered by the inherent structural complexity of silicate melts (see Bottinga et al. (1981) for a brief review). Spectroscopic (Mysen, 1983; 1986), X-ray (Taylor and Brown, 1979a; 1979b), and NMR (Stebbins et al., 1986) studies are extremely important in providing constraints on the kinds and composition of molecules (species) present, but as yet they are not quantitative. In addition, these techniques have been applied for the most part to glasses so that extrapolations are needed to reach conclusions regarding melt speciation (see discussion below of the glass-melt transition). The actual choice of melt species upon which to base a speciation model is therefore open because the present extent of our detailed structural knowledge of silicate melts is still limited to somewhat qualitative inferences. In the geologic literature a variety of species have been proposed for the purposes of thermodynamic modelling including polymers (Masson, 1972; Hess, 1977), ionic species (Bockris et al., 1952; Flood and Knapp, 1968), and neutral species mimicking the stoichiometry of liquidus minerals (Burnham, 1975; Bottinga and Richet, 1978). The recent proliferation of spectroscopic data on silicate glasses (for a review see Mysen, 1983; 1986) offers the opportunity to define discrete melt species on the basis of their ratios of nonbridging oxygens to tetrahedrally-coordinated cations. The consequences of some of these choices are explored in the last section of this chapter.

Stoichiometric solution models

An alternate to the speciation approach is to describe the thermodynamic properties of the melt in terms of a set of liquid components, that is the minimum number of chemical entities needed to span the entire compositional space of interest. Because the number of chemical formulae ('species') is equal to the number of components, iterative techniques are unnecessary to express liquid compositions in terms of the components. For this reason the stoichiometric approach offers the considerable advantage that calibrations of all thermodynamic parameters (standard state and mixing properties) as well as equilibrium computations with these models are greatly simplified.

In a stoichiometric approach to mineral-melt equilibria, all reactions are written to a basic set of component liquid formulae, balanced on the stoichiometry of the liquidus phases. The thermodynamic components are chosen to have the same formulae as the mixing 'species', and we will use the the terms *component* and *species* interchangeably when discussing *simple* solutions. Component formulae can be simple oxides (e.g. SiO_2, CaO), multiples of oxides (e.g. Si_4O_8, Ca_8O_8) and/or combinations of oxides (e.g. Ca_2SiO_4, $Ca_4Si_2O_8$), and need bear no resemblance to actual species present in the melt. Thus Pelton and Blander (1984) and Berman and Brown (1984) used simple oxides to model phase relationships in binary to quaternary synthetic systems, while Bottinga and Richet (1978) and Ghiorso et al. (1983) used combined oxide and multiple oxide formulae to model synthetic and natural systems, respectively. The components must span the compositional space of interest, and in a strict thermodynamic sense they must span the entire 'accessible' compositional space. The reason for this is that the chemical potentials become infinite as the the melt approaches the composition of any component. For example, if we chose MgO and $MgSiO_3$ as components, then the chemical potential of MgO in the melt would approach minus infinity as the composition of the melt approached $MgSiO_3$ in composition. However, from pragmatic considerations, these 'boundary' effects can be minimized if for a given application the composition of the melt phase never approaches the composition of the components. This pragmatic approach has allowed Ghiorso et al. (1983) to model natural melts with components that for the most part mimic mineral formulae, thus reducing the complexity of their model by only considering a portion of the entire compositional space. If one abuses their model and tries to calculate the chemical potentials, for example, at the Mg_2SiO_4 composition, one would find that all of the chemical potentials are infinite, and that discontinuities in the solution properties are encountered in moving to compositions outside the positive component space.

Simple oxides will be used as our liquid components in the examples that follow, but other components could be chosen with similar results. With a stoichiometric approach, equilibrium between a mineral and an *nc* component melt is written as

$$\text{Mineral} = \text{Melt} = \sum_{i}^{nc} \nu_i O_i \quad , \tag{13}$$

where ν_i is the stoichiometric coefficient of the ith liquid oxide component O_i. For example, equilibrium between liquid and forsterite is written

$$\text{Forsterite} = \text{Melt} = 2\text{MgO}_{(\ell)} + \text{SiO}_{2(\ell)} \quad . \tag{14}$$

At equilibrium, the Gibbs free energy of reaction is equal to zero, so that in general

$$\mu^{\circ}_{cr} + RT \ln a_{cr} = \sum_{i=1}^{nc} \nu_i [\mu^{\circ}_i + RT \ln a_i] \quad , \tag{15}$$

where the subscript *cr* refers to the crystalline phase. Substituting the definition of the activity coefficient and assuming that the liquidus mineral is pure gives

$$\mu^{\circ}_{cr} = \sum_{i=1}^{nc} \nu_i [\mu^{\circ}_i + RT \ln X_i + RT \ln \gamma_i] \quad . \tag{16}$$

The example of the forsterite-melt equilibrium is then written

$$\begin{aligned} \mu^{\circ}_{Fo} = \; & 2[\mu^{\circ}_{MgO_{(\ell)}} + RT \ln X_{MgO_{(\ell)}} + RT \ln \gamma_{MgO_{(\ell)}}] \\ + \; & [\mu^{\circ}_{SiO_{2(\ell)}} + RT \ln X_{SiO_{2(\ell)}} + RT \ln \gamma_{SiO_{2(\ell)}}] \quad . \end{aligned} \tag{17}$$

For the sake of illustration, we will use a third degree polynomial to represent G_{excess} and assume that forsterite is in equilibrium with a binary MgO-SiO_2 melt. With substitution of Equation (9) for the activity coefficients of MgO and SiO_2 (labelled as components 1 and 2, respectively), Equation (17) can then be expanded to

$$\begin{aligned} \mu^{\circ}_{Fo} = & [2\mu^{\circ}_1 + \mu^{\circ}_2] + [2RT \ln X_1 + RT \ln X_2] + \\ & W_{112}(X_1^2 + 4X_1X_2 - 6X_1^2X_2) + W_{122}(2X_2^2 + 2X_1X_2 - 6X_1X_2^2) \quad . \end{aligned} \tag{18}$$

Notice that, if all X's are known as in a set of experimental data, Equation (18) remains linear in the Margules parameters, a fact that greatly facilitates their calibration from phase equilibrium data (see 'Methodology' below).

EXPERIMENTAL CONSTRAINTS

Inspection of Equations (1-3) shows the basic thermodynamic data needed to construct a thermodynamic model for silicate melts: data constraining standard state properties ($\Delta_f H^{T_r,P_r}, S^{T_r,P_r}, Cp, V$) and activities. An important practical consideration in choosing between a stoichiometric and speciation model is the amount, type, and quality of available data with which to constrain the different model parameters. In a stoichiometric approach, standard state properties are needed for a minimum number of chemical entities (components) in the melt, and the emphasis is placed on representation of component activities. In a speciation approach, standard state properties must be defined for more melt compositions (species), with need for correspondingly less complex activity coefficient expressions.

In the following paragraphs, we review the available experimental data that can be used to define thermodynamic properties of melts. Because of space limitations, and in order that the reader may easily assess and utilize the constraints imposed by these data, our emphasis here is on the various models that have been proposed to represent, interpolate between, and to some extent extrapolate beyond these data. Complete references to relevant experimental data may be found in the cited studies aimed at representation of these data. We will not attempt to incorporate a critique of the experimental data, nor interpretations of the systematics of such data, instead referring the reader to the informative reviews by Navrotsky (1986) and Richet and Bottinga (1986). We also include a brief review of the experimental constraints on mineral properties because of their obvious importance in modelling liquidus relations.

Thermodynamic properties of melts

Glass-liquid relationships. For many years the inherent experimental difficulties in obtaining thermodynamic data on the melt phase were circumvented by using the properties of glasses to approximate those of their isochemical melts. Fortunately, our knowledge of the thermodynamic properties of melts has grown enormously over the last decade due to the innovative efforts at several experimental laboratories around the world. Before briefly reviewing the state of this knowledge, it is important to recognize the differences between liquids and glasses and the implications of the glass-melt transition on thermodynamic properties. More detailed discussions of this topic with adequate reference to previous work can be found in the reviews presented by Richet and Bottinga (1983; 1986) and Navrotsky (1986).

In rapid cooling of most silicate melts, particularly those with more than about 50 mol % SiO_2, the melt supercools without crystallization at its fusion temperature. At some lower temperature second-order properties decrease drastically (e.g. Cp shown in Fig. 1b) as the melt loses configurational degrees of freedom and becomes a glass. The glass transition temperature is operationally defined as that temperature at which the differing thermodynamic properties of melt and glass intersect (e.g. the heat content curves shown in Fig. 1a). Not only is the glass transition temperature a function of cooling rate, being higher for faster cooling rates, but the thermodynamic properties of the glass are also a function of thermal history (compare the heat content curves of the two glasses shown in Fig. 1a).

It is of utmost importance to recognize these differences between glass and liquid in thermodynamic modelling of melts. As can be seen from Figure 1, the heat capacity of the glass (Cp_g) is quite similar to that of the crystal, but is considerably smaller than that of the melt which for many compositions appears to be independent of temperature (see below). The greater Cp of the liquid compared to the crystal is reflected in the greater slope of its heat content curve (Fig. 1a), illustrating that heats of fusion are very much a function of temperature. What can also be gleaned from this figure is the large difference between the heat of vitrification (the difference between the enthalpy of the glass and crystal at T_g) and the heat of fusion for diopside. This difference is largest for compositions which are least polymerized and have large differences between glass transition and fusion temperatures. Lastly, the differences in Cp between glass and liquid make it necessary to correct heats of mixing measured in glasses to heats of mixing in melts, a process that is further complicated by the nonlinear variation of glass transition temperature with composition (see below).

Heat capacity of glasses. Although the heat capacity of a glass is very different from that of a liquid, Cp_g is often needed in order to compute melt properties such as enthalpies of fusion (see below). The drop calorimetry measurements of White (1919) on $NaAlSi_3O_8$-$CaAlSi_2O_8$ and $MgSiO_3$-$CaSiO_3$ glasses clearly showed that glass heat capacity varied linearly with composition, and the proliferation of recent data obtained by adiabatic calorimetry at low temperature, differential scanning calorimetry between 300 and 1000 K, and drop calorimetry still supports these early observations. These data are indications of ideal mixing which can be expressed by

$$Z_{\text{glass}} = \sum_{i}^{nc} n_i \overline{Z}_i \quad , \tag{19}$$

where Z is some extensive thermodynamic property, in this case Cp or heat content $(H_T - H_{T_r})$, and $\overline{Z}_i$ and n_i are the partial molar properties and number of moles

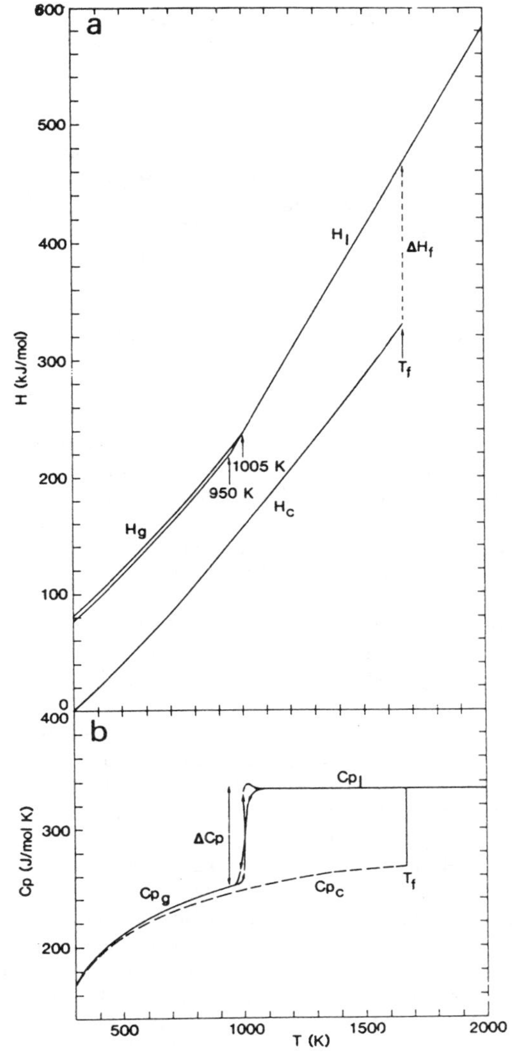

Figure 1. Temperature dependence of the enthalpy (a) and heat capacity (b) of crystalline, liquid, and glassy diopside. The glass transition temperatures of 950 and 1005 K result from different cooling rates. Note the greater heat capacity of liquid compared to glass, which results in a larger heat of fusion than heat of vitrification (the difference between the enthalpy curves for glass and crystal at the glass transition temperature). Taken from Richet and Bottinga (1986).

Table 1: Richet and Bottinga (1982) coefficents for calculation of the heat capacity of glasses ($Jmol^{-1}K^{-1}$): $C_p = \sum X_i(a_i + b_iT + c_iT^{-2} + d_iT^{-1/2})$

	SiO_2	TiO_2	Al_2O_3	FeO	MgO	CaO	Na_2O	K_2O
a_i	127.20	41.71	95.83	26.33	37.34	53.62	64.27	87.62
b_i x10^3	-10.277	41.06	31.61	53.39	19.76	5.298	34.46	0.364
c_i x10^{-5}	4.3127	-0.19	-21.03	-0.13	-3.84	-13.00	-0.08	-11.82
d_i	-1463.9							

of the ith glass oxide, respectively. The oxide glass heat capacities are functions of temperature, which have been represented by the Maier-Kelley function:

$$\overline{Cp}_i = \overline{a}_i + \overline{b}_i T + \overline{c}_i T^{-2} \quad . \tag{20}$$

In recent models $\overline{Cp}$ coefficients for major oxides consistent with (19) and (20) have been derived by Bacon (1977), Richet and Bottinga (1982), and Stebbins et al. (1984). The latter two models were calibrated using different data sets, both greatly expanded over that used by Bacon (1977), and the results of these models are compared in detail by Richet and Bottinga (1986). An important difference between these models is that Richet and Bottinga (1982) forced $\overline{Cp}_{SiO_2}$ to be equal to that of pure SiO_2 glass, a constraint that provides for better estimation of the Cp of SiO_2-rich glasses. Their coefficients for calculation of Cp_g are reproduced in Table 1.

Heat capacity of liquids. Only recently has the Cp of stable liquids (Cp_ℓ) been the focus of systematic experimental studies, but the large amount of high quality data obtained over the last few years has already made it possible to formulate reasonable Cp_ℓ constraints that can be used for the purpose of thermodynamic modelling. Based on heat content measurements on 8 natural and synthetic compositions in the range 1000-1600 K, Carmichael et al. (1977) proposed that melts mix ideally with respect to heat capacity (Eqn. 19) and that the partial molar oxide heat capacities are independent of temperature ($\overline{b}_i, \overline{c}_i = 0$ of Eqn. 20). This model has since been recalibrated by Stebbins et al. (1984) using much additional drop calorimetry data, and their temperature-independent $\overline{Cp}_\ell$ coefficients consistent with the ideal mixing model are presented in Table 2.

Stebbins et al. (1984) noted several examples of data that indicate deviations from composition independent partial molar oxide heat capacities. For example, Cp_ℓ measurements on ternary melts in the system $CaAlSi_2O_8$-$NaAlSi_3O_8$-$CaMgSi_2O_6$ show complex variations with composition, with a positive excess Cp on the join $NaAlSi_3O_8$-$CaMgSi_2O_6$, and a negative excess Cp on the join $CaAl_2Si_2O_8$-$CaMgSi_2O_6$ (Stebbins et al, 1984). Other examples of nonlinear variation of Cp_ℓ with composition for Al-bearing melts are shown in Figure 2, taken from Richet and Bottinga (1986). These experimental observations of excess heat capacity in molten silicates are extremely important from the point of view of formulating thermodynamic models because they indicate that heats of mixing and activity coefficients in some composition ranges are dependent on temperature. In order to accommodate complexities such as these which are only beginning to come to light, Richet and Bottinga (1985) proposed the following model for Al-free melts:

$$Cp_\ell = \sum_i^{nc} X_i Cp_i^\circ + Cp_{xs} \quad , \tag{21}$$

where X_i and Cp_i° are mole fractions and heat capacity coefficients for 20 oxides, respectively, and the excess heat capacity, Cp_{xs}, is given by:

$$Cp_{xs} = W_{SK} X_S X_K^2 \quad , \tag{22}$$

where X_S and X_K are the mole fractions of SiO_2 and K_2O, and W_{SK} is a constant. In order to allow application of this model to Al-bearing melts, Richet and Bottinga (1985) also computed an average Cp° coefficient for Al_2O_3 which neglects the nonideal interactions summarized above.

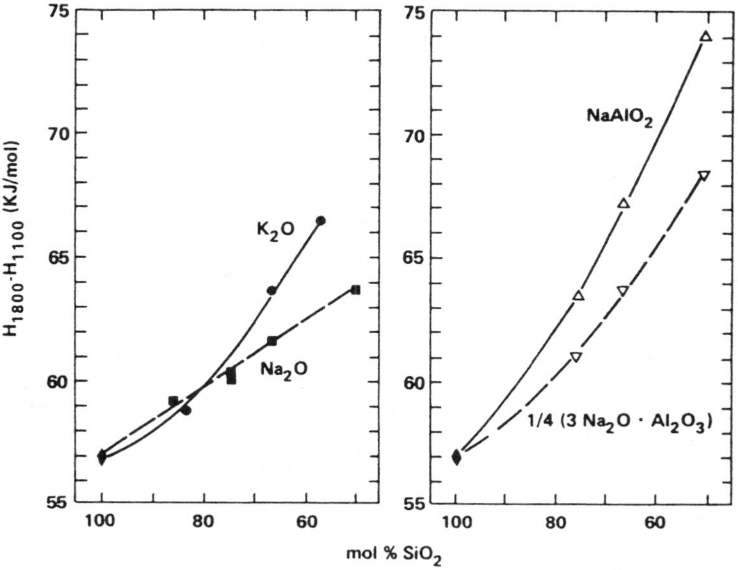

Figure 2. Heat contents of liquids in binary and pseudobinary systems with SiO_2. Note the nonlinear composition dependence of *Cp* (excess heat capacity) on all joins but Na_2O-SiO_2. Taken from Richet and Bottinga (1986).

Table 2: Stebbins et al. (1984) parameters for calculation of silicate liquid heat capacities between 1200-1850 K ($Jmol^{-1}K^{-1}$): $C_p = \sum X_i \overline{C}_{p_i}$

	SiO_2	TiO_2	Al_2O_3	Fe_2O_3	FeO	MgO	CaO	Na_2O	K_2O
$\overline{C}_{p_i}$	80.0	111.8	157.6	229.0	78.9	99.7	99.9	102.3	97.0
	±0.9	±5.1	±3.4	±18.4	±4.9	±7.3	±7.2	±1.9	±5.1

Table 3: Richet and Bottinga (1985) coefficients for calculation of the heat capacity of melts ($Jmol^{-1}K^{-1}$): $C_p = \sum X_i[a_i + b_i T + c_i T^{-2}] + 151.7\, X_S X_K^2$

	SiO_2	TiO_2	Al_2O_3	Fe_2O_3	FeO	MgO	CaO	Na_2O	K_2O
a_i	81.37	75.21	27.21	199.7	78.94	85.78	86.05	100.6	50.13
b_i x10^3	0.0	0.0	94.28	0.0	0.0	0.0	0.0	0.0	15.78
c_i x10^{-5}	0.0	875.3	0.0	0.0	0.0	0.0	0.0	0.0	0.0

Table 4: Bottinga et al. (1982) parameters for calculation of silicate melt volumes (cm^3mol^{-1}):

$$V(T) = (\sum X_i \overline{V}_i^\circ + X_A[V^* + K_V])\, e^{(\sum X_i \overline{\alpha}_i^\circ + X_A[\alpha^* + K_\alpha(V^* + K_V)])(T - T_r)}$$

where $K_V = X_A(\sum X_j K_j / \sum Xj)$ and $X_A = X_{Al_2O_3}$

	SiO_2	TiO_2	Al_2O_3	Fe_2O_3	FeO	MgO	CaO	Na_2O	K_2O
$\overline{V}_i^\circ$	26.75	22.45		44.40	13.94	12.32	16.59	29.03	46.30
$\overline{\alpha}_i^\circ$ x10^{-5}	0.1	37.1		32.1	34.7	12.2	16.7	25.9	35.9
K_j			-20.0		-58.0	-37.8	-8.5	-11.4	-48.1

T_r = *1673 K* $\quad V^* = 41.92 \quad \alpha^* = 87.48x10^{-5} \quad K_\alpha = -2.282x10^{-5}$

The systematics of the temperature dependence of Cp_ℓ are poorly understood at present. Constant Cp_ℓ values have been measured for natural compositions (Carmichael et al., 1977), as well as for many simple systems including SiO_2, $CaSiO_3$, $CaMgSi_2O_6$ for which the data have been gathered up to 2600 K (Stout and Piwinskii, 1982). Increases in Cp_ℓ have been reported for K_2O-SiO_2 liquids (Richet and Bottinga, 1985), aluminosilicate melts (Richet and Bottinga, 1984a; 1984b; Stebbins et al., 1982), and natural basaltic and rhyolitic compositions (Stout and Piwinskii, 1982). Negative temperature dependences have also been measured in many metal systems (e.g. Robie et al., 1979) and synthetic oxide systems such as titanium alkali-silicates (Richet and Bottinga, 1985) and $LiSiO_3$ (Stebbins et al., 1984). The most recent Cp_ℓ model of Richet and Bottinga (1985) accounts for most of these differences by making the $Cp°$ coefficients of Equation (21) dependent on temperature for K_2O, TiO_2, and Al_2O_3. The $Cp°$ coefficients of their model are given in Table 3. For natural melt compositions, the two models summarized above yield very similar results, with predictions generally within the uncertainties (approximately 1%) of the data. The Richet and Bottinga model affords considerably better representation of simple systems (including pure SiO_2). The Stebbins and Carmichael model reproduces the data for aluminosilicates somewhat more closely, but, due to the complexities exemplified in Figure 2, neither model is able to account for these latter data within the precision of the measurements.

The complex behavior of liquid heat capacities with respect both to temperature and composition dependence clearly is an area needing further experimental studies. Part of the problem in interpreting results comes from experimental difficulties which are reflected for example in the poor interlaboratory agreement for Cp_ℓ values for molten diopside and andesine (Stebbins et al., 1983; Richet and Bottinga, 1984a; 1984b). These difficulties are most readily attributed to small amounts of crystallization, and the possibility of nonreproducable glass reference states (Stebbins et al., 1983). The limited temperature range for operation of most calorimeters is another problem that should be somewhat alleviated by the introduction of instrumentation able to measure heat contents up to 2600 K (Stout and Piwinskii, 1982).

Volumetric properties of liquids. Bottinga and Weill (1970) first systematized the fairly extensive set of silicate melt density measurements at 1 atmosphere pressure. They calibrated composition-independent partial molar volumes and constant thermal expansion coefficients for 9 oxides valid between 40 to 80 mol % SiO_2 and 1200 to 1600°C. On the basis of new data, their model was later extended to include TiO_2 (Nelson and Carmichael, 1979) and Fe_2O_3 (Mo et al., 1982), and recalibrated by Stebbins et al. (1984). With these models the volumes of silicate melts can be predicted to within the estimated 1-2% accuracy of most data.

Bottinga et al. (1982; 1983) compiled a much larger data set and proposed a more complex model in which the partial molar volume of Al_2O_3 is a linear function of composition for melts with 35 to 85 mol % SiO_2. Ghiorso and Carmichael (1984) have questioned the validity of the Bottinga et al. model, and Lange and Carmichael (1986) report new measurements on multicomponent melts that support their arguments for a composition-independent partial molar volume for Al_2O_3 and yet also suggest excess volume of mixing between CaO-SiO_2 and CaO-TiO_2 components. Although compositionally dependent partial molar volumes are also clearly evident in other systems (e.g. Li_2O-SiO_2) and from the difference between the computed partial molar volume and the volume of pure SiO_2 liquid (Bottinga et al., 1983), further experimental work is necessary to resolve the nature and magnitude of this dependence for other compositional ranges, as well as to define thermal expansivities which have experimental uncertainties of up to 40% (Bottinga et al., 1983). From a purely empirical perspective, the Bottinga et al. (1982) model leads to markedly smaller residuals in reproducing available density measurements in synthetic systems than alternate models, and their coefficients for calculation of melt volumes are reproduced in Table 4. An alternate model based both on more recent and critically evaluated data is given by Lange and Carmichael (1987).

Relatively few data have been gathered regarding the compressibility of silicate melts. One atmosphere compressibilities have been calculated from ultrasonic measurements on a number of synthetic and natural compositions. Based on these data, an ideal mixing model for prediction of one atmosphere compressibilities is presented by Rivers and Carmichael (1987), and updated by Lange and Carmichael (1987).

Fujii and Kushiro (1977) developed a technique for high pressure density measurements utilizing falling spheres of different densities, and used it to obtain data for an olivine tholeiite. Scarfe et al. (1979) used the same technique to measure densities up to 15 kbar of diopside and sodium silicate melts. Bottinga (1985) extracted compressibilities for molten forsterite, enstatite, diopside, pyrope, jadeite, and albite from the pressure dependence of their experimentally determined fusion curves and auxiliary thermodynamic data. The curvature observed or inferred in the melting curves, particularly of aluminosilicate melts, clearly demonstrates the importance of the pressure dependence of the compressibility of silicate melts. Shock wave measurements up to 230 kbar of Rigden et al. (1984; 1985) on the compositions $An_{36}Di_{64}$ and Di_{100} are the first experiments to directly address this important problem.

Enthalpy and entropy of formation and fusion. The enthalpy of melt components or species are a necessary input to any thermodynamic model, and the compositions for which these data are available place important constraints on the choice of components and species for thermodynamic models. In contrast to minerals for which enthalpies of formation usually are tabulated at 298.15 K, the enthalpy of a melt is linked to that of a mineral by determination of the enthalpy difference ($\Delta_{fs}H$) between the melt and crystal at the fusion temperature (T_{fs}). From this difference, the entropy of fusion can also be calculated from $\Delta_{fs}S = \Delta_{fs}H/T_{fs}$, since the change in free energy of the reaction is zero ($\Delta_{fs}G = 0$) at the fusion temperature. For models which use a reference state for melts of 298.15 K, enthalpies and entropies of formation can be computed from the corresponding fusion properties using the (extrapolated) melt Cp and the relations:

$$H^{298} = H^{T_{fs}} - \int_{298}^{T_{fs}} Cp dT \quad , \tag{23}$$

$$S^{298} = S^{T_{fs}} - \int_{298}^{T_{fs}} \frac{Cp}{T} dT \quad . \tag{24}$$

For very few minerals can $\Delta_{fs}H$ be determined directly because the melting points of most silicate compositions are above the operating temperature of solution calorimeters. Instead $\Delta_{fs}H$ can be derived through calculations involving the glass phase from (see Fig. 1):

$$\Delta_{fs}H^{T_{fs}} = \Delta_{vit}H^{T_s} + \int_{T_s}^{T_g} (Cp_g - Cp_{cr}) + \int_{T_g}^{T_{fs}} (Cp_\ell - Cp_{cr}) \quad , \tag{25}$$

where T_s is the temperature at which the enthalpies of solution have been measured for the crystal (cr) and glass, and T_g is the glass transition temperature. At present there are still only a small number of minerals for which enthalpies of fusion have been determined by calorimetric means. These values are compiled in the review of Richet and Bottinga (1986). As noted above, heats of vitrification ($\Delta_{vit}H$) can be very different from heats of fusion, and Richet and Bottinga (1986) use the former along with Cp_g and Cp_ℓ values from their predictive equations (Richet and Bottinga, 1982; 1985) to compute additional heats of fusion.

Heats of fusion can also be derived from experimentally determined phase relationships, and many authors have for instance computed values from melting curves using the Clausius-Clapeyron equation:

$$\left(\frac{dP}{dT}\right)_{\Delta_r G=0} = \frac{\Delta_{fs}H}{T\Delta_{fs}V} \quad . \tag{26}$$

Errors in the calculated heats of fusion can be drastic (Bottinga and Richet, 1978; Boettcher et al., 1982; Richet and Bottinga, 1985) and reflect the quality of the experimental constraints on the initial fusion curve slopes and the calculated volume changes on melting. Similarly, $\Delta_{fs}H$ values can be extracted from T-X diagrams if the activities of melt components can be computed. This technique has been used for example to retrieve heats of fusion for lime and periclase (Chang and Howald, 1982), two minerals with melting points beyond the range of accessibility to direct calorimetric measurement. Ghiorso et al. (1983) and Berman and Brown (1984) also show that enthalpies of fusion can be estimated for many minerals by calibration of stoichiometric models in multicomponent systems.

Mixing properties of liquids. Mixing relationships in silicate melts can be determined by a variety of techniques, principally including electrochemical measurements, phase equilibration studies, vapor pressure determinations, and calorimetric studies. Activities of various components in silicate melts with relevance to the metallurgical industry have been determined for a number of binary to quaternary systems, and Spencer (1973) provides a good summary of these data. In general, there is only fair agreement in the results between these studies. More recently, activities have been measured by Knudsen cell mass spectrometry for a number of systems including K_2O-SiO_2 (Eliezer et al., 1978), CaO-Al_2O_3 (Allibert et al., 1981), $KAlSi_3O_8$-$NaAlSi_3O_8$ (Rammensee and Fraser, 1982; Rogez et al., 1983), $KAlSi_2O_6$-$NaAlSi_2O_6$ (Fraser et al., 1983), and (K-Na)$AlSi_4O_{10}$ and (K-Na)$AlSi_5O_{12}$ (Fraser et al., 1985).

Julsrud and Kleppa (1986) modified a Calvet-type calorimeter to operate at temperatures up to about 1100°C which is in the temperature range to measure heats of mixing directly in some multicomponent systems. For many systems of geologic interest, however, liquidus temperatures are too high for direct measurements on molten samples. Consequently a large amount of heat of mixing data on geologically interesting compositions has been obtained recently by solution calorimetry of glasses (Navrotsky et al., 1980; 1983; 1985; Hervig and Navrotsky, 1984; Hervig et al., 1985). This technique has widespread applicability, but for derivation of mixing properties of melts one needs to know Cp_g, Cp_ℓ, and the glass transition temperature for each composition that has been calorimetrically measured. The nonlinear and largely unknown variation of glass transition temperatures with composition presents the most difficulty, as discussed in detail by Richet and Bottinga (1986). The importance of these points is well illustrated by the difference between heats of mixing in glasses which show negative excess enthalpies (Hervig and Navrotsky, 1984) and melts which show near-ideal mixing (Rammensee and Fraser, 1982; Rogez et al., 1983) on the $KAlSi_3O_8$-$NaAlSi_3O_8$ join. Fraser and Bottinga (1985) have shown however that, at least for one composition where the requisite data are available to make the conversion, the glass data are entirely consistent with the melt data.

An innovative but very time-intensive technique for determining heats of mixing in geologically relevant melts has been reported by Eliášová et al. (1978). In their study, heat contents of melts across the join $Ca_2MgSi_2O_7$-$CaMgSi_2O_6$ were measured by drop calorimetry. These results were then combined with heats of solution measured on each of the quenched samples to yield the enthalpy of mixing of melts across this join.

By far the most abundant source of data constraining mixing properties of silicate melts is phase equilibrium data which have been studied in synthetic systems since the beginning of the twentieth century. Although very accurate in themselves, many assumptions and much auxiliary data are needed to extract mixing properties from these data. The overall strategy can be appreciated by again considering the example of forsterite-melt equilibrium for which we will assume we have experimental data indicating liquidus compositions and temperatures. Inspection of Equation (18) shows that excess solution properties (expressed through the Margules parameters) are the only unknowns if the standard state properties of the mineral and all melt components are known. Due to the importance of this type of data, the methodology for extracting all thermodynamic parameters from phase equilibrium data is discussed in detail below. The disadvantage of using phase equilibrium data alone is that the enthalpy and entropy contributions to the free energy of mixing cannot be determined separately . In contrast, if heat of mixing measurements are combined with phase equilibrium data, various entropy of mixing models can be tested, as shown by Weill et al. (1980) and Hon et al. (1981).

Thermodynamic properties of minerals

Thermodynamic properties of minerals ($\Delta_f H^{Tr,Pr}, S^{Tr,Pr}, Cp, V$) can be determined by direct calorimetric or physical measurements which yield 'best' values and associated standard deviations for thermodynamic parameters. Various compilations of such data are available, the most widely used for geologic materials being that of Robie et al. (1979). However, stability relationships computed with such tabulated thermodynamic properties often do not coincide with direct phase equilibrium measurements, and it has been pointed out repeatedly (e.g. Zen, 1972; Gordon, 1973; Haas and Fisher, 1976) that this process can be inverted by using phase equilibrium data to derive and/or refine some or all thermodynamic parameters of phases that participate in these equilibria. Considerable disagreement remains on how best to accomplish this refinement but a discussion of this point is deferred until the 'Methodology' section below.

What is clear from analyses of large bodies of data (Helgeson et al., 1978; Robinson et al., 1982; Berman et al., 1985; Holland and Powell, 1985; Berman, 1987) and is worth emphasizing here is that it is the exception for enthalpies of formation refined with phase equilibrium data to agree with nominal calorimetric values, and many refined values fall outside of the uncertainty brackets of calorimetric measurements. Particularly large discrepancies occur with many data determined by low temperature HF solution calorimetry (Fig. 3). Results of high temperature dissolu-

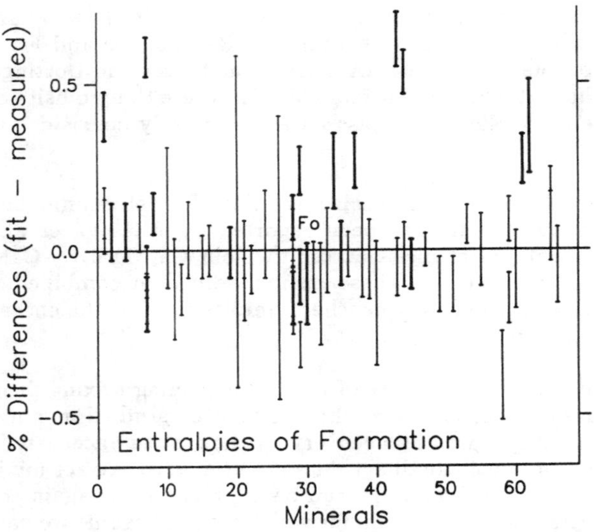

Figure 3. Comparison of calorimetric enthalpies of formation of selected minerals and associated 2σ uncertainties with values refined by MAP analysis of phase equilibrium data (Berman, 1987). Note the large number of inconsistencies with calorimetric results obtained by low temperature HF dissolutions (thick brackets), as well as the amount of scatter among high temperature measurements (thin brackets) on the same mineral (e.g. forsterite).

tions in general yield much better agreement with phase equilibrium studies, with the magnitude of discrepancies being of the same order as those between various calorimetric measurements for the same mineral (e.g. forsterite in Fig. 3). These comparisons suggest that the uncertainties reported for calorimetrically determined enthalpies of formation are somewhat underestimated probably due to small systematic errors related to sample purity.

The above observations dictate that, for the purposes of calibrating melt models, thermodynamic properties of minerals should only be taken from compilations of refined data. The importance of this point cannot be overemphasized because, if liquidus data are used to calibrate a model for melts, any inconsistencies among high temperature mineral properties will be translated directly into the derived melt properties. It is for this reason that some authors have chosen to adjust and refine mineral properties at the same time as deriving melt parameters (e.g. Bale and Pelton, 1983; Ghiorso et al., 1983; Berman and Brown, 1984).

Available compilations of refined thermodynamic data (Helgeson et al., 1978; Robinson et al., 1982; Holland and Powell, 1985; Berman et al., 1985; Berman, 1987) differ in some important regards, notwithstanding the actual minerals in each compilation and the different techniques used to derive their properties. In most of these studies, only enthalpies of formation were optimized, other properties being adopted from directly measured values. The Robinson et al. and Berman databases refined all standard state properties including *Cp* and volume parameters, leading to better overall agreement among all types of experimental data. Other important differences between these data sets are the various equations of state used, and of particular interest in the context of formulating melt models are the type of *Cp* and volume equations used, and the treatment of order-disorder phenomena.

The heat capacity of many minerals has been determined directly by differential scanning calorimetry or by differentiation of drop calorimetry data. For many important minerals liquidus temperatures in synthetic systems are considerably higher than the temperature of calorimetric data, and some technique must be used ensure reasonable high temperature extrapolation of Cp functions. Various methods have been proposed for this purpose (Holland, 1981; Fei and Saxena, 1986), but perhaps the most general is to use the empirical function suggested by Berman and Brown (1985):

$$Cp = k_0 + k_1 T^{-0.5} + k_2 T^{-2} + k_3 T^{-3} (k_1, k_2 < 0) \quad . \tag{27}$$

With the k_1 and k_2 parameters constrained to be negative, this function approaches a constant high temperature limit (k_0) which in practice is in reasonable accord with the Petit and Dulong limit for C_V (converted to Cp). In contrast, fitting Cp data with the Maier-Kelley (1932) Cp function leads to severe overestimation of Cp at high temperature and therefore large errors in computed high temperature enthalpies, entropies, or Gibbs free energies.

Third law entropies at 298.15 K are determined from low temperature heat capacity measurements by integrating Cp/T between 0 and 298.15 K. Any disorder which is 'frozen' into the mineral at low temperature or which occurs at higher temperature must therefore be added to the calorimetrically determined S^{T_r,P_r}. In the simplest treatment, a constant entropy increment can be calculated from the ideal configurational effect of disordering cations on equivalent sites (Ulbrich and Waldbaum, 1976) or from analysis of detailed structural refinements (e.g. Westrich and Holloway, 1981). More sophistocated treatments take account of the temperature dependence of the disordering process. Thus, using Landau theory, Salje (1985) and Salje et al. (1985) have derived functions for representation of the thermodynamic consequences of temperature dependent disordering in albite. Where requisite data are not available to apply this model, it is convenient to define an empirical function to describe the temperature dependence of the disordering process (Helgeson et al., 1978; Berman et al., 1985; Berman, 1987) and calibrate it using data on the enthalpy differences between ordered and disordered phases. Both types of formulation produce very different effects from the maximum configurational entropy assumption, and permit much better agreement to be achieved with phase equilibrium data (Berman et al., 1985; Berman, 1987). Given the importance of high temperature mineral properties in modelling magmatic processes, more work is needed to extend these treatments to additional common minerals of geologic interest (e.g. spinel, anorthite).

Although at low pressures and temperatures it may be reasonable to ignore expansivity and compressibility terms for minerals, this simplifying assumption is not valid under magmatic conditions especially because of the contrasting properties of minerals and melts. Measurements of isobaric expansivities (Skinner, 1966) and isothermal compressibilities (Birch, 1966; Sumino and Anderson, 1982) have been made for many common minerals, but very few data have been obtained that show the pressure and temperature dependence, respectively, of these properties. The general paucity of such data clearly points to the need for improved equations of state for minerals that will allow prediction of volumes at elevated P and T from a limited number of measurements. Work aimed at this end is currently in progress (T. H. Brown, in preparation).

METHODOLOGY

Calibration of thermodynamic models

At the present time phase equilibrium experiments represent the most abundant source of data constraining the thermodynamic properties of geologically relevant melts. In the case of liquidus experiments, knowledge is gained regarding the relative stabilities of minerals and melt. Solid-solid and solid-vapor phase equilibrium data also serve to refine thermodynamic properties of minerals to which the melt properties are linked. In this section we review the fundamental concepts of fitting these and other types of data discussed above, and present a preferred methodology for optimization of thermodynamic parameters in general. The reason we emphasize this aspect of the problem is because successful thermodynamic modelling depends as much on the calibration of model parameters as on formulation of the model itself.

The most important point to recognize in the context of calibrating thermodynamic models is that phase equilibrium data differ from other types of measurements of thermodynamic properties in so far as they do not yield 'best' values with well characterized uncertainties. A successful phase equilibrium experiment shows growth (stability) of one assemblage with respect to another chemically equivalent assemblage, and thus is indicative of the sign of the change in free energy of the reaction. No information is conveyed on where, within a reversed bracket, the equilibrium might lie, and no 'central limit' condition exists to favor the midpoint of a bracket. This 'non-central' characteristic of phase equilibrium data was first recognized by Gordon (1973), who introduced the technique of linear programming (LIP) to the literature of experimental petrology. LIP is a mathematical method for solving systems of linear inequalities which, in the case of phase equilibrium data, express the sign of the free energy change of an experimentally observed reaction. The first step of the solution determines the existence of a 'feasible region' that is consistent with all inequality constraints representing the P-T-X positions of each experimental half-bracket (adjusted for uncertainties). The final step involves minimization of an objective function in order to optimize some linear combination of variables. Following Gordon's lead, the LIP technique has been applied in several studies aimed at deriving thermodynamic properties for minerals (Day and Halbach, 1979; Day and Kumin, 1980; Halbach and Chatterjee, 1982; 1984; Day et al., 1985) and melts (Berman and Brown, 1984).

In contrast to the linear programming method, others have used various types of regression analysis, in which the statistical 'central tendency' was controlled by various schemes of data weighting (Haas et al., 1981; Lindsley et al., 1981). A major problem with regression analysis is that the unique solution produced can be extremely sensitive to small changes in input data. A good example of this sensitivity is the large variation in the values of interaction parameters produced by very small changes in the estimated position of a solvus (see for example the discussion of Tewhey and Hess, 1979). In recognition of this problem, Lindsley et al. (1981) introduced a calibration scheme whereby the experimental data were iteratively adjusted (within the experimental uncertainties) between each regression cycle in order to achieve results that would best represent all data. The reader should recognize the similarity in intent between this scheme and the LIP technique.

Table 5 sets out the major differences between these two mathematical techniques for optimization of thermodynamic parameters. In an effort to incorporate the advantages of both techniques, Berman et al. (1986) proposed using the method of mathematical programming (MAP) which treats phase equilibrium data as linear constraints in $\Delta_r G$ in exactly the same way as LIP, but utilizes a nonlinear 'least-squares' objective function to produce a unique solution that is weighted towards the most precise of the direct measurements of phase properties. This objective function can be formulated as

$$F = \sum_i (U_i - Y_i)^2 / s_i^2 \quad , \tag{28}$$

where the U_i represent fit parameters or some function of the fit parameters, Y_i are experimental values, the s_i are their associated standard deviations, and the summation is taken over all values that have been determined directly and are assumed to have well characterized, normal probability distributions.

The MAP technique for optimization of thermodynamic parameters of minerals has been discussed in detail by Berman et al. (1986). Mathematical programming methods, which include a variety of optimization techniques of which LIP is one, will not be discussed here as numerous texts cover this topic (e.g. Himmelblau, 1972; Gill et al., 1981). Instead we will set up the equations used to analyze phase equilibrium data, and show how to incorporate into the analysis the different types of experimental data constraining thermodynamic parameters of minerals and melts.

In a successful phase equilibrium experiment, liquidus or otherwise, one assemblage is observed to grow at the expense of another chemically equivalent assemblage. The basic thermodynamic information gained from such an experiment may be cast into an inequality in the Gibbs free energy of reaction:

$$\Delta_r G^{P,T} \lessgtr 0 \quad , \tag{29}$$

where the 'greater than' and 'less than' signs indicate whether the reactant or product assemblage, respectively, is stable, and

$$\Delta_r G^{P,T} = \sum_i^{np} \nu_i \mu_i \quad , \tag{30}$$

where np is the number of chemical entities in the reaction. Expanding (29) in terms of Equations (3) and (9) gives:

$$\sum_i^{np} \nu_i [\Delta_f H_i^{Tr,Pr} - T \cdot S_i^{Tr,Pr} + \int_{T_r}^{T} Cp_i dT - T \int_{T_r}^{T} \frac{Cp_i}{T} dT + \int_{P_r}^{P} V_i dP + RT \ln x_i] + \sum_i^{ns} \nu_i W_i^! x_i^! \lessgtr 0 \quad , \tag{31}$$

where we have substituted $W_i^! x_i^!$ for the $RT \ln \gamma_i$ terms, with the $W_i^!$ and $x_i^!$ representing the collection of both the Margules parameters and their respective coefficients (see Eqn. 18) for each of the ns solution phases. Every phase equilibrium experiment provides an inequality which, when expressed through Relation (31), involves a rather large number of parameters, consisting of the experimental variables (P, T, X) and the thermodynamic properties of all phases or phase components/species. LIP or MAP can be used to solve a system of such inequalities for all thermodynamic parameters (both standard state and mixing properties) that are consistent with all phase equilibrium brackets. In practice, it is convenient and less demanding computationally to fix some of these parameters that are thought to be known with sufficient confidence (e.g. all gas properties, Cp, expansivity, compressibility parameters) by regression analysis prior to the MAP analysis. In this case these terms can be evaluated at the temperature and pressure of each experiment and grouped together as a constant, C, that, along with the $RT \ln x_i$ terms, modifies the right hand side (RHS) of the inequality Relation (31):

$$\sum_i^{np} \nu_i[\Delta_f H_i^{T_r,P_r} - T \cdot S_i^{T_r,P_r} + (P-1)V_i^{T_r P_r}] + \sum_i^{ns} \nu_i W_i^! x_i^! \lessgtr C \quad . \qquad (32)$$

The nominal conditions reported for an experiment may not coincide with the actual conditions because of unavoidable uncertainties in control and measurement of run conditions. In order to ensure certainty that each equilibrium lies between the inequality conditions used in the data analysis, it is essential to use in Relations (31) and (32) not the nominal experimental values of P, T, and X, but adjusted values that account for the experimental uncertainties in defining the conditions of the experiment. In general these adjustments are made so that the resulting P, T, X values form the widest possible bracket width (Fig. 4). A detailed discussion of this point in relation to specific data types is given by Berman et al. (1986) who stress that these uncertainties should incorporate careful assessments of both experimental precision (e.g. P, T fluctuations, analytical reproducibility) and accuracy (reflecting calibration procedures, temperature gradients, etc.).

Relation (32) remains linear in all thermodynamic parameters because the coefficients for the individual Margules parameters are combinations of mole fraction terms that are known from each experimental result (Eqn. 18 shows this most clearly). Berman and Brown (1984) exploited this linearity by using LIP to calibrate their solution model for silicate melts. For other functional forms chosen to describe the excess free energies of solutions, the constraint relations may become nonlinear in the W parameters. For example, the quasi-chemical model (Guggenheim, 1952) leads to constraints with nonlinear terms because the interaction parameters cannot be factored so as to be solely functions of compositional variables. In such cases, solution of the problem would require MAP rather than LIP techniques to analyse these nonlinear constraints. The reader should note that the mole fraction terms in Relation (32) are not explicitly known if they apply to species which are greater in number than the thermodynamic components. In this case, an iterative method is necessary which couples a distribution of species calculation to the MAP analysis.

The power of the approach outlined above is that solution properties as well as standard state properties can be efficiently derived that are consistent within the uncertainties of all phase equilibrium data. In this context, this methodology offers a large improvement over techniques of trial and error manual fitting (Saxena and Chatterjee, 1986) or manual adjustment of input data combined with regression analysis (Lindsley et al., 1981). Several other advantages are also noteworthy. As pointed out by Bottinga et al. (1981), adequate calibration of a melt model must also satisfy the conditions that the melt is stable with respect to (a) all non-liquidus minerals, and (b) phase separation in P-T-X regions for which there is no evidence of immiscibility. With MAP/LIP these constraints can be placed directly on the solution by formulating additional inequality relations that express (a) $\Delta_r G < 0$ for all nonliquidus minerals, and (b) $(\frac{\partial^2 G}{\partial x^2}) > 0$ in P-T-X regions without phase

Table 5: Comparison of LIP/MAP and regression techniques for the analysis of phase equilibrium data.

Linear programming analysis	Regression analysis
Treats phase equilibrium data as statements of inequalities in $\Delta_r G$	Treats weighted midpoints of brackets as positions where $\Delta_r G = 0$
Analyses individual half- brackets -provides for constaints from unreversed experiments -every half-bracket can be analysed with different assumptions (eg. solid solution effects, different starting materials)	Analyses pairs of experimental half-brackets
Ensures consistency with all data (if consistency is possible)	Minimizes sum of squares of residuals, but does not ensure consistancy with all data
Provides a range of solutions (feasible region) from which a unique solution is obtained with a suitable objective function	Provides a 'unique' solution (dependent on weighting factors)
Uncertainties approximated by the range of values consistent with all data	Uncertainties computed from variance/covariance matrix

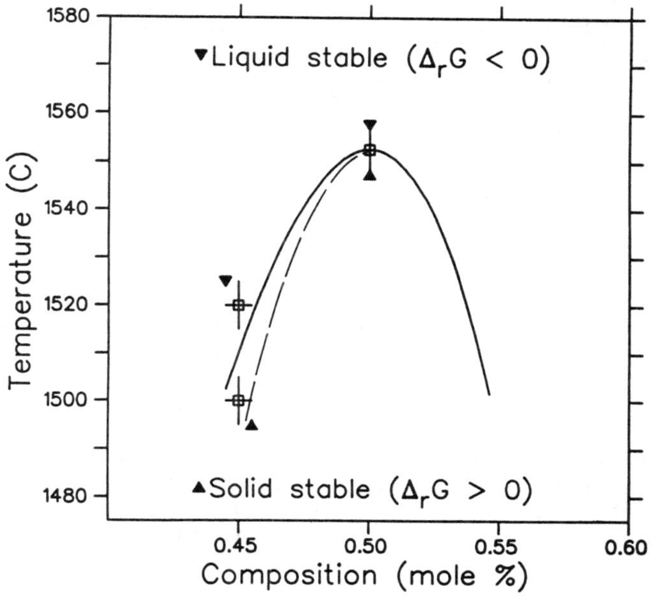

Figure 4. Method used to derive the temperature and composition of data points for LIP/MAP analysis (triangles) from nominal experimental half-brackets (squares) with uncertainties shown by error bars. Note that data adjustment to account for experimental uncertainties must take into account the slope of the equilibrium curve. Note also that, for nonstoichiometric liquidus minerals, the low temperature half-bracket must be analyzed in conjunction with a solution model for the mineral, whereas the high temperature half-bracket is a valid constraint applied to the stoichiometric mineral. This is true because the reduced activity of the nonstoichiometric mineral must displace its liquidus (solid curve) to higher temperature than the liquidus for the stoichiometric mineral (dashed curve).

separation. Lastly, as discussed in detail by Berman and Brown (1984), phase equilibrium data in which the composition of a liquidus mineral is unknown can also be used to some advantage in the calibration by imposing only the 'liquid stable' constraint relation (Fig. 4).

In the above discussion we have placed the most emphasis on the analysis of liquidus data because this is the most abundant source of data for calibration of geologically relevant models. Immiscibility data can also be analyzed by the MAP/LIP method, and details of the constraint formulations are given by Berman and Brown (1984). Analysis of other diverse types of experimental data can be incorporated in a variety of ways: (a) by adding additional constraints among thermodynamic parameters to problem, (b) by placing 'bounds' on individual parameters, or (c) by optimization of these properties through the objective function discussed above. In relation to mineral properties, these techniques are dealt with in detail by Berman et al. (1986) who suggest that bounds reflecting two sigma uncertainties be placed on S^{T_r,P_r} and V^{T_r,P_r}. These values can then be optimized by including them in the objective function which produces the smallest 'least-squares' residuals with the measured values. When it is necessary to estimate or refine *Cp*, expansivity, or compressibility terms, these parameters remain as variables in the problem which are optimized with this objective function.

Thermodyamic properties of melts can be handled in the same way. The experimental data and predictive models summarized in the previous section indicate that heat capacities and volumes have been measured or can be predicted with reasonable confidence for most melt compositions. These data could thus be treated as constants (moved to the RHS), or two sigma bounds could be placed on these values. Bounds could also be put on enthalpies and entropies of fusion or formation, but for many compositions these values will have to remain as fit parameters since they have not been determined directly. When activity data (a_i^{obs}) are available, relations can be added like

$$RT \ln x_i + W_i^! x_i^! < RT \ln a_i^{obs} + a_e \quad (33)$$

$$RT \ln x_i + W_i^! x_i^! > RT \ln a_i^{obs} - a_e \quad (34)$$

to place constraints among the mixing parameters, $W_i^!$. If the uncertainties of these data, a_e, are not well known or if systematic errors are suspected, the data are better utilized as part of the objective function which is minimized:

$$F = \sum [RT \ln a_i^{obs} - (RT \ln x_i + W_i^! x_i^!)]^2 \quad . \quad (35)$$

In this case consistency is not forced with these data by the constraints (33-34), but agreement with them is optimized while ensuring consistency with all other constraint relations (usually phase equilibrium data). Relations similar in form to (33-35) can be written to account for enthalpy of mixing data of melts.

Testing of calibrations

One of the biggest advantages in working with synthetic systems is that model calibrations can be tested for numerical stability by computation and visual inspection of phase diagrams that completely span the composition space of interest. It is convenient therefore to automate these calculations for the purpose of comparison with phase equilibrium data and so that the full implication of model calculations can be quickly appreciated.

Partly with these objectives in mind, we have developed several computer programs which calculate stable phase diagrams in entirety. After the user selects the *P-T-X* range of the diagram and the chemical system, the program reads a thermodynamic database, and calculates all possible reactions among the phases in the selected system. Subsequently, each reaction is followed in turn with all points on the reaction curve tested for stability with respect to all other phases in the system. Finally, metastable extensions are removed and reaction curves with stable assemblages are written to a plot file which is viewed with a separate program. Software for calculation of P-T-X_{CO_2} diagrams among minerals and a vapor phase was developed by Perkins et al. (1986) for use on mainframe facilities, while Brown et al. (1987) have made this software available for microcomputer use. Similarly, programs originally developed for mainframe computation of liquidus diagrams have been converted to run on microcomputers (Theoretical Geochemistry Software Library, Dept. of Geological Sciences, University of British Columbia). The latter programs calculate (a) stable binary liquidus diagrams, (b) stable cotectics in ternary and quaternary sections, and (c) solvus configurations in binary, ternary, or pseudoternary sections. Examples of the graphical output of this software are shown below in relation to the discussion of the Berman and Brown (1984) model.

APPLICATIONS

The number of thermodynamic models proposed for silicate melts is too large to be within the scope of this review. Instead we sketch below the results of some of the models that appear to be the most encouraging in so far as their abilities to (a) represent data adequately, and (b) be extended to multicomponent melts reaching the complexity of magmatic systems. Conspicuously absent from this sketch are most 'structure-based' models which encounter difficulties especially with the latter criterion. It should be appreciated however that these models provide much insight into understanding the thermodynamics of melts on a fundamental level, and the reader is referred to the reviews of such models by Gaskell (1977) and Bottinga et al. (1981).

Speciation models

In their first review of thermodynamic treatments of silicate melts, Bottinga and Richet (1978) setup the framework of a speciation model in which neutral melt species were assumed on the basis of the stoichiometry of the endmember liquidus minerals in any given system. They define the chemical potential of species k by

$$\mu_k = \mu_k^\circ - T\Delta_m\overline{S}_k + \Delta_m\overline{H}_k \tag{36}$$

and use the Flory-Huggins and van Laar equations to represent the partial molar entropy of mixing ($\Delta_m\overline{S}_k$) and enthalpy of mixing ($\Delta_m\overline{H}_k$), respectively:

$$\Delta_m\overline{S}_k = -R[\ln\phi_k + \sum_j^{nsp}\phi_j(1-\frac{V_k}{V_j})] \quad , \tag{37}$$

$$\Delta_m\overline{H}_k = V_k\sum_i^{nsp}\sum_j^{nsp}(A_{ik}-\frac{A_{ij}}{2})\phi_i\phi_j \quad , \tag{38}$$

with $A_{ik} = A_{ki}$ and $A_{ii} = 0$. The volume fractions (ϕ_k) are calculated by

$$\phi_k = \frac{n_k V_k}{\sum_j^{nsp} n_j V_j} \quad , \tag{39}$$

which reduces Equation (37) to a random mixing equation if all molar volumes are identical. Equation (38) is a regular solution formalism with temperature-, composition-independent mixing parameters that yield symmetric excess enthalpies of mixing.

In applications to binary and ternary systems at 1 atmosphere pressure, their model reduced to a stoichiometric approach because the number of species considered in each system was equal to the number of components. They found that 32 binary systems (most with combined oxide endmembers) could be modelled well, with a mean deviation between calculated and experimental liquidus compositions of 2.2 wt %. Another 21 binary systems could not be reproduced in entirety, with the discrepancies being attributed primarily to complications introduced by solid solution in the liquidus minerals and to the non-binary nature of some of the systems. Lastly Bottinga and Richet (1978) showed that the derived binary interaction coefficients could be used for reasonable predictions of the portions of four ternary systems that were bounded by well calibrated binary systems.

In the systems that were reasonably well modelled by Bottinga and Richet (1978), the average deviations are somewhat greater than experimental uncertainties (approximately 1 wt % and 5-10°C for reasonably careful studies) and more than would be acceptable for representation and systemization of the existing large body of liquidus data. Because much of the thermodynamic data for melt species, particularly enthalpies of fusion, had to be estimated, it is likely that use of more recent and accurate $\Delta_{fs}H$ and Cp_{ℓ} data would improve the results somewhat. Nevertheless, it should be noted that there are now experimental data (see discussion of excess Cp above) which indicate that for some compositions, heats of mixing can be expected to be temperature dependent. Turning to the systems in which their model was not successful and with this reservation aside, a major question is whether the difficulties they encountered can be alleviated by inclusion of additional melt species, or whether a more complex activity model (asymmetric or temperature dependent) is needed.

Windom and Wood (1986) addressed this question recently in their attempt to use the same model to fit the $CaAl_2Si_2O_8$-SiO_2 system which could not be reproduced by Bottinga and Richet (1978). In this study, all species were placed on an equal oxygen basis as suggested by Burnham (1975), and the reduced activities of anorthite and cristobalite in this system (Longhi and Hays, 1979) were incorporated. In order to rectify the problem that a single interaction coefficient was unable to reproduce the liquidus curves of both endmembers, Windom and Wood considered an additional melt species, defined through the speciation reaction:

$$CaAl_2Si_2O_8 = Ca_{4/3}Al_{8/3}Si_{4/3}O_8 \text{ (Ca-Tschermak's molecule)} + Si_4O_8 \quad .$$

All interactions involving the $Ca_{4/3}Al_{8/3}Si_{4/3}O_8$ species were assumed to be ideal. As a result of this additional species, the experimental liquidus data could be reproduced to within 10°C, and the equilibrium constant for the speciation reaction was calculated. Spectroscopic and viscosity data summarized by Windom and Wood (unpublished manuscript, 1987) support the proposed speciation model. Difficulty was encountered in extending the model to the ternary system $CaAl_2Si_2O_8$-SiO_2-$CaMgSi_2O_6$ because the SiO_2 liquidus could not be reproduced in the SiO_2-$CaMgSi_2O_6$ binary using a single interaction coefficient. These results suggest that the Bottinga and Richet (1978) model will need to be expanded to include either (a) species that do not have the stoichiometry of liquidus minerals in the given system (such as $CaSiO_3$ or $MgSiO_3$ which have stability fields adjoining those of diopside and tridymite in the CaO-MgO-SiO_2 system), or (b) a more complex (e.g. asymmetric, temperature dependent) heat of mixing equation in order to account for highly nonideal mixing properties in silica-rich portions of systems like this that show liquid immiscibility.

The approach taken by Wood and Windom is similar to the latest version of the Burnham model for granitic systems (Burnham, 1981; Burnham and Nekvasil, 1986). A complete review of this model is beyond the scope of this chapter, and the reader is referred to recent reviews of Bottinga et al. (1981) and Navrotsky (1986). Here we will only highlight the newer points of this model. The crux of Burnham's quasi-crystalline model is that melt species are identified on the basis of the stoichiometry of minerals that appear on the liquidus. All species are normalized to an 8 oxygen basis so as to increase the liklihood of ideal mixing. Based on observations of the equivalent solubilities (expressed as 'albite mole equivalents') of H_2O in molten albite, orthoclase, anorthite, and several natural melt compositions, Burnham (1975) showed that the activity of H_2O was dependent only on its composition, and therefore that the feldspar components of these melts mix ideally (Burnham et al., 1978). Analysis of experimentally determined fusion curves and binary eutectic compositions led Burnham (1981) to extend the ideal mixing model to include the Si_4O_8 component. Differences between eutectic compositions and computed activities of these components $(a_i - x_i)$ are accounted for by postulating the presence of additional melt species rather than assuming (as with a stoichiometric approach) that component activity coefficients are less than unity. The observed $(a_i - x_i)$ differences are fit to empirical equations that are functions of pressure and composition. These equations can be used to compute the activity of any of the four components at any P, T, and X.

The results reported by Burnham and Nekvasil (1986) show good agreement between experimentally determined phase relations in synthetic systems and those computed with this model. Moreover, comparisons between predicted and observed phase relations for natural pegmatite compositions are particularly impressive, and clearly show that the model can be usefully applied to geologic problems. It should be noted as well that, with the exception of the model of Ghiorso et al. (1983), the Burnham model is the only model for multicomponent melts that accounts for pressure and variable H_2O contents. These significant advances are counterbalanced by the loss of generality in the present formulation of empirical $(a_i - x_i)$ relations which do not in general quantify the extent and nature of the hypothesized speciation reactions and through which it is not possible to derive internally consistent thermodynamic properties for the hypothesized species. This would seem to be an important point to pursue in the light of the results reported by Windom and Wood (1986) and discussed above in which phase relations in the system $CaAl_2Si_2O_8$-Si_4O_8 could not be reproduced without nonideal species interactions.

Few complete speciation models have been reported in which thermodynamic properties of all species have been derived explicitly, thereby allowing the free energy of a melt to be computed for any composition. Several examples will be discussed here which illustrate some of the freedom in defining speciation models. Hastie et al. (1982) present a model similar to Burnham's in which melt complexes, most with the stoichiometry of liquidus minerals, mix ideally. The thermodynamic functions for minerals and melt species were adopted from compilations which rely heavily on calorimetric data, occassionally adjusted within the uncertainties of these data, or estimated where experimental data were lacking. Hastie and Bonnell (1985) summarize some of these data for species in the system K_2O-Na_2O-CaO-MgO-Al_2O_3-SiO_2

and use a free energy minimization algorithm (Eriksson, 1975) to show that computed activities of melt components compare favorably with experimental values in various portions of this multicomponent system. Although the present calibration of their model does not reproduce phase equilibria accurately (Hastie and Bonnell, 1985), the predictive merits of the model are well demonstrated by the ability to compute the free energy of the melt in this six component system. Further calibrations utilizing phase equilibrium data are needed in order to assess the validity of the present model assumptions regarding the assumed stoichiometry of the melt complexes and the ideal interactions among these complexes.

Bottinga et al. (1981) used the same model as proposed by Bottinga and Richet (1978), but performed a complete speciation calculation for melts in the system MgO-SiO_2. The melt species assumed in this study were those corresponding to liquidus phases: MgO, Mg_2SiO_4, $Mg_2Si_2O_6$, and Si_4O_8. These authors describe in some detail the general calibration procedures whereby the six 'regular solution' interaction parameters among these species could be derived from the experimentally determined phase diagram. For these calculations, they adopted the Cp_ℓ values of Carmichael et al. (1977), mineral properties of Helgeson et al. (1978), and estimated enthalpies of formation for these species. Bottinga et al. do not provide many details of the results of this analysis other than the average deviation (540 J/mol) between the chemical potentials of the mineral and liquid species at the temperatures and compositions of each of 10 liquidus experiments. These residuals are the same order of magnitude as produced by Bottinga and Richet (1978) in their fit of binary systems using only endmember species, and thus are probably somewhat larger than experimental uncertainties. Considering, however, the significant differences between current Cp_ℓ and enthalpies of fusion and the values adopted by Bottinga et al. (1981), their results appear quite reasonable.

An alternate speciation model, generalized from that proposed by Flood and Knapp (1963), was developed by Bjorkman et al. (1984) and Bjorkman (1985) who analyzed the systems PbO-SiO_2 and Fe-O-SiO_2 without assuming the stoichiometries of the melt species. Instead a calibration procedure was used in which the stoichiometries of melt species were computed. For the latter system with FeO, Fe_2SiO_4, and $FeO_{1.5}$ as components, speciation reactions are written as:

$$pFeO + qFe_2SiO_4 + rFeO_{1.5} = Fe_{p+2q+r}Si_qO_{p+4q+1.5r} \quad , \qquad (40)$$

where p, q, and r are integers. In order that these species be composed of corner-shared silicon tetrahedra, the constraints are imposed that $p \leq -1$, $q \geq 2$, and p varies from $-p = q - 1$ to $-p = 2q$. All species interactions are assumed to be ideal ($a_i = x_i$). Initially, the identity of the melt species are assumed, and free energy changes for all speciation reactions given by (40) are estimated. The mass balance and mass action equations are then iteratively solved to determine the equilibrium distribution of these melt species. Comparison of the resulting values of x_{FeO} with experimental a_{FeO} measurements allows calculation of improved free energy values. This procedure continues until the set of species and free energy values is found that best reproduces the experimental data.

Figure 5 shows the results of these calculations on the join $Fe_{total}O$-SiO_2. The silica-rich portion of the system which shows liquid immiscibility was not included in the analysis. Bjorkman reports that the experimental activity measurements at $X_{FeO} > 0.65$ could only be reproduced within their estimated uncertainties by including the species $Fe_3Si_2O_7$, which together with FeO dominates the species distribution. At $X_{FeO} < 0.60$, the best fit of experimental activities was accomplished

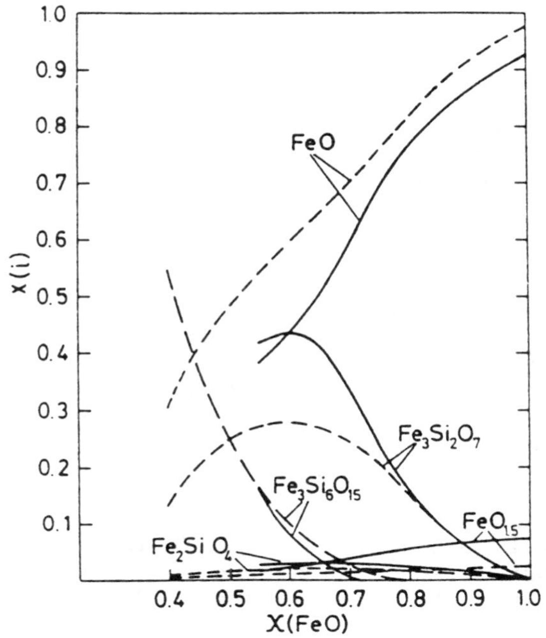

Figure 5. Calculated distribution of melt species in equilibrium with iron as a function of bulk composition in the pseudobinary system FeO-SiO_2. Solid and dashed curves correspond to distribution curves at 1673 and 2153 K, respectively. Taken from Bjorkman et al. (1985).

by incorporation of a $Fe_3Si_6O_{15}$ species. Besides affording excellent representation of the experimental activity data, the derived free energy functions for these species result in good reproduction of the liquidus data in this system. The predicted low abundance (Figure 5) of the orthosilicate species, Fe_2SiO_4, is somewhat unexpected given that the minimum negative excess enthalpy of mixing occurs near the orthosilicate composition in binary metal oxide-SiO_2 systems (e.g. Navrotsky et al., 1985). For this reason, one might question the uniqueness of the final set of adopted species, even though this minimum is much less pronounced in the FeO-SiO_2 system (Pelton and Blander, 1986) than in other systems with more basic metal oxides. In contrast, Bjorkman et al. 's (1984) results for the PbO-SiO_2 system show that the orthosilicate species is most abundant at the composition $X_{PbO} = 0.67$, and that the identities of the other melt species agree well with inferences based on Raman spectroscopy of PbO-SiO_2 glasses.

Comparison of the results of Bottinga et al. (1981) for the MgO-SiO_2 system with the calculations of Bjorkman for the system FeO-SiO_2 clearly shows that activity-composition relations are highly correlated with the choice of species. In the former case, non-ideal interactions were utilized with an assumed choice of melt species, while experimental data at $X_{FeO} > 0.4$ in the latter system could be well represented with ideal mixing among species whose stoichiometries were treated as adjustable parameters. Without detailed independent evidence regarding the melt structures on these joins or studies aimed at testing the predictive capabilities in higher component systems, it is difficult to choose between these different models.

At the present time, the structural data (which are largely limited to studies of glasses not liquids) are not sufficiently detailed to provide this information, primarily because most techniques yield information on short range interactions in the melt. Without more detailed long-range pictures of melt structure, the attractiveness of speciation models in providing insights into melt structure will be somewhat reduced by the need to treat them as empirical models with derived parameters and species of debatable significance.

Stoichiometric models

The first comprehensive attempt to apply a stoichiometric approach to silicate melts was made by Barron (1972), who utilized a symmetric, temperature-independent Margules equation (regular solution) to represent activity relations in several ternary systems of geologic importance. Very precise representation of liquidus relations with this model was not possible in all the systems studied, indicating the need for more complex activity-composition models. Below we review two models with which it is possible to represent multicomponent melt data within the uncertainties of the experimental measurements.

Berman and Brown (1984) presented a thermodynamic model for silicate melts which they applied to the system CaO-Al_2O_3-SiO_2 at one atmosphere pressure. This particular synthetic system provides an extremely stringent test for a melt model because the liquidus fields have been determined for 15 minerals which show minimal complications due to solid solution effects (with the exception of mullite which was excluded from the study). In addition the stable and metastable liquid immiscibility fields on the CaO-SiO_2 and Al_2O_3-SiO_2 joins have also been studied in some detail. In order to reproduce these data within reasonable estimates of their uncertainties, it was necessary to express G_{excess} with a fourth degree, temperature dependent Margules equation. This equation leads to 3 Margules parameters (W_G) per binary and 3 ternary parameters (see Eqn. 6), for a total of 12 W_G parameters (24 parameters in total since $W_G = W_H - T \cdot W_S$) in the ternary system.

Figures 6a and 6b, computed with the software described above, show that this model results in excellent agreement with most of the experimental data, particularly in the lower temperature range where the experimental uncertainties are much smaller. In order to achieve this degree of consistency with the phase equilibrium data, it was necessary to use the calibration techniques described above to derive melt properties simultaneously with those for liquidus minerals using a Cp equation similar to Equation (27) in its ability to provide reliable extrapolations to high temperature. Although Cp_ℓ values for the 3 component oxides were only loosely constrained by the limited experimental data available at that time, the derived melt properties compare well with experimental results. For example, predicted enthalpies of fusion of anorthite (140.4 kJ/mol) and pseudowollastonite (49.9 kJ/mol) are in quite good agreement with the respective calorimetric values: 135.6 ± 8.8 and 133.0 ± 4 for anorthite (Weill et al., 1980; Richet and Bottinga, 1984b) and 57.3 ± 2.9 for pseudowollastonite (Richet and Bottinga, 1984b). Computed heats of mixing on the join $CaAl_2O_4$-SiO_2 are similar in magnitude to those measured in glasses by Navrotsky et al. (1982), but the results are not directly comparable because of the unknown variation of glass transition temperatures across this join (see discussion above). The heat of fusion comparisons suggest, however, that the model provides a reasonable source for first order estimates of melt properties which can be expected to improve when the recent calorimetric or model Cp_ℓ values (see above) are used in future calibrations.

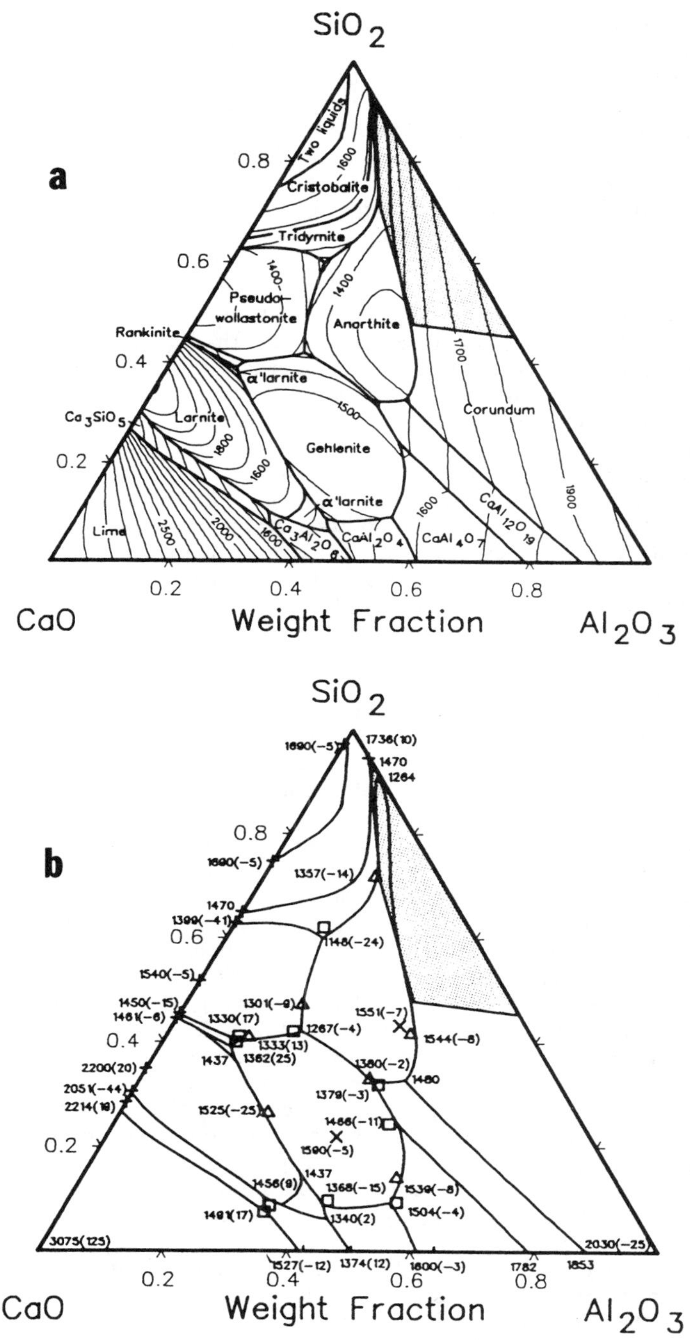

Figure 6. Computer calculated liquidus diagram for the system CaO-Al_2O_3-SiO_2. (a) Liquidus phases, phase boundaries, and isothermal sections at 100°C intervals. The shaded area represents the experimentally determined liquidus field of mullite, which could not be fitted due to extensive nonstoichiometry. (b) Comparison of computed with experimentally constrained invariant points (squares) and maxima (triangles). Temperatures in degrees Celsius, with differences (calculated-experimental) in parentheses. Taken from Berman and Brown (1984).

The study described above was extended by Berman (1983) who applied the same model to the quaternary system CaO-MgO-Al_2O_3-SiO_2 at one atmosphere pressure. Again, the large number of adjustable parameters in this model (31 W_H and 31 W_S parameters for the quaternary system) made it possible to accurately represent the phase equilibrium data in this system. For 69 invariant or piercing points on the joins used to calibrate the liquidi of 24 minerals, the average differences between calculated and experimentally determined invariant points are 0.24 ± 14.5°C and 0.52 ± 0.55 oxide weight percent (1σ errors). In addition computed melt activities compare favorably to those measured in slag systems (Rein and Chipman, 1965).

Figure 7 shows a stereographic pair of the 1 atmosphere quaternary cotectics in the CaO-MgO-Al_2O_3-SiO_2 system, calculated with the liquidus program described above. The computed compositions of the quaternary invariant points are given by Berman (1983), who also shows comparisons with available experimental data. The main point that can be appreciated from Figure 7 is that the model provides a very precise tool for interpolation among the existing experimental data in this system. Because of the highly complex mixing equation used to empirically describe activity-composition relations in the melt, extrapolations cannot be made to temperatures removed from the liquidus and immiscibility surfaces used in the calibration. This point has been emphasized by Barron (1985; 1986) who calculated extraneous solvi both below and above the liquidi in different portions of the ternary CAS system. Barron (1981) showed that extraneous solvi could be expected when the two interaction parameters of an asymmetric binary solution were of opposite sign, and, following the lead of Haller et al. (1974), suggested that molecular weights of melt components be adjusted so that solvi were close to symmetrical on a mole fraction basis. This suggestion would in general serve to reduce the number of mixing parameters needed to reproduce the solvi and therefore add to the predictive merits of the model. Further work is needed, however, to explore the effects of this proposed modification when liquidus as well as liquid immiscibility data are considered.

Another technique to simplify this solution model so that it can be more easily extended to magmatic systems is to choose a different component set which spans only the geologically relevant composition space (e.g. Ghiorso et al., 1983). Not only is the composition space much more restricted and therefore easier to model, but the energetics of mixing are much smaller between combined oxide melt compositions than between oxides because the strongly exothermic heats of mixing between metal oxides and silica are already accounted for (Navrotsky, 1986). It would therefore be instructive to see how much more complex a solution model would have to be than the strictly regular solution model of Ghiorso et al. (1983) in order to reproduce phase equilibrium data in the bounding binary and ternary synthetic systems as well as in natural compositions. Such a model would offer considerable improvement in the ability to predict phase relationships over a wider composition range, since all magmatic compositions would be accessible by interpolation rather than extrapolation. In addition, the interaction parameters among all melt species would be expected to have greater physical significance.

An alternate model which combines some of the advantages of the stoichiometric and speciation approaches has been developed by Blander and Pelton (Pelton and Blander, 1986; Blander and Pelton, 1987) in work aimed at systematizing the extensive body of data regarding the thermodynamic properties of melts in synthetic systems. In this model, simple oxides are also used as components but the formation of a dominant species (ordering) in the melt is explicitly accounted for.

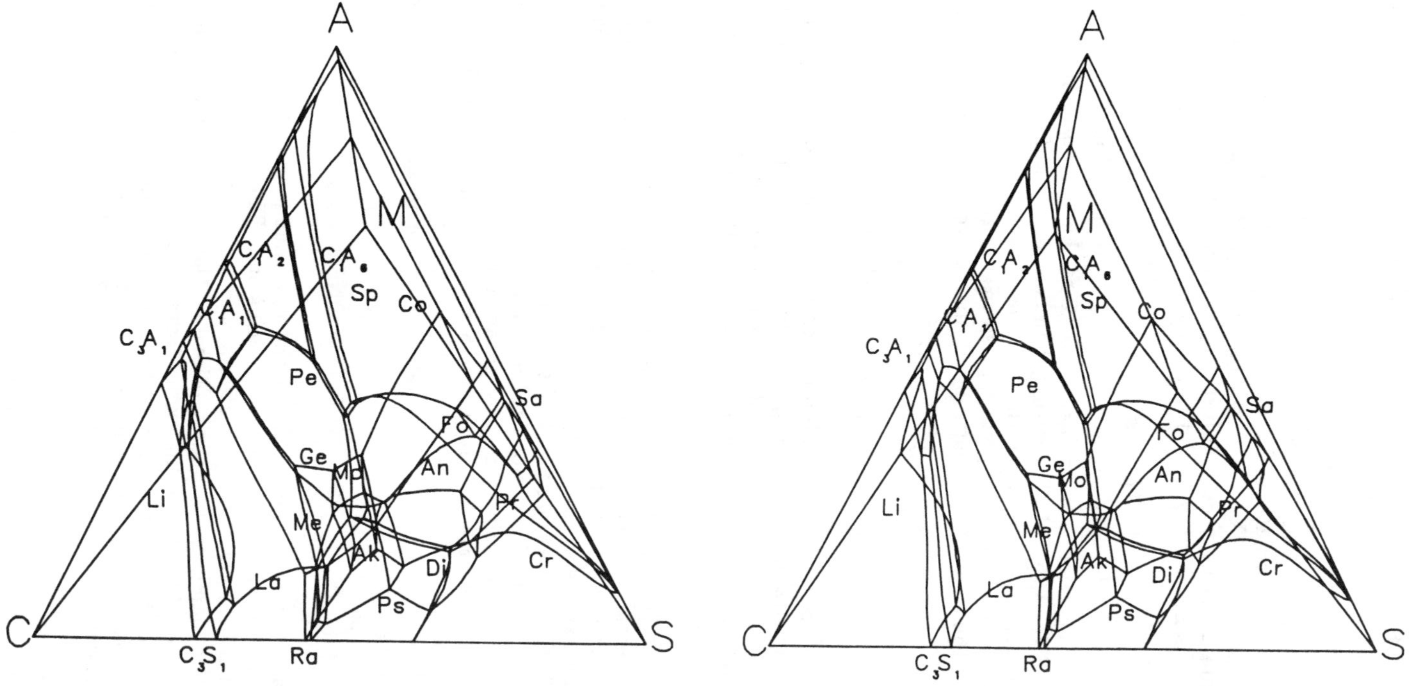

Figure 7. Stereographic pair of the computer calculated liquidus diagram for the system CaO-MgO-Al_2O_3-SiO_2 at one atmosphere. All minerals were assumed to be stoichiometric. Taken from Berman (1983).

This is achieved by considering the formation of 1-2 nearest neighbor pairs from 1-1 and 2-2 pairs in a binary system:

$$[1\text{-}1] + [2\text{-}2] = 2[1\text{-}2] \quad .$$

When the free energy change of this equilibrium (defined as $\Delta G = \omega - T\eta$) is strongly negative, the formation of 1-2 pairs causes ordering which reaches a maximum for the composition $X_1 = X_2 = 0.5$. In order to make the composition at which maximum ordering occurs adjustable, Guggenheim's (1952) quasichemical mixing equation is modified with the introduction of equivalent fractions, Y_1 and Y_2, where:

$$Y_1 = \frac{b_1 X_1}{b_1 X_1 + b_2 X_2} \qquad Y_2 = \frac{b_2 X_2}{b_1 X_1 + b_2 X_2} \tag{41}$$

and b_1 and b_2 are chosen so that $Y_1 = Y_2$ =0.5 at this composition ($X_{SiO_2} = 0.33$ for metal oxide-SiO_2 systems). The enthalpy and entropy of mixing are then defined in terms of ω and η, respectively, and the fractions of each type of pair (x_{11}, x_{22}, x_{12}) in the melt. Derivation of the complete expressions are given by Pelton and Blander (1986). Finally, the enthalpy and entropy of mixing are cast as complex functions of composition:

$$\omega = \omega_0 + \omega_1 Y_2 + \omega_2 Y_2^2 + \omega_3 Y_2^3 + \ldots \quad , \tag{42}$$

$$\eta = \eta_0 + \eta_1 Y_2 + \eta_2 Y_2^2 + \eta_3 Y_2^3 + \ldots \quad , \tag{43}$$

where the ω_i and η_i coefficients are calibrated empirically by least-squares regression of diverse experimental data (phase equilibrium, activity) in these systems. With a variety of assumptions, this modified quasichemical model can be extended to ternary and quaternary systems, as discussed by Pelton and Blander (1986) and Blander and Pelton (1987).

Applications to binary systems result in excellent reproduction of the various types of experimental data which is not surprising in that 5 and 6 adjustable ω_i and η_i parameters were retained in their fitting of the systems CaO-SiO_2 and MgO-SiO_2, respectively (6 parameters were used to fit these systems with the model of Berman and Brown, 1984). One strong point of this model is that it extrapolates well to temperatures and compositions not used in calibrations. For example, predictions of ternary activities from binary mixing parameters are very reasonable (Blander and Pelton, 1987). Nevertheless, in order to represent phase equilibrium data in ternary systems such as CaO-Al_2O_3-SiO_2 with the same fidelity as in binary systems, empirical extensions of the binary calibrations are necessary (Pelton, personal communication, 1987).

CONCLUSIONS

From these various attempts to define thermodynamic models for multicomponent melts we can draw certain general conclusions. The first and most obvious is that there is still a great deal more work that needs to be done in the area of devising thermodynamic melt models. None of the proposed models has all the attributes one would like: generality, some degree of correspondence with the physical structure of melts, and the abilities to represent all available data accurately, to be extended easily to magmatic systems, and to extrapolate beyond calibration conditions and provide predictive power.

The stoichiometric approach appears to offer the most direct and immediate method for systematization of and interpolation among the existing large body of experimental data, particularly phase equilibrium data. Calibration techniques are straightforward, and these models are flexible enough to represent data within the uncertainties of their measurements. The main problem with this approach is that empirical models may lose their predictive power as the mixing equations become increasingly complex in response to the need to faithfully reproduce data as precisely as possible. In this regard, an important area for study is the development and careful testing of strategies for predicting ternary and higher order systems from binary and ternary data (Bertrand et al., 1983; Acree, 1984; Fei et al., 1986; Blander and Pelton, 1987).

The speciation approach is the more fundamentally sound approach because the macroscopic thermodynamic properties can potentially correspond to physical melt structures. The configurational entropy of the melts will thus be accurate, provided that the melt species are chosen realistically. The mathematical treatment of speciation models, in particular the iterative techniques needed for calibration, is more complicated and thus provides a barrier in terms of time, effort, and cost to using this technique for multicomponent systems. The advantage that the speciation approach may offer is in predictive ability. If we take minerals as analogues of melt species, it seems likely that most melt species in anhydrous systems can be accounted for with binary and ternary interactions, and therefore can be defined by studies of binary and ternary systems. To the extent that no new species form in quaternary and higher systems and that interactions are ideal, phase relationships could be predicted in systems of any compositional complexity. In order to accomplish this however, general speciation models must be derived in which the thermodynamic properties of all species are retrieved explicitly.

Finally, there is also a great need for additional experimental work which provides the constraints for thermodynamic models. Although many of the large gaps in the experimental data have been filled over the last few years, more direct measurements of heats of fusion are needed in addition to data addressing some of complexities (nonideality, temperature dependence of Cp_ℓ) that are beginning to be recognized. Studies of melt properties at elevated pressures and particularly related to hydrous systems are urgently needed. The experimental work must provide data not only on macroscopic properties such as heat capacity, phase stability, and mixing properties, but also on microscopic properties important for deciphering the systematics of silicate melt structure.

ACKNOWLEDGMENTS

The authors thank K.E. Windom for making available preliminary results of his work, J.K. Russell for his critical review, and Jennifer Geddes for her enduring patience in helping to prepare this manuscipt. Financial support received by RGB through Geological Survey of Canada sponsorship of DSS contract # 1ST85-00380 is gratefully acknowledged.

REFERENCES

Acree, W.E. (1984) Thermodynamic properties of nonelectrolyte solutions. Academic Press, New York, 308 pp.

Allibert, M., Chatillon, C., Jacob, K.T., and Lourtau, R. (1981) Mass-spectrometric and electrochemical studies of thermodynamic properties of liquid and solid phases in the system CaO-Al_2O_3. J. Amer. Ceram. Soc. 64, 307-314.

Andersen, D.J., and Lindsley, D.H. (1981) A valid Margules formulation for an asymmetric ternary solution: Revision of the olivine-ilmenite thermometer, with applications. Geochim. Cosmochim. Acta 45, 847-853.

Bacon, C.R. (1977) High temperature heat content and heat capacity of silicate glasses: experimental determination and a model for calculation. Amer. J. Sci. 277, 109-135.
Bale, C.W., and Pelton, A.D. (1983) Optimization of binary thermodynamic and phase diagram data. Metal. Trans. B 14B, 77-83.
Barron, L.M. (1972) Thermodynamic multicomponent silicate equilibrium phase calculations. Amer. Mineral. 57, 809-823.
Barron, L.M. (1981) The calculated geometry of silicate liquid immiscibility. Geochim. Cosmochim. Acta 45, 495-512.
Barron, L.M. (1985) Comment on 'A thermodynamic model for multicomponent melts, with application to the system CaO-Al_2O_3-SiO_2' by Berman and Brown. Geochim. Cosmochim. Acta 49, 611-612.
Barron, L.M. (1986) Some thermodynamic properties of the Berman and Brown model for CaO-Al_2O_3-SiO_2. Geochim. Cosmochim. Acta 50, 2727-2733.
Benson, S.W. (1968) Thermochemical kinetics. John Wiley and Sons, New York, 223 pp.
Berman, R.G. (1983) A thermodynamic model for silicate melts, with application to the system CaO-MgO-Al_2O_3-SiO_2. Ph.D. Thesis , University of British Columbia, 178 pp.
Berman, R.G. (1987) Internally-consistent thermodynamic data for minerals in the system K_2O-Na_2O-CaO-MgO-FeO-Fe_2O_3-Al_2O_3-SiO_2-TiO_2-H_2O-CO_2. J. Petrol. (in press).
Berman, R.G., and Brown, T.H. (1984) A thermodynamic model for multicomponent melts, with application to the system CaO-Al_2O_3-SiO_2. Geochim. Cosmochim. Acta 45, 661-678.
Berman, R.G., and Brown, T.H. (1985) Heat capacity of minerals in the system Na_2O-K_2O-CaO-MgO-FeO-Fe_2O_3-Al_2O_3-SiO_2-TiO_2-H_2O-CO_2: representation, estimation, and high temperature extrapolation. Contrib. Mineral. Petrol. 89, 168-183.
Berman, R.G., Brown, T.H., and Greenwood, H.J. (1985) An internally-consistent thermodynamic data base for minerals in the system Na_2O-K_2O-CaO-MgO-FeO-Fe_2O_3-Al_2O_3-SiO_2-TiO_2-H_2O-CO_2. Atomic Energy Canada Ltd. Tech. Rep. 377, 62 pp. (available from SSDO, AECL, Chalk River, Ontario, K0J 1J0.
Berman, R.G., Engi, M., Greenwood, H.J., and Brown, T.H. (1986) Derivation of internally-consistent thermodynamic data by the technique of mathematical programming, a review with application to the system MgO-SiO_2-H_2O. J. Petrol. 27, 1331-1364.
Bertrand, G.L., Acree, W.E. Jr., and Burchfield, T.E. (1983) Thermochemical excess properties of multicomponent systems: representation and estimation from binary mixing data. J. Solution Chem. 12, 327-346.
Birch, F. (1966) Compressibility: elastic constants. In: Handbook of Physical Constants. S.P. Clark, ed., Geol. Soc. Amer. Memoir 97, 97-173.
Björkman, B. (1985) An assessment of the system Fe-O-SiO_2 using a structure based model for the liquid silicate. CALPHAD 9, 271-282.
Björkman, B., Eriksson, G., and Rosén, E. (1984) A generalized approach to the Flood-Knapp structure based model for binary liquid silicates: application and update for the PbO-SiO_2 system. Metal. Trans. B 15B, 511-516.
Blander, M., and Pelton, A.D. (1987) Thermodynamic analysis of binary liquid silicates and prediction of ternary solution properties by modified quasichemical equations. Geochim. Cosmochim. Acta 51, 85-95.
Boettcher, A.L., Burnham, C.W., Windom, K.E., and Bohlen, S.R. (1982) Liquids, glasses, and the melting of silicates to high pressures. J. Geol. 90, 127-138.
Bottinga, Y. (1985) On the isothermal compressibility of silicate liquids at high pressure. Earth Planet. Sci. Lett. 74, 350-360.
Bottinga, Y., and Richet, P. (1978) Thermodynamics of liquid silicates, a preliminary report. Earth Planet. Sci. Lett. 40, 382-400.
Bottinga, Y., Richet, P., and Weill, D.F. (1983) Calculation of the density and thermal expansion coefficient of silicate liquids. Bull. Mineral. 106, 129-138.
Bottinga, Y., and Weill, D.F. (1970) Densities of liquid silicate systems calculated from partial molar volumes of oxide components. Amer. J. Sci. 269, 169-182.
Bottinga, Y., Weill, D.F., and Richet, P. (1981) Thermodyanamic modelling of silicate melts. In: Thermodynamics of Minerals and Melts. R.C. Newton, A. Navrotsky, B.J. Wood, eds., Springer-Verlag, New York, 207-246.
Bottinga, Y., Weill, D., and Richet, P. (1982) Density calculations for silicate liquids. I. Revised method for aluminosilicate compositions. Geochim. Cosmochim. Acta 46, 909-919.
Brockris, J.O'M., Kitchner, J.A., Ignatowicz, S., and Tomlinson, J.W. (1952) Electric conductance in liquid silicates. Trans. Faraday Soc. 48, 75-91.
Brown, T.H., Berman, R.G., and Perkins, E.H. (1987) GEO-CALC: Software package for calculation and display of pressure-temperature-composition phase diagrams using an IBM or compatible personal computer. Computers & Geosciences (in press).
Burnham, C.W. (1975) Water and magmas: a mixing model. Geochim. Cosmochim. Acta 39, 1077-1084.
Burnham, C.W. (1981) The nature of multicomponent aluminosilicate melts. Phys. Chem. Earth 14, 197-229.
Burnham, C.W., Darken, L.S., and Lasaga, A.C. (1978) Water and magmas: application of the Gibbs-Duhem equation: a response. Geochim. Cosmochim. Acta 42, 277-280.
Burnham, C.W., and Nekvasil, H. (1986) Equilibrium properties of granite pegmatite magmas.

Amer. Mineral. 71, 239-263.
Carmichael, I.S.E., Nicholls, J., Spera, F.J., Wood, B.J., and Nelson, S.A. (1977) High-temperature properties of silicate liquids: applications to the equilibration and ascent of basic magma. Philos. Trans. Royal Soc. London A286, 373-431.
Chang, Do R., and Howald, R.A. (1982) Calculation of the heats of fusion of CaO and MgO. High Temp. Sci. 15, 209-218.
Day, H. W., Chernosky, J.V. Jr., and Kumin, H. J. (1985) Equilibria in the sytem $MgO-SiO_2-H_2O$: a thermodynamic analysis. Amer. Mineral. 70, 237-248.
Day, H.W., and Halbach, H. (1979) The stability field of anthophyllite: the effect of experimental uncertainty on permissible phase diagram topologies. Amer. Mineral. 64, 809-823.
Day, H.W., and Kumin, H.J. (1980) Thermodynamic analysis of the aluminum silicate triple point. Amer. J. Sci. 280, 265-287.
de Capitani, C., and Brown, T.H. (1987) The computation of chemical equilibrium in complex systems containing non-ideal solutions. Geochim. Cosmochim. Acta (in press).
Eliášová, M., Proks, I.I., and Zlatovsky, I. (1978) Thermodynamic analysis of the system 2CaO.MgO.-$2SiO_2$- CaO.MgO.$2SiO_2$. Silikaty Prague 22, 97-108.
Eliezer, N., Howald, R.A., Marinkovic, M., and Eliezer, I. (1978) Vapor pressure measurements, thermodynamic paramenters, and phase diagram for the system potassium oxide-silicon oxide at high temperatures. J. Phys. Chem. 82, 1021-1026.
Engi, M. (1983) Equilibria involving Al-Cr spinel: Mg-Fe exchange with olivine. Experiments, thermodynamic analysis, and consequences for geothermometry. Amer. J. Sci. 283-A, 29-71.
Eriksson, G. (1975) Thermodynamic studies of high temperature equilibria. Chem. Scripta 8, 100-103.
Fei, Y., and Saxena, K. (1986) A thermodynamic data base for phase equilibria in the system Fe-Mg-Si-O at high pressure and temperature. Phys. Chem. Minerals 13, 311-324.
Fei, Y., Saxena, S.K., and Eriksson, G. (1986) Some binary and ternary silicate solution models. Contrib. Mineral. Petrol. 94, 221-229.
Flood, H., and Knapp, W.J. (1963) Acid-base equilibria in the system $PbO-SiO_2$. J. Amer. Ceram. Soc. 46, 61-65.
Flood, H., and Knapp, W.J. (1968) Structural characteristics of liquid mixtures of feldspar and silica. J. Amer. Ceram. Soc. 51, 259-263.
Fraser, D.G., and Bottinga, Y. (1985) The melting properties of melts and glasses in the system $NaAlSi_3O_8$-$KAlSi_3O_8$: Comparison of experimental data obtained by Knudsen-cell mass spectroscopy and solution calorimetry. Geochim. Cosmochim. Acta 49, 1377-1381.
Fraser, D.G., Rammensee, W., and Hardwick, A. (1985) Determination of mixing properties of molten silicates by Knudsen-cell mass spectroscopy, II. The systems (Na-K)Si_4O_{10} and (Na-K)$AlSi_5O_{12}$. Geochim. Cosmochim. Acta 49, 349-359.
Fraser, D.G., Rammensee, W., and Jones, R.H. (1983) The mixing properties of melts in the system $NaAlSi_2O_6$-$KAlSi_2O_6$ determined by Knudsen-cell mass spectrometry. Bull. Mineral. 106, 111-117.
Fujii, T., and Kushiro, I. (1977) Density, viscosity, and compressibility of basaltic liquid at high pressures. Carnegie Inst. Wash. Year Book 76, 419-425.
Gaskell, D.R. (1977) Activities and free energies of mixing in binary silicate melts. Metal. Trans. B 8B, 131-145.
Ghiorso, M.S. (1985) Chemical mass transfer in magmatic processes. I. Thermodynamic relations and numerical algorithms. Contrib. Mineral. Petrol. 90, 107-120.
Ghiorso, M.S., and Carmichael, I.S.E. (1984) Comment on density calculations for silicate liquids. I. Revised method for aluminosilicate compositions by Bottinga, Weill, and Richet. Geochim. Cosmochim. Acta 48, 401-408.
Ghiorso, M.S., and Carmichael, I.S.E. (1985) Chemical mass transfer in magmatic processes. II. Applications in equilibrium crystallization, fractionation and assimilation. Contrib. Mineral. Petrol. 90, 121-141.
Ghiorso, M.S., Carmichael, I.S.E., Rivers, M.L., and Sack, R.O. (1983) The Gibbs free energy of mixing of natural silicate liquids: an expanded regular solution approximation for the calculation of magmatic intensive variables. Contrib. Mineral. Petrol. 87, 107-145.
Gill, P.E., Murray, W., and Wright, M.H. (1981) Practical Optimization. Academic Press, New York, 401pp.
Glazner, A.F. (1984) Activities of olivine and plagioclase components in silicate melts and their application to geothermometry. Contrib. Mineral. Petrol. 88, 260-268.
Gordon, T.M. (1973) Determination of internally consistent thermodynamic data from phase equilibrium experiments. J. Geol. 81, 199-208.
Grieg, J.W. (1927) Immiscibility in silicate melts. Amer. J. Sci. 13, 1-44.
Grover, J. (1977) Chemical mixing in multicomponent solutions: and introduction to the use of Margules and other thermodynamic excess functions to represent nonideal behavior. In: Thermodynamics in Geology, D.G. Fraser, ed., D. Reidel Publ. Co., Dordrecht, Holland, 67-97.
Guggenheim, E.A. (1952) Mixtures. Clarendon Press, London, 270 pp.
Haas, J.L. Jr., and Fisher, J.R. (1976) Simultaneous evaluation and correlation of thermodynamic data. Amer. J. Sci. 276, 525-545.
Haas, J.L., Jr., Robinson, G.R., Jr., and Hemingway, B.S. (1981) Thermodynamic tabulations for

selected phases in the system, $CaO-Al_2O_3-SiO_2-H_2O$ at 101.325 kPa (1 atm) between 273.15 and 1800 K. J. Phys. Chem. Ref. Data 10, 576-669.
Halbach, H., and Chatterjee, N.D. (1982) The use of linear parametric programming for determining internally consistent thermodynamic data for minerals. In: High-Pressure Researches in Geoscience. W. Schreyer, ed., E. Schweizerbartsche Verlagsbuchhandlung, 475-491.
Halbach, H., and Chatterjee, N.D. (1984) An internally consistent set of thermodynamic data for twenty-one $CaO-Al_2O_3-SiO_2-H_2O$ phases by linear parametric programming. Contrib. Mineral. Petrol. 88, 14-23.
Haller, W., Blackburn, D.H. Simons, J.H. (1974) Miscibility gaps in alkali silicate binaries-data and thermodynamics interpretation. J. Amer. Ceram. Soc. 57, 120-126.
Hastie, J.W., and Bonnell, D.W. (1985) A predictive phase equilibrium model for multicomponent oxide mixtures. Part II. Oxides of Na-K-Ca-Mg-Al-Si. High Temp. Sci. 19, 275-306.
Hastie, J.W., Horton, W.S., Plante, E.R., and Bonnell, D.W. (1982) Thermodynamic models of alkali-metal vapor transport in silicate systems. High Temp. High Pres. 14, 669-679.
Helgeson, H.C., Delany, J.M., Nesbitt, H.W., and Bird, D.K. (1978) Summary and critique of the thermodynamic properties of rock-forming minerals. Amer. J. Sci. 278A, 229pp.
Hervig, R.L., and Navrotsky, A. (1984) Thermochemical study of glasses in the system $NaAlSi_3O_8$-$KAlSi_3O_8$-Si_4O_8 and on the join $Na_{1.6}Al_{1.6}Si_{2.4}O_8 - K_{1.6}Al_{1.6}Si_{2.4}\ O_8$. Geochim. Cosmochim. Acta 48, 513-522.
Hervig, R.L., Scott, D., and Navrotsky, A. (1985) Thermochemistry of glasses along joins of pyroxene stoichiometry in the system $Ca_2Si_2O_6$-$Mg_2Si_2Si_2O_6$ -Al_4O_6. Geochim. Cosmochim. Acta 49, 1497-1501.
Hess, P.C. (1977) Structure of silicate melts. Can. Mineral. 15, 162-178.
Himmelblau, D.M. (1972) Applied Nonlinear Programming. McGraw-Hill, New York, 498 pp.
Holland, T.J.B. (1981) Thermodynamic analysis of simple mineral systems. In: Thermodynamics of Minerals and Melts. R.C. Newton, A. Navrotsky, and B.J. Wood, eds., Springer-Verlag, New York, 19-34.
Holland, T.J.B., and Powell, R. (1985) An internally consistent thermodynamic dataset with uncertainties and correlations: 2. Data and results. J. Metam. Geol. 3, 343-370.
Hon, R., Henry, D.J., Navrotsky, A., and Weill, D.F. (1981) A thermochemical calculation of the pyroxene saturation surface in the system diopside-albite-anorthite. Geochim. Cosmochim. Acta 45, 157-161.
Julsrud, S., and Kleppa, O.J. (1986) A thermochemical study of the liquid system $NaBO_2$-SiO_2 at 1394 K. Geochim. Cosmochim. Acta 50, 1201-1204.
Lange, R., and Carmichael, I.S.E. (1986) Densities of K_2O-Na_2O-CaO-MgO-Al_2O_3-TiO_2-SiO_2 liquids. Trans. Amer. Geophys. Union 67, 1273.
Lange, R., and Carmichael, I.S.E. (1987) Densities of Na_2O-K_2O-CaO-MgO-FeO-Fe_2O_3-Al_2O_3-TiO_2-SiO_2 liquids: new measurements and derived partial molar properties. Geochim. Cosmochim. Acta (in press).
Langmuir, C.H., and Weaver, J.S. (1982) A general method of calculating phase equilibria applied to complex crystallization processes in oceanic basalts. Trans. Amer. Geophys. Union 63, 475.
Lindsley, D.H., Grover, J.E., and Davidson, P.M. (1981) The thermodynamics of the $Mg_2Si_2O_6$-$CaMgSi_2O_6$ join: A review and an improved model. In: Thermodynamics of Minerals and Melts, R.C. Newton, A. Navrotsky, B.J. Wood, eds., Springer-Verlag, New York, 304 pp.
Longhi, J., and Hays, J.F. (1979) Phase equilibria and solid solution along the join $CaAl_2Si_2O_8$-SiO_2. Amer. J. Sci. 279, 876-890.
Maier, C.G., and Kelley, K.K. (1932) An equation for the representation of high-temperature heat content data. J. Amer. Chem. Soc. 54, 3243-3246.
Masson, C.R. (1972) Thermodynamics and constitution of silicate slags. J. Iron Steel Inst. 210, 89-96.
Mo, X., Carmichael, I.S.E., Rivers, M., and Stebbins, J. (1982) The partial molar volume of Fe_2O_3 in multicomponent silicate liquids and the pressure dependence of oxygen fugacity in magmas. Mineral. Mag. 45, 237-245.
Mysen, B.O. (1983) The structure of silicate melts. Annual Rev. Earth Planet. Sci. 11, 75-97.
Mysen, B.O. (1986) Structure and petrologically important properties of silicate melts relevant to natural magmatic liquids. In: C.M. Scarfe, ed., Mineral. Assoc. Canada Short Course in Silicate Melts 12, 180-209.
Navrotsky, A. (1986) Thermodynamics of silicate melts and glasses. In: C.M. Scarfe, ed., Mineral. Assoc. Canada Short Course in Silicate Melts 12, 130-153.
Navrotsky, A., Hervig, R.L., Roy, B.N., and Huffman, M. (1985) Thermochemical studies of silicate, aluminosilicate, and borosilicate glasses. High Temp. Sci. 19, 133-150.
Navrotsky, A., Hon, R., Weill, D.F., and Henry, D.J. (1980) Thermochemistry of glasses and liquids in the systems $CaMgSi_2O_6$-$CaAl_2Si_2O_8$-$NaAlSi_3O_8$, SiO_2-$CaAl_2Si_2O_8$-$NaAlSi_3O_8$ and SiO_2-Al_2O_3-CaO-Na_2O. Geochim. Cosmochim. Acta 44, 1409-1423.
Navrotsky, A., Peraudeau, G., McMillian, P., and Coutures, J.P. (1982) A thermochemical study of glasses and crystals along the joins silica-calcium aluminate and silica-sodium aluminate. Phys. Chem. Minerals 46, 2039-2047.
Navrotsky, A., Zimmermann, H.D., and Hervig, R.L. (1983) Thermochemical study of glasses in the system $CaMgSi_2O_6$-$CaAl_2SiO_6$. Geochim. Cosmochim. Acta 47, 1535-1538.
Nelson, S.A., and Carmichael, I.S.E. (1979) Partial molar volumes of oxide components in silicate

liquids. Contrib. Mineral. Petrol. 71, 117-124.
Nicholls, J., Russell, J.K., and Stout, M.Z. (1986) Testing magmatic hypotheses with thermodynamic modelling. In: C.M. Scarfe, ed., Mineral. Assoc. Canada Short Course in Silicate Melts 12, 210-235.
Nielsen, R.L., and Dungan, M.A. (1983) Low pressure mineral-melt equilibria in natural anhydrous mafic systems. Contrib. Mineral. Petrol. 84, 310-326.
Pelton, A.D., and Blander, M. (1984) Computer-assisted analysis of the thermodynamic properties and phase diagrams of slags. In: Second Inter. Symp. Metal. Slags and Fluxes. H.A. Fine and D.R. Gaskell, eds., 281-294.
Pelton, A.D., and Blander, M. (1986) Thermodynamic analysis of ordered liquid solutions by a modified quasichemical approach-application to silicate slags. Metal. Trans. B 17B, 805-815.
Perkins, E.H., Brown, T.H., and Berman, R.G. (1986) PT-SYSTEM, TX-SYSTEM, PX-SYSTEM: three programs which calculate pressure-temperature-composition phase diagrams. Computers & Geosciences 12, 749-755.
Rammensee, W., and Fraser, D.G. (1982) Determination of activities in silicate melts by Knudsen-cell mass spectroscopy, I. The system $NaAlSi_3O_8$-$KAlSi_3O_8$. Geochim. Cosmochim. Acta 46, 2269-2278.
Rein, H., and Chipman, J. (1963) The distribution of silicon between Fe-Si-C alloys and SiO_2-CaO-MgO-Al_2O_3 slags. Trans. Metallurgical Soc. AIME 227, 1193-1203.
Richet, P., and Bottinga, Y. (1982) Modèles de calcul des capacités calorifiques des liquides et des verres silicatés. C.R. Acad. Sc. Paris 295, 1121-1124.
Richet, P., and Bottinga, Y. (1983) Verres, liquides, et transition vitreuse. Bull. Mineral. 106, 147-168.
Richet, P., and Bottinga, Y. (1984a) Glass transitions and thermodynamic properties of amorphous SiO_2, $NaAlSi_nO_{2n+2}$ and $KAlSi_3O_8$. Geochim. Cosmochim. Acta 48, 453-470.
Richet, P., and Bottinga, Y. (1984b) Anorthite, andesine, wollastonite, diopside, cordierite and pyrope: Thermodynamics of melting, glass transitions, and properties of the amorphous phases. Earth Planet. Sci. Lett. 67, 415-432.
Richet, P., and Bottinga, Y. (1985) Heat capacity of aluminium-free liquid silicates. Geochim. Cosmochim. Acta 49, 471-486.
Richet, P., and Bottinga, Y. (1986) Thermochemical properties of silicate glasses and liquids: a review. Rev. of Geophys. 24, 1-25.
Rigden, S.M., Ahrens, T.J., and Stolper, E.M. (1984) Densities of liquid silicates at high pressures. Science 226, 1071-1074.
Rigden, S.M., Ahrens, T.J., and Stolper, E.M. (1985) Density of molten diopside at high pressure. Trans. Amer. Geophys. Union 66, 395.
Rivers, M.L., and Carmichael, I.S.E. (1987) Ultrasonic studies of silicate melts. J. Geophys. Res. (in press).
Robie, R.A., Hemingway, B.S., and Fisher, J.R. (1979) Thermodynamic properties of minerals and related substances at 298.15 K and 1 bar (10^5 Pascals) pressure and at higher temperatures. U.S. Geol. Surv. Bull. 1452, 456 pp.
Robinson, G.R. Jr., Haas, J.L. Jr., Schafer, C.M., and Haselton, H.T. Jr. (1982) Thermodynamic and thermophysical properties of selected phases in the MgO-SiO_2- H_2O-CO_2, CaO-Al_2O_3-SiO_2-H_2O-CO_2, and Fe-FeO-Fe_2O_3-SiO_2 chemical systems, with special emphasis on the properties of basalt. U.S. Geol. Surv. Open-file Rep. 83-79, 429 pp.
Rogez, J., Chastel, R., Bergman, C., Brousse, C., and Castanet, R., and Mathieu, J.-C. (1983) Etude thermodynamique du système albite-orthose par calorimétrie de dissolution et effusion de Knudsen couplée à un spectrométre de masse. Bull. Mineral. 106, 119-128.
Salje, E. (1985) Thermodynamics of sodium feldspar I: order parameter treatment and strain induced coupling effects. Phys. Chem. Minerals 12, 93-98.
Salje, E., Kuscholke, B., Wruck, B., and Kroll, H. (1985) Thermodynamics of sodium feldspar II: Experimental results and numerical calculations. Phys. Chem. Minerals 12, 99-107.
Saxena, S.K., and Chatterjee, N. (1986) Thermochemical data on mineral phases: the system CaO-MgO-Al_2O_3-SiO_2. J. Petrol. 27, 827-842.
Scarfe, C.M., Mysen, B.O., and Virgo, D. (1979) Changes in viscosity and density of melts of sodium disilicate, sodium metasilicate, and diopside composition with pressure. Carnegie Inst. Wash. Year Book 78, 547-551.
Skinner, B.J. (1966) Thermal expansion. In: Handbook of Physical Constants. S.P. Clark, ed., Geol. Soc. Amer. Memoir 97, 75-96.
Spencer, P.J. (1973) The thermodynamic properties of silicates. Nat. Physics Lab. Res. Chem. 21, 1-36.
Stebbins, J.F., Carmichael, I.S.E., and Moret, L.K. (1984) Heat capacities and entropies of silicate liquids and glasses. Contrib. Mineral. Petrol. 86, 131-148.
Stebbins, J.F., Carmichael, I.S.E., and Weill, D.E. (1983) The high temperature liquid and glass heat contents and the heats of fusion of diopside, albite, sanidine and nepheline. Amer. Mineral. 68, 717-730.
Stebbins, J.F., Murdoch, J.B., Schneider, E., Carmichael, I.S.E., and Pines, A. (1986) A high-temperature high-resolution NMR study of ^{23}Na, ^{27}Al and ^{29}Si in molten silicates. Nature 314, 250-252.

Stout, N.D., and Piwinskii, A.J. (1982) Enthalpy of silicate melts from 1520 to 2600 K under ambient pressure. High Temp. Sci. 15, 275-292.
Sumino, Y., and Anderson, O.L. (1984) Elastic constants of minerals. In: Handbook of Physical Properties of Rocks. R.S. Carmichael, ed., CRC Press Inc., 39-138.
Taylor, M., and Brown, G.E. (1979a) Structure of silicate mineral glasses I. Geochim. Cosmochim. Acta 43, 61-75.
Taylor, M., and Brown, G.E. (1979b) Structure of silicate mineral glasses II. Geochim. Cosmochim. Acta 44, 109-118.
Tewhey, J.D., and Hess, P.C. (1979) The two phase region in the CaO-SiO_2 system: experimental data and thermodynamic analysis. Phys. Chem. Glasses 20, 41-53.
Thompson, J.B., Jr. (1967) Thermodynamic properties of simple solutions. In: Researches in Geochemistry II. P.H.Abelson, ed., John Wiley and Sons, New York, 340-361.
Tomiska, J. (1986) A simple proceedure for the algebraic conversion among the various representations of the thermodynamic excess functions of binary systems. CALPHAD 10, 239-252.
Ulbrich, H.H., and Waldbaum, D.R. (1976) Structural and other contributions to the third-law entropies of silicates. Geochim. Cosmochim. Acta 40, 1-24.
van Zeggeren, F., and Storey, S.H. (1970) The computation of chemical equilibrium. Cambridge University Press, London, 176 pp.
Weill, D.F., Hon, R., and Navrotsky, A. (1980) The igneous system $CaMgSi_2O_6$-$CaAl_2Si_2O_8$-$NaAlSi_3O_8$: variations on a classic theme by Bowen. In: Physics of Magmatic Processes. R.B. Hargraves, ed., 49-92.
Westrich, H.R., and Holloway, J.R. (1981) Experimental dehydration of pargasite and calculation of its entropy and Gibbs energy. Amer. J. Sci. 281, 922-934.
White, W.P. (1919) Silicate specific heats. Amer. J. Sci. 47, 1-21.
Windom, K.E., and Wood, B.J. (1986) Calculations of melting relations among diopside-anorthite-silica using a Flory-Huggins model. Trans. Amer. Geophys. Union 67, 414.
Wood, B.J., and Nicholls, J.W. (1978) The thermodynamic properties of reciprocal solid solutions. Contrib. Mineral. Petrol. 66, 389-400.
Zen, E-An (1972) Gibbs free energy, enthalpy, and entropy of ten rock-forming minerals: calculations, discrepancies, implications. Amer. Mineral. 57, 524-553.

Chapter 12 Mark S. Ghiorso

MODELING MAGMATIC SYSTEMS: THERMODYNAMIC RELATIONS

INTRODUCTION

This is the first of two chapters concerned with modeling homogeneous and heterogeneous equilibria in magmatic systems. We will focus our attention exclusively on melt models that have been formulated for liquids of magmatic composition and will concentrate on the application of these models to the calculation of phase equilibria in magmas. The application of solution theory to complex silicate melts will first be treated and from there the discussion will move on to calibration of melt models and the calculation of heterogeneous equilibria in both closed and open magmatic systems. Methods of extending the equilibrium models to the calculation of magmatic evolution along a variety of reaction paths, such as polythermal-isobaric crystallization and isenthalpic-isobaric assimilation, will be discussed. Petrologic examples of the modeling methods described here are presented in the subsequent chapter.

General constraints on the formulation of melt models

In recent years a variety of models of magmatic melt behavior have been proposed that provide a means of approximating melt component activities (and therefore chemical potentials of melt components) from experimental phase equilibrium data. These melt models are generally utilized to evaluate solubility products for magmatic minerals in the context of geothermometry, geobarometry or the calculation of liquid lines of descent. The advantage of expressing the experimental phase equilibrium results in a thermodynamic context is twofold: (1) solution theory provides the best means of interpolating and extrapolating the experimental data to other less thoroughly investigated bulk compositions, and (2) a thermodynamic treatment allows for interlaboratory comparison of experimental results and assessment of the internal consistency of data collected over a wide range of temperature, pressure and bulk composition.

The "thermodynamic advantage" is realized as a consequence of the ability to express the equilibrium proportions and compositions of solid and liquid phases in the magma in terms of the constrained minimum of a suitable potential function. Generally, this function is taken to be the Gibbs free energy (G) at a specific temperature (T), pressure (P) and fixed system bulk composition. In order for this function to exist (in the sense of a path independent state variable) the generalized Maxwell relations must be satisfied. These relate the second cross partial derivatives of the Gibbs function with respect to the number of moles of system components (n_1, n_2, etc.):

$$\left(\frac{\partial^2 G}{\partial n_i \partial n_j}\right)_{T,P,n_{k\neq i,j}} = \left(\frac{\partial^2 G}{\partial n_j \partial n_i}\right)_{T,P,n_{k\neq i,j}} . \quad (1)$$

Utilizing the standard definition of the chemical potential,

$$\left(\frac{\partial G}{\partial n_i}\right)_{T,P,n_{k\neq i}} = \mu_i = \mu_i^o + RT \ln a_i \quad , \tag{2}$$

where the superscript ° denotes the standard state, R the universal gas constant, and a_i the activity of the i-th component, Equation (1) may be written:

$$\left(\frac{\partial \ln a_i}{\partial n_j}\right)_{T,P,n_{k\neq j}} = \left(\frac{\partial \ln a_j}{\partial n_i}\right)_{T,P,n_{k\neq i}} \quad , \tag{3}$$

which is often simplified with the substitution $a = \gamma X$ to yield:

$$\left(\frac{\partial \ln \gamma_i}{\partial n_j}\right)_{T,P,n_{k\neq j}} = \left(\frac{\partial \ln \gamma_j}{\partial n_i}\right)_{T,P,n_{k\neq i}} \quad . \tag{4}$$

Equation (3) establishes constraints on the formulation of activity composition relations which express the compositional dependence of melt activities in magmatic systems. In the calibration of activity models from experimental phase equilibrium data, violation of Equation (3) renders the Gibbs function non-existent. Consequently, activity models which violate Equation (3) are completely empirical and cannot be used to estimate the activities of melt components not explicitly calibrated; i.e., describe the overall shape of the Gibbs function surface. Such models do not provide a thermodynamic description of the melt phase.

The easiest way to incorporate the constraint of Equation (3) into the construction of a melt model is to propose a functional form for the Gibbs potential at the onset and differentiate to obtain expressions for the melt activities. This is the approach utilized by Ghiorso et al. (1983) and is the standard procedure in the thermodynamic description of solid solutions (see chapter by Wood, this volume) and aqueous electrolytes (see chapter by Pitzer, this volume). The development of arbitrary expressions for individual melt activities can easily lead to violations of Equation (3). As an example, we might consider the melt model of Glazner (1984). Glazner fits the natural logarithms of "melt activities" (obtained from phase equilibrium data in a manner discussed below) to arbitrary polynomial functions of the mole fractions of melt components. Manipulation of his expressions for "forsterite" and "fayalite" (Glazner, 1984, Tables 3 and 4) reveal that they cannot satisfy Equation (3). The cross partial derivatives of Glazner's activity expressions are inconsistent as a consequence of neglecting to insure that the postulated activity-composition relations represent valid expressions for the derivatives of some hypothetical Gibbs function.

Melt models can be devised which are inconsistent with Equation (3) even if they are based on a conceptually pleasing formulation of "species interactions" in the melt. As an example the model of Burnham and Nekvasil (1986; Nekvasil, 1986) might be considered. They propose a melt speciation model for hydrous-"granitic" melts (the system $NaAlSi_3O_8$-

$CaAl_2Si_2O_8$-$KAlSi_3O_8$-SiO_2-Al_2O_3-$LiAlSi_2O_6$-H_2O-F-B) involving a variety of melt components which resemble mineral species that have been recalculated to an 8-oxygen basis. These are chosen to reflect the stoichiometry of solid phases which appear on the liquidus in this idealized granitic system. The method is an extension of one originally proposed for anhydrous multicomponent melts by Burnham (1981) and independently for simple systems by Bottinga and Richet (1978). Burnham and Nekvasil (1986) write the activity of the i-th component in the melt as a function of the form:

$$a_i = X_i^{am} (1 - f_i)(1 - X_w)^2 \qquad \text{for } X_w \leq 0.5 \quad ,$$

$$a_i = X_i^{am} (1 - f_i) \exp\left\{\left(6.52 - \frac{2667}{T}\right)\left[\ln(1 - X_w) + X_w + 0.193\right]\right\}$$

$$\text{for } X_w > 0.5 \quad ,$$

where X_w refers to the mole fraction of water in the melt and the superscript ^{am} indicates that mole fractions are recalculated on an anhydrous basis. Burnham and Nekvasil express the f_i as quite general functions of pressure, the mole fraction of water, and the remaining melt mole fractions recalculated to an anhydrous basis. From Equation (3), it is clear that a functional relationship must exist between the f_i and the f_j corresponding to the i-th and j-th "anhydrous" melt components. This relation is given by:

$$(1 - f_j)\left(\frac{\partial f_i}{\partial n_j}\right)_{n_{k \neq j}} = (1 - f_i)\left(\frac{\partial f_j}{\partial n_i}\right)_{n_{k \neq i}} \quad .$$

Inspection of the various definitions of f_i provided by Burnham and Nekvasil (1986, their Tables 2a,b,c and d) reveal that the cross-partial derivative relationships are not satisfied. In addition, the derivatives in the various f_i possess discontinuities at unrelated pressures which indicate that Burnham and Nekvasil's model demands the existence of first order phase transitions in the melt (a discontinuity in $\partial\mu/\partial P$ implies a discontinuity in $\partial V/\partial n$ and an overall ΔV where the pressure dependence of the f_i change). Such first order phase transitions imply the formation of immiscible liquids and are experimentally unsupported. It is clear that the melt speciation concept itself (i.e., Burnham, 1981; Bottinga and Richet, 1978; Bottinga et al., 1981) is <u>not</u> what causes Burnham and Nekvasil's activity expressions to violate Equation (3). If equilibrium constants between melt species were extracted from the phase equilibrium data and the melt "activity coefficients" set to unity, the violation of Equation (3) would never occur. The problem arises solely as a result of choosing an inconsistent mathematical form for the "activity coefficient" functions in the model.

Review of models that satisfy the thermodynamic requirements

Three models of "magmatic"-melt behavior currently in use satisfy the Maxwell criteria embodied in Equation (3). These are the ideal site mixing model for mafic magmas proposed by Doyle and Naldrett (1986,

1987), the modified two-lattice ideal site mixing model described most recently by Nielsen and Dungan (1983), and the regular solution model of Ghiorso et al. (1983).

Doyle and Naldrett (1986, 1987) construct a melt model by projecting the compositions of natural liquids into pseudoternaries, where the ternary "components" (e.g. FeO, MgO and matrix) are assumed to mix ideally. By equilibrating various synthetic analogues of magmatic liquids with metallic iron and metallic nickel, they calculate the activities of FeO and NiO in the melt and find them to be proportional to the molar ratios of FeO/(FeO+MgO) and NiO/(NiO+MgO). By extension, they infer activities of NiO and FeO in mafic magmas, and use these relations to analyze data on the compositional dependence of liquid-solid distribution coefficients. Doyle and Naldrett (1986, 1987) have not attempted to demonstrate that their model predicts consistent solubility products for solid solutions in mafic magmatic systems.

Nielsen and Dungan (1983) have extended the approach of Nielsen and Drake (1979) to formulate a "two-lattice" model of melt behavior. This model is best understood by thinking of the melt as consisting of two non-interacting but completely miscible phases. The first is described as a solution of the network forming components (taken as SiO_2, $NaAlO_2$ and $KAlO_2$) and the second by a solution of network modifying components (such as MgO, FeO, CaO, etc.). The Gibbs free energy of the network forming "lattice" is taken to be independent of that of the network modifying "lattice" such that the mole fraction of a network forming component is not a function of the mole fraction of any network modifying component and vice versa. The components on each lattice are assumed to mix ideally. One consequence of the two-lattice model that seems counter-intuitive is that the silica activity in the melt is not a function of the abundance or identity of network modifying components. Thus, the addition of MgO, FeO or CaO to a mafic magma will not change the activity of silica. In addition, the two lattice model demands that the activities of both network formers (e.g. SiO_2) and network modifiers (e.g. MgO) be unity, independent of melt composition, throughout a limiting binary such as $MgO-SiO_2$.

Nielsen and Dungan (1983) have applied their melt model to the calibration of equilibrium constants for liquid-solid reactions between mafic melts and end-member components of olivine, pyroxene, plagioclase, rhombohedral oxides and spinel. This has allowed them to model successfully low-pressure phase relations in natural anhydrous mafic systems. However, from a thermodynamic perspective, their model suffers from two drawbacks. Equilibrium constants obtained from Nielsen and Dungan's calibration are not constrained by the available standard state thermodynamic data for the solid and liquid phases. As an example, the log K for the reaction

$$\underset{\text{enstatite}}{Mg_2Si_2O_6} = \underset{\text{forsterite}}{Mg_2SiO_4} + \underset{\text{liquid}}{SiO_2}$$

can be derived from combining Nielsen and Dungan's equilibrium constants for enstatite- and forsterite-melt equilibria. The value obtained, 2701/T-2.03, predicts melt silica activities associated with the pure end-member solid phases that disagree strongly with available thermodynamic data (see Fig. 3, Chapter 13, this volume). The predicted and calculated silica activity curves are essentially orthogonal, intersecting at 1200°C, but above 1200° Nielsen and Dungan predict enstatite to be stable in the presence of leucite, nepheline and perovskite, and below 1200° forsterite to be stable in the presence of quartz. The difficulty exhibited here probably stems in part from the assumption of ideality. Nielsen and Dungan's extracted "equilibrium constants" are forced to account for any non-ideal behavior the system may exhibit, including the decoupling of melt mole fractions between the two lattices. This point highlights the second drawback of the Nielsen and Dungan approach and is a general criticism of all thermodynamic descriptions of the melt phase that assume ideal mixing of melt components. Silicate melts exhibit liquid immiscibility and consequently substantial positive deviations from ideality over much of the natural composition range.

One way out of this dilemma is to adopt a mathematical form for a solution model that has the potential to account for non-ideal behavior, yet is consistent with available thermodynamic measurements on multicomponent silicate systems. The trick is to satisfy all available thermodynamic data for the pure component solid and liquid phases (the standard state) in addition to those measurements of the mixing properties of multicomponent melts. Ghiorso et al. (1983) have attempted to construct such a melt model by applying multicomponent regular solution theory to the melt phase. Their expression for the Gibbs free energy of the melt is an extended, multicomponent version of that proposed by Carmichael et al. (1977). It may be written:

$$G = \sum_i n_i \mu_i^o + RT \sum_i n_i \ln X_i + n \tfrac{1}{2} \sum_i \sum_j W_{ij} X_i X_j$$

$$+ n RT \left[X_w \ln X_w + (1-X_w) \ln (1-X_w) \right] , \qquad (5)$$

where n_i denotes the moles of the i-th component in the liquid, X_i the associated mole fraction, X_w the mole fraction of water in the melt, n the sum of liquid moles, and the W_{ij} the regular solution binary interaction parameters. The superscript ° refers to the standard state of unit activity of the pure melt component at any T and P. The last term in Equation (5) represents an extension of the configurational entropy term to hydrous melts and insures that the activity of water in the melt will be proportional at low water concentrations to the square of the mole fraction (Stolper, 1982a,b). If the liquid components are chosen to be linear combinations of the simple oxides, it has been shown by Lange and Carmichael (1987) that to a level of precision estimated to be 0.6 %, anhydrous silicate melts (containing less than 20 wt % TiO_2) exhibit no excess volume of mixing. Lange and Carmichael also demonstrate that the first pressure and temperature derivatives of the excess volume are zero at atmospheric pressure. There are no experimental data available from elevated pressures to modify the conclusions drawn from the 1 bar results. These observation require the W_{ij} in Equation (5) to be pressure independent. To a relative precision of 4.4 %, ideal mixing relations are exhibited by the heat capacities of oxide components in the melt (Stebbins et al., 1984), which requires the temperature dependence of the excess entropy of mixing to be zero (dW_{ij}/dT must be a constant). Ghiorso et al. (1983) further simplify their formulation by assuming that the W_{ij} are independent of temperature. Compositional derivatives of Equation (5) yield the chemical potential of the k-th "anhydrous" liquid component:

$$\mu_k = \mu_k^o + RT \ln X_k + \sum_i X_i W_{ik} - \tfrac{1}{2} \sum_i \sum_j W_{ij} X_i X_j + RT \ln (1-X_w) \quad (6a)$$

and the chemical potential of water in the melt:

$$\mu_w = \mu_w^o + 2\, RT \ln X_w + \sum_i X_i W_{iw} - \tfrac{1}{2} \sum_i \sum_j W_{ij} X_i X_j \quad . \quad (6b)$$

Adopting a set of liquid components, for which the standard state data are either known or can be estimated, fixes the equilibrium constants for solid-liquid reactions in the magma and allows the interaction parameters in Equations (5) and (6) to be calibrated from experimental data (e.g. Bottinga et al., 1981; Ghiorso et al., 1983). In the next section we review the types of experimental data and calibration methods that allow thermodynamic models to be constructed for magmatic systems.

CALIBRATION METHODS

The majority of experimental data available for the calibration of melt models involves equilibration of the liquid with a solid phase. Ideally, such experiments provide knowledge of coexisting solid and liquid compositions, the temperature and pressure of the experiment and the oxidation state under which equilibrium was achieved. Often, less information is available, but the results are still usable.

Consider a solid phase whose composition can be described in terms of a set of end-member components (ϕ_p) which we will index on p. If thermodynamic components, c_k, are chosen to express the composition of the liquid which coexists with ϕ, then a set of p-chemical reactions between the liquid and the end-member components of the solid can be symbolized by:

$$\phi_p = \sum_k \nu_{pk} c_k \quad , \quad (7)$$

where the ν_{pk} denote the stoichiometric amount of the k-th liquid component in the p-th solid end-member. We can write the mass action expression corresponding to Equation (7) as:

$$-A_{\phi_p} = \Delta G_{\phi_p} = \Delta G^o_{\phi_p} + RT \sum_k \nu_{pk} \ln a_k - RT \ln a_{\phi_p} \quad , \quad (8)$$

where A and ΔG denote respectively the chemical affinity and Gibbs free energy change of the reaction in Equation (7). If the liquid chemical potentials are modeled by Equation (6a), and if the liquid and solid are assumed to be in equilibrium, then Equation (8) may be rewritten to yield:

$$-\Delta G^{o}_{\phi_p} + RT \ln a_{\phi_p} - \sum_k \nu_{pk} \left[RT \ln X_k + RT \ln (1-X_w) \right]$$

$$= \sum_k \nu_{pk} \sum_i W_{ki} X_i - \tfrac{1}{2} \left(\sum_k \nu_{pk} \right) \sum_i \sum_j W_{ij} X_i X_j \quad , \qquad (9)$$

where those quantities gathered on the left are "known", and the right hand side of the expression contains the, as yet, undetermined coefficients of the model. In order for the solid component activities (ϕ_p) to be calculated, thermodynamic models for the mineral phases must be adopted. If multiple statements of Equation (9) are assembled for a variety of solid phases and liquid compositions, then the binary interaction parameters can be extracted from the data using linear least-squares techniques rooted in generalized-inverse theory (Ghiorso, 1983).

The methods discussed in the previous paragraph have been employed by Ghiorso et al. (1983) to calibrate their expression (Eqn. 5) for the Gibbs free energy of natural silicate liquids. Additionally, they have utilized water solubility measurements in natural composition melts to calibrate binary interaction parameters between water and the "anhydrous" melt components. The interaction parameters so obtained are tabulated in Ghiorso et al. (1983, Table A4-3). Through Equation (5), they provide a description of the Gibbs energy of silicate melts in the range of "naturally occurring" compositions. It should be borne in mind that because of assumptions regarding the configurational entropy of the melt phase (cf. Eqn. 5), these parameters are unlikely to describe the excess free energy in the limiting binary and ternary systems that are far removed from "natural" melt compositions. The Ghiorso et al. (1983) regular solution model is a thermodynamically consistent, but local, approximation to the Gibbs surface that is applicable only to magmatic compositions and temperatures.

Experimental data are often available where the identity of some solid phase present on the liquidus at a specified T and P is known, but the specific composition of the solid has not been determined. Such data are still usable for calibration purposes if it is assumed that the bulk composition of the system is the same as that of the liquid at the instant when the first solid phase appears. To see this we reorganize Equations (8) and (9) to obtain:

$$-A_{\phi_p} - \Delta G^{o}_{\phi_p} - \sum_k \nu_{pk} \left[RT \ln X_k + RT \ln (1-X_w) \right] = \qquad (10)$$

$$- RT \ln X_{\phi_p} - RT \ln \gamma_{\phi_p} + \sum_k \nu_{pk} \sum_i W_{ki} X_i - \tfrac{1}{2} \left(\sum_k \nu_{pk} \right) \sum_i \sum_j W_{ij} X_i X_j$$

where as before the quantities to the left of the equality are "known" (the chemical affinity is zero if the phase is on the liquidus). As additional unknowns on the right-hand side of Equation (10) we have the p-mole fractions in the solid phase. It should be recalled that the activity coefficients in the solid are implicitly functions of these mole fractions so that the γ's do not represent additional parameters. Equation (10) may be used to establish a least-squares constraint on the coefficients W_{ij} (or the parameters of equivalent model formulations) if it is recognized that the sum of the mole fractions in the solid phase must be unity. Thus, if there are three end-members in the solid solution forming on the liquidus, only two end-member mole fractions may be obtained independently from the three statements of Equation (10). The "extra equation" is available to constrain the interaction parameters. The least-squares problem corresponding to multiple statements of Equation (10) is non-linear, and results in an estimate of the composition of the solid phase present on the experimental liquidus as well.

As a final calibration technique, it is possible to utilize information regarding the absence of a solid phase in order to constrain the liquid solution properties. Such constraints allow petrologic experience and intuition to play a quantitative role in the development of magmatic solution models. For instance, leucite will never crystallize from a tholeiitic liquid nor will low-Ca pyroxene ever form from an alkali-olivine basalt. If, at a given T and P, a solid phase is absent from a liquid of specified bulk composition, then the chemical affinity associated with the formation of that phase will be strictly greater than zero, i.e. the phase will be undersaturated in the melt. This affinity is defined as the weighted sum over all the end-member reactions:

$$A_{\phi} = \sum_{p} X_{\phi_p} A_{\phi_p} \quad . \tag{11}$$

If we assume that the compositions of the hypothetical solid end-members can be defined in terms of exchange equilibrium with the melt (Ghiorso, 1987), then "exchange reactions" of the form:

$$\phi_p + \sum_k \nu_{P+1,k}\, c_k = \phi_{P+1} + \sum_k \nu_{pk}\, c_k \quad , \tag{12}$$

provide relationships between the affinities of the end-member reactions:

$$A_{\phi_{p+1}} - A_{\phi_p} = 0 \quad . \tag{13}$$

Combining Equations (11) and (13) results in an equivalence of the end-member reaction affinities and the overall affinity for undersaturation:

$$A_{\phi_p} = A_{\phi} \quad \text{for all p.} \tag{14}$$

Insertion of Equation (14) into Equation (10) demonstrates how knowledge of the absence of a solid phase from the equilibrium assemblage adds but one additional unknown, A_{ϕ}, to the modified mass-action expression. A value for the overall affinity may be obtained in the course of the least-squares calibration, and the restriction that this affinity be positive translates to a constraint on the binary interaction model parameters.

To make this discussion of "phase-absent" constraints a bit more concrete, let us consider a magmatic liquid, which at a given T and P does <u>not</u> coexist with olivine. Let $Mg_4Si_2O_8$ and $Fe_4Si_2O_8$ be two liquid components and Mg_2SiO_4 and Fe_2SiO_4 be components in the olivine. The potential dissolution reaction between solid and liquid then becomes:

$$Mg_2SiO_4 \;=\; \tfrac{1}{2}\, Mg_4Si_2O_8$$

solid liquid

$$Fe_2SiO_4 \;=\; \tfrac{1}{2}\, Fe_4Si_2O_8 \quad ,$$

from which two statements of Equation (10) may be constructed

$$- \Delta G^{o}_{Fo} - \tfrac{1}{2} RT \ln X_{Mg_4Si_2O_8} - \tfrac{1}{2} RT \ln (1-X_w) = A_{01} - RT \ln X^{01}_{Fo} - RT \ln \gamma^{01}_{Fo} + \tfrac{1}{2} \sum_i W_{Mg_4Si_2O_8,i} X_i - \tfrac{1}{4} \sum_i \sum_j W_{ij} X_i X_j$$

$$- \Delta G^{o}_{Fa} - \tfrac{1}{2} RT \ln X_{Fe_4Si_2O_8} - \tfrac{1}{2} RT \ln (1-X_w) = A_{01} - RT \ln X^{01}_{Fa} - RT \ln \gamma^{01}_{Fa} + \tfrac{1}{2} \sum_i W_{Fe_4Si_2O_8,i} X_i - \tfrac{1}{4} \sum_i \sum_j W_{ij} X_i X_j$$

To the above mass action expressions may be added the constraint that

$$X_{Fo} + X_{Fa} = 1 \quad ,$$

which gives us three equations and the ability to uniquely define values for X_{Fo}, X_{Fa} and A_{ol}, and apparently provides no added information concerning the W_{ij}'s. Recognizing that the reactions must proceed to the right, however, allows us to write

$$A_{ol} > 0 \quad ,$$

which clearly demonstrates that the constraint of undersaturation influences the calibration of the W_{ij} to force A_{ol} to be positive. As additional information and in the course of solving the least squares problem, we obtain the composition of the olivine on the saturation surface at a point which is "closest" to equilibrium with this liquid at the indicated T and P.

CALCULATING SOLID-LIQUID EQUILIBRIA

Although the formulation and calibration of thermodynamic models for magmatic melts is a stimulating exercise in and of itself, the real impetus for the creation of such a model is the prediction of phase equilibria in systems that have not been investigated experimentally. With further assumptions, melt models can be used to predict the stoichiometry of irreversible reaction paths of petrological interest. The rest of this chapter will be devoted to a review of methods of predicting phase equilibria and reaction paths in magmatic systems. The methods described below are applicable for any thermodynamically consistent description of the solution properties of the liquid phase.

Notation and mathematical statement of the problem

Let us write the total Gibbs free energy of the magma as G, and consider contributions to G divided according to the liquid (L) and solid (ϕ) phases in the system:

$$G = G_L + \sum_{\phi} G_{\phi} \quad . \tag{15}$$

Although Equation (15) is written assuming the presence of one liquid phase, it can be readily generalized to include two or more coexisting immiscible liquids. We will express the composition of the system by choosing a set of components for each phase. These components will be arranged as the elements of a number of vectors, one vector for each phase in the system. Thus, the vector $\mathbf{n}_L$ contains elements whose values represent the mole numbers of each of the liquid components, the vector $\mathbf{n}_{\phi=1}$ contains elements whose values represent the mole numbers of the end-member components in the first solid phase, etc. The lengths of the vectors $\mathbf{n}_L$, $\mathbf{n}_{\phi=1}$, $\mathbf{n}_{\phi=2}$, etc. need not be equal, as the number of thermodynamic components necessary to describe the composition of each phase in the system need not be the same. A vector that describes the composition of the system can now be constructed from the individual "component"-vectors for each phase. We will call this vector $\mathbf{n}$, and construct it by stacking all those vectors previously defined on top of each other:

$$\mathbf{n} = \begin{bmatrix} \mathbf{n}_L \\ \mathbf{n}_{\phi=1} \\ \mathbf{n}_{\phi=2} \\ \cdot \\ \cdot \\ \cdot \end{bmatrix} \qquad \text{i.e.} \tag{16}$$

The picture on the right is meant to illustrate the process involved in constructing $\mathbf{n}$, and to emphasize the fact that the individual vectors that contribute to $\mathbf{n}$ need not be equal in length. Equation (16) allows us to define a new vector, $\boldsymbol{\mu}$, as the derivative of G with respect to $\mathbf{n}$:

$$\boldsymbol{\mu} = \left(\frac{\partial G}{\partial \mathbf{n}}\right)_{T,P} = \begin{bmatrix} \mu_L \\ \mu_{\phi=1} \\ \mu_{\phi=2} \\ \cdot \\ \cdot \\ \cdot \end{bmatrix} \qquad \text{i.e.} \tag{17}$$

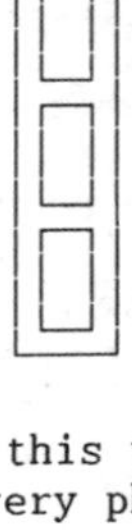

We use the symbol μ for this quantity because this vector contains the chemical potentials of every component in every phase in the system.

Note that it too has a segmented structure like $\mathbf{n}$. The bulk composition of the system, which we will denote by the vector $\mathbf{b}$, may be expressed in terms of the components of any phase we choose. For convenience, let us use the same compositional variables for the system as we do for the liquid phase. It then remains to define a set of transformation matrices, $\mathbf{T}_\phi$, which stoichiometrically convert "solid-phase" components into "liquid-phase" components. These definitions allow us to write:

$$\mathbf{b} = \mathbf{n}_L + \sum_\phi \mathbf{T}_\phi \mathbf{n}_\phi \tag{18}$$

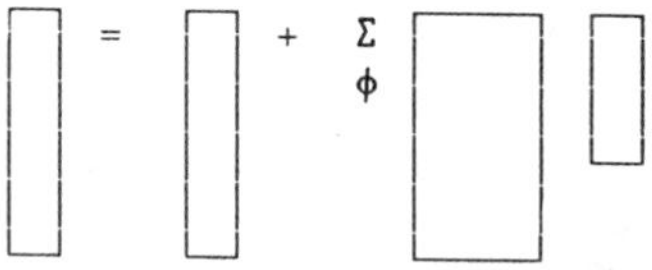

The picture on the right lets us visualize the $\mathbf{T}_\phi$ as projection operators that transform one component (coordinate) space into another. Just as it is convenient to group the $\mathbf{n}_L$, $\mathbf{n}_{\phi=1}$, $\mathbf{n}_{\phi=2}$, etc. to form the vector $\mathbf{n}$, it is convenient to assemble the matrices $\mathbf{T}_{\phi=1}$, $\mathbf{T}_{\phi=2}$, etc. to form the matrix $\mathbf{C}$. We write

$$\mathbf{C} = \left[\mathbf{I} \mid \mathbf{T}_{\phi=1} \mid \mathbf{T}_{\phi=2} \mid \cdots\right] \tag{19}$$

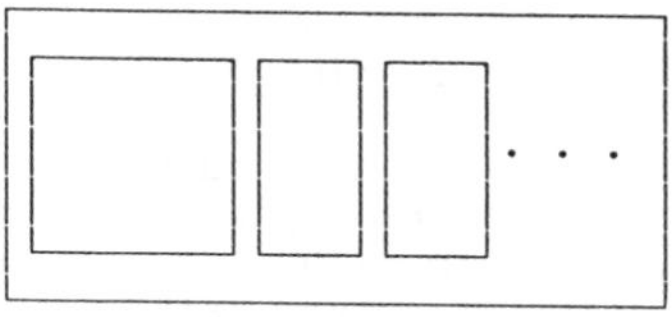

where the picture on the right demonstrates that we are forming a "block-partitioned" matrix with the number of rows equal to the number of liquid components and the number of columns equal to the sum of all components in every phase. The left-hand most block of $\mathbf{C}$ is the identity matrix ($\mathbf{I}$) which transforms, trivially, the stoichiometric description of the liquid components into the compositional variables for the system. The definitions in Equations (16)-(19) allow us to state mathematically the problem of characterizing the equilibrium state in a multiphase thermodynamic system. There are two alternate formulations:

$$\begin{aligned} \mathbf{T}^T_{\phi=1}\, \mu_L &= \mu_{\phi=1} \\ \mathbf{T}^T_{\phi=2}\, \mu_L &= \mu_{\phi=2} \\ &\ \vdots \\ \mathbf{b} &= \mathbf{C}\,\mathbf{n} \end{aligned} \qquad \Bigg| \qquad \begin{aligned} &\text{minimize } G \\ &\text{with respect to } \mathbf{n} \\ &\text{such that } \mathbf{b} = \mathbf{C}\,\mathbf{n} \end{aligned} \tag{20}$$

Most readers will recognize the equilibrium criteria embodied in the system of non-linear equations to the left of the vertical line in Equation (20) (the system is non-linear because the chemical potentials are non-linear functions of the component mole numbers). The system must be solved in terms of the elements of the vector $\mathbf{n}$, and the equations constitute the usual set of liquid-solid mass action expressions coupled with a statement of conservation of mass for the system ($\mathbf{b} = \mathbf{Cn}$). This method of solving for the equilibrium state of a magmatic system has been used by Russell and Nicholls (1985) with great success. It also forms the core of most calculational algorithms based upon "distribution" coefficients and mass balance constraints (Langmuir and Hanson, 1981; Nielsen and Dungan, 1983). The "integrated form" of this non-linear system is provided by the equilibrium criteria stated to the right of the vertical line in Equation (20). This equivalence may become clearer if it is recalled that to minimize a function (G), you take the derivative with respect to the unknowns ($\mathbf{n}$), set the result equal to zero (which generates the mass action expressions), and solve for the optimal values ($\mathbf{n}$ at equilibrium), subject to any constraints on the minimum ($\mathbf{b} = \mathbf{Cn}$). Thus, the right-hand side is the integrated form of the left-hand side in Equation (20). From a computational perspective, it turns out to be a little easier to minimize G directly by a method other than equating its derivative to zero, rather than the alternative of solving the non-linear system of mass-action/mass-balance expressions. It also proves to be conceptually easier to modify the integrated form of the equilibrium criteria in order to characterize chemical equilibrium in open magmatic systems, and in systems subject to very general non-linear constraints (such as isenthalpic, isochoric constraints). In the next section we will summarize an algorithm suitable for finding the minimum of G in a multiphase system.

An algorithm for finding the minimum of G

Considerable attention has been focused in the chemical engineering (the classic paper by White et al., 1958; more recently the review by Smith and Missen, 1982), mathematical (cf. Gill et al., 1981), and geochemical literature (Karpov and Kaz'min, 1972; Karpov et al., 1973; Shvarov, 1976; Saxena, 1982; Ghiorso, 1985; Harvie et al., 1987) on the construction of numerical algorithms that are capable of finding rapidly the minimum of a complex potential function like G subject to simply defined linear constraints (such as those embodied in mass conservation equations and implied by the constancy of temperature and pressure). We will outline the algorithm used by Ghiorso (1985) for calculation of magmatic phase relations.

Whenever faced with understanding the behavior of a complex non-linear function, the usual method is to expand the function in a Taylor series and truncate after a suitable (preferably small number) of terms. This mathematical procedure requires that we know the properties of the function at some suitable reference point, "close" to the point of interest. In the context of minimizing G, the reference point, which we will denote $\tilde{\mathbf{n}}$, will provide an initial guess to the equilibrium distribution of material between the solid and liquid phases in the system. It will turn out that $\tilde{\mathbf{n}}$ need not be chosen very precisely in the context of numerical values for the mole numbers of all the system components, but that the identity of phases that will eventually constitute the equilibrium assemblage must be known in order to construct a suitable guess for this vector. Therefore, prior to solving the minimization problem, we must have a method for determining, at any specified T, P and liquid composition, the degree of supersaturation of any solid phase which might potentially precipitate from the liquid. Fortunately, this problem has already been solved.

In the previous section on calibration methods, we discussed the case of obtaining information on liquid solution properties by knowing that a solid phase is present on the liquidus at a specified T, P and liquid composition, without a specific knowledge of the composition of the solid. The relevant argument centered around discussion of Equation (10). We now do the calibration "backwards". Knowing the liquid solution properties (the W_{ij}), and assuming the potential solid is on the liquidus ($A_\phi = 0$), we write one statement of Equation (10) for each end-member in the solid-solution and solve these equations simultaneously for the mole fractions of each end-member. The sum of these calculated mole fractions we call the saturation index (Σ_ϕ, after Reed, 1982). If the saturation index is equal to unity, then the solid phase of interest is on the liquidus (just saturated), if Σ_ϕ is greater than unity, the solid phase is supersaturated in the magma, and if less than unity, the solid phase in undersaturated. By reviewing the saturation indices of all potential solid phases, the identity of the potential equilibrium phase assemblage is indicated. The vector $\tilde{\mathbf{n}}$ may be constructed by assigning trivial masses to the potential solid phases and taking the liquid composition as essentially that of the bulk composition of the system. It is important to realize that all initially supersaturated solid phases may not eventually appear in the equilibrium description of the system, just as some solid phases may supersaturate only after others have precipitated from the liquid. Our numerical algorithm must be robust enough to detect these eventualities, and specific methods for dealing with these problems are discussed below and in Ghiorso (1985).

Having chosen $\tilde{\mathbf{n}}$, we expand G in a multivariable Taylor series:

$$G \simeq G\Big|_{\tilde{\mathbf{n}}} + \mu^T\Big|_{\tilde{\mathbf{n}}}(\mathbf{n}-\tilde{\mathbf{n}}) + \tfrac{1}{2}(\mathbf{n}-\tilde{\mathbf{n}})^T \mathbf{H}\Big|_{\tilde{\mathbf{n}}}(\mathbf{n}-\tilde{\mathbf{n}}) \tag{21}$$

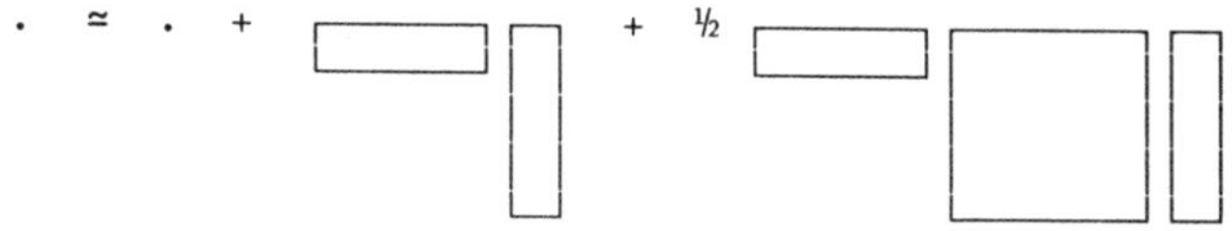

and truncate the series after the third term. The second derivative of G with respect to $\mathbf{n}$ (i.e. $\partial\mu/\partial\mathbf{n}$) is a matrix of second cross partial derivatives which we abbreviate as $\mathbf{H}$. The name for this matrix in the mathematical literature is the Hessian matrix. In Equation (21) the function (G) and its derivatives (μ and $\mathbf{H}$) are numerically evaluated at the point about which G is expanded ($\tilde{\mathbf{n}}$). The symbol $|_{\tilde{\mathbf{n}}}$ indicates the point of evaluation and the superscript ^{T} indicates that the transpose of the vector must be taken. Examination of the picture associated with Equation (21) shows that the transpose operator merely lays the vector on its side and that the vector-vector and vector-matrix-vector multiplications are so arranged as to maintain each term in the Taylor series as a scalar quantity.

The advantage of Equation (21) is that the minimum in G can be expressed as the analytical solution of a system of linear equations because G has been rendered as a quadratic function of $\mathbf{n}$. There are two problems, however. The solution vector $\mathbf{n}$ so obtained, represents the minimum of G only if $\tilde{\mathbf{n}}$ is "close" to $\mathbf{n}$. This can be dealt with by repeatedly choosing $\tilde{\mathbf{n}}$ until the calculated value of $\mathbf{n}$ which minimizes Equation (21) is identical to it. The second problem is that no knowledge of the equality constraints ($\mathbf{Cn} = \mathbf{b}$) is present in Equation (21), so that the value of **n** which minimizes Equation (21) will most likely violate the mass balance constraint on the system. The only way out of this dilemma is to incorporate the mass balance constraint directly into Equation (21). There is a very simple way to do this.

Think of the elements of the vector **n** as defining numerical values along the axes of some multidimensional orthogonal coordinate system - one perpendicular direction for each element of **n**. Thus, as a simple example we might visualize a three variable system with axes corresponding to the amount of MgO in the liquid, the amount of SiO_2 in the liquid, and the amount of forsterite in the system. The vector **n** points to a particular spot in this multidimensional space. The mass balance constraints tell us that all points (or vectors) in this space are not physically permissible solutions to the minimum of G. In particular if the amount of MgO and SiO_2 is fixed in our simple system example, then the set of physically permissible solutions forms a line in this three variable space. We are free to choose any solution vector whose tip touches that line of permissible solutions but not any other vector pointing to an arbitrary amount of liquid MgO, liquid SiO_2 and forsterite. To maintain feasibility of our solution vector we must restrict our attention to choosing vectors which terminate on the feasible line, or in mathematical terms, we must project our solution vector into the "null space" of the constraints. This is most easily done by rotating our coordinate system so that the "feasible line" is parallel to one of the rotated axes. The intersection of the "feasible line", which is referred to as the null space, with the plane formed by the second and third rotated axes, which we call the range space or constraint space, fixes the value of this coordinate and embodies the mass constraint on the system. We are now free to chose the coordinates of any point along the feasible line which minimizes G without worrying about the system mass constraint. In general terms, our multidimensional space which contains the vector **n** must be reduced in dimension by the number of mass constraints so that we are free to minimize G by choosing vectors arbitrarily in the null space. In the general problem, we can rotate the coordinate system so that the null space is orthogonal to the constraint space by constructing a projection matrix (**K**) which renders the constraint matrix (**C**) block upper triangular (Lawson and Hanson, 1974):

$$\mathbf{C}\,\mathbf{K} = \mathbf{R} \tag{22}$$

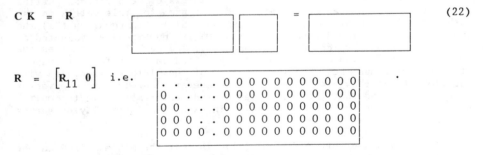

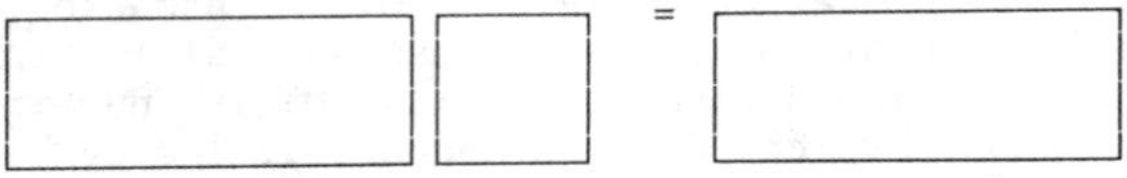

$$\mathbf{R} = \begin{bmatrix} \mathbf{R}_{11} & \mathbf{0} \end{bmatrix} \quad \text{i.e.} \quad \begin{bmatrix} . & . & . & . & . & 0 & 0 & 0 & 0 & 0 & 0 & 0 & 0 & 0 & 0 & 0 \\ 0 & . & . & . & . & 0 & 0 & 0 & 0 & 0 & 0 & 0 & 0 & 0 & 0 & 0 \\ 0 & 0 & . & . & . & 0 & 0 & 0 & 0 & 0 & 0 & 0 & 0 & 0 & 0 & 0 \\ 0 & 0 & 0 & . & . & 0 & 0 & 0 & 0 & 0 & 0 & 0 & 0 & 0 & 0 & 0 \\ 0 & 0 & 0 & 0 & . & 0 & 0 & 0 & 0 & 0 & 0 & 0 & 0 & 0 & 0 & 0 \end{bmatrix} .$$

We will not pursue here algorithms for constructing **K**, but will only mention the relevant facts which will be used later, namely that **K** is square and orthogonal (which means that $\mathbf{K}^T\mathbf{K} = \mathbf{I}$). As the picture associated with Equation (22) shows, the resulting matrix **R** has an interesting partitioned structure. The left partition is an upper triangular matrix (only the upper triangle contains non-zero elements) which we will symbolize as R_{11}, and the left partition consists entirely of zeroes. In order to incorporate this supposedly useful decomposition of the constraint matrix into Equation (21) we partition the matrix $\mathbf{K}$ into two parts according to the following definition and picture:

$$\mathbf{K} = \left[\mathbf{K}_1 \mid \mathbf{K}_2\right] \qquad (23)$$

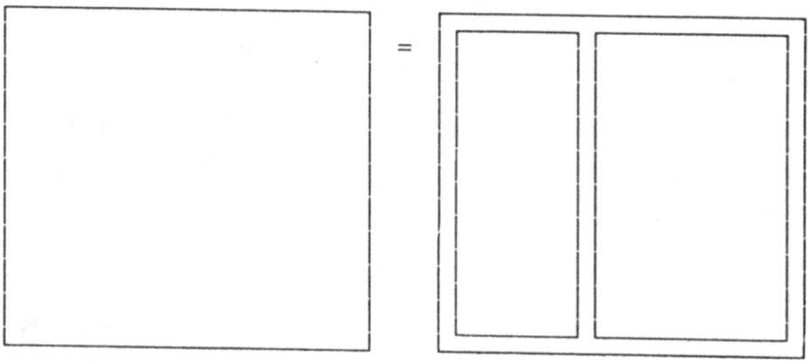

The matrix $\mathbf{K}_1$ has column dimension equal to the number of mass balance constraints and the matrix $\mathbf{K}_2$ has column dimension equal to that of the null space. The columns of $\mathbf{K}_1$ can be thought of as a set of vectors which define the rotated set of orthogonal axes of the constraint space, and the columns of $\mathbf{K}_2$, vectors which define the orthogonal axes of the null space. By specifying coordinates, or solutions to the minimization problem, in the constraint space and null space reference frame, we simplify the problem by mathematically decoupling the two aspects of the solution. Let $\mathbf{n}_1$ denote a vector of coordinates in the constraint space, and let $\mathbf{n}_2$ denote a vector of coordinates in the null space. We define these vectors according to the following formula:

$$\mathbf{n} = \mathbf{K}_1 \mathbf{n}_1 + \mathbf{K}_2 \mathbf{n}_2 \qquad (24)$$

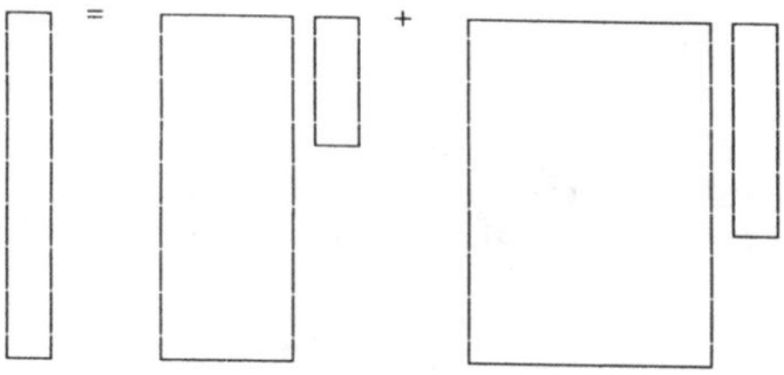

where the associated picture emphasizes the various lengths and widths of the vectors and matrices involved. We proceed now by rewriting Equation (22) as:

$$\mathbf{C} = \mathbf{R}\,\mathbf{K}^T \quad ,$$

which allows us to express the mass constraint on the system as:

$$\mathbf{C}\,\mathbf{n} = \mathbf{R}\,\mathbf{K}^T\,\mathbf{n} = \mathbf{b} \quad . \qquad (25)$$

The various definitions adopted in Equations (22), (23) and (24) permit Equation (25) to be rewritten as:

$$\left[\mathbf{R}_{11} \mid \mathbf{0}\right]\left[\mathbf{K}_1 \mid \mathbf{K}_2\right]^T \mathbf{n} = \mathbf{b} \quad . \qquad (26)$$

Careful examination of Equation (26) reveals that the column dimension of $\mathbf{R}_{11}$ and $\mathbf{K}_1$ are identical so that the matrix of zeroes multiplies elements of $\mathbf{K}_2$ on the left-hand side of the equation. This observation tells us that $\mathbf{K}_2$ will not contribute to the specification of $\mathbf{n}$ in Equation (26). Since the columns of $\mathbf{K}_1$ and $\mathbf{K}_2$ are orthogonal, we have from Equation (24):

$$\mathbf{K}_1^T \mathbf{n} = \mathbf{K}_1^T \mathbf{K}_1 \mathbf{n}_1 + \mathbf{K}_1^T \mathbf{K}_2 \mathbf{n}_2 = \mathbf{n}_1 \quad ,$$

which allows us to rewrite Equation (26) in a form which uniquely defines the coordinates of the constraint space:

$$\mathbf{n}_1 = \mathbf{R}_{11}^{-1} \mathbf{b} \quad . \tag{27}$$

The projection matrix $\mathbf{K}$ has permitted us to transform our mass balance constraints into a form (Eqn. 27) which fixes the coordinates of the equilibrium description of the system in the constraint space. The mass balance constraints put no restriction on $\mathbf{n}_2$, the coordinates of the solution in the null space. To find these coordinates we minimize the Gibbs free energy with respect to $\mathbf{n}_2$. Taking our definition for $\mathbf{n}$ (Eqn. 24), we rewrite Equation (21) to define the equilibrium condition in terms of the unconstrained minimum with respect to $\mathbf{n}_2$:

$$\text{minimize} \quad G\Big|_{\tilde{\mathbf{n}}} + \mu^T\Big|_{\tilde{\mathbf{n}}}(\mathbf{K}_1\mathbf{n}_1 + \mathbf{K}_2\mathbf{n}_2 - \tilde{\mathbf{n}})$$

$$+ \tfrac{1}{2}\,(\mathbf{K}_1\mathbf{n}_1 + \mathbf{K}_2\mathbf{n}_2 - \tilde{\mathbf{n}})^T\, \mathbf{H}\Big|_{\tilde{\mathbf{n}}}\, (\mathbf{K}_1\mathbf{n}_1 + \mathbf{K}_2\mathbf{n}_2 - \tilde{\mathbf{n}}) \tag{28}$$

with respect to $\mathbf{n}_2$.

The value of $\mathbf{n}_2$ which minimizes the quadratic approximation to the Gibbs function in Equation (28), is obtained by setting the first derivative equal to zero. A bit of manipulation yields:

$$\mathbf{K}_2^T \mathbf{H}\Big|_{\tilde{\mathbf{n}}} \mathbf{K}_2 \mathbf{n}_2 = -\mathbf{K}_2^T \mu\Big|_{\tilde{\mathbf{n}}} - \mathbf{K}_2^T \mathbf{H}\Big|_{\tilde{\mathbf{n}}} \mathbf{K}_1 \mathbf{n}_1 \tag{29}$$

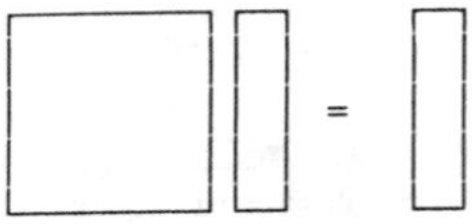

where all quantities are known except for the vector $\mathbf{n}_2$. The picture in Equation (29) reveals that a solution for the vector $\mathbf{n}_2$ involves solving a system of <u>linear</u> equations. Once $\mathbf{n}_2$ has been determined, that is the coordinates in the null space have been fixed, the constrained solution vector $\mathbf{n}$ can be reconstructed using Equation (24).

Although the above procedure may appear unnecessarily complicated, the simplicity arises through the fact that the final solution is obtained via a linear system of equations. Though additional numerical enhancements help speed convergence (Ghiorso, 1985), essentially setting the derived $\mathbf{n}$ equal to $\tilde{\mathbf{n}}$ and repeating the above process four to five times generally determines the equilibrium state of the system within computational precision. At each iteration the matrices $\mathbf{K}$ and $\mathbf{R}$ and the vector $\mathbf{n}_1$ need not be redetermined, as the mass constraint on $\mathbf{n}$ doesn't change. The only complication that can occur is when a phase (usually one of the solids) decides to no longer participate in the equilibrium assemblage. At this point the mole numbers of each of its end-members will fall below some established tolerance and the phase will be deleted from the equilibrium configuration. This necessitates recalculating the **K**, **R** and $\mathbf{n}_1$ and iterating anew on $\mathbf{n}_2$. After the "final" equilibrium proportions of solids and liquid are obtained, the saturation indices for all solid phases not present in the system must be reevaluated in order to determine if some supersaturation has developed as a consequence of the precipitation of some other phase. If one is detected, a new phase must be added to the system and its proportions and composition determined through the minimization procedure. The whole process continues until no residual supersaturations are detected. This is the equilibrium state of the system.

Modeling irreversible reactions

At the onset it should be recognized that thermodynamics is insufficient to model the details of any interesting process associated with magma generation or crystallization. This is because all processes are irreversible, and to model their progress necessitates understanding their kinetics. Despite this, however, thermodynamics does give us a means of simulating an irreversible process. This can be done by discretizing the evolution of the system into a series of steps in temperature, pressure, enthalpy, bulk composition, or the like, and determining the equilibrium state of the system at each step. Linking the steps together gives us a net approximation to the overall system evolution. What we assume in this approximation is that the kinetics is rapid enough, with respect to the time available for the process to be accomplished, so that the system essentially maintains a continuous equilibrium state. This assumption is obviously better justified the higher the temperature. What we lose as a consequence of this assumption are the results of processes that depend on transient spatial gradients in the system. We cannot model the interplay of heat loss and crystal growth for example, nor make any comment concerning the time it takes the system to evolve from state A to B. Whether or not this modeling approximation is useful depends largely on the petrologic utility of the results. Calculations presented in the subsequent chapter will demonstrate that for magmas the step-wise equilibrium approach is quite informative despite its inherent simplicity.

The implementation of step-wise equilibrium modeling techniques does provide additional conceptual challenges, in that the "equilibrium path" along which the system is forced to evolve may not be one best characterized by explicit changes in temperature, pressure or system bulk composition (i.e. states characterized by a minimum in the Gibbs free energy). We may wish to model the evolution of the system along an oxygen fugacity buffer, or trace the temperature of the system by modeling the withdrawal of known amounts of heat from the magma. These scenarios demand that we be able to characterize equilibrium in the evolving model system as the minimum of a suitable potential function of quite general thermodynamic variables. For example, suppose we wanted to model the crystallization of a magma body emplaced high enough in the crust, such that the walls of the magma chamber behave as a rigid box. Clearly, the independent variables that characterize equilibrium in such a system should be temperature, bulk composition and volume rather than pressure. Thus, we seek a thermodynamic potential which is minimal in a system at equilibrium under specified temperature, volume and bulk composition. This is the Helmholtz free energy. The evolution of the system in our example could be calculated by first minimizing the Gibbs free energy at the initial temperature, lithostatic pressure and magma composition to determine the correct phase assemblage and system volume. Then to simulate crystallization, the temperature of the system could be decremented, and an equilibrium assemblage calculated at each step by minimizing the Helmholtz potential subject to specified (fixed) volume and bulk composition. The internal pressure of the magma would be calculated at each step as a derived rather than specified variable (just as volume is calculated as a derived quantity when G is minimized at a given T and P).

The four commonly used thermodynamic potentials do not offer sufficient possibilities for modeling the myriad reaction paths of petrological interest. For this reason serious modelers of magmas require serious potentials. We discuss a method for the design of these potentials in the next section.

GENERALIZED THERMODYNAMIC POTENTIALS

Legendre transformations

Generalized thermodynamic potentials have found their greatest utility in the description of equilibria in open systems. Korzhinskii (1949, 1956, 1959) was perhaps the first to discuss how equilibrium in a system open to a particular component could be characterized by minimizing a suitably constructed thermodynamic potential, as long as the corresponding chemical potential of that component was fixed by external constraints. Such components he termed perfectly mobile components. Korzhinskii demonstrated the construction of these thermodynamic potentials by the use of Legendre transforms. We outline this method below, and show how it can be used to construct quite general functions.

Let Ψ be a thermodynamic potential which is minimal in a system at thermodynamic equilibrium for specified values of its independent variables, x_i, indexed on i. Ψ may be the Gibbs free energy of the system, which is minimal at specified T, P and bulk composition. New potentials can be readily constructed from Ψ, if we transform the independent variable x_i to $(\partial\Psi/\partial x_i)$. The new potential, Φ, is defined by the Legendre transform of Ψ:

$$\Phi\left(x_1, x_2, .., \left(\frac{\partial\Psi}{\partial x_i}\right)_{x_j, i\neq j}, ..\right) = \Psi(x_1, x_2, .., x_i, ..) - x_i \left(\frac{\partial\Psi}{\partial x_i}\right)_{x_j, i\neq j} \tag{30}$$

As an example, if we begin with the Gibbs free energy, we can "replace" the independent variable P with the variable $(\partial G/\partial P)$, the system volume, to form the new potential A, the Helmholtz free energy:

$$A(T,V,\mathbf{n}) = G(T,P,\mathbf{n}) - \left(\frac{\partial G}{\partial P}\right)_{T,\mathbf{n}} P = G(T,P,\mathbf{n}) - V\,P \quad .$$

Perhaps a slightly more interesting exercise, is to replace one of the system components, say oxygen, with the chemical potential of oxygen. This can be done singly, or in conjunction with other independent variables as the following table demonstrates:

$G = G(T,P,\mathbf{n})$	$L(T,P,\mathbf{n}^*,\mu_{O_2}) = G - n_{O_2}\mu_{O_2}$	(31)
	$A(T,V,\mathbf{n}) = G - PV$	(32)
	$A^*(T,V,\mathbf{n}^*,\mu_{O_2}) = G - PV - n_{O_2}\mu_{O_2}$	(33)

Equation (31) provides a definition of an appropriate "Korzhinskii" potential for the calculation of thermodynamic equilibrium in magmatic systems crystallizing along a fixed oxygen buffer. The $\mathbf{n}^*$ in Equation (31) refers to a subset of the original system components, save oxygen. The minimum in L is calculated for a system open to O_2, at constant T, P and chemical potential of O_2. Examples of the use of this function in reaction path modeling applications in magmas is demonstrated in the subsequent chapter.

Thermodynamic potentials which are minimal for specified heat content can be constructed using similar techniques applied to a slightly modified form of the Gibbs free energy. These functions were derived by Ghiorso and Kelemen (1987) and are defined in the following table:

$\frac{G}{T} = \frac{G}{T}\left(\frac{1}{T},P,\mathbf{n}\right)$	$-S(H,P,\mathbf{n}) = \frac{G}{T} - \frac{1}{T}H$	(34)
	$-S^*(H,P,\mathbf{n}^*,\mu_{O_2}) = \frac{G}{T} - \frac{1}{T}H - n_{O_2}\frac{\mu_{O_2}}{T}$	(35)

Equation (34) establishes that equilibrium in a system of specified heat content (enthalpy, H), pressure and bulk composition is given by a maximum in entropy, a result which allows isenthalpic reaction paths to predict temperature as a function of the degree of assimilation in magmatic systems.

Minimization of generalized thermodynamic potentials

With some modifications the numerical methods described in the section entitled, "An algorithm for finding the minimum of G," may be used to minimize the potentials given in Equations (31) through (35). We will not discuss these modifications, but rather give an indication of the difficulties via the following example. For details, the reader is referred to Ghiorso and Kelemen (1987) or Gill et al. (1981).

Suppose we wish to calculate the equilibrium state of a system given a specified heat content, Θ, pressure and bulk composition. Equation (34) tells us that the mathematical problem is:

$$\begin{aligned} &\text{minimize} && -S(H,P,\mathbf{n}) \\ &\text{such that} && \mathbf{C}\,\mathbf{n} = \mathbf{b} \\ &\text{and} && H(T,P,\mathbf{n}) = \Theta\ . \end{aligned} \tag{36}$$

As before (Eqn. 20), the mass constraint in Equation (36) is a linear function of **n**, whereas the new numerical twist is that H (the enthalpy) is a non-linear function of **n** and the new dependent variable T. As might be imagined, this makes it harder to apply the equality constraints in the minimization process. The problem can be visualized by referring back to our geometric construction of two rotated coordinate systems defining the null space and range space of constraints. In our previous discussion, the null space and range space were fixed; independent of the vector **n** and the scalar T. Since H is non-linear, the null space corresponding to Equation (36) will change orientation as a function of **n** (or T), which means any projection matrix equivalent to **K** will only apply locally "close" to a reference point, ($\tilde{\mathbf{n}}$,T°), where the constraints are satisfied. The feasible line, in the simple system example we used previously, now becomes a "deformed plane" or feasible surface in 4-space. The problem is one of staying on the feasible surface when choosing (**n**,T) in the process of minimizing -S.

The constraints in Equation (36) may be linearized by expansion in a Taylor series about some initial guess to the minimum of -S, ($\tilde{\mathbf{n}}$,T°), which itself satisfies the constraints. By truncating after the second term, the Taylor series approximation to the constraints is linear in **n** and T. This linearized system can be decomposed into a projection operator which locally defines an orthogonal range and null space for the linearized constraints. The function, -S, is expanded in a Taylor series about ($\tilde{\mathbf{n}}$,T°) and truncated, as before, after the quadratic term. It is minimized in the null space and the approximation to the true minimum, (**n**,T), reconstructed as in Equation (24). The problem is that (**n**,T) may not satisfy the heat content constraint, because (**n**,T) may be sufficiently different from ($\tilde{\mathbf{n}}$,T°) to render the first order Taylor series expansion of this constraint inadequate. In practice the situation is most easily corrected by adjusting T until the enthalpy constraint is satisfied. Then the whole minimization process is repeated, including in this case the construction of new projection operators, until the derived values of (**n**,T) are numerically equivalent to those of the presumed minimum ($\tilde{\mathbf{n}}$,T°). Thus the phase proportions and compositions as well as the temperature of the system are determined from the specified heat content, bulk composition and pressure.

SUMMARY

In this chapter methods for calibrating thermodynamic models to phase equilibrium data have been described and criteria for the correct formulation of these models have been reviewed. Thermodynamic formulations for the mixing properties of magmatic liquids allow heterogeneous equilibria to be modeled as a function of a variety of convenient independent variables. The generality and robustness of potential minimization techniques makes this numerical method best suited for the task at hand. The simulation of irreversible processes as a sequence of linked steps in reaction progress, with each step characterized by heterogeneous equilibrium, is a fruitful geochemical technique at temperatures and pressures where reaction kinetics are relatively rapid. The algorithms described in this chapter for the modeling of magmatic processes are based on this approximation. Software which implements the methods discussed here has been described previously (Ghiorso, 1985; Ghiorso and Carmichael, 1985; Ghiorso and Kelemen, 1987) and is available from the author. Modeling examples of petrologic interest are presented in the following chapter.

REFERENCES

Bottinga, Y., Richet, P. (1978) Thermodynamics of liquid silicates, a preliminary report. Earth Planet. Sci. Letts. 40, 382-400.

Bottinga, Y., Weill, D.F., Richet, P. (1981) Thermodynamic modeling of silicate melts. In: R.C. Newton, A. Navrotsky, B.J. Wood, eds., Thermodynamics of Minerals and Melts, Springer-Verlag, New York, 207-302.

Burnham, C.W. (1981) The nature of multicomponent aluminosilicate melts. In: Chemistry and Geochemistry of Solutions at High Temperatures and Pressures, D.T. Rickard and F.E. Wickman, eds., Phys. Chem. Earth 13/14, 197-229.

Burnham, C.W., Nekvasil, H. (1986) Equilibrium properties of granite pegmatite magmas. Amer. Mineral. 71, 239-263.

Carmichael, I.S.E., Ghiorso, M.S. (1986) Oxidation-reduction relations in basic magma: a case for homogeneous equilibria. Earth Planet. Sci. Letts. 78, 200-210.

Carmichael, I.S.E., Nicholls, J., Spera, F.J., Wood, B.J., Nelson, S.A. (1977) High-temperature properties of silicate liquids: applications to the equilibration and ascent of basic magma. Phil. Trans. R. Soc. London A 286, 373-431.

Doyle, C.D., Naldrett, A.J. (1986) Ideal mixing of divalent cations in mafic magma and its effect on the solution of ferrous oxide. Geochim. Cosmochim. Acta 50, 435-443.

Doyle, C.D., Naldrett, A.J. (1987) Ideal mixing of divalent cations in mafic magma. II. The solution of NiO and the partitioning of nickel between coexisting olivine and liquid. Geochim. Cosmochim. Acta 51, 213-219.

Ghiorso, M.S. (1983) LSEQIEQ: A FORTRAN IV subroutine package for the analysis of multiple linear regression problems with possibly deficient pseudorank and linear equality and inequality constraints. Computers and Geosciences 9, 391-416.

Ghiorso, M.S. (1985) Chemical mass transfer in magmatic processes. I. Thermodynamic relations and numerical algorithms. Contrib. Mineral. Petrol. 90, 107-120.

Ghiorso, M.S. (1987) Chemical mass transfer in magmatic processes. III. Crystal growth, chemical diffusion and thermal diffusion in multicomponent silicate melts. Contrib. Mineral. Petrol. 96:291-313.

Ghiorso, M.S., Carmichael, I.S.E. (1980) A regular solution model for met-aluminous silicate liquids: Applications to geothermometry, immiscibility, and the source regions of basic magma. Contrib. Mineral. Petrol. 71, 323-342.

Ghiorso, M.S., Carmichael, I.S.E. (1984) Comment on "Density calculations for silicate liquids. I. Revised method for aluminosilicate compositions" by Bottinga, Weill and Richet. Geochim. Cosmochim. Acta 48, 401-408.

Ghiorso, M.S., Carmichael, I.S.E. (1985) Chemical mass transfer in magmatic processes. II. Applications in equilibrium crystallization, fractionation and assimilation. Contrib. Mineral. Petrol. 90, 121-141.

Ghiorso, M.S., Carmichael, I.S.E., Rivers, M.L., Sack, R.O. (1983) The Gibbs free energy of mixing of natural silicate liquids; an expanded regular solution approximation for the calculation of magmatic intensive variables. Contrib. Mineral. Petrol. 84, 107-145.

Ghiorso, M.S., Kelemen, P.B. (1987) Evaluating reaction stoichiometry in magmatic systems evolving under generalized thermodynamic constraints: Examples comparing isothermal and isenthalpic assimilation. In: Magmatic Processes: Physicochemical Principles. B.O. Mysen, ed., Geochemical Society, Spec. Publ. No. 1, p. 319-336.

Gill, P.E., Murray, W., Wright, M.H. (1981) Practical Optimization. Academic Press, New York, 401 pp.

Glazner, A.F. (1984) Activities of olivine and plagioclase components in silicate melts and their application to geothermometry. Contrib. Mineral. Petrol. 88, 260-268.

Harvie, C.E., Greenberg, J.P., Weare, J.H. (1987) A chemical equilibrium algorithm for highly non-ideal multiphase systems: Free energy minimization. Geochim. Cosmochim. Acta 51, 1045-1057.

Karpov, I.K., Kaz'min, L.A. (1972) Calculation of geochemical equilibria in homogeneous multicomponent systems. Geochem. Int'l. 9, 252-262.

Karpov, I.K., Kaz'min, L.A., Kashik, S.A. (1973) Optimal programming for computer calculation of irreversible evolution in geochemical systems. Geochem. Int'l. 10, 464-470.

Korzhinskii, D.S. (1949) The phase rule and systems with fully mobile components [in Russian]. Dokl. Acad. Nauk SSSR 64, 361-364.

Korzhinskii, D.S. (1956) Deduction of thermodynamic potentials for systems with perfectly mobile components [in Russian]. Dokl. Acad. Nauk SSSR 106, 295-298.

Korzhinskii, D.S. (1959) Physiochemical Basis of the Analysis of the Paragenesis of Minerals. Consultants Bureau, New York, 141 p.

Lange, R.A., Carmichael, I.S.E. (1987) Densities of Na_2O-K_2O-CaO-MgO-FeO-Fe_2O_3-Al_2O_3-TiO_2-SiO_2 liquids: new measurements and derived partial molar properties. Geochim. Cosmochim. Acta (in press).

Langmuir, C.H., Hanson, G.N. (1981) Calculating mineral-melt equilibria with stoichiometry, mass balance and single component distribution coefficients. In: R.C. Newton, A. Navrotsky, B.J. Wood, eds., Thermodynamics of Minerals and Melts (Advances in Physical Geochemistry, v.1), Springer-Verlag, New York, 247-272.

Lawson, C.L., Hanson, R.J. (1974) Solving Least Squares Problems. Prentice-Hall, New Jersey, 340 pp.

Nekvasil, H. (1986) A Theoretical Thermodynamic Investigation of the System Ab-Or-An-Qz(H_2O) and Implications for Melt Speciation. Ph.D. thesis, Pennsylvania State Univ., University Park, PA, 267 pp.

Nielsen, R.L., Drake, M.J. (1979) Pyroxene-melt equilibria. Geochem. Cosmochim. Acta 43, 1259-1272.

Nielsen, R.L., Dungan, M.A. (1983) Low pressure mineral-melt equilibria in natural anhydrous mafic systems. Contrib. Mineral. Petrol. 84, 310-326.

Reed, M.H. (1982) Calculation of multicomponent chemical equilibria and reaction processes in systems involving minerals, gases and an aqueous phase. Geochim. Cosmochim. Acta, 46, 513-528.

Russell, J.K., Nicholls, J. (1985) Application of Duhem's theorem to the estimation of extensive and intensive properties of basaltic magmas. Canadian Mineral. 23, 479-488.

Saxena, S.K. (1982) Computation of multicomponent phase equilibria. In:

Saxena, S.K., ed., Advances in Physical Geochemistry, v. 2, Springer-Verlag, New York, 225-242.
Shvarov, Yu. V., (1976) Algorithm for calculation of the equilibrium composition in a multicomponent heterogeneous system. Dokl. Akad. Nauk SSSR, 229[5], 1224.
Shvarov, Yu. V., (1978) Minimization of the thermodynamic potential of an open chemical system. Geochem. Int'l. 15[6], 200-203.
Smith, W.R., Missen, R.W. (1982) Chemical Reaction Equilibrium Analysis. John Wiley and Sons, New York, 364 p.
Stebbins, J.F., Carmichael, I.S.E., Moret, L.K. (1984) Heat capacities and entropies of silicate liquids and glasses. Contrib. Mineral. Petrol. 86, 131-148.
Stolper, E. (1982a) Water in silicate glasses: an infrared spectroscopic study. Contrib. Mineral. Petrol. 81, 1-17.
Stolper, E. (1982b) On the speciation of water in silicate melts. Geochim. Cosmochim. Acta 46, 2609-2620.
White, W.B., Johnson, S.M., Dantzig, G.B. (1958) Chemical equilibrium in complex mixtures. J. Chem. Phys. 28, 751-755.

Chapter 13

Mark S. Ghiorso and Ian S. E. Carmichael

MODELING MAGMATIC SYSTEMS: PETROLOGIC APPLICATIONS

INTRODUCTION

In this chapter we present several petrologic applications that arise from thermodynamic modeling of magmatic systems. The calculations presented here are based on the techniques established in the previous chapter and on the regular solution model for magmatic melts published by Ghiorso et al. (1983) and by Ghiorso (1985; Appendix 1). Enhancements to the published model have been used in the preparation of Figures 1-3 and these modifications are discussed in the appendix to this chapter and in greater detail in a forthcoming publication (Ghiorso and Carmichael, in preparation). The applications presented here demonstrate in some detail how complex petrologic hypotheses can be tested to yield quantitative and occasionally surprising results.

The first two examples will utilize the ability to model the shape of the Gibbs free energy surface in composition-temperature space. The remaining examples will focus on the modeling of several irreversible reaction paths corresponding to the evolution of magmas under a variety of geologic conditions.

THERMODYNAMIC CLASSIFICATION OF IGNEOUS ROCKS

It is appropriate that we should begin this series of examples with a section on the thermodynamic classification of igneous rocks. Since the turn of the century controversy has raged concerning the merits of classification schemes based predominantly on rock composition and those based essentially on mineral content (Carmichael et al., 1974). The arguments center largely on which scheme best describes the genetic relationships among rock types and illuminates most accurately the critical aspects of their petrogenesis. Shand (1927, 1943) was perhaps the first to point out that either scheme, alone, is inadequate to the task. The bulk composition of the rock must be viewed in light of its mineral content and vice versa. A thermodynamic classification has the perspective to describe igneous rocks in just such a fashion.

The most comprehensive and well used scheme to connect an igneous rock's chemical composition to its mineral assemblage is that devised by Cross, Iddings, Pirsson and Washington (1902). These authors developed an intricate procedure to derive a series of idealized anhydrous mineral components, called the norm, from the chemical analysis of the rock. Although this procedure, and the calculated CIPW norm, is still widely used today, the CIPW quantitative classification of igneous rocks, with its classes, subclasses, and orders derived from the norm, and rangs and subrangs reflecting various oxide ratios derived from the chemical analysis, has never been widely adopted.

Of the classifications emphasizing the mineralogical assemblage of an igneous rock, Shand's (1913) saturation concept, although not the earliest of this type (see Johannsen, 1939), has become the most widely known. This is partly because of the emphasis it places on significant compositional variation. For example, Shand divided the essential and accessory minerals found in igneous rocks into two groups: the first, or saturated group of minerals, occur with quartz or tridymite, and the second, or undersaturated group, never occur with quartz or tridymite. From this division of the rock-forming minerals and the common accessories, came the threefold classification of igneous rocks into (1) oversaturated, those containing quartz or tridymite, (2) saturated, those without quartz or tridymite but with saturated minerals, and (3) undersaturated, those comprised of undersaturated minerals. This concept of oversaturation, saturation and undersaturation, emphasizing the SiO_2 content of the rock, is conveniently represented quantitatively by the thermodynamic affinity for saturation with quartz. This topic will be taken up below.

At about the same time that Shand was developing his mineralogical classification, later fully described in his book Eruptive Rocks (Shand, 1927, 1942), Harker (1909) was representing igneous rock suites and petrographic provinces, as series on oxide variation diagrams with SiO_2 as abscissa: the Harker diagram. Although he advocated a mineralogical basis for the classification of igneous rocks (op. cit., p. 376), some of his contemporaries chose the weight percentage of SiO_2 as a means of classification (e.g., Hatch, 1891, and later editions). Of this, Shand (1913), with typical prescience, said, "a chemical classification, to be of any value, must have regard to all the molecules present in the rock, not to one alone."

Little has changed from the beginning of the century as we approach its close. Streckeisen (1976) has advocated, on behalf of the International Union of Geological Sciences (IUGS) Subcommission on the Systematics of Igneous Rocks, a classification based on the modal abundances of minerals in plutonic rocks. This mineralogical construct was modified to a chemically equivalent one for volcanic rocks (Streckeisen and Le Maitre, 1979), and more recently condensed to a classification of volcanic rocks based entirely on weight percentages of SiO_2 and $Na_2O + K_2O$ (Middlemost, 1980; Le Maitre, 1984; Le Bas et al., 1986).

To connect a stable igneous mineral assemblage to the liquid that gives rise to it is a challenge that was posed over seventy years ago after the CIPW procedure was published (Shand, 1914, p. 488). "Anybody who can produce a system in which chemical and mineralogical characters can be correlated in a satisfactory manner will have solved the great problem of petrography" (op. cit.). The underlying basis for such a system has its foundation in thermodynamics.

To see this let us imagine ourselves to be multidimensional creatures capable of visualizing Gibbs free energy surfaces for liquid and solid phases in magmatic systems. Let us assume we position ourselves on the liquid free energy surface at a "point" that corresponds to a tholeiitic melt. "Looking" around in Gibbs energy-composition space, we see various curved surfaces corresponding to the Gibbs energy of different solid phases. Depending on the temperature and pressure, some of these surfaces cut through that of the liquid, while others are suspended far above it, possessing a significantly higher free energy than that of the melt. Many of the solid Gibbs free energy surfaces are broad and flattened corresponding to extensive solid solution between fixed end members. Others are pointed and sharply define a minimum value at a particular coordinate in our composition space. These of course, are the solids which form with fixed stoichiometry. From our "tholeiitic" vantage point, which for illustrative purposes we will assume is at a temperature and pressure just above the liquidus, we look out and see various Gibbs surfaces, corresponding to the solid phases olivine, plagioclase, clinopyroxene, nepheline, kalsilite, perovskite, etc.

We could position (or if preferred, classify) our tholeiite by simply measuring the distance to the closest approach of any or all of the solid Gibbs surfaces. A better discriminant, however, is to construct a measure of how close a particular solid phase is from achieving equilibrium with the melt. We do this by forming the tangent surface to the liquid Gibbs function which intersects at the composition of our tholeiite. In two dimensions the tangent surface would be a line, in three dimensions a plane, in n-dimensions an n-1-dimensional hypersurface. A perpendicular vector to this tangent surface can be made to intersect the Gibbs functions of the various solids. If we position the origin of the vector on the tangent surface such that its length is minimized with respect to intersection with the Gibbs function of a solid phase of interest, then the length of that vector is a measure of the departure of the solid phase from equilibrium with the melt. This is a geometric interpretation of the chemical affinity for the formation of this solid phase from the liquid. The composition of the "potential" solid is given by the intersection of the tip of this same vector with the Gibbs surface for the solid phase.

Igneous rocks may be classified by tabulating the chemical affinity for the formation of all potential solid phases for a particular bulk composition. In Chapter 12 the chemical affinity[1], A, associated with a chemical reaction is defined as (Eqn. 8):

$$A = -\sum_i \nu_i \left(\frac{\partial G}{\partial n_i}\right)_{T,P,n_{j\neq i}} = -\sum_i \nu_i \mu_i ,$$

where the ν_i are the stoichiometric reaction coefficients of the products (defined to be positive) and reactants (defined to be negative). If we take as an example the affinity for quartz saturation, we have for the crystallization reaction liquid ⇒ solid (caution; in Chapter 12 we considered solid ⇒ liquid reactions):

$$\Delta G = -A = \mu^{qtz}_{SiO_2} - \mu^{liquid}_{SiO_2} ,$$

so that if quartz is stable, ΔG will be a negative quantity, and the affinity positive. The magnitude of the affinity represents the extent of supersaturation or undersaturation, that is the direction in which the reaction of interest will proceed. The affinity will be dependent upon the temperature, pressure and composition of the liquid. Equation (6a) of Chapter 12, or an equivalent expression appropriate to a thermodynamic formulation other than that of Ghiorso et al., (1983), may be used to estimate the chemical potential of liquid silica in the lava.

1. Affinity has also been used in quite a different way in treating the the interrelationship of igneous rock series, where it is taken to denote kinship (Coombs, 1963).

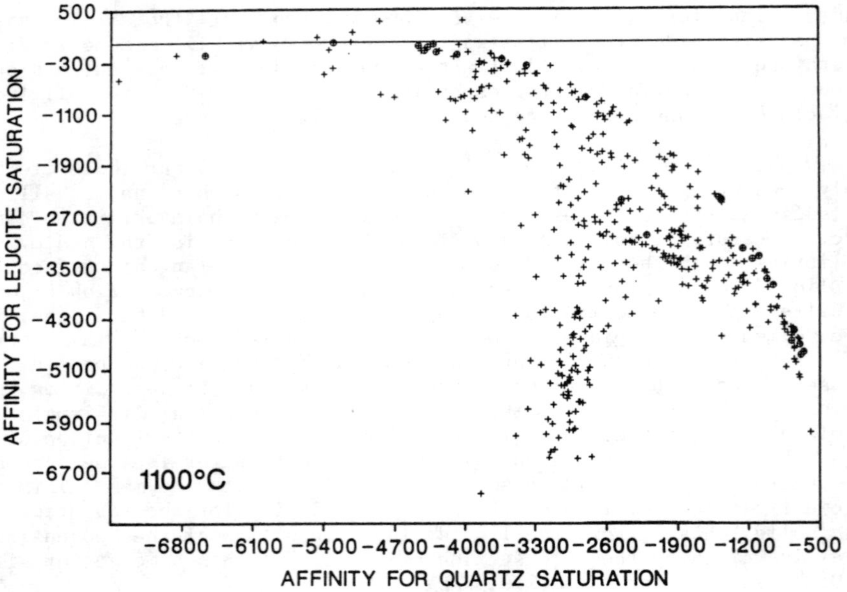

Figure 1. The chemical affinity (negative molar Gibbs free energy) for the reaction:

$$\underset{\text{leucite}}{KAlSi_2O_6} = \tfrac{1}{2}\,\underset{\text{liquid}}{K_2Al_2Si_2O_8} + \tfrac{1}{4}\,\underset{\text{liquid}}{Si_4O_8}$$

plotted versus the affinity for the reaction:

$$\underset{\text{quartz}}{SiO_2} = \tfrac{1}{4}\,\underset{\text{liquid}}{Si_4O_8}$$

for a comprehensive spectrum of lava compositions. Affinities are evaluated at 1100°C. Circled crosses indicate those lava types which possess phenocrysts of quartz and/or leucite. The solid line denotes an axis of zero affinity.

The 1100°C affinities for saturation with pure quartz for over 400 analyses of diverse lava types, have been plotted in Figure 1 against the calculated affinities for saturation with pure leucite. The affinities have been estimated using the thermodynamic formulation of Ghiorso et al. (1983), modified as described in the Appendix to this chapter. The lavas shown in Figure 1 were selected simply by culling the files of wet chemical analyses performed by the second author. Those lavas which contain either phenocrysts of quartz or leucite are indicated by circled crosses. The variety of lava types includes almost all known terrestrial examples with the exception of komatiite and boninite.

There are only a few examples of supersaturation of leucite, and the extent of this is at most a few hundred calories per mole of $KAlSi_2O_6$, possibly within error of the data and the solution model. Otherwise, the saturation curve (A = 0) at 1100°C virtually intersects the cluster of lavas with leucite phenocrysts, and suggests that among these compositions, namely leucite basanites and nephelinites, there are none which are accumulative or strongly oversaturated (A > 1000 cals/mol), in leucite.

No silicic compositions approach saturation with quartz at 1100°C, and those which contain phenocrysts have negative affinities of between 500 and 1500 cals/mol. Clearly, the affinity would approach zero with decreasing temperature, and for the representative silicic lava types shown in Figure 1, saturation for anhydrous liquids is achieved at about 800-900°C.

There are a concentration of points in Figure 1 with affinities for quartz saturation between -2600 and -3600 cals/mol, and with affinities for leucite saturation of -4000 to -6000 cals/mol. These are a group of mid-ocean ridge tholeiitic glasses, and they typify the position of alkali-poor tholeiitic magma in this plot. In general terms, the normal basaltic liquid line of descent would lead either to a siliceous residuum with A_{quartz} approaching zero, or to silica-poor derivatives which have correspondingly large and negative affinities for leucite crystallization, and yet are highly undersaturated with respect to quartz. The points defining the curved edge of the array in Figure 1 are lavas in petrogeny's residua system, namely those rich in the components $KAlSiO_4$, $NaAlSiO_4$ and SiO_2.

The affinity for saturation with a solid solution can also be calculated, except that unlike quartz or leucite, the composition of the solid is another variable(s). In Chapter 12 (discussion centering around Eqns. 11-14) it is shown how this affinity can be defined uniquely given the assumption of exchange equilibrium between solid end-members and the melt. We illustrate here an alternative procedure which does not involve this assumption but at the same time does not tell us the most likely solid composition to form from the melt. We calculate the 1200°C affinity for plagioclase saturation, stipulating the composition of the plagioclase to be An_{85}, and also for saturation with olivine of Fo_{90} composition. These particular solid compositions can be considered to be convenient reference points to evaluate the saturation states of most mafic lavas.

In Figure 2 the results are plotted for the same array of lava compositions as shown in Figure 1, with those lavas containing phenocrysts of either olivine or plagioclase being represented by circled crosses. Just saturated ($A = 0$) or just oversaturated ($A > 0$) with respect to An_{85} are a cluster of mid-ocean ridge tholeiitic glasses. These basalts contain both kinds of phenocrysts within a short distance towards the center of their pillows. It should be remembered that supersaturation ($A > 0$) of a solid solution phase of a particular composition may be reduced by either elevating the temperature slightly, or changing the composition of the solid solution towards the high-melting component. The two lavas represented by solid squares in Figure 2 are rich in plagioclase phenocrysts, and their affinities for plagioclase crystallization become negative only at very high temperatures at 1 bar. These lavas are accumulative varieties, and were never entirely liquid.

Olivine-rich lavas (picritic varieties) are also slightly supersaturated at 1200°C and 1 bar. However, the cluster of lavas with an affinity for Fo_{90} of about 1600 cals/mol and an affinity for An_{85} of approximately -2500 cals/mol are a group of minettes that are not particularly rich in olivine phenocrysts (1-4%), but have abundant phlogopite phenocrysts. These lavas, not obviously accumulative, are clearly supersaturated in the anhydrous equivalent of phlogopite, namely olivine. This point illustrates the fact that the chemical affinity for a phase may indicate supersaturation, but it does not mean that the phase in ques-

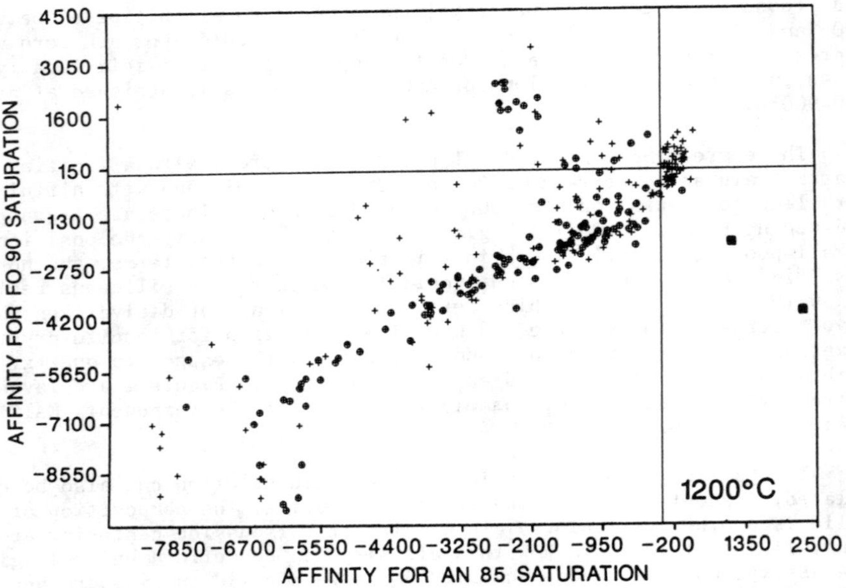

Figure 2. The chemical affinity (negative molar Gibbs free energy) for the reaction:

$$\underset{\text{Plagioclase}}{Ca_{0.85}Na_{0.15}Al_{1.85}Si_{2.15}O_8} = \frac{17}{80}\underset{\text{liquid}}{Ca_4Si_2O_8} + \frac{9}{320}\underset{\text{liquid}}{Na_{16/3}Si_{8/3}O_8} + \frac{111}{320}\underset{\text{liquid}}{Al_{16/3}O_8} + \frac{43}{80}\underset{\text{liquid}}{Si_4O_8}$$

plotted versus the affinity for the reaction:

$$\underset{\text{Olivine}}{Mg_{1.8}Fe_{0.2}SiO_4} = \frac{9}{20}\underset{\text{liquid}}{Mg_4Si_2O_8} + \frac{1}{20}\underset{\text{liquid}}{Fe_4Si_2O_8}$$

for a comprehensive spectrum of lava compositions. Affinities are evaluated at 1200°C. Circled crosses indicate those lava types which possess phenocrysts of olivine and/or plagioclase. Solid squares denote cumulative lavas. Solid lines denote axes of zero affinity.

tion is the thermodynamically stable one at that P, T and bulk composition. Gibbs free energy minimization methods such as those described in Chapter 12, must be used to determine the equilibrium phase assemblage, and <u>all</u> potential solid phases must be considered in the minimization process.

The lavas with the most negative affinities (undersaturated) in Figure 2 are those types which at lower temperatures would precipitate both plagioclase and olivine, but both phases will be enriched in their low-melting components (i.e., Ab and Fa). These are rhyolites, trachytes and phonolites, and basaltic liquid lines of descent generally will terminate at these compositions.

The spectrum of lavas plotted in Figures 1 and 2 demonstrate quite nicely the continuum of compositions found in nature. The distribution of data emphasizes the mineralogical control on the bulk compositions of common lava types and, in turn, betrays classification schemes which "pigeonhole" these lavas without regard to their genetic relationships.

Eventually, with the appropriate experimental data, it will be possible to include amphiboles and micas among the phases for which affinities may be calculated. These calculations however, await the development of a more comprehensive solution model for hydrous magmatic melts. Clearly, the potential exists to evaluate the degree of under- or oversaturation of any phase of any stipulated composition with any known lava type. A classification scheme based upon these various levels of "mineral saturation" is both quantitative and comprehensive in that it offers the means of taking the complete composition of a liquid into account, rather than selecting as significant, or revealing, just a few oxide components.

THE ACTIVITY OF SILICA AND THE DEPTH OF ORIGIN OF MAFIC MAGMAS

Thermodynamic models for magmatic liquids allow us to predict the activity of silica in the melt phase as a function of temperature and pressure. The accuracy of this prediction depends on our ability to model the shape of the Gibbs function surface (G), in that the activity of silica is directly related to a particular compositional derivative of G. In this example we explore controls on the activity of silica in magmatic systems and evaluate predictions of silica activity with regard to hypotheses concerning the depth of origin of mafic magmas.

Silica activity and magmas

Many of the common igneous rock-forming minerals are related to one another by a silication reaction of the general form:

$$\underset{\text{solid}}{\sum_r \nu_r \,\text{Mineral}_r} \;=\; \underset{\text{liq}}{SiO_2} \;+\; \underset{\text{solid}}{\sum_p \nu_p \,\text{Mineral}_p} \;,$$

where Mineral_r and Mineral_p represent reactant (r) and product (p) solid phases with stoichiometric reaction coefficients ν_r and ν_p respectively. If the solids involved in these silication reactions stably coexist and are assumed to be in the standard state, which is defined by unit activity of the pure substance at any temperature and pressure, then the variation of the melt silica activity with temperature or pressure is given by:

$$\log a_{SiO_2} = \frac{\Delta G^o}{\ln(10)\; R\; T} \;,$$

where ΔG^o indicates the standard state free energy change of the reaction of interest, R is the universal gas constant, ln(10) represents the natural logarithm of ten and T denotes the absolute temperature. If the solids are taken to be outside the standard state, the melt silica activity will vary as a function of the extent of solid solution. For example, decreasing the activity of a solid product will increase the activity of silica in equilibrium with the assemblage, whereas decreasing the activity of a solid reactant will decrease the silica activity in the melt.

Curves are plotted in Figure 3a which correspond to the 1 bar tem-

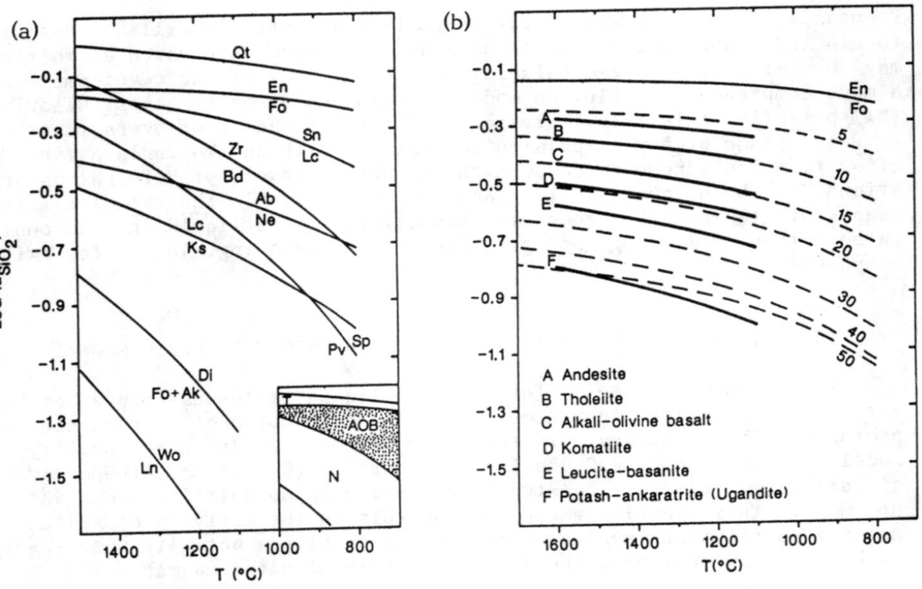

Figure 3. log a_{SiO_2} versus temperature plotted for a variety of silica buffers and represenstitive lava types. The thin solid curves in figures (a) and (b) indicate buffer reactions evaluated at one bar pressure and correspond to:

Qz	Quartz = SiO_2
En-Fo	2 Enstatite = Forsterite + SiO_2
Sn-Lc	Sanidine = Leucite + SiO_2
Zr-Bd	Zircon = Baddeleyite + SiO_2
Ab-Ne	½ Albite = ½ Nepheline + SiO_2
Sp-Pv	Sphene = Perovskite + SiO_2
Lc-Ks	Leucite = Kalsilite + SiO_2
Di-Fo+Ak	$\frac{4}{3}$ Diopside = $\frac{1}{3}$ Forsterite + $\frac{2}{3}$ Akermanite + SiO_2
Wo-Ln	2 Wollastonite = Larnite + SiO_2.

The dashed curves on figure (b) indicate values of silica activity consistent with the enstatite-forsterite buffer at the indicated pressure (in kbar). The heavy solid curves in (b) indicate calculated silica activities for the following rock types:

	Andesite	Tholeiite	Olivine Basalt	Komatiite	Leucite Basanite	Ugandite
SiO_2	61.61	52.29	50.20	47.96	44.54	36.71
TiO_2	0.60	1.17	2.43	0.36	2.16	5.54
Al_2O_3	17.82	14.75	16.65	7.44	15.50	9.30
FeO_T	4.85	12.25	10.28	11.39	9.15	12.73
MnO	0.10	0.22	0.24	0.20	0.19	0.26
MgO	2.54	5.30	3.62	24.35	9.12	6.34
CaO	5.70	9.89	7.53	7.46	11.14	14.08
Na_2O	4.77	2.60	5.27	0.64	4.08	2.40
K_2O	1.43	0.33	2.16	0.06	1.82	6.05
P_2O_5	0.20	0.16	0.82	0.03	0.61	1.11

The inset in figure (a) demonstrates the range of silica activities applicable to the rock types Tholeiite (T; bounded by the buffers Q and En-Fo), Alkali-olivine basalt (AOB; bounded by the buffers En-Fo and Sp-Pv), and Nephelinities (N; bounded by the buffers Sp-Pv and Wo-Ln).

perature dependence of the logarithm of the activity of silica for a number of silication reactions which are useful for understanding the variation of silica activity in magmas. The solid phases are assumed to be pure and the stoichiometric relations are provided in the figure legend. These silication reactions are of four types: (1) reactions involving end-member alkali-aluminosilicates, (2) reactions involving the desilication of an end-member metasilicate into a corresponding orthosilicate, (3) reactions involving desilication to an oxide, and (4) reactions involving more than two solid phases. The slopes of the curves shown on Figure 3a are related to the reaction enthalpy, the magnitude of which is proportional to the steepness of the slope.

Casual inspection of Figure 3a reveals evidence of mineral incompatibilities. One example is the lack of overlap in the stability fields of forsterite and quartz. Another is the mutual incompatibility of enstatite with either leucite or nepheline. It is apparent that neither kalsilite nor larnite are stable with feldspar, and in addition that the presence of perovskite is incompatible with that of a feldspar (at least below 1200°C). Within Figure 3a lies the thermodynamic representation of Shand's (1913) saturation concept, as discussed in the previous section. It also represents a thermodynamic basis for the desilication steps in the CIPW norm algorithm (Johannsen, 1939). It is evident that the activity, or the chemical potential, of silica in a multicomponent liquid will have a profound effect on the crystalline phases which precipitate at equilibrium and 1 bar. There is a parallel, not to be forced, between the role of silica activity in magmas and that of the activity of the hydrogen ion (pH) in aqueous solutions.

The equilibria plotted in Figure 3a demonstrate the range of silica activity exhibited by magmatic liquids. The presence of a silica mineral constrains the upper limit. The appropriate curve shown in Figure 3a corresponds to the activity of silica in equilibrium with β-quartz. It should be recognized that at high temperatures quartz will be metastable, its place being taken by tridymite or cristobalite. It is because of this metastability that the curve extends above zero (melt silica activities in excess of unity) in the figure. The lower limit of silica activity in natural liquids is more difficult to estimate, but there is no record of larnite in uncontaminated igneous rocks (Carmichael et al., 1970), so that terrestrial magmas do not reach this limit. As the assemblage of diopsidic augite, olivine, and melilite is found in some silica-poor lavas, the curve defined by Di ⇒ Fo + Ak (Fig. 3a) will not be the lower limit of silica activity. Terrestrial lavas will therefore, lie between this reaction and that of larnite stability; the range in the activity of silica covering just about two orders of magnitude at one bar.

Two of the silication reactions intermediate within this range are of considerable significance, for the congeries of magmas may be conveniently subdivided by them. Although sphene and perovskite have only been found infrequently to coexist in lavas (Tilley, 1953; Smith, 1970), unlike sphene, perovskite has never been reported to occur with feldspar. Therefore, the activity of silica defined by the perovskite ⇒ sphene reaction (Fig. 3a) separates all those natural liquids which precipitate feldspar at some stage of their cooling interval, from those magmas that never do so (note that this incompatibility occurs only below 1200°C in Fig. 3a). On the high silica activity side are the alkali olivine basalts, basanites, tephrites, and all their feldspathic derivatives including phonolites, whereas on the low silica activity side are all the feldspar-free rock and lava types, with their arresting array of names and titles, all conveniently called nephelinites here, and labeled as such in Figure 3a.

The other silication reaction of particular interest is that involving forsterite and enstatite, for it models the reaction relationship of olivine to liquid which characterizes the tholeiitic magma type and its derivatives (Carmichael et al., 1974). The forsterite-enstatite reaction thereby distinguishes this great group of basaltic magmas from the alkali olivine basalt magma series, which displays neither this reaction relation, nor its mineralogical representation in the form of a calcium-poor pyroxene (Carmichael et al., 1970) (Fig. 3a). It is interesting to note that the baddelyite-zircon reaction falls close to the forsterite-enstatite one at high temperature, so that at these temperatures zircon will only be found in the tholeiitic magma type, whereas at low temperatures zircon will and does occur in nepheline-bearing assemblages, such as nepheline-syenites or phonolites (Deer et al., 1982).

The activity of silica in magmatic liquids can be correlated with other broad trends in the igneous economy. It is inversely proportional, in general, to the volume of volcanic eruption. The eruptive volume of the most silica-poor lavas is tiny in comparison to the huge volumes of a single tholeiitic lava flow (e.g., Columbia River Plateau), or the vaster eruptions of silica-rich pyroclastic flows. The activity of silica is also inversely proportional to the concentration of trace metals, particularly the incompatible elements (e.g., U, Th, Ta, REE, Zr, P) and to anticipate the next section, it is inversely proportional to the depth to the respective mantle source regions. Silica activity is also inversely proportional to the iron redox state of lavas (Carmichael and Ghiorso 1986). Lastly, and perhaps to the dismay of those who relish celebrating petrographic variety with a plethora of names and titles, almost all of which seem quite uninformative, the activity of silica is inversely proportional to the proliferation of rock names.

The melt solution model of Ghiorso et al. (1983, 1985, Appendix) may be used to calculate silica activity in a wide variety of lava types. In the context of Figure 3a, such calculations provide little more than another means of classifying rocks. To acquire a deeper genetic understanding of silica activities in terrestrial lavas, the effect of pressure on the various end-member reactions shown in Figure 3a must be evaluated; i.e., we must begin to appreciate what equilibria buffered the silica activity for the pressure and temperature conditions under which the magma formed. In the next section we explore one aspect of this problem and develop a petrogenetic grid for basic magmas.

Silica activity of basic magmas and a petrogenetic grid

As a consequence of our standard state definition, the pressure dependence of the activity of silica in a magmatic liquid is given by:

$$\left(\frac{\partial \ln a_{SiO_2}}{\partial P}\right)_{T,n_j} = \frac{\bar{v}_{SiO_2} - v^o_{SiO_2}}{R\,T} \quad ,$$

where $\bar{v}_{SiO_2}$ is the partial molar volume of SiO_2 in a multicomponent silicate liquid, and $v^o_{SiO_2}$ is the molar volume of SiO_2 in pure SiO_2 liquid (perhaps metastable), calculated at the temperature and pressure of interest. The best experimental evidence suggests that this difference does not significantly depart from zero (Lange and Carmichael, 1987). Therefore, the activity of silica in a natural silicate liquid will not be a function of pressure but only of temperature and composition. If magmas ascend isochemically, which for the moment we will assume to be the case, then the only variation in silica activity is that caused by temperature drop during ascent[2].

Building on this simple assumption, a petrogenetic grid can be constructed which depicts the conditions of equilibration of various magma types with an idealized mantle or source assemblage. If it is granted that olivine and orthopyroxene could coexist in equilibrium in the source regions of all basic magmas, then the activity of silica in the liquid will be buffered by the reaction:

$$\underset{\text{olivine}}{Mg_2SiO_4} + \underset{\text{liq}}{SiO_2} = \underset{\text{orthopyroxene}}{2\ MgSiO_3}$$

where

$$\log a_{SiO_2} = \frac{\Delta G^o(T,P)}{\ln(10)\ R\ T} + 2 \log a^{opx}_{MgSiO_3} - \log a^{olivine}_{Mg_2SiO_4} .$$

To a good approximation, the log terms on the right hand side of this equation cancel. This can be appreciated by considering that at the temperatures and pressure of interest, the forsterite component of mantle olivine ($\sim Fo_{90}$) has an activity approximately equal to $(X_{Fo})^2$. This olivine typically coexists with an orthopyroxene of composition En_{90}. Assuming that under the P-T conditions corresponding to mafic magma genesis, the concentration of Al in the M1 site of the orthopyroxene is small, the activity of the enstatite component in this pyroxene is essentially equal to its mole fraction. The stoichiometric coefficient of two preceding the pyroxene log activity term therefore cancels the contribution from the olivine. Assuming this cancellation of log terms, calculated isobars of silica activity corresponding to the equilibrium of olivine and orthopyroxene are shown in Figure 3b. Also plotted in Figure 3b are the variation in calculated $\log a_{SiO_2}$ for six lava types whose compositions are provided in the figure legend. These melt activities were determined using the solution model of Ghiorso et al. (1983) modified as discussed in the Appendix.

Under the assumptions outlined above, the conditions of anhydrous equilibration are represented by the intersection of the curves drawn for each lava type with the mantle source assemblage isobars. As these

---

2. To standardize our calculations we shall assume that during their ascent to the surface, magmas have their Fe_2O_3/FeO ratios determined by an oxygen fugacity which follows the QFM buffer (Carmichael and Ghiorso, 1986).

isobars are essentially co-parallel with the silica activity curves, the equilibration pressures are defined within narrow limits. For example, the tholeiitic lava (see figure legend for analysis) would be in equilibrium with its mantle source assemblage at 10 kbars. The temperature of this equilibration would necessarily be above the 1 bar liquidus temperature (ca. 1200°C), and would depend on the geothermal gradient. The indicated equilibration conditions for anhydrous andesitic magma, 6-7 kbar (Fig. 3b), would suggest that the mantle source region of this magma type lies in the pressure regime of the continental crust, requiring an impossibly steep geothermal gradient and quite different lithology to that of the lower crust. Either andesitic magmas of the composition used in the construction of Figure 3b are not generated by partial fusion of shallow mantle material, or if they are, then their activities of silica are significantly reduced. This lowering of silica activity could be caused by water dissolved in the liquid (Carmichael et al., 1974). A "wet" andesitic magma would therefore correspond to equilibration conditions deeper in the mantle.

The equilibration conditions of a komatiite, albeit a rather magnesian-poor variety, lies in the liquid metastable region, but nevertheless indicates that at about 20 kbar, a mantle residue of orthopyroxene and olivine could equilibrate with a komatiite magma above its liquidus temperature (ca. > 1600°C).

The silica-poor potash ankaratrite from Uganda has about the lowest silica activity of any terrestrial silicate magma, and this would suggest that its source region, if it contains olivine and orthopyroxene, will be very deep, well in excess of 100 km. As Ghiorso et al. (1983) have shown, this could be a region in which diamonds are stable, as evidenced by their olivine and orthopyroxene inclusions. Indeed, sporadic diamonds have been found in this part of the Uganda rift related to explosion-craters of this lava type, and it is this general type of potassium-rich, silica-poor magma that many petrologists believe to be related to the lamproites of Western Australia and to the kimberlites of the world's diamond fields.

The general conclusion of the relations displayed in Figure 3b is that the more silica-rich lavas such as tholeiites would have a more shallow source than the more silica-poor, alkali-rich lavas. The greatest difficulty concerned with interpreting these relations is in regard to the assumption of isochemical ascent. Crystal fractionation, assimilation and devolatilization will clearly modify the melts' silica activity. Such considerations necessitate the examination of specific cases, and in the remainder of this chapter we demonstrate examples of the thermodynamic modeling of magmatic processes that alter the bulk composition of the melt.

FRACTIONAL CRYSTALLIZATION OF OLIVINE THOLEIITIC MAGMA

This is our first example of the simulation of an irreversible magmatic process using the modeling technique discussed in the previous chapter. We will consider the fractional crystallization of the olivine tholeiitic magma whose composition is reported in the following table:

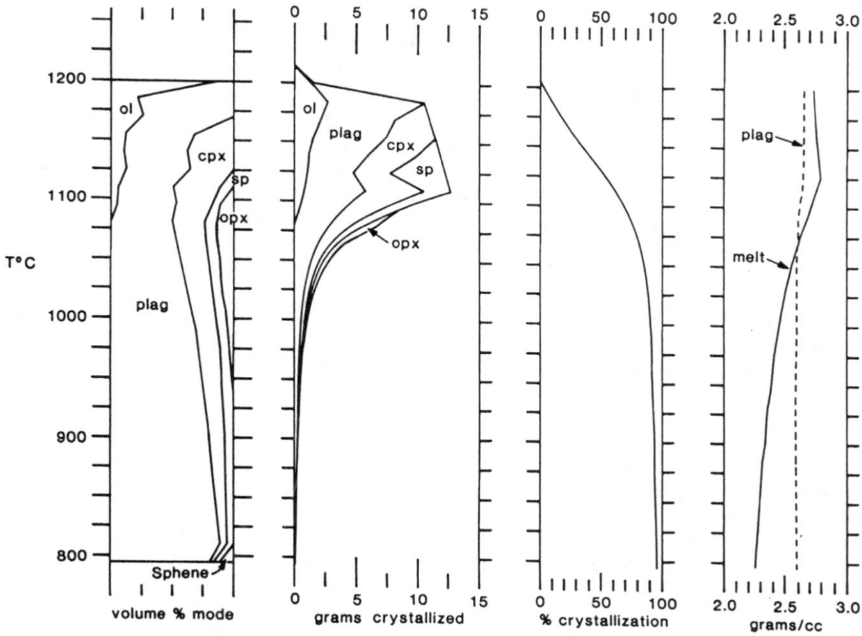

Figure 4. Calculated mineral proportions and melt and plagioclase densities for the fractional crystallization of an olivine tholeiite along the QFM buffer at 1 bar. Far left: Instantaneous solid mode. Middle left: Instantaneous mass crystallized. Initial liquid mass was 100 grams. Middle right: Cumulative per cent crystallized. Far right: Instantaneous melt and plagioclase density. ol, olivine; plag, plagioclase; cpx, clinopyroxene; sp, ulvöspinel-magnetite solid solutions; opx, orthopyroxene. From Ghiorso and Carmichael (1985).

Table 1. Initial composition of olivine tholeiitic liquid (wt %)
From Ghiorso and Carmichael (1985)

SiO_2	48.47	Al_2O_3	15.31	FeO	9.90	CaO	11.57	K_2O	0.20
TiO_2	1.71	Fe_2O_3	1.66	MgO	8.77	Na_2O	2.30	H_2O	0.10

This lava represents the most primitive composition of the Thingmuli volcanic series, Iceland (Carmichael, 1964) a sequence of rocks ranging from olivine tholeiites through tholeiites, basaltic andesites and icelandites to rhyolites. The chemical and mineralogical trends of this volcanic series have been documented in great detail (Carmichael, 1964; 1967) and the evidence suggests that these lavas have been derived by periodic eruption of a shallow magma chamber undergoing crystal fractionation. This interpretation allows us to compare our model calculations with field and petrographic evidence.

Crystal fractionation is simulated for the bulk composition in Table 1 by lowering the temperature in 15°C increments and subtracting precipitated mass at each stage. The liquid is maintained on the QFM buffer by opening the system to oxygen and calculating heterogeneous equilibrium using an appropriate Korzhinskii potential for fixed oxygen chemical potential (Eqn. 31 of the previous chapter). Crystallization along the QFM buffer is chosen to insure consistency between calculated olivine and spinel compositions and those observed in the derived lavas

(Ghiorso and Carmichael, 1985). In Figure 4 a summary of calculated phase proportions for crystal fractionation at 1 bar is provided. The natural crystallization sequence is (Carmichael, 1964) olivine, plagioclase, augite, spinel (Fe-Ti oxide), then olivine-out replaced by pigeonite with essentially the simultaneous appearance of minor amounts of ilmenite. Apatite crystallizes from the basaltic andesites and icelandites and fayalitic olivine returns in the icelandites, replacing Ca-poor pyroxene in the more acid rocks. In the latter stages of crystallization magnetite phenocrysts become shrouded with sphene. With the exception of ilmenite and apatite (whose saturation surfaces cannot be modeled due to inadequate thermodynamic data and liquid activity-composition relations), the only major inconsistency between the calculated and observed trend concerns the persistence of orthopyroxene without the reappearance of an iron-rich olivine. This points to difficulties in modeling the saturation surfaces for mafic minerals in silica-rich hydrous melts.

Figure 4 shows that the vast majority of crystallization takes place in the first 100°C. The calculated solid mode in this early stage is dominated by plagioclase with subordinate augite. This is corroborated by petrographic evidence from the tholeiites of the Thingmuli series (Carmichael, 1964). The more evolved lavas contain largely plagioclase phenocrysts and this is also indicated by the calculations shown here. The spike in spinel precipitation at 1125°C is reflected in the large numbers of Fe-Ti oxide phenocrysts found in the more evolved tholeiites and the preponderance of magnetite-rich glass which dominates the groundmass of these lavas (Carmichael, 1967). During the whole of the first 80% of crystallization, the calculated density of the precipitated plagioclase is less than that of the liquid (Fig. 4). The decrease in calculated liquid density which begins at 1120°C is related to the precipitation of spinel.

Harker variation diagrams of residual liquids produced from fractionation calculations at 1 bar and 2 kbar are shown in Figure 5. Compositions of the Thingmuli lavas reported by Carmichael (1964, Table 9) are plotted for comparison. Percent crystallization is indicated at the top of the figure and cumulative oxygen (grams of O_2 per 100 grams of magma) released (+) or absorbed (-) by the system is indicated at the base. Agreement between calculated and observed liquid compositions is quite good, especially when one considers that the most primitive lava has been assumed to be a primary melt and that the degree of crystal fractionation is extreme. For example, a 0.05 wt % error in the determination of any oxide in the composition of the olivine tholeiite could magnify itself into a 1.00 wt % error in the derived liquid at 95% crystallization. Where the calculated curves in Figure 5 dramatically depart from the natural trends the discrepancy can usually be ascribed to the neglect of a phase known to form in the sequence. In the case of TiO_2, the suppression of ilmenite probably accounts for the high titania values in the residual liquids.

Figure 5 clearly shows the initial increase and subsequent decrease in silica concentration which is characteristic of fractionating tholeiitic liquids. The turnover to higher silica contents, which is accentuated by sharp breaks in all the variation curves, denotes the first appearance of spinel on the liquidus. The negative slopes, prior to this event, reflect the precipitation of clinopyroxene.

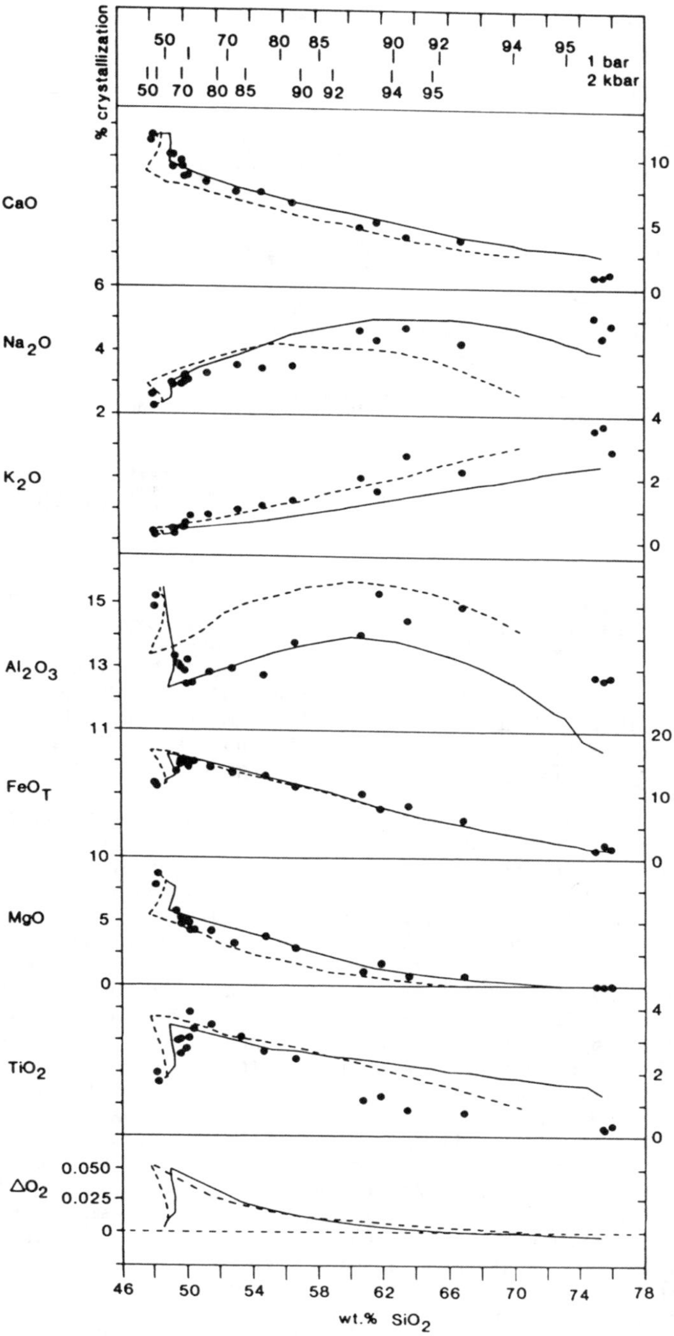

Figure 5. Calculated liquid variation diagram for the fractional crystallization of an olivine tholeiite along the QFM buffer at 1 bar (solid curves) and two kilobars (dashed curves). The per cent crystallization as a function of silica content is indicated at the top of the diagram. At the bottom, the cumulative amount of assimilated (-) or expelled (+) oxygen is reported as a function of wt % SiO_2 in the melt. All compositional units are in wt %. Solid circles denote data on the Thingmuli fractionation series taken from Carmichael (1964). From Ghiorso and Carmichael (1985).

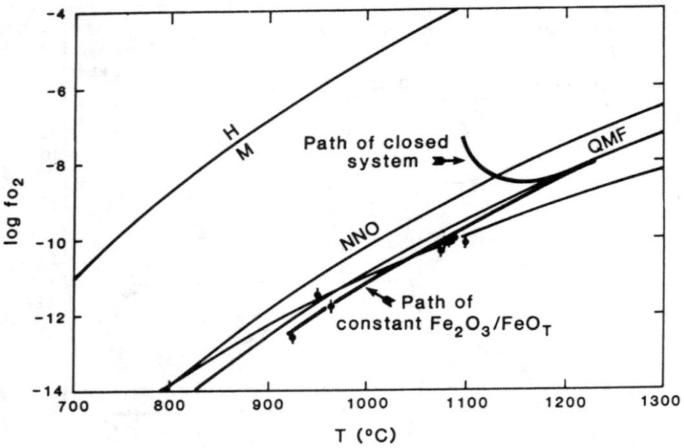

Figure 6. Equilibration values of oxygen fugacity derived from coexisting Fe-Ti oxides of the Thingmuli lava series of tholeiite to rhyolite (Carmichael, 1967). Also shown is the calculated closed system line of descent. After Carmichael and Ghiorso (1986).

The effects of open versus closed system crystal fractionation are readily seen in Figures 6 and 7. Figure 6 shows oxygen fugacity as a function of temperature for several common oxygen buffers as well as calculated temperature-oxygen fugacity pairs for Fe-Ti oxide assemblages found in the icelandites and rhyolites of the Thingmuli series (Carmichael, 1967). The oxygen fugacity-temperature path of a closed system crystal fractionation calculation is shown as well. This path was calculated by minimizing the Gibbs free energy at each step in reaction progress while maintaining constant bulk composition. It is clear that fractionation along the QFM buffer is an excellent approximation to the natural Thingmuli trend and that the system must be buffered, either externally or internally (e.g. Carmichael and Ghiorso, 1986), to maintain this trend; the natural course of crystallization drives the oxygen fugacity up toward the hematite-magnetite buffer. The explanation of these trends is best understood by examining Figures 7a and 7b.

The open system crystal fractionation path for the Thingmuli olivine tholeiite is displayed in Figure 7a, where a comparison is made to lavas analyzed by Carmichael (1964). The open system path is replotted on Figure 7b along with the path of the closed system simulation. The reason for the divergence can be understood in terms of the ferric/ferrous ratios of the residual liquids. In the initial stages of fractionation the melt is crystallizing olivine, plagioclase and clinopyroxene which creates a tendency to increase the ferric/ferrous ratio of the liquid. Increasing the ferric/ferrous ratio raises the oxygen fugacity of the melt (Kilinc et al., 1983). This effect can be seen in the closed system path in Figure 6. If the system is buffered, oxygen is expelled from the melt by transforming at each step "excess" ferric to ferrous iron. On the AFM projection (Fig. 7b) the paths of the closed and open system calculations coincide until spinel appears on the liquidus. The spinel which saturates is approximately the composition Uv_{70} and at this temperature the closed system has risen to about one log unit above QFM in oxygen fugacity. The ferric/ferrous ratio of the spinel is such that precipitation from the "open" system consumes more ferric iron relative to ferrous than is available, and the system, in order to maintain an

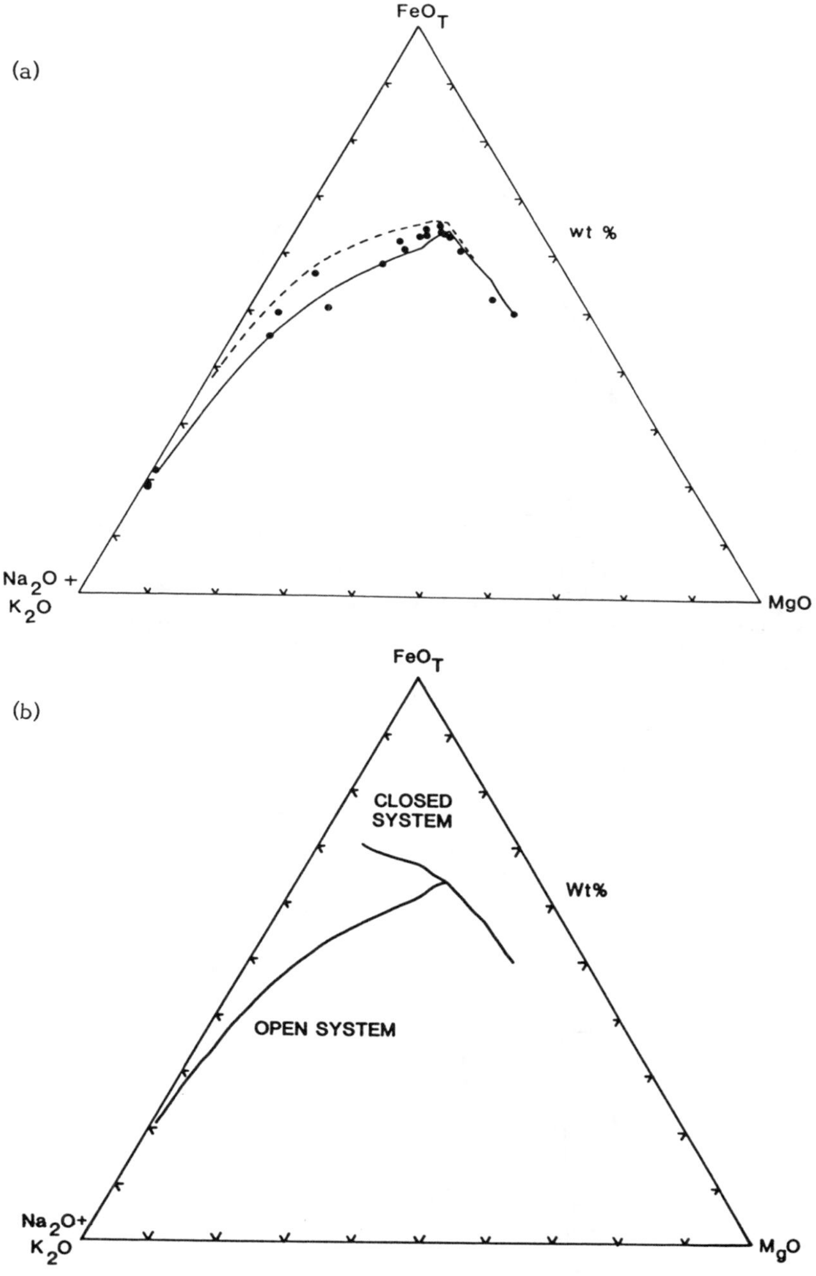

Figure 7. Calculated liquid path for the fractional crystallization of an olivine tholeiite. (a) Calculations performed along the QFM buffer at 1 bar (solid curves) and 2 kbar (dashed curves). Solid circles denote data on the Thingmuli fractionation series taken from Carmichael (1964). (b) Calculations performed at 1 bar along the QFM buffer (labeled OPEN SYSTEM) and in a system closed to oxygen exchange (labeled CLOSED SYSTEM). From Ghiorso and Carmichael (1985).

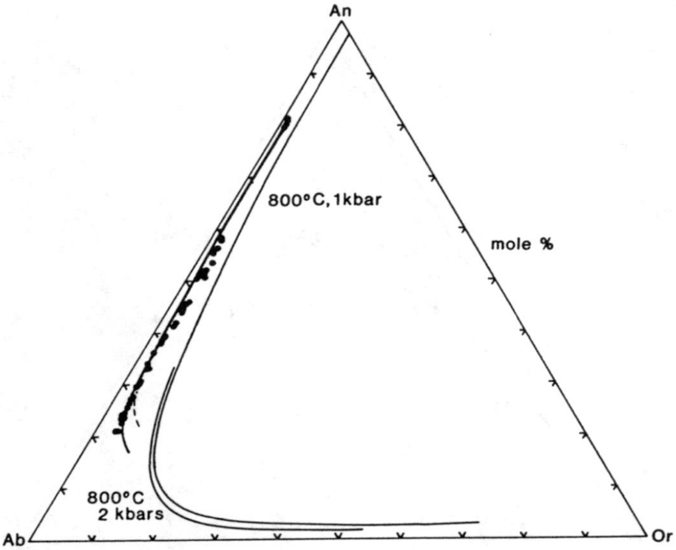

Figure 8. Calculated compositions of plagioclase precipitated during the fractional crystallization of an olivine tholeiite along the QFM buffer at 1 bar (solid curve) and 2 kbar (dashed curve). Solid circles denote data on plagioclase phenocryst compositions from lavas of the Thingmuli fractionation series (Carmichael, 1967). The 800°C ternary solvii were computed using the feldspar solution model of Ghiorso (1984). From Ghiorso and Carmichael (1985).

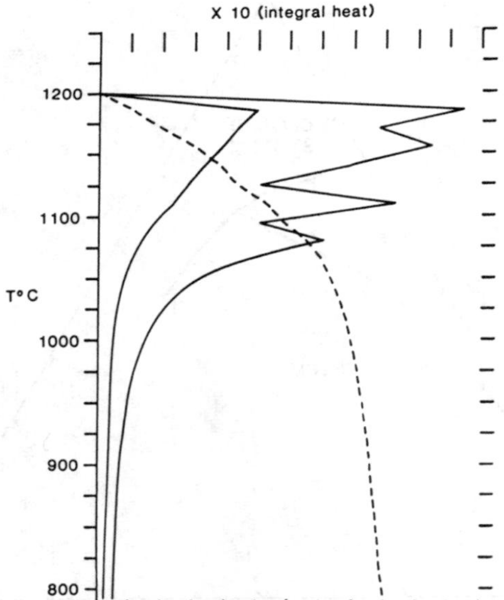

Figure 9. Calculated heat production due to the fractional crystallization of an olivine tholeiite along the QFM buffer at 1 bar. The outer solid curve represents the instantaneous heat output and the inner solid curve that portion released purely to cool the system (heat capacity contribution). The difference between the two is the instantaneous latent heat of fusion. The dotted curve corresponds to the integral heat production. At 1215°C the system consists of 100 g of stable liquid. From Ghiorso and Carmichael (1985).

oxygen fugacity buffered on QFM, absorbs oxygen. The "closed" system however has quite the opposite response! The ferric/ferrous ratio of the spinel relative to the more oxidized melt causes the melt ferric/ferrous ratio to increase as a consequence of spinel formation. This results in a rapid increase in melt oxygen fugacity. The continued formation of spinel induces the melt ferric/ferrous ratio to increase even further (the spinel is not changing composition fast enough to overcome the trend in the melt) until a temperature of 1095°C is reached, when the melt ferric/ferrous ratio has increased to the point where the precipitation of pure magnetite would be insufficient to reverse the oxidizing trend. As these events take place, there develops a tendency to reduce the degree of spinel supersaturation with each temperature decrease, consequently preventing the removal of large amounts of iron from the residual liquid. The net effect is to drive the liquid towards the FeO_T apex in Figure 7b while simultaneously advancing in composition toward the alkali-iron join. It is surprising how this seemingly subtle distinction in reaction path has such a profound effect on the nature of highly fractionated liquids in this system.

A comparison between calculated and measured feldspar compositions is provided in Figure 8. The agreement is quite good over the entire range (An_{81} to An_{17}) and the prediction of a single feldspar, zoned to anorthoclase in the more acidic melts, is borne out by petrographic observation (Carmichael, 1967). In general, the accuracy of the prediction of solid phase compositions in mafic to intermediate melts is on the order of ±5 mole % for plagioclase, olivine, and Fe-Ti oxide spinels and considerably greater for pyroxene compositions. In this modeling scheme the pyroxenes are approximated as quadrilateral solid solutions and the orthorhombic phase is assumed to be a binary enstatite-ferrosilite solid solution. Orthopyroxene proxies for pigeonite. Despite these approximations there is excellent agreement between the open system modeling of the fractionation of the Thingmuli olivine tholeiite and the observed mineral phases in the derived lavas for about the first 85 % of crystallization. This is to be expected from the observed match of the residual liquid compositions (Figs. 6 and 7a).

Before turning to a new example involving a different reaction path simulation, it is useful to examine the behavior of several thermodynamic state functions which characterize the fractionation of the Thingmuli olivine tholeiite. In Figure 9 the instantaneous and cumulative heats of crystallization are indicated for 100 grams of initial liquid crystallizing, at a pressure of 1 bar, to 95.6 % solid. Spikes in the latent heat of crystallization reflect the appearance of new phases on the liquidus. These spikes persist regardless of the size of the temperature increment used in the modeling scheme. Heat flow models constructed using this record of variable heat production (Ghiorso, in press) indicate that instantaneous heat output is roughly proportional to time spent at a given temperature. If eruption events are periodic, then the results presented in Figure 9 indicate there is a greater likelihood of sampling magmas which have recently acquired a new phase on the liquidus. The volumetric abundance of lavas of a particular range in composition may consequently have a direct relationship to the latent heat released in the source chamber.

In Figure 10 the Gibbs free energy differences between the supercooled liquid and the stable liquid+solid assemblage are plotted for each interval of the open system fractionation of the Thingmuli olivine

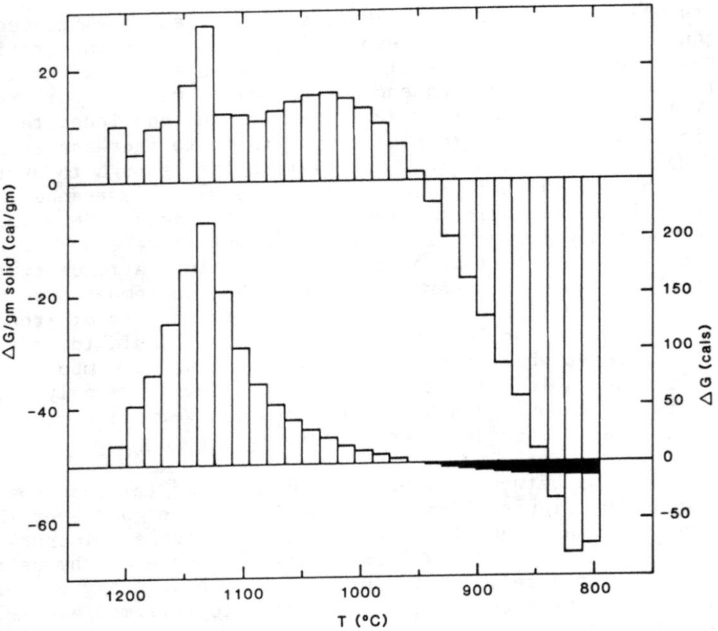

Figure 10. Bargraphs of the calculated difference in the Gibbs free energy of the multiphase system and the Gibbs free energy of a metastable liquid of equivalent bulk composition at each stage in the fractional crystallization of an olivine tholeiite along the QFM buffer at 1 bar. The lower graph should be read according to the right hand scale, the upper according to the left. The apparently metastable nature of the solid-liquid assemblages at lower temperatures is a consequence of the open system calculations. From Ghiorso and Carmichael (1985).

tholeiite. These free energy differences are small, and consequently significant, in that their magnitude demonstrates that the rate of crystal growth in magmatic systems will be essentially proportional to the free energy of crystallization (Ghiorso, 1987). The results of Figure 10 emphasize that an appropriate kinetic theory of crystallization and dissolution in magmatic systems must acknowledge that such systems rarely depart substantially from the equilibrium state.

The results presented in Figure 10 demonstrate that at advanced stages of fractionation, the supercooled liquid has a lower Gibbs free energy than the calculated stable liquid+solid assemblage. In Figure 10, this point in the cooling history occurs where the free energy change goes to zero and subsequently becomes negative. It corresponds exactly to a net increase in the oxygen content of the system (where ΔO_2 passes through zero in Fig. 5). The thermodynamic stability of the calculated solid+liquid assemblage under these conditions can be understood if it is realized that the open system Korzhinskii potential, not the Gibbs free energy, is minimal at each step of this reaction path. Figure 10 demonstrates that the minimum in these two functions is not necessarily coincident.

ASSIMILATION OF PELITE INTO THOLEIITIC MAGMA

In this example we will consider simulation of reaction paths involving assimilation under both isothermal and isenthalpic conditions.

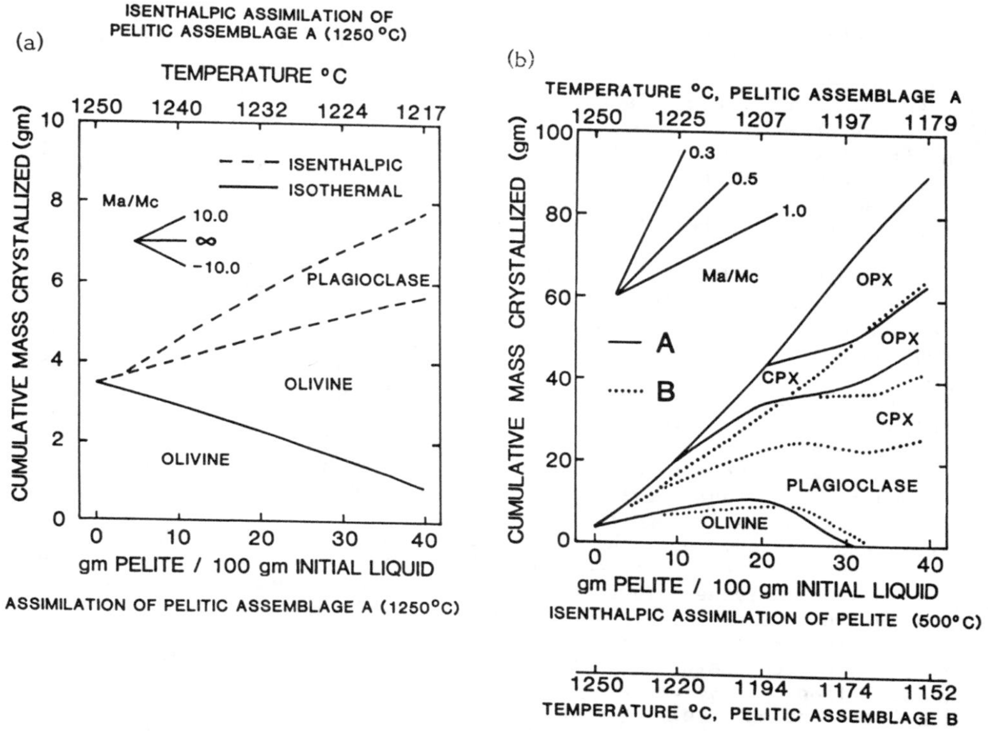

Figure 11. Calculated total solid composition for isothermal and isenthalpic assimilation of pelitic rock in an anhydrous magnesian MORB (FAMOUS 527-1-1) at 3 kbar, along the QFM oxygen buffer. Phase assemblages A and B have identical bulk composition:

SiO_2	64.46	Al_2O_3	17.11	MgO	2.75	Na_2O	1.59
TiO_2	0.73	FeO_T	6.69	CaO	3.11	K_2O	3.56

B includes 1.35 wt % H_2O, but A is composed of Qtz-Ilm-Kspar-Opx-Spinel-Plag, whereas B is composed of Qtz-Ilm-Musc-Bio-Garnet-Plag. (a) Initial temperature of the liquid, and the temperature of the assimilate throughout the calculation, was set at 1250°C. M_a/M_c refers to the ratio of mass assimilated over mass crystallized. (b) Input parameters as in (a), but with the temperature of the assimilate set at 500°C throughout the calculation. From Ghiorso and Kelemen (1987).

The isenthalpic reaction paths will be modeled by "step-wise" minimization of the potential given by Equation (35) of the previous chapter, where the enthalpy content of the system is fixed, at each step in reaction progress, by the combined heat contents of the melt and the added solid assimilate. The example is taken from Ghiorso and Kelemen (1987) and involves the interaction of pelitic country rock with a magnesian MORB (FAMOUS 527-1-1). The reaction path is constrained to follow the QFM buffer at a total pressure of 3 kbar, with the initial magma taken at 1250°C, close to its dry liquidus. Three different reaction paths are modeled: (1) isothermal assimilation, with the solid assimilate at 1250°C, (2) isenthalpic assimilation, with the assimilate at 1250°C, and (3) isenthalpic assimilation, with the assimilate at 500°C. The composition of the assimilate is taken from Ghiorso and Kelemen (1987) and is reported in the legend to Figure 11. In order to model realistically the isenthalpic reaction paths, presumed pelitic mineral assemblages are adopted for this bulk composition. These parageneses are reported in turn in the figure legend.

Solid phase proportions as a function of the extent of assimilation are displayed in Figures 11a and 11b. It should be noted that the vertical scale (cumulative mass crystallized) is ten times as large for reaction with the assimilate at 500°C than for reaction with the assimilate at 1250°C. As an example, isenthalpic assimilation of 40 grams of pelite at 1250°C produces less than 10 grams of olivine and plagioclase, whereas almost 100 grams of solid are produced after the assimilation of 40 grams of pelite at 500°C. This difference results from the need in the latter case to heat the assimilate to magmatic temperatures.

Figure 11a demonstrates quite dramatically the distinction between isothermal and isenthalpic reaction paths. The isothermal constraint causes olivine to be consumed as pelite is assimilated, largely as a consequence of altering the liquid bulk composition in a fashion which lowers the liquidus temperature. When the extent of assimilation is carried beyond the degree shown, olivine disappears completely and no solid phases are present in the system. This reaction path demands, however, that the system absorb heat to melt the pelite and the existing olivine. When the heat content of the system plus assimilated pelite is maintained constant, the liquid responds by precipitating solid as pelite is assimilated. The amount of solid formed is not as great as that assimilated, but the temperature of the system drops 33°C to accommodate these first 40 grams of assimilate. If the assimilate was forsteritic olivine rather than pelite quite a different scenario would unfold. In this case (Ghiorso and Kelemen, 1987) more solid would precipitate than that assimilated and the heat balance would be such as to raise the temperature of the system as a function of the degree of assimilation. These calculations emphasize that assimilation should not be thought of as a melting event which dilutes (or perhaps contaminates) the magma, but instead as an overall chemical reaction which takes solid plus liquid and forms a new solid-liquid assemblage. The overall enthalpy change associated with such a reaction is not always positive!

Isenthalpic assimilation of two isochemical pelitic phase assemblages are compared in Figure 11b. In both cases the pelite is at 500°C which is perhaps a more reasonable temperature contrast for wall rock assimilation into a magma body. Phase assemblage A might be thought of as a granulite facies rock which cooled to 500°C after the peak of metamorphism. It contains no hydrous minerals. Assemblage B includes muscovite and biotite and might be characteristic of amphibolite facies metamorphism. The results shown in Figure 11b demonstrate that assimilation of the hydrous assemblage results in a more substantial temperature decrease and less solid produced when compared to the anhydrous assemblage. The addition of water to the melt suppresses, in part, the crystallization of the solid phases and causes the more pronounced temperature drop.

The reaction paths modeled in Figures 11a and 11b result in quite distinct melt compositions. A few key variables are plotted in Figure 12. Assimilation of pelitic material, previously heated to 1250°C, results in a continuous increase in the silica and alkali contents of the melt with little distinction between reaction paths. The magnesium number ($MgO/[MgO+FeO_T]$ in wt %) of the liquid decreases as a function of added assimilate with some divergence of the isothermal and isenthalpic paths at extreme levels of contamination (40%). Solid phases change composition in the expected directions: olivine moves from Fo_{87} to Fo_{82} and plagioclase has a composition of An_{72} when the melt has reached

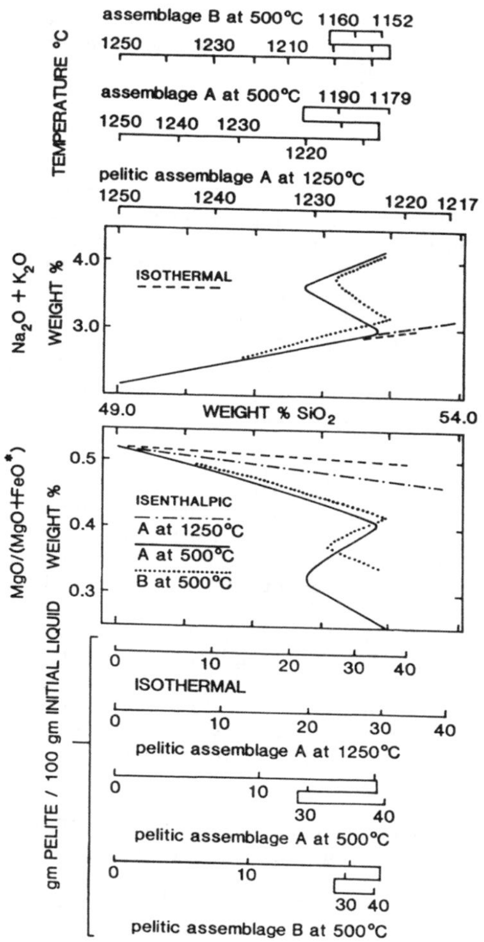

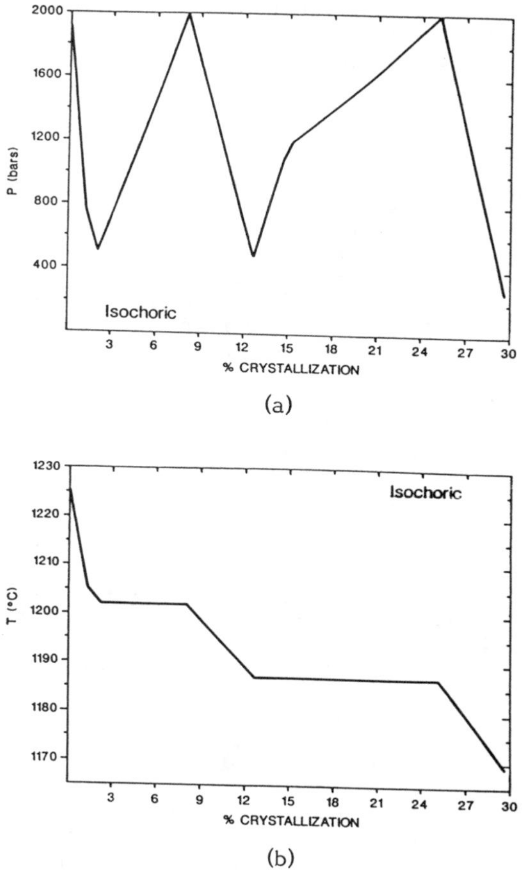

Figure 12 (left). Calculated liquid composition for isothermal and isenthalpic assimilation of pelitic rock in a magnesian MORB. Input parameters as for Figure 11. From Ghiorso and Kelemen (1987).

Figure 13 (right). Changes in intensive variables during the isochoric-isobaric equilibrium crystallization of olivine tholeiitic magma. Solid triangles indicate transition points between the isochoric-isobaric intervals of crystallization. (a) Pressure versus per cent crystallization. (b) Temperature versus per cent crystallization.

1217°C. These observations are in direct contrast to the calculations made using the low temperature assimilate. Examination of Figure 12 reveals that for these reaction simulations the silica content of the melt at first increases but then decreases at the onset of the precipitation of orthopyroxene. When the olivine consuming reaction has gone to completion silica again increases in the melt as assimilation of the pelite proceeds. In contrast, the alkali contents of the contaminated melts increase continuously with each added increment of assimilate as does the iron-magnesium ratio of the liquid phase. As in Figure 11b, the distinction between the "wet" and "dry" pelitic assimilates is profound and underscores the effect on the phase relations due to the addition of small amounts of water to the dry MORB liquid. The plagioclase composition predicted to coexist with these contaminated melts (after 40 % assimilation) is approximately An_{66}, the orthopyroxene composition is about En_{77}, and the clinopyroxene, approximately $Di_{83.5}Fs_{15}$.

ISOCHORIC CRYSTALLIZATION

When a magma body is emplaced into the upper crust and is then allowed to cool, the surrounding country rock must respond by deforming in order to accommodate the change in volume which accompanies crystallization and cooling. If the deformational response lags behind the rate of magmatic crystallization, then the internal pressure of the magma will depart from that of the surrounding country rock. The extent of this pressure difference will depend on the relative rates of the processes involved as well as the maximum level of stress that the crust can accommodate before fracturing. One could envisage the fracturing that results from this pressure drop as a mechanism for the instigation of volcanic eruptions. The reaction path which characterizes crystallization in such a magma body is one expressed in terms of the independent variables volume and temperature. An idealized scenario might involve isochoric (constant volume) crystallization down to an internal pressure which represents a cracking threshold, followed by reestablishment of the external pressure on the magma. Such a crystallization path can be modeled as a series of isochoric steps, from which pressure is deduced as a dependent variable. This reaction path can be calculated by minimization of the Helmholtz free energy. The next step in the evolution of the system is computed by incrementing the pressure in a stepwise fashion (isothermally) back to its initial value, minimizing the Gibbs free energy to compute the equilibrium phase relations at each step. The process is followed by another episode of isochoric crystallization, etc. The consequences of this idealized reaction scenario will be pursued in this example.

We choose an initial magma composition corresponding to the olivine tholeiitic liquid used in the crystal fractionation example reported above. The magma is assumed to be emplaced at a depth corresponding to 2 kbar just at its liquidus temperature of 1225°C. Crystallization is assumed to take place along the QFM oxygen buffer, which necessitates minimization of the potentials given by Equation (31) (isothermal path) and Equation (33) (isochoric path) of the previous chapter. An internal pressure of 500 bars was taken to correspond to the onset of fracturing and two complete isochoric-isothermal "cycles" were computed. Crystallized solids were not allowed to back react with the residual liquids (fractional crystallization).

Pressure and temperature as a function of the extent of magmatic crystallization are plotted in Figures 13a and 13b. The rapid decrease in pressure as a consequence of the relatively minor degree of crystallization results from the fact that the bulk melt compressibility, β, is of the order of 10^{-6}. Relatively minor differences in the volume of melt and crystals translate as a result into large pressure drops in the system ($\Delta P \simeq \Delta V/\beta$). The kinks in the pressure-temperature curves displayed in Figures 13a and 13b correspond to the appearance of new phases on the liquidus. A summary of the phase relations for this reaction path are displayed in Figure 14.

Figure 14a provides a "normal" reference frame in which to view the results of the isochoric-isothermal simulation which are displayed in Figure 14b. In Figure 14a the reaction path corresponds to that of an isobaric system with the pressure fixed at 2 kbar. The differences in phase proportions displayed in the two figures are dramatic. The first interval of isochoric crystallization causes precipitation of olivine from the melt. The density of olivine is such that its crystallization results in a rapid decrease in pressure. Plagioclase saturates just as the system achieves 500 bars. During the subsequent isothermal crystallization interval, olivine and plagioclase are produced in about the same proportion as that in the isobaric simulation (Fig. 14a) and the system attains its original pressure at about 8 % crystallization. An isochoric interval follows during which olivine and plagioclase precipitate, but unlike the isobaric path, clinopyroxene does not appear at the same stage of crystallization. The stability of the pyroxene is enhanced at higher pressure and the isochoric reaction path heads the system away from clinopyroxene saturation. The second isochoric interval corresponds to an extent of crystallization nearly twice as great as the first. If it is recalled that the plagioclase which forms from this tholeiitic liquid is actually less dense than the coexisting melt (Fig. 4) then it becomes clear that more olivine must crystallize in order to compensate this effect and in addition, lower the internal pressure of the magma. Disregarding for the moment the effect of the thermal coefficient of expansion, if this magma had only plagioclase on the liquidus, isochoric crystallization would raise the internal pressure above that on the country rocks! The second isothermal stage of the isochoric-isothermal reaction path leads to extensive crystallization of clinopyroxene. Figure 15 demonstrates that this has the usual effect of lowering the silica content of the melt. In the isobaric reaction scheme the progressive decrease in the melt concentration of silica accompanying pyroxene precipitation continues until saturation with spinel is reached. By contrast, during the final isochoric path displayed in Figure 14b, clinopyroxene leaves the liquidus (and presumably does not return until the pressure is elevated during the next cycle). This results in the increasing silica contents shown in Figure 15.

The compositions of the precipitated solid phases also dramatically reflect the differences in modeled reaction paths. In Figure 16 instantaneous olivine and plagioclase compositions are compared for the isobaric (Fig. 16a) and the isochoric-isothermal models (Figs. 16b and 16c). In the isobaric model both the olivine and plagioclase become progressively more fayalitic and albitic, respectively. These trends could be interpreted as giving rise to normally zoned crystals. Olivine crystallizing during the isochoric-isothermal path also becomes progressively more fayalitic but plagioclase shows markedly different behavior (Fig. 16b). During the isochoric intervals the plagioclase increases in Ca-content as the pressure in the magma drops. This is a consequence of the large partial molar volume of soda in the melt, which at higher pressure forces soda into the coexisting plagioclase in preference to CaO, and renders the plagioclase more albitic. The trends shown in Figure 16b indicate normal plagioclase zoning during the isothermal legs of the crystallization path and reversed zoning during the isochoric intervals. The net result could be an oscillatory zoned crystal.

ISOBARIC VESICULATION

In our final example we will treat a simple isothermal-isobaric

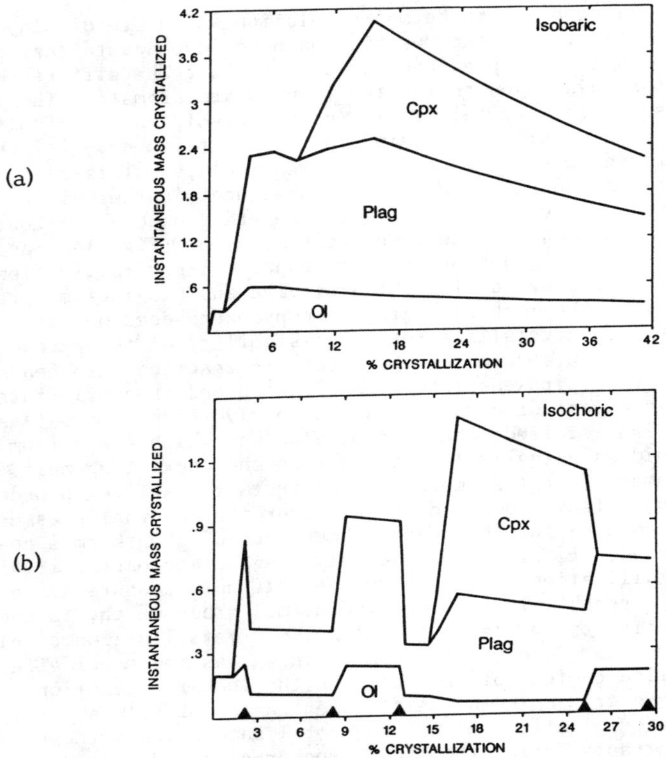

Figure 14. Instantaneous mass crystallized plotted versus per cent crystallization for two different reaction paths of the equilibrium crystallization of olivine tholeiitic magma. (a) Isobaric cooling at two kilobars. (b) Combined isochoric-isobaric reaction path corresponding to Figure 13. Solid triangles indicate transition points between the isochoric-isobaric intervals of crystallization. Symbols refer to plagioclase (plag), olivine (ol) and clinopyroxene (cpx).

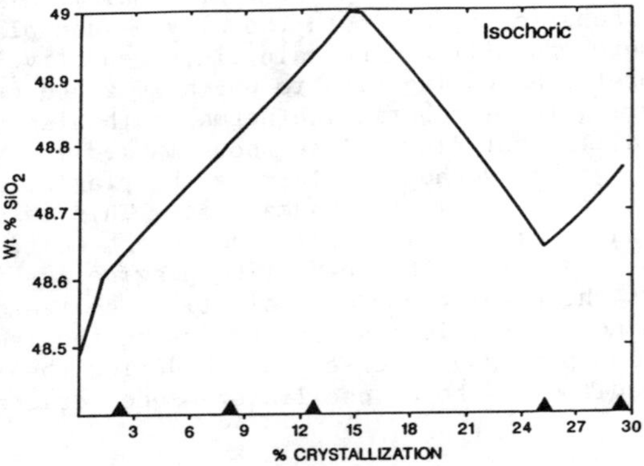

Figure 15. Wt % SiO_2 in the melt plotted versus per cent crystallization for the isochoric-isobaric equilibrium crystallization of olivine tholeiitic magma. The reaction path corresponds to Figure 13 and the solid triangles indicate transition points between the isochoric-isobaric intervals of crystallization.

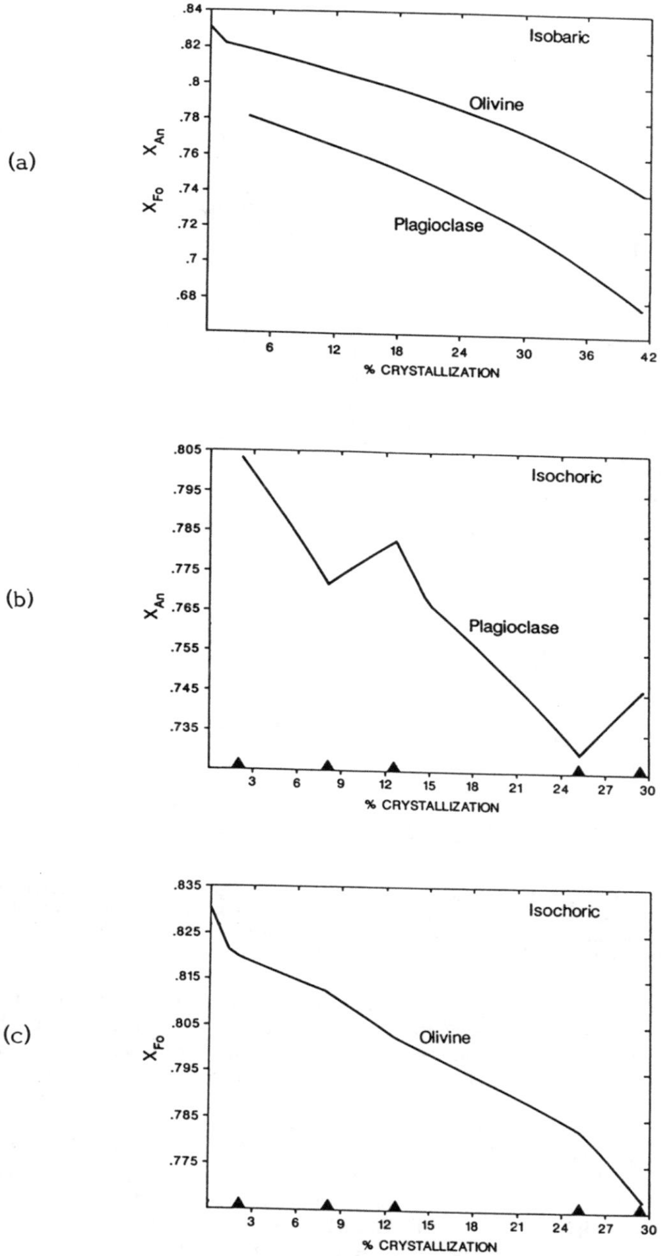

Figure 16. Olivine and plagioclase compositions plotted versus per cent crystallization for two different reaction paths of the equilibrium crystallization of olivine tholeiitic magma. X_{Fo} refers to the mole fraction of forsterite in olivine and X_{An} to the mole fraction of anorthite in plagioclase. (a) Isobaric cooling at two kilobars. (b) and (c) Combined isochoric-isobaric reaction path corresponding to Figure 13. Solid triangles indicate transition points between the isochoric-isobaric intervals of crystallization.

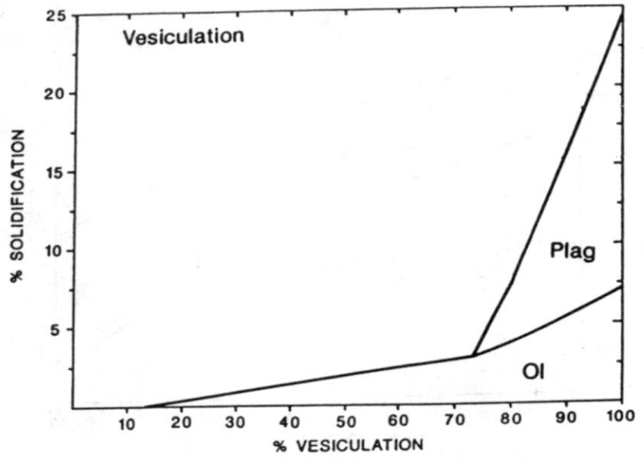

Figure 17. Per cent solidification plotted versus per cent vesiculation for an isothermal (1175°C)-isobaric (1 bar) reaction path involving the devolatilization of olivine tholeiitic magma. Symbols refer to plagioclase (plag), olivine (ol) and clinopyroxene (cpx).

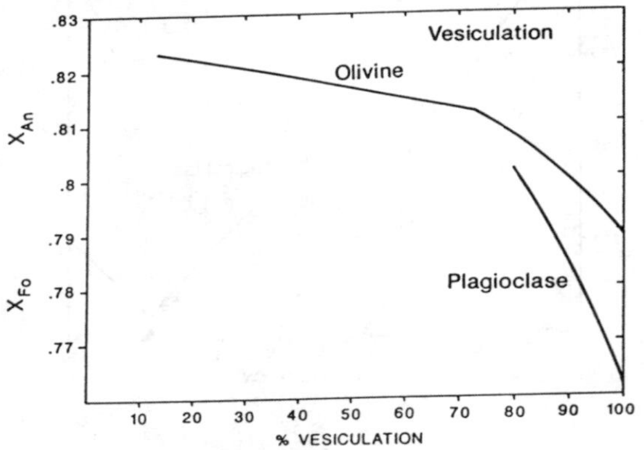

Figure 18. Olivine and plagioclase compositions plotted versus per cent vesiculation for an isothermal (1175°C)-isobaric (1 bar) reaction path involving the devolatilization of olivine tholeiitic magma. X_{Fo} refers to the mole fraction of forsterite in olivine and X_{An} to the mole fraction of anorthite in plagioclase.

reaction path involving the degassing of olivine tholeiitic magma. Once again we consider the lava whose composition is reported in Table 1, but this time we specify the water content as 1.5 wt %. The lava is assumed to have been brought to the surface rapidly enough so that vesiculation takes place irreversibly at 1 bar pressure. For this simple example, heat loss is neglected and the temperature is held constant at 1175°C. Phase proportions are reported in Figure 17 as a function of the extent of vesiculation. With 1.5 wt % dissolved water, the magma is initially superliquidus at 1175°C. Substantial crystallization (from 2.5 to 25 %) takes place only within the last 25% of the vesiculation path. This interval corresponds to a melt water content decreasing from 0.38 wt %. Figure 17 underscores the fact that small amounts of water have a more sensitive impact on magmatic phase proportions than higher water contents in the melt.

A more mechanistic way to view the results shown in Figure 17 is to consider the vesiculation process as effectively raising the liquidus temperature of the magma and consequently creating an undercooled condition at low water contents. The extent of undercooling would amount to a maximum of roughly 50°C. Presumably, this undercooling could result in the rapid nucleation and growth of groundmass crystallites. An indication of the compositions of these potential secondary crystals is provided in Figure 18. As might be expected, solid phase compositions are predicted to change much more rapidly at extremely low melt water contents. Plagioclase becomes considerably more sodic and olivine more fayalitic. In the context of rapid degassing, the petrographic expression of these results should be a population of small crystallites of much more evolved composition than any phenocrysts that might be present in the lava. The phenocrysts may, in turn, be abruptly rimmed by more sodic feldspar and more fayalitic olivine.

This example of the vesiculation of water from basaltic magma is evidence of the ability to extrapolate known phase relations using thermodynamic modeling of magmatic systems. The calibration of the Ghiorso et al. (1983) regular solution model, upon which Figures 17 and 18 are based, included no explicit information on the effect of water content on solid phase proportions and compositions in mafic melts. Information was available on the solubility of water in granitic liquids, the effect of small amounts of water on plagioclase, olivine and clinopyroxene phase compositions in rhyolitic melts, and the plagioclase and olivine saturation surface in a wide variety of anhydrous mafic liquids. Though the model fails to predict accurately the <u>saturation surface</u> for water in basaltic liquids, it does seem to have captured the essential features of the shape of the Gibbs free energy surface for water undersaturated conditions. Many more phase equilibrium experiments are necessary in order to verify these predictions in detail.

SUMMARY

The examples presented in this chapter demonstrate that thermodynamic solution models for magmatic systems provide an effective means of testing reaction and crystallization hypotheses of petrologic interest. We have shown that reaction paths of considerable complexity can be modeled in the context of a step-wise equilibrium approximation to the overall irreversible process. Such calculations require no preconceived notions regarding the identity, compositions or proportions of solid phases in the magma, yet provide comprehensive and internally consistent predictions of the major element evolution of all the phases in the system. In this context, thermodynamic modeling of magmatic systems places additional constraints on petrologic hypotheses suggested by trace element or isotopic data.

Thermodynamic modeling of magmatic systems paves the way toward a quantitative understanding of dynamic processes such as crystal growth and transport. The goal of this modeling is to provide a basis from which to build a comprehensive non-equilibrium theory of magmatic evolution.

APPENDIX

A revised version of the regular solution model of Ghiorso et al. (1983) was used for construction of Figures 1-3. The new model extends the capabilities of the former in predicting crystal-liquid equilibria in alkalic magmas. New liquid components were chosen, some entropies of fusion were estimated and new standard state data were adopted for the liquid phase according to the following table:

Liquid component	Sources of data
Si_4O_8	Richet et al. (1982); Lange and Carmichael (1987)
$Ca_{8/3}Ti_{8/3}O_8$	Robie et al. (1978); Berman and Brown (1985); Lange and Carmichael (1987); Stebbins et al. (1984)
$Al_{16/3}O_8$	Robie et al. (1978); Berman and Brown (1985); Lange and Carmichael (1987); Stebbins et al. (1984); Stull and Prophet (1971)
Fe_6O_8	Robie et al. (1978); Berman and Brown (1985); Lange and Carmichael (1987); Stebbins et al. (1984)
$Mg_2Cr_4O_8$	Robie et al. (1978); Kelley (1960)
$Fe_4Si_2O_8$	Robie et al. (1982a); Berman and Brown (1985); Lange and Carmichael (1987); Stebbins et al. (1984)
$Mn_4Si_2O_8$	
$Mg_4Si_2O_8$	Robie et al. (1978, 1982b); Berman and Brown (1985); Lange and Carmichael (1987); Stebbins et al. (1984); Ghiorso and Carmichael (1980)
$Ca_4Si_2O_8$	Robie et al. (1978); Berman and Brown (1985); Lange and Carmichael (1987); Stebbins et al. (1984);
$Na_{16/3}Si_{8/3}O_8$	Stull and Prophet (1971); Berman and Brown (1985); Lange and Carmichael (1987); Stebbins et al. (1984)
$K_2Al_2Si_2O_8$	Robie et al. (1978); Berman and Brown (1985); Lange and Carmichael (1987); Stebbins et al. (1984)
$P_{16/5}O_8$	
H_2O	

These changes reflect the adoption of the Berman and Brown (1985) solid heat capacity formulation and the much more precise liquid volume measurements of Lange and Carmichael (1987). Additionally, the standard state for the Si_4O_8 liquid component was taken to be that of amorphous silica (Richet et al., 1982). This choice effectively renders the standard state that of silica glass throughout the magmatic temperature range. If this standard state choice is not made, and the properties of metastable silica liquid simply extrapolated down in temperature, one has the logical inconsistency that the liquid is more stable than the solid at 1200°C. This sensitivity to choice of standard state results from the fact that the temperature integral of the heat capacity difference between glass and supercooled liquid is of the same order as the enthalpy of fusion.

Thermodynamic data for the remaining solid phases were generally adopted from calorimetric measurements, with the exception of clinoenstatite, clinoferrosilite, hedenbergite, leucite and the end-member spinels. Heat capacity expressions were taken from, or refitted to the form of, Berman and Brown (1985). The enthalpy of formation of clinoenstatite and clinoferrosilite were fixed by matching the exchange equilibria (Fo + 2Fs = Fa + 2En) tabulated by Miyano and Klein (1986) and by forcing consistency between silica activity buffers displayed in Figure

3 and petrographic data on natural phase assemblages in volcanic rocks. The enthalpy of formation of hedenbergite was set in order to minimize the residuals for the fit of this phase to the model (this enthalpy has not been determined calorimetrically). Data for leucite was adjusted as described in the following paragraph. Enthalpies of formation and third law entropies of spinel end members were adjusted within reported calorimetric uncertainty to be consistent with the reciprocal exchange potentials determined by Sack (1982). Data for ulvospinel and chromite were fixed by Sack's reciprocal terms. Further details on the choice of standard state data, the selection of additional solution models for the solid phases, and the degree of internal consistency of the revised database are provided in Ghiorso and Carmichael (in preparation).

The regular solution interaction parameters used in the construction of Figures 1-3 were calibrated from solid-liquid equilibria involving olivine (forsterite-252 cases and fayalite-247), plagioclase (albite-150, anorthite-153, sanidine-80), potassium-feldspar (albite-1, anorthite-1, sanidine-1), orthopyroxene (enstatite-35, ferrosilite-35), clinopyroxene (diopside-108, hedenbergite-108), leucite (53), nepheline (3), spinel (magnesioferrite-53, magnetite-38, spinel-150, hercynite-128, ulvospinel-49, di-magnesium titanate-4, picrochromite-93, chromite-93), rhombohedral oxides (hematite-17, ilmenite-8, geikielite-8), pseudobrookite (7). These data represent a significant extension over the 1983 database. The new model parameters have been calibrated incorporating the solid-phase-absent constraints discussed in the preceding chapter. These were imposed in order to clear up an existing problem involving saturation surface calculations for leucite and quartz. Analysis revealed that the thermodynamic data for leucite is most probably in error. This probability gains some support from petrologic observations (cf. Fig. 3a). In the course of calibration of the model we have found the following data for leucite to be optimal: ΔH_f(298.15 K, 1 bar) = -728,645 cals/mol (a correction of 2.4 kcals/mol) and S° (298.15 K, 1 bar) = 42.35 cals/mol-K (a correction of 5.5 cals/mol-K).

REFERENCES

Berman, R.G., Brown, T.H. (1985) Heat capacity of minerals in the system Na_2O-K_2O-CaO-MgO-FeO-Fe_2O_3-Al_2O_3-SiO_2-TiO_2-H_2O-CO_2: representation, estimation, and high temperature extrapolation. Contrib. Mineral. Petrol. 89, 168-183.

Carmichael, I.S.E. (1964) The petrology of Thingmuli, a Tertiary volcano in Eastern Iceland. J. Petrol. 5, 435-460.

Carmichael, I.S.E. (1967) The mineralogy of Thingmuli, a Tertiary volcano in Eastern Iceland. Amer. Mineral. 52, 1815-1841.

Carmichael, I.S.E., Ghiorso, M.S. (1986) Oxidation-reduction relations in basic magmas: a case for homogeneous equilibria. Earth Planet. Sci. Lett. 78, 200-210.

Carmichael, I.S.E., Nicholls, J., Smith, A.L. (1970) Silica Activity in igneous rocks. Amer. Mineral. 55, 246-263.

Carmichael, I.S.E., Nicholls, J., Spera, F.J., Wood, B.J., Nelson, S.A. (1977) High-temperature properties of silicate liquids: applications to the equilibration and ascent of basic magma. Phil. Trans. R. Soc. London A. 286, 373-431.

Carmichael, I.S.E., Turner, F.J., Verhoogen, J. (1974) Igneous Petrology. McGraw-Hill Book Company, New York, 739 pp.

Coombs, D.S. (1963) Trends and affinities of basaltic magmas and

pyroxenes as illustrated on the diopside-olivine-silica diagram. Min. Soc. Amer. Spec. Pap. 1, 227-250.
Cross, C.W., Iddings, J.P., Pirsson, L.V., Washington, H.S. (1902) A quantitative chemico-mineralogical classification and nomenclature of igneous rocks. J. Geol. 10, 555-690.
Deer, W.A., Howie, R.A., Zussman, J. (1982) Rock-forming minerals, Vol. 1A, Orthosilicates, 2nd Ed., Longmans, New York, 919 pp.
Ghiorso, M.S. (1984) Activity/composition relations in the ternary feldspars. Contrib. Mineral. Petrol. 87, 282-296.
Ghiorso, M.S. (1985) Chemical mass transfer in magmatic processes. I. Thermodynamic relations and numerical algorithms. Contrib. Mineral. Petrol. 90, 107-120.
Ghiorso, M.S. (1987) Chemical mass transfer in magmatic processes. III. Crystal growth, chemical diffusion and thermal diffusion in multi-component silicate melts. Contrib. Mineral. Petrol. 96:291-313.
Ghiorso, M.S. (in press) Temperatures in and around cooling magma bodies. In Perchuk, L.L., ed., Advances in Physical Geochemistry, Springer, New York.
Ghiorso, M.S., Carmichael, I.S.E. (1980) A regular solution model for met-aluminous silicate liquids: Applications to geothermometry, immiscibility, and the source regions of basic magmas. Contrib. Mineral. Petrol. 71, 323-342.
Ghiorso, M.S., Carmichael, I.S.E. (1985) Chemical mass transfer in magmatic processes. II. Applications in equilibrium crystallization, fractionation and assimilation. Contrib. Mineral. Petrol. 90, 121-141.
Ghiorso, M.S., Carmichael, I.S.E. (in preparation) Chemical mass transfer in magmatic processes. IV. A recalibration of the regular solution model and examples of crystallization calculations for alkalic magmas.
Ghiorso, M.S., Carmichael, I.S.E., Rivers, M.L., Sack, R.O. (1983) The Gibbs free energy of mixing of natural silicate liquids; an expanded regular solution approximation for the calculation of magmatic intensive variables. Contrib. Mineral. Petrol. 84, 107-145.
Ghiorso, M.S., Kelemen, P.B. (1987) Evaluating reaction stoichiometry in magmatic systems evolving under generalized thermodynamic constraints: Examples comparing isothermal and isenthalpic assimilation. In: Magmatic Processes: Physicochemical Principles. B.O. Mysen, ed., Geochem. Soc., Spec. Publ. No. 1, 319-336.
Harker, A. (1909) The Natural History of Igneous Rocks, MacMillan, New York, 384 pp.
Hatch, F.H. (1891) Petrology of the Igneous Rocks, Thomas Murby, London, 128 pp.
Johannsen, A. (1939) A Descriptive Petrography of the Igneous Rocks, Vol. 1, Univ. of Chicago Press, 318 pp.
Kelley, K.K. (1960) Contributions to the data on theoretical metallurgy. XIII. High-temperature heat-content, heat-capacity, and entropy data for the elements and inorganic compounds. U.S. Bureau of Mines. Bull. 584, 232 pp.
Kilinc, A., Carmichael, I.S.E., Rivers, M.L., Sack, R.O. (1983) The ferric-ferrous ratio of natural silicate liquids equilibrated in air. Contrib. Mineral. Petrol. 83, 136-140.
Lange, R.A., Carmichael, I.S.E. (1987) Densities of Na_2O-K_2O-CaO-MgO-FeO-Fe_2O_3-Al_2O_3-TiO_2-SiO_2 liquids: new measurements and derived partial molar properties. Geochim. Cosmochim. Acta 51, in press.
Le Bas, M.J., Le Maitre, R.W., Streickeisen, A., Zanettin, B. (1986) A chemical classification of volcanic rocks based on the total

alkali-silica diagram. J. Petrol. 27, 745-750.
Le Maitre, R.W. (1984) A proposal by the IUGS Subcommission on the systematics of igneous rocks for a chemical classification of volcanic rocks based on the total alkali-silica (TAS) diagram. Austral. J. Earth Sci. 31, 243-255.
Middlemost, E.A.K. (1980) A contribution to the nomenclature and classification of volcanic rocks. Geol. Mag. 117, 51-57.
Miyano, T., Klein, C. (1986) Fluid behavior and phase relations in the system Fe-Mg-Si-C-O-H: Application to high grade metamorphism of iron-formations. Amer. J. Sci. 286, 540-575.
Nicholls, J., Carmichael, I.S.E. (1972) The equilibration temperature and pressure of various lava types with spinel- and garnet-peridotite. Amer. Mineral. 57, 941-959.
Nicholls, J., Carmichael, I.S.E., Stormer, J.C. Jr. (1971) Silica Activity and $\underline{P}_{total}$ in igneous rocks. Contrib. Mineral. Petrol. 33, 1-20.
Nicholls, J., Stout, M.Z. (1982) Heat effects of assimilation, crystallization, and vesiculation in magmas. Contrib. Mineral. Petrol. 84, 328-339.
Richet, P., Bottinga, Y., Denielou, L., Petitet, J.P., Tequi, C. (1982) Thermodynamic properties of quartz, cristobalite and amorphous SiO_2: Drop calorimetry measurements between 1000 and 1800 K and a review from 0 to 2000 K. Geochim. Cosmochim. Acta 46, 2639-2658.
Robie, R.A., Finch, C.B., Hemingway, B.S. (1982a) Heat capacity and entropy of fayalite (Fe_2SiO_4) between 5.1 and 383 K; comparison of calorimetric and equilibrium values for the QFM buffer reaction. Amer. Mineral. 67, 463-469.
Robie, R.A., Hemingway, B.S., Takei, H. (1982b) Heat capacities and entropies of Mg_2SiO_4, Mn_2SiO_4, and Co_2SiO_4 between 5 and 380 K. Amer. Mineral. 67, 470-482.
Robie, R.A., Hemingway, B.S., Fisher, J.R. (1978) Thermodynamic properties of minerals and related substances at 298.15 K and 1 bar (10^5 Pascals) pressure and at higher temperature. U.S. Geol. Surv. Bull. 1452, 456 pp.
Sack, R.O. (1982) Spinels as petrogenetic indicators: activity-composition relations at low-pressures. Contrib. Mineral. Petrol. 79, 169-186.
Shand, S.J. (1913) On saturated and unsaturated igneous rocks. Geol. Mag. Series 5, 10, 408-514.
Shand, S.J. (1914) The principle of saturation in petrography. Geol. Mag. Series 6, 1, 485-493.
Shand, S.J. (1927) Eruptive Rocks, Wiley, New York, 360 pp.
Shand, S.J. (1943) Eruptive rocks, their genesis, composition, classification, with a chapter on meteorites. Thomas Murby and Co., London, 460 pp.
Smith, A.L. (1970) Sphene, perovskite and coexisting Fe-Ti oxide minerals. Amer. Mineral. 55, 264-269.
Stebbins, J.F., Carmichael, I.S.E., Moret, L.K. (1984) Heat capacities and entropies of silicate liquids and glasses. Contrib. Mineral. Petrol. 86, 131-148.
Streckeisen, A. (1976) To each plutonic rock its proper name. Earth Sci. Rev. 12, 1-33.
Streckeisen, A., Le Maitre, R.W. (1979) A chemical approximation to the modal QAPF classification of the igneous rocks. N. Jb. Mineral. Abh. 136, 169-206.
Stull, D.R., Prophet, H. (1971) JANAF Thermochemical Tables, 2nd ed. Nat. Stand. Ref. Data Ser., Nat. Bur. Stand. (U.S.), 37, 1141 pp.
Tilley, C.E. (1953) The nephelinite of Etinde, Cameroons, West Africa. Geol. Mag. 90, 145-151.